Sensors

Volume 2
Chemical and Biochemical Sensors

Sensors

A Comprehensive Survey

Edited by
W. Göpel   (Universität Tübingen, FRG)
J. Hesse   (Zeiss, Oberkochen, FRG)
J. N. Zemel (University of Pennsylvania,
              Philadelphia, PA, USA)

Published
Vol. 1 Fundamentals and General Aspects
    (Volume Editors: T. Grandke, W. H. Ko)
Vol. 2 Chemical and Biochemical Sensors, Part I
    (Volume Editors: W. Göpel, T. A. Jones †, M. Kleitz,
    J. Lundström, T. Seiyama)
Vol. 4 Thermal Sensors
    (Volume Editors: T. Ricolfi, J. Scholz)
Vol. 5 Magnetic Sensors
    (Volume Editors: R. Boll, K. J. Overshott)

Remaining volumes of this closed-end series:

Vol. 3 Chemical and Biochemical Sensors, Part II (scheduled for 1991)
Vol. 6 Optical Sensors (scheduled for 1991)
Vol. 7 Mechanical Sensors (scheduled for 1992)
Vol. 8 Cumulative Index and Selective Topics (scheduled for 1993)

© VCH Verlagsgesellschaft mbH, D-6940 Weinheim (Federal Republic of Germany), 1991

Distribution

VCH, P.O. Box 10 1161, D-6940 Weinheim (Federal Republic of Germany)

Switzerland: VCH, P.O. Box, CH-4020 Basel (Switzerland)

United Kingdom and Ireland: VCH (UK) Ltd., 8 Wellington Court, Wellington Street,
    Cambridge CB1 1HZ (England)

USA and Canada: VCH, Suite 909, 220 East 23rd Street, New York, NY 10010-4606 (USA)

ISBN 3-527-26768-9 (VCH, Weinheim)   ISBN 0-89573-674-8 (VCH, New York)

Sensors

A Comprehensive Survey

Edited by
W. Göpel, J. Hesse, J. N. Zemel

Volume 2

Chemical and Biochemical Sensors
Part I

Edited by
W. Göpel, T. A. Jones †, M. Kleitz,
J. Lundström, and T. Seiyama

VCH

Weinheim · New York · Basel · Cambridge

Series Editors:

Prof. Dr. W. Göpel
Institut für Physikalische und
Theoretische Chemie der Universität
Auf der Morgenstelle 8
D-7400 Tübingen, FRG

Prof. Dr. J. Hesse
Carl Zeiss,
ZB „Entwicklung"
Postfach 1380
D-7082 Oberkochen, FRG

Prof. Dr. J. N. Zemel
Center for Sensor Technology
University of Pennsylvania
Philadelphia, PA 19104-6390, USA

Volume Editors:

Prof. Dr. W. Göpel
see above
Dr. T. A. Jones †
Health and Safety
Executive
Sheffield, UK

Dr. M. Kleitz
L.I.E.S./E.N.S.E.E.G
Domaine Universitaire,
B.P. 75
F-38402 Saint-Martin
d'Hères, France

Prof. J. Lundström
Linköping Institute of
Technology
Dept. of Physics and
Measurement Technology
S-58183 Linköping, Sweden

Prof. T. Seiyama
Tokuyama Soda Co., Ltd.
Tenjin 1-10-24
Chuo-ku, Fukuoka-shi,
Japan 810

Published jointly by
VCH Verlagsgesellschaft mbH, Weinheim (Federal Republic of Germany)
VCH Publishers Inc., New York, NY (USA)

Editorial Directors: Dipl.-Phys. W. Greulich, Dipl.-Chem. Dr. M. Weller, N. Banerjea-Schultz
Production Manager: Dipl.-Wirt.-Ing. (FH) H.-J. Schmitt

Library of Congress Card No.: applied for

British Library Cataloguing-in-Publication Data:
Sensors: a comprehensive survey: Vol 2. Chemical
and biochemical sensors, Part I. − (Sensors)
I. Goepel, W.   II. Jones, T. A.   III. Kleitz, M.
502.8

ISBN 3-527-26768-9

Deutsche Bibliothek Cataloguing-in-Publication Data:
Sensors : a comprehensive survey / ed. by W. Göpel ... –
Weinheim ; Basel (Switzerland) ; Cambridge ; New York, NY :
VCH.
NE: Göpel, Wolfgang [Hrsg.]
Vol. 2. Chemical and biochemical sensors. – Part 1. Ed. by W.
   Göpel ... – 1991
   ISBN 3-527-26768-9 (Weinheim ...)
   ISBN 0-89573-674-8 (New York)

Preface to the Series

The economic realities of productivity, quality, and reliability for the industrial societies of the 21st century are placing major demands on existing manufacturing technologies. To meet both present and anticipated requirements, new and improved methods are needed. It is now recognized that these methods must be based on the powerful techniques employing computer-assisted information systems and production methods. To be effective, the measurement, electronics and control components, and sub-systems, in particular sensors and sensor systems, have to be developed in parallel as part of computer-controlled manufacturing systems. Full computer compatibility of all components and systems must be aimed for. This strategy will, however, not be easy to implement, as seen from previous experience. One major aspect of meeting future requirements will be to systematize sensor research and development.

Intensive efforts to develop sensors with computer-compatible output signals began in the mid 1970's; relatively late compared to computer and electronic measurement peripherals. The rapidity of the development in recent years has been quite remarkable but its dynamism is affected by the many positive and negative aspects of any rapidly emerging technology. The positive aspect is that the field is advancing as a result of the infusion of inventive and financial capital. The downside is that these investments are distributed over the broad field of measurement technology consisting of many individual topics, a wide range of devices, and a short period of development. As a consequence, it is not surprising that sensor science and technology still lacks systematics. For these reasons, it is not only the user who has difficulties in classifying the flood of emerging technological developments and solutions, but also the research and development scientists and engineers.

The aim of "Sensors" is to give a survey of the latest state of technology and to prepare the ground for a future systematics of sensor research and technology. For these reasons the publishers and the editors have decided that the division of the handbook into several volumes should be based on physical and technical principles.

Volume 1 (editors: T. Grandke/Siemens (FRG) and W. H. Ko/Case Western Reserve University (USA)) deals with general aspects and fundamentals: physical principles, basic technologies, and general applications.

Volume 2 and 3 (editors: W. Göpel/Tübingen University (FRG), T. A. Jones †/Health and Safety Executive (UK), M. Kleitz/LIE-NSEEG (France), L. Lundström/Linköping University (Sweden) and T. Seiyama/Tokuyama Soda Co. (Japan)) concentrate on chemical and biochemical sensors.

Volume 4 (editors: J. Scholz/Sensycon (FRG) and T. Ricolfi/Consiglio Nazionale Delle Ricerche (Italy)) refers to thermal sensors.

Volume 5 (editors: R. Boll/Vacuumschmelze (FRG) and K. J. Overshott/Gwent College (UK)) deals with magnetic sensors.

Volume 6 (editors: E. Wagner and K. Spenner/Fraunhofer-Gesellschaft (FRG) and R. Dändliker/Neuchâtel University (Switzerland)) treats optical sensors.

Volume 7 (editors: N. F. de Rooij/Neuchâtel University (Switzerland), B. Kloeck/Hitachi (Japan) and H. H. Bau/University of Pennsylvania (USA)) presents mechanical sensors.

Each volume is, in general, divided into the following three parts: specific physical and technological fundamentals and relevant measuring parameters; types of sensors and their technologies; most important applications and discussion of emerging trends.

It is planned to close the series with a volume containing a cumulated index.

The series editors wish to thank their colleagues who have contributed to this important enterprise whether in editing or writing articles. Thank is also due to Dipl.-Phys. W. Greulich, Dr. M. Weller, and Mrs. N. Banerjea-Schultz of VCH for their support in bringing this series into existence.

W. Göpel, Tübingen   J. Hesse, Oberkochen   J. N. Zemel, Philadelphia, PA

August 1991

Preface to the Volumes "Chemical and Biochemical Sensors"

Planning "Sensors", it soon became clear that chemical and biochemical sensors would have to be treated in two volumes to appropriately present the wealth of material.

Thus, these volumes present for the first time a comprehensive description of chemical and biochemical sensors with emphasis placed upon both, technical and scientific fundamentals and applications. The aim is to offer well-funded knowledge to scientists and technicians and to show todays technical capabilities in this sensor field. Furthermore, both volumes together are intended to foster the future developments and applications of sensors and at the same time serve as a useful reference work.

The arrangement of the material presented here deviates in some way from that of the other volumes in the series "Sensors", which are devoted to physical sensors (mechanical, thermal, magnetic, optical sensors). With those sensors the internal structure of each single volume follows a classification by the input signal of the first transduction principle, which in most cases is identical with a classification according to the measurand. (For a complete discussion of sensor definitions and classifications see Volume 1, Chapter 1.). The number of measurands in chemical and biochemical systems, however, is many orders of magnitude larger than in physical systems because of the huge number of different compounds which can occur in gaseous, liquid, and solid media. Therefore another structuring criterion had to be used. As described in detail in Chapter 1 of the present volumes, different types of (bio-)chemical sensors may be classified according to the different sensor properties used for the detection of chemical state, ie, of concentrations, partial pressures, or activities of particles. We adopted this classification to organize the "core" of the present volumes which consists of a description of "basic sensors" (like liquid and solid electrolyte sensors, etc.). For the sake of completeness, the core should be surrounded by a "shell" of articles devoted to other important aspects of the field:

- physical and chemical parameters and measurands;
- the theoretical physical or physico-chemical background underlying the sensing mechanisms (selected textbook knowledge on mechanics, optics, thermodynamics, kinetics, statistics, etc.);
- the technology to produce sensor elements or components (thin-film, thick-film, ceramics technologies, etc.);
- applications (car engine regulation, environmental control, etc.).

We tried to cover all these aspects by organizing the books in the following way:
First volume:

- Definitions, typical examples for chemical and biochemical sensors, and some historical remarks are given in Chapters 1 and 2.
- Chemical sensor technologies and interdisciplinary tasks to design chemical sensors are described in Chapter 3.

— Physical and physical chemistry basics of different detection principles and also pattern recognition approaches for multicomponent analysis with sensor arrays are described in Chapters 4–6.
— The major part of the volumes consists of a careful description of basic sensors in Chapters 7–13. They include liquid electrolyte sensors, solid electrolyte sensors, electronic conductivity and capacitance sensors, field effect sensors, calorimetric sensors, optochemical sensors, and mass sensitive sensors.

Second volume:

— Biosensors often make use of transducer properties of the basic sensors mentioned above and usually have additional biological components. They are therefore described in a separate Chapter 14.
— Application aspects are dealt with in Chapters 15–25. Here, the possibilities and limitations of sensors if compared with the conventional instrumentation in analytical chemistry and calibration aspects are described first. Specific facettes of certain fields of applications are then presented by specialists from different fields including environmental, biotechnological, medical, or chemical process control.

A major input to the present books originally came from Dr. T. A. Jones from the National Health and Safety Executive in Sheffield, U. K. He was a distinguished scientist in chemical sensor basic research and an expert in the particular field of combustable gas sensors. In 1989 he passed away. We could like to take the opportunity here to thank him for his input and enthusiastic support in the planning phase of the volumes on chemical and biochemical sensors.

After all this effort, the editors would now like to thank all participating authors for their contributions and their help in structuring the book by coordinating their individual chapters to the overall guidelines. The editors would also like to thank Dr. Klaus Schierbaum and the staff of VCH, particularly the editorial staff Mrs. N. Banerjea-Schultz and Dipl.-Phys. W. Greulich for their professional input and patience.

Wolfgang Göpel, Michel Kleitz, Ingemar Lundström, Tetsuro Seiyama

Tübingen, Grenoble, Linköping, Fukuoka

August 1991

Contents

Volume 2: Chemical and Biochemical Sensors, Part I

Volume 3: Chemical and Biochemical Sensors, Part II
(Contents given as an overview)

List of Contributors

List of Contributors

Volume 2: Chemical and Biochemical Sensors, Part I

Dr. Mårten Armgarth
Sensistor AB
Box 76
S-58102 Linköping, Sweden
Tel.: (0046-13) 113422
Tfx: (0046-13) 123422

Dr. Gilbert E. Boisde
CEA-CEN-Saclay
DEIN-SAI
F-91191 Gif-sur-Yvette Cedex, France
Tel.: (0033-1) 69088543
Tfx: (0033-1) 69087819

Prof. Dr. Karl Cammann
Westfälische Wilhelms-Universität
Anorg. Chem. Institut
Lehrstuhl für Analyt. Chemie
Wilhelm-Klemm-Str. 8
D-4400 Münster, FRG
Tel.: (0049-251) 833141
Tfx: (0049-251) 833169

Dr. Pierre Fabry
L.I.E.S.G./E.N.S.E.E.G.
Domaine Universitaire, B. P. 75
F-38402 St. Martin d'Heres, France
Tel.: (0033-76) 826557
Tfx: (0033-76) 826630

Dr. Jacques Fouletier
L.I.E.S.G./E.N.S.E.E.G.
Domaine Universitaire, B. P. 75
F-38402 St. Martin d'Heres, France
Tel.: (0033-76) 826557
Tfx: (0033-76) 826630

Prof. Dr. Günter Gauglitz
Universität Tübingen
Inst. f. Physik u. Theor. Chemie
Auf der Morgenstelle 8
D-7400 Tübingen, FRG
Tel.: (0049-7071) 296927
Tfx: (0049-7071) 296910

Prof. Dr. Wolfgang Göpel
Universität Tübingen
Inst. f. Physik u. Theor. Chemie
Auf der Morgenstelle 8
D-7400 Tübingen, FRG
Tel.: (0049-7071) 296904
Tfx: (0049-7071) 296910

Dr. T. Alwyn Jones †
Health and Safety Executive
Broad Lane
Sheffield S3 7HQ/UK

Dr. Michel Kleitz
L.I.E.S.G./E.N.S.E.E.G.
Domaine Universitaire, B. P. 75
F-38402 St. Martin d'Heres, France
Tel.: (0033-76) 826557
Tfx: (0033-76) 826630

Prof. Ingemar Lundström
Linköping Inst. of Technology
Dept. of Physics and Measurement
Technology
S-58183 Linköping, Sweden
Tel.: (0046-13) 281200
Tfx: (0046-13) 137568

Dr. Maarten S. Nieuwenhuizen
Prins Maurits Laboratory TNO
P. O. Box 45
NL-2280 AA Rijkswijk,
New Netherlands
Tel.: (0031-15) 843519
Tfx: (0031-15) 843991

Dr. Claes I. Nylander
Sensistor AB
Box 76
S-58102 Linköping, Sweden
Tel.: (0046-13) 113422
Tfx: (0046-13) 123422

Dipl.-Chem. Friedrich Oehme
Hühnerbühl 34
D-7883 Görwihl, FRG
Tel.: (0049-7754) 7358

Dr. Klaus-Dieter Schierbaum
Universität Tübingen
Inst. f. Physik. u. Theoret. Chemie
Auf der Morgenstelle 8
D-7400 Tübingen, FRG
Tel.: (0049-7071) 295282
Tfx: (0049-7071) 296910

Prof. Tetsuro Seiyama
Tokuyama Soda Co., Ltd.
Tenjin 1-10-24
Chuo-ku, Fukuoka-shi, Japan 810
Tel.: (0081-92) 7516566
Tfx: (0081-92) 7111089

Dr. Elisabeth Siebert
L.I.E.S.G./E.N.S.E.E.G.
Domaine Universitaire, B. P. 75
F-38402 St. Martin d'Heres, France
Tel.: (0033-76) 826557
Tfx: (0033-76) 826630

Stefan Vaihinger
Universität Tübingen
Inst. f. Physik. u. Theoret. Chemie
Auf der Morgenstelle 8
D-7400 Tübingen, FRG
Tel.: (0049-7071) 296933
Tfx: (0049-7071) 296910

Dr. Albert van den Berg
CSEM SA
Rue de la Maladiere 71
CH-2007 Neuchatel, Switzerland
Tel.: (0041-38) 205387
Tfx: (0041-38) 254078

Dr. Hendrik H. van den Vlekkert
Priva B. V.
Zijlweg 3/P. O. Box 18
2678 LC/2678 ZG De Lier,
The Netherlands
Tel.: (0031-1745) 13921
Tfx: (0031-1745) 17195

Dr. Bartholomeus H. van der Schoot
Université de Neuchatel
Institut de Microtechnique
Rue A.-L.-Brequet 2
CH-2000 Neuchatel, Switzerland
Tel.: (0041-38) 205387
Tfx: (0041-38) 254078

Prof. Dr. Adrian Venema
Delft University of Technology
Electrical Engineering Faculty
4 Mekelweg
NL-2629 CD Delft, The Netherlands
Tel.: (0031-15) 786466
Tfx: (0031-15) 785755

Dr. Peter Walsh
Health and Safety Executive
Res. and Lab. Service Division
Broad Lane
Sheffield S3 7HQ, UK
Tel.: (0044-742) 768141 x 3179
Tfx: (0044-742) 755792

Dr. habil. Hans-Dieter Wiemhöfer
Universität Tübingen
Inst. f. Physik. u. Theoret. Chemie
Auf der Morgenstelle 8
D-7400 Tübingen, FRG
Tel.: (0049-7071) 296753
Tfx: (0049-7071) 296910

Prof. Dr. Otto Wolfbeis
Joanneum Research
The Optical Sensor Institute
Steyrergasse 17
A-8010 Graz, Austria
Tel.: (0043-316) 8020222
Tfx: (0043-316) 8020181

Prof. Jay N. Zemel
University of Pennsylvania
Center for Sensor Technologies
Philadelphia, PA 19104-6390, USA
Tel.: (001-215) 8988545
Tfx: (001-215) 8981130

Volume 3: Chemical and Biochemical Sensors, Part II

Dr. Hansjörg Albrecht
Laser-Medizin-Zentrum GmbH
Krahmerstr. 6-10
W-1000 Berlin 45, FRG
Tel.: (0049-30) 8344002
Tfx: (0049-30) 8344004

Dr. Hiromichi Arai
Kyushu University
Materials Science & Technology
6-1 Kasugakoen Kasuga-shi
Fukuoka 816, Japan
Tel.: (0081-92) 5739611 x 310
Tfx: (0081-92) 5752318

Dr. Friedrich G. K. Baucke
Schott Glaswerke
Hattenbergstr./Postfach 2480
D-6500 Mainz 1, FRG
Tel.: (0049-6131) 333239
Tfx: (0049-6131) 333341

Dr. Karen Colbow
Prof. Konrad Colbow
Simon Fraser University
Dept. of Physics
Barnaby, British Columbia V5A156, Canada
Tel.: (001-604) 2913162
Tfx: (001-604) 2913592

Dr. Martin Gerber
Boehringer Mannheim Co.
Biochemistry R & D Division
9115 Hague Road/P. O. Box 50100
Indianapolis, IN 46250-0100, USA
Tel.: (001-317) 5767589
Tfx: (001-317) 5767525

Dr. Michael Hofer
Joanneum Research
The Optical Sensor Institute
Steyrergasse 17
A-8010 Graz, Austria
Tel.: (0043-316) 8020222
Tfx: (0043-316) 8020181

Dr. Klaus Kaltenmaier
Dr.-Ing. Häfele Umweltverfahrens-
technik GmbH & Co.
Gerwigstr. 69
D-7500 Karlsruhe, FRG
Tel.: (0049-721) 616093
Tfx: (0049-721) 621599

Dr. Petra Krämer
University of California
Dept. of Entomology
Davis, CA 95616, USA
Tel.: (001-916) 7525109
Tfx: (001-916) 7521537

Prof. Dr. Hans-Heinrich Möbius
Ernst-Moritz-Arndt-Universität
Inst. f. Physikalische Chemie
Soldtmannstr. 16
DO-2200 Greifswald, FRG
Tel.: (0037-822) 75479
Tfx: (0037-822) 63260

Prof. Dr. Michael Oehme
Norwegian Inst. f. Air Research
Dept. Organic Analyt. Chemistry
Postboks 64
N-2001 Lillestrom, Norway
Tel.: (0047-6) 814170
Tfx: (0047-6) 819247

Dr. Kenneth F. Reardon
Colorado State University
Dept. of Agricultural and
Chemical Engineering
Fort Collins, CO 80523, USA
Tel.: (001-303) 4916505
Tfx: (001-303) 4917369

Prof. Dr. Friedrich Scheller
Zentralinstitut für Molekularbiologie
Robert-Rössle-Str. 10
DO-1115 Berlin-Buch
Tel.: (0037-2) 3463681/-2918
Tfx: (0037-2) 3494161

Dr. Thomas Scheper
Universität Hannover
Inst. f. Technische Chemie
Callinstr. 3
D-3000 Hannover, FRG
Tel.: (0049-511) 7622509
Tfx: (0049-511) 7623456

Prof. Dr. Rolf D. Schmid
GBF
Mascheroder Weg 1
D-3300 Braunschweig, FRG
Tel.: (0049-531) 6181300
Tfx: (0049-531) 6181303

Prof. Dr. Hanns-Ludwig Schmidt
TU München
Allgemeine Chemie und Biochemie
D-8050 Freising-Weihenstephan, FRG
Tel.: (0049-8161) 713253/-54
Tfx: (0049-8161) 713583

Dr. Florian Schubert
Physikalisch-Technische Bundesanstalt
Abbestr. 2-12
D-1000 Berlin 10, FRG
Tel.: (0049-30) 3481235
Tfx: (0049-30) 3481490

Dr. Wolfgang Schuhmann
TU München
Allgemeine Chemie und Biochemie
D-8050 Freising-Weihenstephan, FRG
Tel.: (0049-8161) 713319
Tfx: (0049-8161) 713583

Dr. Wolfgang Trettnak
Joanneum Research
The Optical Sensor Institute
Steyrergasse 17
A-8010 Graz, Austria
Tel.: (0043-316) 8020222
Tfx: (0043-316) 8020181

Dr. Karl Wulff
Boehringer Mannheim GmbH
Abtlg. GT-GL
Sandhofer Str. 116
D-6800 Mannheim, FRG
Tel.: (0049-621) 7594019
Tfx: (0049-621) 7593087/7502890

1 Definitions and Typical Examples

WOLFGANG GÖPEL and KLAUS-DIETER SCHIERBAUM, Universität Tübingen, FRG

Contents

1.1   Definitions and Classifications

(Bio-)chemical sensors are (miniaturized) devices which convert a chemical state into an electrical signal (Figure 1-1) [1-4]. A chemical state is determined by the different concentrations, partial pressures, or activities (see Table 1-1) of particles such as atoms, molecules, ions, or biologically relevant compounds to be detected in the gas, liquid, or solid phase. The chemical state of the environment with its different compounds determines the complete analytical information. Several more general or more specific definitions have been given for sensors, some of which are summarized in Table 1-2. Most scientists agree that sensors should operate reversibly in the thermodynamic sense. A variety of biosensors, however, does not fulfill this requirement.

The different types of (bio-)chemical sensors may be classified according to the different sensor properties used for the particle detection. Most commonly used properties are conductivities, potentials, capacities, heats, masses, or optical constants which change upon variations in the composition of chemical species interacting with the sensor (Table 1-3).

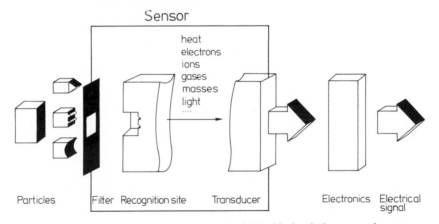

Figure 1-1. Key-lock arrangement of a typical chemical or biochemical sensor to detect atoms, molecules, or ions ("particles") in the gas or liquid phase. Some details of real sensor setups are also indicated. As an example, a filter or membrane may be used to separate the particles to be monitored from interfering particles. The detection process may include a heat or chemical pretreatment which transforms the analyte into detectable particles.

The highly selective and quantitative detection of chemical species is no principal problem. It is possible in well-established analytical chemistry laboratories by using their classical approach of monitoring quantitative specific chemical reactions and/or applying sophisticated spectrometers (see Section 1.4 and Chapter 15). In contrast to this, the detection of chemical species with chemical and biochemical sensors is usually less selective but these devices can have small dimensions and are comparatively cheap. Usually, sensors operate continuously. Examples for discontinuous operation are those (bio-)chemical sensors for the control of biotechnology processes which require repeated sampling and subsequent analyzing.

Many scientist agree that the definition of chemical sensors also includes the classical *detectors,* such as flame ionization detectors in chromatographic setups which detect those invidual

Table 1-1. Common definitions of amounts of chemical species in different environments [3], [5].

Concentrations in gases

Particle concentration

number of particles of gas "i" per total number of particles, eg,

$1\% \quad = 1:10^2$

$1 \text{ ppm} = 1:10^6$

$1 \text{ ppb} = 1:10^9$

Mass concentration $\rho_i = m_i/V$ $\qquad\qquad\qquad\qquad\qquad\qquad\qquad$ (mg/m^3)

mass of gas "i" per $V = 1 \text{ m}^3$ of the gas phase

Partial pressure $\qquad\qquad\qquad\qquad\qquad\qquad\qquad\qquad\qquad$ (Pa, mbar)

the partial pressure p_i of the component "i" is $p_i = (n_i/\sum n_i) \cdot p_{tot}$ with n_i as the number of moles of component "i", $\sum n_i$ as the total number of moles of all components, and p_{tot} the total pressure

Concentrations in solutions

Molarity $c_i = n_i/V$ $\qquad\qquad\qquad\qquad\qquad\qquad\qquad\qquad\qquad$ (mol/L)

number of moles n_i of component "i" per $V = 1 \text{ L}$ of the solution

Normality $c_{eq,i} = z \cdot n_i/V$ $\qquad\qquad\qquad\qquad\qquad\qquad\qquad\qquad$ (mol/L)

number of equivalent moles $z \cdot n_i$ of component "i" per $V = 1 \text{ L}$ of the solution

(an equivalent mole is a mole multiplied by the number z of reacting particles (molecules, atoms, or ions) in the corresponding specific reaction)

Molality $c_{m,i} = n_i/m$ $\qquad\qquad\qquad\qquad\qquad\qquad\qquad\qquad\qquad$ (mol/kg)

number of moles of component "i" per $m = 1 \text{ kg}$ of the solvent

Mass concentration $\rho_i = m_i/V$ $\qquad\qquad\qquad\qquad\qquad\qquad\qquad\qquad$ (kg/L)

mass m_i of component "i" per $V = 1 \text{ L}$ of the solution

(1 mg/L = 1 µg/mL; 1 µg/L = 1 ng/mL)

Volume fraction $\varphi_i = v_i/V$ $\qquad\qquad\qquad\qquad\qquad\qquad\qquad\qquad$ (dimensionless)

volume v_i of component "i" per $V = 1 \text{ L}$ of the solution

(1 µL/L = 1 nL/mL; 1 nL/L = 1 pL/mL)

Mole fraction $x_i = (n_i/\sum n_i)$ $\qquad\qquad\qquad\qquad\qquad\qquad\qquad\qquad$ (dimensionless)

number of moles n_i of component "i" per total number of moles $\sum n_i$ of the solution

Mass fraction $w_i = m_i/\sum m_i$ $\qquad\qquad\qquad\qquad\qquad\qquad\qquad\qquad$ (dimensionless)

mass m_i of component "i" per total mass $\sum m_i$ of the solution

Activities

Some sensors monitor *concentrations,* other thermodynamic *activities* in solutions with

activity a = concentration $c \cdot$ activity coefficient y

The activity describes the effective concentration or active masses of the component at a stoichiometric concentration c. The activity coefficient y varies with the concentration. In an ideal solution, $y = 1$ holds. The activity is defined by the formal Nernst equation to describe the chemical potentials of the species i by $\mu_i = \mu_i^0 + RT \cdot \ln(a_i/a_{i0})$ with $\mu_i^0 = \mu_i$ for $a_i = a_{i0}$ as the standard activity. Often $a_{i0} = c_{i0}$ is chosen. Different diluted ionic solutions have identical activity coefficients if the ion strength of the solution is the same. The ion strength I_i is defined as $I_i = 0.5 \cdot \sum c_i \cdot z_i^2$ with c_i as the concentration of the ion "i" in moles per liter and z_i its charge.

Table 1-2. Common definitions of chemical sensors.

DIN 1319 [6]
The sensor is the first part of a measuring instrument.

VDI/VDE 2600 [7]
The adaptor, sensor, or monitor is the primary part of the transducer, which converts the measuring parameter directly and reacts sensitively to it.

IEC-draft 65/84 [8]
The sensor is the the primary part of a measuring chain which converts the input variable into a signal suitable for measurement.

Mackintosh-report [9]
A sensor is defined as a device which converts a physical or chemical parameter into an electrical signal. With the rapidly falling costs of microelectronic circuitry the need for lower cost sensors is highlighted.

after Tränkler [10]
The sensor is an adaptor which can be prepared in large quantities by means of common microtechnologies. The sensor has a limited accuracy and low price.

after Wartmann [11]
Sensors are setups to determine physical and chemical quantities which are prepared by a variety of technologies and structures.

after Prognos [12]
A sensor is a compact measuring unit which converts binary, physical, chemical, or biological state parameters by means of a transducer into an electrical output. The sensor evaluation electronics for signal processing is incorporated in the same housing. Capsulated sensor elements as well as measuring adaptors are also called sensors even if they don't have an adapting electronics. Finally, also those measuring units are called sensors which include in a compact unit intelligent data evaluation for, eg, linearizing the measuring curve, automatic calibration, or compensation for noise. These may be called intelligent or smart sensors. Sensors for monitoring complex quantities like optical pattern recognition may be called sensor systems.

---

components that have been separated from a mixture in a preceding separation process. These detectors should be non-specific and sensitive to all *components* to be detected. Examples will be discussed in Chapter 15. Other examples are electrodes of liquid electrolyte sensors (Chapter 7). The term *analyzer* usually describes a sensor for the detection of a specific chemical compound. Oxygen analyzers, for instance, may be based upon the paramagnetic oxygen detection (see Chapter 23.6.3) or upon the zirconia-based solid electrolyte cell (see Chapter 8). A few other terms often used in sensor research are given in Table 1-4.

In the general description of sensors for physical, chemical or biological quantities, usually a distinction is made between sensor elements, sensors without special intelligent features ("integrated sensors"), and smart sensors with special intelligent features (see Volume 1, Chapter 12). The latter may include the measuring window, transducer, signal conditioning, intelligent special electronics, and power supply [15] (see Figure 1-2). In this context, chemical sensors may act as the primary part of a measuring chain and as the interface between the analytical informations in the chemical environment and the microprocessors or computers.

In a systematic treatment of sensors, a five-dimensional coordinate system should be envisaged (Figure 1-3). The coordinates describe the quantities to be monitored, the detection

Table 1-3. Characteristic sensor properties monitored by different types of chemical sensors [13] to be treated in the following chapters.

Liquid state electrolyte sensors (Chapter 7):
  Voltages V, currents I, conductivities σ

Solid-state electrolyte sensors (Chapter 8):
  Voltages V, currents I

Electronic conductance and capacitance sensors (Chapter 9):
  Resistances R or conductances G, real (ohmic R) and imaginary part (capacitance C) of frequency-dependent conductances (complex admittances) $\tilde{Y}$

Field-effect sensors (Chapter 10):
  Potentials ΔV, work function changes $\Delta\Phi$

Calorimetric sensors (Chapter 11):
  Heat of adsorption or reaction Q_{ads} or Q_{react}

Optochemical and photometric sensors (Chapter 12):
  Optical constants ε as a function of frequency v

Mass-sensitive sensors (Chapter 13):
  Masses Δm of adsorbed particles

---

principle, the physics as well as physical chemistry background of the sensor, the technology of the sensor manufacturing principle, and the applications.

In physical systems, the number of independent physical quantities monitored by mechanical, thermal, magnetic, or optical sensors is in the order of hundreds. In chemical and biochemical systems, however, the number of measurands to be monitored is many orders of magnitudes larger because of the huge amount of different chemical compounds that can occur in gases, liquids, and solids. As a result, physical sensors may be classified systematically according to their measuring parameters as it is done, for instance, in the corresponding volumes of the present book series. Because of the huge number of measuring parameters for chemical sensors, however, they may at best be classified according to their detection principles. These principles define the different basic sensors in Chapters 7-13 and the different types of biosensors described in Chapter 14.

Table 1-4. Alternative terms used in describing sensors and sensor systems [14].

Device	Application	Mode of action and operation [a]
Probe	Single shot assay	A or U
Dipstick	Single shot assay	A or U
Dosimeter	Measurement of exposure to chemical species or radiation	I [b], A
Badge	Usually a personal dosimeter	I, A
Alarm device	Gives a signal when a present threshold is exceeded	R or I, U
Monitor, detector, analyzer	Sensor, dosimeter, or alarm device	I, or R, A or U

(a) R, reversible; I, irreversible; A, attended operation; U, unattended operation
(b) Can be made a quasi-sensor, if the sensing area is steadily renewed

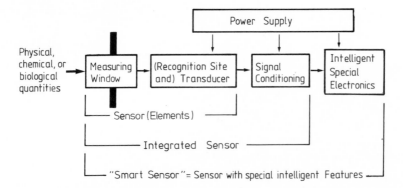

Figure 1-2. Schematic representation of general sensor elements and sensors. The sensor consists of the transducer, which transfers physical, chemical, or biological quantities through the measuring window to the sensor element. The latter transforms non-electric p_i into electric signals x_i' which are conditioned subsequently to x_i. Further processing with intelligent special electronics is done in "smart sensors" [12].

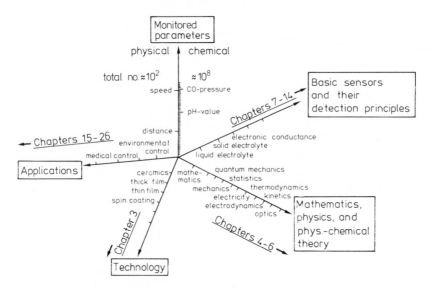

Figure 1-3. Five important aspects to describe sensors. Reference is also given to different chapters in this volume.

1.2 Typical Examples

Research and development of chemical and biochemical sensors are characterized by a huge gap between the large amount of new ideas and prototypes utilizing new principles of chemical sensing on the one hand and a very limited number of practically important sensors which are currently manufactured in large quantities on the other hand.

Typical examples for commonly used chemical sensors are

— the amperometric CO-sensor based upon current measurements in liquid electrolytes to monitor air quality (Figure 1-4),
— the Lambda-probe based upon potential measurements using solid electrolytes to detect oxygen in the car-exhaust (Figure 1-5),
— the Taguchi SnO_2 sensor based upon conductance measurements using oxides to detect reducible gases in gas warning systems (Figure 1-6),
— the pellistor based upon measurements of reaction heats using oxide catalysts to monitor reducible gases (Figure 1-7),
— the pH-sensitive electrode based upon potential measurements using gel-type glasses to monitor hydrogen ions in water (Figure 1-8),
— the ion-sensitive field effect transistor (ISFET) with an oxide gate to monitor hydrogen ions in water (Figure 1-9),
— the Clark electrode to monitor oxygen in water (Figure 1-10), or
— the fluorosensor to monitor NADH in water (Figure 1-11).

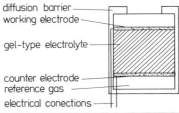

diffusion barrier
working electrode
gel-type electrolyte
counter electrode
reference gas
electrical conections

Figure 1-4.
Schematic setup (top) and photograph (bottom) of an amperometric CO-sensor. (Published with kind permission of Bayer-Diagnostic + Electronic GmbH, Compur Monitors, Munich, FRG).

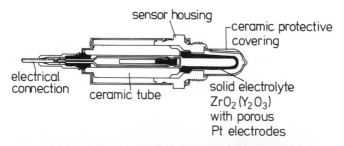

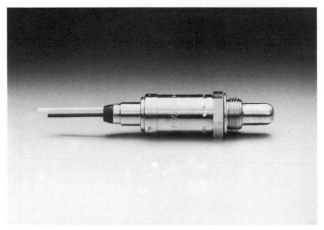

Figure 1-5. Schematic setup (top) and photograph (bottom) of a ZrO_2-based Lambda-probe to measure the oxygen partial pressure. (Published with kind permission of Bosch GmbH, FRG).

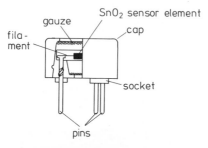

Figure 1-6.
Schematic setup (top) and photograph (bottom) of an SnO_2-based Tagushi sensor to monitor reducible gases in air. (Published with kind permission of Figaro Inc., Japan).

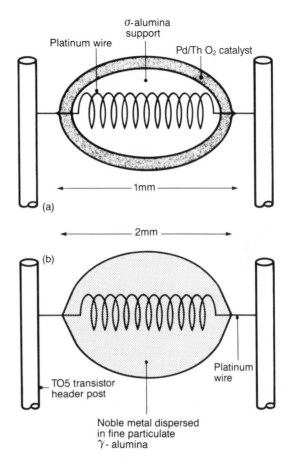

Figure 1-7. Schematic setup (top) and photograph (bottom) of pellistors based upon measurements of reaction heats to monitor reducible gases. (Published with kind permission of Health & Safety Executive, Sheffield, UK).

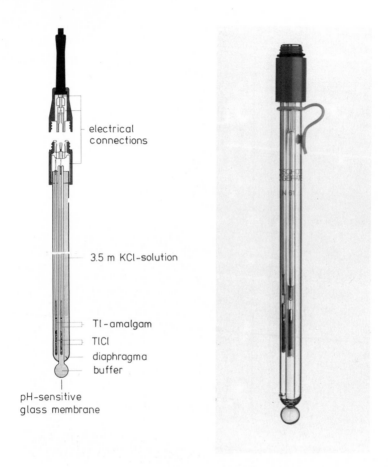

Figure 1-8. Schematic setup (left) and photograph (right) of a pH-sensitive glass electrode to determine H^+-concentrations in water. (Published with kind permission of Schott GmbH, FRG).

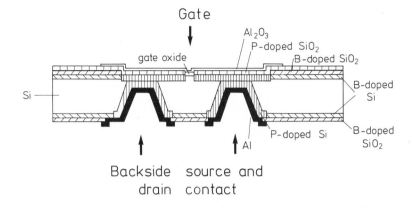

Gate

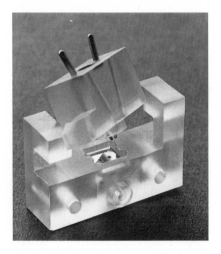

Figure 1-9. Schematic setup (top) and photograph (bottom) of an ionsensitive field effect transistor (ISFET) to monitor the concentration of hydrogen ions in water in a flow-through cell. ISFETs can also be used as basic structures for biosensors. (Published with kind permission of Sentron, NL).

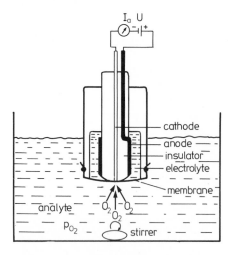

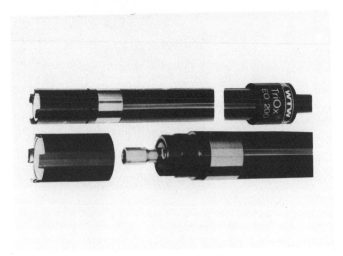

Figure 1-10. Schematic setup (top) and photograph (bottom) of a Clark electrode to monitor oxygen in water. (Published with kind permission of Wissenschaftlich-Technische Werkstätten GmbH, FRG).

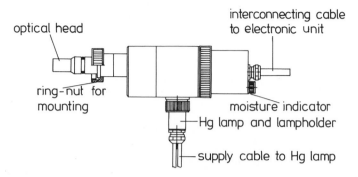

Figure 1-11. Schematic setup (top) and photograph (bottom) of a fluorosensor to monitor NADH (nicotinamide-adenine-dinucleotid) in bioprocesses. (Published with kind permission of Dr. W. Ingold AG, Switzerland).

In order to provide mechanical and to some extent chemical ruggedness of the sensor system, the sensing element is usually encapsulated and is also provided with membranes as diffusion barriers, and/or dust protections. These additional components might enhance the selectivity. The very least requirement is, however, that they do not suppress the required detection of the particular chemical species.

1.3 Application Fields and Sensor Markets

Typical application fields of (bio-)chemical sensors are summarized in Table 1-5. Future markets are extremely difficult to predict carefully. Following one of the many recent predictions on future markets for the two specific examples "environmental protection" and "medical applications", the worldwide increasing demand (in million DM) is listed for the years 1988–2000 in Tables 1-6 and 1-7. For further details, see eg, [12].

Table 1-5. Typical application fields for (bio-)chemical sensors [16], [17]

Environmental and immission control (air, water, soil, ...)
Working area measurements (workplace, household, car, ...)
Emission measurements (car, car exhaust, waste water, ...)
Fire warning and safety control (household, mining, laboratory, tunnel, hotel, ...)
Breathing gases: control and regulation (living rooms, diving or medical equipment, airplanes, ...)
Household appliances: control and regulation
Car engines: control and regulation
Process control and regulation (biotechnological and chemical plants, fermentation processes, general chemical processes, drying, ...)
Chemical and biochemical analysis
Medical applications (clinical diagnostics, prosthetics, anesthetics, veterinary, ...)
Agriculture (analysis in agriculture and gardening, detection of pesticides, ...)

Table 1-6. Worldwide demand of sensors for medical applications 1988–2000 [12]. Numbers denote units of 10^6 DM

	1988	1990	1995	2000
Diagnostics:				
ultrasound sensors	5.0	6.0	9.5	13.0
optical sensors	8.0	11.5	20.0	35.0
electrochemical sensors	40.0	57.0	110.0	170.0
biosensors	64.0	305.0	490.0	710.0
total	117.0	379.5	629.0	932.0
Control				
transcutaneous electrodes	102.0	123.0	181.0	266.0
breathing gas sensors*	135.0	166.0	261.0	390.0
optical sensors	140.0	163.0	218.0	278.0
pH-sensors (electrodes)	18.0	23.0	40.0	67.0
pressure sensors	40.0	48.0	77.0	110.0
temperature sensors	25.0	27.0	43.0	55.0
electrochemical* and fluorescence sensors*	18.0	21.0	30.0	39.0
ionsensitive sensors (ISFETs)	8.0	11.5	23.0	40.0
flow-through sensors	2.0	3.0	5.0	7.0
biosensors	30.0	334.0	388.0	450.0
total	496.0	819.5	1266.0	1702.0

* for O_2 and CO_2

Table 1-7. Worldwide demand of sensors for environmental protection 1988–2000 [12]. Numbers denote units of 10^6 DM

	1988	1990	1995	2000
Emission and immission measurent in gases and air				
electrochemical sensors	30	40	86	162
semiconducting sensors	8	15	80	200
phys.-chem. sensors	83	105	150	180
biosensors	–	–	4	8
total	121	160	320	550
Waste water and environmental control				
electrochemical sensors	20	24	65	130
semiconducting sensors	–	5	15	70
phys.-chem. sensors	9	11	20	35
biosensors	–	10	30	55
total	29	50	130	290
total both	150	210	450	840

1.4 Chemical Sensing and Analytical Chemistry

The detection of particles is a particular case of the more general analytical task of chemists to obtain qualitative or quantitative time- and spatial-resolved information about specific chemical components (Figure 1-12 and Chapter 15). Examples for quantitative informations are concentrations, activities, or partial pressures (for details, see Table 1-1) of specific components or in some cases of classes of chemical compounds with similar properties such as their flammability in air. Examples for qualitative informations are the presence or absence of certain molecules, the presence of particles exceeding a certain threshold-limited value (TLV), or the lower explosive limits (LEL) of combustible gases.

All of these informations can be obtained from a complete chemical analysis system. It generally requires [19], [20]

– sampling (step 1),
  (selecting increments of the bulk material and combining them (gross sample), dividing sub-sample and analysis sample)
– sample transport (step 2),
– sample pretreatment (step 3),
– separation of the components (step 4)
  (chemical separation, ion exchange, electrolysis and electrophoresis, destillation, crystallisation, sublimation extraction, liquid and gas chromatography)
– detection (step 5),
  (titrimetry and gravimetry, electroanalytical methods such as electro-gravimetry, conductimetry, potentiometry, voltammetry, thermoanalytical methods such as calorimetry and

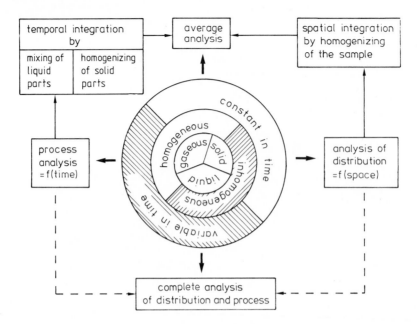

Figure 1-12. Basic tasks in analytical chemistry to obtain time- and spatial-resolved and/or -integrated analytical informations from gaseous, liquid, or solid samples for different sample states. After [18].

thermal analysis, and spectroanalytical methods using photons, electrons, neutrons, ions, and neutral particles as probes such as colorimetry and spectrophotometry, spectrofluorometry, infared spectrophotometry, atomic emission spectroscopy, atomic absorption and flame emission spectroscopy, mass spectrometry, ultraviolet spectroscopy, nuclear magnetic resonance spectroscopy)
— data treatment (step 6)
(statistical evaluation to determine errors, accuracy, precision)
— data interpretation (step 7).

An overview of typical detection ranges of several different instrumental analytical methods is shown in Figure 1-13 [18].

Obviously, specific chemical sensors may perform sucessfully the step 5. If specificity for a particular component is realized (compare Figure 1-1), these sensors do not require additional separation (step 4) of the different interfering chemical components which are usually present in the sample. In some applications chemical sensors may operate directly in the environment and these general considerations are not as crucial. An example is gas monitoring in alarm systems of households.

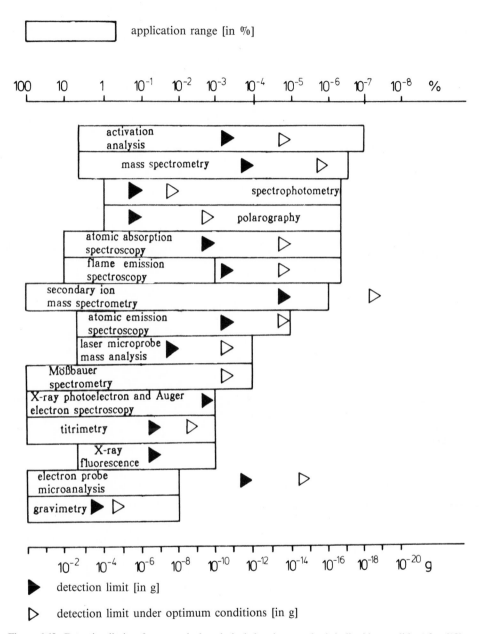

Figure 1-13. Detection limits of some typical analytical chemistry methods in liquids or solids. After [18].

1.5   Sensor Parameters

To simplify the following discussion, we generally use "p_i" to describe any analytical information (some of which are listed in Table 1-1) of the specific compound "i".

We first focus at the measuring signals x' of invidual sensors and at the corresponding information deduced therefrom. Practically important sensor parameters are summarized in Table 1-8.

Table 1-8. Practically important sensor parameters [16]. Examples of parameters are given in parentheses.

General operating data
Measuring parameter [O_2 concentration or partial pressure, ...]
Measuring medium [air, water, ...]
Measuring principle [electrochemical, amperometric, ...]
Technology and experimental set-up of sensor or sensor system
Measuring range [ppm]
Measuring accuracy [± ppm]
Reproducibility [± ppm]
Resolution [± ppm]
Warm up time [s]
Response time [s]
Decay time [s]
Maximum measurement frequency [Hz]
Linearity of output signal [%]
Selectivity and cross-sensitivity towards other substances and their influence on the measuring parameter
[eg, ppm of disturbance variable related to 1% of measuring parameter]
Possibility to sterilize (specific characteristics of biochemical sensors)

Operation conditions
Maximum and minimum values for:
− relative humidity [from/to % r.h.]
− temperature [from/to °C]
− range of concentration for component detection [from/to ppm]
− range of concentration for other components [from/to ppm]
− volume of flow and amount of gas or solvent to use
− turbidity and coloration of solution (specific characteristics of biochemical sensors)

Requirements regarding
− power supply/battery stability [± %]
− accuracy of flow [± %]
− temperature of flow [°C]

Stability
Drifts of zero point [%/h]
Drifts of sensitivity [%/h]
Lifetime [h]
Contamination
Regeneration after contamination
Regeneration after pollution
Time between maintenance checks [h]

Table 1-8. Continuation

Calibration
Type of calibration
Calibration drift [%/h]
Calibration intervals [h]
Required equipment for calibration (eg, standard gases or buffering solutions)

Connections
Display
Error signal
Required equipment for complete sensor system (eg, standard gases or buffering solutions)
Possibility of measurement of several components
Measuring connections
Calibration connections
Electrical connections
Power consumption
Other consumption materials (gases, solvents, …)

Measuring outputs
Type of measuring outputs
Maximum contact power
Alarm signal
Output for multiple component measuring
Sensitivity (% of full deflection)
Linearity

General Data
Dimensions of the sensor system ($l \times w \times h$ in mm)
Dimensions of the sensor ($l \times w \times h$ in mm)
Weight
Mode of protection
Registration
Price (per unit)
Delivery time (week)
Delivery size

For an unequivocal characterization and test of their static and dynamic properties, the sensors have to be calibrated and tested in experimental setups which make it possible to perform fast changes of "p_i" ie, of the concentration, partial pressure, etc (see Chapter 16). Typical response signals $x'(t)$ of chemical sensors during their exposure to certain amounts $p_1(t)$ and $p_2(t)$ of atoms, molecules, or ions are indicated in Figure 1-14. Characteristic static and dynamic sensor parameters may be deduced from these measurements. They are described in more detail in Chapter 6 and will therefore only be summarized here. We assume fast stepwise increases and decreases of the partial pressures p_1 and p_2 of two different components "1" and "2" (Figure 1-14a, during times "A" and "B"). Figure 1-14b characterizes the reponse of a "reliable" sensor which shows a constant zero level without drifts ("C") in the absence of detectable components, a fast rise $(dx'/dt)_{t_0}$ upon gas exposure at t_0, a short overall response time Δt or 90% response time $\Delta t_{90\%}$ upon exposure to p_1 ("D"), a stable measuring signal x' ("E"), which is an unequivocal function of p_1, and a fast signal decay to the same starting signal after removing p_1 ("F"). Response times are generally defined as the times to achieve a certain percentage of the final change in the sensor signal, eg, 90%. In the case of nonspecific sensors, the sensor also yields a signal in the presence of another component "2" ("G") introduced at t_0'. It should be mentioned that signals of non-specific sensors may also be evaluated to determine gases specifically. This, however, requires a more sophisticated signal evaluation by means of pattern recognition methods as it is used in multicomponent analysis. This will be described in Chapter 6. Drifts of the baseline after switching on the sensors ("H") and during sensor operation over longer times ("I") as shown schematically in Figure 14c indicate typical practical problems which have to be minimized and/or compensated for.

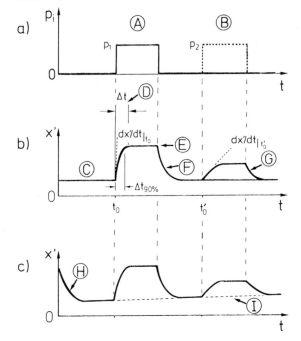

Figure 1-14.
a) Fast stepwise increases and decreases of the partial pressures p_1 and p_2 of two different components "1" and "2"; b) and c) typical measuring signals deduced from different chemical sensors as the result of changes according to a). Further explanations are given in the text.

Mathematical transformations of the sensor signals x' to "adapted" sensor signals x make possible an adaption of the data evaluation to the specific measuring problem (for details, see Chapter 6.2). This signal preprocessing aims at faster, more reliable, and more direct data evaluations. As a simple example different transformations may be performed experimentally by adjusting different electrical operation modes to the sensor. To be more specific, if the sensor signal is based upon conductance changes, operation in the constant voltage or current mode generates inverse response signals and yields resistances R or conductances G, respectively. As indicated in Figure 1-14b, independent signals may also be deduced from the slopes $(dx'/dt)_{t_0}$ or $(dx'/dt)_{t_0'}$ obtained after exposure of the sensor to component "1" or "2" at t_0 or t_0' or from the time-dependent relaxation of the sensor signal $x'(t)$. This requires, however, the sensor to be operated in a pumped flow setup with the gas switched stepwise between the unknown gas and a reference gas.

For further details of experimental and mathematical methods to transform signals, the reader is referred to an extensive literature [21].

Calibration Curve and Partial Sensitivity

In the simplest case, the calibration curve of a sensor with respect to the monitored species "1" is unequivocally given by

$$x' = f(p_1) \tag{1-1}$$

as indicated in Figure 1-15. In this particular case, the sensor signal for $p_1 = 0$, ie, in the absence of component "1", is x'_0 (zero level). The partial sensitivity is defined by the slope

$$\gamma_1 = \frac{\partial x'}{\partial p_1}. \tag{1-2}$$

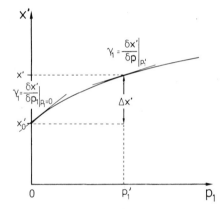

Figure 1-15.
Typical calibration curve of a chemical sensor indicating the measuring signal x' after exposure of different concentrations or partial pressures p_1 of a component "1". Further explanations are given in the text.

For non-linear functions $x' = f(p_1)$, the value γ_1 depends on p_1 as shown in Figure 1-15 for $p_1 = 0$ and $p'_1 > 0$. This is the case for most practical sensor devices. An example of great practical importance is the logarithmic response described by the Nikolsky equation

$$x'_i = V_i = RT/nF \; [\ln \, (a_i + \sum_j k_{ij}^{m/n} \, a_j)] \tag{1-3}$$

[2] for potentiometric electrolyte sensors (see Chapter 7 and 8). Here, V_i denotes the voltage-signal of the potentiometric electrolyte sensor, R the gas constant, T the temperature, k_{ij} the selectivity to monitor ions "i" in the presence of ions "j" (see Equation 1-8), $n(m)$ and $a_{i(j)}$ the number of elementary charges and activities of the $i(j)$-th ion component and F the Faraday constant. In this case, partial sensitivities increase with decreasing activities.

In simple cases, the calibration curve of a sensor is linear. This holds, eg, for amperometric liquid or solid electrolyte sensors, and allows for a simple calibration procedure. For specific sensors, the linear response requires

$$x' = \gamma_1 p_1 \tag{1-4}$$

with constant partial sensitivities $\gamma_1 = (\partial x'/\partial p_1)$ independent of the concentration or partial pressure p_1 and not influenced by other partial pressures $p_{j \neq 1}$.

In the more general case, the measuring signal x' is a multi("n")-dimensional function of the analytical information p_i of all detectable components "i" with $1 \leq i \leq n$. For the simple extension from one to two components this is illustrated in Figure 1-16a. From the calibration curve, all important data for the evaluation of sensor parameters can be deduced provided that the signals are reproducible.

Reversibility

For a simple data evaluation, we require time-independent and stable calibration curves. This implies the measuring signal to be a state function in the thermodynamical sense of *reversibility* [7]. Over large ranges of pressures and temperature this is usually not fulfilled. For limited ranges and restricted combinations of partial pressures, however, and for a certain temperature range the values x' are often found to be independent of the history of the sensor. This situation is shown in Figure 1-16a. Under these conditions variations of x' are described by

$$dx' = \left(\frac{\partial x'}{\partial p_1} \right)_{p_{j \neq 1}, T} dp_1 + \left(\frac{\partial x'}{\partial p_2} \right)_{p_{j \neq 2}, T} dp_2 + \ldots + \left(\frac{\partial x'}{\partial p_i} \right)_{p_{j \neq i}, T} dp_i$$

$$\ldots + \left(\frac{\partial x'}{\partial T} \right)_{p_j} dT \tag{1-5}$$

and hence

$$\oint dx' = 0 \tag{1-6}$$

must hold. The integral extends over the different ranges of p_i and the sensor operation temperature T of reversible sensor response. The empirical determination of these ranges of complete reversibility of a sensor is a very important first task to test its potential future applications.

Specificity

If only one of the different partial sensitivities $\partial x'/\partial p_i$ is large if compared with all the others, we have a *specific* sensor which makes possible to quantitatively determine the concentration or partial pressure p_i of one component "i".

Sensitivity and Cross-sensitivity

In contrast to the dimension-dependent partial sensitivities $\gamma_i = (\partial x'/\partial p_i)$, the sensitivity S_i is usually defined as a dimensionless measuring signal referred to a reference value x'_0.

$$S_i = x'/x_0' \quad \text{or} \quad S_i = \Delta x'/x_0' = 1 - x'/x_0' \tag{1-7}$$

for a given concentration or partial pressure p_i as indicated for $i = 1$ in Figure 1-15. The reference value x_0' may be defined by the sensor signal in the absence of the component to be monitored or by the signal for a certain standard concentration of this component. As indicated in Figure 1-16b cross-sensitivity results from the lack of specificity of the sensor and leads to changes of the calibration curve $x' = f(p_1)$ in the presence of the interfering component "2". In this particular case, the partial sensitivities γ_1 are also affected (Figure 1-16c).

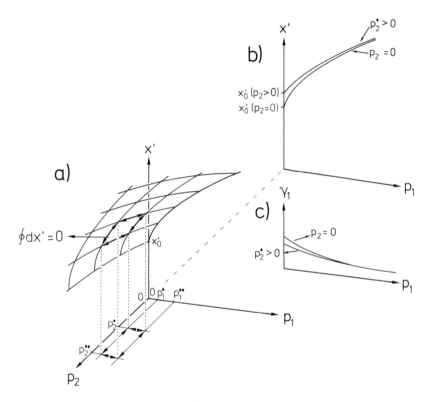

Figure 1-16. a) The measuring signal x' of a non-specific chemical sensor as a function of concentrations or partial pressures p_1 and p_2 of two components "1" and "2" in a three-dimensional plot. b) Projection of x' in the x'-p_1-plane indicating the calibration curve $x' = f(p_1)$ in the absence of component "2", ie, for $p_2 = 0$, and in the presence of "2", ie, for $p_2^* > 0$. c) Partial sensitivities $\gamma_1 = \partial x'/\partial p_1$ for $p_2 = 0$ and for a selected value $p_2^* > 0$ obtained from the calibration curve $x' = f(p_1, p_2)$.

Selectivity

Selectivity of a single sensor is usually defined by the ratio of the different (partial) sensitivities related to the species to be monitored and to the interfering components. As an example, the selectivity of potentiometric electrolyte sensors with their logarithmic response (see Equation 1-3) is determined by

$$k_{ij} = f(V_i \, [a_i = \text{const}, \, a_j] \tag{1-8}$$

with V_i as the sensor signal (voltage) for the ion "i" with activity a_i in the presence of interfering ions "j". For highly specific potentiometric electrolyte sensors, $k_{ij} = 0$ holds. Details are described in Chapters 5 and 7.

General definitions of the terms *specificity, sensitivity,* and *selectivity* will be introduced unequivocally for sensor arrays in Chapter 6.

Instabilities and Drifts

Instabilities and drifts are serious problems in almost every application of chemical sensors affecting largely the accuracy in the determination of the analytical information. Non-cumulative drifts denote statistical variations of the sensor signal. Cumulative drifts lead to irreversible changes of the calibration curves and can result for example from a "poisoning" of the active sensor area with undesired chemical compounds. In contrast, short-term drifts may be observed after switching-on (eg, Figure 1-14c, "H"). They are caused by the time required to establish steady-state conditions, such as a constant operation temperature of the sensor. Thermal drifts are related to changes of the sensor signal upon variations of the ambient temperature and may be suppressed by keeping the sensor operation temperature constant. Long-term drifts are usually related to the stability of the zero level or the base line. A particular example is indicated in Figure 1-14c ("I") which is independent of the presence of the component "1" or "2". In this situation, the reference value x_0' has to be recalibrated between different detection cycles. One experimental method to suppress these drift effects in the subsequent data evaluation involves reference measurements by comparing signals of an active sensor with those of an inactive sensor with both having the same drift properties.

Error and Accuracy

The accuracy and precision of analytical procedures are determined by systematic and statistical errors. In contrast to systematic errors (errors of the measuring procedure itself) which in principal can be corrected for or even be avoided, the statistical errors are caused by interferences in the information transfer during the analysis. Depending on their influence on the analytical information, we distinguish mathematically between constant, multiplicative, and non-linear errors.

The distribution of the measuring signals x' in a series of measurements is often a Gaussian distribution function as shown in Figure 1-17 with

$$P(x') = \frac{1}{\sigma \sqrt{2\pi}} \cdot \exp\left[-\frac{(x' - \mu)^2}{2\sigma^2}\right]. \tag{1-9}$$

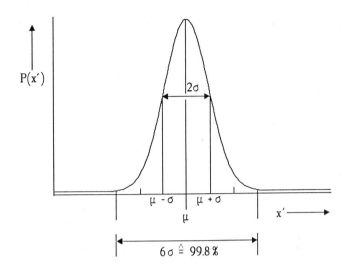

Figure 1-17. Gaussian function $P(x')$ to describe normal distributions of x' [18].

For large numbers of measuring signals ($N \to \infty$), the maximum of the function becomes exactly the average value μ and corresponds to the most probable value $P_{max}(x')$. The scattering of the measuring signals x' is described by the standard deviation σ of the distribution. For statistical reasons, μ and σ cannot be calculated exactly from a limited number of measuring signals and

$$\bar{x}' = \frac{1}{N} \sum_n x' \approx \mu \text{ and} \tag{1-10}$$

$$s = \sqrt{\frac{1}{f} \sum_n (x' - \bar{x}')^2} = \sigma \tag{1-11}$$

holds with $f = N - 1$ as statistical degree of freedom. Minima in the absolute (s) or relative standard deviations ($s_r = s/\bar{x}'$) characterize optimum performance of the analytical procedures for a given range of concentrations. They are determined by the sum of all individual errors.

Analytical results in the determination of the average value $\bar{x}'$ are only valid within a specific confidence range $\varDelta$

$$\varDelta (\bar{x}') = \frac{s \cdot t (P, f)}{\sqrt{N}} . \tag{1-12}$$

This value depends on the statistical probability P (usually chosen as 99.8% for $|x' - \mu| < 3\sigma$). The parameter $t(P, f)$ is calculated for the degree of freedom f and the probability P and is summarized in statistical tables [18].

Limits of Detection and Determination

In analytical chemistry different definitions are used for the minimum concentration or partial pressure of a component "k" detectable with different statistical probabilities as demonstrated in Figure 1-18.

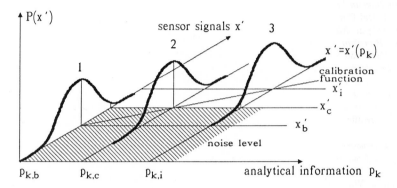

Figure 1-18. Schematic representation of different statistical definitions of minimum detectable concentrations or partial pressures p_k of a component "k" with a sensor in the case of a linear calibration function. The probabilities $P(x_i')$ are Gaussian functions centered at three different levels x_b', x_c', x_i' of the component "k"[18].

The component "k" with a concentration or partial pressure "$p_{k,b}$" is analyzed as "absent" with a statistical probability of 99.8%. At the detection limit "$p_{k,c}$", the component is detected only with a probability of 50%, whereas the determination at "$p_{k,i}$" is characterized by a high probability of 99.8%.

Dynamic Range

The dynamic range or operation range of the sensor is characterized by the lowest detectable amount of a component, which is given by the noise of the sensor signal, and by the maximum detectable amount which is for example restricted by saturation effects of the sensor. Sensors with logarithmic response functions show particularly high dynamic ranges. Their partial sensitivities and hence their resolution decrease with increasing concentration.

1.6   References

[1] Moseley, P. T., Tofield, B. C., *Solid State Gas Sensors,* Bristol: Adam Hilger, 1987.
[2] Janata, J., *Principles of Chemical Sensors,* New York and London: Plenum Press, 1989.
[3] Oehme, F., *Chemische Sensoren,* Braunschweig: Vieweg, 1991.

[4] Madou, M. J., Morrison, S. R., *Chemical Sensing with Solid State Devices,* San Diego: Academic Press, 1989.

[5] Willard, H. H., et al., *Instrumental Methods of Analysis,* Belmont California:, Wadsworth Publishing Comp., 1981.

[6] DIN 1319, *Grundbegriffe der Meßtechnik,* Berlin: Beuth, 1985.

[7] VDI/VDE 2600, Blatt 3, *Gerätetechnische Begriffe,* Düsseldorf: VDI-Verlag, 1973.

[8] *Terms and Definitions in Industrial Process Measurement and Control,* (IEC-draft 65/84), International Electrotechnical Committee, 1982.

[9] "Low cost sensors", in: *Mackintosh State of the Art Series,* Luton, UK: Benn Electronics,

[10] Tränkler, H. R., *Proceedings of the VDI-Congress,* Bad Nauheim, FRG, 19. – 21. 3. 1984.

[11] Wartmann, G., Hard and Soft, Fachbeilage Mikroperipherik, Düsseldorf: VDI-Verlag, 9 (1989).

[12] Schröder, N., Zerressen, A., *Sensortechnik 2000.* Prognos Weltreport, Technologieanalyse und -prognose der Sensoren bis zum Jahr 2000, Basel, Switzerland: Prognos, 1988.

[13] Göpel, W., Solid-State Chemical Sensors: Atomistic Models and Research Trends, *Sensors and Actuators* **16** (1989) 167–193.

[14] Wolfbeis, O. S., Chemical Sensors – Survey and Trends, *Fresenius J. Anal. Chem.* **337** (1990) 522–527.

[15] Schröder, N., *Eurosensors IV,* Karlsruhe (FRG) 1990 and *Sensors and Actuators,* to be published.

[16] Infratest Industria, *Chemische und Biochemische Sensoren,* Marktübersicht, Infratest Industria, München, 1989.

[17] Oehme, F., Göpel, W., *Chemische Feldeffekt-Sensoren,* Hard and Soft, Fachbeilage Mikroperipherik, Düsseldorf: VDI-Verlag, 1987.

[18] Danzer, K., et al. *Analytik,* Stuttgart: Wissenschaftliche Verlagsgesellschaft, 1987.

[19] Graber, N., Lüdi, H., Widmer, H. M., The Use of Chemical Sensors in Industry, *Sensors and Actuators,* **B1** (1990), 239–243.

[20] Jeffery, G. H., et al., *Textbook of Quantitative Chemical Analysis,* Essex, England: Longman Scientific & Technical, 1989.

[21] Profos, P., *Handbuch der industriellen Meßtechnik,* Essen: Vulkan-Verlag, 1987, pp. 142.

2 Historical Remarks

Section 2.1: WOLFGANG GÖPEL, Universität Tübingen, FRG
Section 2.2: T.A. JONES†, Health and Safety Executive, Sheffield, UK,
        WOLFGANG GÖPEL, Universität Tübingen, FRG
Section 2.3: JAY N. ZEMEL, University of Pennsylvania, Philadelphia, PA,
        USA
Section 2.4: TETSURO SEIYAMA, Tokuyama Soda Co., Fukuoka, Japan

Contents

2.1   Overview of the History and Early Electrochemistry Development

Chemical and biochemical sensors have a long history which can be traced back to the turn of the century. Their research and development (R&D) has really boomed during the past ten years.

Determining parameters have been:

- the discovery of new sensor principles, new sensor materials, and new sensor preparation technologies,
- the evolution of different environmental problems and subsequent major national funding programs and protective laws,
- mid-term strategies of major companies to support research and development.

This chapter is not aimed at a comprehensive description of historical developments and the present state of R&D activities. For details on both topics, the reader is referred to a variety of recent textbooks or market surveys.

In an attempt to give a brief overview of the history including the early electrochemistry development Table 2-1 shows a few milestones. It covers the invention of the sensor principles as well as data on the first practical sensors.

Over 30 years have passed since the appearance of "real" chemical sensors. The devices, which in the beginning were called by a variety of names, gradually came to be known by the single term "sensor" and a classification of sensors into physical and chemical types has been established. The chemical sensor has made great advances and has taken root in human life and industry as a feature of modern technology. Moreover, remarkable progress is expected in the future. Numerous chemical sensors of various types were proposed in the early 1960s. Then, in the 1970s, new sensing devices such as the ion-sensitive field effect transistors (ISFET) were proposed and at the same time some of the sensors proposed earlier began to be produced commercially. This period may be considered to be the first stage of the chemical sensors development which culminated in the first International Meeting on Chemical Sensor held in Fukuoka in 1983. Success in this stage still required the effective combination of basic research, high technology, and strong effort in basic science.

The worldwide current activities in chemical and biochemical sensor development and applications may be characterized in many ways. In an attempt to give a brief overview, scientific papers in the currently most popular journal *Sensors and Actuators* published between 1988 and 1990 and in recent conference proceedings of international sensor conferences [49] in 1990 are listed in Table 2-2. These papers cover those different basic sensor principles according to which the following Chapters 7 to 14 of these books are organized.

Characteristic of the present situation in chemical and biochemical sensors R&D is the huge range between extremely optimistic predictions about potential sensor markets in the future and more pessimistic predictions. In all surveys and studies reliable sensors are considered as the bottleneck for future applications of information technologies to control chemical and biochemical environments.

Because of different traditions, environmental problems, and technological developments, different evolutionary waves could be observed in the development of chemical sensors in Europe, in the United States and Canada, and in Japan. With the aim of characterizing briefly these developments, different authors from Europe, the US, and Japan will describe in Sections 2.2 to 2.4 their specific view of significant historical developments in the different countries.

Table 2-1. History of chemical and biochemical sensors (survey on significant inputs).

Year	Type of sensor	Investigator	Reference
1885	2-Electrode cell with platinized platinum electrodes	Kohlrausch, F.	[1]
1888	Metal electrodes in their salt solutions	Nernst, W.	[2]
1897	Hydrogen electrode for pH-monitoring	Böttger, W.	[3]
1904	Hot wire sensor for gas analysis (heat conductance)	MAN company	[4a]
1906	Glass electrode	Cremer, M.	[4b]
1909	pH-determination based upon color indicators	Sörensen, S.	[5]
1913	2-electrode cell Cu/Pt for measuring Cl in solution	Rideal, E.K., Evans, U.R.	[6]
1922	Mercury drop electrodes in polarography	Heyrovsky, J.	[7]
1923	Catalytic combustion-type sensor	Jonson	
1925	Antimony electrode for pH-monitoring	Kolthoff, I.M., Hartong, R.D.	[8]
1928	2-electrode cells Zn/Ag for monitoring dissolved oxygen	Tödt, F.	[9]
1929	Practical use of glasselectrodes for pH-monitoring	MacInnes, D. A., Cole, M.	[10]
1933	Type-2 electrodes for monitoring anions	LeBlance, M., Harnapp, O.H.	[11]
1938	Humidity sensor using LiCl films	Dunmore	
1941	Paramagnetic oxygen sensor	Klauer, F., Turowski, E., Wolff, V.,	[12]
1946	Glassbar-refractometer	Karrer, E., Orr, S.	[13]
1952	Galvanic cell-type gas sensor	Hersch	
1957	Membrane coated 2-electrode cell for monitoring dissolved oxygen	Clark, L. C.	[14]
1957	Theory of electromotive forces of a solid electrolyte cell	Wagner, C.	
1958	Membrane-coated pH glass electrode for CO monitoring in blood	Severinghaus, W., Bradley, A. F.	[15]
1958	Flame ionization detector for gas-chromatography	Harley, J., Nel, W., Pretorius, V.	[16]
1959	Thermocatalytical sensor for combustable gases (wire)	Sieger, J.	[17]
1960	Electron capture detector for gas-chromatography	Lovelock, J.E., Lipsky, S.R.	[18]
1960	Photoionization detector for gas-chromatography	Lovelock, J. E.	[19]
1961	Ion-conducting solid electrolyte sensor for gaseous oxygen	Weissbarth, J., Ruka, R. and Peters, H., Moebius, H. H.	[20] [21a]
1961	Ion electrode sensor	Pungor, E.	[21b]
1962	Thermocatalytical sensor for combustable gases (pellistor)	Baker, A. R.	[22]
1962	Basic concept of biosensors: enzyme electrodes	Clark, L. C., Lyons, C.	[23]
1962	Oxide semiconductor-type gas sensor	Seiyama, T.	[24]
1964	Ion-selective electrodes with Silicon/rubber-membranes	Pungor, E., Toth, K.	[25]
1964	Piezoelectric quartz crystal sensor	King, W. H.	[26]

Table 2-1. continued

Year	Type of sensor	Investigator	Reference
1965	Ion-selective electrodes with solid state electrolytes (ceramic pellets)	Riseman, J., Wall, R. A.	[27]
1965	Practical use of catalytic combustion-type sensor	Riken-Keiki Comp.	
1966	Fluoride-selective electrodes based upon LaF_3	Frant, M., Ross, J.	[28]
1967	Glucose sensor	Updike, S. J., Hicks, G. P.	[29]
1967	Ion-selective electrodes with liquid gel membranes	Ross, J.	[30]
1967	Practical use of oxide semiconductor-type gas sensor	Figaro Eng. Inc.	
1967	Practical use of ion electrode sensor	Metrimpex Co.	
1967	SnO_2-based conductance sensor	Taguchi, K.	
1970	Practical use of semiconductor gas sensors based upon SnO_2	Tagushi, K.	[31]
1970	3-electrode cell with catalytic working electrode for CO-monitoring	Energetic Sciences Comp.	[32]
1970	Chemically sensitive field effect transistor	Bergveld, P.	[33]
1970	Optical fiber gas sensor	Harsick	
1972	Gassensitive electrodes for monitoring dissolved gases	Frant, M., Ross, J., Riseman, J.	[34]
1972	Piezo-immunotest	Shons, A., Dorman, F., Najarian, J.	[35]
1974	Practical use of electrochemical gas sensor (potentiostatic electrolysis type)	Belanger	
1974	Enzyme thermistor	Mosbach, K., Danielson, B.	[36]
1975	Pd gate field effect transistor (FET) hydrogen sensor	Lundström, I.	[37]
1975	Enzyme-FET (ENFET)	Janata, J.	[38]
1976	Practical use of oxygen ZrO_2-based sensors for automobiles	Bosch Co.	
1976	Practical use of the $MgCr_2O_4$-TiO_2-based humidity sensor	Mathushita Comp.	
1976	Solid electrolyte oxygen amperometric sensor	L. Heyne	[39]
1977	Potentiometric immunosensor	Aizawa, M., Kato, S., Suzuki, S.	[40]
1979	Immuno-FET (IMFET)	Janata, J., Huber, R. J.	[41]
1979	Amperometric immunosensor	Mattiason, B., Nilsson, H., Olsson, B.	[42]
1979	Tissue electrode	Rechnitz, G.A., Arnold, M.A., Meyerhoff, M. E.	[43]
1982	Surface plasmon resonance	Nylander, C. et al.	[44]
1979	Surface acoustic wave device	Woltjen, H., Dessy, R.	[45]
1980	Fibre optic sensors for pH-monitoring	Peterson, J. I., Goldstein, R. S., Fitzgerald, R. V.	[46]
1984	Integrated optical sensors	Tiefenthaler, K., Lukosz, W.	[47]
1986	Neuronal receptrode	Belli, S. L., Rechnitz, G. A.	[48]

Table 2-2. Numbers of scientific papers published in *Sensors and Acutators* in 1988–1990 and in the conference proceedings of the "3rd International Meeting on Chemical Sensors" (Chicago, USA, 1990) and of the "Eurosensors IV" (Karlsruhe, FRG, 1990).

Basic Sensors	UK	France	Germany	Holland	Sweden	Italy	Finland	Switzerland	Austria	Other European	USA	Canada	Japan	Other	total
Liquid Electrolyte	1		6	1				3	1	2	10		4		28
Solid State Electrolyte	1	5	14			4			1		6	2	6	3	42
Electronic Conductivity	20	13	29			6	11	3		8	8	4	32	2	136
FET (Ions)	1	6	12	8		1		7		1			3		39
FET (Gases)	1		3	1	3	2				2	8		2		22
Calorimetric	1	2	2	2				3			1	1	2		14
Optochemical	1	1	6	2				3	2		1		5		21
Mass Sensitive		1	4	3				1			10		2		21
Biosensors	3		7	3	3	4		3	1		11		15		50
Others	2			2	1	2					3		1	1	12
total	31	28	83	22	7	19	11	23	5	13	58	7	72	6	385

2.2   Chemical Sensor Research and Development in Europe

2.2.1   Introduction

In attempting to overview the research and development effort in chemical sensors in Europe there are a number of problems: firstly, the field itself covers such a multiplicity of disciplines, interests, and applications that, working in one area one can only have limited knowledge of other areas of work; secondly Europe itself is still very much a collection of very independent states with languages, traditions, and means of running and funding research of their own. This makes it difficult to ascertain, in detail, what is taking place in the various countries. There will therefore be omissions and possibly errors in this section.

2.2.2   The Need for Sensors

The need for chemical sensors is multifold but the interest, in recent years, has grown from an awareness of the need in the future for sensors of all types in information technology, robotics, and automatic processing systems. With the need the awareness has grown that sensor technology, in general, and chemical sensors technology, in particular, lags behind both electronic device and computer technology both of which have made major strides forward recently. Within this framework the applications of chemical sensors may be identified as being in the following six major areas.

2.2.2.1   *Industrial Hygiene*

There has been an increase in awareness of the actual and potential damage to health of exposure at the workplace to a variety of toxic chemicals. Some effects are immediate, others may be very long term. In Europe a framework of Health and Safety at work legislation has grown aimed at improving and controlling the working environment. The European Economic Community (EEC) has been instrumental in adding impetus to this by issuing EEC directives to the member nations which they, in turn, have to implement through their own legislative processes. Such legislation currently in draft as consultative form together with associated approved codes of practice is, for example, the legislation concerning the control of substances hazardous to health (COSHH) in the UK. Similar legislation is under consideration in the other European States. These regulations and codes of practice place emphasis on monitoring both the workplace and personnel. Three stages are identified in which monitoring has a role to play; firstly to identify the hazard, secondly to assess the seriousness of the hazard, and thirdly to control the hazard. It is fairly obvious that sensors could play a major part in all three steps, always providing that suitable sensors are available.

2.2.2.2   *Pollution Measurement and Control*

This is a major area of concern throughout Europe, both in terms of atmospheric and water pollution. The problems of acid rain, resulting in defoliation and destruction of plant life in some parts of the continent, have been well published, as have some incidents of severe pollution of waterways and heavy metal pollution of some coastal waters. The problems of acid rain, particularly in terms of tracing sources and mapping plume dispersal may be more effectively addressed using remote monitoring laser techniques. However, monitoring the effluent levels at and in stacks as a control of filtration systems and monitoring in urban and industrial environments requires a range of sensors capable, in some instances, of operating in very severe conditions. Real time monitoring of water pollutants, as opposed to sampling and consequent analysis, has advantages. Sensors for monitoring both total toxic pollutants and specific pollutants such as heavy metals, pesticides, and fertilizers are required.

2.2.2.3 Hazard Monitoring

The monitoring of the explosive gas hazard has largely been solved through the availability of the calorimetric catalytic and the metal oxide sensors. In Europe, particularly in applications requiring quantitative or semiquantitative measurement, the calorimetric device has been preferred to the metal oxide. This is largely because it is a true hazard measuring sensor, giving a measure of the explosibility of the atmosphere, irrespective of the nature of the gas. A hazard area to which chemical sensors have not been fully applied is that of early fire detection, either by selective detection of CO or through detection of a variety of gaseous products evolved when materials are burnt. Another major hazard area, in which chemical sensors will play a part in the future, is in detection of hidden explosives.

2.2.2.4 Process Control

Across the whole of manufactoring industry, from chemical plant and refineries to high technology electronic component production, a wide range of gas phase and liquid phase chemical measuring techniques as well as means of measuring physical parameters are required for process control, quality control, and safety monitoring in complex processes. The scope for specialized sensors in this area is enormous but the demands on sensor reliability and integrity are very high.

2.2.2.5 Combustion Control

The major applications in combustion control are the control of gas and oil fired boilers in industrial, commercial, and domestic premises and in the control of automobile engines. The method usually used is to monitor the exhaust gases and to control the fuel to air mixture on the basis of this. Both applications have grown from the requirement to conserve fuel and to minimize pollution emission. Europe lies well behind both the USA and Japan in its legislation on pollution emission from motor vehicles. A further complication for in-engine sensors is the fact that leaded petrol accounts for a high portion of all the fuel used in motor vehicles in Europe.

2.2.2.6 Medicine

In medicine, the demand for sensors, for monitoring certain chemical species in the blood and other body fluids and also in exhaled breath have been increasing (see Chapters 20 to 24), as medical treatment gets more complex resulting in progressively stricter requirements for real-time monitoring of the patient's condition. The demand for sensors in this field is reflected in the increased proportion of the chemical sensor literature and conferences which is devoted to biosensor work and in the increasing number of Biotechnology Laboratories which are being established.

2.2.3   Coordination of Research and Development

2.2.3.1 European Efforts

There is active European participation in several aspects of chemical sensor research. The effort is not, however, well coordinated or controlled either in Europe as a whole, among the members of the EEC or even at individual nation level. There is, therefore, some duplication between different groups, although it is a measure of the complexity and diffuseness of this field that despite the lack of coordination, even on a world wide basis, direct duplication is, in fact, rare. This problem is not confined to sensor research and development and there have been, within the EEC, attempts at coordinating and supporting research in certain fields. This is done, eg, through EEC initiatives which provide funds for research on selected general topics. A typical program funded by the EEC was the European Strategic Program for Research in Information Technology (ESPRIT) which was started in 1984, with a five year budget of 750 million ecu (approx 750 million $), a small portion of which was available for sensor research. The role of sensors in information technology is now well established and some of the funds from the follow-up ESPRIT program have gone into sensor research. As an example, the International Conference on "Advanced Materials" 1991 in Straßbourg, for the first time, had a symposium on "New materials, physics and technologies for micronic integrated sensors" in which materials aspects of sensors were stressed. Most of the EEC funding so far is aimed at advanced materials technologies, design methodology, assurance for products and processes, application of manufacturing technologies, and technologies for manufacturing processes.

The EEC also supports the most important information source concerning European activities in the field of sensors, ie, the annual "Eurosensors" conferences. The first one was held 1987 in Cambridge (GB) followed by 1988 in Twente (NL), 1989 in Montreux (CH), 1990 in Karlsruhe (FRG) and 1991 in Rome (I).

The EEC also supports research efforts through schemes which aim at stimulating European cooperation and scientific and technical interchange between leading laboratories in the different member countries. These schemes provide grants for travel and short to medium term stays at the collaborating laboratories.

The European Coal and Steel Community, founded in 1951, still exists within the EEC framework and administers its own program of research contracts financed by annual levies on the coal and steel industries in member states. Under three major headings, ie, first Safety in Mining, second Industrial Hygiene in Mines, and third Technical Control of Nuisances and Pollution at the Place of Work and in the Environment of Iron and Steel Works, a number of sensor-related R&D projects relevant to these particular industries have been funded. The funding usually amounts to some 60% of the total cost of the project.

A major problem generally identified within Europe is that of slowness in transferring technology from the laboratory to the commercial market place, particularly in comparison with both the USA and Japan. The quality of the research work is comparable, the exploitation and marketing is not.

2.2.3.2   National Efforts

In addition to EEC and other international schemes, individual nations in Europe have their own schemes for supporting work on sensors. Of course, it is an extremely difficult task to summarize all these activities adequately here. We can, therefore, give only a few typical data.

Chemical and biochemical sensors cover a wide range of different measuring principles, fabrication technologies, and applications. In the field of general sensor research they represent the largest area of activity with many strong groups in universities and research institutes and more scientific publications than for any other advanced sensors. A few selected topics will be summarized in Section 2.4 below.

In Europe, there are more than 700 companies and institutions involved in the research and development of advanced sensors. This activity embraces more than 2000 qualified scientists and engineers. Total spending on advanced sensor R&D by governments and industry in Europe is in the order of 250 mill. ecu with private companies financing around 150 mill. ecu. In comparing the advanced sensor work in different countries supported today, it is important to keep in mind the varying motivations for the stimulation of advanced sensor R&D and the sources of financial funding, ie,

1. government funding of basic research, mainly to universities and research institutes, to encourage academic work and multidisciplinary sensor research as a good training ground for engineers, scientists etc.;

2. government funding for applied research to stimulate industry with company financial involvement on a 50/50 shared cost basis;

3. development programs with sensor manufacturing companies;

4. development programs with user companies, generally large companies such as automotive and major manufacturing companies that have continuously changing sensing needs and that feel that their future interests are inadequately safeguarded by sensor manufacturers;

5. defense program on specialized sensors.

In the major European countries, funding for advanced sensors R&D can come from all of the sources and, in general, users spending is the largest. In the smaller countries there are fewer large users. Their needs are often satisfied by imported sensor products and government budgets are limited. A comparison of advanced sensor research in different European countries shows up dramatic differences due to government funding policies and company interests and also reflects fundamental characteristics in the nature of the sensor industry and sensor markets in each country.

Germany

Germany has by far the most activities in Europe and leads advanced-sensor research in several areas. Reasons for this are the structure of the market, the needs of industry, and government policy. The sensor market in Germany is twice as large as in other countries of comparable size. Sensors are critical for German industry, particularly in areas dominated by German companies such as machine tool/factory automation, advanced industrial manufacturing systems and metrology/analytical/scientific equipment. In these areas the sensor is a critical part of the machine or assembly in that the superior functioning of the machine depends heavily on the characteristics of the individual sensor and the total control network. In addition, companies in consumer markets such as automobile and domestic appliances do integrate microelectronics intelligence with their products and need appropriate sensors. In Germany, some 400 companies invest around 100 mill. ecu per year in advanced sensor research and development. These include many sensor manufacturers involved in advanced product development but the largest sums of money in sensor research are invested by the users and system manufacturers. Another reason for German's domination in sensor R&D is government policy in sensor research. The German Ministery of Research and Technology (BMFT) invests a major amount per year in well-structured and coordinated research programs. The coherent policy of treating all the technology disciplines including solid state and optical sensors, signal processing and network actuators lends weight to considering the sensor industry as a whole and not divided into individual subjects. Strongly represented in this program are silicon technology and micro-actuators, chemical sensors, gas sensors including solid electrolyte and thin film devices materials development, and integrated optic sensors.

France

In France, there is a thriving sensor industry of about 150 companies, specializing in areas related to strength, in the French economy, such as avionics and defence technology. 75 companies are involved in advanced sensor R&D, many of them users or system manufacturers. Government funding in the order of 10 mill. ecu per year basically comes from the sensor committee of the Ministry for Research and Technology mainly for university work, from the major government research laboratories (LETI, LAAS, ESIEE) and also defence spending. France is well represented in silicon, thin film, thick film and fiber optic sensor technology.

In the chemical sensor field, 38 joint university or industry programs have been supported by the sensors committee between 1983 and 1989. A microchemical sensor society coordinates the research activities of abont 20 industrial firms and 15 university laboratories. It holds several meetings per year.

United Kingdom

There is a rigorous industrial base of chemical sensor and instrument manufacturers in the UK.

Important market sectors are in the oil, gas and mining industries, and lately in health-care. Notable success in the translation of products from research to market include the pellistor and the electrochemical, disposable glucose sensor based on glucose oxidase with a ferrocene mediator. Industrial investment in new biosensor and immunosensor developments has been very heavy.

Considerable and very sucessful efforts have been made to coordinate sensor R&D. In this context, the science and engineering research council (SERC) which administers the government support for research at universities in the UK initiated a major program of support for work in chemical sensors in 1984. This has benefitted a number of the leading academic groups in the UK, primarily those working in the biosensor field. Concern over the lack of coordination and the need for comprehensive information relating to this field of work particularly with regard to "what work was going on where" also resulted in the Department of Trade and Industry through one of its constituent organizations, The Laboratory of the Government Chemist, setting up what is now known as the UK Chemical Sensors Club, in 1985. The club holds general meetings every year and disseminates information relating to work carried out on chemical sensors and to current events and developments in the sensor field. It does, in fact, provide a very good forum through which anybody can acquaint himself with other's work and needs. The UK government has a priority in the promotion of collaborative research work involving industry with the acedemic community and research institutes. Examples of this have been the Harwell work on gas sensor materials, voltage effect gas sensors, and biosensors, all three of which have been jointly funded by the Department of Trade and Industry, a consortium of industrial sponsors. The UK Optical Sensors Collaborative Association (OSCA) is similarly funded to suggest pre-competitive research on the use of optical fibers in sensor technology and to promote information exchange. In 1988, the UK government announced a major new initiative, named LIMK and involving government departments together with the SERC, whose objective was the promotion of academic-industrial partnerships. Two of the program areas featuring in this initiative involve chemical sensors.

Italy

Italy has only a small R&D effort in advanced sensors. This stems from the nature of the economy which is centered on manufacturers of consumer goods such as cars, domestic appliances, clothes, shoes etc. and not in the machine industries. Thus, as users, the Italian companies import most of their sensors and instrumentation and do not generate them internally. The government sponsors a small national program on sensors but this goes almost exclusively to universities. Some additional funding occurs directly through the Government Laboratories (IROE, IESS). Within research institutes and universities, there is essentially no research on silicon micromachining, actuators, and integrated optics. Some interesting work has been done on thick film sensors such as mechanical or chemical sensors, and there are a few highspots in thin film and fiber optic sensors, too.

Switzerland

Switzerland, as one of the smaller countries in Europe holds a strong position in sensor research. The sensor industry is a traditional precision industry serving the machine tool manufacturers, ie, with sensors for mechanical variables such as position, force, pressure, and vibration sensors, and the pharmaceutical industries with their need for chemical sensors. Scaled according to the size of the country, Switzerland maintains an effort in advanced sensor research almost equivalent to Germany. 35 companies invest 15 million ecu per year, employing 140 qualified research staffs, but government spending on research is proportionally lower when compared to Germany. Swiss companies are strong in silicon research and operate several silicon foundries. Switzerland also has a traditional strength in thin film technology. A powerful effort is dedicated to chemical and biochemical sensors including those for pH, electrochemical and pollution monitoring and more recent research is directed to microchemical analysis systems and biosensors, eg, based on optical grating couplers.

Holland

Holland has, to some extent, a similar profile to Switzerland. There are two centres of excellence in advanced sensor research, the Universities of Twente and Delft. These have encouraged small companies to take up the new technology. Government funding in Holland goes exclusively to universities and there is no special government scheme to encourage industrial participation in research on a 50/50 shared costs basis. Holland is strong in silicon micromachined sensors and actuators, and chemical and biosensors.

Sweden

Sweden invests around 6 mill. ecu per year in advanced sensor research with 50 qualified researchers. Government funding is very modest, around 0.75 mill. ecu which supports mainly university research. Some of the university teams have strong track records, particularly in the chemical and biochemical area, with MOSFET sensors at Linkoping University, and thin film thermistors for thermal assay probes at Lund University. This has produced several small spin-off companies now active in chemical, biochemical, and medical areas. At one time *ASEA* had the largest industrial fiber optic group in the world, but this was sold off in 1988. Recently, however, a Swedish Company has developed a biosensor technology based on surface plasmon resonance and introduced it to the market.

Finland

In Finland there is research on sensors based on metal oxides. A Finn company has also been successful with a humidity sensor based on capacity changes.

2.2.4   Research and Development of a Few Selected Topics

2.2.4.1   Liquid Electrolyte Sensors

In this category (see also Chapter 7) are included amperometric, polarographic, and galvanic gas sensors, pH-electrodes, ion-selective electrodes including both, those applied to biosystems and those using biological systems such as enzyme electrodes and membranes, and the more conventional techniques of coulometry and conductivity measurements. Interest in this general field is prevalent throughout Europe with a number of strong academic and industrial groups involved in the UK where much of the SERC initiative support has been channelled into this area and the allied field of use of FET sensing devices in liquids with strong emphasis on biosensor work. France, Sweden, Italy, Holland, and Germany also have significant efforts aimed at this area, much of the work having a biosensor bias.

2.2.4.2   Solid Electrolyte Sensors

The literature on the use of solid state ionic conductors in chemical sensing (see also Chapter 8) is dominated by studies of O_2 effects in ZrO_2. With the ZrO_2 based sensors becoming accepted as working devices, although retaining effort aimed at improving such devices, most groups particularly in France, Germany, and the UK are shifting emphasis to the search for low temperature conducting electrolytes, investigating the protonic conductors and the utilization of these types of materials to detecting a range of other chemical species with "micro-ionic devices".

2.2.4.3   Electronic Conductivity Devices

Although there is a large effort in France, Germany, Switzerland, and the UK with some work in most other countries, it is unlikely that the total European effort in this area compares with that in Japan (see also Chapter 9). The bulk of the work is aimed at the various aspects of gas adsorption on metal oxide semiconductors. It has been realized that fundamental understanding of the mechanisms of gas adsorption and consequent conductivity changes are all important in the development of this type of sensor and a number of academic groups, particularly in Germany, are looking to a range of modern techniques to provide new insight. A number of European workers have become interested in organic materials as gas sensors and are studying sublimed and Langmuir-Blodgett films of macromolecules and polymers. The LB-film aspect could be interesting both in this and other types of sensor. The countries where, to date, most general investigations of these films have taken place are the UK, France and Germany. Work on conducting polymers is also prominent in Europe and several groups have demonstrated the possibility to use such polymers, prepared by numerous methods, as chemical sensors.

2.2.4.4   FET Devices for Gas Phase Measurement

Following the pioneering and continuing work of Lundstrom in Sweden on the catalytic metal gate FETs, a number of organizations in most European countries have devoted considerable effort to the development of this type of sensor (see also Chapter 10). The extension of this technique to a significant number of gases is to date mainly applied to gases like H_2, H_2S and NH_3, amines, alcohols, and unsaturated hydrocarbons.

2.2.4.5 FET Devices for Liquid Phase Measurement

This area of development is very closely aligned with the conventional liquid electrolyte devices with the development of ion selective FETs, immunochemical FETs and others. These systems are widely studied in most European countries with considerable effort in the UK, in France, in Germany, Sweden, Switzerland and Holland. Both liquid phase and gas phase FETs are in principle attractive devices for major electronic component manufacturers because of the ease with which they can be fabricated, using conventional electronic industry techniques. Much of the work on Chem FETs is now aimed at the development of selective membranes and coatings in order to improve the selectivity of the device, rather than of improvement of the basic device.

2.2.4.6 Calorimetric Chemical Sensors

Following the development of calorimetric gas sensors of the pellistor type with long term resistance to most catalyst poisons, there is little R&D effort now devoted to this type of device. Other systems based on measurement of heat of reaction, either using pyroelectric materials or utilizing the Seebeck effect have received only limited attention. Biosensors based on enzyme coated thermistors have also been developed. The main future development in this type of sensor could be the utilization of microelectronic fabrication techniques to provide miniaturization, low power, and single chip design (see also Chapter 11).

2.2.4.7 Optochemical Sensors

Optochemical techniques are becoming increasingly popular (see also Chapters 12, 17, and 18 as well as Volume 6); they have been identified as an area of high technical and commercial potential, particularly those devices and systems involving the use of fiber optic techniques. The major area of investigation is the development of chemically sensitive coatings. Another application of fiber optic technology, which will inevitably find use in monitoring gases, is the dispersed infra-red-system using the fibers for radiation transmission. A number of commercial organizations have also been readdressing the problems of miniaturizing and increasing the robustness of conventional infra-red absorption techniques for selective low-level measurement of gases. Other optochemical sensors use phenomena in thin layers, such as surface plasmon resonance and optical wave propagation. Sensors, based on the so-called photoacoustic or optothermal effect, are developed and marketed mainly by Scandinavian companies.

2.2.4.8 Mass Sensitive Sensors

Although investigation of piezo effect systems and, in particular, the development of associated adsorptive coatings have been taking place in most countries in Europe for a number of years there has not been a major impact on the gas sensing field outside the laboratory in the past. A Swedish Company, however, developed and sells a piezoelectric sensor for anaesthetic gases. This might explain why workers in Europe have not pursued the SAW type devices with the alacrity of others, particularly in the USA. There is, however, growing interest and work going on particularly in the UK, Italy, Germany, Belgium, and Holland.

2.2.4.9 Other Sensors

Various phenomena such as surface plasmon resonance and surface voltage/potential measurement have been and are studied, usually as individual efforts or as part of a multi-application research group. Conventional chemical detection systems such as FIDs and Electron Capture Devices as well as other detectors for chromatographic systems are continuously redesigned and reengineered by manufacurers of analytical equipment.

2.2.4.10 Biosensors

This is probable the biggest single area of growth at present (see Chapters 14 and 22). Virtually all the techniques mentioned can be utilized in some way to make measurements on biological systems. The use of biologically based molecules, enzymes, aminoacids, etc. as systems for improving selectivity of devices such as ion selective electrodes and ion selective FETs is increasing. The theoretical lock-and-key approach often described for such systems in the literature is proving attractive to many workers. The use of arrays of chemical sensors for the identification of odors and prediction of the composition of multicomponent gas mixtures is an interesting concept under extensive study both in USA and Japan. The electronic nose which combines sensor arrays with pattern recognition routines is one of the more interesting possibilities for chemical sensors. In Europe, activities in that area are found in many countries, notably in UK, Germany, and Sweden.

2.2.5   Trends and Areas of Future Development

A need has been universally identified for much improved understanding of the mechanisms involved in existing sensor systems; this is particularly true of the electronic conductivity type gas sensors, in which a number of parameters require elucidation. A number of avenues are now being explored including the engineering of well-defined sensor structures in terms of both microstructure and macrostructure and careful experimentation aimed at defining the critical parameters and properties. In carrying out such investigations full use is now being made of the modern analytical techniques, including XPS, Auger spectroscopy, EELS, SIMS and others. Although some of these techniques require that the studies are carried out under vaccum, the information obtainable provides data giving insight into the processes taking place in these devices. On the basis of the greater understanding of these devices it may become possible to predict useful material/additive combinations for specified applications (for details, see Chapter 3, for example).

In the near future, there will be much work aimed at the development of special coatings, eg, for mass effect and optochemical sensors and the development of membrane and barrier materials for liquid based electrolytic and FET systems. In all these areas, biologically based materials can be expected to play a wider role.

There will inevitably be improvements in fabrication techniques; the use of microelectronic techniques in miniaturizing sensors and in making integrated sensors as well as sample handling systems will provide the major impetus (see also Volumes 1 and 7). The use of such technique to provide multisensor arrays on single chips have not, to date, received the same attention in Europe as in the USA and Japan, but a number of organizations with the necessary expertise are becoming involved. The increased interest in molecular electronics will result in the use of new materials and new forms of materials in a number of areas. These will include Langmuir-Blodgett films, materials deposited by chemical vapor deposition techniques, materials into which active additives are inserted by ion implantation techniques, complex mixed oxide systems and again, an increased range of biological materials.

There will also be intensified use of existing sensors, despite their limitations, particularly in the gas detection area. A greater emphasis on defining applications and selecting sensors or combinations of sensors, together with computerized pattern recognition techniques to match the application, is likely. It has to be recognized, perhaps belatedly in some parts of Europe, that sensors which are not perfect and not the ideal answer are usually better than nothing, in the absence of anything better, and should, with care, be used while progress to better sensors is being pursued. It should be realized that major leaps forward are unlikely and that progress in chemical sensors, as a whole, is going to be achieved by careful experimentation and painstaking data collection − together with an element of inspiration [50–57].

2.3   History of Microfabricated Chemical Sensors in North America

Chemical sensors that transduce chemical concentration or activity into electrical signals have a long and complex history. One of the major events was the recognition that a new class of sensor, the microfabricated chemical sensors, might evolve from semiconductor research which became apparent with the development of the transistor. Electrochemical structures, devices based on the quasi-reversible change in electrical impurities and a wide range of photonic structures also evolved during this time. In the past two decades, field effect based semiconductor structures have received considerable attention. During this time, work began to appear on other types of devices such as acoustic wave structures and thermal devices. While commercially successful chemical sensors of classical design are readily available, a practical microfabricated chemical sensor using contemporary microelectronic batch processing technology is still awaited.

The beginnings of chemical sensor research dates from the turn of the 20th Century when Cremer [58] and Haber and Klemensiewicz [59] first described pH-sensitive glasses. The role of defects in determining the electrical properties of solids was suggested by Frenkel in 1926 [60]. Later work by Wagner and Schottky confirmed this suggestion [61]. Kröger has extensively reviewed this subject and one can see from the material there that compounds have considerable potential as chemically sensitive materials [62]. The next major event was the observation of chemical sensitivity in germanium and silicon "cat-whisker" microwave detector diodes during the Word War II era. With the anticipated growth of microwave communication systems, the stabilization of these essential devices became an important task. The first step toward the solution of the problem was the recognition that there are electronic states associated specifically with the surface of semiconductor materials that could hold electronic charge [63]. The presence of these states gave a working explanation for the deviations of the current flow under bias for these semiconductor "cat-whisker" diodes. The next step in the evolution of contemporary solid state electronics was the study of how different gases and chemical environments influenced the electronic surface state density. To test the surface environment in the vicinity of a "cat-whisker", Bardeen and Brattain introduced another „cat-whisker" probe to monitor the surface conditions. It was on this basic structure, called the type A transistor, that Bardeen and Brattain observed transistor action on germanium in 1948 [64]. In 1953, Bardeen and Brattain reported that a chemical cycle, consisting of wet and dry oxygen, wet and dry nitrogen, and ozone, could be used to modify the surface potential of a free standing semiconductor surface [65]. This chemical process became known as the Bardeen-Brattain cycle. Many other gaseous combinations were developed and were reviewed by Kingston [66]. The existence of electronic consequences to a modification of the ambient of a semiconductor has been a major driving force in the development of intensive studies on surface phenomena.

For the most part, the investigations that have evolved over the past four decades have clarified many aspects of gas-solid interactions. However, the early chemical response of semiconductor materials was found to be so erratic and non-reproducible that the semiconductor industry opted to eliminate the problem by hermetic sealing of the electronic components in well-defined environments. The possible use of the semiconductor surface for chemical sensing was generally appreciated but the drive toward microelectronic applications left little resources for this direction.

At the same time that the transistor was announced, Shockley and Pearson reported on an experiment where a capacitive structure was used for modifying the surface potential of a semiconductor directly with an electric field [67] and the stage was set for the eventual evolution of the metal-oxide-semiconductor field effect transistor (MOSFET); see also Volume 1, Chapters 3, 4, 7, and 10. Here too, the presence of large electronic surface state densities were observed. In this case, the high surface state densities precluded the use of the field effect in microelectronic applications. It would not be until the development of the silicon MOS structure that the field effect would begin to impact heavily on microelectronics [68]. The early MOS experiments were directed at understanding the origin and character of interfacial electronic states on semiconductor materials. The chemical sensitivity of the MOSFET structures was universally viewed as a problem to be eliminated rather than a potential opportunity for new classes of measurements. As a consequence, chemical sensitivity acquired a pejorative character in the semiconductor industry and efforts to eliminate or stabilize the surface dominated the attention given to chemical sensitivity. Nevertheless, microfabricated chemical sensor research is intimately connected with the evolution of microfabricated electronic circuitry as will be discussed below.

The sensitivity of other semiconductor materials to their gaseous environment was a well known phenomenon even in the 1950's. In 1956, the sensitivity of the lifetime of cleaved lead sulfide single crystals to oxygen and water vapor was noted [69]. While the focus of semiconductor research in the United states was on electronic application driven by commercial and military requirements, the situation was less constrained in both Japan and Europe.

The next major development in the microfabricated chemical sensor area was the seminal paper of Bergveld from the Netherlands in 1970. Bergveld introduced the concept of the ion sensitive field effect transistor (ISFET) based on the then rapidly developing MOSFET [70]; see also Chapter 10. By the late 1960's, it has become apparent that silicon manufacturing technology could have important consequences for microfabricated sensors such as pressure devices [71] and Hall effect elements [72]. By introducing the concept of the ISFET, Bergveld opened up the possibility that new classes of chemical sensors could be designed and fabricated in silicon using the rules and procedures developed by the microelectronic device processors. In 1974, Matsuo and Wise described the pH sensitivity of an ISFET structure using a reference electrode and with this, the activity in the United States on ISFET structures began. Shortly thereafter, Janata and co-workers at Utah began an extensive study of thick polymeric coatings as ion sensitive membranes in 1975 [73]. The large body of work by Janata and co-workers since then has provided much of the electrochemical basis for the evolution of contemporary chemically sensitive semiconductor devices (CSSD). Other classes of CSSDs, including some that had not as yet been developed, were reviewed by Zemel [74]. In that paper, the possibility of using Pd as the gate material in a chemically sensitive device was discussed. However, it was Lundstrom in Sweden who first described a working hydrogen sensitive Pd-MOS device [75]. Buck, whose work had been directed primarily toward macroscopic ion selective electrodes, initiated a program on Faradaic ISFETs and succeeded in producing anionic sensors with AgBr chemically sensitive membranes [76]. The work of Freiser and his group [77, 78] on coated wire sensors was intimately related to the non-Faradaic sensors of Janata, Bergveld and others.

The automobile industry became an important factor in the early 70's as they came face to face with the need to control emissions and improve automobile efficiency. The result was a focus on electrochemical sensors with particular emphasis on oxygen sensors. The devices that emerged were electrochemical units that operated at elevated temperature. Through this period, the academic community divided into those dealing with electrochemical sensors of macroscopic dimensions (>5 mm in minimum dimension) and those that focussed on microelectronic style, photolithographic batch processing.

From these beginnings, the use of silicon based manufacturing technology in CSSD emerged. There is litte question that when properly employed, planar photolithographic fabrication of silicon based devices offers the prospect of a highly cost effective manufacture. Alternates to the MOSFET were proposed by Wen et al. that employed the gate controlled diode [79, 80]. Thus the three classes of MOS structures were available for application. The one advantage of the gate controlled diode configuration used by Wen et al. was the introduction of a thermally migrated connection path from one side of the wafer to the other. This development demonstrated that it would be possible to isolate the chemical processes on one side of the silicon chip and the electronic functions on the other. While this specific technology did not give rise to a commercial device, the principle of chemical isolation of the electronics had been demonstrated by McConnell and co-workers who introduced an ingeneous twist on the MOS capacitor using the surface photovoltage to measure changes in surface potential induced by chemical reactions [81].

Another major stumbling block to the wide spread application of silicon microelectronic fabrication technology was the difficulty in developing a compatible planar photolithography for defining the area containing the chemically sensitive materials. The major advantage of small size and high density can only be achieved if more than one chemically sensitive layer can be deposited on the ISFET chips. In order to combine the advantages of separating the chemical sensitive regions with their exposure to operational fluids with the now well developed silicon microfabrication methods Lauks and co-workers devised an extended gate structure which consisted of an integrated shielded gate line that allowed the electronics to be laterally displaced away from the chemically sensitive region of the chip [82]. Katsube and co-workers introduced sputtered IrO_x films as chemically sensitive layers for field effect based pH measurements. This spatial separation could be implemented in standard CMOS technology [83]. The procedure proposed initially by Van der Spiegel et al. was to use a conventional design procedure based on a silicon foundry's design rules up to the point where the chemically sensitive layer would be applied. The application of the layer to the silicon wafer using proprietary photolithographic methods was the final step in the processing prior to final electrical test and device packaging [84]. This was in contrast to the techniques developed by Janata, Huber and co-workers who employed solvent casting methods on the individual field effect elements [85, 86].

In the intervening years since Bergveld alerted the community to this class of sensors, other physiochemical principles have been applied to the development of microfabricated chemical sensors. Some of this work began before 1970 as was the case with Sauerbrey in Germany [87], King [88], and Stockbridge [89] who employed high quality quartz oscillator crystals as mass monitors in various applications. While Sauerbrey's work was seminal, it was King at Esso who first demonstrated that quartz oscillators with appropriate chemically sensitive coatings could be employed as vapor detectors in the late 60's. During the past few years, the work of Guilbault deserves particular notice for bringing the enzyme detector [90] to the foreground. The mass sensitivity is quite respectable, being of the order of fractions of a nanogram. An extension of this concept was proposed by T. Hirschfeld whereby either a crystal with several nodes on its surface or a device with multiple cantilevered or ribbon elements is used [91]. Different selective membranes are deposited on the various surface nodal maxima or the discrete cantilevered or ribbon oscillating elements and the mass uptake on each material measured simultaneously. This data is used to obtain a "fingerprint" pattern for the identification of different gases in mixtures. "Fingerprint" measurements appear to be of increasing importance in all areas of chemical sensing. Both silicon and quartz have been suggested as substrate materials for preparing these multi-element devices.

The development of extremely high frequency surface acoustic devices for signal processing led a number of investigators to explore their possible use as gravimetric or mass based chemical sensors. Wohltgen has played a seminal role in the development of the surface acoustic wave (SAW) class sensor at the Naval Research Laboratory [92, 93]. Recent work well illustrates the potential of this class of devices [94]. Surface acoustic wave research under the direction of White at the University of California, Berkely, is also exploring chemical sensor device applications [95].

The SAW based devices present an interesting prospect. There is the argument about their inherent sensitivity. Like almost all chemically sensitive devices, they require a coating to act as the transducer interface phase between the external local universe and the sensor element. The characteristics of the interface phase have to be as well established for the operation of the SAW as for the chemical uptake or loss. This has been a continuing source of research interest as illustrated in reviews that appeared in the March, 1987 issue of the IEEE Transactions on Ultrasonics, Ferroelectrics and Frequency Control. While not a serious limitation, it is nevertheless correct that at the higher operational frequencies in excess of 100 MHz, the cost of the associated electronics and amplifiers rises more steeply than the frequency itself. Whether the inherent sensitivity of the SAW more than compensates for the added system cost is an unresolved question.

Preliminary work has suggested that microcalorimetric measurements may be conducted with ferroelectric materials like $LiTiO_3$. These pyroelectric materials have a high thermal heat flow sensitivity and have been widely developed as infrared sensors [96]. Sensitivities to heat flows as small as several nanowatts/cm^2 have been reported. A pyroelectric micro-calorimeter has been constructed from $LiTaO_3$ and some interesting results obtained [97, 98]. However, the results make it quite clear that far more work will have to be done before these devices reach commercial character. An integrated ZnO based pyroelectric element deposited on silicon has been used to detect chemical reactions [99]. In general, microfabricated thermally based chemical sensors have not been generally studied.

The breadth of activity of fiber optic chemical sensors (see also Chapters 12, 17, and 18) is one of the major developments in the past decade. Some of the most important initial work was carried out by Peterson and co-workers [100–102]. In recent years the Fiber Optic Sensors conference has illustrated the remarkable growth of this field. The fiber optics sensors group at the Naval Research Laboratory has demonstrated that fiber optic systems can be used to detect a variety of chemical species in solutions [103]. In this case, a dye was coated on the fiber and an evanescent wave interaction provides the coupling to the absorbing outer layer. While the optical fiber is unquestionably a small structure, it is far more difficult to use standard planar photolithographic methods on a small cylindrically symmetric structure than a planar structure. However, the work of Butler and Ginley is particularly noteworthy in that it illustrates the potential advantage of a specialized optical fiber element in certain environments [104]. These authors employed a layer of palladium as a strain sensitive hydrogen detecting material. The strain resulting from the uptake of the hydrogen modifies the optical properties and dimensions of the fiber, thereby changing its optical path length. This change is measured interferometrically. This type of structure was also used by these authors to examine the strain induced by an electrodeposited film as it grows in the bath. This is quite a unique example of the potential value of these types of sensors. Surprisingly, there does not seem to be too much research on the use of planar waveguide elements. Part of the reason is that most planar waveguide elements are constructed with $LiNbO_3$ substrates. These substrates are expensive and not easily designed to be three dimensional. There is an urgent need for a three dimensional polymeric material that can be deposited on glass or silicon substrates but could still be photolithographically micromachined. This material should have properties that would allow it to survive immersion in various solutions, especially aqueous solutions, for reasonable periods of time and could be processed so that optically active molecular species can be either grafted to the polymer or physically entrapped in the polymeric matrix. These molecular grafts could be designed to interact either chemically or optically with their surroundings. The photoformability is needed to insure that the planar fabrication technology can be applied.

Electrochemical devices based on both thick and thin film technologies are being studied in a large number of laboratories (see also Volume 1, Chapter 6) even though they are not made by microfabricated technologies yet. The solid electrolyte work on the yttria and calcia stabilized zirconia at Ford reviewed by Logothetis and Hetrick are excellent examples of the research that has been conducted in this area [105]. Chang and Hicks have demonstrated interesting Hall effect results on SnO_x [106, 107]. Similar work has been conducted in both

Europe and Japan. Amperometric measurements also appear to hold considerable promise [108]. Stetter has demonstrated in an impressive fashion that amperometric electrodes have a valuable role to play in the detection of environmental pollutants [109]. Furthermore, the system developed by Stetter has reached the point where commercial development is beginning to arise [110]. The University of Washington has a strong effort on sensors through another NSF co-sponsored University-Industry Research Center which operates under the rubric of the Center for Process Analysis and control (CPAC). One of the less standard chemical sensors is based on the charge flow transistor as a dielectrometer [111, 112]. This program of Senturia and co-workers has led to the commercial development of a microdielectrometer [113].

There also has been a rapid evolution of research efforts in Canada. With its long history of biomedical research, the University of Toronto led the way toward a Canadian presence in chemical sensor research. Cobbold and co-workers initiated an in-depth study of the electrolyte-SiO_2-Si system which led them to apply the site binding model of Yates et al. [114] to their system [115, 116]. Barabash have been investigating the noise and thermal properties of ISFETs and have demonstrated that in the site binding model approximation, the electrochemical noise associated with the exchange currents between the bulk electrolyte and the surface sites is well below the noise levels associated with the operation of the ISFET itself [117]. Also at the University of Toronto, M. Thompson has headed a group that has been exploring the use of Langmuir-Blodgett as chemically and specifically biologically sensitive membranes for some time [118]. The device structures used in their studies have included both bulk and surface acoustic wave elements. There has been a recent review of the utility of bulk acoustic wave devices as immunosensors by this group [119]. S. R. Morrison has headed a long standing effort on gas surface interactions at Simon Fraser University in British Columbia.

The evolution of microfabricated chemical sensor research in North America has included most of the contemporary trends in this field. The development of silicon as a substrate for sensors is extremely well advanced both in North America and other parts of the world. There is every reason to expect continued growth around the world on various types of chemical sensors. It is obvious that the key problems will not be the manufacture of silicon structures, but rather the incorporation by planar technology of chemically sensitive layers onto previously prepared electronic, or other micromachined structures. In turn, the decision as to which chemical species are to be detected will be a technology rather than a scientific issue. Some effort is needed to address the general issue of which classes of chemical species are best measured in the microsensor area by a given class of phenomena. Nevertheless, it has become quite apparent that in the United States, the economic issue will determine the future directions for microfabricated sensor development. The development of specialized surface coatings is a general problem in all areas of sensor science and technology. Much of this work is proprietary and it may be some time before the scientific community will be appraised of the details of this research. It will be interesting to see how the emerging specialized technologies will modify this current view of sensor research.

2.4   History of Chemical Sensors in Japan

2.4.1   Introduction

There can be no question about the view that Japan is one of the leading countries for research and development of chemical sensors. Especially in Japan, it is noted that extensive efforts are devoted to practical application of chemical sensors, which have triggered interests in other Asian countries. In this review, research and development in the field of chemical sensors in Japan is briefly summarized. Some overviews have been also published from the different aspects [120, 121].

2.4.2   Historical Progress on Chemical Sensors in Japan

The historical progress on chemical sensors is summarized in Table 2-3 from the view of practical use. Since 1960, onwards studies or proposals on detection or sensing of gaseous components had begun to appear in literature, though the commercial needs of sensing devices are still ambiguous. In Japan, Seiyama [122] and shortly afterwards Taguchi [123] proposed semiconductor oxide type gas sensors in 1962. In 1967 Figaro Engineering Inc. succeeded in

Table 2-3. Research and development of chemical sensors in Japan.

1959	Catalytic combustion type gas sensor	Komyo Rigaku Kogyo K.K.
1962	Proposal of oxide semiconductor type gas sensor	Seiyama, Taguchi
1964	Thermistor type gas sensor	Denshisokki Co.
1967	ZrO_2-based solid electrolyte type oxygen analyzer	NGK Insulator Ltd.
1967	Oxide semiconductor type gas sensor	Figaro Eng. Inc.
1967	Galvanic cell type oxygen monitor	Riken Keiki Co., Ltd.
1968	Leakage alarm for LP gas	New COSMOS Electric Co., Ltd.
1976	Humidity sensor using $MgCr_2O_4$-TiO_2 system	Matsushita Electric Ind. Co., Ltd.
1977	ZrO_2-based oxygen sensor for automotive exhaust	Toyota Motor Corp.
1980	SnO_2-based oxide semiconductor type sensor for detection of town gas	Figaro Eng. Inc.
1982	Oxide semiconductor type oxygen sensor for automotive exhaust using TiO_2-based system	Nissan Motor Co., Ltd.
1983	The First International Meeting on Chemical Sensors in Fukuoka	
1984	Microbial sensor for BOD determination	Tokyo Institute of Technology Nisshin Electric Co, Ajinomoto Co., Ltd.
1984	Solid electrolyte type lean-burn sensor using ZrO_2-based oxide	Toyota Motor Corp.
1984	ISFET for diagnosis examination	Kuraray Co.
1988	Foul breath checker, semiconductive oxide type	Figaro Eng. Inc.
1989	TiO_2-based NO_x sensor	Tokuyama Soda Co.
1989	Fragrance meter using lipid coated quartz oscillator	Okahata, Sogo Pharmaceutical Co.

the commercial production of this type of gas sensors. It may be said that this was the first R&D on devices which were later called chemical sensors. Since then, a variety of chemical sensors have started to be investigated and research speeded up after 1970.

Distinguished researchers in Japan contributed much to the developments. The leaders in the respective fields, are Suzuki in biosensors, Ishibashi in ion sensors, Matsuo in FET sensors, and Seiyama in gas sensors and humidity sensors.

Semiconductor oxide type sensors mostly using tin oxide began to be used practically and produced commercially as sensors for LPG and town gas in Japan. They were put to practical use by Figaro Engineering Inc., New Cosmos Electric Co., Ltd., and Yazaki Meter Co., Ltd., etc. Since the first study in 1962, Seiyama has accumulated a wide range of information about this kind of semiconductor oxide sensors including their sensitivity, selectivity, and sensitization by noble metals [124]. Although he and his co-workers, Yamazoe et al., have studied the physico-chemical properties intensively of these sensors [125], the sensing mechanisms and kinetics are still unclear in part because of complicated cooperative effects, such as structural defects, adsorption properties, space charge layer, surface reactivity, spillover phenomena, metal-semiconductor junctions, and the particle size effect.

Selectivity of semiconductor oxide sensors can be modified by investigation of an adequate selection of nobel metal species as sensitizers and varying the working temperature of the sensor. The sensors for H_2, CO, and alcohols have been developed and commercialized, though their selectivities are still insufficient. These sensors are incorporated in combustion devices, air conditioning systems, air cleaning installations, household electric appliances, and so on. A considerable number of these sensors have been developed by Ihokura and his co-workers at Figaro Engineering Inc. Catalytic combustion type sensors using a supported platinum catalyst are also commercialized for detection of flammable gases. The total production of these gas sensors is estimated to be 10 million devices.

Recently, odor or olfactory sensors are also becoming more important. Egashira et al. [126] have studied ruthenium-loaded titanium sensors which show high selectivity to amines, and proposed an application as freshness sensors for foods. A sensor for organic sulfur compounds as mercaptans developed by Figaro Engineering Inc. [127] has a practical application as a foul breath sensor.

Clark's proposal in 1962 has triggered development of biosensors. Updike and Hicks did their first research in glucose sensors using immobilized enzymes in 1966. Since then, Suzuki of the Tokyo Institute of Technology, previously engaged in biological or microbial electrode reactions, took up the study on biosensors actively. He and his co-workers Aizawa and Karube have produced a number of contributions to enzyme sensors, microbial sensors, immunosensors etc. [128]. Interesting and useful among them are two particular proposals [129]. One is a new type of one-chip FET biosensor with working and reference electrodes, and the other is a multi-sensor system for monitoring the freshness of fish or meat.

Glucose sensors are used practically for processing and diagnostic examination, whereas others still have some problems for commercialization in spite of active research. However, a wide variety of new biosensors have been proposed recently. Some of them utilize electron mediators together with enzymes; and some others aim at development of highly functionalized electrodes. Application of monoclonal antibodies is also one of the recent topics.

For practical use of biosensors, miniaturization of electrodes is required. For example, highly sensitive micro-enzyme electrodes with a chip size of 50–500 µm have been reported by Yamauchi et al., in which the enzyme is immobilized on a Pt wire by electro-deposition

[130]. A FET type multi-biosensor is also being studied and seems to have bright prospects for practical utilization [131]. Furthermore, application of biosensors to the detection of flow injection analysis (FIA) system and to real-time measurement are in progress. Although biosensors have an inevitable weak-point in their stability, as mentioned above, the practical utilization of biosensors progresses steadily for diagnostic examination and other applications. Shichiri et al. [132] have developed long term stability of the embedded-type glucose sensor. The disposable-type biosensor has been studied by Suzuki et al. [133].

Stimulated by the original research of ISFET by Bergveld in 1970, Matsuo of Tohoku University, who had been studying amplifier actions of MOS type transistor, started to study of FET type sensor with his co-worker Esashi. Since then, they have done excellent work [134] on the pH sensor, pNa sensor and so on. In particular, they fabricated a micro ISFET, which is 10 μm in chip size and therefore useful to place in a living body. From 1980's, FET type sensors have been widely studied in Japan on gas sensors and biosensors as well as ISFET. Stabilities of these sensors are, however, still unsatisfactory, and stability is urgently required for ISFET for practical applications. One such problem is to improve firm contact between the ion sensing membrane and the gate substrate. As a result of these efforts, a new integrated sensor for simultaneous detection of K^+, Na^+, and Cl^- was reported to be stable for more than two months [135].

In the 1960's decade, analytical chemists were stimulated by the studies of Pungor on ion electrodes (ion sensors). In Japan, Ishibashi of Kyushu University, who had been working previously on ion-exchange membranes or liquid membranes, started to study ion sensors. His interesting contributions are sensing mechanisms of liquid-membrane type ion-sensors and development of medical ion-sensors for vitamin B, acetylcholine, and ephedrines [136].

Ion sensors for Na^+, K^+, etc. have already been commercialized by some companies, for example, Hitachi Ltd. and Shimadzu Ltd. Also, selective and high performance sensors have been developed, using crown ethers, by Kuraray Co. and Tokuyama Soda Co. for diagnostic examination. Further, ion sensors have been extending their detection power to various fine-chemicals, for example, vitamin B series [137]. Detection at extremely low concentration, utilization of chemically modified electrodes, and simplification of the sensor structure, eg, a coated wire electrode, are also being researched extensively.

One marked trend of the research activity in Japan is the strong contribution of industry. In the field of chemical sensors, a number of efforts for practical application has occurred, as listed in Table 2-3. For quality control in the steel industry, measurement of the oxygen concentration in molten iron has been studied since 1950. In 1966, Matsushita and Goto [138] reported the application of a solid electrolyte (stabilized zirconia) cell to the oxygen concentration measurement. These devices had already been considered useful for this purpose by Kiukhola and Wagner in 1957. Since then, in Japan, this type of oxygen sensor for steel industry has been actively investigated and extended to commercial production. On the other hand, the need to severely regulate exhaust gases from automobiles has also speeded the development of oxygen sensors. The companies which responded to this requirement were Nippon Gaishi, Toyota Motors, and others. The commercially produced sensors have been called λ-sensors, used to keep the air-fuel ratio at the equivalent value. For this purpose, sensors of stabilized zirconia type have certain preferred characteristics. Accordingly, the production increased year by year, and now in Japan, it is estimated to be 10 million items.

In the next stage, automobile companies requested a "lean-burn type" oxygen sensor, which works in the lean-burn engine. Since the conventional λ-type sensor is ineffective in this region, Takeuchi et al. of Toyota Motors coated the electrode of the sensor with a porous ceramic layer to form an oxygen diffusion layer [139]. The limiting-current type sensor thus assembled exhibited an excellent response in the lean-burn region. This film fabrication and miniaturization of this type of oxygen sensor is now under investigation [140]. It is noteworthy that, the limiting-current type oxygen sensor exhibits linear response in the range of oxygen concentrations from 0 to 100%, when Knudsen diffusion is dominant in the pores [141]. Solid electrolyte sensors are generally operative only at elevated temperatures. However, a series of oxygen sensors operative at ambient temperature were designed using proton-conductive membranes attached to a gas diffusion layer [142].

In Japan, humidity sensors find many uses in air conditioning systems and other installations because of hot temperatures and high humidity. In 1976, Nitta et al. developed a humidity sensor using $MgCr_2O_4 - TiO_2$ ceramics [143] and it was produced by Matsushita Electric Industry Co. Ceramic humidity sensors have also been commercialized by Toshiba and Mitsubishi Electric Co. The sensing mechanism of humidity sensors using porous ceramics has been proposed and elucidated by Seiyama et al. [144]; see also Chapter 20.

Another type of commercialized humidity sensor is based on the electrical conductivity change caused by water vapor absorption into a hydrophilic polymer film. Fundamental studies of hydrophilic polymer membranes by Sakai et al. are also notable for this type of humidity sensor [145]; see also Chapter 20. Hijikigawa [146] developed an FET type humidity sensor by applying silicon planar technology using the cross-linked cellulose, acetate butylate as the sensing element and incorporating it in insulated-gate field effect transistor; thus, a microchip sensor has been constructed. The chip dimension is about 1.5 mm square. At present, total production of humidity sensors of various kind amounts to millions.

2.4.3   Activities of Academic Meetings and Publishing in Japan

As mentioned above, the expansion of research and practical development in the chemical sensor field have increasingly promoted several symposium organizations and publication planning. These activities have become international in character. Research Association for Chemical Sensor Development (present name, Japan Association of Chemical Sensors) was called for by Seiyama and was organized in 1977, with the support of the Electrochemical Society of Japan. This was the first academic association on chemical sensors in Japan. It now consists of about 100 active researchers from universities and institutes and 45 companies. With the cooperation of members of this association, the first International Meeting on Chemical Sensors was held in Fukuoka in 1983. Another association is the Technical Committee of Electron Devices in the Institute of Electrical Engineers in Japan. A sensor symposium, which covers physical and chemical sensors, has been held annually by the Institute since 1981, and the English proceedings have been published. The 4th International Conference on Solid-State Sensors and Actuators was organized mainly by the Institute members and held in Tokyo in 1987.

In addition, a commercial journal on sensor technology entitled "Sensor Gijutsu (Sensor Technology)" has been published monthly by Technical Research Center since August 1981. In recent months, Kodansha/Elsevier and Scientific Publishing Division of MYU started an English annual review, "Chemical Sensor Technology", and a new journal, "Sensor and Materials", respectively. As for books, "Kagaku Sensor (Chemical Sensor)" edited by

Seiyama et al. was the first (1982). Since then many books have been published. Among them, "Kagaku Sensor Jitsuyo Binran" (1986) is the most comprehensive handbook on chemical sensors. The former was published by Kodansha Scientific and the latter by Fuji Technosystem.

The research activity on sensors in Japan stimulated those in China and Korea. Although commercial sensors in these countries are often based on Japanese technologies, recently, several distinguished researchers have been active on chemical sensors: in China, Ren-Yu Fang (Changchun Inst.), Luo Weigen (Shanghai Ceramic Inst.), Sun Liongyan (Jirin Univ.), Shen Yu Sheng (Univ. Sci. Tech.); in Korea, Soon Ja Park (Seoul Univ.), K. S. Yun (KIST), Duk-Dong Lee (Kyngpook Univ.); and in Taiwan, Chien-Min Wang (Ind. Tech. Inst.). In this situation, it is expected that an East Asia Sensor Conference will be organized in the near future.

2.4.4   Future Scope of Chemical Sensors in Japan

In 1981–1986, 296 papers were presented at the periodical domestic meetings organized by the two research associations described above. The largest contribution were those of semiconductor metal oxide sensors for combustible gases, mostly using SnO_2 and other oxides, the next one were the humidity sensors, then followed by biosensor and ISFET studies. It is noted that the research on chemical sensors is increasing and flourishing.

In conclusion, we will point out several aspects which will become important in future.

(1) Microfabrication of sensor element is advantageous in several aspects, and much required for in-vivo sensing. Accordingly, related papers are increasing in number. For example, development of micro oxygen electrodes [147], micro SnO_2 sensors [148], micro enzyme sensors [130] etc. have been reported.

Research into sensors of thick film type and thin film type together with FET type have been actively continued in Japan. But the commercial production of these sensors have not been realized succesfully yet because of unsatisfactory reproducibility and stability of the sensor elements. It is expected that intensive studies will overcome these problems in the near future.

Under such circumstances, it appears that studies on sensors of optical fiber type and SAW type are not so vigorous. However, there is one remarkable advance in miniaturization of infrared absorption gas sensor, which is reported by San-yo Electric Inc. [149]. The progress of light sources and detectors is expected so that the chemical sensors using optical methods are likely to be popular in the next generation.

(2) Gas sensors for environmental sensing, in particular, sensors for global environment are going to be subjected to active research and development. For example, research on limiting current type oxygen sensors [141], CO sensors [150], and NO_x sensors using SnO_2 [151], and NO_2 sensors using lead phthalocyanine [152] is also in progress aiming at practical utilization. Importantly, some of them are reported to be operational at room temperature [150, 153]. Moreover, sensors for CO_2, ozone [154], and chlorofluorocarbon compounds [155] have already begun to be studied intensively.

(3) One of the important directions is to make an effort for biomimetic approach to sensing functions of the living body. Umezawa et al. are trying to make sensors for K^+ and Ca^{2+} that have an amplification function by ion channels across a synthetic membrane [156]. Here studies of olfactory sensors using lipid bi-molecular layer membranes reported by Okahata et

al. are interesting not only as a new application of the quartz oscillator sensor, but also as an approach to olfactory action by using biomimetic materials [157].

(4) The fourth approach is intelligent sensors. Ikegami et al. [158], as well as Müller in Germany, took the initiative in the study of the multisensor system and the intelligent sensor (see also Volume 1, Chapters 11 and 18). This is one of the important directions for advances and utilization of sensors in future. For the multi-sensor system, Igarashi et al. [159] reported clinical monitoring by multi-ISFET, on which sensor elements for pH, Na$^+$, and K$^+$ were constructed into one chip. As the latter example, Morizumi et al. [160] reported an olfactory sensor system using seven quartz oscillator arrays with a neural network. Studies on olfactory sensors are also more active so that the progress in this field is expected.

2.5   References

[1] Kohlrausch, F., *Wied. Ann.* **26** (1885) 161.
[2] Nernst, W., *Z. phys. Chemie* **2** (1888) 613.
[3] Böttger, W., *Z. phys. Chemie* **24** (1897) 253.
[4a] DRP 165349 (1904), Masch. Fabrik Augsburg–Nürnberg (MAN).
[4b] M. Cremer, *Z. Biol. (Munich)* **47** (1906) 562.
[5] Sörensen, S., *Biochem. Z.* **21** (1909) 131.
[6] Rideal, E. K., Evans, U. R., *J. Soc. Publ. Anal. Chem.* **1** Aug. (1913).
[7] Heyrovský, J., *Pilos. Mag.* **45** (1923) 313.
[8] Kolthoff, I. M., Harting, B. D., *Rev. Trav. Chim.* **44** (1925) 113.
[9] Tödt, F., *Z. Elektrochemie* **34** (1928) 586.
[10] MacInnes, D. A., Dolge, M., *Ind. Eng. Chem.,* Anal. Ed. **1** (1929) 57.
[11] LeBlance, M., Harnapp, O. H., *Z. phys. Chemie* **A 166** (1933) 322.
[12] Klauer, F., Turowski, E., Wolff, V., *Angew. Chemie* **54,** 494 (1941).
[13] Karrer, E., Orr, R. S., *J. Opt. Soc. Am.* **36** (1946) 42.
[14] Clark, L. C., *US-Pat. 2913386*, 1958.
[15] Severinghaus, W., Bradley, A. F., *J. App. Physiol.* **13** (1958) 515.
[16] Harley, J., Nel, W., Pretorius, V., *Nature* **181** (1958) 177.
[17] Sieger, J., *Brit. Pat. 864293* (1959).
[18] Lovelock, J. E., Lipsky, S. R., *J. Am. Chem. Soc.* **82** (1960) 431.
[19] Lovelock, J. E., *Nature* **188** (1960) 401.
[20] Weissbarth, J., Ruka, R., *Rev. Sci. Instr.* **32** (1961) 593.
[21a] Peters, H., Moebius, H. H., *Pat. 21673 Ger (East)*, 1961.
[21b] Pungor, E., Hollós-Rosinyi, E., *Acta Chim. Acad. Sci. Hung.* **27** (1961) 63.
[22] Baker, A. R., *Brit. Pat. 892530*, 1962.
[23] Clark, L. C., Lyons, E., *Ann. N. Y. Acad. Sci.* **102** (1962) 29.
[24] Seiyama, et al., *Anal. Chem.* **34** (1962) 1502.
[25] Pungor, E., Toth, K., *Acta Chim. Acad. Sci. Hung.* **41** (1964) 239.
[26] King, W. A., *Anal. Chem.* **36** (1964) 1735.
[27] Riseman, J., Wall, R. A., *US-Pat. 3306837* (1967).
[28] Frant, M., Ross, J., *Science* **154** (1966) 1553.
[29] Updike, S. J., Hicks, G. P., *Nature* **214** (1967) 986.
[30] Ross, J., *Science* **156** (1967) 1378.
[31] Tagushi, K., *US-Pat. 3631436* (1970).
[32] Energetic Sciences Inc., New York, USA.
[33] Bergveld, P., *IEEE Trans. Biomed. Eng.,* **BME-19** (1970) 70.
[34] Frant, M., Ross, J., Riseman, J., *Anal. Chem.* **44** (1972) 10.

[35] Shons, A., Dorman, F., Najarian, J., *J. Biomed. Mater. Res.* **6** (1972) 565.
[36] Mosbach, K., Danielson, B., *Biophys. Acta* **364** (1974) 140.
[37] Lundstrom, M. S., Shivaraman, M. S., Svenson, C., *J. Appl. Phys.* **46** (1975) 3876.
[38] Janata, J., *J. Am. Chem. Soc.* **97** (1975) 2914.
[39] Heyne, L., in: *Measurement of Oxygen,* Degn, H., Baslev, I., Brook, R. J. (eds.); Amsterdame: Elsevier, 1976 pp. 65–88.
[40] Aizawa, M., Kato, S., Suzuki, S., *J. Membrance Sci.* **2** (1977) 125.
[41] Janata, J., Huber, R. J., *Ion Selective Electr. Rev.* **1** (1979) 32.
[42] Mattiason, B., Nilsson, H., Olsson, B., *J. Appl. Biochem.* **1** (1979) 377.
[43] Rechnitz, G. A., Arnold, M. A., Meyerhoff, M. E., *Nature* (London) **278** (1979) 466.
[44] Nylander, C., Liedberg, B., Lind, T., *Sensors and Actuators* **3** (1982) 79.
[45] Wohltjen, H., Dessy, R., *Anal. Chem.* **51** (1979) 1458.
[46] Peterson, J. I., Goldstein, R. S., Fitzgerald, R. V., *Anal. Chem.* **52** (1980) 864.
[47] Tietonthaler, K., Lukosz, W., *Opt. Lett.* **9** (1984) 137.
[48] Belli, S. L., Rechnitz, G. A., *Anal. Lett.* **19** (1986).
[49] International Meeting on Chemical Sensors, Cleveland, USA, Sept. 90, and Eurosensors IV, Karlsruhe, FRG, Okt. 90.
[50] *Proc. 1st Int. Meeting on Chemical Sensors, Fukuoka, Japan, 1983.*
[51] *Proc. 2nd Int. Meeting on Chemical Sensors, Bordeaux, France, 1986.*
[52] *Proc. 3rd Int. Conf. on Solid-State Sensors and Actuators (Transducers '85), Philadelphia, PA, 1985.*
[53] *Proc. 4th Int. Meeting on Solid-State Sensors and Actuators (Transducers '87), Tokyo, 1987.*
[54] Göpel, W., "State and Perspectives of Research on Surfaces and Interfaces", *CEC-Status Report in Physical Sciences DGXII,* Rep. No EUR 13108 EN, ISBN 92-826-1795-5.
[55] *Transducers '89, The 5th Int. Conference on Solid-State Sensors and Actuators & Eurosensors III, Montreux, Switzerland 1989.*
[56] *Proc. 3rd Int. Meeting on Chemical Sensors, Cleveland, USA, 1990.*
[57] *Conf. Proc. Eurosensors IV, Karlsruhe, FRG, 1990.*
[58] Cremer, M., *Z. Biol.,* **47** (1906) 562
[59] Haber, F., Klemensiewicz, Z., *Z. Physik. Chem.* (Leipzig), **67** (1909), 385.
[60] Frenkel, J., *Z. Physik.* **35** (1926) 652.
[61] Wagner, C., Schottky, W., *Z. Physik Chem.,* **B 11,** 163 (1930).
[62] Kröger, F. A., *The Chemistry of Imperfect Crystals* (3 Volumes), North Holland Publishing Co., Amsterdam, (1974).
[63] Bardeen, J., *Phys. Rev.,* **71** (1947) 717.
[64] Bardeen, J., Brattain, W., *Phys. Rev.,* **74** (1948) 230.
[65] Brattain, W. H., Bardeen, J., *Bell Syst. Tech. J.,* **32** (1953) 1.
[66] Kingston, R. H., *J. Appl. Phys.* **27** (1956) 101.
[67] Shockley, W., Pearson, G. L., *Phys. Rev.* **74** (1948) 232.
[68] Nicollian, E. H., Brews, J. R., *MOS (Metal-Oxide-Semiconductor) Physics and Technology,* John Wiley and Sons, New York, 1982. This monograph contains an excellent summary of the evolution of the Si-MOSFET.
[69] Scanlon, W. W., in *Semiconductor Surface Physics* Kingston, R. H. (ed.); University of Pennsylvania Press, Philadelphia, 1957 p. 239.
[70] Bergveld, P. *IEEE Trans. on Biomed. Eng.,* **BM-17** (1970) 70.
[71] Tufte, O. N., Chapman, P. W., Long. D., *J. Appl. Phys.,* **33** (1962) 3322.
[72] Maupin, J. T., Geske, M. L., in *The Hall Effect and its Applications, Proc. of the Comm. Symp.,* Chien, C. L., Westgate, C. R. (eds.), Plenum Press, New York, 1980, p. 421. The history and application of the Hall effectis discussed in other portions of this book.
[73] Moss, S. D., Janata, J., Johnson, C. C. *Anal. Chem.,* **47** (1975) 2238.
[74] Zemel, J. N., *Anal. Chem.,* **47** (1975) 224A.
[75] Lundström, I. et al., *Appl. Phys. Lett.,* **26** (1975) 55.
[76] Buck, R. P., Hackleman, D. E., *Anal. Chem.* **49** (1977) 2315.
[77] James, H., Carmack, G., Freiser, H., *Anal. Chem.,* **44** (1972) 856.
[78] Catrall, R. W., Tribuzio, S., Freiser, H., *Anal. Chem.,* **46** (1974) 2223.
[79] Wen, C. C., Chen, T. C., Zemel, J. N., *IEEE Int. El. Dev. Meet.,* (1978) Washington DC, p. 108.
[80] Wen, C. C., Chen, T. C., Zemel, J. N., *IEEE Trans., El. Dev.,* **ED-26** (1979) 1959.

[81] Hafeman, D. G., Parce, J. W., McConnell, H. M., *Science,* **240** (1988) 1182.

[82] Van der Spiegel, J. et al., *Sens. and Actuators,* **4** (1983) 291.

[83] Lauks, I., et al. *Tech. Digest, 3rd Int. Conf. on Solid State Sensors and Actuators,* Philadelphia, PA, (1985) p. 122.

[84] Lauks, I. R., Istat Inc., Dover NJ, Private communication.

[85] Brown, R. B. et al., *Technical Digest, 3rd Int. Conf. on Solid State Sensors and Actuators,* Philadelphia, PA (1985) p. 125.

[86] Janata, J., "Chemically Sensitive Field Effect Transistors", in *Solid State Chemical Sensors* Janata, J., Huber, R. J. (eds.), Academic Press, New York, 1985, p. 66. References to the earlier work by Janata and co-workers may be found here.

[87] Sauerbrey, G., *Z. f. Phys.,* **155** (1959) 206.

[88] King, W. H., et al., *Anal. Chem.,* **40** (1968) 1330.

[89] Stockbridge, C. D., in *Vacuum Microbalance Techniques,* Behmdt, K. H. (ed.), Plenum Press, New York, p. 193.

[90] Guilbault, G. G., *Anal. Chem.,* **55** (1983) 1682.

[91] Hirschfeld, T., Invited Talk, University of Pennsylvania, 1985.

[92] Wohltgen, H., Dessy, R., *Anal. Chem.,* **51** (1979) 1458.

[93] Wohltgen, H., *Sens. and Actuators,* **5** (1984) 307.

[94] Wohltgen, H., et al., *IEEE Trans. On Ultrasonic, Ferroelectrics and Frequency Control,* **UFFC-34** (1987) 172.

[95] White, R. M., et al., *IEEE Trans. On Ultrasonics, Ferroelectrics and Frequency Control,* **UFFC-34** (1987) 162.

[96] Putley, E. H., in *Semiconductors and Semimetals,* **5,** Willardson, R. K., Beer, A. C. (eds.), Academic Press, New York, 1970, p. 259.

[97] Zemel, J. N. et al. "Recent Advances in Chemically Sensitive Electronic Devices" in *Fundamentals and Applications of Chemical Sensors,* Schuetzle, D., Hammerle, R., ACS Symposium Series 309, American Chemical Society, Washington DC, 1986 p. 2.

[98] Zemel, J. N., in *Solid State Chemical Sensors,* Janata J. Huber, R. J. (eds.), Academic Press, New York, 1985, p. 163.

[99] Polla, D. L., White R. M., Muller, R. S., *Technical Digest, 3rd Int. Conf. on Solid State Sensors and Actuators,* Philadelphia PA, 1985, p. 33.

[100] Peterson, J. I., Goldstein, S. R., Fitzgerald, R. V., *Anal. Chem.,* **52** (1980) 864.

[101] Peterson, J. I., et al., *Anal. Chem.* **56** (1984) 16 A.

[102] Peterson, J. I., Yurek, G. G., *Science,* **224** (1980) 123.

[103] Giuliani, J. F., et al., *Technical Digest, 3rd Int. conf. on Solid State Sensors and Actuators,* Philadelphia PA, 1985 p. 74.

[104] Butler, M. A., Ginley, D. S., *Proc. Symp. on Chemical Sensors,* Turner, D. R., (ed.), Proc. Vol. 87–9, (Electrochemical Society, Pennington NJ, 1987), p. 502.

[105] Logothetis, E. M. Hetrick, R. E., in *Fundamentals and Applications of Chemical Sensors,* Schuetzle, D., Hammerle, R. (eds.), ACS Symposium Series 309, American Chemical Society, Washington DC, 1986. p. 136.

[106] Chang, S. C., Hicks, D. B., in *Fundamentals and Applications of Chemical Sensors,* Schuetzle, D., Hammerle, R. (eds.), ACS Symposium Series 309, American Chemical Society, Washington DC, 1986 p. 71.

[107] Chang, S. C., Hicks, D. B., *Technical Digest, 3rd Int. Conf. on Solid State Sensors and Actuators,* Philadelphia PA (1985) p. 381.

[108] Durst, R. R., Blubaugh, E. A., in *Fundamentals and Applications of Chemical Sensor,* Schuetzle, D., Hammerle (eds.), ACS Symposium Series 309, American Chemical Society, Washington DC, 1986, p. 245.

[109] Stetter, J. R., in *Fundamentals and Applications of Chemical Sensors,* Schuetzle, D., Hammerle, R. (eds.), ACS Symposium Series 309, American Chemical Society, Washington DC, 1986, p. 299.

[110] Transducer Research, Inc. Chicago IL.

[111] Garverick, S. L., Senturia, S. D., *IEEE Trans. El. Dev.,* **ED-29** (1982) 90.

[112] Sheppard, N. F. et al., *Sens. and Actuators,* **2** (1982) 263.

[113] Micromet, Cambridge Mass. 02139.

[114] Yates, D. E., Levine, S., Healy, T. W., J. Electrochem. Soc. Faraday Trans. 1, **70** (1974) 1807.

[115] Siu, W. M., Cobbold, R. S. C., *IEEE Trans. El. Dev.,* **ED-26** (1979) 1805.
[116] Barabash, P. R., Cobbold, R. S. C., *Technical Digest, 3rd Inf. Conf. on Solid State Sensors and Actuators,* Philadelphia PA, (1985), p. 144.
[117] Barabash, P. R., Ph. D. Dissertation, University of Toronto (1987).
[118] Krull, U. L., Thompson, M., *IEEE Trans., El. Dev.,* **ED-32** (1985) 1180.
[119] Thompson, M., et al., *IEEE Trans. On Ultrasonics, Ferroelectronics and Frequency Control,* **UFFC-34** (1987) 127.
[120] Seiyama, T., Yamazoe, N., in: *Fundamentals and Applications of Chemical Sensors,* ACS Symposium Series 309, Schuetzle, D., Hammerle, R. (eds.), Washington, D. C.: American Chemical Society, 1986, pp. 39–55.
[121] Seiyama, T., *Chemical Sensor Technology;* Tokyo: Kodansha-Elsevier, 1988; Vol. 1, pp. 1–2. Yamazoe, N., *Proc. Symposium on Chemical Sensors, Honolulu, Hawaii, 1987;* p. 1.
[122] Seiyama, T., Kato, A., Fujiishi, K., Nagatani, M., *Anal. Chem.* **34** (1962) 1502. Seiyama, T., Kagawa, S., *Anal. Chem.* **38** (1966) 1069.
[123] Taguchi, N., *Jpn. Patent 45-38200, 1962.*
[124] Seiyama, T., Kagawa, S., *Kagaku no Ryoiki* **73** (1965) 73.
[125] Yamazoe, N., Fuchigami, J., Kishikawa, M., Seiyama, T., *Surf. Sci.* **86** (1979) 335. Yamazoe, N., Kurokawa, Y., Seiyama, T., *Proc. International Meeting on Chemical Sensors, Fukuoka, Japan, September 1983;* Tokyo: Kodansha-Elsevier, 1983; p. 33. Yamazoe, N., Seiyama, T., *Transducers '85, International Conference of Solid-State Sensors and Actuators, Philadelphia, PA, June 1985, Digest of Technical Papers;* Piscataway, NJ: The Institute of Electrical and Electronics Engineers, Inc., 1985, p. 376. Tamaki, J. et al., *Shokubai* **31** (1989) 393.
[126] Egashira, M., Shimizu, Y., Takao, Y., *Transducers '89, The 5th International Conference of Solid-State Sensors and Actuators & Eurosensors III, Montreux, Switzerland, June 1989, Abstracts;* Lausanne: Conference Organizers in Medicine, Science and Technology, 1989, A4.2.
[127] Nakahara, T., Takahata, K., Matsuura, S., *Proc. Symposium on Chemical Sensors, Honolulu, Hawaii, 1987;* p. 55.
[128] Suzuki, S., Aizawa, M., *Dekinkagaku* **27** (1972) 1026. Aizawa, M., Kato, S., Suzuki, S., *J. Membrane Sci.* **2** (1977) 1254. Karube. I., Mitsuda, S., Suzuki, S., *Eur. J. Appl. Microbiol. Biotechnol.* **7** (1979) 343.
[129] Miyahara, Y., et al., *J. Chem. Soc. Jpn.* (1983) 823. Karube, I., Matsuoka, H., in: *Biosensor:* Suzuki, S. (ed.): Tokyo: Kodansha, 1984, p. 143.
[130] Ikariyama, Y., Yamauchi, S., Uchida, H., *54th Meetin of the Electrochemical Society of Japan, Osaka, Japan, 1987, Abstracts;* Tokyo: The Electrochemical Society of Japan, 1987, F 102.
[131] Nakamoto, S., Ito, N., Kuriyama, T., Kimura, J., *Sensors and Actuators* **13** (1988) 165.
[132] Ueda, N., et al., *Jinko Zoki* **17** (1988) 196.
[133] Suzuki, H., et al., *Transducers '89, The 5th International Conference of Solid-State Sensors and Actuators & Eurosensors III, Montreux, Switzerland, June 1989, Abstracts;* Lausanne: Conference Organizers in Medicine, Science and Technology, 1989, D 11.3.
[134] Matsuo, T., et al., *IEEE Trans. Biomed. Eng.* **BEE-17** (1974) 70 Shoji, S., Esashi, M., Matsuo, T., *Proc. The International Meeting on Chemical Sensors, Fukuoka, Japan, 1983;* p. 473.
[135] Tsukada, K., Sebata, M., Miyahara, Y., Miyagi, H., *Sensors and Actuators* **18** (1989) 329.
[136] Ishibashi, N., Kohara, H., *Anal. Lett.* **4** (1971) 785. Jyo, A., Torikai, M., Ishibashi, N., *Bull. Chem. Soc. Jpn.* **47** (1974) 2862. Ishibashi, N. et al., *Anal. Sci.* **4** (1988) 527.
[137] Ishibashi, N., et al., *Anal. Sci.* **4** (1988) 527.
[138] Matsushita, Y., Goto, K., *Trans. Iron & Steel Inst. Jpn.* **6** (1966) 132.
[139] Takeuchi, T., *Proc. 2nd International Meeting on Chemical Sensors, Bordeaux, July 1986;* Bordeaux: Bordeaux Chemical Sensors, 1986, p. 69.
[140] Takahama, T., Masuda, M., Sugiyama, Y., *Fuji Jiho* **61** (1988) 503.
[141] Usui, T., Nuri, K., Nakazawa, M., Osanai, H., *Jpn. J. Appl. Phys.* **26** (1987) L2061.
[142] Yoshida, N., et al., *Proc. 9th Kagaku Sensor Kenkyu-kai, 1989;* p. 73.
[143] Nitta, T., Terada, Z., *National Tech. Rep.* **22** (1976) 885. Nitta, T. Terada, Z., Hayakawa, S., *J. Am. Ceram. Soc.* **63** (1980) 295.
[144] Seiyama, T., Arai, H., Yamazoe, N., *Sensors and Actuators* **4** (1983) 85.
[145] Sakai, Y., Sadaoka, Y., Matsuguchi, M., *J. Electrochem. Soc.* **136** (1989) 171.

[146] Hijikigawa, M., *Proc. 2nd International Meeting on Chemical Sensors, Bordeaux, July 1986;* Bordeaux: Bordeaux Chemical Sensors, 1986, p. 101.

[147] Karube, I., Tamiya, E., *Proc. 2nd International Meeting on Chemical Sensors, Bordeaux, July 1986;* Bordeaux: Bordeaux Chemical Sensors, 1986, p. 588.

[148] *Nikkei New Materials* (Japanese), No. 16 (1986) **85;** Tokyo Nikkei McGraw-Hill, Inc.

[149] Yokoo, T., et al., *Jpn. J. Appl. Phys.* **26** (1987) 1082–1087. Shibata, K., et al., *Jpn. J. Appl. Phys.* **26,** (1987) 1898–1902.

[150] Haruta, M., Sano, H., Nakane, M., *Sensors and Actuators* **13** (1988) 339.

[151] Satake, K., Kobayashi, A., Nakahara, T., Takeuchi, T., *9th Meeting on Chemical Sensors in Japan, 1982, Abstracts;* p. 97.

[152] Sadaoka, Y., Jones, T. A., Göpel, W., *Transducers '89. The 5th International Conference of Solid-State Sensors and Actuators & Eurosensors III, Montreux, Switzerland, June 1989, Abstracts;* Lausanne: Conference Organizers in Medicine, Science and Technology, 1989, p. 24.

[153] Miura, N., Kaneko, H., Yamazoe, N., *J. Electrochem. Soc.* **134** (1987) 1875.

[154] Takada, T., Komatsu, K., *Transducers '87, Digest;* Tokyo: 1987, p. 693.

[155] Komiya, H., *Sensor Gijutsu* **9** (1989) 56.

[156] Sugawara, M., Kojima, K., Sazawa, H., Umezawa, Y., *Anal. Chem.* **59** (1987) 2842.

[157] Okahata, Y., Ebato, H., Ye, Y., *J. Chem. Soc. Chem. Commun.* (1988) 1037.

[158] Ikegami, A., Kaneyasu, M., *Transducers '85, International Conference of Solid-State Sensors and Actuators, Philadelphia, PA, June 1985, Digest of Technical Papers;* Piscataway, NJ: The Institute of Electrical and Electronics Engineers, Inc., 1985, pp. 136–139.

[159] Igarashi, I., et al., *Transducers '89, The 5th International Conference of Solid-State Sensors and Actuators & Eurosensors III, Montreux, Switzerland, June 1989, Abstracts;* Lausanne: Conference Organizers in Medicine, Science and Technology, 1989, p. 3.

[160] Nakamoto, T., Fukunishi, K., Morizumi, T., *Transducers '89, The 5th International Conference of Solid-State Sensors and Actuators & Eurosensors III, Montreux, Swsitzerland, June 1989, Abstracts;* Lausanne: Conference Organizers in Medicine, Science and Technology, 1989, p. 298.

3 Chemical Sensor Technologies: Empirical Art and Systematic Research

Wolfgang Göpel, Universität Tübingen, FRG

Contents

3.1 Sensor Materials

The development of sensors and sensor systems is an interdisciplinary task which requires the detailed knowledge of experts from completely different fields. In general, computer scientists, electrical and electronic engineers, physicists, chemists, biochemists, biologists, or medical engineers may be involved [1–9].

Picking three examples: excellent empirical knowledge is required since many years in the proper choice and development of glasses for conventional pH-electrodes [10, 11]. Semiconductor device technology with new materials must be handled in the development of ion-sensitive field effect transistors based upon Ta_2O_5 [12–14]. Enzyme-based biosensors require an understanding of covalent linking of macromolecules to inorganic or organic substrates [15].

Evidently the many different technological aspects of designing sensors or complete sensor systems cannot be treated in detail here. We will merely focus at discussing the most critical part which currently requires most R + D work, ie, the proper choice of *materials for individual sensor elements,* some of which are listed in Table 3-1. We have selected a choice of materials which had been discussed at recent international conferences on chemical and biochemical sensors [16–18].

Table 3-1. Typical materials for chemical and biochemical sensors, (i) inorganic, (o) organic materials.

1. Electron-conductors
   i: oxides (SnO_2, TiO_2, ZnO, Ta_2O_5, IrO_x, Ga_2O_3, ...), semiconductors (Si, GaAs, ...), metals (Pt, Pd/Cu, Pt/Cu, ...), ...
   o: phthalocyanines (PbPc, CuPc, H_2Pc, ...), ...

2. Mixed conductors
   i: perovskites ($SrTiO_3$, $CaMO_3$, ...), ...

3. Solid ion-conductors
   i: ZrO_2, CeO_2, LaF_3, AgJ, AgCl, β-alumina's, Fe thiocyanate, ...
   o: Nafion, ...

4. Cage compounds
   i: zeolites, ...
   o: β-cyclodextrines, calixarenes, valinomycine, ...

5. Optimized heterogeneous catalysts
   Substrates: Al_2O_3, SiO_2, TiO_2, MoO_x, VO_x, ...
   Chemical modifications: oxides of Ce, Zr, Cr, Co, ...
   Promotors: Pt, Rh, Ru, Ni, Pd, ...

6. Langmuir-Blodgett films
   phthalocyaninato-polysiloxanes, ...

7. Polymers
   polypyrrole, polyindole, polysiloxanes, polysilanes, Nafion, ...

8. Membranes
   polysiloxanes, ...

9. Enzyme systems and mediators
   glucose-oxidase, ferrocene, choline esterase, dehydrogenase, ...

10. Antibodies, receptors, organelles, microorganisms, animal and plant cells

The different inorganic, organic, and biological materials require different technologies to design chemical and biochemical sensor structures, some of which will be discussed in Section 3.3.

The most important parameter of many sensor materials is their *conductivity* (Figure 3-1) because the majority of sensors is based upon interactions involving electrical charges (see "Basic Sensors" in Chapters 7-10). Figures 3-2 and 3-3 indicate the importance of appropriate *contacts* to make use of ionic and electronic conductivities. Table 3-2 indicates for the particular example of electron-blocking electrodes that *temperature ranges* of specific conduction mechanisms have to be adjusted carefully. The influence of different interfaces may lead to *non-ohmic behavior* of devices with some typical examples shown in Figure 3-4. Whenever one of these characteristics or properties changes selectively and reversibly upon particle interactions, this phenomenon can be utilized in a chemical sensor.

Optical properties determine the second important parameter of a variety of chemical sensors. In an overview, the dielectric constant ε (v) characterizes typical frequency ranges for possible chemical sensor operations. Examples in Figure 3-5 indicate the low frequency sensor operation in complex impedance spectroscopy ($v < 10^5$ Hz) with electrical responses deter-

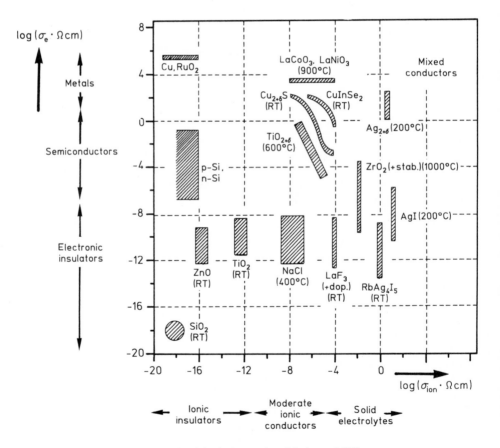

Figure 3-1. Ionic and electronic conductivity in inorganic solids (survey) [19].

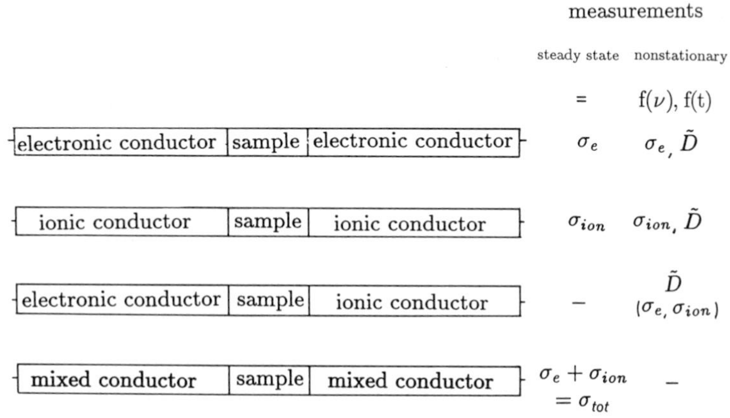

measurements

steady state   nonstationary

			$=$	$f(\nu), f(t)$
electronic conductor	sample	electronic conductor	σ_e	$\sigma_e,\ \tilde{D}$
ionic conductor	sample	ionic conductor	σ_{ion}	$\sigma_{ion},\ \tilde{D}$
electronic conductor	sample	ionic conductor	$-$	$\tilde{D}$ (σ_e, σ_{ion})
mixed conductor	sample	mixed conductor	$\sigma_e + \sigma_{ion}$ $= \sigma_{tot}$	$-$

Figure 3-2. Schematic representation of different electrodes to measure electronic (σ_e), ionic (σ_{ion}), or mixed ($\sigma_e + \sigma_{ion}$) conduction.
$=$ denotes DC-measurements, ν is the frequency, t the time, $\tilde{D}$ the chemical diffusion coefficient

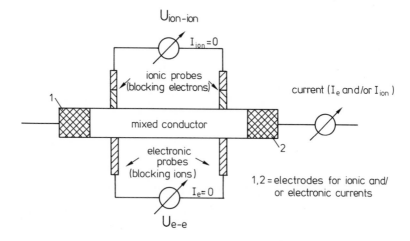

Figure 3-3. Four probe set-ups to measure ionic as well as electronic conductivity of a mixed conductor which is contacted with the mixed conducting contacts 1 and 2. For details, see text and [19].

mined by bulk, grain boundary, three phase boundary, or contact properties. Also indicated is the infrared, visible, and UV range as the domain of optical sensors and of classical optical spectrometers. The latter monitor characteristic excitations responsible for the different maxima in $\varepsilon(\nu)$. If $\varepsilon(\nu)$ changes reversibly upon sensor/particle interaction at any frequency this change may be made use of to design a sensor. Besides the huge amount of analytical spectroscopic tools acting as "sensors", new approaches particularly in fibre optics aim at cheaper "non-spectrometer" solutions.

Table 3-2. Typical electron-blocking electrodes to be used as ionic probes in Figure 3-3 [19]. Examples are given for solid and liquid ion conductors.

Electrode interface (electron ↔ ion conduction)	Exchangable Ion	Temperature range
Ag/AgI	Ag^+	$> 145 \,°C$
$Ag/RbAg_4I_5$	Ag^+	$>\ \ 0 \,°C$
$Cu/CuBr$	Cu^+	$> 300 \,°C$
$Cu/Rb_4Cu_{16}J_7Cl_{13}$	Cu^+	$>\ \ 0 \,°C$
$Cu/CuPbBr_3$	Cu^+	$> 150 \,°C$
$Cu/CuSO_{4(aq)}$	Cu^{++}	$< 100 \,°C$
$Ag/Nafion\ (Ag^+)$	Ag^+	$< 100 \,°C$
$Na\text{-}Hg/Nafion\ (Na^+)$	Na^+	$< 100 \,°C$
$Ag,O_{2(g)}/ZrO_2(+Y_2O_3)$	O^{2-}	$> 300 \,°C$
$Pt,O_{2(g)}/ZrO_2(+Y_2O_3)$	O^{2-}	$> 500 \,°C$
$Pb/PbCl_2$	Cl^-	$> 150 \,°C$
$Ag/AgCl_{(s)}/Cl^-_{(aq)}$	Cl^-	$< 100 \,°C$
$Ag/AgF/LaF_3$	F^-	$<\ \ 80 \,°C$
$Ag/AgCl_{(s)}/Cl^-_{(aq)},\ F^-_{(aq)}/LaF_3$	F^-	$< 100 \,°C$

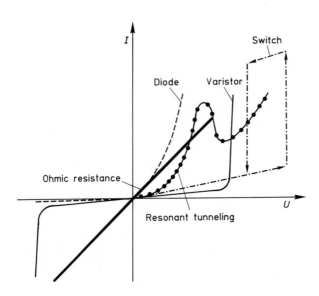

Figure 3-4. Current/voltage characteristics of typical bulk and interface structures.

The combined use of *optical and conductivity properties* leads to other types of sensors. Many basic concepts in the development of chemical and biochemical sensor materials, technologies, and techniques of investigation make use of results obtained in other fields of materials science [20–29]. They will therefore be discussed briefly in the next chapter.

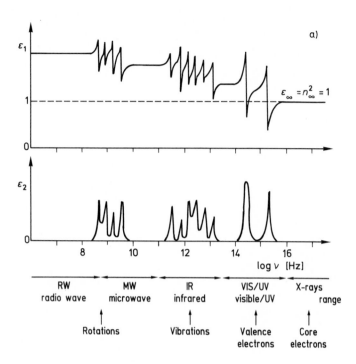

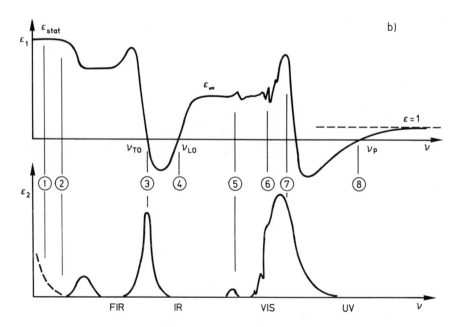

Figure 3-5, Part I. Optical properties of materials: frequency-dependent real (ε_1) and imaginary (ε_2) part of the dielectric constant for a) free molecules and b) semiconductors.

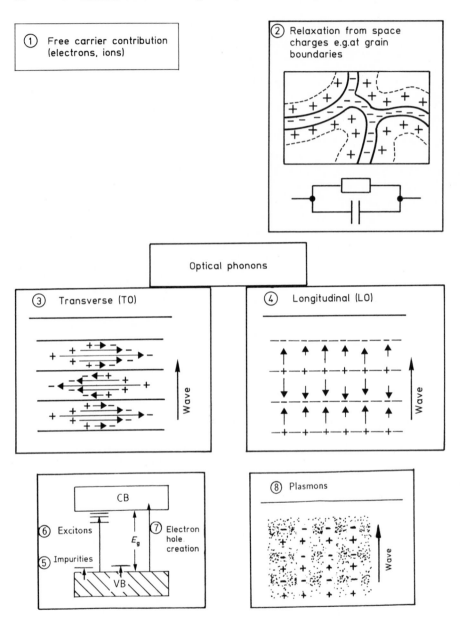

Figure 3-5, Part II. Atomistic origin of the frequency dependence of the dielectric constant of semiconductors. The numbers 1 to 8 correspond to the numbered points of the $\varepsilon-\gamma$ curves in Figure 3-5, Part I, b).

3.2 Sensor Technologies in the Context of General Materials Science

3.2.1 The Key Role of Controlled Interface Properties

The typical setup of practically important sensors (see Chapter 1), the different sensor materials (see Table 3-1), the survey of different contact materials (see Figures 3-2 and 3-3) and of optical properties (see Figure 3-5) indicate clearly the importance of a complete control of interface phenomena to design a chemical sensor [7, 8]. Different surfaces and interfaces may be distinguished, some of which must show reversible sensor/particle interactions and others must not show any chemical reactions in order to avoid drift effects. This complete control of interface phenomena is a well-known task in many other fields of current materials sciences.

Over the last ten years, major key technologies in materials science were developed based upon new ideas in solid-state physics. With increasing demands to reduce the size of solid-state devices, to reduce energy costs, to consider environmental aspects, or to optimize material's performances, the research of surfaces and interfaces now becomes of primary interest for the development of new key technologies in material science [30].

By definition, an interface is the junction of two different substances or of two phases of the same substance. A survey on possible interfaces is given in Figure 3-6. Examples for properties determined by interface structures averaged over different distances are shown in Figure 3-7.

The properties of interfaces as quasi two-dimensional entities are often remarkably different from the properties of bulk matter of the same composition. Concepts to treat interfaces, surfaces, and thin films are linked closely: Surfaces are interfaces between the bulk and the gas phase, thin films are systems whose interior can be strongly influenced by the close proximity of its interfaces.

Current interest in this field becomes evident, eg, from recent Nobel prizes in Physics and Chemistry, from the large amount of new journals, new experimental techniques to study or process interfaces, or from the various national material science research programs in Europe, Japan, or USA. The following typical examples characterize different types of current activities with implications for future sensor developments [2, 6, 21, 32]:

- Scientific concepts are developed for the basic relations between the macroscopic properties of matter (eg, conductivity or optical properties) and its atomic level structure. Because the detailed structures of interfaces in thin films can be manipulated with greater control than those of bulk solids or liquids, they provide particularly attractive systems for studying these basic relations.
- The structure, properties, and reactivity of matter and hence the sensor properties at an interface can be very different from those of matter in the bulk because of the close proximity of the interfacial matter to matter of different composition or of the interfacial matter to vacuum.
- A number of interesting phenomena only appear in systems with extremely small dimensions. Thin films, wires, clusters, and particles can be produced of a small size, so that the quantized states of confined electrons lead to specific size effects and completely new phenomena, which may even be independent of the materials chosen [33].

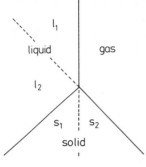

Figure 3-6.
Schematic classification of phases and their interfaces.
l_1, l_2 and s_1, s_2 denote two different liquids and solid materials, respectively.

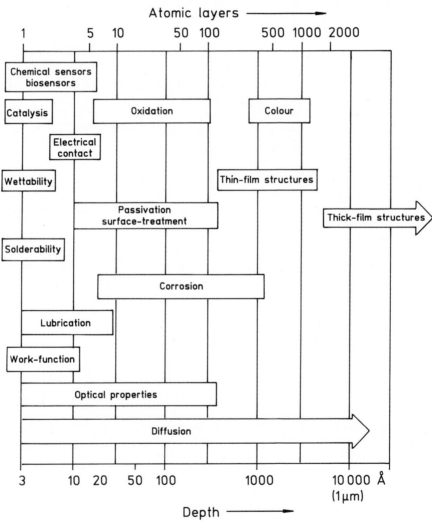

Figure 3-7. Examples for interface phenomena with their characteristic depths [31].

3.2.2 Four Main Directions of Research and Development

The rapidly growing research and development in this field is based upon four main directions, ie, search for

— new physical chemistry concepts for an unequivocal description of the macroscopic phenomenological properties in relation to the atomic structure of interfaces including their dynamic structure,
— new or upgraded materials,
— new experimental tools to characterize the behavior and to observe these structures with different spatial resolution down to the atomic scale (Figure 3-8), and
— new techniques to prepare and tailor interface properties with a control down to the atomic scale.

The first approach to understand unequivocally the structure of surfaces and interfaces usually starts from the characterization of corresponding homogeneous volume structures. Surface and interface properties are then defined by deviations from these ideal volume structures in terms of "excess quantities".

3.2.3 Man-Made and Biological Structures

— Current research and development (R + D) is characterized on the one side by significant progress *to synthesize stable inorganic and organic structures* with the aim of obtaining long-term stabilities and minimum failure rates. As an example for different requirements to optimize interface structures, Figure 1-1 of Chapter 1 shows characteristic surfaces and interfaces of a chemical sensor with (1) chemically active material with controlled grain boundaries and/or bulk doping, (2) thin film/substrate, (3) thin film/contact, (4) thin film/gas phase interfaces, and (5) interfaces involving protecting overlayers. Other typical examples are interfaces in current microelectronics.

— The alternative R + D approach is *to extract and utilize well-defined biological systems.* Compared with man-made synthetic structures these show the inherent differences to have an inbuilt repair mechanism for failures, to have interface properties based upon cooperative dynamic effects, and to require a controlled (bio-)chemical environment. Dynamic structures determine their physical and chemical behavior. Fluctuations and self-organization of matter have optimized these structures for specific functions. As an example, Figure 1-2 in Chapter 1 showed a schematic representation of a biomembrane with different functional units to provide stability, selective diffusion, adsorption, reaction sites, etc.
— Many key-questions in biology and medicine can only be answered by adapting microelectronic and micromachining technologies to biological systems. In this context, large R + D efforts are required *to develop stable interfaces of hybrid systems* to, eg, couple man-made electronic and/or optical devices to "biological devices" such as enzymes, receptors, organelles, membranes, cells, or organisms. As an example, Figure 3-9 illustrates different

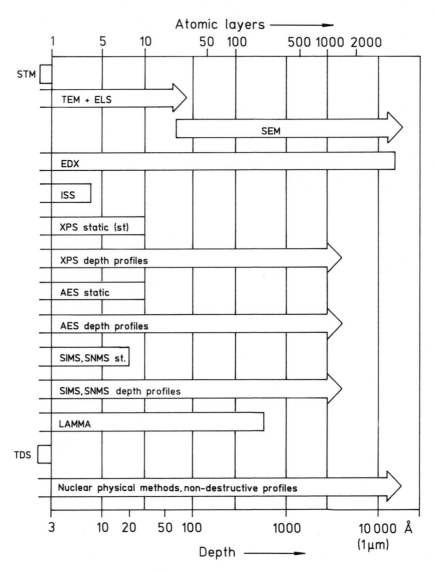

Figure 3-8. Examples for experimental tools of interface analysis utilizing different probes (such as photons, electrons, or ions) and their characteristic information depths [31]. For details of the techniques see Section 3.4.2.

experimental approaches to develop biosensors based upon modified natural or artificial membranes which at the molecular recognition site produce electrical or optical signals utilized for the specific detection of ions, substrates, hormons, antigens etc. The aim is to control the interconnection between electronic, ionic, and general mass transports on the molecular level.

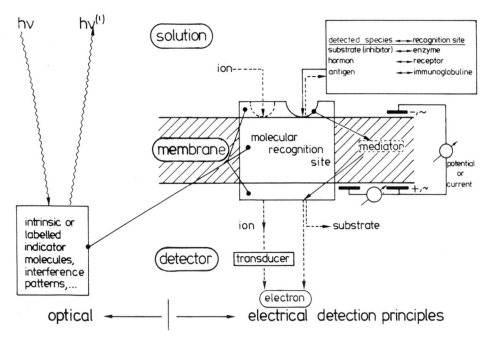

detected species	⟶ recognition site
substrate (inhibitor)	⟶ enzyme
hormon	⟶ receptor
antigen	⟶ immunoglobuline

Figure 3-9. Characteristic interfaces of hybrid systems. This example schematically shows experimental approaches to characterize the functions of modified biological membranes for their use as biosensors. Suitable electrical contacts make possible to monitor electron and/or ion transport as well as interface charges [15].

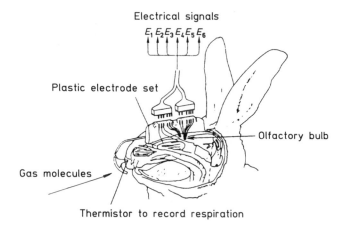

Figure 3-10. "Interfacing" biological sensors to microelectronics. This simple example shows the schematic setup to deduce electrical signals E_1 ,..., E_6 from the olfactory system of a rabbit [34].

Another typical example is the direct deduction of electrical signals from living organisms. As an illustration, Figure 3-10 shows signals E_i deduced from the olfactory system of a rabbit. Monitoring the response upon exposure of the nose to different odors and applying subsequent data evaluation makes possible to study pattern recognition and to simulate parallel processing of neuronal networks in the brain (see Chapter 6).

3.2.4   Empirical Knowledge and Systematic Research

In many fields of practical sensor applications, surfaces and interfaces have so far been optimized empirically. Alternatively the control of surface and interface properties on the atomic level is the key issue and the general aim of their theoretical description, their experimental characterization with high spatial, time, and spectroscopic resolution (Section 3.4), and their preparation technologies (Section 3.3).

For a variety of applications, this control can be obtained in comparative studies performed under ultra-high-vacuum (UHV) conditions on both, well-defined "prototype" structures and structures of practical importance, respectively. In this R + D approach, ideal thin film structures may be prepared and optimized systematically (see Figure 3-11) with respect to specific physical or chemical properties needed for their application [7, 12, 35–40].

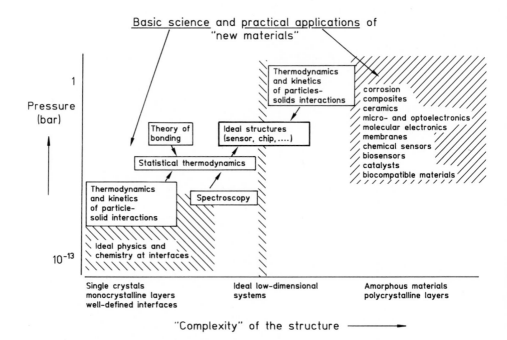

Figure 3-11. Control of surfaces and interfaces on the atomic level: Systematic approach for new materials which can be treated under UHV conditions [8, 9]. A few examples from different fields of applications are given in Section 3.2.6.

In many cases, UHV conditions are not directly suitable to prepare or characterize interfaces. Also, there is an increasing interest in studying burried interfaces non-destructively. This cannot be achieved with many conventional surface spectroscopies. Therefore a variety of new experimental approaches and techniques is now being developed to account for these problems [41].

As can be seen in Figure 3-11 the traditional disciplines of Physical Chemistry offer suitable concepts to coordinate this interdisciplinary R + D work by answering the basic questions:

- Under which conditions do we expect stability or reactions at surfaces and interfaces (→ *thermodynamics*)?
- If reactions are expected, how fast are they (→ *kinetics*)?
- How can we characterize unequivocally static or dynamic structures in both approaches, experimentally (→ *spectroscopy*) and theoretically (→ *theory of bonding*)?
- What is the correlation between spectroscopically and/or theoretically available information and macroscopic phenomena such as electrical, optical, or chemical properties (→ *statistical thermodynamics*)?

3.2.5 Interface Phenomena in Typical Fields of Application

The current research and development of "new materials" concerns

- the upgrading and systematic improvement of existing materials and
- the search for new materials with new chemical compositions.

Table 3-3. Functional properties and applications of some selected ceramic materials which may be used as components in chemical sensors [49].

Class of material	Functional properties	Material	Application
Insulators	Dielectricity	SiO, SiO_2, Si_3N_4, Al_2O_3, SiC, AlN	Diffusion masks, surface passivation, impurity insulation, selected area masks, passivation with high thermal conductivity
	Ferroelectricity	$BaTiO_3$, TiO_2	Capacitors, electrooptics
	Piezoelectricity	$Pb(Zr,Ti)O_3$ [PZT], perovskite-type materials in general, quartz	High voltage generators, ultrasonic generators, mechanical sensors, resonators, relays, pumps, motors, fans, positioners, printers, touch control
	Electrooptics	$(Pb,La)(Zr,Ti)O_3$	Shutters, modulators, color filters, goggles, displays, memories, image storages, holographic recording, optical waveguides
	Pyroelectricity	PZT, $PbTiO_3$ $LiTaO_3$, $SrNbO_3$	Infrared radiation detectors, temperature sensors

(continued on next page)

Table 3-3. (continued).

Class of material	Functional properties	Material	Application
Electrical conductors	**Volume properties**		
	Ionic conductivity	ZrO_2, β, β''-alumina NASICON	Solid electrolytes, gas sensors
	Semiconductivity	TiO_2, SnO_2, ZnO, Pervoskite-type oxides (eg, $SrTiO_3$, $BaTiO_3$, $DrSnO_3$)	Gas sensors
	Negative temperature coefficient of resistance (NTC)	Ion-conducting (eg, ZrO-Y_2O_3) and semiconducting (eg, NiO-TiO_2) thermistor materials	Temperature sensors, temperature compensations
	Grain boundary properties		
	Positive temperature coefficient of resistance (PTC)	Doped $BaTiO_3$	Selfregulating heating elements, temperature compensations
	Varistor effect	ZnO, $BaTiO_3$	Overvoltage protection, surge current absorbers, electronic sensors
	Surface properties		
	Surface ionic conductivity	SiO_2, $ZnCr_2O_4$	Humidity sensors

In both cases, an atomistic understanding of interface properties is crucial. A few characteristic examples for typical fields of application in which recent "materials science" led to high performance are

- control of corrosion, friction, and wear [42, 43],
- development of composites [44, 45],
- ceramics for electronics and electrochemical reactors with typical materials shown in Table 3-3 [46–48],
- inorganic materials for micro- and optoelectronics [33],
- organic materials for molecular electronics and optics with some typical organic devices including chemical sensors shown in Figure 3-12 and a list of molecular electronic materials in Tables 3-4 and 3-5 [50, 51],
- membranes with a survey of ranges of various separation processes shown in Figure 3-13 and a survey of materials, structures, preparations and applications given in Table 3-6 [52–54],
- heterogeneous catalysts [55, 56], and
- biocompatible materials [57, 58].

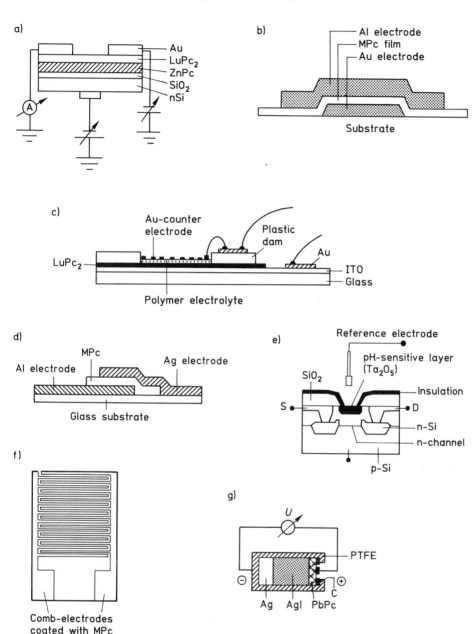

Figure 3-12. Typical thin-film devices illustrating recent developments in the use of molecular materials in (opto-)electronics and chemical sensing: (a) thin-film organic transistor, (b) switching device, (c) electrochemically driven optical display, (d) diode for chemical sensing or solar energy conversion, (e) ion-sensitive field-effect transistor (ISFET), (f) comb structure for thin- film chemical sensors, and (g) sandwich structure with solid electrolyte to monitor gases like O_2 and NO_2. Here, M(Pb, Lu-)Pc denotes metal (lead-, luthetium-)phthalocyanines as typical organic materials [41].

Table 3-4. Examples for applications of selected molecular electronic materials which may be used as components in chemical sensors.

1. Molecular wires, coatings, and shieldings
  a) $(-SN-)_x$ chains
  b) conjugated polymers with ions of AsF_5, K, Li, I_2 as dopants, eg, trans PA, polyphenylene, polyphenylene sulfide, PPy, PMP, PMPS, PTh, PPPO, PPPV, PPP, PHD, PMA
    $\Rightarrow$ Conductivity can be varied by selectively doping the material
  c) donor-acceptor-systems:
    – donors such as TTF, TMTTF, HMTTF, TMTSF, HMTSF, HMTTeF, TTT, TST, BEDT-TTF, perylene
    – acceptors such as TCNQ, DMTCNQ, TNAP, I_2, Cl_2

2. Switches
  a) photochromic systems: fulgide, aberochrome
  b) piezopyroelectric systems: PVDF
  c) liquid crystals: cyanobiphenylene, phenylcyclohexane, phenylbenziate

3. Memory devices
  a) Ag and Cu salts of TCNQ, TNAP and derivatives
    $\Rightarrow$ Exhibit reversible memory switching, threshold switching voltage can be altered by changing donor or acceptor
  b) mass-storage: PMMA, PC, PVC, CAB, BPAPC, PMP, PVDF, PVF, dimethoxy-1,2,dipenylethanone

4. Electronic devices (diodes, transistors, capacitors)
  a) Poly(N,N′-dibenzyl-4,4′-bipyridium), $LuPc_2$, ZnPc
    $\Rightarrow$ Demonstration of molecule-based rectifier; threshold voltage varied by altering molecular materials used
  b) Na^+ implanted polyacetylene
    $\Rightarrow$ Stable in O_2 atmosphere
  c) 3,4,9,10-perylene tetracarboxylic dianhydride on n-Si or p-Si
    $\Rightarrow$ High reverse breakdown voltage
  d) Polypyrrole on Si, $LuPc_2$, ZnPc
    $\Rightarrow$ organic transistor elements
  e) Quinolinium-TCNQ, N-methyl acridinium-TCNQ
    $\Rightarrow$ electrolytes in solid electrolyte capacitors

5. (Rechargeable) Batteries
  a) Perchlorate doped polyacetylene-lithium
    $\Rightarrow$ High energy capacity; lightweight rechargeable system
  b) Electrochemically doped poly(n-vinyl carbazole)
    $\Rightarrow$ Voltage remains constant during discharge; rechargeable lightweight system

c) Poly(vinyl pyridine)-I_2
  ⇒ Comprises cathode in LiI battery
d) Doped Polypyrrole

6. Transparent conductive thin films
  Polypyrrole/polymer complexes
  ⇒ Antistatic packaging for semiconductor and electronics industries; high optical transmission and electrical conductivity

7. Microphones and speakers (transducers)
  Polyvinylidine
  ⇒ Versatile organic piezoelectric material used in sonar hydrophones, microphones, and medical imaging

8. Selected chemical sensor materials already in use
  a) Phthalocyanines (Pc's)
  ⇒ Thermally evaporated PbPc films or Langmuir-Blodgett films of asymmetrically substituted Cu-phthalocyanine detect NO_2
  b) Poly(ethylene maleate) cyclopentadiene
  ⇒ Detects acetone, methylene chloride, benzene, MeOH, pentane in a Surface Acoustic Wave (SAW) device
  c) Triethanolamine on Si
  ⇒ Detects SO_2 in a SAW device

9. Resists
  Polymers, aromatic bisacido components, PMMA

Abbreviations:

TTF	=	tetrathiafulvalene	PMPS	=	poly m-phenylensulfide
TMTTF	=	tetramethyltetrathiafulvalene	PTh	=	poly 2.5-thienylene
HMTTF	=	hexamethyltetrathiafulvalene	PPPO	=	poly p-phenylenoxide
TMTSF	=	tetramethyltetraselenfulvalene	PPPV	=	polyphenylenvinyl
HMTSF	=	hexamethyltetraselenfulvalene	PPP	=	poly p-phenylene
HMTeF	=	hexamethyltetratellurfulvalene	PHD	=	polyheptadienyl
TTT	=	tetrathiatetracene	PMA	=	polymethylacetylene
TST	=	tetraselentetracene	PMMA	=	polymethylmethacrylate
BEDT-TTF	=	bis(ethylendithiolo)tetrathiafulvalene	PC	=	polycarbonyl
TCNQ	=	tetracyanochinodimethane	PVC	=	polyvinylchloride
DMTCNQ	=	dimethyltetracyanochinodimethane	CAB	=	celluloseacetobutyrate
TNAP	=	tetracyanonaphthochinodimethane	BPAPC	=	bisdiallylpolycarbonate
PA	=	polyacetylene	PMP	=	poly-4-methylpentene
PPy	=	polypyrrole	PVDF	=	polyvinyldifluoride
PMP	=	poly m-phenylene	PVF	=	polyvinylfluoride

Table 3-5. Examples for applications of some typical molecular optical materials which may be used as components in chemical sensors.

1. **Optical display devices**
   a) Conducting polythiopene complexes
   ⇒ Produces colored displays with fast switching times and better viewing geometry (no polarizing elements) than LCD devices
   b) Tetrahiafulvalene or pyrazoline in polymethacrylonitrile doped with $LiClO_4$,
   ⇒ Poly(3-bromo-N-vinylcarbazole)
   c) Very fast (<200 msec) photochromic changes observed
   d) N,N'-di(n-heptyl)-4,4'-bypyridinium dibromide
   ⇒ High speed, reversible, exhibits memory effects

2. **Electrophotography**
   a) Poly(vinyl carbazole)-trinitrofluorenone
   ⇒ High quantum efficiency
   b) CuTCNQ + poly(N-vinyl carbazole)-trinitrofluorenone
   ⇒ Charged by halogen lamp, multiduplication up to 50 copies
   c) Polycarbonate-triphenylalanine
      exchange lines
   d) N,N'-diphenyl-N,N'-bis(3-methyl phenyl)-[1,1'-biphenyl]-4,4'-diamine in polycarbonate

3. **Optical information storage (optical recording)**
   a) Metal-TCNQ complexes
   ⇒ Erasable, high contrast media
   b) Photochromic dyes
   ⇒ High sensitivity, stable, and erasable
   c) Liquid crystalline polymers
   ⇒ Provide high contrast and high resolution with high sensitivity
   d) 1,4-dihydroxyanthraquinone
   ⇒ Used in amorphous SiO_2 matrix as frequency domain storage media; requires cryogenic temperatures

4. **Photovoltaics**
   a) Polyacetylene/n-ZnS
   ⇒ Used in Schottky barrier configuration; polyacetylene bandgap matches solar spectrum well
   b) Electrochemically doped Poly(N-vinyl carbazole) merocyanine dyes
   ⇒ Conversion efficiencies 0.015 to 2.00%

5. **Optoelectronic devices**
   Metal-TCNQ complexes
   ⇒ Electrical switching behavior occurs in conjunction with applied optical field

Tab. 3-5. (continued).

6. Photonic devices (solid state optical devices)

a) Electrochemical doped polythiophenes

⇒ Optoelectronic switching accompanied by optically induced doping; subsequent undoping occurs electrochemically

b) Urea

⇒ Used in optical parametric oscillators; conversion efficiencies approach 20%; tunable through visible and IR

c) Methylnitroaniline

⇒ Used as frequency doubling material and non-linear optical material

d) 1 : 1 copolymer of methylmethacrylate and glycidilmethylmethacrylate polyphenylsiloxane, polydiacetylene, polymethylmetacrylate, polystyrrole, polycarbonate

⇒ Used as optical waveguide in integrated optics and waveguide material for lowcost local networks

e) Liquid crystalline polymers

⇒ Used as light valve in optical logic networks

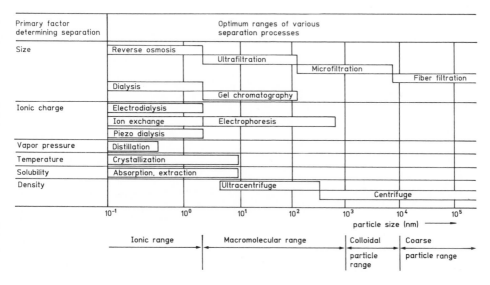

Figure 3-13. Ranges of various separation processes of those technical membranes which are important components of chemical sensor systems [53].

Table 3-6. Synthetic membranes, their structure, preparation, and typical application [53]. All of these membranes may be used as important components (eg, filters) of sensor systems.

Membrane type	Membrane material	Membrane structure	Preparation procedure	Applications
Ceramic and metal membranes	Clay, silica, aluminum oxide, graphite, silver, tungsten	Microporous 0.05 to 20 μm pore diameter	Molding and sintering of ceramic or metal powders	Filtration at elevated temperatures, gas separations
Glass membranes	Glass	Microporous 10 to 100 μm pore diameter	Leading of an acid soluble phase from a two component glass mixture	Filtration of molecular mixtures
Sintered polymer membranes	Polytetrafluoroethylene, polyethylene, polypropylene	Microporous 0.1 to 20 μm pore diameter	Molding and sintering of a polymer powder	Filtration of suspensions, air filtration
Stretched polymer membranes	Polytetrafluoroethylene, polyethylene	Microporous 0.1 to 5 μm pore diameter	Stretching of a partial crystalline polymer film	Air filtration, filtration of organic solvents
Track-etched membranes	Polycarbonate, polyester	Microporous 0.02 to 20 μm pore diameter	Irradiation of a polymer film and acid leading	Filtration of suspensions, sterile filtration of biological solutions
Symmetric microporous phase inversion membranes	Cellulosic esters	Microporous 0.1 to 10 μm pore diameter	Casting of a polymer solution and precipitation of the polymer by a non solvent	Sterile filtration, water purification, dialysis
Asymmetric membranes	Cellulosic ester, polyamide, polysulfone	Homogeneous or microporous, "skin" on a microporous substructure	Casting of a polymer solution and precipitation of the polymer by a non-solvent	Ultrafiltration and reverse osmosis separation of molecular solutions
Composite membranes	Cellulosic esters, polyamide, polysulfone	Homogeneous polymer film on a microporous substructure	Deposition of a thin polymer film on microporous substructure	Reverse osmosis, separation of molecular solutions
Homogeneous membranes	Silicon rubber	Homogeneous polymer film	Extrusion of a homogeneous polymer film	Gas separations
Ion exchange membranes	Polyvinylchloride, polysulfone, polyethylene	Homogeneous or microporous polymer film with positively or negatively charged fixed ions	Immersing an ion exchange powder in a polymer, or sulfonation and amination of a homogeneous polymer film	Electrodialysis, desalination

3.2.6   The Key Role of Physical Chemistry to Coordinate Interdisciplinary Research and Development

The key role of physical chemistry to design "new materials" in general was already pointed out in Section 3.2.4 (Figure 3-11). The design of chemical and biochemical sensors in particular is characterized by the following concepts and aims:

— First, *long-term stability* of the devices is highly desired. This requires negligible interface reactions for corrosion protection of biocompatible materials, electronic devices, ceramics, membranes etc. except at the chemically sensitive sites of sensors and biosensors (see Chapter 4).
— Second, completely *reversible changes* of interface properties are required to detect or convert molecules or ions sensitively, selectively, and reproducibly in sensors and biosensors. The changed interface properties used as (bio)sensor signals are usually monitored as phenomenological parameters, such as frequency-dependent electronic or ionic conductivities, potentials, or optical properties.
— Third, a large variety of *different materials* may be chosen, such as organic and inorganic semiconductors, conductors, ionic and mixed conductors, polymers, membranes, enzymes, antibodies, receptors, micro-organisms, or cells.
— Fourth, all of these structures are developed by *combining empirical knowledge and systematic spectroscopic data.* In a first approach they are optimized empirically by varying systematically certain materials, structures, and preparation techniques, and, if necessary, by subsequently characterizing them in a success/failure analysis. Alternatively they may be optimized systematically by determining spectroscopically the atomistic structure of surfaces and interfaces during their synthesis and test under controlled conditions.
— Fifth, an *atomistic theoretical understanding of interface phenomena* is crucial. Comparison between the atomistic structures as determined spectroscopically and the phenomenological parameters as determined under practical operation conditions makes possible to understand structures, stabilities, and reactivities of surfaces and interfaces on the atomic scale and, as a result, to develop sensor structures with controlled chemistry down to the monolayer range ("interface engineering").

3.3   Preparation Technologies

A large variety of sensor preparation procedures is available or currently being developped

— to produce controlled surfaces of ideal bulk material (Section 3.3.1),
— to modify these surfaces by thermodynamically controlled surface, subsurface, and/or bulk reactions (Section 3.3.2),
— to modify these surfaces irreversibly by overlayer formation (Section 3.3.3), and
— to tailor interface properties on the atomic scale (Section 3.3.4).

Because of the huge amount of R + D work in this field, the different approaches can only be characterized briefly with references given to more detailed and specific literature. Preparation procedures are usually quite different for inorganic if compared to organic materials. Silicon is by far the most important inorganic material often used as base material or substrate, and compatibility of general sensor materials with silicon is usually highly desired. SnO_2 is another important inorganic sensor material, which is used as single crystal, thin film, or thick film material.

3.3.1 Preparation of Clean Surfaces from Bulk Material

This approach is perfect for basic studies of sensor properties but usually too expensive for commercial mass production.

Starting from single crystal bulk or from ideal film materials, three different methods are commonly used to prepare well-ordered clean surfaces:

— Cleavage of single crystals or scraping of single crystals and thick films under UHV conditions without further processing. In certain cases and for certain orientations of single crystals this may immediately lead to atomically clean and geometrically ideal surfaces.
— Cutting of single crystals along well-defined orientations or scraping of single crystals and thick films, subsequent polishing either mechanically, chemically, or electrochemically and then annealing under ultra-high-vacuum conditions to obtain atomically and geometrically clean surfaces. For compounds, care has to be taken to obtain stoichiometric surfaces with their ideal atomic composition within the unit cell. A common problem in this context is the preferential evaporation of volatile elements or formation of thermal decomposition products during heat treatment of the sample. In addition, careful investigations are usually required to study segregation phenomena of bulk impurities in order to avoid surface contaminations at elevated temperatures which may even occur under UHV conditions.
— Cutting of single crystals or scraping of single crystals and thick films and ion bombardment. This procedure is necessary if surface impurities cannot be removed by the heat treatment alone.

For all preparation procedures deviations from geometrically ideal surfaces may occur which may be characterized by the dimension zero, one, or two.

Avoiding these surface atom irregularities is often an extremely difficult task. Surface irregularities, on the other hand, are sometimes prepared intentionally, too. As an example, regular step arrays make possible to introduce lower dimensional defects or to taylor the substrate for subsequent preferential orientation of overlayer molecules.

3.3.2 Thermodynamically Controlled Modifications of Clean Surfaces

Clean surfaces prepared by any of the techniques described above in Section 3.3.1 or below in Section 3.3.3 make possible systematic modifications which at best may be performed under thermodynamically controlled conditions. In these reversible experiments thermodynamic variables are changed repeatedly such as

— temperatures,
— chemical potentials, partial pressures or concentrations of particles in the gas or liquid phase,
— electrochemical and electrical potentials, or
— light intensity.

Examples for film structures depending on thermodynamic conditions of the environment are given in Figure 3-14. Other examples are listed in Table 3-7.

Complete reversibility of these surface modifications has to be checked carefully, eg, by applying the "cheap" or anyone of the other electrical, optical, or spectroscopic techniques to be mentioned below in Section 3.4. The "cheap" techniques are of particular importance for their use as "work horse" techniques in systematic sensor screening tests and for practical sensor applications.

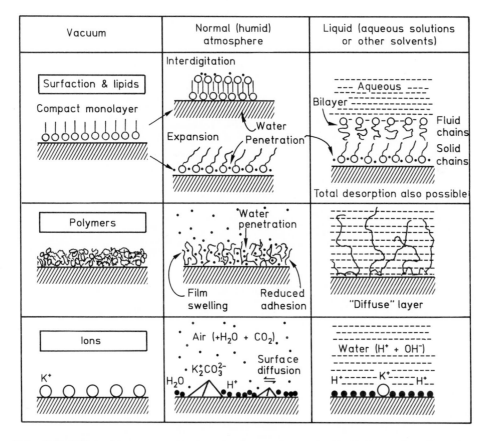

Figure 3-14. Effects of exposure to vacuum, normal atmosphere, and a liquid shown schematically for surfactants and lipids, polymers and ions [59].

Table 3-7. Thermodynamic approach to controlled modification and characterization of surfaces.

---

1. **Identify spectroscopically and search for thermodynamically controlled**
    1.1 Structures of clean surfaces
    1.2 Physisorption
    1.3 Chemisorption
    1.4 Formation of intrinsic defects
        (a) surface defects only
        (b) bulk defects in equilibrium with surface defects
    1.5 Bulk phase transitions
2. **Study the influence of dopants**
    2.1 Surface doping only
    2.2 Bulk doping in equilibrium with surface doping (segregation equilibria)
3. **Study coadsorption phenomena**
4. **Study the kinetics of general solid/gas (liquid) interactions after various sample pretreatments**
5. **Choose different forms of substrates**
    5.1 Single crystals
    5.2 Thin films
    5.3 Thick films
    5.4 Ceramics
    5.5 Polycrystalline and amorphous materials
    5.6 Structured electronic and ionic thin- and thick-film devices
6. **Compare results from 'cheap' techniques of practical importance with results from spectroscopic techniques**
    6.1 'Cheap': V, I, σ, Φ, C, Q_{ads}, Q_{react}, ε, m (see Section 3.4.1, Table 3-9 and Chapter 1, Table 1-2).
    6.2 Spectroscopic: XPS, TDS, AES, EELS, SIMS, ... (see Section 3.4.2 and Tables 3-10 and 3-11.

Several experimental approaches may be chosen for the thermodynamically controlled modification of single-crystal, thin-film, thick-film, ceramic, polycrystalline, and amorphous materials, or of structured electronic and ionic thin- or thick-film devices interacting with the gas or liquid phase. Kinetic steady-state conditions may also lead to reproducible results. They are therefore also mentioned in Table 3-7 as a generalization of the thermodynamic approach to produce controlled surface modifications under continuous flow conditions as they are applied, for instance, in catalytic calorimetric sensors.

All of these reversible modifications may in principle be utilized to build a chemical sensor.

3.3.3 Preparation of Overlayers and Thin Films

Overlayer formation is possible by a huge variety of different preparation procedures with a survey given in Table 3-8. For details, see also [60–63].

– In the first five preparation procedures (PVD, CVD, plasma-polymerization, electrochemical deposition, and chemical deposition) the overlayer is formed by atoms, molecules, ions, or clusters.

Table 3-8. Techniques of overlayer formation [60].

Physical vapor deposition (PVD)
- thermal evaporation
- sputtering
  diodes, triodes, magnetrons, ion beams
- ion plating
  DC-glow-discharge, HF-glow-discharge, magnetron-discharge, cathode arc discharge, low-voltage arc discharge, thermal arc discharge, ioncluster beam deposition
- reactive modifications of these processes

Chemical vapor deposition (CVD)
- thermal CVD
- plasma-activated CVD
- photon-activated CVD
- laser-induced CVD

Plasma polymerization

Electrochemical deposition
- cathodic deposition
  aqeous electrolytes, non-aqeous electrolytes, melt electrolysis, dispersion coatings, laser-activated deposition
- anodic oxidation
  passivation layers of Al, Ta, Nb, ..., duplex layers on Al, oxide layers on Al
- electrophoresis
  electro-dipcoating, enamel processing

Chemical deposition
- electroless plating
- displacement reaction
- homogeneous precipitation
- spray pyrolysis
- chromate formation
- phosphate formation

Thermal spray processes
- flame
- explosion
- arc
- plasma
- plasma in vacuum

Weld deposition
- flame
- arc
  tungsten inert gas, metal inert gas, metal active gas
- plasma
  plasma powder, plasma metal/inert gas, plasma hot wire
- laser beam

Plating processes
- cast plating
- mill plating
- explosion plating

(continued on next page)

Table 3-8. (continued).

  − point plating
  − friction plating
  − aluminothermic plating

Deposition from metallic melts
  − melt dipping
  − liquid quenching

Deposition of organic polymers, emulsions, and pastes
  − mechanical
    painting, printing, screen printing, dip coating, Langmuir-Blodgett coating, spin coating,
    casting, laminating
  − thermal
    melt extrusion, melt floating, dip coating
  − spray process
    mechanical, mechanical-electrostatic, electrostatic, flame spraying

---

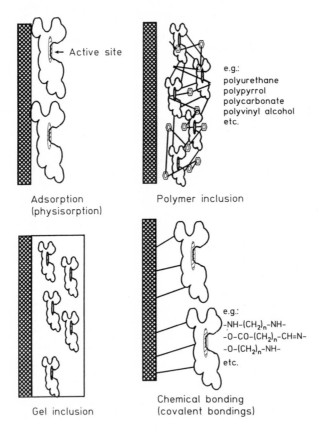

Figure 3-15. Different possibilities for the immobilization of enzymes at electrodes. The active part of the enzymes is indicated by ellipses [65].

— In the next two deposition techniques (thermal spraying and weld deposition) solid or liquid particles of macroscopic dimensions with diameters of tens of μm and more are added.
— In the following two deposition techniques (plating and deposition from metallic melts) the overlayer material is added as a liquid or solid material.
— In the last group, deposition techniques for organic polymer emulsions and pastes are summarized [64].

Specific problems occur in the preparation of biosensors. To illustrate two specific examples, Figure 3-15 shows different immobilization procedures for fixing enzymes of amperometric biosensors by adsorption, immobilization, inclusion, or direct chemical bonding. Specific problems also arise in the proper choice of a mediator with typical structures shown in Figure 3-16.
In the following we focus at some recent techniques which make possible the tayloring of interface properties down to the atomic scale. Some of these techniques were listed in Table 3-8.

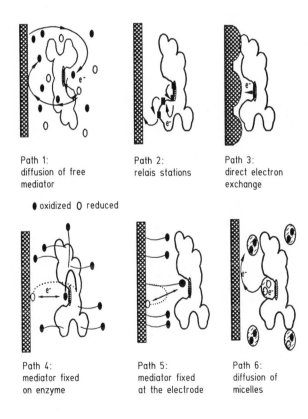

Path 1:
diffusion of free
mediator

● oxidized ○ reduced

Path 2:
relais stations

Path 3:
direct electron
exchange

Path 4:
mediator fixed
on enzyme

Path 5:
mediator fixed
at the electrode

Path 6:
diffusion of
micelles

Figure 3-16. Different possibilities for the mediated electron transfer between the active sites of enzymes and the electrode [65].

3.3.4 Future Trends

3.3.4.1 Monolayer Films

Inorganic monolayer thin films have been studied intensively in the past. Their controlled preparation down to the submonolayer range is possible under UHV conditions with, eg, the molecular beam epitaxy (MBE), ion cluster beam deposition (ICB), or electron beam deposition (EBD) techniques. A variety of small molecule/substrate systems is of interest to study and make use of two-dimensional gases, liquids, and solids with particular emphasis on thermodynamically controlled systems. Organic monolayer thin films may at best be prepared by the Langmuir-Blodgett (LB) technique [66].

As-prepared LB films may subsequently be modified for their practical use as sensors by inserting synthetic or biological multifunctional groups with a few examples illustrated in Figure 3-17.

Strategies are shown in the examples of Figure 3-18 to stabilize lipid membranes by making use of specific membrane/molecule interaction mechanisms, which can also be made use of for LB films [67, 52].

Recent UHV spectroscopic studies at prototype LB films indicate the potential of this technique to prepare, characterize, and modify well-ordered overlayers, particularly if isotope-labelled functional groups are inserted at specific sites [68]. A specific problem of current in-

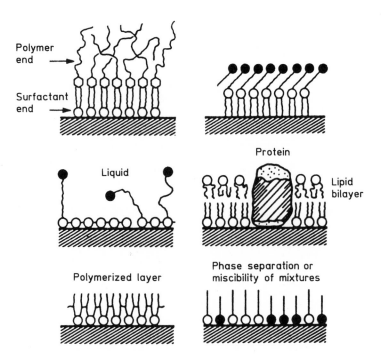

Figure 3-17. Examples for different types of thin-film design with multifunctional groups and multicomponent films [59].

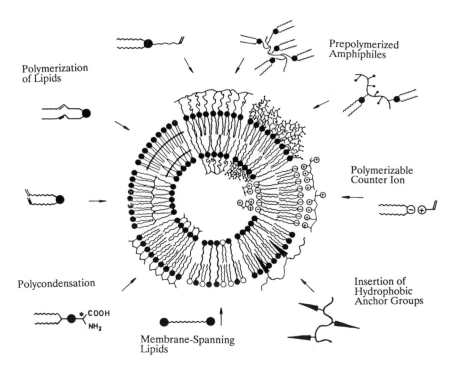

Figure 3-18. Strategies to stabilize lipid membranes [52].

terest is to apply electrical contacts on top of nonconducting LB films without short-circuiting the dielectric film. Recent investigations indicate evaporated Al and Mg atoms to form suitable contacts because of the formation of a dense metal hydroxide in the first surface layer [69]. These investigations point out the need of the preparation to be checked by surface spectroscopic investigations if control of interface structures is desired down to the atomic scale (see Section 3.4.2).

3.3.4.2 Atomically Abrupt Interfaces

So far, the best available inorganic systems with atomically abrupt interfaces have been prepared by the molecular beam epitaxy (MBE) technique [33].

The preparation may at best be done with an in-situ control of the chemical composition of both, substrate and evaporation material, of the evaporation rate, and of geometric structures. Usually the latter is checked subsequently by transmission electron microscopy. This technique requires a very careful sample preparation.

Abrupt interfaces between organic and inorganic material can be prepared for "non-reactive" inorganic substrates. Atomically clean Si, for example, is a highly "reactive" substrate for most organic materials. Saturation of the dangling surface bonds or covalent linking is therefore necessary to avoid organic overlayers to be decomposed in the first monolayer. In this context, empirical knowledge from a variety of other fields of materials science (such as chemical modifications of SiO_2-derived high pressure liquid chromatography (HPLC) materials or catalysts, eg, used for the immobilization of enzymes) is extremely useful to start with a proper choice of materials.

Data on the bulk kinetics and the bulk thermodynamics of possible reactions at the interfaces can usually be used as a first approximation to find combinations for "non-reactive" lower molecular interface materials.

3.3.4.3   From Micromechanics to Nanotechnology

So far we have controlled the composition down to the monolayer range only *perpendicular to the film plane.* Controlled structures *in the film plane* require patterning or micromachining.

Lithographic patterning is a common step in semiconductor device technology with typical procedures as characterized in Figure 3-19. Controlled polymerization with ion beams, electron beams, laser or synchrotron light is used for the production of lateral structures in the micrometer range.

Anisotropic wet etching of silicon (usually in a mixture of ethylenediamine, brenzcatechine and water or KOH in water) is an alternative approach to produce ordered structures in the micrometer range (Figure 3-20).

This makes it possible to produce a variety of microsensor transducers and also micromechanic devices such as resonators, springs, valves, scanning tunneling microscope (STM) components, tip arrays for future STMs etc.

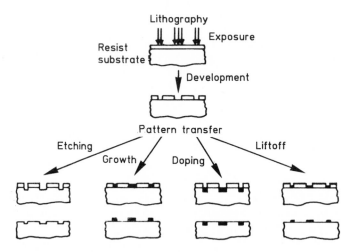

Figure 3-19. The planar process is the most effective means of producing a relief structure, a pattern of selective doping, or a pattern of one material on top of another. In the lithographic step a radiation-sensitive film known as a resist is exposed to some pattern of radiation which can be electrons, ions, or photons. The radiation alters the resist which is usually a polymer enabling the pattern to be developed as a relief image in the resist. This image can be transferred to the substrate by etching, growth, doping, or lift-off [70].

Etching rate

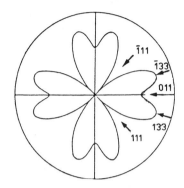

Surfaces

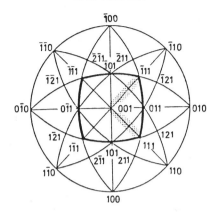

Basic structures

20 nm

Figure 3-20. Etching rate as a function of surface orientation of a Si(100) wafer [71].

In *microsystem technology* a general miniaturization and integration is aimed at in which different electronic, optical, and mechanical properties of a variety of different materials are utilized in miniaturized systems with controlled dimensions in the (sub-)micrometer range.

The next miniaturization step is sometimes called *"nanofabrication"* which requires electron beam or scanning tunneling microscope lithography [70–72].

Current interest in this area is to produce quantum effect devices with their potential use for optical sensors and to contact artificial or synthetic macromolecules for sensor applications. An optimum strategy to produce these structures is to employ electrons, ions, or X-rays whenever they are most effective in solving the particular problem at hand. The resist most widely used for nanometer lithography is PMMA (polymethylmetacrylate).

Recent investigations indicate the potential of using the scanning tunneling microscope not only for the detection but also for the production of nanostructures. An example is shown in Figure 3-21. Ion-conduction of Ag^+ through a NAFION membrane occurs in the high electrical field of the tunneling tip. Discharging of the silver ions leads to silver atom deposition at the surface [73].

For high applied voltages the local generation of resistive heat by a STM tip may produce high temperatures and therefore a structuring on the nanoscale [72].

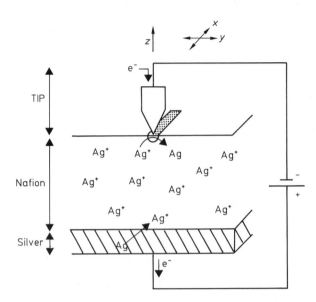

Figure 3-21.
Schematic presentation of a scanning tunneling microscope (STM) technique to deposit microstructured silver on Ag$^+$-conducting Nafion films [73].

Systematic spectroscopy in certain electronic and vibrational states and studies of subsequent energy transfer as well as heat losses will be an active field of research in the near future to utilize manipulation of sensor materials with STM or related scanning techniques.

3.3.5   Control and Manipulation of Individual Molecules

The last step in the controlled miniaturization is to manipulate and modify individual molecules on well-prepared substrates.

Controlled *chemical synthesis of individual molecules* or supramolecular structures is one approach to prepare well-defined systems. Starting from extremely pure materials at best under clean room conditions makes possible to prepare inorganic or organic clusters with controlled outer or inner surfaces, guest/host interfaces etc. As-prepared synthetic molecular structures may subsequently be modified, deposited, embedded etc.

On the other hand a variety of preparation techniques makes possible to extract organelles, cells, mytochondria, membranes, enzymes, hormons etc. from their biological environment. As a simple example, natural bilayer lipid membranes (BLM) are often used as experimental systems for biomolecular electronic device developments.

Examples for *visualizing and manipulating of organic molecules* are shown in Figures 3-22 and 3-23.

Avoiding sample damage during imaging if atomic manipulation is not wanted requires further systematic studies.

The controlled deposition of *individual molecules* on well-prepared, mostly inorganic "prototype" substrates is the first step towards contacting or embedding these molecules in future molecular sensors.

The problem of *individual addressing of molecules* was already pointed out above. Currently available structures from microelectronics have controlled dimensions in the submicron

Figure 3-22.
STM manipulation: this example shows
two individual 2-Ethyl-Hexyl-Phthalate
molecules on graphite which can be fixed
to the substrate by electrical pulses [74].

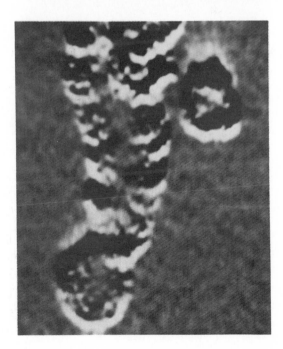

Figure 3-23.
STM picture of a DNA molecule [75].

range. Only thicknesses of extended thin films can be controlled down to the atomic scale. A variety of geometric arrangements for microcontacts is possible with different degrees of realized technologies to produce structures in atomic dimensions. The prerequisite for controlled attachment is an understanding of covalent linking of molecules to the substrate.

3.4   Characterization of Sensor Properties

In the past, most chemical and biochemical sensors have been optimized only empirically. With an increasing demand for thin film fabricated microstructurized sensors and sensor arrays and with an increased number of different materials combined in chemical and biochemical sensors there is an increasing interest in systematic research replacing the purely empirical approach to design an optimized chemical sensor. Also, failure analysis of existing sensors and sensor structures usually requires an atomistic understanding about chemical reactions involved. In addition, there is a variety of electrical, electrochemical, and optical measurements which may be used to characterize sensor structures, to understand the sensing mechanism, and which also may be used for practical applications. We therefore distinguish in the following application-oriented techniques from basic-science-oriented techniques. The latter are usually orders of magnitudes more expensive and usually require an operation by experts.

3.4.1   Application-Oriented Techniques

A survey on electrical, electrochemical, and optical techniques is given in Table 3-9.

The electrical and optical characterization of materials concerns a wide area of theoretical topics and many aspects of applications. The optical principles usually follow classical approaches of optical spectrometers, which in the design of optochemical sensors are simplified, reduced in size or spectroscopic resolution etc. Details are discussed in Chapter 12. Many sensor principles aim at a direct conversion of chemical information into electronic signals. This requires a detailed understanding of electrical properties including:

– conductivities of ions and electrons,
– dielectric constants, both, under static and frequency-dependent conditions,
– current/voltage behavior,
– capacity, and
– relations between these quantities including their dependence on structure, morphology, composition, and on thermodynamic quantities like partial pressures or concentrations.

The various techniques in this field make possible to analyze:

– bulk electrical properties of homogeneous materials, single crystals, thin layers, polycrystalline materials,

Table 3-9. Characteristic techniques to test sensors for practical applications.

1. **Electrical and electrochemical steady-state techniques (without or with photons)**
   Current-voltage characteristics
   Capacitance-voltage characteristics
   Point contact measurements
   2- and 4-point conductivity measurements
   Hebb-Wagner polarization techniques
   Determination of transference numbers
   Permeation measurements
   Thermoelectronic power measurements
   Work function and band bending measurements
   Field effect transistor characteristics
2. **Electrical and electrochemical resonance and relaxation techniques (without or with photons)**
   Time-dependent polarization measurements
   Complex impedance spectroscopy and related Fourier transform techniques
   Pressure modulation spectroscopy
   Thermally stimulated depolarization measurements
   vibrating oscillator measurements,
   surface acoustic wave measurements

— electrical properties of surfaces, their dependence on surface orientation, adsorption, defects, segregation, reconstruction, roughness, adsorption etc.,
— conductivities and dielectric properties of heterogeneous materials, of two-phase materials, ceramics and multiphase materials, materials with dispersions, etc.,
— current/voltage relations and capacities of interfaces and two-phase boundaries of thin-film/substrate, of membrane/solution, or of electrode/electrolyte interfaces,
— electrical and dielectric properties of contacts on various sensor materials, and
— charge flow, electrical fields, charge storage, and stability in microstructures such as gates of gas- or ion-sensitive field effect transistors, or in biological microstructures like cell membranes or cell organelles.

3.4.1.1 Characterization of Bulk Properties: Homogeneous Materials

Homogeneous materials may be characterized phenomenologically by static and frequency-dependent conductivities and dielectric constants. A variety of standardized methods are available for these measurements: Conductivities can be measured by direct current (DC) and alternating current (AC) methods. The choice of materials for the required contacts and the choice of a suitable sample geometry are practical problems to be solved first. In many cases, this may not be trivial.

Electronic and ionic conductivities of a specific sensor can in principle be distinguished by using electronically and/or ionically conducting contacts (see Figure 3-2) [76].

Usually applied setups to compensate the electrode's influence on the measured data are two- or four-point arrangements (Valdes, Van-der-Pauw and other methods) [77–79].

Static bulk properties are measured at low and medium frequencies. At relatively high frequencies additional information on the dynamics of charge carrier transport is available on the microscopic scale from the frequency dependence of dielectric constants or conductivities (Figure 3-5) [80].

Additional information is obtained by monitoring mobilities and diffusion coefficients of charge carriers.

3.4.1.2   Characterization of Heterogeneous Materials, Interfaces, and Electrodes by Relaxation Methods

Heterogeneous materials such as ceramics, multi-phase mixtures, porous materials, layer structures, or structurized membranes play an important role in many applications. Most important in this context is an understanding of the mechanism of charge transport along and through the various discontinuities of the sensor system as they are represented by electrodes, microcontacts, grain boundaries, interfaces, conducting channels and layers, conducting pores and channels in membranes, etc. [81–83]. The analysis of electrical data has to include the influence of geometric factors like grain size and pore size, roughness at contacts, volume fraction of dispersed materials etc., and non-linear current/voltage relations at inner surfaces and interfaces, as well as additional effects on the frequency dispersion of electrical quantities due to the inhomogeneities of the material down to the atomic scale [84].

Useful techniques developped extensively during the last years to characterize electrical response properties concern relaxation methods such as *impedance spectroscopy* and equivalent time-dependent techniques (Fourier transform techniques) [84–87]. The determination and evaluation of real and imaginary parts of the complex impedance as a function of frequency usually shows characteristic ranges of frequencies that are determined by the bulk, electrodes, grain boundaries, etc. (see, eg, Figure 3-5). Impedance spectroscopy has gained particular importance in characterizing sensor properties of ceramic materials or electrochemical interfaces. Data of complex impedances may be evaluated with respect to both, electronic as well as ionic contributions, the detailed procedure depending on the choice of electrode materials. The fundamentals of these techniques are formulated within the framework of linear response theory. Therefore the investigations have to be restricted to small amplitudes.

New and fascinating topics in this area concern effects on charge transport arising from

- fractal geometry and percolation paths in heterogeneous media [88, 89],
- grain size distributions and degree of porosity [81, 84].
- compositions of grain surfaces [84], or
- enhanced grain boundary transport [84].

Impedance measurements with heterogeneous materials and interfaces used to analyze such microstructural effects will be of increasing importance for material characterization and device testing in the future. Also of growing interest are measurements with microcontacts. Further details about particular applications are discussed in Section 3.4.1.9.

3.4.1.3   Separation of Surface, Interface, and Bulk Conductivities

A simple experimental approach to separate surface and interface contributions to the total conductivity consists in measuring the sample conductivity as a function of film thickness for different substrate and overlayer (or residual gas) compositions. Surface excess conductivities of larger samples can be measured by choosing a suitable contact geometry as described for example by ASTM standards [77–79]. Alternative approaches concern the use of microcontacts and the corresponding measurements of local conductivities. A well established technique to characterize inhomogeneous conductivities is the spreading resistance technique based on resistance measurements with a small point contact. This often provides a good probe for local conductances.

3.4.1.4   Characterization of Nonlinear Current-Voltage Relationships

Nonlinear current/voltage curves can be used to characterize nonequilibrium properties and contacts of electronic and ionic conductor sensors [90, 91]. The I/V-curves obtained for metal/semiconductor contacts and semiconductor/semiconductor heterocontacts are usually analyzed with different models describing the injection of electrons or holes, space charge limited currents, tunneling currents, thermionic emission, or high field effects such as avalanche-breakdown (see, eg, Figure 3-4) [91].

Another important area in sensor applications is the electrochemistry of liquid electrolyte/solid interfaces [90]. Investigations of electrode kinetics in that field are mainly based on evaluating current/voltage curves at different scan rates. A successful new approach is to combine and discuss these data along with independent results from spectroscopic techniques and techniques of microanalysis (see below).

Further details can be derived from current/voltage curves if hysteresis and switching effects occur which do or do not depend on the previous history of the sample. These topics are of increasing interest, eg, in molecular sensor devices. Interpretation of these results is usually only possible with additional detailed knowledge about the atomistic structures of involved interfaces, traps in the band-gap of low-conductivity organic compounds etc.

3.4.1.5   Local Electrical Properties

Interest in local electrical properties has led to the development of a variety of techniques using microcontacts, microelectrodes, and recently to the application of the tunneling microscope to measure electrical properties with atomic resolution [92, 93]. The latter measurements make possible the characterization of non-ohmic I/V behavior involving occupied and empty states around the Fermi level for a given specific molecular sensor site at the solid/gas or electrolyte interface. This makes possible to derive detailed pictures about the physical origins of I/V curves based upon occupied and empty electronic states of sensing molecules.

Microcontacts with a lateral resolution down to 0.5 μm are applied in electrochemistry and semiconductor physics to monitor ionic motion along membranes, ion activities in living cells, changes of stoichiometry in solids by diffusion, and local electronic conductivities in semiconductor devices [94, 19]. The study of biological membranes with their inhomogeneous structure characterized by pores and gates and channels is another increasing area of future application for microcontacts and microelectrodes for chemical sensing. Further developments also concern the use of arrays of microelectrodes [95].

3.4.1.6   Measurements of Electrode Potentials, Membrane Potentials, and Electromotive Forces

In electrochemical sensor systems potentiometric measurements make possible the determination of thermodynamic data from electrode equilibria at surfaces, at electrodes, and in multiphase systems. The fundamental properties of the involved interfaces become of increasing importance, eg, for the optimization of electrochemical sensors and microelectronic devices [89].

3.4.1.7   Measurements of Work Functions

The widely used *vibrating capacitor-technique* or microstructured *split-gate devices* monitor sensitively work function changes induced by charge transfer reactions at metal or semiconductor surfaces, by surface dipole moments, and/or by changes in the bulk Fermi level.

The disadvantage of these methods is that the three different contributions to the overall work function cannot be separated and optimized independently. This is possible, however, in *ultraviolet photoemission experiments* (UPS) (see below, Section 3.4.2).

3.4.1.8   Optical Spectroscopy and Photoconductivity

Most of the experiments mentioned above may also be performed with additional light illumination in different frequency ranges. This leads to new sensor principles and makes possible the use and the characterization of electronic excited states or of characteristic relaxation times of localized or delocalized electronic states. Irreversible changes make possible to characterize photochemical reactions.

The interaction of light with matter monitored by typical frequency-dependent absorption, transmission, or reflection effects with unchanged or changed frequency of the light offers a variety of light spectrometer-type arrangements for chemical sensors. These may be simplified by replacing expensive light sources, monochromators, and/or detectors by corresponding cheap and simple components for their use in opto-chemical sensors (see Chapter 12).

3.4.1.9   Other Techniques

For practical sensor applications and a phenomenological characterization, a variety of other experimental techniques is available to monitor not only electronic and ionic conductivity, potentials, or EMF values but also optical, magnetic, or catalytic properties as functions of the partial pressures or concentrations or of the temperature under both, steady state or time-dependent conditions. A new approach to characterize phenomenologically the kinetics of interface reactions under atmospheric pressure conditions utilizes frequency-dependent variations in both, the partial pressure above and the electrical potentials across elec-trochemical cells. Results from the many other techniques to study time-dependent sensor responses or results from sensor arrays make it possible to determine parameters which characterize multi-component mixtures. By means of pattern recognition, these parameters may then be evaluated to identify the individual components in mixtures (see Chapter 6).

3.4.2 Basic-Science Oriented Techniques

3.4.2.1 Systematic Concepts to Treat Interface Properties

An atomistic understanding of molecular recognition in chemical sensors can be deduced by applying interface spectroscopic tools to the sensor structures.

We will briefly characterize typical informations which can be obtained from spectroscopic results. They usually require an understanding of bonds and bands in molecules and solids as they are related to geometric structures and to chemical reactivities.

a) Systematics to Characterize Geometry and Bonding

In a systematic approach to characterize surface and interface properties one usually starts from

— individual atoms and molecules (Figures 3-24a and b) or
— semi-infinite solids (Figure 3-24c to f).

In both cases the most detailed information can be obtained from those materials which form crystalline solids. In this case the following list indicates systematic steps to synthesize complex systems by starting from:

— ideal solids and their quasiparticles,
— bulk defects,
— ideal surfaces,
— surface defects, ie, lattice irregularities at the surface and foreign atoms,
— interfaces of two solid or a solid and a liquid phase,
— sandwich structures between two interfaces at a distance
    a) large enough to neglect any mutual influence,
    b) small enough to allow for mutual interaction, eg, via electrical fields, and
    c) of atomic dimensions with quantum effects of confined particles,
— controlled structures in two and three dimensions,
— low-order structures in mixed systems with
    a) microcrystals and clusters,
    b) nanocrystals, glassy subphases and mesoscopic structures,
    c) fractal subphases.

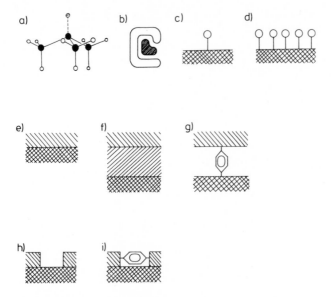

Figure 3-24. Simple molecular and solid-state pictures to characterize geometry and bonding at surfaces
and interfaces (a-i) and to characterize chemical sensing at surfaces (a-d, g, i) [32]:
a) cluster to simulate atomic adsorption at a surface, b) receptor/substrate interaction in a
supramolecule with specific interface bonding, c) "classical" solid-state picture of adsorption at a free surface, d) monolayer adsorption, e) "classical" picture of interfaces,
f) "classical" picture of layer structures, g) molecule interfacing two solids, h) structured
thin films, i) molecule contacted by a structured thin film.

Individual molecules or molecular units which do not crystallize may be used, eg, as independent particles, glasses, polymers, liquids, defects, or subphases in a host lattice.

*b) Spectroscopic, Electrical, and Optical Properties: Quantum Structures as Their Common
   Origin*

An atomistic understanding of the correlation between structure, bonding, and the physical
properties conductivity and optical behavior is of primary interest in current sensor science
and applications. As an example, Figure 3-25 schematically shows bonds and bands in
molecules and solids, respectively, which can be monitored directly by various spectroscopic
tools (see Section 3.4.2.2). The position of the Fermi level E_F (ie, the electrochemical potential of electrons) relative to the valence and conduction band edges E_V and E_C determines the
electrical behavior of

- insulators ($E_C - E_F$ and $E_F - E_V \gg k_B T$),
- semiconductors ($E_C - E_F$ or $E_F - E_V$ up to several $k_B T$), or
- metals ($E_F - E_C \gg k_B T$).

A schematic overview of *optical properties* in a wide frequency range was already given in
Figure 3-5 with its characteristic real (ε_1) and imaginary (ε_2) parts of the dielectric constant

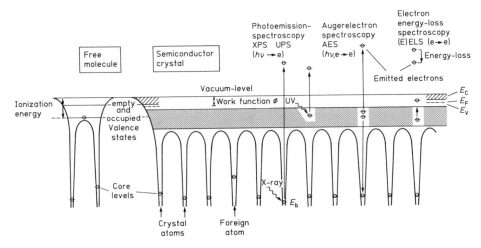

Figure 3-25. Schematic presentation of electron levels in free molecules and solids. Also indicated are some spectroscopic approaches to identify elements and electronic orbitals by means of XPS, UPS, AES, and EELS. For details, see Section 3.4.2.2.

of free molecules, semiconductors, and metals. The microscopic understanding in terms of motions or energy transitions on the atomic scale, ie, the "quantum structure" was also indicated there.

Figure 3-1 showed the (temperature-dependent) *electrical conductivity* of typical bulk materials with their absolute values ranging over more than 24 orders of magnitudes. Interfaces between different materials may lead to typical deviations from the linear current/voltage behaviour (Ohm's law) with a few characteristic examples already shown in Figure 3-4.

In the study of electrical properties, the proper choice of electrical contacts and an atomistic understanding of their interfaces is of general interest for many practical applications with the extreme limit of contacts mediating charge transfer to individual molecules (see Section 3.3.5). In this context, recent developments in scanning tunnelling microscopy even make possible to determine current/voltage curves of individual molecules and to correlate these electrical properties with the quantum structure of this "interface".

3.4.2.2 Information Available from Experiments

a) Survey

As already mentioned above, experimental approaches can be divided in two categories,

- techniques to measure macroscopic phenomenological parameters characterizing mechanical, optical, or electrical properties (see Chapter 3.4.1), and
- techniques to detect geometric structures on the microscopic down to the atomic scale and to monitor energy levels of atoms, molecules, solids etc. attributed to the allowed quantum states of electrons, vibrations, rotations, etc. These techniques are summarized in Tables 3-10 and 3-11. They will be discussed briefly now.

Table 3-10. Common experimental methods of surface and interface analysis to study chemical and biochemical sensors.

1. Geometric arrangement of atoms	
SEM	scanning electron microscopy
(SPA) LEED	(spot profile analysis in) low-energy electron diffraction
SAM	scanning Auger microscopy
(S)SIMS	(scanning) secondary ion mass spectrometry
(S)EDX	(scanning) energy dispersive X-ray analysis
(S)ESD	(scanning) electron stimulated desorption
ISS	ion scattering spectroscopy
STM	scanning tunneling microscopy
AFM etc.	atomic force microscopy and related techniques
2. Elemental composition, contaminations	
AES	Auger electron spectroscopy
SAM	
(M)XPS	(monochromatic) X-ray photoemission spectroscopy
(S)EDX	
TDS	thermodesorption spectroscopy
(S)SIMS	
ISS	
3. Electronic structure of core and valence band levels	
(M)XPS	
(PAR)UPS	(polarization and angle-resolved) ultraviolet photoemission spectroscopy
ELS	energy loss spectroscopy
AES	
UV-VIS	optical spectroscopy
ELL	ellipsometry
4. Dynamic structure, vibrations	
HREELS or	high resolution electron energy loss spectroscopy
EELS	
(FT)IR	(Fourier transform) infrared spectroscopy
RAM	Raman spectroscopy
5. Coverage of adsorbed particles, sticking coefficients	
Δm	mass differences
TDS	
(S)ESD	
AES	
(M)XPS	
6. Bond stabilities of particles at sensor surfaces, heats of reaction, energies and entropies of desorption	
TDS	
(M)XPS	
HREELS	
$Q_{ads, react}$	measurements of adsorption and reaction heats

Table 3-11. Matrix of excitation and detection processes. (For an explamation of the abbreviations see List of Symbols and Abbreviations at the end of this volume).

Detection by ＼ Excitation by	Electrons	Photons	Neutral particles	Ions	Phonons	Electric or magnetic fields
Electrons	AES SAM SEM LEED RHEED EM HEED SES EELS TEM IETS EBIC	(X)CIS (X)APS EMA EDX	ESD	ESD		
Photons	XPS (ESCA) UPS PARUPS CIS CFS X-AES $\Delta\varphi$	SES OS LM FER XD ESR NMR ELL XRF HOL	PD	LAMMA	PAS	PC PDS MPS PVS
Neutral particles			PM AIM DM NIS	MBT FAB—MS	HAM	
Ions	INS	IMXA IEX PIXE	SNMS IID	SIMS IMPA IID ISS RBS		CDS
Phonons		TL FER	TDS		SAM	SDS
Electric or magnetic fields	FES FEM IETS STM $\Delta\varphi$			FIM FIMS FIAP FDM		MPS SDS PDS PC CPD FEC PVS HE CDS CM

The only technique covering both, the determination of geometric structures and the characterization of electrical properties (or "electronic structures") *on the atomic scale* is the scanning tunnelling microscope (STM) [92, 93]. For the same approach on the mesoscopic scale, several additional techniques are available or under development now.

For the determination of structures and properties of individual molecules and supramolecules a variety of techniques from analytical chemistry can be applied successfully such as NMR, ESR, MBS, FTIR, UV-VIS, DTA, XRD etc. (for details see Chapter 15). Most of these techniques, however, are too insensitive to study surfaces and interfaces of well-ordered sensor systems. In many cases, these techniques can therefore only be applied to study bulk materials, structures in solution, or high-surface area materials.

For the investigation of well-controlled surfaces and interfaces with their typical atom densities of 10^{15} cm^{-2}, a large variety of different experimental techniques has been developped now and is constantly increasing. Electrons, photons, ions, atoms, electric and magnetic fields as well as microcontacts are used to probe surfaces and interfaces by specific interactions with schematic experimental arrangements shown in Figure 3-26.

A list of the different methods and an explanation of acronyms is added in Table 3-11.

Recent developments in spectroscopies for sensors and their interfaces aim at

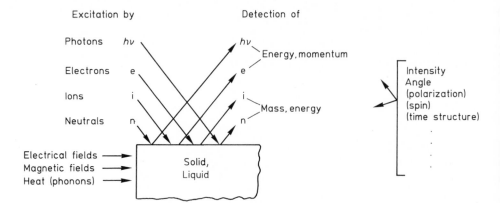

Figure 3-26. Schematic presentation of typical spectroscopic arrangements to study sensor surfaces and interfaces.

— the combined use of several independent tools of interface analysis,
— higher surface sensitivity,
— smaller detection volumes,
— better spatial resolution down to the atomic scale,
— more specific determination, eg, of electronic structures in a given energy and momentum range,
— faster probing,
— lower damage,
— in-situ measurements, eg, under electrochemically controlled conditions,
— study of buried interfaces, and
— on-line preparation control of synthetically prepared or tailored interfaces.

Some details of the different topics:

The *use of several independent tools of interface analysis* applied to the same sample is highly desired. This leads to sophisticated constructions of multi-method interface analysis systems. These usually combine several spectroscopies to characterize the geometric, electronic, dynamic, and chemical structure, electrical and optical properties of the sample, and to study conditions under which sample damage can be avoided (the damage might be introduced by probing with electrons, ions, etc.). These multi-method analysis systems must contain fast sample entry facilities and sample transfer systems linked with independent high pressure or electrochemical cells. An example is shown in Figure 3-27.

The preparation and characterization of ceramic, thick-, or thin-film structures in general and of chemical sensors in particular may at best also be done under controlled ultra-high-vacuum (UHV) conditions. As-prepared prototype devices with well-defined interfaces may then be compared with practical devices by testing both types of devices under realistic measuring conditions and by characterizing them by means of electron, photon and ion spectroscopies using UHV transfer facilities with a typical experimental set-up shown in Figure 3-27 to prepare, modify, and study solid-state gas sensors or electrochemical sensors. For further details, the reader is referred to an extensive literature on this subject [96, 97].

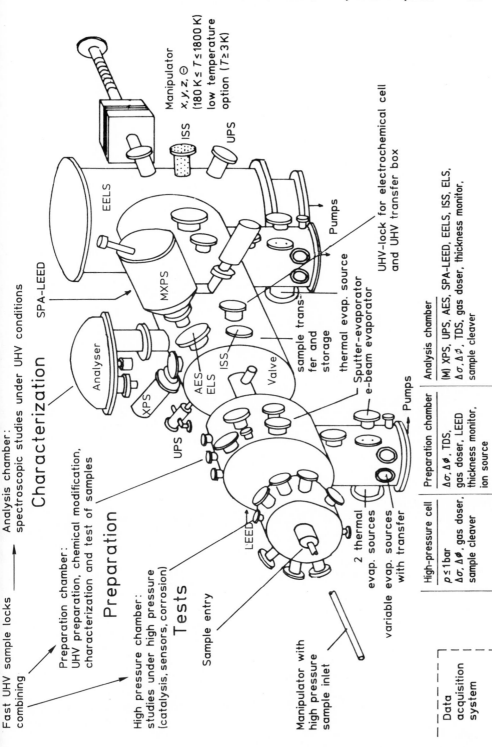

Figure 3-27. Experimental set-up to prepare, modify, test, and study solid-state gas sensors thermodynamically or kinetically controlled under high or atmospheric pressure, under electrochemical and under ultrahigh vacuum (UHV) conditions [8].

If, in addition the specific influence of grain boundaries or surface segregation has to be studied and taken into consideration, high spatial resolution is desirable for all spectroscopies even though the energetic resolution in the determination of quantum structures (electronic, vibrational, rotational, ... states) may be poor if compared to the resolution of most of the instruments shown in Figure 3-27. This compromise to gain high spatial resolution requires additional experimental set-ups with other spectrometers such as STM, AFM, SEM, SAM, SIMS, etc. Grain boundaries play an important role, eg, in temperature and pressure dependent segregation phenomena of ceramic sensors. Organic molecules and biological structures may at best be studied by using optical spectrometers. The detection of hydrogen as well as the characterization of larger molecules in biosensor devices is possible with a variety of optical spectroscopies or with secondary ion mass spectrometry (SIMS) which may also be used in the scanning mode as SSIMS.

High surface sensitivity is usually gained by applying modulation techniques which are sensitive to surface properties or by utilizing probes with a small information depth. The latter is characterized by a small mean free path or a large capture cross section. Two extreme examples are small mean free paths (in the order of atomic dimensions) of electrons in solids at kinetic energies around 80 eV (Figure 3-28) if compared with large mean free paths (in the order of centimeters) of neutrons of the same kinetic energy. The latter can therefore only be used for the characterization of bulk structures of large samples or of surface structures of high-surface area powders.

Small detection volumes are determined by capture cross sections if the detected size is not limited by the experimental set-up itself.

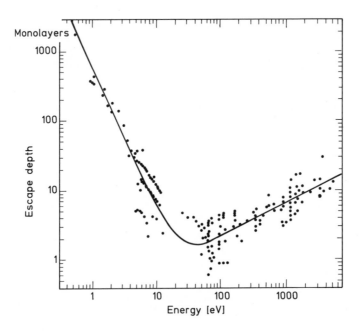

Figure 3-28. Inelastic mean free path (escape depth) Λ of electrons in matter.

Better spatial resolution down to the atomic scale can in principle be achieved by different approaches (Figure 3-29). In scattering techniques it is determined by the wavelength of the probes chosen. For atomic resolution in the order of 1 Å a certain energy of these probes is required (Figure 3-30). The resolution for direct pictures of matter requires even smaller wavelengths. The resolution of beam scanning techniques is usually restricted by the beam diameter and minimum detectable current. The scanning tunnelling microscope also shown in Figure 3-29 offers unique advantages in small-scale sample probing.

For the more *specific determination* of electronic, vibronic etc. levels a variety of approaches are possible, such as the use of monochromatic probes with high energy resolution, spin or angle dependent measurements etc.

Fast probing and low damage during the experiment are often contradicting goals and compromises have to be found. As an example, sample damage cannot be avoided during ion bombardment in SIMS and related techniques (Figure 3-31).

In-situ-measurements to, eg, study catalysts under high pressure conditions or electrochemical reactions in liquid solutions require mean free paths of probes not to be influenced by the air or solution, respectively. In this context techniques to determine optical, electrical, or magnetic properties are usually favorable whereas spectroscopies utilizing photoemitted electrons fail.

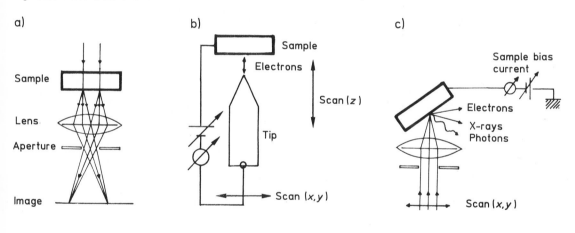

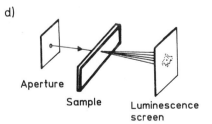

Figure 3-29. Different measuring principles to characterize geometric structures down to the atomic scale: a) direct imaging (TEM, ...), b) scanning tunneling microscope and related techniques (STM, ...), c) general scanning techniques (SEM, SAM, ...), d) scattering techniques (XRD, LEED, ...)

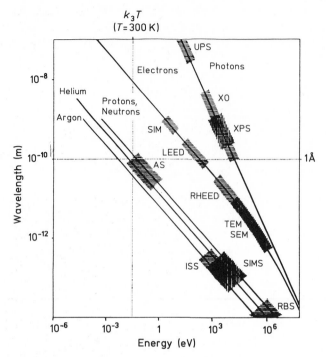

Figure 3-30. Wavelength and energy of different probes for specific techniques of surface and bulk analysis.

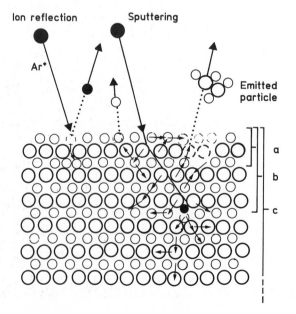

Figure 3-31.
Ion-induced modifications at the surface of a solid; a) emission zone, b) implantation zone, c) lattice perturbation zone.

The study of buried interfaces becomes of increasing importance. A variety of different approaches altogether are still unsatisfactory. Electrical studies such as measurements of voltage and/or frequency dependent conductivities with microcontacts may be performed (Section 3.4.1). Optical measurements may be performed which make use of changed symmetries or dielectric constants at the inner surfaces. For these techniques layers with escape depths of the photons below or in the order of the film thickness and/or angle dependent spectroscopies may be applied. Formation of buried interfaces may also be studied spectroscopically during overlayer formation starting from the submonolayer range.

The *on-line preparation control* of synthetically prepared or tailored interfaces is usually highly desired but requires sophisticated experimental setups. Some details have been discussed above.

b) Geometric Arrangement of Atoms and Morphology

Geometric arrangements of atoms and morphologies may in principle be determined down to the atomic scale by direct imaging (eg, TEM [98]), scanning techniques with particle beams (eg, SEM [99] and SAM) or tunnelling tips (eg, STM [92]), or scattering techniques (eg, LEED [100, 101]) with principles shown in Figure 3-29 and a few details of commonly used techniques summarized in Table 3-12. Recent developments, eg, in scanning tunnelling microscopy or transmission electron microscopy indicate clearly the importance of a physical understan-

Table 3-12. Characteristic features to determine the geometric arrangement of atoms.

	Long-range order required	Spatial resolution	Surface sensitivity	Elemental analysis	Conductivity required	UHV required
LEED	x		1. layer	–	x	x
RHEED	x			–	x	x
PED	–			x	x	x
SEXAFS	–			x	x	x
NEXAFS	–			x	x	–
He Diffr.	x		1. layer	–	–	x
ISS	–	30–100 μm	1. layer	x	x	x
Grazing X-Ray	x		1. layer		–	–
FIM	–		1. layer	x	x	x
EM	–	SEM: ≤0.005 μm	–		x	vacuum
		TEM: ≤0.001 μm	–		x	vaccum
STM	–	>0.01 nm	1. layer		x	–
STM-related	–	>0.2 nm	1. layer		–	–
ESD	–	–	1. layer	x	x	x

of the detection principle for an unequivocal interpretation of the results because of possible sample damage during the experiment. A critical analysis of the influence of the probe chosen and of the physics underlying these experiments is therefore required in all experiments to determine geometric arrangements of atoms. The particular emphasis is now put on studying buried interfaces by – if at all possible – non-destructive methods. Usually destructive methods had to be chosen so far.

c) Elemental Composition

A variety of techniques makes possible the determination of elemental compositions [102, 103] all of which, however, fail so far on the atomic resolution scale. Characteristic features of commonly used techniques are summarized in Table 3-13.

Table 3-13. Characteristic features of techniques to determine elemental compositions.

	Spatial resolution (µm)	Surface sensitivity (nm)	Conductivity required	Quantitative analysis	Detection limit at.% per monolayer	ppm	Chemical analysis	Sample damage
XPS	≥ 20	0.5–10	x	x (10–30%)	0.1	$3 \cdot 10^3$	x	seldom
AES	<0.1	0.5–10	x	x (10–30)%	<0.1	10^3	(x)	seldom
SIMS	1	0.3–100	x	(x) (20–1000%)	1–10^{-3}	1–10^{-4}	(x)	yes
SNMS		0.3–100	x	x	10^{-2}	1	(x)	yes
LAMMA	1–4	10–300	x	–	–	<50	x	yes
TDS	–		–	x	10^{-4}		(x)	no

d) Electronic Structure

A few experimental details of commonly used techniques to study electronic structures [104] of occupied and unoccupied electronic states in the valence band range are summarized in Table 3-14.

Table 3-14. Characteristic features of techniques to determine electronic structures.

	Information about filled bands	Unfilled bands	Surface sensitivity
UPS	x		2–3 layers
IPS	–	x	2–3 layers
PIS	x		1 layer
ELS	"x"	"x"	2–3 layers
$\Delta\varphi$	x		"1. layer"
UV VIS	"x"	"x"	–

e) Dynamic Structure

Experimental details of the most commonly used techniques are summarized in Table 3-15.

f) Improvement of Sensor Performance by Systematic Spectroscopic Studies: Three Simple Case Studies

– For all sensor/gas interactions the mechanism changes significantly with temperature. To adjust reliable sensor operation conditions a detailed understanding about these

Table 3-15. Characteristic features of techniques to determine dynamic structures.

Characteristics	IRAS	SERS	Technique EELS	IETS	NIS	AIS
Resolution (FWHM) cm^{-1}	1–5	1–5	25–100	1–5	50–75	1–5
Spectral range (approx.) cm^{-1}	1500–4000	1–4000	240–5000	240–8000	16–1600	80–500
Sensitivity (% monolayer)	0.1	0.1	0.1	0.1	0.1	0.1
Sample area (mm^2)	10	1	1	1	large	1
Substrate	Crystals, metals and insulators	Roughened Ag, Cu, Au	Crystals, conducting	Oxide-metal	Finely-divided metals and insulators	Crystals, metals and insulators
Adsorbate	Few, mainly CO	Few, CN$^-$, Pyridine	Many	Many	Mainly H and other light atoms	None to date
Ambient pressure	≤1 atm	Sample preparation ≤1 atm	<10^{-6} Torr	Sample preparation ≤1 atm	≤ several atm	≤10^{-6} Torr
UHV compatible	x	x	x	x	x	x
Theoretical situation	Classical	Evolving	Evolving	Evolving	Developed	Evolving

temperature-dependent interaction mechanisms is required. Often, an overview can easily be obtained from thermal desorption spectra. Figure 3-32 shows a typical example of oxygen (O_2) desorption from an oxide sensor material with characteristic ranges in which physisorbed and chemisorbed oxygen are desorbed, point defects are formed at the surface, and bulk defects are involved. Further details about these mechanisms are discussed in Chapter 4.

— The formation of stable ohmic contacts, Schottky barriers, or more generally stable three phase boundaries between metals, oxides, and the gas phase is a typical problem in electronic or ionic conductivity sensors. As an example, comparative spectroscopic and elec-

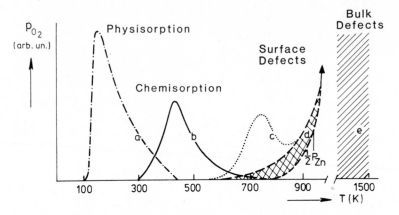

Figure 3-32. Characteristic thermal desorption spectra (TDS) of molecular oxygen on ZnO. Curve c is only obtained during the first heat treatment. Similar results are obtained on SnO_2 and TiO_2 surfaces [7].

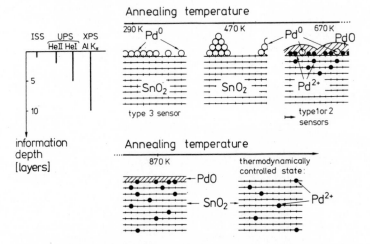

Figure 3-33. Schematic presentation of different tools of interface analysis (ISS, UPS, XPS) to investigate the distribution of those nobel metal atoms (here: Pd) at oxide surfaces (here: SnO_2) which produce different selective interface and sensor properties. Also given is a survey of temperature-dependent interaction mechanisms [105].

trical studies of sensor performances indicate the Pt or Pd metal atoms above the first monolayer of common metal oxide substrates such as TiO_2 or SnO_2 to form Schottky barriers whereas their indiffusion between the first and second layer produces a drastic change in the electrical performance from Schottky-barrier to ohmic behavior. Some typical results and experimental spectroscopic approaches to study diffusion and segregation are shown in Figure 3-33. Evidently, Schottky barrier devices are only stable if indiffusion of metal atoms is thermodynamically or kinetically not favorable. Details about these sensors, the band structure, and conduction mechanisms are discussed in Chapter 4.

– Light sensitivities and drift problems in ion-sensitive field effect transistor (ISFET) structures to monitor pH-values (see Chapter 1, Figure 1-9a and Chapter 5, Figure 5-22) are general problems, which can be solved by characterizing and subsequently optimizing the influence of different preparation steps on the resulting sensor performance. The latter is determined by interfacial bonding of, eg, Ta_2O_5/SiO_2 sandwiches, of Al-treated gates, or of metallic interlayers to protect the electrode from-light generated carriers. In this context, depth profile measurements are essential (Figure 3-34) which are correlated with fabrication steps to produce these ISFET structures.

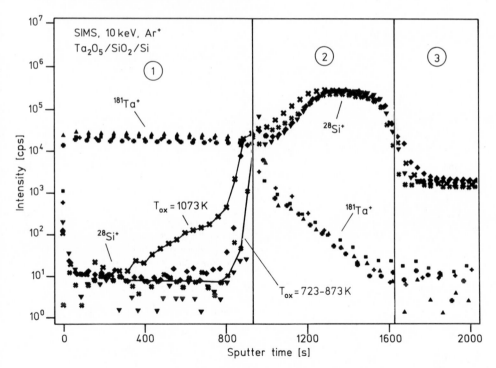

Figure 3-34. Depth profile of the gate of an ion-sensitive Ta_2O_5-based field effect transistor (TOSFET) with different interfaces between Ta_2O_5 and SiO_2 for different oxidation temperatures T_{ox} of the final Ta_2O_5 treatment. Sharp interfaces ($T_{ox} = 723\text{-}873$ K) are required for reversible pH responses in the range between 1 and 10. Note the logarithmic scale of SIMS intensities of the different fragments $^{181}Ta^+$ and $^{28}Si^+$ after different T_{ox}. The region 1 denotes the Ta_2O_5 overlayer, 2 the SiO_2 layer and 3 the Si substrate. The transistor is shown schematically in Figure 5.22 [14].

3.5   References

[1] Gorl, P., McGinley, Ch. (eds.), European Centres of Development on Sensors − *Directory of European Activities 1988*; London: Metra Martech, 1988.

[2] *European Community Research Programmes (Catalog);* Brussels: Commission of the European Communities, 1989.

[3] *Vorschlag für einen Beschluß des Rates über das gemeinschaftliche Rahmenprogramm im Bereich der Forschung und Technologischen Entwicklung (1980–1994);* Brussels: European Communities, 1989.

[4] Kato, S., *Future Technology in Japan:* Tokyo: Institute for Future Technology, 1988.

[5] *Science and Technology in Japan* (journal).

[6] Yano, I. (ed.), *NIPPON − a Charted Survey of Japan 1989/90:* Tokyo: Kokusei-Sha, 1989.

[7] Göpel, W., *Progr. Surf. Sci.* **20** (1985) 9.

[8] Göpel, W., *Sens. Actuators* **16** (1989) 167.

[9] VDI/VDE-Technologiezentrum Informationstechnik GmbH, *Technologietrends in der Sensorik:* Berlin: Hellmich, 1988.

[10] Cammann, K., *Das Arbeiten mit ionenselektiven Elektroden:* Berlin: Springer, 1977.

[11] Freiser, H., *Ion-Selective Electrodes in Analytical Chemistry;* New York: Plenum Press, 1980.

[12] Gimmel, P., et. al., *Sens. Actuators* **17** (1989) 195.

[13] Gimmel, P., et al., *Sens. Actuators* **B1** (1990) 345.

[14] Gimmel, P., et al., "Reduced Light Sensibility in Optimized Ta_2O_5 ISFET Structures", in: *Conf. Proc. Eurosensors IV,* Karlsruhe, FRG, and *Sens. Actuators* **B4** (1991) 135.

[15] Göpel, W., et al., "Biosensor Systems Based upon Receptor Functions", in: *GBF Monographs 13, Biosensors: Applications in Medicine, Environmental Protection and Process Controll,* Schmid, R. D., Scheller, F. (eds.); Weinheim: VCH, 1989.

[16] *Conf. Proc. Eurosensors III,* Montreux, CH, 1989; Lausanne: Elsevier Sequoia.

[17] *Conf. Proc. Eurosensors IV,* Karlsruhe FRG, 1990.

[18] *Proc. Int. Conf. Chemical Sensors,* Cleveland, USA, 1990.

[19] Wiemhöfer, H.-D., *Habilitation Thesis,* University of Tübingen, Tübingen, FRG, 1991.

[20] Sheppard, L. M., *Ceram. Bull.* **68** (1989) 1624.

[21] Thomas, I. L. (ed.), *Material Sciences Programs 1988;* Washington, DC: Department of Energy, 1988.

[22] van Binst, G. (ed.), "Design and Synthesis of Organic Molecules Based on Molecular Recognition", in: *Proc. XVIII Solvay Conf. Chemistry;* Berlin: Springer, 1983.

[23] Yanagida, H., *Angew. Chem.* **100** (1988) 1443.

[24] *Advanced Materials* (journal).

[25] Kreysa, G. (ed.), *Untersuchungen zur Bewertung moderner Entwicklungslinien der chemischen Grundlagenforschung mit hohem technischen Innovationspotential;* Frankfurt: DECHEMA, 1988.

[26] Oshima, K., Saito, K., Hirooka, M. (eds.), *Advanced Materials for Innovations in Energy, Transportation and Communications: Perspectives and Recommendations;* Tokyo: IUPAC CHEMRAWN VI, 1987.

[27] NRC-Committee to Survey Opportunities in the Chemical Sciences, *Opportunities in Chemistry;* Washington, DC: National Academy Press, 1985.

[28] *Frontiers in Chemical Engineering: Research Needs and Opportunities;* Washington, DC: National Academy Press, 1987.

[29] Razim, C., *Angew. Chem.* **100** (1988) 748.

[30] Armstrong, J. A. et al. (eds.), *Research Briefings 1986: Science of Interfaces and Thin Films;* Washington, DC: National Academy Press, 1986.

[31] Stemme, R. (ed.), "Oberflächenanalyse als Dienstleistung", *Technologie Aktuell* **1**; Düsseldorf: VDI Verlag, 1984.

[32] Göpel, W., *State and Perspectives of Research on Surfaces and Interfaces − Review Study for DG XII, CEC;* Brussels: Commission of the European Community, 1990.

[33] Ploog, K., *Angew. Chem.* **100** (1988) 611.

[34] Ikegami, A., Kaneyasu, M., "Olfactory Using Integrated Sensor", *Conf. Proc. Transducers* 85, p. 136.

[35] Harke, S., Wiemhöfer, H.-D., Göpel, W., *Sens. Actuators* **B1** (1990) 188.

[36] Schierbaum, K. D., Wiemhöfer, H.-D., Göpel, W., *Solid State Ionics 28-30* (1988) 1631.

[37] Schindler, K. et al., *Sens. Actuators* **17** (1989) 555.

[38] Mockert, H., Schmeißer, D., Göpel, W., *Sens. Actuators* **19** (1989) 159.

[39] Weiß, W. et al., *J. Vac. Sci. Technol.* **B8** (1990) 715.

[40] Ziegler, Ch. et al., "Interface Analysis of the System Si/YBa$_2$Cu$_3$O$_7$", *Conf. Proc. Angew. Oberflächenanalytik,* Kaiserslautern, FRG and *Fres. Z. Analyt. Chem.,* in press.

[41] Göpel, W., "Chemical Sensing, Molecular Electronics, Nanotechnology: Interface Technologies Down to the Molecular Scale", *Conf. Proc. Eurosensors IV,* Karlsruhe, FRG, 1990 and *Sens. Actuators* **B4** (1991) 7.

[42] *DECHEMA Corrosion Handbook,* Vols. 1-12; Weinheim: VCH, 1989.

[43] LeGressus, C., Maire, Ph., *Surf. Interf. Anal.* **11** (1988) 219.

[44] Brown, A., Gorne, P., Gyulai, G., *European Centers of Development on Advanced Composite Materials − Directory of European Activities 1988;* London: Metra Martech, 1988.

[45] Chou, T.-W., McCullough, R. L., Pipes, R. B., *Scientific American* **255,** No. 4 (1986) 166.

[46] Ismail, M. I. (ed.), *Electrochemical Reactors;* Amsterdam: Elsevier, 1989.

[47] Levinson, L. M. (ed.), *Electronic Ceramics;* New York: Marcel Dekker, 1988.

[48] Buchanan, R. C. (ed.), *Ceramic Materials for Electronics;* New York: Marcel Dekker, 1986.

[49] Arndt, J., in: *Sensors Vol. 1,* Grandke, T., Ko, W. H. (eds.); Weinheim: VCH, 1989, pp. 247-278.

[50] Aviram, A., *Adv. Mat.* **1** (1989) 124.

[51] Carter, F. L. (ed.), *Molecular Electronic Devices;* New York: Marcel Dekker, 1982.

[52] Ringsdorf, H., Schlarb, B., Venzmer, J., *Angew. Chem.* **100** (1988) 117.

[53] Strathmann, H., *J. Mem. Sci.* **9** (1981) 121.

[54] Strathmann, H., *Trends in Biotechnology* **3** (1985) 112.

[55] Anderson, J. R., Boudart, M. (eds.), *Catalysis − Science and Technology;* Berlin: Springer, 1981.

[56] King, D. A., Woodruff, D. P. (ed.), *The Chemical Physics of Solid Surfaces and Heterogeneous Catalysis;* Amsterdam: Elsevier, 1981.

[57] Heimke, G., *Adv. Mat.* **1** (1989) 234 and 345.

[58] Leonhard, E. F., Turitto, V. T., Vroman, L., "Blood in Contact with Natural and Artificial Surfaces", *Annals of the New York Academy of Sciences 516;* New York: New York Academy of Science, 1987.

[59] Swalen, J. D. et al., *Langmuir* **3** (1987) 932.

[60] Haefer, R. A., *Oberflächen- und Dünnschichttechnologie;* Berlin: Springer, 1987.

[61] Frey, H., Kienel, G. (eds.), *Dünnschichttechnologie;* Düsseldorf: VDI Verlag, 1987.

[62] Herman, H., *Scientific American* **259,** No. 3 (1988) 78.

[63] Maissel, L. I., Glang, R., *Handbook of Thin Film Technology;* New York: McGraw-Hill, 1970.

[64] Feast, W. J., Munro, H. S., *Polymer Surfaces and Interfaces;* Chichester: Wiley, 1987.

[65] Löffler, U., *PhD Thesis,* University of Tübingen, Tübingen, FRG, 1991.

[66] Agarwal, V. K., *Physics Today* **41,** No. 6 (1988) 40.

[67] Bain, C. D., Whitesides, G. M., *Adv. Mat.* **1** (1989) 110.

[68] Schreck, M. et al., "Interaction of Slow Electrons with Organic Solids: Comparative HREELS and IR-Studies on Selectively Deuterated Molecules", *Surf. Sci.,* in press.

[69] Schreck, M. et al., *Thin Solid Films* **175** (1989) 95.

[70] Smith, H. I., Craighead, H. G., *Physics Today,* **43** (1990) 24.

[71] Benecke, W., "Mikromechanik für Sensoren", *Mikroperipherik IV* (1988).

[72] Staufer, U. et al., *Appl. Phys. Lett.* **51** (1987) 244.

[73] Craston, D. H., Lin, C. W., Bard, A. J., *J. Electrochem. Soc.* **135** (1988) 785.

[74] Foster, J. S., Frommer, J. E., Arnett, P. C., *Nature* **331** (1988) 324.

[75] Kobbe, B., *Bild der Wissenschaft* **4** (1989) 12.

[76] Rickert, H., *Electrochemistry of Solids;* Berlin: Springer, 1982.

[77] Wieder, H. H., Laboratory Notes on Electrical and Galvano-Magnetic Measurements, *Mat. Sci. Monographs* **2**; Amsterdam: Elsevier, 1979.

[78] van der Pauw, L. J., "A Method of Measuring Specific Resistivity and Hall Effect of Diks of Arbitrary Shape", *Philips Research Reports* **13** (1958).

[79] "Standard Test Methods for DC Resistance or Conductance of Insulating Materials", *ASTM-Standards, D 257–78, 1983.*

[80] Funcke, K., Riess, I., *Z. Phys. Chem.* **NF 140** (1984) 217.

[81] De Levie, R., *Electrochim. Acta* **8** (1963) 751; and *Electrochim. Acta* **10** (1965) 131.

[82] McLaughlin, S., *Current Topics in Membranes and Transport* **9** (1977) 71.

[83] Neher, E., Stevens, C. F., *An. Rev. Biophys. Bioeng.* **6** (1977) 345.

[84] McDonald, J. R., *Impedance Spectroscopy – Emphasizing Solid Materials and Systems;* New York: Wiley, 1987.

[85] Bäuerle, J. E., *J. Phys. Chem. Solids* **30** (1969) 2657.

[86] McDonald, J. R., *Solid State Ionics* **25** (1987) 271.

[87] Hodge, J. M., Ingram, N. D., West, A. R., *J. Electro Anal. Chem.* **74** (1976) 125.

[88] Avnir, D., *The Fractal Approach to Heterogeneous Chemistry;* Chichester: Wiley, 1989.

[89] Kirk-Patrick, S., *Rev. Modern Phys.* **45** (1973) 574.

[90] Bard, A. J., Faulkner, L. R., *Electrochemical Methods;* New York: Wiley, 1980.

[91] Henesch, H. K., *Semiconductor Context;* Oxford: Clarendon Press, 1984.

[92] Hansma, P. K., Tersoff, J., "Scanning Tunneling Microscopy", *J. Appl. Phys.* **61** (1987) R1.

[93] Wickramasinghe, H. K., *Scientific American* **261**, No. 4 (1989) 74.

[94] Ammann, D., *Ion-Selective Microelectrodes;* Berlin: Springer, 1986.

[95] Madou, M. J., Morrison, S. R., *Chemical Sensing with Solid State Devices;* Boston: Academic Press, 1989.

[96] Wutz, M., Adam, H., Walcher, W., *Theorie und Praxis der Vakuumtechnik;* Braunschweig: Vieweg, 1982.

[97] Roth, A., *Vacuum Technology;* Amsterdam: Elsevier, 1983.

[98] Reimer, L., "Transmission Electron Microscopy", *Springer Series in Optical Science;* Berlin: Springer 1989.

[99] Reimer, L., *Scanning Electron Microscoppy:* Berlin: Springer, 1985.

[100] Clarke, L. J., *Surface Chrystallography: An Introduction to Low Energy Electron Diffraction;* Chichester: Wiley, 1985.

[101] Van Hove, M. A., Weinberg, W. H., Chan, C.-M., "Low Energy Electron Diffraction", *Springer Series in Surface Science* **6**; Heidelberg: Springer 1986.

[102] Briggs, D., Seah, M. P. (eds.), *Practical Surface Analysis by Auger- and X-Ray Photoelectron Spectroscopy;* Chichester: Wiley, 1983.

[103] Rivière, J. C., *Surface Analytical Techniques;* Oxford: Oxford University Press, 1990.

[104] Henzler, M., Göpel, W., *Oberflächenphysik;* Stuttgart: Teubner, 1991.

[105] Geiger, J. F., Schierbaum, K.-D., Göpel, W., *Vacuum* **41** (1990) 1629.

4 Specific Molecular Interactions and Detection Principles

WOLFGANG GÖPEL and KLAUS-DIETER SCHIERBAUM, Universität Tübingen, FRG

Contents

4.1   Survey on Basic Sensor Principles

4.1.1   Chemical Sensing with Natural and Artificial Structures

Molecular recognition in chemical and biochemical sensing may be achieved by specific key-lock interactions as they are known from both,

— natural biological systems, eg, for the detection of tastes and odors by receptor proteins incorporated in the biomembrane with the latter shown schematically in Figure 4-1 [1], or from
— microstructurized man-made devices, eg, for the detection of small molecules by electronic charge transfer in thin film layer structures with an example shown in Figure 4-2 [2].

Up to now, details about the molecular structure of receptors shown schematically in Figure 4-1, about their signal transfer and data processing in the brain, ie, details about the molecular-scale understanding of tasting or smelling are not known yet. In addition, our human senses are insensitive to certain hazardous molecules such as CO (see Figure 4-3), they generally do not make possible quantitative estimations of particle concentrations. Man-made chemical and biochemical devices, however, may solve both problems.

Todays man-made (bio-)chemical sensors make use of relatively simple molecule/sensor interaction principles and data evaluations. Most chemical sensors of great practical importance have been developed and optimized empirically without an atomistic understanding of their functions. Some details of the historical development of chemical sensors have been discussed in Chapter 2. Typical examples of widely used chemical sensors including the pH-electrode, the Lambda-probe, the Taguchi-sensor, or the Clark-cell have been presented in Chapter 1.

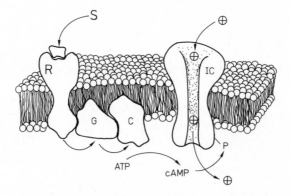

Figure 4-1. Schematic presentation of a biomembrane with characteristic principles of biological detection, transduction, and amplification: Binding of an effector molecule S at a receptor protein R initiates the formation of a second messenger (here cycloadenosinemonophosphate cAMP) by adenylatcyclase C which subsequently induces a catalytic amplification that finally triggers the ion channel (IC) opening. The activation of C involves the coupling G-protein (G). P is the phorphorylation site.

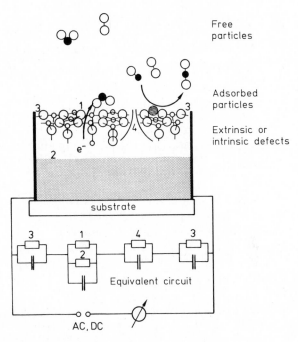

Figure 4-2. Characteristic examples of surfaces and interfaces utilized for chemical sensing in inorganic devices. The detection principles illustrated here are based upon conductance changes of a semiconductor attributed to the surface (1), bulk (2), three phase boundaries or contacts (3), and grain boundaries (4). These processes are associated with charge transfer of electrons (e^-). Further explanations are given in the text.

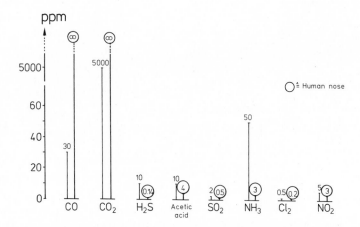

Figure 4-3. Comparison of detection limits of the human olfaction sense (numbers in circles) with the TWA-values, ie, the eight-hour time weighted average of allowed values, of some inorganic and organic compounds in parts per million at one atmosphere total pressure. In contrast to CO and CO_2 which cannot be smelled directly, other hazardous gases such as H_2S, SO_2, and Cl_2 can be recognized extremely sensitively by the human nose.

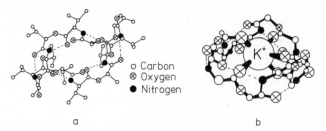

o Carbon
⊗ Oxygen
● Nitrogen

a   b

Figure 4-4. Valinomycin as example for a fully synthesized receptor molecule to detect K^+ ions by their selective embedding in a molecular cage [3]. (a) shows the conformation before complexation, (b) after the complexation [5].

The different types of (bio-)chemical sensors may be classified according to the different sensor properties used for molecular detection such as conductances, potentials, capacitances, heats, masses, or optical constants (Table 1-1).

To illustrate some specific aspects of our current understanding of chemical sensors, Figure 4-2 shows schematically, that specific detection can be achieved by specific reactions at the surface (1), in the bulk (2), at the three phase boundary or contacts (3), or at grain boundaries (4) [3, 4].

All of these reactions might involve molecules, ions, or electrons and all of them require a chemically inert substrate and/or coating of the sensor. The physical and physical chemistry background to understand sensor properties on the basis of specific molecular interactions will be described below in this Chapter.

Evidently there is a huge gap concerning both, size and complexity, between the world of man-made and of biological molecular recognition structures with an overlap between the two worlds in the nanometer range. Future "nanotechnology" is therefore hoped to bridge this gap. Some details about the necessarily interdisciplinary research and development of chemical sensors had been discussed in Chapter 3.

One possible approach is to synthesize receptor molecules for the specific detection of small particles with an example shown in Figure 4-4. An alternative approach is to extract and utilize well-defined biological systems in man-made hybrid devices, ie, in a specific type of biosensor (see, eg, Figure 3-9).

There is, however, also a close similarity between man-made and biological molecular recognition. Mixtures of gases or odors may be identified with sensor arrays (eg, with about ten receptors of the human nose) coupled to a sophisticated data evaluation (eg, with the human brain to identify in the order of 1000 odors [6]). Current trends in utilizing this concept for man-made sensor arrays will be discussed in Chapter 6.

4.1.2 Basic Sensors

The following Figures 4-5 to 4-11 show a few typical examples of schematic setups of those basic sensors which from the application point of view will be discussed in the Chapters 7 to 13. Some typical phenomenological sensor responses x_i' upon changed analytical information p_i (see also Table 1-1) are listed in Table 4-1. A survey on our current physical and chemical understanding of these responses will be given in the following and in Sections 4.4 and 4.5.

Table 4-1.   Survey on sensor responses x' of basic sensors

Liquid state electrolyte sensors (Chapter 7)

Potentiometry *(Section 7.1)*

Voltages V

$$V = E \pm E_0 = \pm (RT/nF) \cdot \ln a_i$$

a_i activity of ion i (see also Table 1-1)

F Faraday constant ($F = 9.6485399 \cdot 10^4 \; C \cdot mol^{-1}$)

n number of electrons involved in the potential-determining reaction

R gas constant ($R = 8.31451 \; J \cdot mol^{-1} \cdot K^{-1}$)

T temperature

Amperometry *(Section 7.2)*

currents I_D

$$I_D = (n \cdot F \cdot A \cdot D \cdot a_i)/\delta$$

A electrode area

D diffusion coefficient

δ thickness of diffusion layer at the surface of the working electrode

Conductometry *(Section 7.3)*

conductivities σ

$$\sigma = \sum_i \sigma_i = \sum_i c_i (z_i e) u_i$$

σ_i partial conductivity of ion i

c_i concentration of ion i

z_i number of charges of the ion i

e elementary charge ($e = 1.60217733 \cdot 10^{-19} \; C$)

u_i mobility of the ion i

Solid-state electrolyte sensors (Chapter 8)

Voltages V $V = (RT/nF) \cdot \ln a_i$

currents I $I = (n \cdot F \cdot A \cdot D \cdot a_i)/\delta$

Electronic conductance and capacitance sensors (Chapter 9)

Resistances R, conductances G $R^{-1} = G = f(p_i)$,

eg, $G \sim p_{O_2}^{-1/m}$, $G \sim p_i^{n_i}$

p_i partial pressure of component i (see also Table 1-1)

p_{O_2} partial pressure of oxygen

m characteristic parameter of defects to be involved (see Table 4-2)

n_i characteristic parameter (see Section 9.5.1)

Capacitances C $C = f(p_i)$

Frequency-dependent conductances $\tilde{Y}$ $|\tilde{Y}| = |G + i\omega C| = f(p_i)$

$i = \sqrt{-1}$

$\omega = 2\pi v$ with v as frequency

(continued on next page)

Table 4-1. (continued).

Field-effect sensors (Chapter 10)

Potentials ΔV $\Delta V = f(\rho_i)$
Source-drain current I_D $I_D = f(\rho_i)$
Work function changes $\Delta\Phi$ $\Delta\Phi = f(p_i)$
               eg, $\Delta\Phi \sim n_i \ln p_i$

Calorimetric sensors (Chapter 11)

Heat of reactions Q_{react} per time $\Delta V \sim \dot{Q}_{react} \sim \Delta P \sim r \cdot \Delta H = f(p_i)$

          ΔV change of balance voltage
          ΔP change in electrical power
          r reaction rate
          ΔH heat of reaction per mole

Optochemical and photometric sensors (Chapter 12)

Optical constants ε as a
function of frequency v $\varepsilon(v) = f(p_i)$
Intensity of light I

Mass-sensitive sensors (Chapter 13)

Masses Δm of ad- or absorbed particles $\Delta f = -C_f \cdot \dfrac{f_0^2}{A} \cdot \Delta m = f(p_i)$

          Δf frequency change
          C_f mass sensitivity
          A coated area
          f_0 resonance frequency of the oscillating quartz

Anode reaction:

$$CO + H_2O \longrightarrow CO_2 + 2H^+ + 2e^-$$

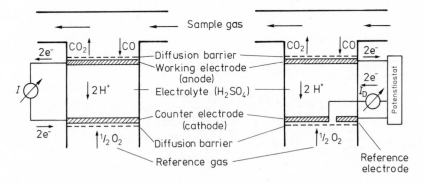

Cathode reaction

$$\tfrac{1}{2} O_2 + 2H^+ + 2e^- \longrightarrow H_2O$$

Figure 4-5. Examples for liquid electrolyte sensors with two-electrode (left side) and three-electrode arrangement (right side) to monitor CO utilizing the amperometric detection principle. The sensor signal is $x' = I_D$ (see Table 4-1).

— The detection of CO in air with two- and three-electrode *liquid electrolyte sensors* utilizing an amperometric operation mode (ie, with currents I_D as sensor response x' at constant voltage V) is illustrated in Figure 4-5. The principle is based upon the electrocatalytic reaction between CO and O_2 molecules to form CO_2. The oxidation of CO occurs at the anode and the reduction of O_2 at the cathode:

$$
\begin{array}{lll}
CO + H_2O & \rightarrow CO_2 + 2H^+ + 2e^- & \text{anode reaction} \\
1/2\ O_2 + 2H^+ + 2e^- & \rightarrow H_2O & \text{cathode reaction} \\
\hline
CO + 1/2\ O_2 & \rightarrow CO_2 & \text{total reaction} \quad (4\text{-}1)
\end{array}
$$

The rate-determining step in the overall reaction is controlled by the diffusion of CO molecules from the diffusion barrier to the surface of the anode where the fast oxidation of CO molecules takes place. Since the activity gradient between diffusion barrier and anode itself is proportional to the partial pressure of CO, the reaction rate and hence the current $I_D = x'$ is a linear function of p_{CO}. In principle, this polarographic principle can be utilized for the detection of all electrochemically active species which can be oxidized or reduced at a certain potential. Since this potential is characteristic for a specific electrode and chemical component its precise adjustment in a three-electrode arrangement (Figure 4-5, right side) improves the selectivity if compared to the cheaper two-electrode-arrangement (Figure 4-5, left side).

— The potentiometric detection principle is demonstrated in Figure 4-6 for the ZrO_2-based λ-probe as *solid-state electrolyte sensor* to measure the oxygen partial pressure p_{O_2}, eg, in the car exhaust. The solid electrolyte shows high bulk conductivity of the O^{2-}-ions. The porous Pt-electrodes make possible the dissociation of molecular O_2 from the gas phase and the electron transfer between the cathode and anode:

$$
\begin{array}{lll}
O_{2,gas} + 4e^- & \rightleftharpoons 2O^{2-} & \text{cathode reaction} \\
2\ O^{2-} & \rightleftharpoons O_{2,gas} + 4e^- & \text{anode reaction} \quad (4\text{-}2)
\end{array}
$$

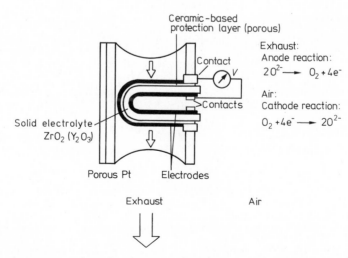

Figure 4-6. Solid-state electrolyte sensor to determine oxygen partial pressures p_{O_2} (λ-probe) [7]. The sensor signal is $x' = V$ (see Table 4-1).

At a constant temperature and a constant oxygen partial pressure p_{O_2} in the reference gas (usually air), a potential V is generated in the concentration gradient between anode and cathode. The potential V is determined by the difference in the chemical potentials of O_2 between exhaust and the constant reference phase (Nernst equation, see Table 4-1). Liquid and solid electrolyte sensors may be operated as potentiometric ($I \approx 0$), amperometric (V = const), or conductometric devices. Details will be discussed in Chapters 5, 7, and 8.
– As an example of an *electronic conductance sensor,* Figure 4-7 shows a metal-oxide based polycrystalline sensor to monitor reducing gases. The response signal x' is the electronic conductance G (in contrast to the ionic conductance of conductometric electrolyte sensors). It is usually measured between two ohmic contacts (Figure 3-4). Its value is determined by competitive electronic charge transfer reactions between negative oxygen species at the surface or at grain boundaries (formed by chemisorption and/or dissociation of O_2 from the gas phase at the n-type semiconductor) and the reducing molecules. As a result of this reaction, a change occurs in the electron density and hence in the conductance G of the oxide. The definition of electronic conductance sensors also includes Schottky-diode sensors (see Figure 3-4) with their voltage-dependent conductances and capacitance (dielectric) sensors with their capa-

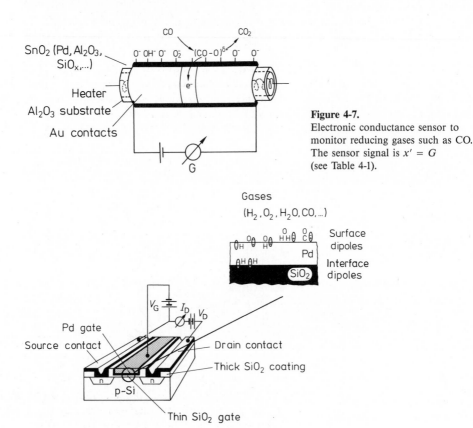

Figure 4-7.
Electronic conductance sensor to monitor reducing gases such as CO. The sensor signal is $x' = G$ (see Table 4-1).

Figure 4-8. Chemical field-effect sensor to monitor H_2 [8]. The sensor signal is $x' = \Delta V = \Delta V_D$ or I_D (see Table 4-1).

citances C as sensor signals. The latter are monitored with alternating currents (AC). The general sensor responses of these types of sensors may be characterized by a change of the frequency-dependent complex conductance (or admittance $\tilde{Y}$, see Chapter 3, Section 3.4.1 and Chapter 9). This concept may also be applied to characterize MOS-diodes with their metal-oxide-semiconductor sandwich structure.

− The detection principle of *field-effect chemical sensors* is illustrated in Figure 4-8 for a typical gas-sensitive device (Gas-FET). It is based upon hydrogen-atom induced changes in the electrical double layer at the Pd/SiO_2 interface which can be monitored by a changed drain-source current I_D through the n-conducting channel. The value of I_D is usually adjusted at a certain level by an electrical field perpendicular to the channel which is generated by a positive voltage V_G at the Pd gate. Adsorption and dissociation of H_2 at the Pd surface and subsequent diffusion of H atoms to the Pd/SiO_2 interface lead to the formation of surface and interface dipoles which then modify the conductance of the n-type channel at a given voltage. Ion-sensitive field-effect transistors (ISFETs) monitor the concentration of ions or adsorbed oriented dipoles. In these devices the metallic gate of Gas-FETs is replaced by the liquid and a reference electrode is added. Details will be discussed in Chapters 5 and 10.

− Typical *calorimetric sensors* are pellistors (Figure 4-9) which are based upon the measurement of reaction heat fluxes $\dot{Q}_{react}$ resulting from the catalytic oxidation of reducing gas species in air at the surface of the heated pellistor. They are usually operated in a Wheatestone bridge arrangement with an inactive reference sensor. The sensor response x' may be chosen as the change in the temperature ΔT monitored by the changed resistance of a platinum filament. The latter leads to a deviation in the difference voltage ΔV in the balanced bridge. In a second operation mode x' may be chosen as the changed power ΔP required to keep the pellistor temperature at a constant value. Oxidation of reducing gaseous species leads to a decrease in the electrical power consumption ΔP. Diffusion barriers in front of the pellistor may lead to a linearization of the sensor signal ($\Delta P \sim p_i$).

− As example for an *optochemical sensor*, the spectroscopic detection of NADH-molecules by a "fluorosensor" is shown in Figure 4-10. Absorption of light with the wavelength $\lambda = 360$ nm leads to excited NADH* molecules which are desactivated by emission of a characteristic fluorescence radiation at $\lambda = 450$ nm. The latter is monitored by a photodetector.

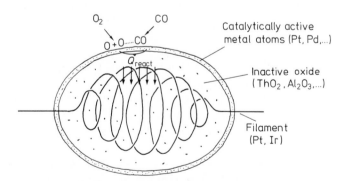

Figure 4-9. Calorimetric sensor to detect reducing gases [9]. The sensor signal is $x' = \dot{Q}_{react}$ (see Table 4-1).

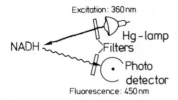

Figure 4-10.
Optochemical sensor ("fluorosensor") to determine the concentration of NADH (nicotinamide adenine dinucleotide) in
culture broths [10]. The sensor signal is $x' = I$ (intensity of
fluorescence (see Table 4-1)).

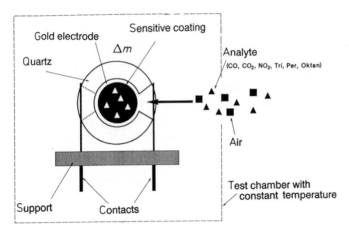

Figure 4-11. Mass-sensitive sensor to detect particles by their selective reaction with the quartz coating.
The mass increase leads to a shift of the resonance frequency [11]. The sensor signal is
$x' = \Delta f$ (see Table 4-1).

– Typical bulk *mass-sensitive sensors* are quartz microbalance sensors (Figure 4-11) which
are coated with a chemically sensitive layer. These devices make possible a sensitive weighing
of mass changes upon adsorption of particles from the gas or liquid phase at the chemically
sensitive layer. The latter lead to a shift of the resonance frequency f_0. The shift may be
monitored as difference signal x' if compared with an oscillating uncoated quartz. Alternatives are surface acoustic wave devices (SAWs).

4.2   Equilibrium and Rate-Determined Sensor Responses:
Thermodynamics and Kinetics

4.2.1   Overview

In the last chapter we introduced basic sensors and their characteristic setups. Their different principles to detect particles can be described phenomenologically in a uniform way.
For this purpose, we apply thermodynamic and kinetic concepts of physical chemistry to
describe chemical reactions in general. Three different types may thus be distinguished:

1. Needed: sensor property x' = partition function

$$\Rightarrow \quad dx' = \left(\frac{\partial x'}{\partial p_1}\right)_{p_{i\neq 1},T} dp_1 + \left(\frac{\partial x'}{\partial p_2}\right)_{p_{i\neq 2},T} dp_2 + \dots \left(\frac{\partial x'}{\partial T}\right)_{p_i} dT$$

and $\oint dx' = 0$ (compare Chapter 1, Equations (1-1) to (1-6))

2. Why sensor/molecule interaction?
How fast?

$\Rightarrow$ driving forces

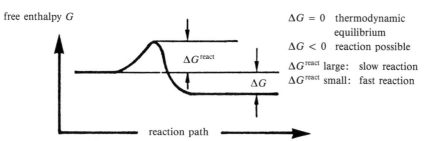

free enthalpy G

ΔG^{react}

ΔG

reaction path

$\Delta G = 0$ thermodynamic
equilibrium

$\Delta G < 0$ reaction possible

ΔG^{react} large: slow reaction

ΔG^{react} small: fast reaction

3. Why G and not simply *energy*?

$\Rightarrow$ Thermodynamics: $\Delta G = \Delta H - T\Delta S = \Delta(U + p_i V) - T\Delta S$ (compare Equation (4-3))

Examples:

Adsorption $\Delta U(\Delta H) < 0$ $\Delta S < 0$
$\rightarrow$ favored at $T = 0$ K
desorption at $T > 0$, if $|T\Delta S| > |\Delta H|$

Point defects $\Delta U(\Delta H) > 0$ $\Delta S > 0$
$\rightarrow$ negligible at $T = 0$ K
favored at high T, if $|T\Delta S| > |\Delta H|$

4. $G \leftrightarrow$ energies E_i of electrons, photons, plasmons, …?

$\Rightarrow$ Statistical thermodynamics

$$E_i \leftrightarrow Z = \sum_i e^{-\frac{E_i}{k_{B}\cdot T}} \leftrightarrow G = k_B T \left(\ln Z - \frac{\partial \ln Z}{\partial \ln V} \right) \leftrightarrow \theta = f(p_i)_{T=\text{const}} \quad \text{etc.}$$

(compare Equations (4-7), (4-8))

Figure 4-12. Survey on the prerequisites for reproducible sensor action (1), driving forces and kinetics of sensor/molecule interaction (2), phenomenological thermodynamics of general sensor/molecule interactions (3), and link between coverages Θ, free enthalpies G and spectroscopic data E_i characterizing the atomistic structure of sensor surfaces and interfaces (4).

- equilibrium sensors (described by thermodynamic equilibria, Section 4.2.2.),
- rate-determined sensors (described by kinetic steady-state flow conditions, Section 4.2.3), and
- one-way sensors (Section 4.2.4).

If reversibility is required in the general definition of a sensor (see Chapter 1), the last type should be excluded. This strict requirement, however, would restrict, eg, the general treatment of biosensor principles. We will therefore not rigorously exclude the discussion of one-way sensors in this book, particularly in those cases where future developments aim at regenerable one-way sensors.

In practical applications, all three types of sensors usually show problems because of drifts or uncontrolled memory effects which depend on the previous history of the sensor and which often restrict potential fields of applications very seriously.

As we described formally in Chapter 1, sensors respond with signals x_i' to the analytical information p_i (ie, to concentrations, partial pressures, ion activities etc.) of certain species i. The "driving force" required for an interaction between the particle i and the sensor is described thermodynamically by a negative change in the total free enthalpy ($\Delta G < 0$) during particle/sensor interaction along the "reaction path" [12]. In the simplest approximation, this path is considered as the distance of the particle from the surface. This is shown schematically in Figure 4-12. In reality, the general sensor/particle interactions, the reaction paths, and the energy minima are more complex if compared to this simplified picture.

The energy minima for instance correspond to minima in a $3n$-dimensional representation of the total free enthalpy as a function of all $3n$ coordinates of the sensor/particle system with its n atoms [13]. As it is well known from classical thermodynamics, the free enthalpy change

$$\Delta G = \Delta H - T\Delta S = \Delta (U + p_i V) - T\Delta S =$$

$$= \Delta \left[E_{\text{pot}}^{T=0} + \int_{T=0}^{T} C_V \, dT + \sum \Delta E_{i,\text{ph}} (T_{\text{ph}}) + p \, V_i \right] - T\Delta S \tag{4-3}$$

is a compromise between two terms:

- The enthalpy change ΔH ("generation of heat") which comprises changes in the inner energy ΔU at $T = 0\,\text{K}$ ($E_{\text{pot}}^{T=0}$), its specific heat (C_V) term at $T > 0\,\text{K}$, possibly energy changes $\Delta E_{i,\text{ph}}$ from phase transitions at T_{ph}, and changes in the mechanical work $\Delta p_i V$.
- The entropy change ΔS ("generation of chaos").

Here, p_i and V denote the partial pressure and the volume, respectively. Reactions are fast and hence have high reaction rates r if activation free enthalpies ΔG_{act} are small. Thermodynamic equilibrium is characterized by $\Delta G = 0$.

The two Δ-terms in Equation (4-3) explain the generally observed drastic change in the mechanism of sensor response with temperature:

- *Adsorption of particles* is always characterized by
  - a decrease in the energy ΔU or $\Delta H < 0$ (because reaction energy ΔH is released and total energies are reduced upon particle bonding) and

— a decrease in entropy $\Delta S < 0$ (because particles have a higher order and show less "chaos" in the adsorbed state as compared to their free gas or liquid environments).

As formally described by Equation (4-3), adsorption is therefore favored at low temperatures and leads to low-temperature adsorption equilibria of particles. At higher temperatures, however, these particles desorb because of the predominant contribution of an entropy increase $\Delta S > 0$ to ΔG which drives the reverse reaction because of the influence from the term $-T\Delta S$.

— Particularly at elevated temperatures, entropy effects evidently become important. Under these conditions, we have to generalize the possible particle/sensor interactions by also including *desorption, defect formation, incorporation of particles,* or *bulk reactions.* Some of these reactions show the opposite temperature dependence if compared with adsorption reactions. As an example, the formation of point defects such as oxygen vacancies V_O in an ideal oxide sensor material requires energy ($\Delta H > 0$) and increases the entropy $\Delta S > 0$. The latter results from an increased disorder of the previously perfect lattice in the absence of defects. Because of the energy required to produce point defects, eg, in the reaction

$$O_1 \rightleftharpoons 1/2 \ O_2 + V_O^{\cdot\cdot} + 2e' \tag{4-4}$$

the point defect concentration is negligible at low temperatures and increases at higher temperatures. In Equation (4-4) O_1 denotes lattice oxygen, $V_O^{\cdot\cdot}$ positively charged oxygen vacancies, and e' free electrons in the oxide. The equilibrium constant for this reaction

$$K = \frac{\sqrt{p_{O_2}} \ [V_O^{\cdot\cdot}] \ [e']^2}{[O_1]} \tag{4-5}$$

is influenced by the oxygen partial pressure p_{O_2} in the gas phase, and by the concentrations of both, oxygen vacancies $[V_O^{\cdot\cdot}]$ and electrons $[e']$. For low defect concentrations, $[O_1]$ is constant. Changes in p_{O_2} may be detected sensitively by conductivity changes because of changes in

$$[e'] = 2[V_O^{\cdot\cdot}] \sim p_{O_2}^{-1/6}. \tag{4-6}$$

This particular conductivity sensor will be described in more detail in Section 4.7.

In conclusion, because of the two opposing temperature dependences in adsorption and defect formation steps, we expect strong temperature effects in the sensor response behavior with temperature-dependent relative contributions from different adsorption and point defect reactions, all of which may determine the overall sensor response signal.

In principle, ie, from an academic point of view, spectroscopic and theoretical calculations make it possible to calculate the driving force for sensor operation ΔG provided that all the possible energies E_i of the system are known. The E_i values are determined by all involved atoms and their electronic states, vibrations, rotations and translations. The key role in this calculation is the determination of the partition function Z of the total sensor/particle-system

$$Z = \sum_i e^{-\frac{E_i}{k_B T}} \tag{4-7}$$

with E_i as all possible energetic Eigenstates of the system. From this, the driving force G of a specific sensor/particle situation with its E_i values is deduced from

$$G = -k_{B}T\left[\ln Z - \left(\frac{\partial \ln Z}{\partial \ln V}\right)_{T}\right]. \tag{4-8}$$

Often, the energies E_i of the system contain additive contributions from electronic, vibronic, or rotational energies of individual molecules or the solid. These energies may be determined spectroscopically. They are of importance in the systematic optimization of sensor structures as it was described in Chapter 3, Section 3.4.2.1.

Theoretical calculations or spectroscopic experimental determinations of E_i and G may be done for sensor conditions before and after particle/sensor interactions. The difference yields ΔG. Theoretically one may also determine reaction paths during particle/sensor interactions as indicated schematically in Figure 4-13. This yields ΔG^{react}. For details of these statistical thermodynamics concepts, the reader is referred to corresponding textbooks [14].

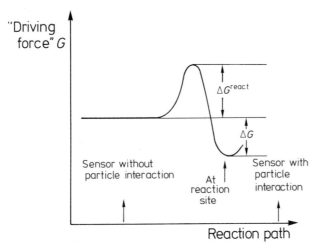

Figure 4-13. Simple thermodynamic and kinetic description the sensor/particle interaction: the driving force is the change in the free enthalpy ΔG between the sensor without particle interaction and the sensor/particle system after interaction. For $\Delta G = 0$, we observe an equilibrium between free and bonded particles. For $\Delta G < 0$, sensor and particles react spontaneously if the free enthalpy of activation ΔG^{react} is small.

4.2.2   Equilibrium Sensors

Figure 4-14 schematically shows the thermodynamic condition for equilibrium sensors. The total free enthalpy G_{tot} which consists of contributions from both, the selective sensor and the chemical component with its partial pressure p_i in the analyte, has two minima. In the ideal case of a completely selective sensor, G_{tot} does not depend on any other component than i. (Cross interference effects make G_{tot} depend on other components). Under equilibrium conditions the free enthalpy per mole v_i of particles, ie, the chemical potential of the component i

$$\mu_i = \left(\frac{\partial G_{tot}}{\partial v_i} \right)_{T, p, v_j \neq i, s} \tag{4-9}$$

is identical in all states "s" of the component i (here: in two states, the state of the free particles and of the interacting particles at the sensor). In the gas phase

$$\mu_{i, gas} = \mu_{i, gas}^0 + RT \ln (p_i / p_i^0) \tag{4-10}$$

holds with $\mu_{i, gas}^0$ as standard chemical potential at standard pressure p_i^0. Pressure and temperature changes therefore determine changes in the concentration of particles i in the different sensor interaction steps. This in terms determines the changes in the sensor properties some of which are indicated schematically in Figure 4-14.

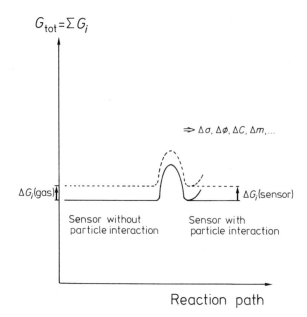

Figure 4-14.
Equilibrium sensors: The free enthalpy G_{tot} of the system "particles + sensor" is identical with the system "particle/sensor". Changes in the partial pressure of component i lead to changed ΔG_i(gas) and hence to corresponding changes $\Delta \sigma$, $\Delta \Phi$, ΔC, Δm ... (see Table 4-1).

Simple interaction steps as they may occur in gas sensors are physisorption, chemisorption, surface-, and bulk-defect reactions (see also Figure 3-32).

A specific example is shown in Figure 4-15. Here, general O_2/oxide interactions are described which involve different possible steps. We have plotted the total potential energies $E_{pot} = E_{pot}^{T=0}$ (inner energies at $T = 0$ K) instead of free enthalpies G because these potential energies only are completely independent of the partial pressure and temperature (see Equation (4-3)) and therefore make possible a general schematic drawing of all interaction mechanisms in one picture. Because of the partial-pressure and temperature dependence of the chemical potentials in the gas phase and in all interaction steps, we can change the interaction mechanisms and adjust different values of the minima in the different interaction stages.

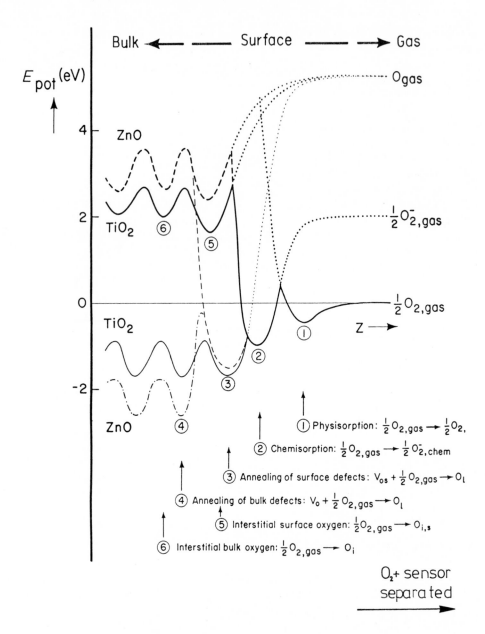

Figure 4-15. Characteristic interaction steps between oxygen and the thermodynamically most stable surfaces of ZnO and TiO$_2$ as characterized by the potential energy $E_{pot}^{T=0}$ attributed to the different reactions listed in the lower part as a function of distance z from the surface. O$_{l(s)}$ denotes oxygen at ideal lattice (surface) sites [15].

A prerequisite, however, is that the activation barrier between different interaction steps is not too high if compared with the thermal energy $k_B T$. Otherwise, the reactions are kinetically hindered. For complete kinetic hindrance, reversible sensor/gas interactions are also expected. Inbetween-situations lead to drifts.

4.2.3 Rate-Determined Sensors

Figure 4-16 shows a simple example of the free enthalpy G_{tot} as a function of reaction path for rate-determined sensors. The CO oxidation on oxide surfaces, for instance, is possible in the presence of O_2 in the gas phase and it usually involves chemisorbed or lattice oxygen. We find intermediate positively charged species $(CO + O)^{\delta+}$ at the surface of n-type semiconductors. This species leads to a conductivity increase and determines the sensor response to CO in the gas phase. The species finally reacts to form CO_2 which desorbs into the gas phase. Under ideal conditions the CO_2 is pumped away quickly. For thermodynamic reasons the back reaction is completely negligible at typical sensor operation temperatures around 350 °C. The steady-state concentration of intermediate $(CO + O)^{\delta+}$ species in a contineous flow of CO in air is determined by the partial pressures of CO and O_2 in the gas phase and therefore makes possible to monitor both gases. For constant O_2 in air we monitor CO.

$$G_{tot} = \Sigma G_i$$

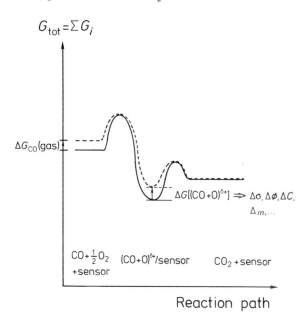

Figure 4-16.
Rate-determined sensors: The overall decrease of the free enthalpy is negative and makes possible a faster oxidation of CO by O_2 molecules if the activation barrier is lowered because of increased p_{co} and hence ΔG_{co} (gas). In this case, the sensor acts as a catalyst. The concentration of intermediate complexes $(CO + O)^{\delta+}$/sensor determines the sensor responses $\Delta\sigma$, $\Delta\Phi$, ΔC, Δm, ... (see Table 4-1).

In reality, different other intermediate steps may also be involved. The steady-state concentrations of these intermediately formed species may determine the sensor signal of conductivity sensors if charge transfer reactions are involved. The generation of heat at the catalytically active surface determines the sensor signal of calorimetric sensors. Other changes may be used for other sensor principles. In the most general case, the overall rate r of catalytic CO_2 formation may involve surface, bulk, interface, or even three phase boundary reactions of CO and O_2. The overall rate r is often determined by the highest intermediate activation barrier E_A and is usually described by

$$r = k_0 \, e^{-E_A/k_B T} \cdot p_1^{n_1} \cdot p_2^{n_2} \cdots p_i^{n_i} \qquad (4\text{-}11)$$

with the pre-exponential factor k_0, activation energy E_A, and reaction orders n_1, n_2, ... n_i with respect to the different partial pressures p_1, p_2, ... p_i.

As a simple example for characteristic adsorption complexes formed during the (low-temperature) CO-oxidation, Figure 4-17 shows typical surface species which occur as intermediates in the CO oxidation at the thermodynamically most stable surface of ZnO. The Coulomb interaction between acceptor-type surface species O_2^- formed by chemisorption of O_2 and donor-type species $(CO_2 \cdot V_{Os})^+$ formed by reaction of CO with lattice oxygen determines the oxidation kinetics in the rate-limiting step.

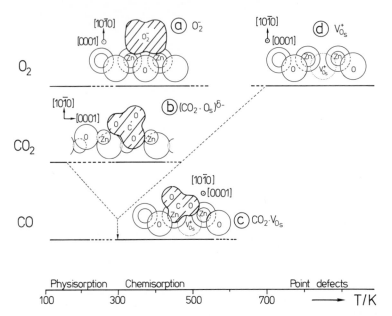

Figure 4-17. Characteristic adsorption complexes at ZnO surfaces formed by O_2, CO_2, and CO interaction at different temperatures. Zn and O denote ideal ZnO lattice atoms, V_{Os} is an oxygen vacancy (point defect) [16].

4.2.4   One-Way Sensors

The sensors discussed in Sections 4.2.2 and 4.2.3 responded reversibly to increasing or decreasing partial pressures p_i. For one-way sensors the $\Delta G < 0$ change is so large that it cannot be reversed under experimental conditions of the setup. A typical example is an antigene/antibody reaction in biochemical sensors as shown schematically in Figure 4-18 (see also, Section 4.12).

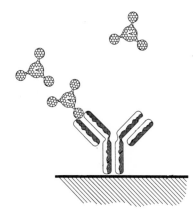

Figure 4-18.
Schematic representation of a key-lock reaction between
an adsorbed antibody molecule and the antigene A [17].
Molecular weights of involved species are in the order of
10^5 and more.

4.3   Survey on Temperature- and Concentration-Dependent Sensing Mechanisms

As discussed above, three operation principles describe all the basic sensors of Table 4-1
with their practical setups for real applications to be discussed in the following Chapters 7
to 14. For all of these sensors, we may either obtain equilibrium (thermodynamic), steady-state
flow (kinetic), or static (irreversible) values of their responses during exposure of the sensor
to an environment with time-independent specific analytical information. Alternatively, we
may evaluate time constants or rate changes by which the sensor adjusts to systematically
changed analytical informations or to systematically changed dynamical sensor operation
conditions (compare Figures 6-1 and 6-2).

Some of the basic sensor principles are understood in great detail today. Solid-state gas sensors in particular can be treated easily because of well-established theories and experiments
to describe gases and solids.

This does not hold for liquids. Therefore, liquid or solid electrolyte sensors with their time-
dependent structures and cooperative phenomena have to be treated more phenomenologically. This will be done in a separate Chapter 5.

In the following Sections 4.4 to 4.10, we will illustrate with a few examples the present
"state-of-the-art" of an atomistic understanding of simple sensor principles with particular
emphasis on solid-state gas sensors.

Chemical gas sensors may in principle monitor reactions involving physisorption, chemisorption, surface defects, grain boundaries, three phase boundaries, or bulk defects
(Figure 4-2). Because reactions are predominately energy-driven at low temperatures or en-
tropy-driven at high temperatures, we expect adsorption or defect reactions to dominate at low
or at high temperatures, respectively. All of these processes may change upon pressure variations in the gas phase. Reversible changes are highly desirable for reliable sensors. A careful
choice of temperature and partial pressure ranges is therefore extremely important for reversible sensor operation. The aim is usually to utilize the predominant influence of only one type
of solid/gas interaction step.

Binary and ternary oxides represent the most important class of gas-sensor materials operating under ambient air conditions (see Table 3-1).

They are always non-stoichiometric compounds $(N_l)M_nO_{m-\delta}$ $(l, m, n =$ integers) with the deviation δ of the ideal stoichiometry $(l):n:m$ usually adjusted thermodynamically at high temperatures within the stability ranges of the corresponding oxide phases [18]. These ranges are shown in Figure 4-19 for various binary compounds. Constant free enthalpies of formation ΔG^0 in the two-oxide phase regions are characterized by a constant partial pressure p_{O_2} of oxygen. In contrast, ΔG^0 is a function of p_{O_2} within the single-oxide phase regions (because of Gibbs rule [18]) and the partial pressure of oxygen determines their nonstoichiometry at constant temperature.

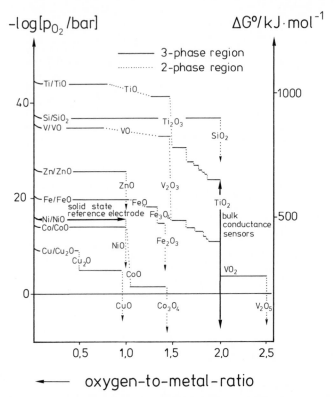

Figure 4-19. The stability ranges of different binary oxides at 1000 K as a function of the oxygen partial pressure p_{O_2} or free enthalpy of formation ΔG^0 characterized by the oxygen-to-metal ratio. The specific stability range of a bulk conductance sensor (see Figure 4-24 and Chapter 9, Section 9.5.3) based upon TiO_2 to monitor O_2 and of a NiO/Ni solid state reference electrode (see Chapters 5 and 8) are indicated by arrows.

This nonstoichiometry results from ionic point defects, eg, oxygen vacancies V_O, interstitials O_i, metal vacancies V_M, and/or interstitials M_i in both, the oxygen and metal bulk sublattice. As a result of their ionization, additional electronic defects, ie, electrons and/or defect electrons are present in the conduction and/or valence band. The concentrations of all defects may be adjusted thermodynamically by adjusting the different defect equilibria between the bulk and the gas phase. The latter can be treated by the law of mass action including charge neutrality (see Equation (4-4)) with a particular example of an acceptor-doped oxide shown in Table 4-2 [19].

Systematic studies of oxides for sensors usually start from undoped bulk materials and then introduce different bulk or surface intrinsic and extrinsic (foreign atom) dopants to optimize their electronic and/or ionic conductance by adding donors or acceptors. Elementary steps of oxide/oxygen interactions must be understood first before the interaction with other gases can be studied successfully (Figure 4-15). Detailed results are available for the well-investigated "prototype" materials TiO_2 [20], SnO_2 [21], ZnO [22] and ZrO_2 [23] but also for organic materials such as lead phthalocyanine PbPc [24].

Table 4-2. Defect concentrations for pure and doped Frenkel-like oxides [19] with e' as electron, h' as hole, $V_O^{\cdot\cdot}$ as double positively charged oxygen vacancy, O_i'' as double negatively charged oxygen interstitial, N_M' as single negatively charged acceptor, and the equilibrium constants $K_F = [O_i''][V_O^{\cdot\cdot}]$, $K_i = [e'][h^\cdot]$, and $K_R = [V_O^{\cdot\cdot}][e'] p_{O_2}^{1/2}$.

Predominant electroneutrality condition of defects	P_{O_2} range	[e']	[h']	[$V_O^{\cdot\cdot}$]	[O_i'']
$[e'] = 2[V_O^{\cdot\cdot}]$	Low	$(2K_R)^{1/3} P_{O_2}^{-1/6}$	$\dfrac{K_i P_{O_2}^{1/6}}{(2K_R)^{1/3}}$	$\dfrac{[e']}{2}$	$K_F\left(\dfrac{4}{K_R}\right)^{1/3} P_{O_2}^{1/6}$
$[O_i''] = [V_O^{\cdot\cdot}]$	Intermediate	$\left(\dfrac{K_R}{K_F^{1/2}}\right)^{1/2} P_{O_2}^{-1/4}$	$K_i\left(\dfrac{K_F^{1/2}}{K_R}\right)^{1/2} P_{O_2}^{1/4}$	$K_F^{1/2}$	$K_F^{1/2}$
$2[V_O^{\cdot\cdot}] = [N_M']$	Intermediate	$\left(\dfrac{2K_R}{[N_M']}\right)^{1/2} P_{O_2}^{-1/4}$	$\dfrac{K_i[N_M']^{1/2}}{(2K_R)^{1/2}} P_{O_2}^{1/4}$	$\dfrac{[N_M']}{2}$	$\dfrac{2K_F}{[N_M']}$
$[h^\cdot] = [N_M']$	High	$\dfrac{K_i}{[N_M']}$	$[N_M']$	$\left(\dfrac{K_R[N_M']^2}{K_i^2}\right) P_{O_2}^{-1/2}$	$\left(\dfrac{K_F K_i^2}{K_R[N_M']^2}\right) P_{O_2}^{1/2}$
$[h^\cdot] = 2[O_i'']$	High	$\left(\dfrac{K_R K_i}{2K_F}\right)^{1/3} P_{O_2}^{-1/6}$	$\left(\dfrac{2K_F K_i^2}{K_R}\right)^{1/3} P_{O_2}^{1/6}$	$\left(\dfrac{4K_R K_F^2}{K_i^2}\right)^{1/3} P_{O_2}^{-1/6}$	$\dfrac{[h^\cdot]}{2}$

* The third and fourth rows apply only if $[N_M'] > K_F^{1/2}$, as the apparent intrinsic defect concentration. Otherwise row 2 is used.

4.4 Physisorption Sensors

Physisorption describes weak van-der Waals sensor/particle interaction similar to the intermolecular forces between two molecules observed in non-ideal gases [13]. Low-temperature physisorption sensors typically monitor reversible changes in masses or dielectric constants if chemisorption bonding is either not possible, or if the activation barrier between physisorption and chemisorption is too high so that chemisorption is kinetically hindered. Because of the rather unselective intermolecular forces involved in physisorption, cross interference with other gases is usually a problem and may be avoided by, eg, monitoring changes in the sensor signals upon temperature variations around the desorption temperature of the species to be detected. Humidity sensors are the most popular sensors of this type which monitor the physisorption or the multilayer condensation of water at a certain temperature.

4.5 Chemisorption Sensors

Selective chemisorption bonds may lead to very specific changes in a variety of sensor responses x_i' such as their optical or electrical parameters.

Examples to characterize selective chemisorption bonds and changed geometric and electronic surface properties are shown in Figure 4-20. They may be calculated for surface atom clusters in simple quantum chemistry approaches for free surfaces and surfaces after exposure to small molecules.

In the simplest case of electronically conducting sensors, chemisorption may lead to charge transfer from the particle to the sensor (donor-type) or vice versa (acceptor-type) (Figure 4-21). This causes characteristic changes in sheet conductivities $\sigma_\square$ and work functions Φ which can be monitored in simple experiments.

Their theoretical understanding is possible in the following band scheme picture: A specific example of an acceptor-type interaction with an n-type semiconductor is shown in Figure 4-22. Chemisorption leads to a decreased surface conductivity $\Delta\sigma$, an increased band bending $e\Delta V_s$ at the surface (index s), and generally also to a change in the electron affinity $\Delta\chi$. Assuming flat band condition before chemisorption, ie, $e\Delta V_s = 0$ and hence $\Delta\sigma = 0$, the sheet conductance of a rectangular homogeneous thin film sample of resistance R is given by

$$\sigma_\square = R^{-1}\, l^2\, A^{-1} = \sigma_b \cdot d \qquad (4\text{-}12)$$

with the sensing outer surface A, the distance l between the two electrodes covering the two opposite sides of the sensor, and the thickness d. The specific conductivity σ_b is related to the bulk concentration n_b and/or p_b of electrons and/or defect electrons by

$$\sigma_b = e \cdot (\mu_{b,e} \cdot n_b + \mu_{b,p} \cdot p_b) \qquad (4\text{-}13)$$

where e denotes the elementary charge and $\mu_{b,e}$ ($\mu_{b,p}$) denotes the bulk mobilities of the electrons (defect electrons) in the conduction (valence) band.

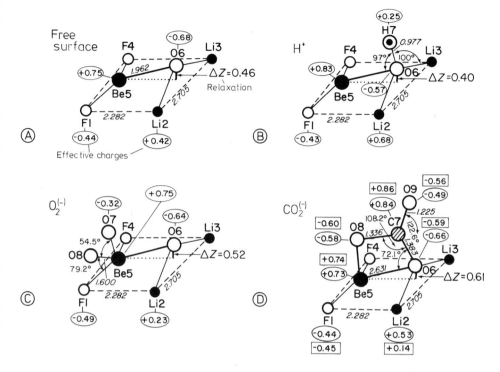

Figure 4-20. Free surface cluster (A) and characteristic adsorption complexes (B–D) simulating chemisorption at ZnO (10 $\bar{1}$0) with (B) hydrogen (donor-type) chemisorption, (C) oxygen (acceptor-type) chemisorption, and (D) CO_2 (weak acceptor-type) chemisorption [25].

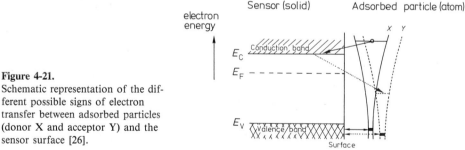

Figure 4-21.
Schematic representation of the different possible signs of electron transfer between adsorbed particles (donor X and acceptor Y) and the sensor surface [26].

The surface excess conductivity $\Delta\sigma$ upon chemisorption is then determined by the deviation of the sheet conductance $\sigma_\square$ as measured under flat-band conditions ($\sigma_\square = \sigma_b d$)

$$\Delta\sigma = \sigma_\square - \sigma_b d .$$

(4-14)

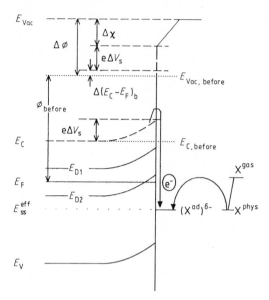

Figure 4-22.
Schematic diagram of chemisorption with electronic charge transfer with surface and bulk electronic levels, the vacuum level E_{Vac}, work function changes $\Delta\chi$ upon surface reactions, electron affinity changes $\Delta\chi$, band bending $-e\Delta V_s$, changes in the bulk Fermi level position $\Delta(E_C - E_F)_b$, minimum of conduction band E_C, and maximum of the valence band E_V, and effective surface state E_{ss}^{eff} attributed to chemisorbed species $(X_{ad})^{\delta-}$. The latter is formed by charge transfer from the physisorbed species X^{phys} [15]. For further details, see text.

The value $\Delta\sigma$ contains contributions from mean mobilities $\mu_{s,e}$ ($\mu_{s,p}$) of electrons and defect electrons with excess concentrations ΔN and ΔP per unit area in the space charge layer formed upon chemisorption

$$\Delta\sigma = e \cdot (\mu_{s,e} \cdot \Delta N + \mu_{s,p} \cdot \Delta P) . \tag{4-15}$$

To a first approximation, the values $\mu_{s,e}$ ($\mu_{s,p}$) are often assumed to be identical to the corresponding bulk values $\mu_{b,e}$ ($\mu_{b,p}$). The surface excess concentrations ΔN (ΔP) are defined by

$$\Delta N = \int_0^d [n(z) - n_b] \, dz . \tag{4-16}$$

Generally the variations in the sheet conductance $\sigma_\square$ upon gas exposure may contain both, changes in bulk as well as in surface contributions. If bulk contributions are negligible and defect electron concentrations can be ignored as it is the case for many n-type semiconductor materials and $\mu_{s,e}$ ($\mu_{s,p}$) is known or estimated from $\mu_{b,e}$ ($\mu_{b,p}$), ΔN (ΔP) can be deduced from $\Delta\sigma$.

If the electronic structure of the bulk with its concentrations of free electrons, holes, and charged point defects is known, it is possible to solve the Poisson equation and hence to determine the band bending $-e\Delta V_s$ and the charge per unit area Q_{sc} in the space charge layer [15]. The latter is identical to the negative charge Q_{ss} per unit area in surface states E_{ss}^{eff} (compare, Figure 4-22). From this, the partial charge δ can be calculated, which is formally attributed to the chemisorption complex

$$\delta = Q_{ss}/e n_{(s)}^{ad} . \tag{4-17}$$

The surface concentration of adsorbed species $n_{(s)}^{ad}$ may be determined in independent thermal desorption spectroscopy (TDS) experiments which start at a given temperature and surface coverage (adjusted by the pressure, see Chapter 3, Section 3.4.2.2). Assuming the value δ to be given by Fermi statistics, we can calculate the position of the electronic level E_{ss}^{eff} relative to the position of the Fermi level at the surface ($E_{ss}^{eff} - E_F$). This level is formally attributed to the chemisorption state.

A comparison of experimentally determined work function changes $\Delta\Phi$ (eg, from Kelvin-probe measurements) with band bending changes $e\Delta V_s$ (eg, from UPS, see Chapter 3, Section 3.4.2.2) leads to an estimation of electron affinity changes $\Delta\chi$ from

$$\Delta\Phi = -e\Delta V_s + \Delta\chi + \Delta(E_C - E_F)_b \qquad (4\text{-}18)$$

provided that variations in the bulk Fermi level position $\Delta(E_C - E_F)_b$ can be neglected. The latter holds, if bulk diffusion of atoms or ions can be neglected. An estimation of the dipole moment μ^{ad} formally attributed to electron affinity changes $\Delta\chi$ is possible by using the equation

$$\Delta\chi = e(\varepsilon_s \varepsilon_0)^{-1} \mu^{ad} n_{(s)}^{ad} . \qquad (4\text{-}19)$$

From thermal desorption experiments starting at different temperatures and pressures we may also determine the adsorption isotherms $n_{(s)}^{ad}(p, T)$, ie, the number of adsorbed particles $n_{(s)}^{ad}$ per unit area or their coverage Θ and from this the heat of adsorption per mole

$$Q_{ad} = -R\,[\partial \ln p/\partial(1/T)]_{n_{(s)}^{ad}} \qquad (4\text{-}20)$$

provided that equilibrium sensors are investigated. The value Q_{ad} determines the sensitivity of calorimetric chemisorption sensors. The TDS experiments also make possible the determination of molar activation energies of desorption $\tilde{E}^{des}$, corresponding entropies of desorption $\tilde{S}^{des}$ and reaction orders of desorption m (the tilde ($\sim$) indicates molar quantities). These values determine characteristic temperature ranges of reversible sensor/gas interactions. The initial sticking coefficient S_0 is determined, eg, from conductivity changes $d\Delta\sigma/dt$ extrapolated to zero coverage. This value determines the response time of the sensor.

The parameters δ, μ^{ad}, S_0, Q_{ad}, $\tilde{E}^{des}$, $\tilde{S}^{des}$, and m characterize specific sensor-gas interactions for one specific molecule; for details, see Ref. [15]. In coadsorption experiments corresponding excess parameters can be defined. They are determined from changed parameters with changed partial pressures of one component for different but constant compositions of all other components.

As already mentioned above, theoretical cluster calculations make possible an atomistic understanding of these parameters in terms of partial charges, dipole moments, and binding energies attributed to the chemisorption complex [15]. In thin-film or polycrystalline materials, drastic changes in the sensor response occur for thicknesses or grain sizes below the characteristic attenuation length (Debye length, ie, the "non-flat" band range in Figure 4-22) of trapped surface charges (for details, see Chapter 9).

4.6   Surface Defect and Catalytic Sensors

These defects are often involved in chemisorption or surface reactions including catalytic conversion reactions because the binding energy of surface lattice atoms is usually of the same order as chemisorption energies. The most important intrinsic surface defect in n-type conducting oxides is a donor-type oxygen vacancy (Figure 4-23, see also eg, Figure 4-15). For entropy reasons, its formation is thermodynamically favored at low p_{O_2} and high T, or at high p_{CO}. Extrinsic surface defects, on the contrary, involve atoms not present in the ideal lattice. They may result from segregated bulk impurities or dopants. Their electronic or geometric influence on sensor properties is usually correlated with their influences on catalytic properties of the surface (see eg, Figures 3-33 and 4-17) [27].

Field effect structures with catalytic metal gates constitute another type of catalytic sensors, where chemisorbed species and/or reaction intermediates on the sensing surface change the work function of the metal. These changes are either due to hydrogen atoms at the metal/insulator interface and for thin discontinuous metal films also due to adsorbates on the metal and insulator surface. The size of the work function change, which is detected by the sensor, depends in an oxidizing atmosphere generally on the rate of chemical reactions on the surface. This sensor type is thus another example of a rate-limited sensor similar to that described in Fig. 4-16. Further details are found in [8].

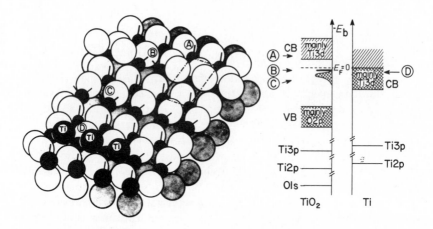

Figure 4-23. Geometric and electronic structure of different intrinsic surface defects at TiO$_2$ (110) with ideal oxygen positions (A), oxygen vacancies (B), oxygen vacancy line-defects, ie, one-dimensional surface defects (C), and metallic titanium deposition at the surface (D).

4.7   Bulk Defect Sensors

At low temperatures, bulk defects are often undesirable or at least should remain unchanged because bulk defects in the subsurface region may lead to irreversible drifts of sensor parameters if first a defect concentration had been adjusted before during the sample preparation at high temperatures, if second this low-temperature defect concentration is far away from the thermodynamic equilibrium, and if third the thermal energy $k_B T$ is comparable to activation barrier heights for the diffusion between the chemisorbed state and the bulk (see Figure 4-15).

At high temperatures, however, these defects can be essential for a specific and reversible sensor operation. Reversibility of bulk defect sensors usually requires operation at temperatures high enough to make use of either mixed (electron and ion, see Figure 4-24) or fast ion conduction. The activation barrier of the first reaction at the surface occurring in this solid/gas interaction usually determines the response kinetics of bulk defect sensors in the lower temperature range.

A further typical example for a bulk defect sensor is shown in Figure 4-25. It also demonstrates their separation from the influence of surface defects including chemisorbed species: From the thickness-dependence of sheet conductances, we conclude that NO_2 molecules are chemisorbed whereas O_2 molecules form bulk defects. In this particular example, a detailed study indicates that incorporation of O_2^- in PbPc forms highly mobile defect electrons h^+ in the mixed $(O_2^- + h^+)$ conducting bulk of PbPc [28]. This process is reversible only above 470 K. Chemisorption of NO_2 on the other side is reversible around 300 K and negligible at 470 K.

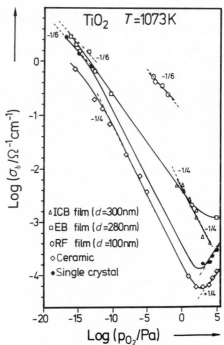

Figure 4-24.
Specific bulk conductivities σ_b of TiO_2 single crystals, ceramics, ion-cluster-deposited (ICB), electron-beam evaporated (EB), and rf-sputtered (RF) films with thicknesses d between 100 and 300 nm as a function of oxygen partial pressure p_{O_2} at constant temperature $T = 1073$ K in a log/log-plot [3]. Slopes are indicated for characteristic ranges p_{O_2} between $+\frac{1}{4}$ and $-\frac{1}{4}$.

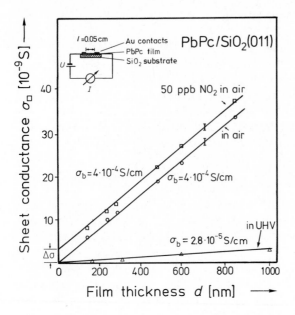

Figures 4-25.
Separation of surface and bulk effects in chemical sensing: sheet conductance $\sigma_\square$ as a function of film thickness for lead phthalocyanine (PbPc) thin films under ultra high vacuum (UHV) conditions, in air without and with NO_2.

4.8   Microcrystal and Grain Boundary Sensors

Small particles offer unique advantages to adjust high surface sensitivities because of a large surface-to-volume atom ratio, to adjust different surface orientations and by changing their size to adjust systematically chemical or electronic properties ranging from individual atoms to the infinite solid.

Typical structures of clusters and the development of single crystal surfaces with an increasing number of atoms for "magic numbers of atoms" in highly symmetric and energetically preferred particle shapes are shown in Figure 4-26.

These clusters may either be deposited at a ceramic or monocrystalline material or they may be embedded in a matrix.

Electrical conductance between these clusters in a polycrystalline matrix may be described by statistical percolation paths along which the overall conductance occurs and by bootlenecks at the interface between two clusters. In the simplest case this interface may be considered as two surfaces of the type discussed above in Figure 4-22. The conductance across this interface is usually determined by the concentration of free carriers at the interface. This concentration is sensitively influenced by donor- and acceptor-type of chemisorption. Therefore it is possible to develop acceptionally sensitive semiconducting sensors with small particles. Drastic sensitivity enhancements and surprising reproducibility is obtained for cluster diameters below the characteristic screening length of electrical charges of the surface (Debye-length). In this case the opposing surfaces of a single cluster influence each other. Details will be discussed in Chapter 9.

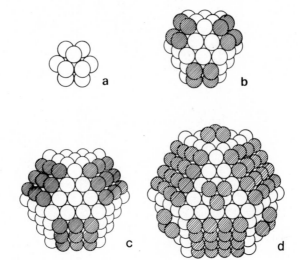

Figure 4-26.
Typical structures of clusters and the development of single crystal surfaces with increasing number of atoms per cluster [14].

4.9   Interface and Three-Phase Boundary Sensors

— As a first example of a sensor monitoring potential variations at interfaces, the chemically sensitive field-effect sensor was already shown schematically in Figure 4-4 and further discussed in Section 4.6 above.

— A second example is illustrated in more detail in Figure 4-27. The mixed conduction of PbPc and the ion conduction of AgI may be utilized to measure changes in the overall voltages

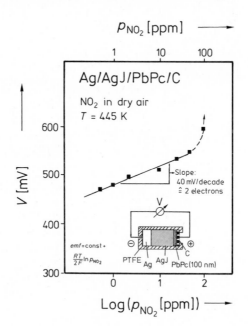

Figure 4-27.
Sensor response V of a potentiometric Ag/AgI/PbPc/C solid-electrolyte sensor as a function of the NO$_2$ partial pressure in air.

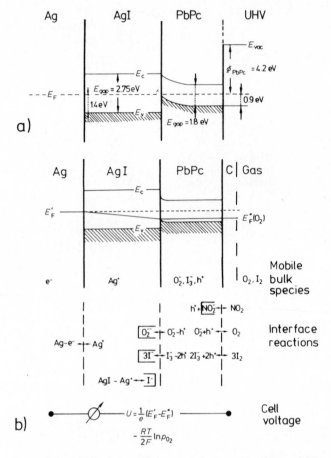

Figure 4-28. Schematic band diagram of the sensor (see Figure 4-27) to monitor O_2 and NO_2 in the gas phase. The sandwich-structure makes use of electron conduction (e) across the contacts Ag and C, ionic conduction (Ag^+) in AgI and mixed conduction (O_2^-, I_3^-, h^+) in PbPc. For further details, see text.

of the electrochemical cell upon changes in O_2 or NO_2 partial pressures. Details about the chemical composition, geometric structure, mobile bulk species, interface reactions with trapped species (O_2^{2-}, I^-, NO_2^- indicated by frames in Figure 4-28, lower part), about positions of occupied and empty electronic states including the valence band E_V, conduction band E_C, the bandgap E_{gap}, work function Φ and Fermi level E_F for different operation modes have been determined under UHV (Figure 4-28a) and O_2 as well as NO_2 gas atmosphere conditions (Figure 4-28b) by means of XPS, UPS, HREELS and electrical measurements [30]. For simplification, a few other mobile bulk species are not mentioned in this drawing (such as I_x^- with $x = 5,7, \ldots$ in PbPc with the relative concentrations of all the different I_x^- depending on the applied voltage and external I_2 partial pressure). The sandwich structure operates as an electrochemical potentiometric cell with the cell voltage V determined by p_{O_2} or p_{NO_2} in the Nernst equation.

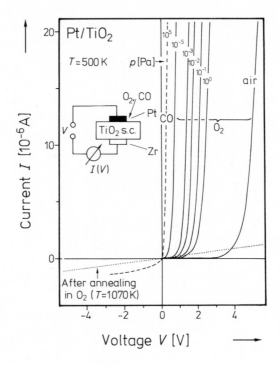

Figure 4-29.
Current-voltage (*I-V*) curves of Schottky barrier sensors (solid and dashed lines) with Schottky Pt/TiO$_2$ contacts and ohmic Zr back contacts at low temperatures. The dotted line shows the ohmic *I-V*-curve of bulk TiO$_2$ sensors. (Compare Figure 4-24).

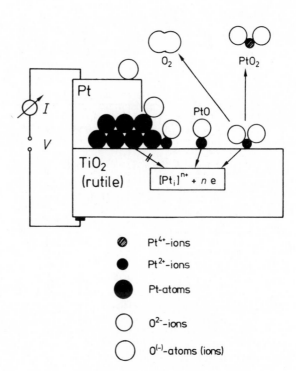

Figure 4-30.
Schematic representation of Pt/TiO$_2$-contacts with Schottky barrier properties at low temperatures. At the three-phase boundary, oxidation of Pt atoms occurs at high temperatures with O$_2$ from the gas phase. This leads to the formation of Pt^{n+} ions which diffuse into the TiO$_2$ subsurface thereby forming irreversibly donor states and hence ohmic Pt/TiO$_2$ contacts [30].

— A third example of a typical three-phase boundary sensor is a Schottky diode device shown in Figure 4-29. It is based upon changes in the Pt/TiO_2 interface conductance produced by changes in the O_2 or CO partial pressure. Because of the higher work function of Pt if compared with the n-type semiconductor TiO_2, a Schottky barrier is formed at the interface. The acceptor-type of O_2 interaction increases the barrier at the three-phase boundary. It is catalyzed by Pt. At elevated temperatures, however, Pt diffuses into the TiO_2 subsurface layers and produces ohmic contacts (compare also Figure 3-33 for the system Pd/SnO_2 and Figure 4-30) [31]. The same device may then be used as a high-temperature bulk defect sensor to monitor O_2 by mixed conduction.

— A fourth example is the Pt/ZrO_2/gas three-phase boundary which is essential for reversible O_2 monitoring in ion-conducting sensors. At this boundary, the conversion from O_2 to O^{2-} must take place rapidly, with the latter species subsequently being incorporated in bulk ZrO_2. If this conversion is kinetically hindered, other gases may also be monitored. Mixed-conductivity electrodes instead of Pt lead to new sensors with interface reactions similar to the ones illustrated in Figure 4-28 at the AgI/PbPc interface [32].

All the examples illustrate the great potential of using materials with different transport properties (electronic, ionic, or mixed conduction) and interface reactivities (formation of ions, dissociation, etc.) to design new chemical sensors with well-defined three phase boundaries.

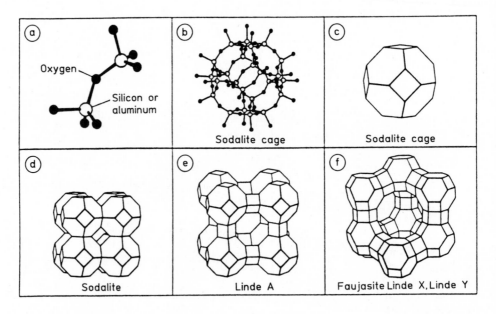

Figure 4-31. The framework of zeolites is constructed of tetrahedral building blocks. In some cases there is a characteristic "sodalite cage". The four tetrahedral vertexes are filled by oxygen atoms and a silicon or aluminum atom lies at the center (a). Each oxygen is shared by two tetrahedrons In prepresentations of the sodalite cage (b), the ball-shaped atoms are usually omitted (c). Sodalite cages are found in all of the following structures depicted here: sodalite itself (d), the synthetic Linde A (e), and a group of like-structured crystals namely the mineral faujasite and the synthetic zeolites Linde X and Linde Y (f) [33].

4.10   Cage Compound Sensors

In all examples discussed so far a clear distinction could be made between surface and bulk properties. Inorganic or organic cage compounds however with cages of molecular dimensions cannot be treated with this concept because of cooperative phenomena determining the interactions between these compounds and molecules. The most prominent inorganic cage compounds are zeolites (Figure 4-31, see preceding page).

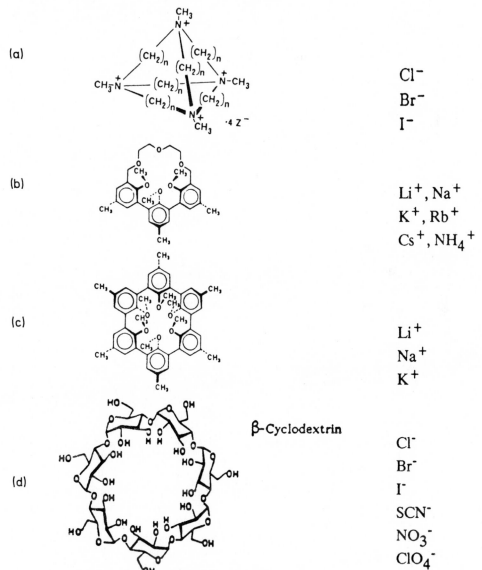

Figure 4-32. Typical organic cage compounds [34–39].
(Continued on next page).

Dibenzo-18-Krone-6

(e)

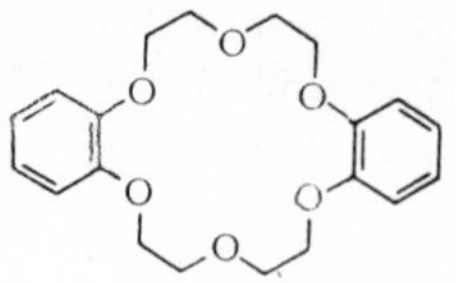

Na$^+$

K$^+$

Cs$^+$

(f)

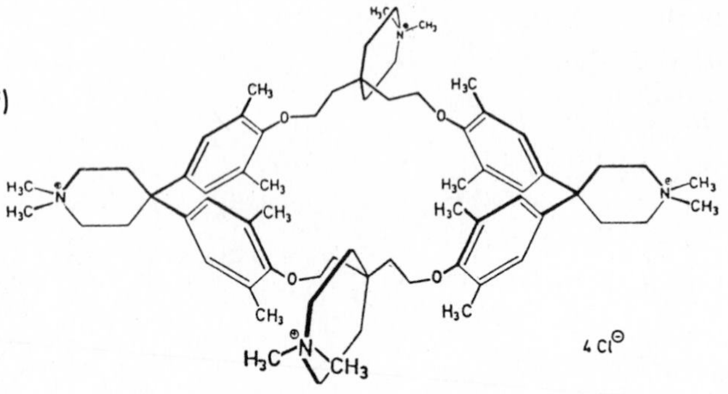

Perylene

Fluoranthene

Pyrene

4 Cl$^\ominus$

Figure 4-32 (continued). Typical organic cage compounds [34–39].

Typical organic cage compounds are shown in Figure 4-32. The chemical synthesis of these structures was stimulated by structures in biological systems with their specific key-lock molecular recognition centers.

Usually their practical use in chemical sensors requires to solve the signal deduction during the key-lock event. This may be done optically, by mass-sensitive sensors, or by incorporation of cage compounds in membranes (see below). It usually requires a very careful choice of solvent or dry deposition techniques, choice of appropriate matrixes, covalent bonding etc. The same problems are encountered in the stabilisation of enzymes in bioelectrodes of biosensors (compare Chapter 3 and the following Section 4.11).

4.11   Modification of Electrodes in Biosensors

Characteristic key-lock interactions as discussed above may be used in biosensors with recognition sites as they had been listed already in Chapter 3 (Figure 3-9). A typical example for a complete biosensor based upon selective metabolism is shown in Figure 4-33. This often applied amperometric principle shown here makes use of several independent principles to gain selectivity in the determination of compounds in a complex environment such as the human blood.

– The first selectivity is obtained by choosing a proper outer membrane which prevents large molecules above a certain size from indiffusion.

− The second selectivity is obtained from the highly selective enzyme reaction. In this case the glucose oxidase is used which converts glucose according to

$$d\text{-glucose} + O_2 \xrightarrow{\text{glucose oxidase}} \text{gluconolactone} + H_2O_2 \qquad (4\text{-}21)$$

at the catalytic active site of the enzyme.

− Signal transduction from the enzyme to the electrode may be obtained by choosing different mediators with concepts already mentioned in Chapter 3 (Figure 3-16). The third step to gain selectivity is therefore a careful choice of a proper mediator system.

− The fourth selectivity is obtained by to adjusting the electrode material and electrode potential to the specific enzyme and mediator system.

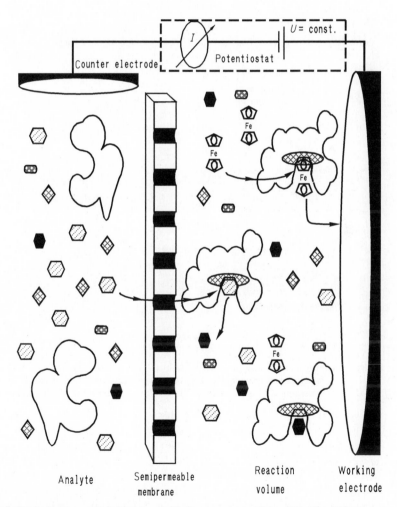

Figure 4-33, Part I. Typical amperometric biosensor with ferrocene as mediator [40]. The different molecular structures are explained in Part II (next page).

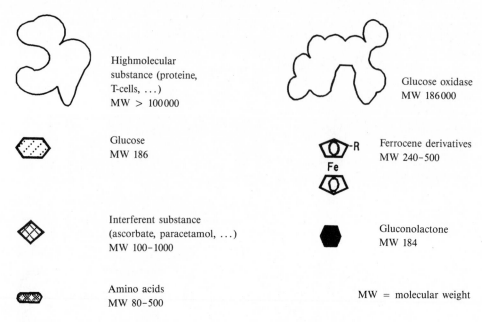

Figure 4-33, Part II. The different molecular structures involved in Figure 4-33, Part I.

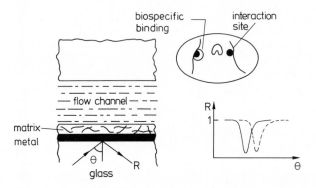

Figure 4-34. Schematic drawing of a coupling matrix on a gold surface used for biosensing based on surface plasmon resonance. The drawing illustrates how one of the molecules of a biospecific interaction pair is immobilized in the matrix and how biospecific binding affects the reflectance R (Θ).

4.12   Biospecific Interaction Sensors

Molecular recognition occurs in many important biochemical environments. Antigen-antibody interaction is a good example (see Fig. 4-18). Several biosensor concepts have been developed to measure such interactions. With optical techniques like ellipsometry [41], wave propagation in integrated optical wavequides [42] and surface plasmon resonance [43] it is possible to monitor such interactions directly. Recently a sensing surface consisting of one of the molecules in the interaction pair covalently bound to a polymer (dextran) matrix on a gold surface has been developed for biosensing using surface plasmon resonance [44, 45] (Fig. 4-34). This surface is possible to be used several times for, eg, antigen detection by decreasing the pH outside the sensing surface to break the antigen-antibody binding by keeping the covalently bound antibody in the polymer matrix [44, 45].

4.13   References

[1] Singer, S. J., Nicolson, G. L., *Science* **175** (1972) 720; Winter, R., "Struktur und Dynamik von Modell-Biomembranen", *Chem. in unserer Zeit* **2** (1990) 71–81.

[2] Schierbaum, K. D. et al., "Prototype Structure for Systematic Investigations of Thin-film Gas Sensors", *Sens. Actuators B* **1** (1990) 171–175.

[3] Schierbaum, K. D., Kirner, U., Geiger, J., Göpel, W., "Schottky-Barrier and Conductivity Gas Sensors based upon Pd/SnO$_2$ and Pt/TiO$_2$", *Sens. Actuators B* **4** (1990) 87–94.

[4] Göpel, W., Schierbaum, K. D., Wiemhofer, H. D., "Three Phase Boundaries in Gas Sensing: The Prototype Systems Pd/SnO$_2$, Pt/TiO$_2$, and Pt/ZrO$_2$", *Proc. of the 3rd Int. Meeting on Chemical Sensors, Cleveland (USA), 1990,* and *Sens. Actuators,* in press.

[5] Rothschild, K. J., Asher, J. M., Stanley, H. E., Anastassakis, E., "Raman Spectroscopy of Uncomplexated Valinomycin, 2. Nonpolar and Polar Solution", *J. Amer. Chem. Soc.* **99** (1977) 2032.

[6] Edwards, P. A., Jurs, P. C., "Correlation of Odor Intensity with Structural Properties of Odorants", *Chemical Senses* **14** (1989) 281–291 and references given there.

[7] Technical information, Bosch GmbH, Stuttgart (FRG).

[8] Lundstrøm, I., Söderberg, S., "Hydrogen Sensitive MOS-Structures Part 1: Principles and Applications", *Sens. Actuators* **1** (1981) 403–426; Lundstrøm, I., Söderberg, S., "Hydrogen Sensitive MOS-Structures, Part 2: Characterization", *Sens. Actuators* **2** (1981/1982) 105–138; Lundström, I., Armgarth, M., Spetz, A., Winquist, F., Gas Sensors Based on Catalytic Metal-Gate Field Effect Devices, *Sens. Actuators* **10** (1986) 399–421, and Lundström, I., Armgarth, M., Petersson, L.-G., Physics with Catalytic Metal Gate Chemical Sensors, *CRC Crit. Rev. Solid State Mater. Sci.* **15** (1989) 201–278.

[9] Gentry, S. J., Walsh, P. T., "Poison-Resistant Catalytic Flammable-Gas Sensing Elements", *Sens. Actuators* **5** (1984) 239–251.

[10] Technical information, Dr. W. Ingold AG, Urdorf (Switzerland).

[11] King Jr., W. H., "Piezoelectric Sorption Detector", *Anal. Chem.* **36** (1964) 1735–1739; Sauerbrey, G., "Verwendung von Schwingquarzen zur Wägung dünner Schichten und Mikrowägung", *Z. Physik* **155** (1959) 206–222.

[12] Göpel, W., "Solid state chemical sensors: Atomistic models and research trends", *Sens. Actuators* **16** (1989) 167–193.

[13] Atkins, P. W., *Physical Chemistry*; Oxford: Oxford University Press 1990.

[14] McClelland, B. J., *Statistical Thermodynamics;* London: Chapman and Hall, Science Paperbacks 1973.

[15] Göpel, W., "Chemisorption and Charge Transfer at Ionic Semiconductor Surfaces: Implications in Designing Gas Sensors", *Progr. in Surf. Sci.* **20**, No. 1 (1985) 9.

[16] Esser, P., Feierabend, R., Göpel, W., "Comparative Study on the Reactivity of Polycrystalline and Single Crystal ZnO Surfaces: Catalytic Oxidation of CO", *Ber. Bunsen Ges. Phys. Chem.* **85** (1981) 447–455.

[17] Schasfoort, R., PhD thesis, University of Twente (Netherlands), 1989.

[18] Sørensen, O. T., "Thermodynamics and Defect Structure of Nonstoichiometric Oxides", in: *Nonstoichiometric Oxides*, Sørensen, O. T. (ed.); New York: Academic Press 1981.

[19] Tuller, H. L., "Mixed Conductivity in Nonstoichiometric Oxides", in: *Nonstoichiometric Oxides,* Sørensen, O. T. (ed.); New York: Academic Press 1981.

[20] Rocker, G., Göpel, W., "Chemisorption of H_2 and CO on Stoichiometric and Defective TiO_2 (110)", *Surf. Sci.* **175** (1986) L675–L680; Göpel, W., Rocker, G., Feierabend, R., "Intrinsic Defects of TiO_2 (110): Interaction with chemisorbed O_2, H_2, CO, and CO_2", *Phys. Rev. B* **28** (1983) 3427–3438; Kirner, U. et al., "Low and High Temperature TiO_2 Oxygen Sensors", *Sens. Actuators B* **1** (1990) 103–107.

[21] Kohl, D., "Surface Processes in the Detection of Reducing Gases with SnO_2-based Devices", *Sens. Actuators* **18** (1989) 71–113; Schierbaum, K. D., Wiemhöfer, H. D., Göpel, W., "Defect structure and sensing mechanism of SnO_2 gas sensors: Comparative electrical and spectroscopic results", *Solid State Ionics* **28–30** (1988) 1631–1636; Geiger, J. F., Schierbaum, K. D., Göpel, W., "Surface spectroscopic studies on Pd-doped SnO_2", *11th Int. Vacuum Congr., 7th Int. Conf. on Solid Surfaces and Vacuum,* in press.

[22] Hotan, W., Göpel, W., Haul, R., "Interaction of CO_2 and CO with Nonpolar Zinc Oxide Surfaces", *Surf. Sci.* **83** (1979) 162–180; Göpel, W., Bauer, R. S., Hansson, G., "Ultraviolet Photoemission Studies of Chemisorption and Point Defect Formation on ZnO Nonpolar Surfaces", *Surf. Sci.* **99** (1980) 138–158; Göpel, W., Lampe, U., "Influence of Defects on the Electronic Structure of Zinc Oxide Surfaces", *Phys. Rev. B* **22** (1980) 6447–6462.

[23] Subbarao, E. C., Maiti, H. S., "Oxygen sensors and pumps", in: *Advances in Ceramics,* Vol. 24b: *Science and Technology of Zirconia III,* Somiya, S. N., Yamamoto, N., Yaagida, H. (eds.), Colombus, OH: American Ceramics 1988, 731–747; Badwal, S. P. S., Ciacchi, F. T., "Performance of Zirconia Membrane Oxygen Sensors at Low Temperatures with Nonstoichiometric Oxide Electrodes", *J. Appl. Electrochem.* **16** (1986) 28–40; Schindler, K. et al., "Spectroscopic and Electrical Studies of Yttria-Stabilized Zirconia for Oxygen Sensors", *Sens. Actuators* **17** (1989) 555–568.

[24] Rager, A. et al., "Stability of Organic Thin Films on Inorganic Substrates: Prototype-Studies Using Metal Phthalocyanines", *J. Mol. Electronics* **5** (1989) 227; Mockert, H., Schmeisser, D., Göpel, W., "Lead Pthalocyanine as Protoype Organic Material for Gas Sensors: Comparative Electrical and Spectroscopic Studies to Optimize O_2 and NO_2 Sensing", *Sens. Actuators* **19** (1989) 159.

[25] Göpel, W., Pollmann, J., Ivanov, I., Reihl, B., "Angle-resolved photoemission from polar and nonpolar zinc oxide surfaces", *Phys. Rev. B* **26** (1982) 3144.

[26] Göpel, W., "Entwicklung chemischer Sensoren: Empirische Kunst oder systematische Forschung", *Techn. Messen* **52** (1985) 47; 92; 175.

[27] Seiyama, T., "Surface Reactivity of Oxide Materials in Oxidation Reduction Environments", in: Novotny, J., et al. (ed.) *Surface and Nearsurface Chemistry of Oxide Materials;* Amsterdam: Elsevier, 1988

[28] Mockert, H., Schmeisser, D., Göpel, W., "Lead Phthalocyanine as Prototype Organic Material for Gas Sensors: Comparative Electrical and Spectroscopic Studies to Optimize O_2 and NO_2 Sensing", *Sens. Actuators* **19** (1989) 159.

[29] Perez, O. L., Romen, D., Yacaman, M. J., "Distribution of Surface Sites on Small Metallic Particles", *Appl. Surf. Sci.* **13** (1982) 402.

[30] Göpel, W., "Phthalocyanines as Prototype Materials for Chemical Sensors and Molecular Electronic Devices", *Conf. Proc. of the Int. Conf. on Synthetic Metals, Tübingen (FRG), 1990;* and *Synth. Metals,* in press.

[31] Kirner, U., Schierbaum, K. D., Göpel, W., *Angewandte Oberflächenanalytik, Kaiserslautern (FRG), 1990;* and *Fres. Z. Anal. Chem.,* in press.

[32] Wiemhöfer, H. D., Schmeisser, D., Göpel, W., "Lead Phthalocyanine as a Mixed Conducting Oxygen Electrode", *Conf. Proc. Solid State Ionics, Hakone (Japan), 1989;* and Solid State Ionics.

[33] Göpel, W., *State and Perspectives of Research on Surfaces and Interfaces,* Luxembourg: Office for Official Publications of the European Communities, 1990.

[34] Schmidtchen, F. P., "Molecular Catalysis by Polyammonium Receptors", *Topics Curr. Chem.* **132** (1986) 101.

[35] Cram, D. J., Trueblood, K. N., "Concept, Structure, and Binding in Complexation", *Topics Curr. Chem.* **98** (1981) 43.

[36] Cram, D. J., "Von molekularen Wirten und Gästen sowie ihren Komplexen (Nobelvortrag)", *Angew. Chem.* **100** (1988) 1041.

[37] Saenger, W., "Cyclodextrin-Einschlußverbindungen in Forschung und Industrie", *Angew. Chem.* **92** (1980) 343.

[38] Pedersen, C. J., Frensdorf, H. K., „Makrozyklische Polyäther und ihre Komplexe", *Angew. Chem.* **84** (1972) 16.

[39] Diederich, F., "Cyclophane zur Komplexierung von Neutralmolekülen", *Angew. Chem.* **100** (1988) 372.

[40] Löffler, U., PhD thesis, University of Tübingen, Tübingen (FRG) 1991.

[41] Arwin, H., Lundström, I., "Surface-Oriented Optical Methods for Biomedical Analysis", *Methods in Enzymology* **137** (1988) 366–381.

[42] Nellen, Ph. M., Tiefenthaler, K., Lukosz, W., "Integrated Optical Input Grating Couplers as Biochemical Sensors", *Sens. Actuators* **15** (1988) 285–295.

[43] Liedberg, B., Nylander, C., Lundström, I., "Surface Plasmon Resonance for Gas Detection and Biosensing", *Sens. Actuators* **4** (1983) 299–304.

[44] Löfås, S., Malmquist, M., Rönnberg, I., Stenberg, E., Liedberg, B., Lundström, I., Bioanalysis with surface plasmon resonance, *Sens. Actuators,* submitted.

[45] Löfås, S., Johansson, B., "A Novel Hydrogel Matrix on Gold Surfaces in Surface Plasmon Resonance Sensors for Fast and Efficient Covalent Immobilization of Ligands", *Chemical Comm.* (1990) 1526–1528.

5 Specific Features of Electrochemical Sensors

HANS-DIETER WIEMHÖFER, Universität Tübingen, FRG
KARL CAMMANN, Universität Münster, FRG

Contents

5.1 Introduction

Electrochemical sensors based upon potentiometry, amperometry, coulometry, AC conductivity and capacitance measurements are already well-established in analytical and clinical chemistry laboratories. These sensors require little complex equipment and are relatively easily calibrated or need no calibration in case of coulometry. They sense and measure concentrations and activities of ions and neutral species in liquid solutions, solids and gases. In this chapter we will discuss some fundamental detection principles of electrochemical sensors. For more details on electrochemistry we refer to the extensive literature [see eg, 1–7].

Electrochemical sensors are characterized by a direct relation of measured electrical currents and potential differences to concentrations, activities or partial pressures of chemical species. Just as chemical sensors based upon semiconductors and semiconductor devices, electrochemical sensors make possible a direct coupling of electrical signals to the status of a chemical system.

The terms electrochemical sensor, electrochemical device or electrochemical cell are used for ionic conductors (ie, electrolytes) between at least two electrodes, although more sophisticated devices may involve additional electrodes and more complicated sequences of conducting phases, interfaces and membranes. In a broader sense, devices with semiconductors or mixed conductors with electronic and ionic conductivity might also be termed electrochemical sensors, if chemical and electrochemical equilibria with molecules or ions determine their impedance, voltage or current-voltage relation. In a general classification one may distinguish the following two classes.

One *class of electrochemical sensors* is based upon conductivity changes and changes of dielectric constants as measured by AC- or DC-techniques. The sensor response is governed predominantly by surface or bulk impedance changes of the respective electrolyte or conductor due to its interaction with ions and molecules of the environment. These types of sensors are treated in detail in Chapters 9 for electronic conductors and in Chapter 7 for liquid electrolytes.

A second, more important *class of electrochemical sensors* is based upon electrochemical equilibria and reactions at interfaces of two electrically conducting phases. These types of sensors are presented in Chapter 7 for the use of liquid electrolytes and in Chapter 8 for the use of solid electrolytes. In any case, such a sensor is designed to possess an interface at which the generated potential, the mass-transport-controlled current, the frequency-dependent AC-responses or the time-dependent DC-responses can be related directly and quantitatively to concentrations or activities of chemical species to be monitored. The actual device performance depends on several factors including thermodynamic, kinetic, and the mass transport parameters that determine observables such as response range, response stability, noise, drift, response selectivity, response time, temperature coefficient, lifetime, pretreatment requirements, and others.

As sensors systems may become more elaborate by making use of multiple interfaces of passive membranes and thin surface coatings, there is a great need for recognizing how all of these factors can be optimized. A particular question is, what type of sensor behavior should be optimized by choosing certain potential-generating or potential-controlling interfaces. The latter may be termed unblocked, blocked or kinetically controlled (eg, with responses between blocked and unblocked interfaces) depending on the nature of the electrochemical coupling processes.

Unblocked interfaces between two electrical conductors with ionic or electronic conductivity allow charge transfer through the interface and continuous charge flow without degradation. Electrochemical equilibria control the electrical properties at these interfaces. Ions or molecules that take part in these equilibria may in principle be determined by means of suitable electrochemical sensors containing such an interface. Blocked interfaces, on the other hand, can be found for instance in ion-sensitive field-effect-transistors between the sensitive gate layer and the transistor base. It is, however, a generally accepted conclusion that unblocked reversible interfaces always provide the best sensors [8].

5.2 Galvanic Cells

Potentiometric electrochemical sensors are based upon galvanic cells. Figure 5-1 shows a simple example of a galvanic cell with the solid silver ion conductor AgCl. The open-circuit voltage E, also called electromotive force (emf) in electrochemistry, multiplied by Faraday's constant F determines the maximum electrical work ΔW_{el} for the transport of electrons from the negative to the positive electrode. It is given by:

$$\Delta W_{el} = -FE .$$ (5-1)

The cell in Figure 5-1 gives the following cell reaction:

$$Ag_{(s)} + \frac{1}{2} Cl_{2(g)} \rightarrow AgCl_{(s)} .$$ (5-2)

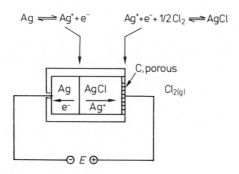

Figure 5-1.
Example for a galvanic cell with AgCl as solid electrolyte (a silver ion conductor). E denotes the open-circuit voltage (which corresponds to the electromotive force). The electrode reaction equilibria at the Ag/AgCl and AgCl/C, $Cl_{2(g)}$ interfaces are indicated. The cell may be used as a potentiometric gas sensor for chlorine gas operating above 150 °C.

The change of Gibbs free energy of the cell reaction under reversible conditions correponds to:

$$\Delta G_{cell} = \mu^0_{AgCl} - \mu^0_{Ag} - \frac{1}{2} \mu^0_{Cl_2} - \frac{RT}{2} \ln \left(\frac{p_{Cl_2}}{p^0} \right)$$

$$= \Delta_f G^0_{AgCl} - \frac{RT}{2} \ln \left(\frac{p_{Cl_2}}{p^0} \right) .$$ (5-3)

$\mu_{Cl_2}^0$, μ_{AgCl}^0 and μ_{Ag}^0 denote the standard chemical potentials of gaseous Cl_2, solid AgCl and Ag. p_{Cl_2} and p^0 denote partial pressure of Cl_2 and standard pressure (usually $p^0 = 1.013$ bar). $\Delta_f G_{AgCl}^0$ is the standard value of Gibbs free energy for formation of AgCl.

For constant values of temperature and total pressure the change of Gibbs free energy is equivalent to the maximum electrical work W_{el} corresponding to the given cell reaction:

$$\Delta G_{cell} = W_{el} = -nFE \qquad (p, T = \text{const}) \tag{5-4}$$

where n denotes the number of electrons involved in the cell reaction (here $n = 1$). Comparison of Equations (5-3) and (5-4) yields the following expression for the emf E:

$$E = E^0 + \frac{RT}{2F} \ln \left(\frac{p_{Cl_2}}{p^0} \right). \tag{5-5}$$

The constant E^0 contains the standard value of the Gibbs free energy of formation of solid AgCl:

$$E^0 = -\left(\frac{\Delta_f G_{AgCl}^0}{F} \right). \tag{5-6}$$

Galvanic cells like the example in Figure 5-1 provide the basis of a series of well-known *potentiometric electrochemical gas sensors* as they are treated in detail in Chapters 7.1 for liquid electrolyte sensors and Chapter 8 for solid electrolyte sensors.

Figure 5-2 shows another example of a galvanic cell with a liquid electrolyte. The net cell reaction in that case is:

$$H_{2(g)} + Cl_{2(g)} \rightarrow 2H_{aq}^+ + 2Cl_{aq}^-. \tag{5-7}$$

The corresponding change of Gibbs free energy is:

$$\Delta G_{cell} = 2\mu_{H^+} + 2\mu_{Cl^-} - \mu_{H_2} - \mu_{Cl_2}. \tag{5-8}$$

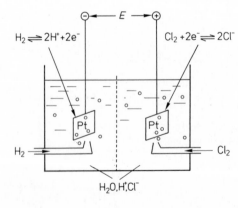

Figure 5-2.
Example for a galvanic cell with aqueous HCl as liquid electrolyte, E denotes the open-circuit voltage (corresponding to the electromotive force). The two half-cells may contain different concentrations of HCl. Then, they have to be separated by a diaphragm, as indicated in the figure.

The result for the emf E is:

$$E = E^0 + \frac{RT}{2F} \ln \left[\frac{\left(\dfrac{p_{H_2}}{p^0}\right)\left(\dfrac{p_{Cl_2}}{p^0}\right)}{a_{H^+}^2 \, a_{Cl^-}^2} \right] . \tag{5-9}$$

The emf may always be written as the difference of two electrode potentials. The usual convention to scale electrode potentials in electrochemistry with aqueous electrolytes is to define the electrode potential of an electrode as the emf of a cell consisting of this electrode and the normal-hydrogen-electrode (NHE). A normal-hydrogen-electrode consists of a solution containing protons with activity $a_{H^+} = 1$ and inert platinum in contact with hydrogen gas under normal pressure ($p_{H_2} = 1.013$ bar). Electrode potentials defined in this way are denoted by ε in the following. For a chlorine electrode the electrode potential is therefore given by:

$$\varepsilon_{Cl_2/Cl^-} = E\,(Cl_2/Cl^- \;\; vs. \;\; H_2/H^+ \; \text{with } a_{H^+} = 1, \;\; p_{H_2} = p^0)$$

$$= \varepsilon^0_{Cl_2/Cl^-} + \frac{RT}{2F} \ln \left[\frac{\left(\dfrac{p_{Cl_2}}{p^0}\right)}{a_{Cl^-}^2} \right] . \tag{5-10}$$

The explicit expression for $\varepsilon^0_{Cl_2/Cl^-}$, the standard electrode potential of the chlorine electrode, is:

$$\varepsilon^0_{Cl_2/Cl^-} = \frac{1}{2F} \left(\mu^0_{Cl_2} - 2\mu^0_{Cl^-} + \mu^0_{H_2} - 2\mu^0_{H^+} \right) . \tag{5-11}$$

Accordingly, the standard electrode potential of the hydrogen electrode is set zero. Every emf E between any pair of electrodes in aqueous electrolytes can be written as a difference of the two electrode potentials ε_1 and ε_2:

$$E = \varepsilon_2 - \varepsilon_1 . \tag{5-12}$$

Standard electrode potentials (usually for 25 °C) for a large number of electrode reactions are tabulated [see eg, 9, 10]. ε_{Cl_2/Cl^-} and the corresponding standard value $\varepsilon^0_{Cl_2/Cl^-}$ ($= \varepsilon_{Cl_2/Cl^-}$ for $p_{Cl_2} = p^0$ and $a_{Cl^-} = 1$) characterize an electrode potential being determined by the electrode reaction or electrode equilibrium:

$$Cl_2 + 2e^- \rightleftharpoons 2Cl^- . \tag{5-13}$$

If we consider a general electrode equilibrium between oxidized and reduced species involving n electrons:

$$ox + ne \rightleftharpoons red \tag{5-14}$$

we obtain the following general expression for the electrode potential:

$$\varepsilon_{ox/red} = \varepsilon_{ox/red}^0 + \frac{RT}{nF} \ln \frac{a_{ox}}{a_{red}} . \tag{5-15}$$

This is the well-known Nernst equation with a_{red} and a_{ox} as a short-hand notation for products of the activities of reacting species on the left and right hand sides of Equation (5-14) with suitable exponents given by the stoichiometric coefficients.

The basic idea of potentiometric electrochemical sensors is to use the emf as a sensor signal, if it depends on the chemical activity or partial pressure of the species to be analyzed. In order to have a definite signal depending only on one single chemical species, the activities of all other species in the corresponding electrode reactions have to be kept constant. The electrode with its electrode potential changing with the activity of the species to be measured is called the indicator or measuring electrode.

The electrode potential of the second electrode needed in any galvanic cell, ie, of the reference electrode has to be kept constant. In electrochemical experiments with liquid electrolytes the most popular reference electrodes are Hg/Hg_2Cl_2, Cl^- ("calomel") and $Ag/AgCl$, Cl^--electrodes. The electrode reactions correspond to the formation of the solid chlorides from the metal and dissolved chloride ions in the electrolyte. The chemical activities of the solids, the metal and the solid metal chloride, are constant at a given temperature. If the activity of the dissolved chloride ions is held constant, too, the electrode potential must be constant.

5.3   Thermodynamic View of Electrode Equilibria

Electrodes in galvanic cells as described above provide a definite coupling between the electrochemical potentials $\tilde{\mu}_e$ of electrons in the electron-conducting electrode and $\tilde{\mu}_{ion}$ of ions in the electrolyte. The coupling is determined by the electrochemical equilibrium at the electrode/electrolyte interface. In this case, we speak of unblocked reversible electrodes. A galvanic cell thus corresponds to an electrochemical cell with two unblocked reversible electrode interfaces coupled by an ionic conductor. The difference of the electrochemical potentials of electrons at both electrodes provides the open-circuit voltage or emf E. Therefore, the properties of galvanic cells, in particular the value of the emf, are determined by the single electrode equilibria at the electrode interfaces. As at least two interfaces are involved, each electrode interface can be analyzed separately. At the electrode interface Ag/AgCl in Figure 5-1 we observe the equilibrium electrode reaction:

$$Ag \rightleftharpoons Ag^+ + e^- , \tag{5-16}$$

which gives the following relation between the electrochemical potentials of electrons in the metallic $Ag(\tilde{\mu}_e')$ and of Ag^+ ions in the solid electrolyte $(\tilde{\mu}_{Ag^+})$ and the chemical potential of silver in the metal (μ_{Ag}^0)

$$\mu_{Ag}^0 = \tilde{\mu}_{Ag^+} + \tilde{\mu}_e' . \tag{5-17}$$

At the second electrode interface (C/AgCl) we observe the following electrochemical equilibrium between silver ions, gaseous chlorine, electrons and AgCl:

$$\tilde{\mu}_{Ag^+} + \tilde{\mu}_e'' + \frac{1}{2}\mu_{Cl_2} = \mu_{AgCl}^0 . \tag{5-18}$$

The emf E corresponds to the difference of the electrochemical potentials of electrons at the two electrodes:

$$E = -\frac{1}{F}(\tilde{\mu}_e'' - \tilde{\mu}_e') . \tag{5-19}$$

Replacing $\tilde{\mu}_e'$ and $\tilde{\mu}_e''$ by the preceeding two equations finally results in Equation (5-5), too, because $\tilde{\mu}_{Ag^+}$ cancels. The treatment of single electrode reactions is a different way of deriving the emf of a galvanic cell.

Oxygen sensors based on oxide ion conducting ZrO_2-ceramics are classic examples for potentiometric gas sensors (see also Chapter 8):

$$O_2\,(p_{O_2}'),\ Pt\,|\,ZrO_2\,(stab.)\,|\,Pt,\ O_2\,(p_{O_2}'') . \tag{5-20}$$

The electrochemical potentials of electrons at the electrode/electrolyte boundary are determined by the equilibrium between the electrons of platinum, oxide ions of the electrolyte and oxygen in the gas phase:

$$\frac{1}{2}O_{2(g)} + 2e^- \rightleftharpoons O^{2-} , \tag{5-21}$$

which relates $\tilde{\mu}_e$ to the chemical potential μ_{O_2} of oxygen in the gas phase according to:

$$\tilde{\mu}_e\,(Pt) = -\frac{1}{4}\mu_{O_2} + \frac{1}{2}\tilde{\mu}_{O^{2-}}\,(ZrO_2) . \tag{5-22}$$

The electrochemical potential of oxygen ions, $\tilde{\mu}_{O^{2-}}\,(ZrO_2)$, is constant throughout the electrolyte for zero current, and therefore it has the same value at the two electrodes.

Taking the explicit expression for μ_{O_2}:

$$\mu_{O_2} = \mu_{O_2}^0 + RT\ln\frac{p_{O_2}}{p^0} \tag{5-23}$$

we obtain the emf from Equations (5-19) und (5-22) as:

$$E = -\frac{1}{F}(\tilde{\mu}_e'' - \tilde{\mu}_e') = \frac{RT}{4F}\ln\frac{p_{O_2}''}{p_{O_2}'} . \tag{5-24}$$

If p_{O_2}' at the reference electrode is kept constant, the voltage at constant temperature is then directly related to p_{O_2}''.

In general, any electrode equilibrium at an electrode/electrolyte interface involves electrons of the electronic conductor, ions of the ionic conductor and one or more neutral chemical species. The thermodynamic equilibrium condition for the general case can be written as:

$$\tilde{\mu}_e = \text{const} - \frac{1}{z_{ion}} \tilde{\mu}_{ion} \tag{5-25}$$

z_{ion} is the charge number of the ions in the electrolyte (may be positive or negative). The constant includes the chemical potentials of the other chemical species involved in the electrode reaction. The neutral species may be molecules from the gas phase adsorbed at the interface, neutral components of the ionically conducting phase or of the electronic conductor, or they may be components from additional solid phases in contact with the components at the electrode interface. This forms the general basis for the use of electrode potentials in potentiometric sensors to identify neutral components. If only one chemical potential in Equation (5-25) varies, and the others are fixed by the phase equilibria, a one-to-one correspondence between electrode potential and variable chemical potential is expected. This is required for unequivocal sensing.

As an example, the following cell can be used to measure partial pressures of sulfur in the gas phase [7]:

$$\text{Ag} | \text{AgI} | \text{Ag}_2\text{S}, \, \text{S}_{x \, (g)} \, (2 \leq x \leq 8) \,. \tag{5-26}$$

Ag_2S is a mixed conductor for electrons and silver ions, AgI is a good silver ion conductor at temperatures above 145 °C [7]. The reaction at the indicator electrode Ag_2S is when formulated with gaseous S_x-species:

$$\frac{1}{x} \text{S}_x + 2\text{e}^- + 2\text{Ag}^+ \rightleftharpoons \text{Ag}_2\text{S} \,. \tag{5-27}$$

The equilibrium condition corresponds to:

$$\tilde{\mu}_e (\text{Ag}_2\text{S}) = \frac{1}{2} \mu^0_{\text{Ag2S}} - \frac{1}{2x} \mu_{\text{S}_x} - \tilde{\mu}_{\text{Ag}^+} = \text{const} - \frac{RT}{2x} \ln \left(\frac{p_{\text{S}_x}}{p^0} \right) - \tilde{\mu}_{\text{Ag}^+} \,. \tag{5-28}$$

The emf of the cell, Equation (5-26), is given by:

$$E = -\frac{1}{F} [\tilde{\mu}_e (\text{Ag}_2\text{S}) - \tilde{\mu}_e (\text{Ag})] = \text{const}' + \frac{RT}{2xF} \ln \left(\frac{p_{\text{S}_x}}{p^0} \right) \,. \tag{5-29}$$

In electrochemistry, electrodes such as Pt, O_2/ZrO_2 are called electrodes of the first kind, because only one chemical potential of a neutral component is involved in addition to the electrochemical potentials of ions and electrons. Accordingly, $\text{AgI}/\text{Ag}_2\text{S}, \, \text{S}_{x(g)}$ is an electrode of the second kind, because the electrode reaction involves an additional component which is coupled to the electrochemical potentials of mobile silver ions and electrons by a secondary chemical equilibrium (ie, of $\text{S}_{x(g)}$ with the compound Ag_2S).

In this way one may use additional equilibria by introducing further phases that couple partial pressures, ie, chemical potentials of various chemical species, to the electrochemical potentials of electrons and mobile ions. This concept has been applied to a variety of chemical species and solid electrolytes (see Chapter 8).

5.4   Electrochemical Potentials and Band Schemes

The use of electrochemical potentials of electrons and ions is essential for the thermodynamics of equilibria at electrodes and more general, also at interfaces of any two conducting phases. The electrochemical potential of electrons is called Fermi energy or Fermi level (E_F) in physics and electronics (usually taken per particle). With N_A for Avogadro's number, $\tilde{\mu}_e$ and E_F are related by:

$$\tilde{\mu}_e = N_A \cdot E_F . \tag{5-30}$$

As Fermi-Dirac-statistics holds also for ions in electrolytes (no ionic position can be occupied by more than one ion), one could use the term Fermi energy of ions for the electrochemical potential of ions as well.

Electrochemical equilibrium within an electron conductor under zero current implies that the Fermi energy or electrochemical potential of electrons is constant everywhere. An ionic conductor, on the other hand, must show a constant electrochemical potential of ions for thermodynamic equilibrium. The electrochemical potential E_F of electrons, however, usually shows a gradient in the ionic conductor of a galvanic cell. This is illustrated in Figure 5-3 for a cell with ZrO_2.

Changes of electrode potentials with oxygen partial pressures are related to distinct changes of oxygen concentration, bonding and surface energies at the Pt/ZrO_2 interface [11]. The oxygen partial pressure may be varied over more than 30 orders of magnitude. This provides well-defined shifts of the Fermi energy E_F with regard to the band edges of ZrO_2 at the electrode interfaces as explained in the following.

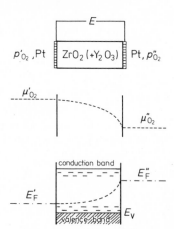

Figure 5-3.
Galvanic cell $O_2(p')$, $Pt|ZrO_2$ (Y_2O_3-stabilized)$|Pt$, $O_2(p'')$: schematic band scheme. The differences $E_F - E_V$ at the electrode interfaces are determined by the partial pressures of oxygen in the surrounding gas phase. E_F denotes the Fermi energy and E_V the valence band edge of the solid electrolyte ZrO_2. Localized electronic levels are indicated in the band gap which may give rise to electron or hole conduction, if E_F shifts towards the band edges (for very low or very high p_{O_2}).

Stabilized zirconia has a band gap of 5.2 eV [12] and shows a small variable deviation of the oxygen content from an ideal stoichiometric $(1:2)$ composition. The electronic conductivity and this small non-stoichiometry depend on the partial pressure of oxygen and thus on the chemical potential μ_{O_2} of oxygen. The Fermi energy or its equivalent, the electrochemical potential of electrons $\tilde{\mu}_e$ and the chemical potential of oxygen are defined everywhere within the electrolyte and coupled to the electrochemical potential $\tilde{\mu}_{O^{2-}}$ of ions as described by Equation (5-22). Evaluation of Equation (5-22) and using:

$$\tilde{\mu}_e = \mu_e - F\varphi \tag{5-31}$$

$$\tilde{\mu}_{O^{2-}} = \mu_{O^{2-}} - 2F\varphi \tag{5-32}$$

with φ as the electrical potential and μ_e and $\mu_{O^{2-}}$ as the chemical potentials of electrons and oxygen ions yields:

$$\mu_{O_2} = 2\mu_{O^{2-}} - 4\mu_e . \tag{5-33}$$

The chemical potential of oxygen ions, $\mu_{O^{2-}}$, is constant because of the high point defect concentration [7]. Therefore, the following equation holds for partial pressure changes:

$$\Delta\mu_e = -\frac{1}{4}\Delta\mu_{O_2} = -\frac{RT}{4}\Delta \ln p_{O_2} . \tag{5-34}$$

For changes $\Delta\mu_e$, we expect from Fermi-Dirac statistics that:

$$\Delta\mu_e = -N_A \cdot \Delta(E_C - E_F) = N_A \cdot \Delta(E_F - E_V) \tag{5-35}$$

where E_C and E_V denote the conduction and valence band edge.
This results in:

$$\Delta(E_F - E_V) = -\Delta(E_C - E_F) = -\frac{k_B T}{4}\Delta \ln p_{O_2} . \tag{5-36}$$

Equation (5-36) shows that changes of electrode potentials due to changes of the oxygen partial pressures can be identified by corresponding shifts of E_F within the electronic band scheme of ZrO_2.

Thus, a fundamental aspect of electrode equilibria is the coupling between the electrochemical potential (or Fermi energy) of electrons $\tilde{\mu}_e$ and the corresponding quantity $\tilde{\mu}_{ion}$ of the mobile ions within the electrolyte. This makes possible the "transduction" of chemical information, ie, of chemical potentials involved in the electrode equilibrium, into voltage or current signals that can be processed in electronic circuits.

5.5   Electrical Potential Differences at Electrodes

Although electrical potential differences between any two conducting phases are not measureable, changes of these differences can be related to changes of electrode potentials and their discussion plays an important role in the electrochemistry of electrodes in liquid electrolytes [see eg, 1, 2, 7]. We take as an example the Cl_2/Cl^--electrode of the cell in Figure 5-2. The electrode reaction and corresponding equilibrium condition is:

$$Cl_{2(g)} + 2e^-_{(Pt)} \rightleftharpoons 2Cl^-_{aq} \tag{5-37}$$

$$\mu_{Cl_2} + 2\tilde{\mu}_e = 2\tilde{\mu}_{Cl^-} . \tag{5-38}$$

The electrochemical potentials may be split into a chemical potential and an electrostatic potential energy according to Equations (5-31) and (5-32). From Equation (5-38) we obtain the electrical potential difference:

$$\Delta\varphi\,(Cl_2/Cl^-) = \varphi\,(Pt) - \varphi\,(electrolyte)$$

$$= \frac{1}{2F}\,[\mu_{Cl_2} + 2\mu_e - 2\mu_{Cl^-}] = const + \frac{RT}{2F}\,\ln\frac{p''_{Cl_2}}{a^2_{Cl^-}} . \tag{5-39}$$

The constant contains the standard values of the chemical potentials. The electrical potential difference $\Delta\varphi$ is called the Galvani-potential difference at that interface.

Figure 5-4 schematically shows the space dependence of the electrical potential in the galvanic cell of Figure 5-2. Figures 5-5 und 5-6 show specific features of electrode/liquid electrolyte interfaces in more detail.

The electrode equilibrium determines the distribution of surface and space charges with a corresponding space-dependent electrical potential φ at the interface. The region along which the electrical potential differences occur is called the electrochemical double layer of the electrode.

The electromotive force as the difference of the electrical potentials between the two platinum electrodes is given by:

$$E = -\frac{1}{F}\,(\tilde{\mu}''_e - \tilde{\mu}'_e) = \varphi_{Pt}\,(Cl_2/Cl^-) - \varphi_{Pt}\,(H_2/H^+)$$

$$= \Delta\varphi'' + \Delta\varphi_{diff} + \Delta\varphi' . \tag{5-40}$$

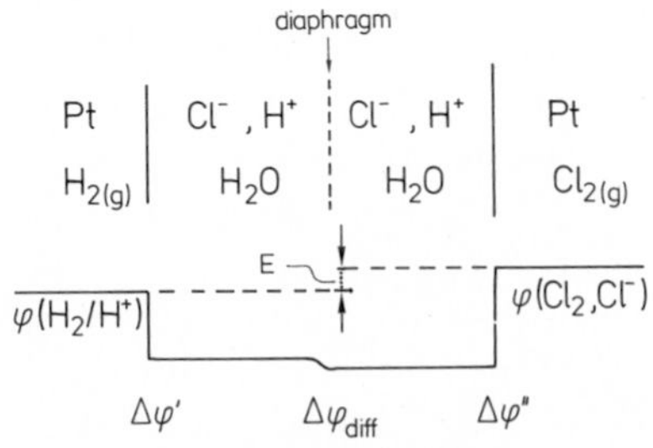

Figure 5-4.
Electrical potential differences (ie, Galvani potential differences) at the various interfaces of a galvanic cell (for the example of the cell in Figure 5-2). $\Delta\varphi_{diff}$ corresponds to an additional diffusion potential developed across the diaphragm, if different concentrations of HCl occur in the two half cells. $\Delta\varphi''$ can, eg, be expressed by:

$$\Delta\varphi'' = const + \frac{RT}{2F}\,\ln p_{Cl_2} .$$

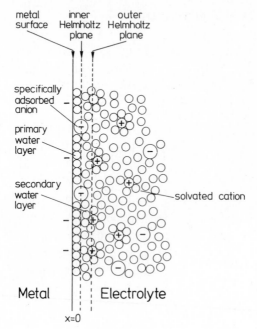

Figure 5-5.
Atomic model for the ionic double layer at the interface metal electrode/aqueous electrolyte. (+ positive ions, − negative ions, water molecules are denoted by empty circles).

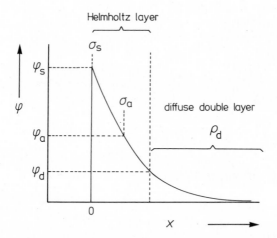

Figure 5-6.
Distribution of electrical potential and charge in the electrochemical double layer of Figure 5-5 (σ_s and σ_a denote the positions of the surface charges on the metal surface and on the outer Helmholtz plane, ρ_d denotes the space charge in the diffuse part of the double layer. The electrical potential is denoted by φ. x is the distance from the metal surface.).

If the half-cells at the two electrodes contain different ionic concentrations, one has to separate the two compartments by a diaphragm which prevents mixing of the two electrolyte solutions. Due to a difference in the electrical mobility of the different ions there may be a diffusion potential difference $\Delta\varphi_{\text{diff}}$ across the diaphragm. It is very important to reduce diffusion potential differences in galvanic cells for chemical sensor applications or at least to prevail constancy between calibration and measurements, since these potential differences can exceed 10 mV in certain cases (eg, strong pH gradients).

For the specific case of two HCl-solutions with different mean ionic activities $a''_{H+} = a''_{Cl-} < a'_{H+} = a'_{Cl-}$ we obtain approximately [see eg, 1-6]:

$$\Delta\varphi_{diff} = -\frac{2.3\,RT}{F}\frac{u_{H+} - u_{Cl-}}{u_{H+} + u_{Cl-}} \cdot \log\left(\frac{a'_{H+}}{a''_{H+}}\right) = A \cdot \log\left(\frac{a'_{H+}}{a''_{H+}}\right). \qquad (5\text{-}41)$$

The prefactor A at 25 °C is $-$ 37.9 mV. This corresponds to the value of the diffusion potential difference for $a'_{H+} = 10 \cdot a''_{H+}$. In order to minimize this effect a salt bridge with a high concentration of ions of equal mobility (eg, KCl) is used.

5.6 Overpotentials and Mixed Potentials

The difference between an actual electrode potential ε (for non-zero current density) to the equilibrium value ε_{eq} is usually defined as the overpotential η:

$$\eta = \varepsilon - \varepsilon_{eq}. \qquad (5\text{-}42)$$

For every electrode we get a typical dependence of the current density j at the interface on the overpotential η. This dependence may be determined by various so-called "polarization effects". Examples for these are [1–6]:

− charge transfer or activation polarization ($\rightarrow$ charge transfer overpotentials),
− diffusion or concentration polarization ($\rightarrow$ diffusion or concentration overpotentials),
− reaction polarization ($\rightarrow$ overpotentials associated with a preceding slow reaction step),
− ohmic polarization ($\rightarrow$ ohmic voltage drop).

Diffusion overpotentials are important for amperometric sensors and will be discussed thereafter in Section 5.7. Ohmic polarization is used in conductivity sensors.

Charge-transfer overpotentials result from an activation barrier for the charge transfer of ionic or electronic charges in the rigid part of the double layer at the electrode/electrolyte interface. This is an inherent characteristic property of the specific type of electrode reaction as well as of the specific type of electrode and electrolyte. Activation polarization gives rise to the so-called charge-transfer impedance and a nonlinear dependence of the current density on the overpotential η. For a given electrode reaction we find the following typical dependence of the current density on the charge-transfer overpotential η [eg, 1–6] known generally as the Butler-Volmer equation:

$$j(\eta) = j_+(\eta) - j_-(\eta) = j_0\left[\exp\left(\frac{\alpha nF\eta}{RT}\right) - \exp\left(-\frac{(1-\alpha)\,nF\eta}{RT}\right)\right]. \qquad (5\text{-}43)$$

Here, we have neglected any interference due to additional polarization effects (for detailed considerations, see eg, [1]).

Figure 5-7 shows the functional relationship in a diagram of current density vs. overpotential. The exponential relation between current density and overpotential results from the activation barrier for the charge transfer reaction step. The parameter α is the transfer coefficient denoting the portion of the overpotential affecting the height of the anodic activation barrier, j_0 denotes the exchange current density. It corresponds to the values of cathodic and anodic partial current densities $j_-(\eta)$ and $j_+(\eta)$ under equilibrium (for $\eta = 0$: $j_+(0) = -j_-(0) = j_0$). It is directly related to the ability of the electrode potential to compensate for disturbing effects of competing electrode processes. Table 5-1 shows typical experimentally obtained exchange current densities for some electrode reactions. An electrode reaction equilibrium with a very high exchange current density leads to a very stable electrode potential. On the other hand, a low exchange current density characterizes electrodes whose potential may be disturbed by small traces of interfering species or even the very small electrical current drawn in every voltage measurement (eg, 10^{-14} A).

In liquid solutions for example, the presence of oxygen may lead to an influence on the electrode potential of an indicator electrode and to deviations in ε from the expected value. A similar problem may occur with solid electrolytes, if several gases react at the electrode of a solid electrolyte electrode interface.

Under these conditions, the measured electrode potential deviates for zero current from the equilibrium values for both electrode reactions giving a *mixed potential* between the two dif-

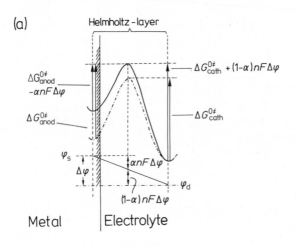

(a)

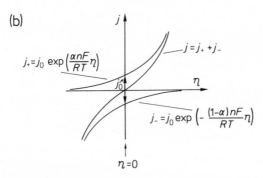

(b)

Figure 5-7.
(a) Gibbs free energy diagram for the charge transfer reaction at a metal electrode/liquid electrolyte interface. $\Delta G^{0\neq}_{anod}$ and $\Delta G^{0\neq}_{cath}$ denote the Gibbs free energies of activation of the anodic and cathodic steps for $\Delta\varphi = 0$. The activation barriers are changed linearly with $\Delta\varphi$ as indicated. (b) The derivation gives an exponential dependence of the current density j on the overpotential η [1–6]. j is given as the sum of the anodic and cathodic partial current densities j_+ and j_-. The constant prefactor j_0 is the exchange current density ($j_0 = j_+ = -j_-$ for $\eta = 0$, $\eta = \Delta\varphi - \Delta\varphi_{eq}$).

Table 5-1. Exchange current density j_0 and transfer coefficient α of several electrode reactions [41]. j_{00} corresponds to the standard exchange current density defined for $c_{ox} = c_{red} = 1$ mol/L.

Redox system (electrode reaction)	Electrolyte	Temperature	Electrode	j_0 [A cm^{-2}]	j_{00} [A cm^{-2}]	α
Fe^{3+}/Fe^{2+} ($5 \cdot 10^{-3}$M)	1 M HClO$_4$	25 °C	Pt	$2 \cdot 10^{-3}$	$4 \cdot 10^{-1}$	0.58
Ag/10^{-3} N Ag$^+$	1 M HClO$_4$	20 °C	Ag	$1.5 \cdot 10^{-1}$	13.4	0.65
Cd/10^{-2} N Cd^{++}	0.8 N K$_2$SO$_4$	20 °C	Cd	$1.5 \cdot 10^{-3}$	$1.9 \cdot 10^{-2}$	0.55
Cd(Hg)/1.4 $\cdot$ 10^{-3} M Cd^{++}	0.5 M Na$_2$SO$_4$	25 °C	Cd(Hg)	$2.5 \cdot 10^{-2}$	4.8	0.8
Zn(Hg)/2 $\cdot$ 10^{-2} M Zn^{++}	1 M NaClO$_4$	0 °C	Zn(Hg)	$5.5 \cdot 10^{-3}$	0.10	0.75
H$_2$/OH$^-$	1 N KOH	25 °C	Pt	10^{-3}	10^{-3}	0.5
H$_2$/H$^+$	1 M H$_2$SO$_4$	25 °C	Hg	10^{-12}	10^{-12}	0.5
H$_2$/H$^+$	1 M H$_2$SO$_4$	25 °C	Pt	10^{-3}	10^{-3}	0.5
O$_2$/OH$^-$	1 N KOH	25 °C	Pt	10^{-6}	10^{-6}	0.3
O$_2$/H$^+$	1 M H$_2$SO$_4$	25 °C	Pt	10^{-6}	10^{-6}	0.25

ferent equilibrium values [1, 14]. Such situations are of interest for the discussion of selectivity, cross sensitivity and disturbance of electrode potentials by impurities. A non-equilibrium electrode potential in general corresponds to a net cathodic or anodic current for that electrode reaction.

Figure 5-8 illustrates the case of a silver electrode in a solution containing silver ions, protons and dissolved oxygen. At least two reactions may occur: the redox-equilibrium between dissolved silver ions, electrons and silver and the redox-equilibrium of oxygen at the silver metal surface as indicated in Figure 5-8. This leads to a shifted electrode potential with a value between the thermodynamic equilibrium values of the two single electrode reactions, ie, to a "mixed potential". The diagram in Figure 5-8 shows how the mixed potential is determined from the two respective current-potential curves of the two single electrode reactions. The reaction of silver ions shows a high exchange current density, whereas the oxygen reaction shows a much lower value. Under conditions of zero net current (potentiometric measurement with an electrometer), the total current at the electrode through the interface must be zero. This is only possible, if the anodic oxidation of silver is accompanied by a steady state cathodic reduction of dissolved oxygen. There is only one value of the potential in Figure 5-8

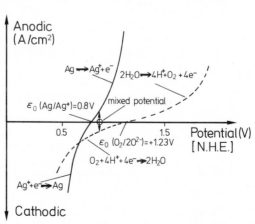

Figure 5-8.
Establishment of a mixed potential at a silver electrode (the influence of oxygen is exaggerated here). At the measured mixed potential ($\varepsilon_{mix} \approx 0.9$ V) the anodic silver oxidation current is exactly equal to that of the cathodic oxygen reduction current.

where this condition is fulfilled: This mixed potential, however, may shift with time, eg, if the oxygen concentration is changing, because of a limited diffusion rate ("concentration polarization") influenced by hydrodynamic parameters (like stirring rate). Therefore, mixed potentials are often unstable in time and show an enhanced dependence on hydrodynamic conditions even at high ion concentration where streaming potentials are absent.

5.7   Concentration Polarization and Amperometric Sensors

With regard to the charge transfer only, there is no limitation for the electrical current with increasing overpotential as can be seen from Equation (5-43). At larger values of the over-potential, however, other sources of polarization effects become important, such as ohmic polarization or concentration polarization. The latter implies a limitation of the current density. For instance, the transport of reactive ions to or from the electrode is described by the following equation for the partial electrical current density of ion i:

$$j_i = -z_i F D_i \operatorname{grad} c_i + \frac{\sigma_i}{\sigma_{\text{tot}}} j_{\text{tot}} , \tag{5-44}$$

where z_i, D_i, c_i and σ_i are charge number, diffusion coefficient, concentration and partial conductivity of the ion i, and j_{tot} is the total electrical current density. If σ_i is of the same order as the total conductivity σ_{tot}, the migration term in Equation (5-44) may predominate.

Current limitation due to a limited diffusion rate of ions, however, is favored, if the total conductivity of the electrolyte is much higher than the conductivity of the species to be reduced or oxidized at the electrode interface. In principle, this effect may occur with dissolved ions or molecules as well as with gaseous species that participate in the electrode reaction. If, for instance, gas molecules are oxidized or reduced, concentration gradients occur in the gas phase, in the pores of the electrode structure, or at the electrode surface. In these cases the current density will be proportional to the concentration gradient $\operatorname{grad} c_i$ at the electrode/electrolyte interface, ie,:

$$j_i = -z_i F D_i \operatorname{grad} c_i . \tag{5-45}$$

Figure 5-9 shows the concentration profile of a reacting species in the vicinity of the electrode surface. For a linear geometry, ie, a plane electrode surface, the length of the depletion zone will increase with time and produce a corresponding decrease of concentration gradients at the electrode. If it is possible to limit the diffusion zone to a defined distance, a steady-state diffusion current will be reached for longer times which depends linearly on the concentration of the reactive species. A large enough overpotential so that arriving species are immediately consumed at the electrode leads to depletion of reactive species at the electrode ($c_i = 0$ at $x = 0$). Assuming a linear concentration gradient, a simple expression for the limiting current density results:

$$j_{i,\text{ lim}} \cong -z_i F D_i \cdot \frac{c_i(\infty)}{\delta} . \tag{5-46}$$

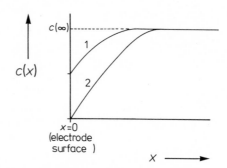

Figure 5-9.
Concentration profiles at an electrode; $x = 0$ corresponds to the electrode surface for (1) a potential where $c(x = 0)$ is about $c(\infty)/2$; (2) a potential where $c(x = 0) \approx 0$ and $j = j_{lim}$.

Here, δ denotes the thickness of the diffusion layer and $c_i(\infty)$ the concentration of ions i in the solution outside the depletion layer. Diffusion-limited currents are essentially used in *amperometric and voltammetric sensors* (or polarographic in case of a mercury working electrode). Figure 5-10 shows typical current-potential curves for large overpotentials illustrating the effect of concentration polarization on the current.

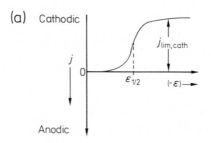

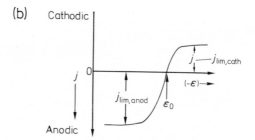

Figure 5-10.
Current limitation due to concentration polarization: (a) Current-potential curve of a Nernstian reaction involving two soluble species with only oxidant present initially ($\varepsilon_{1/2}$-half wave potential). (b) Current-potential curve for a Nernstian system involving two soluble species with both forms initially present (ε_0-equilibrium electrode potential).

A well-defined diffusion zone with constant thickness δ may be achieved by several different ways, ie, by using a dropping mercury electrode, by strongly stirring the solution, by using rotating disk electrodes with a defined diffusion-convection zone, or by using a membrane in front of the electrode which acts as a diffusion barrier with defined thickness for the reacting species [see eg, 6].

Many examples of practical amperometric gas sensors will be discussed in Chapters 7 for liquid and in 8 for solid electrolytes. As a gas permeable membrane must separate the gas phase from electrode and electrolyte, and at the same time the electrode must provide mechanical contact to the electrolyte, the electrode material has to be porous so that the electrolyte phase is in intimate contact with both, the inert working electrode material and the membrane. On the other hand a diffusion layer must be thin enough in order to yield small response times after changing the concentration gradient. Figure 5-11 shows the principle of this fundamental setup. In order to eliminate electrode polarization at the reference or counter electrode and an interfering ohmic voltage drop, one may use a third electrode as a high ohmic potential probe for the precise control of the working electrode.

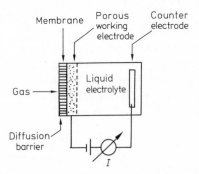

Figure 5-11.
Schematic setup of a simple amperometric gas sensor with liquid electrolyte (see also Chapter 7).

There is a straightforward and simple way of achieving selectivity with amperometrically driven sensor devices. This can be easily understood, eg, by considering a typical steady-state current-voltage curve in a solution, which contains different redox-active components in the presence of an excess of inert conducting ions. Under certain conditions a steady-state current-voltage curve may be observed like the one shown in Figure 5-12 with well-separated current

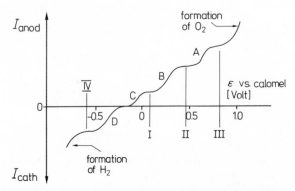

Figure 5-12. The influence of the choice of the electrode potential is demonstrated in this figure. A quasi-stationary current-voltage curve may be obtained for instance with a rotating Pt disk electrode or glassy-carbon electrode. The oxidation of species A, B and C starts at different electrode potentials. Species D can be reduced only. Depending on the applied electrode potential the following selectivities can be obtained:

 I — only C is monitored (and all species with similar standard electrode potential as C),
 II — B and C are monitored,
 III — A, B and C are monitored,
 IV — only D can be monitored.

steps [6, 15]. The position of the current steps is more or less depending on the standard electrode potentials ε_0 of the different electrode reactions. Shifts may occur due to additional polarization effects. Selectivity is achieved by operating with a working electrode potential at which the species of interest is reduced or oxidized (at a corresponding limited current). However, as it can be seen in Figure 5-12, all other compounds with half-wave potentials between the zero current point and the corresponding working electrode potential are electrochemically sensed at the same time, too. Only in certain cases near the zero current point, a single species is selectively oxidized or reduced. In order to gain more selectivity other parameters have to be optimized, too. The choice of the electrode material, for instance, may influence the voltage of the current step for a certain electrode reaction and may therefore be used to change the selectivity of a sensor. Interfering half-wave potentials may also be shifted along the potential scale by selective complexation (masking) of those compounds. This is possible for a number of metal ions. Another way of gaining more selectivity is to use special membranes as diffusion barriers which are predominantly permeable for the species of interest or to build up an amperometric biosensor (see Chapter 14).

5.8 Ion-Selective Electrodes

Ion-selective electrodes are the basic components of electrochemical sensors for the potentiometric determination of the activity of certain ions in the presence of other ions and neutral species [16–24]. The technique is usually applied to aqueous solutions, although application to non-aqueous electrolytes is possible in certain cases, too. As our discussion of electrodes shows, electrode potentials depend on the activities of the ionic species involved in the electrode equilibria. Therefore, in principle, electrode equilibria may be used to measure ionic activities. For instance, calomel and Ag/AgCl, Cl$^-$ electrodes have electrode potentials depending on the activity of chloride ions. A silver wire dipped into a solution of silver ions develops an electrode potential depending on the activity of silver ions and of other interfering redox systems in the solution. In most cases, however, electrode potentials in liquid electrolytes are sensitive to other species that may easily be reduced or oxidized at the electrode/electrolyte interface leading to mixed potentials. In particular, measurements at low ion activities suffer from this problem.

Therefore, most of the available ion-selective electrodes make use of the electrochemical equilibria at interfaces between liquid electrolytes and ion-conducting solids (solid electrolytes) or ion-conducting polymer or liquid membranes. As only transfer of ions is possible at these interfaces, an interference of redox equilibria, eg, resulting from dissolved oxygen gas or other redox systems like Fe^{2+}/Fe^{3+}, Cr^{3+}/CrO_4^{2-} etc., is excluded.

Figure 5-13 shows a schematic potential diagram of a typical setup of an ion-selective sensor cell. It shows the sequence of interfaces and the respective electrical potential differences. To choose the specific example of LaF_3 as ion-selective membrane material, the electrochemical equilibrium between fluoride ions in the aqueous solution and in the solid membrane determines the electrical potential difference ($\Delta\varphi_3 + \Delta\varphi_4$) between the two ion-conducting phases. The usual cell construction makes sure that only the electrical potential difference $\Delta\varphi_3$ between the test solution and the ion-selective membrane is changed, if the concentra-

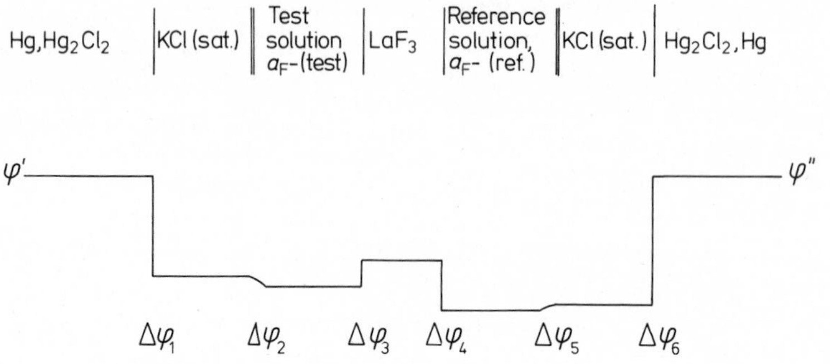

$$\text{Hg,Hg}_2\text{Cl}_2 \left| \begin{array}{|} \text{KCl (sat.)} \end{array} \right| \begin{array}{|} \text{Test} \\ \text{solution} \\ a_{F^-}\text{(test)} \end{array} \left| \text{LaF}_3 \right| \begin{array}{|} \text{Reference} \\ \text{solution,} \\ a_{F^-} \text{ (ref.)} \end{array} \left| \begin{array}{|} \text{KCl (sat.)} \end{array} \right| \text{Hg}_2\text{Cl}_2\text{,Hg}$$

$$E = \varphi'' - \varphi'$$
$$= \underbrace{\Delta\varphi_1 + \Delta\varphi_2 + \Delta\varphi_5 + \Delta\varphi_6}_{E_0} \quad + \quad \underbrace{\Delta\varphi_3 + \Delta\varphi_4}_{\displaystyle -\frac{RT}{F}\ln\frac{a_{F^-}\text{(test)}}{a_{F^-}\text{(ref.)}}}$$

Figure 5-13. Galvanic cell for a typical "ion-selective electrode" with the electrical potential profile showing characteristic differences at the various interfaces. $\Delta\varphi_2$ and $\Delta\varphi_5$ indicate possible diffusion potential differences.

tion of fluoride ions in the test solution is changed. All other electrical potential differences at the remaining interfaces have to be kept constant. Then, the following equation describes the emf of the cell as a function of fluoride ion activity in the test solution:

$$E = E_0 - \frac{RT}{F} \ln \frac{a_{F^-} \text{ (test)}}{a_{F^-} \text{ (ref.)}} \; . \tag{5-47}$$

The constant E_0 contains all constant Galvani-potential differences at the various interfaces except the potential difference between reference and test solution that changes with the fluoride ion activity.

Assuming that ions i are the only species that can be transferred through the interface membrane/solution in Figure 5-14, it follows that in the equilibrium case the electrochemical potential $\bar{\mu}_i$ of these ions has the same value on both sides of the membrane and inside the membrane. The following equation holds:

$$\bar{\mu}_i' = \bar{\mu}_i^m = \bar{\mu}_i'' \; . \tag{5-48}$$

The electrochemical potential in general, may be split into three parts, the standard chemical potential of ions μ_i^0, a second term containing the logarithm of the ion activity, and an electrostatic term $z_i F \varphi'$ (for F^- $z_i = -1$):

$$\bar{\mu}_i' = \bar{\mu}_i^0 + RT \ln a_i' + z_i F \varphi' \; . \tag{5-49}$$

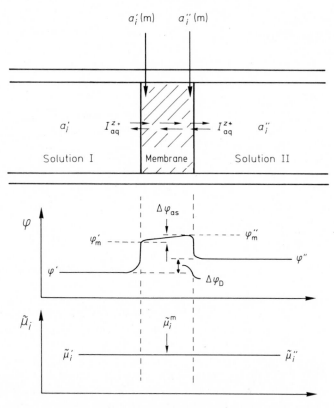

Figure 5-14. Electrochemical equilibrium of ions i along an ion-selective membrane that interacts selectively with i. The electrical potential differences at the two interfaces are determined by the corresponding values of the activity a_i on both sides of each interface.

Replacing the electrochemical potentials in Equation (5-48) by Equation (5-49), we obtain the electrical potential difference between the liquid electrolyte and the membrane at one of the two interfaces. This is the so-called Donnan-potential difference:

$$\Delta\varphi = \varphi'_m - \varphi' = \frac{1}{z_i F}(\mu_i^{0\prime} - \mu_i^{0m}) + \frac{RT}{z_i F}\ln\left(\frac{a'_i}{a_i^{m\prime}}\right). \tag{5-50}$$

In deriving this equation we have to assume that the membrane is thick if compared to the thickness of the double layer at the interfaces. Otherwise the electrical potential differences at the two interfaces influence each other. The total electrical potential difference $\Delta\varphi$ between the two electrolyte solutions separated by the membrane is given by:

$$\varphi'' - \varphi' = \Delta\varphi = \Delta\varphi' + \Delta\varphi_{as} + \Delta\varphi'' = \frac{RT}{z_i F}\ln\left(\frac{a'_i}{a''_i}\right) + \Delta\varphi_{as}. \tag{5-51}$$

The first term on the right-hand side of Equation (5-51) depends on the ratio of the ion activities in the two liquid electrolytes. The second term on the right-hand side (corresponding to $\Delta\varphi_{as}$ in Figure 5-14) describes a possible asymmetry within the membrane due to differences in the chemical composition, which gives rise to a difference in μ_i^{0m} between both interfaces.

The selectivity of proper ion-selective electrodes is mainly determined by the choice of the membrane material. Every ionic species that can be transferred between the membrane phase and an adjacent liquid solution must in principle have an influence on the electrical potential difference $\Delta\varphi_3$ in Figure 5-13 and thus on the signal of the sensor. Therefore, high selectivity means that interaction with the membrane has to be allowed only for the so-called potential-determining ion. Ions with similar size and charge as the potential-determining ion usually cannot be prevented from interacting with the membrane. The selectivity of a membrane material towards an interfering ion j is usually quantified by so-called selectivity coefficients K_{ij}. The selectivity coefficients are defined by the Nicolsky-Equation [25]:

$$E = E_i^0 + \frac{RT}{z_i F} \ln \left[a_i' + \sum_j K_{ij} a_j' \right].$$
(5-52)

The equation holds, if the ion i and j have the same charge number z_i. a_j' corresponds to the activity of interfering ions.

The best selectivities correspond to very low values of K_{ij}. The best ion-selective electrode, the pH-glass-membrane electrode, shows a selectivity coefficient of almost 10^{-12} towards most cations. The second best electrode, the LaF_3-based fluoride selective electrode, still has a coefficient in the 10^{-7} range for many ions except for OH^- ions (which, however, may easily be excluded by applying low pH-values). In general, any ion "selective" electrode should have a selectivity coefficient lower than 10^{-4} in order to justify its name. This means a 10^4-fold excess of an interfering ion just results in an error of 100%, ie, twice of the accurate analyte concentration.

If the counter ions are also able to cross the electrode interface to some extent (breakthrough of the Donnan exclusion principle), the theoretical slope of the E vs. $\ln a_i$ curve, ie, the Nernst factor or Nernstian slope, tends to be considerably less than theoretically expected. On the other hand, neutral species do normally not at all interfere with the electrochemical equilibria which determine the galvanic potential differences at the membrane/electrolyte boundary. Therefore, most dissolved neutral molecules give selectivity coefficients of practically zero.

To explain selectivity and cross sensitivity with regard to different ions, quantitative models have been developed [16, 17, 26–29]. They are based on thermodynamic and kinetic properties of the ion transfer reaction at the ion-selective membrane. The process of equilibration at the membrane/solution interface is accompanied by an exchange of ionic charges. There is an ionic double-layer extending in this case on both sides of the membrane/solution interface containing net positive and negative charge densities. Changes in the ionic activities in the solution or in the membrane correspond to changes in the chemical potential differences. This effect is counter-acted by a change of the charge densities, so that the electrical potential difference cancels the chemical potential difference and gives an equal electrochemical potential in both phases (see Figure 5-14).

The kinetic approach as applied to the electrochemical equilibrium at metallic electrodes is also useful for the electrochemical equilibrium at interfaces between ionically conducting phases. As in the case of metal electrode reactions, the transfer of ions through the interface is assumed to be thermally activated. Activation barriers arise, because the stabilization of ions by hydration or solvation in the case of non-aqueous electrolytes is lost for the ion transfer from aqueous solutions into the membrane phase. On the other hand, transfer of ions from the bulk of an ion-conducting membrane to its surface sites usually involves the addition of energy. Therefore, the kinetic picture for ionic charge transfer with two partial ionic currents that cancel in the equilibrium case is applied here in analogy to metal electrode reactions. The ionic current density j_i at an ion-selective membrane is taken as

$$j_i = j_i^0 \exp\left[\frac{\alpha z_i F \eta}{RT}\right] - j_i^0 \exp\left[-\frac{(1-\alpha)z_i F \eta}{RT}\right] \tag{5-53}$$

with the following definitions:

$$\eta = \Delta\varphi - \Delta\varphi_{eq}$$
$$j_i^0 = z_i F (k_+ a_i)^{1-\alpha} (k_- a_i^m)^\alpha$$

j_i^0 is the ionic exchange current density. k_+ and k_- are rate constants describing the transfer kinetics of ions into and out of the ion-selective membrane at the contact with an electrolyte solution. Other polarization effects such as reaction layers with high resistivity are neglected in Equation (5-53). The ability of a membrane to be selective for a special kind of ion can be expressed quantitatively by the ratio of the exchange currents of interfering and parent ions.

Figure 5-15 shows the situation in the pre-equilibrium time immediately after emersion of an ion-selective membrane electrode into a sample solution. The directed partial ion currents at the interface are plotted versus time. The directed current density $\vec{j}_i$ of the primary ion to be measured starts at a relatively high level depending on the local gradient of the electrochemical potential at time zero and on the activation energy to pass the interface. Because of the interface charging it decreases then with time in approaching equilibrium. At the same time, the oppositely directed current density of this ion species $\overleftarrow{j}_i$ increases, because it is favored by the potential change at the interface, until both are equal and the exchange current density for the primary ion has been established. For the interfering ion species similar pre-

Current

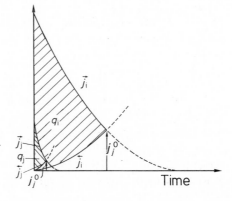

Figure 5-15.
Schematic representation of the directed current densities $\vec{j}$ *and* $\overleftarrow{j}$ across the interface; the selectivity coefficient is given by the ratio of the amount of transported charge due to the interfering ion (q_j) to that due to the measured ion (q_i).

equilibrium partial current densities can be drawn as a function of time as symbolized by $\vec{j}_j$ and $\overleftarrow{j}_j$ in Figure 5-15. The hedged areas denote the net amount of charge q_i and q_j transported across the interface during this equilibration time. The ratio of the two charges is a measure of the ion selectivity of that interface.

With this model in mind, the question concerning a more accurate description of the expression "potential determining" ion is possible: For any "electrified" interface the species which controls the transfer of the greatest number of charges dq transferred across that interface is potential determining, since the change of the electrical potential difference at the interface during equilibration is given by $d(\Delta\varphi') = dq/C$ (with C as the double-layer capacitance).

The ratio of the transferred charges q_j/q_i can also be expressed by the ratio of the two different exchange current densities j_j^0/j_i^0, if the geometric similarity theorem is applied (see Figure 5-15). In many cases this ratio was found to be numerically equal to the selectivity coefficient:

$$K_{ij} \cong \frac{q_j}{q_i} = \frac{j_j^0}{j_i^0} . \qquad (5\text{-}54)$$

This was shown in a series of studies with different kinds of ion-selective membrane materials [26–28]. On a longer time scale, the established electrical potential difference controlled by the activity of the primary ion with a higher exchange current density acts like O_2 in the case of Ag/Ag^+ (cf. Figure 5-8) as an overpotential for other ionic species (with lower exchange currents) in the sample solution as well as in the ion-selective membrane phase. The sign and amount of this overpotential depends on the difference of the corresponding equilibrium potential differences of the other ion species at that interface. In any case, an overpotential for a disturbing ionic species stimulates a corresponding ion transfer at the interface and leads to a mixed potential. This cross-sensitivity can change the activity of the ions in the membrane or at least at the membrane surface. Such a situation occurs for fluoride-sensitive LaF_3 surfaces with respect to OH^- as an interfering ion. In case a positive shift of the potential at the membrane is set-up for the quasi-equilibrium with OH^-, a relatively small current density of OH^- ions is directed into the LaF_3 membrane leading for long times to a lanthanum hydroxide layer $(La(OH)_3)$ on top of the LaF_3.

The partial currents characterizing the steady-state or mixed potential of a potential-determining ion in the presence of an interfering species can be limited on a longer time scale by diffusional processes within the membrane or solution phase. If the current limitation is due to hindered transport on the solution side, an influence of the degree of stirring on the cell potential will be observed. If it is the membrane phase, a behavior like in chronopotentiometry and/or cyclic voltammetry (with current peaks) may result. The effect of potential overshoot after a certain concentration change of the primary ion can originate in those processes. The steady-state mass flow of the primary ion species under the steady-state conditions of a mixed potential should not alter the chemical activity of this ion within the membrane phase. Therefore, as it is also pointed out in Chapter 8, there has to be an "ion-buffering capacity" in the membrane in order to avoid drifting membrane potentials.

Different models have been used by other authors to derive an expression for the selectivity coefficient and the interference of ions at ion-selective membranes without refering to transfer kinetics and mixed potentials [see eg, 6, 17]. For liquid membranes, the selectivity coefficients are often identified with the equilibrium constant K_{ij}^{exch} of the ion-exchange equilibrium of the primary and the interfering ion at the membrane surface [22]:

$$I_{aq}^{z^+} + J_{membrane}^{z^+} \xrightleftharpoons{K_{ij}^{exch}} J_{aq}^{z^+} + I_{membrane}^{z^+} \cdot \qquad (5\text{-}55)$$

This corresponds to the assumption that the ion activities in the membrane change according to the ratio of their activities in the test solution. Compared to the model discussed above this means that the chemical potential differences of the two ions have to coincide after a certain time which is only possible by an adjustment of the chemical composition of the membrane, which needs some time. In principle, this equilibrium cannot be complete as the equilibrium at the membrane/reference electrolyte interface usually involves different activities. Therefore, a diffusion potential difference is expected along the membrane in these cases. A corresponding derivation for the selectivity coefficient then gives (u_i, u_j — electrical mobilities) [6, 17]:

$$K_{ij}^{pot} = \frac{u_j}{u_i} \cdot K_{ij}^{exch} \cdot \qquad (5\text{-}56)$$

Fast response of thick membranes and super-Nernstian slopes (ie, slopes higher than predicted by theory) can be better explained by the kinetic approach. Details on mechanistic models, materials and combinations of materials that have been developed and proposed for ion-selective membranes can be found in a series of monographs and reviews [16-24].

5.9   Microelectrodes

A microelectrode can be defined as an electrode which is microscopic in at least one of its dimensions. A wide range of geometries have been considered and investigated; these include microdisc, microring and microband electrodes as well as microstructurized electrodes [see eg, 30–34, 40]. Ion-sensitive electrodes have been fabricated with tip diameters down to 1 μm. The latter are miniaturized, if very small sample volumes have to be measured such as in living cells, nerves etc. Figure 5-16 shows examples for construction principles of such electrodes.

With respect to voltammetric and amperometric techniques based on planar electrodes, the use of microelectrodes has important experimental advantages, too. Normally, a diffusion-limited current at a large planar electrode is due to planar diffusion of the reacting species. The current changes with time proportional to $t^{-1/2}$, if the extension of the diffusion layer is not limited (eg, by convection, stirring, or a diffusion membrane). Thus, no steady-state value of the current can be reached without additional measures. At a hemispherical microelectrode, however, the current density depends approximately on r^{-2} (r being the distance from the center of the microelectrode). There will be a non-zero steady-state current density of the reacting species after a typical time $t_0 = a^2/D$, where a denotes the radius of the microelectrode and D the diffusion coefficient with charge number z (typically 10^{-5} cm^2/s in aqueous solutions):

$$j(t \gg t_0) = -zFD \cdot \frac{c_\infty}{a} = \text{const} \cdot \qquad (5\text{-}57)$$

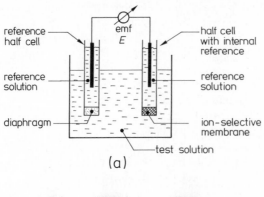

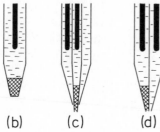

Figure 5-16.
Construction principles of ion-selective microelectrodes: a) schematic principle of the cell assembly, b) miniaturized indicator electrode with the ion-selective membrane at the end of a capillary tube, c), d) combinations of indicator and reference electrodes in a double wall or twin tube (for details, see, eg, [22]).

Therefore, a microelectrode with small radius a gives a much faster response due to the small effective diffusion length (of the order of a) together with a stable steady-state current in case of diffusion limitation without the need for a constant stirring.

Further important advantages concern low capacitive background currents and small ohmic voltage drops in the electrolyte, because small currents flow. Microelectrodes are particularly useful for media of low conductivity. Experiments using ultramicroelectrodes in the gas phase have been reported, too. Figure 5-17 shows a schematic diagram of a corresponding ring ultramicroelectrode assembly [32].

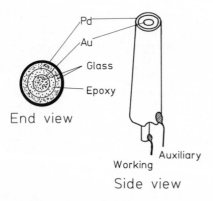

Figure 5-17.
Simplified picture (not to scale) of a ring ultramicroelectrode assembly used as a detector for gas chromatography [35]. The working electrode is a thin gold film deposited on a 5 μm quartz fibre. This is sealed inside a glass tube by epoxy resin. The auxiliary electrode is a palladium film deposited on the outside of this tube.

5.10   Unblocked and Blocked Interfaces

Up to now, we have discussed unblocked interfaces. Electrochemical equilibria determine and stabilize electrical and chemical potential differences between the two conducting bulk phases at these interfaces. Figure 5-18 shows some types of unblocked interfaces admitting a continuous charge transfer. Two types of such unblocked interfaces have been treated above:

– the first (corresponding to (c) in Figure 5-18) was the interface *electronic conductor/ionic conductor (ie, electrolyte)* in the presence of an electrode equilibrium which couples $\tilde{\mu}_e$ and $\tilde{\mu}_{ion}$. Applications concern the measurement of activities or partial pressures of neutral species in gases, liquids, melts, and solids (see Chapters 7 and 8).
– and the second (corresponding to (b) in Figure 5-18) was the interface *electrolyte/ion-selective membrane,* ie, the interface of two different ionic conductors at which the electrochemical equilibrium of an ionic species couples their electrochemical potential $\tilde{\mu}_{ion}$ in both phases. Such interfaces are the basis for ion-selective electrodes usually applied to measure ionic activities in liquid solutions, although ionic activities in solid solutions (solid ionic conductors) could be determined according to the same principles.

Case (a) in Figure 5-18 is the electronic analogue to the contact between two ionic conductors in case (b). Although the equilibrium at interfaces between two electronic conductors ac-

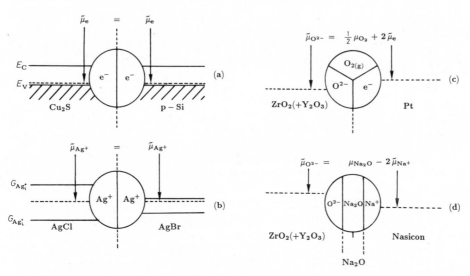

Figure 5-18. Various cases for reversible unblocked interfaces with coupling of the electrochemical potentials of two conducting phases (electronic or ionic conductors): a) Interface of two electronic conductors (as viewed in the band scheme, E_C, E_V denote conduction and valence band edges). b) Interface of two silver ion conductors (viewed in the free energy scheme, $G_{Ag_i^+}$, $G_{Ag_i^+}$, are free energies of silver ions on normal and interstitial lattice sites). c) Interface between an electronic and ionic conductor in equilibrium with a gas phase (reversible electrode). d) Interface between an oxide ion conductor and a sodium ion conductor (Nasicon) in the presence of dissolved or finely dispersed Na_2O.

cording to case (a) does only involve electrons, it may be sensitive to chemical activities of ions or molecules in the surroundings. For instance, the interface impedance for electronic charge transport through the interface is influenced by changes in work function and conductivity due to adsorption or changes in chemical composition.

Case (d) corresponds to an unblocked interface between two ionic conductors with different mobile charge carriers. The electrochemical equilibrium is characterized by participation of neutral components and can in principle also be used for the detection of such components.

In the examples of Sections 5.3 to 5.5, complete thermodynamic equilibrium of the reactants and products involved in the electrode reaction was assumed at the electrode interfaces as well as in the bulk of the solid electrolyte. At low temperatures, however, kinetic effects may prevent the complete equilibration of an electrode reaction, particularly for multistep electrode reactions involving dissociation of strong bonds. Oxygen electrodes like Pt, O_2/ZrO_2 in cells like (5-20) show reversible behavior only at elevated temperatures. Electrode equilibria corresponding to only a partial reduction of oxygen are found at lower temperatures, eg, with LaF_3 [13, see also Chapter 8]. These involve O_2, O_2^-, and O_2^{2-} or their protonated forms HO_2, HO_2^-, H_2O_2 at the solid electrolyte surface. These intermediates are known to be formed also during the complete reduction of O_2 to H_2O at oxygen electrodes in liquid electrolytes and by gas phase interaction of oxygen with oxide surfaces at lower temperatures. In these cases one may speak of kinetically controlled or partially blocked interfaces. It is difficult to use them for electrochemical sensors, although it is possible and has been reported for some examples (cf. Chapter 8).

Interfaces may be termed as *blocked,* if a continuous charge flow is not possible. This does not exclude, however, an interaction and an exchange of charged species with the interface region of both phases at such an interface. For instance, binding sites for ions such as H^+, OH^-, Na^+ etc. may exist at the boundary between an aqueous electrolyte and an oxide like SiO_2, Ta_2O_5, TiO_2, although a transport of these ions is excluded in the bulk oxide phase due to their negligible mobility at ambient temperature. These sites may be located at the oxide surface or may extend a certain distance over the subsurface region. An electrochemical equilibrium between adsorbed ions and ions dissolved in the electrolyte generates and determines the surface charge on the oxide, the electrical capacitance of the interface, and the electrical potential difference between the oxide surface and the bulk of the electrolyte.

The difference between blocked and unblocked interfaces can be seen in the different behavior after applying an external potential difference. Whereas an unblocked interface develops a steady-state current after some time, the current through a blocked interface decays to zero (except very small leakage currents in practical cases due to defects or impurities).

The pH-glass electrode represents the case of a transition between a blocked and unblocked interface. Detailed models for the sensing principle of pH-glass electrodes were developed earlier based on experimental results [36]. Recent theories of surface interactions at blocked interfaces are strongly related to those concepts. The dissociation equilibria of Si-OH units at the interface glass/aqueous electrolyte play the decisive role as potential-determining processes. Similarly, many other oxides MO_x (eg, M = Sb, Ta, Al, . . .) show formation of M-OH groups (at the surface or extended over a hydrated subsurface layer) in contact with liquid electrolytes leading to pH-dependent dissociation equilibria which are also employed in pH-sensors.

Blocked interfaces have found interesting applications in a series of potentiometric devices like coated wire electrodes (CWEs) or ion-selective field effect transistors (ISFETs) (see eg,

[37–40]. Figure 5-19 shows the sequence of interfaces forming an electrochemical cell with an ISFET containing a blocked interface between liquid electrolyte and insulating gate material. Semiconductor and electrolyte are coupled only capacitively in an ideal ISFET. The surface of Ta_2O_5 has binding sites for H^+ and OH^- and an electrochemical equilibrium occurs between dissolved protons and hydroxide ions and adsorbed ions on the Ta_2O_5-surface. If the transistor base and the reference electrode are connected, the surface equilibrium between sensitive layer and electrolyte determines a surface charge together with defined potential differences $\Delta\varphi_2$ (in the insulating layer) and $\Delta\varphi_3$ (between bulk electrolyte and oxide surface). The selectivity of the binding sites in or on the gate material determines the selectivity of this device. The theoretical concepts describing the sensing mechanisms of ISFETs with oxide layers for pH-measurement are often discussed within the so-called site-binding theory [37–40].

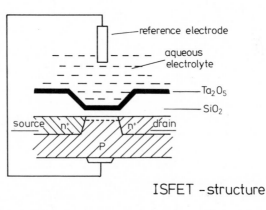

ISFET -structure

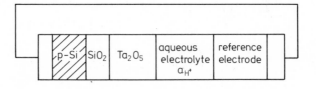

Figure 5-19.
Schematic view of an electrochemical cell with an ISFET. The lower part of the figure shows the electric potential differences along the different interfaces.

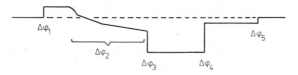

For a good Nernstian response, it is essential that the surface density of sites at the interface is high enough. This means that the chemical potential of adsorbed ions is practically constant. This is in principle analogue to the constant chemical potentials of mobile ions in fast ion conductors or solid electrolytes where a high density of ionic sites in the bulk phase is responsible. There is also an analogy to the Fermi level pinning at the surface of semiconducting compounds, if high electronic surface state densities occur. The electrochemical equilibrium between solution and surface sites is characterized by an equal electrochemical potential of ions. As the chemical potential at the surface is constant, only the change of the chemical potential in the bulk solution determines the electrical potential difference between surface and bulk $\Delta\varphi$ according to:

$$\Delta\varphi = \text{const} + \frac{1}{z_{ion}F}\,\mu_{ion}\,(\text{solution}) = \text{const} + \frac{RT}{z_{ion}F}\,\ln a_{ion}\,. \tag{5-58}$$

The terms *blocked and unblocked interface* are idializations and extreme cases for the electrical interaction at the interface of two phases. As the foregoing discussion has shown, interfaces in electrochemical sensors are characterized by defined electrochemical equilibria regardless wether they involve bulk sites in both phases (unblocked interface) or are restricted to surface or subsurface sites at one of the two phases (blocked interface).

5.11   Outlook

We have discussed important aspects of electrochemical sensing principles. However, we have not covered specific details of conceivable and more sophisticated sensor mechanisms and sensor setups. This is done in the following Chapters 7 and 8.

Also, we have not discussed in detail the wide field of new possibilities by applying new materials such as electronically and/or ionically conducting polymers, crown ethers, biological molecules and compounds, dealt with in the field of supramolecular chemistry etc.

Nevertheless, the few central ideas of electrochemistry in the context of potentiometric or amperometric principles discussed above make possible to understand the great flexibility to construct new sensors with new materials (such as catalysts, membranes, electrolytes, electrodes), with new geometries (such as micro-, ultramicroelectrodes, microstructures), with new measurement techniques (such as time-dependent, frequency-dependent) and with new techniques for signal processing (such as pattern recognition).

5.12   References

[1] Vetter, K. J., *Electrochemical Kinetics;* New York: Academic Press 1967.
[2] O'M Bockris, J., Reddy, A. K. N., *Modern Electrochemistry* Vol 1 und 2; New York: Plenum Press, 1970.
[3] Conway, B. E., *Theory and Principles of Electrode Processes;* New York: Ronald Press, 1965.
[4] Delahay, P., *Double Layer and Electrode Kinetics;* New York: Interscience Publishers, 1965.
[5] Koryta, J., Dvorak, J., Bohackova, V., *Electrochemistry;* London: Methuen, 1970.
[6] Bard, A. J., Faulkner, L. R., *Electrochemical Methods;* New York: J. Wiley, 1980.
[7] Rickert, H., *Electrochemistry of Solids;* Berlin: Springer, 1982.
[8] Buck, R. P., *J. Chem. Soc. Faraday Trans. 1* **82** (1986) 1169.
[9] Latimer, W. M., *The Oxidation States of the Elements and Their Potentials in Aqueous Solutions;* Englewood Cliffs, N J: Prentice Hall, 1952.

[10] Milazzo, G., Caroli, S., *Tables of Standard Electrode Potentials;* New York: Wiley-Interscience, 1977.

[11] Schindler, K., et al., *Sens. Actuators* **17** (1989) 555.

[12] Wiemhöfer, H.-D., Harke, S., Vohrer, U., *Solid State Ionics* **40/41** (1990) 433.

[13] Wiemhöfer, H.-D., Göpel, W., *Sens. Actuators B* **4** (1991) 365.

[14] Wagner, C., Traud, W., *Z. Elektrochem.* **44** (1938) 391.

[15] Cammann, K., "Elektrochemische Untersuchungsmethoden", in: *Moderne Untersuchungsverfahren in der Chemie;* Stuttgart: G. Thieme Verlag, (1990).

[16] Cammann, K., *Working with Ion-Selective Electrodes;* Berlin: Springer-Verlag, (1979)

[17] Morf, W. E., *The Principles of Ion Selective Electrodes and of Membrane Transport;* Amsterdam: Elsevier Scientific Publ., (1981).

[18] Buck, R. P., *Anal. Chem.* **50** (1978) 17R.

[19] Covington, A. K., *Ion-Selective Electrode Methodology* Vol. I and II, Boca Raton, FA: CRC Press, 1979.

[20] Koryta, J., *Ion-Selective Electrodes;* Cambridge: Cambridge University Press, (1975).

[21] Freiser, H. (ed.), *Ion-Selective Electrodes in Analytical Chemistry;* New York: Plenum, (1979).

[22] Ammann, D., *Ion-Selective Microelectrodes;* Berlin: Springer, 1986.

[23] Buck, R. P., *Sens. Actuators* **1** (1981) 197.

[24] Morf, W. E., Simon, W., in: *Ion-Selective Electrodes in Analytical Chemistry* Vol. 1, Freiser, H. (ed.); New York: Plenum 1978.

[25] Nicolsky, B. P. et al., in: *Glass Electrodes for Hydrogen and Other Cations,* Eisenman, G., (ed.): New York: M. Dekker, (1967).

[26] Cammann, K., "Mixed Potential Ion-Selective Electrode Theory", in: *Conference on Ion Selective Electrodes, Budapest 1977,* Amsterdam: Elsevier, 1978, p. 297.

[27] Cammann, K., „Exchange Kinetics at Potassium-Selective Liquid Membrane Electrodes", *Anal. Chem.* **50** (1978) 936.

[28] Cammann, K., "A More Kinetically Oriented Ion-Selective Electrode Theory − An Approach Towards a Better Understanding of Potentiometric Sensors", *Drug. Res.* **1**, No. 4 (1978) 709.

[29] Cammann, K., "Ion-Selective Bulk Membranes as Models for Biomembranes", in: *Topics in Current Chemistry* Vol 128; Berlin: Springer-Verlag, 1985.

[30] Wightman, R. M., *Anal. Chem.* **53** (1981) 1125A.

[31] Pons, S., Fleischmann, M., *Anal. Chem.* **59** (1987) 1391A.

[32] Morita, M., Longmire, M. L., Murray, R. W., *Anal. Chem.* **60** (1988) 2770.

[33] Kittlesen, G. P., White, H. S., Wrighton, M. S., *J. Amer. Chem. Soc.* **106** (1984) 7389.

[34] Wohltjen, H., *Anal. Chem.* **56** (1984) 87A.

[35] Ghoroghchian, J. et al., *Anal. Chem.* **58** (1986) 2278.

[36] Baucke, F. G. K., "Contribution to the Electrochemistry of pH Glass Electrode Membranes", in: *Conference on Ion Selective Electrodes, Budapest 1977,* Amsterdam: Elsevier, 1978, p. 215.

[37] Sibbald, A., *J. Molecular Electronics* **2** (1986) 51.

[38] Lauks, I., *Sens. Actuators* **1** (1981) 261 and 393.

[39] Yates, D. E., Levine, S., Healy, T. W., *J. Chem. Soc. Faraday Trans. 1* **70** (1979) 1807.

[40] Madou, M. J., Morrison, S. R., *Chemical Sensing with Solid State Devices,* New York: Academic Press, 1989.

[41] Hamann, C. H., Vielstich, W., *Elektrochemie* Vol. 2, Weinheim: VCH 1981, p. 28.

6 Multi-Component Analysis in Chemical Sensing

STEFAN VAIHINGER and WOLFGANG GÖPEL, Universität Tübingen, FRG

Contents

6.1   Introduction

So far we have discussed the importance of developing and designing sensors for the specific detection of one component in a mixture of many others. In reality, however, this complete selectivity cannot be realized. Therefore, a variety of different methods have been developed to overcome this problem by using an array of different sensors and evaluating the data in a subsequent data processing step similar to the olfactory system of humans with a few receptors and the subsequent data evaluation in the brain.

This chapter surveys the current "state of the art" in the field of multi-component analysis with (bio-)chemical sensors. We first describe formal concepts of multi-component analysis as defined in analytical chemistry and of pattern recognition as treated in information theory, in Section 6.2. We also give a short survey on processing of measuring signals utilized in (bio-)chemical sensing. We then discuss the few practical examples that have been reported so far in the field of chemical sensing based on pattern recognition, in Section 6.3.

Possibilities and limitations of current multi-component analysis in chemical sensing are pointed out. All mathematical approaches used in the few practical examples are based on sequential computational methods only, in spite of the rapidly growing interest in parallel processing and neural networks. The basic reason is that practical implications of these new mathematical approaches are still negligible in the field of chemical sensing. Therefore, this gap in the degree of mathematical sophistication between "man-made" pattern recognition approaches and biological pattern recognition in nature cannot yet be bridged. The same holds for the gap in the molecular complexity of individual "man-made" sensor elements if compared with biological sensors.

The application of appropriate pattern recognition methods depends on the measuring problem, which can be described from two different points of view:

– first, the operator is interested in the qualitative and quantitative analysis of all components present in the analyte, or
– second, the operator is interested in the quantification of one specific component only without determining other components present in the analyte.

A typical example of the second goal is the quantitative detection of perchloroethylene in dry-cleaning machines, or more generally, of certain hazardous vapors for personal protection.

To solve the problem of a real multi-component analysis or of the quantification of one species in a multi-component mixture, three different principles can be applied (Figure 6-1).

a. Sensors with high selectivity may be used. Each sensor responds to a single component in the mixture. For each component one sensor is needed. If the sample contains more gas components than sensors, the system evidently cannot identify all components in the mixture. Ideally, a sensor responding to one species should not be influenced by another component. The sensor function is then described by the key-lock principle, similar to pheromones used by insects. Examples of sensor-active materials of this kind are enzymes and even larger biological structures, which may be immobilized on substrates or field-effect transistors. Other examples are electrochemical sensors with filters, although they never reach the selectivity of biological sensors.

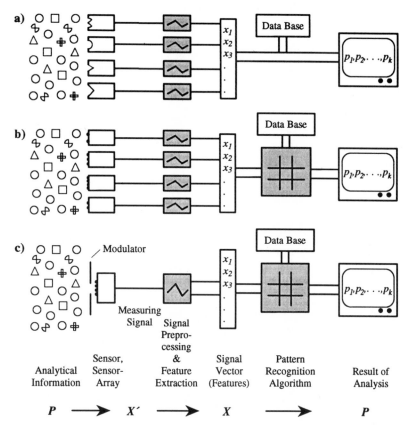

Figure 6-1. Basic principles for applying pattern recognition methods to gas analysis. The sensor or sensor array is exposed to a gas mixture, the measuring signal x_i' is preprocessed and the resulting individual features x_i are extracted. The complete signal vector X containing all the sensor's information is used as input of the pattern recognition algorithm. For details of the three principles a), b), and c), see text.

This principle is the ideal case, where no mathematical pattern recognition algorithm is required. The incoming signal is compared with calibrated standards from a database. As no cross-sensitivity effects occur, each sensor signal corresponds unequivocally to the concentration of a specific gas. The disadvantage of this approach for many practical applications is the small number of really selective sensors with long lifetimes available so far. As an example, different coatings of electrochemical devices or of gates for field-effect transistors may be optimized by immobilizing biological materials such as enzymes, ionophores, or antigens. The latter do show specific reactions with very high selectivity, but have poor long term stability. The enzymes and ionophores usually show cross-sensitivity effects and also instability. The interface between the substrate and the biological part is usually the most critical component concerning drifts.

b. Sensors with partly overlapping selectivity may be used. Most of the commonly used sensors, such as electrochemical sensors or SnO_2-based sensors, show a sensitivity not only to one specific gas, but also to groups of gases, eg, to reducing or oxidizing gases. The sensitivity of such sensors can be shifted by using metal dopants or different sensor temperatures (eg, of semiconductor sensors) or by using different metals and grain sizes of the electrode material (eg, of electrochemical sensors). For analysis of gas mixtures with these sensors having different sensitivities and selectivities, they are operated in an array configuration. A subsequent pattern recognition method is needed [1]. Several practical examples of this second principle are discussed in Section 6.3.

c. The selectivity of sensors may be increased by evaluating dynamic measurements. Dynamic features may be represented by a set of parameters, some of which are specific for a specific gas and independent of the gas concentration even in mixtures. These parameters make possible the quantitative identification of certain species [2, 3]. As an example, the response of the sensor upon stepwise excitation with initial slopes and equilibrium values or with phase shifts and equilibrium values may be used (Figure 6-2). The latter may be evaluated for different frequencies as utilized in complex impedance or pressure modulation spectroscopy.

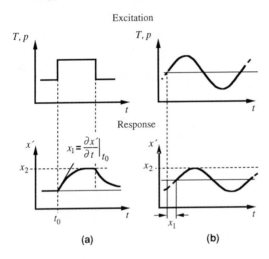

Figure 6-2.
Different sensor signals obtained from (a) stepwise or (b) periodic perturbation of sensor operation conditions; in these examples they are determined by T and p [9].

6.2   Multi-Component Analysis and Pattern Recognition: Approaches from Analytical Chemistry and Information Theory

6.2.1   Measuring Signals and Analytical Information

For simplification, and because of the larger number of available practical examples, we shall present predominantly gas analysis data in the following with the analytical information P and the sensor signals X' and X, respectively (cf, Figure 6-1). All the concepts to be discussed now may, however, be extended to other sensor environments including electrolytes and/or biochemical analytes.

Analytical methods provide general correlations between the analytical information P and the output of the sensors. This leads to a feature vector X after preprocessing of the data X'.

Analytical information such as local and time-dependent concentrations c, activities a, or partial pressures p can be described as a vector. For simplification, the different species of a mixture are indicated by the subscript k in the following and their concentrations are denoted by p_k. The complete analytical information of a system which, for instance, contains n different gases is given by the vector $P = (p_1, p_2, p_3, \ldots, p_k, \ldots, p_n)$. The measuring signal x' as output of the transducer or the sensor (such as voltages V, currents I, and resistances R (see Table 1-3) is the detected change in a sensor property and determines the vector $X' = x'_1, x'_2, x'_3, \ldots, x'_m)$. By a subsequent data processing and feature extraction the measuring signal is transformed to the *signal vector* or *feature vector* $X = (x_1, x_2, x_3, \ldots, x_m)$.

6.2.2 Data Preprocessing and Feature Extraction

Data preprocessing is the link between the sensor output X' (vector of measuring signal) and the input $X = (x_1, x_2, \ldots, x_i, \ldots, x_m)$ to the mathematical algorithm for the identification of a mixture. Most pattern recognition algorithms prefer linear transformations, which have special requirements for the input signals. Often it is not possible in a simple way to consider drift effects of the sensor in the pattern recognition ("PARC") algorithm. Correction for drifts may be done, however, in the data preprocessing step. Very often this also includes the evaluation of difference signals between an active and inactive sensor of the same kind. This difference signal may then be processed or used directly for the identification step.

Below we list typical examples of data preprocessing in static and dynamic measurements:

— *Static measurements:*
  Difference signal between active and inactive sensors:

$$x_i = x'_{a,i} - x'_{b,i} .$$
(6-1)

  Offset substraction, eg, for considering linear drift effects:

$$x_i = x'_i - x'_{i,0} .$$
(6-2)

  Relative signals, eg, for considering drifts in the sensitivity coefficient:

$$x_i = \frac{x'_i}{x'_{i,0}} .$$
(6-3)

  Averaging to reduce noise effects:

$$x_i = \frac{1}{N} \sum_{j=1}^{N} x'_{ij} .$$
(6-4)

  Linearization transformations:

$$x_i = \ln x'_i$$
(6-5)

$$x_i = \sqrt{x'_i} .$$
(6-6)

Normalization [4-7]:

$$x_i = \frac{x'_i}{\sqrt{\sum\limits_{i=1}^{m} x'^2_i}} \qquad\qquad (6\text{-}7)$$

$$x_i = \frac{\log x'_i}{\sum\limits_{i=1}^{m} \log x'_i} . \qquad\qquad (6\text{-}8)$$

Weighting based on a priori information about the sensor behavior [7, 8].
— *Dynamic measurements (see also Figure 6-2):*
Derivative [9] and equilibrium (or 90%) values:

$$x_i = \left.\frac{dx'_i}{dt}\right|_{t_0} , \qquad x_i = x'_i \,(90\%) . \qquad\qquad (6\text{-}9)$$

Evaluation of periodically modulated signals:
— ratio of modulation amplitude to the average of the modulated signal;
— ratio of modulation amplitude to the upper or lower value of the amplitude [10];
— Fourier transformation and comparison of Fourier signal in the range of the exciting frequency;
— modulation in the range of the most pronounced gradient for selective detection [10, 11].
Evaluation of general response functions to stepwise changes:
— approximation by a polynomial of degree *N;*
— non-linear fit algorithm to determine the sensor response with a sum of exponential functions [3].

In most practical cases, combinations of the procedures listed for static and dynamic measurements are applied to obtain a signal vector useful for the subsequent pattern recognition. As an example, offset subtraction is useful in both static and dynamic measurements. A general recipe for feature extraction from data of chemical sensors cannot be given here. The success of a pattern recognition method or of an analysis often depends critically on this procedure. Sophisticated feature extraction aims at minimizing the computing effort of the subsequent PARC algorithm.

6.2.3   Sensitivity, Selectivity, and Specificity of Sensor Arrays

Usually, simple mathematical transformations between the vector of the features x_i derived from the measuring signal x'_i of the ith sensor element by data preprocessing $X = (x_1, x_2, x_3, \ldots, x_i, \ldots, x_m)$ and the analytical information vector $P = (p_1, p_2, p_3, \ldots, p_k, \ldots, p_n)$ are possible only for small variations dX and dP because of the often found non-linearity of the functions $x_i(p_k)$ over a larger range. Therefore, to a first approximation only

$$dX = E\,dP \tag{6-10}$$

holds with small variations $dX = (dx_1, dx_2, \ldots, dx_i, \ldots, dx_m)$, $dP = (dp_1, dp_2, \ldots, dp_k, \ldots, dp_n)$, and E as the functional or calibration matrix:

$$E = \begin{pmatrix} \gamma_{11} & \gamma_{12} & \cdots & \gamma_{1n} \\ \gamma_{21} & \gamma_{22} & \cdots & \gamma_{2n} \\ \vdots & \vdots & & \vdots \\ \gamma_{m1} & \gamma_{m2} & \cdots & \gamma_{mn} \end{pmatrix}. \tag{6-11}$$

In general, the functional matrix has to be calibrated for different concentrations of values $p_1, \ldots, p_n$. The *partial sensitivity*, γ_{ik}:

$$\gamma_{ik} = \left(\frac{\partial x_i}{\partial p_k} \right)_{p_{j \neq k}} \tag{6-12}$$

describes the local variation of x_i with p_k at constant $p_{j \neq k}$. It can be determined geometrically from the calibration graphs $x_i(p_k)$. In the general concept of multi-component analysis to determine the sensitivities, selectivities, and specificities in analytical chemistry, γ_{ik} is the partial sensitivity of the sensor i with respect to changes in the partial pressure of the component k [12]. In general, γ_{ik} may vary with p_k and $p_{j \neq k}$. The latter characterizes non-linear response.

The transformation described in Equation (6-10) is unique provided that the determinant det $(E) \neq 0$. This determinant is called the overall sensitivity of this method of analysis.

In a totally selective analysis, several components can be determined independently of each other. In this case, only elements of the main diagonal of the calibration matrix exist. The *selectivity* $\varXi$ of any not completely selective analysis is usually defined as the smallest value of the *partial specificities* $\varPsi_i$:

$$\varPsi_i = \left(|\gamma_{ii}| \Big/ \sum_{\substack{k=1 \\ k \neq i}}^{n} |\gamma_{ik}| \right) - 1 \tag{6-13}$$

with

$$\varXi = \operatorname*{Min}_{i=1\ldots m} \varPsi_i. \tag{6-14}$$

The specificity for a component a is defined in analogy with the selectivity according to

$$\varPsi_a = \left(|\gamma_{aa}| \Big/ \sum_{\substack{k=1 \\ k \neq a}}^{n} |\gamma_{kk}| \right) - 1. \tag{6-15}$$

In Section 6.3.2.3, these definitions will be applied to a particular practical example, viz, to an SnO_2-based gas sensor array, in order to describe accuracies in the multi-component analysis of gas mixtures.

6.2.4 Pattern Recognition Methods

Pattern recognition methods can be applied if the properties of a sample or a chemical system cannot be determined directly. Physical or chemical methods are used from which a "hidden" property can be calculated. A sample is characterized by a set of features ("pattern"). These features have to be relevant for the properties in which the operator is interested. If the features are chosen well, objects with similar not directly measurable properties should have similar patterns. Pattern recognition methods have to classify different objects by similarities of patterns. The affiliation of a pattern to a specific class corresponds to a specific property of this object. Therefore, pattern recognition is an instrument for qualitative analysis. Quantification requires subsequent methods. In this context two different tasks can be defined:

1. A set of patterns of objects is given with a known affiliation for each object to a specific class. The rules for discrimination have to be found which attach the object to the classes by aid of the patterns. The determination of these rules is called "training". It is desirable that the classification rules are able to classify new objects correctly, if the system is not trained for these objects.

2. A set of patterns of objects is given which cannot be classified a priori. Cluster analysis has to determine which patterns can be grouped together in classes by reason of similarities. Further, it has to determine whether the objects have common properties inside a cluster.

In this section some mathematical methods are presented which may be applied for multi-component analysis in chemical sensing [13]. Further details can be obtained from examples in the literature [14-24].

6.2.4.1 Linear Regression

Most analyses use a single-component, single-sensor model with a linear relationship between sensor response x' and the concentration of the analyte p (direct calibration model):

$$p = ax + b + \varepsilon \tag{6-16}$$

with ε as random variable. A calibration is required to find the sensitivity of the sensor, a, as the slope of the regression line, and the intercept on the ordinate, b. The residual error ε is the portion of the response not described by the model.

For an analysis the response to unknown samples is measured and the concentrations are estimated. Using least-squares regression it is assumed that the calibration plot is linear and the sensor is fully specific for the analyte of interest. Matrix effects and the presence of background responses are the most commonly encountered problems. Matrix effects are due to changes in the sensitivity coefficient a between calibration and quantification. This type of effect is most readily treated by using the method of standard addition. Background or interference correction cannot be included directly in the model.

6.2.4.2 *Multiple Linear Regression (MLR)*

A typical problem in applying the multiple linear regression (MLR) method [25] is the determination of the concentration of n analytes in p samples with an array of m sensors. Two important assumptions have to be made to solve the problem. First, the sensors have to respond linearly, and second, the responses of the sensors to two or more analytes are additive.

The response matrix X is a $p \times m$ matrix with p samples (rows) and m sensor responses (columns):

$$P = X E + \varepsilon \qquad\qquad (6\text{-}17)$$

where P is a $p \times n$ matrix of unknown concentrations with p samples (rows) and n analytes (columns), E is the calibration matrix of regression coefficients, and ε is an error matrix. Before analyzing an unknown sample, first a calibration set with P_0 of known concentrations with the attached sensor responses X_0 has to be performed. Using P_0 and X_0 the calibration matrix E can be determined by linear algebra.

This MLR procedure calculates the best least-squares fit describing the data points for each analyte and each sensor using the criteria given. In many cases this is an appropriate method, but in some instances its application is not appropriate or is impossible. If the number of analytes n is greater than the number of sensors m (under-determined case), MLR gives an infinite number of solutions, for the same number of analytes and sensors there is one unique solution, and, if more sensors than analytes are used, more information is available and a better statistical fit is achieved.

The main problem is collinearities of the sensor responses. If the sensor responses are linearly dependent, the inversion procedure becomes sensitive to errors when calculating E. A small error in the observed sensor response will cause a large error in the result of the analysis.

6.2.4.3 *Background Identification*

For the calibration step of MLR it is assumed that the calibration mixture contains the same number as or more analytes than the unknown mixture. The presence of a background, an unknown analyte, will cause erroneous results. A method proposed by Osten and Kowalski [26] for background identification can be used to test the necessity for further calibration steps.

The major problem with sensor array data is that background detection is possible but not correctable without knowing the unknown components of the background.

6.2.4.4 *Principal Component Regression (PCR)*

An alternative to the linear calibration problem, principal component regression (PCR), couples factor analysis or principal component analysis (PCA) with multi-linear regression [27]. PCA is a method based on the Karhunen-Loeve expansion used in classification models, which yields qualitative results for pattern recognition. In a classification technique, normalized response vectors from the sensor array are used in combination with methods such as this Karhunen-Loeve expansion to locate areas in an m-dimensional space (m is the number of sensors) where classes of the samples are grouped together.

6.2.4.5 Partial Least Squares (PLS)

Partial least squares (PLS) was described first in the mid-1960s by Wold [28] and was subsequently used in econometrics, sociology and psychology during the 1970s [29]. The first use in chemistry was reported by Wold, Gerlach and Kowalski in the late 1970s. Since then, the groups of Wold and Martens have been refining and specializing the method for chemical applications [25, 30, 31]. A tutorial for a complete description was given by Geladi and Kowalski [32].

The PLS method of regression is based on the properties of multiple linear regression (MLR) and of principal component analysis (PCA). The important aspects of PLS method are:

- model building;
- prediction;
- parsimony.

One important field of use of PLS is in multivariate calibration in analytical chemistry. Applications in this field can easily be extended to the calibration of sensors. Most of the chemical applications have used the two-block PLS model, where the response and concentration matrices are considered as blocks of data. One potential advantage of PLS is the use of more than one response block to be regressed with the concentration block. This aspect of PLS has not yet been applied in a chemical experiment.

6.2.4.6 Rank Annihilation Factor Analysis (RAFA)

An important problem for the analytical chemist is the analysis of only a few components in a complex mixture. It would be convenient if quantitative information could be obtained for the analytes of interest without worrying about the other sample components. Second-order bilinear sensors, ie, sensors that give a two-dimensional data matrix of the form

$$M_{ij} = \sum_k \beta_k x_{ik} y_{jk} \tag{6-18}$$

are specially suited for this purpose, and the technique for quantification is known as rank annihilation [33, 34]. So far this method has been applied to excitation-emission fluorescence [33, 35] and to liquid chromatography with UV detection [36] with excellent results. One problem with the calculation has been that an iterative solution requiring many matrix diagonalizations was necessary. Lorber [37] and Sánchez and Kowalski [38] have reported a non-iterative solution, rank annihilation factor analysis (RAFA), presenting the problem as a generalized eigenvalue problem in which a direct solution is found by using a singular value decomposition.

6.2.4.7   *Artificial Neural Networks (ANN)*

A more recent approach for the solution of the multi-valent data analysis problem in multi-component analysis is the use of artificial neural networks (ANN).

An ANN is a network mostly implemented in software that is supposed to simulate the function of the human brain, in our case the function of the biological olfactory system. As an alternative to the software implementation, the modeling of such structures with transputer configurations is possible. Owing to the parallel signal processing, the recognition ability is higher in comparison with the PARC methods described previously. The base element of every neural network is the single neuron. A model of a neural processing element simulated in the computer (perceptron) is shown in Figure 6-3 [39, 40]. The input information with N input elements x_i is weighted each by a factor w_i, which corresponds to the different transmitting resistances of the synapses. After summation of all input functions and offset correction, the activation function f decides the amplitude of the output signal y, which is the new input attraction for the next neuron. An important influence on the function of the neuron model is

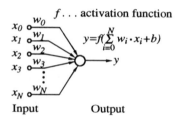

Figure 6-3. Model of a perceptron [40].

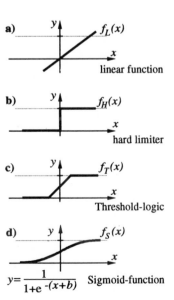

Figure 6-4.
Different activation functions. These functions are important for the teaching process [40].

given by the activation function *f*. Some examples are given in Figure 6-4. For an efficient learning algorithm the non-linear sigmoid function has proved to be useful [41].

A learning algorithm called the *back-propagation algorithm* makes possible the learning procedure also in complicated network structures. Such a network with an input layer, two hidden layers, and an output layer is shown in Figure 6-5. This structure is called *three-layer perceptron*. The input layer has no network function and is the information receiver of the first hidden unit. The number of input and output elements is defined by the task of the network. For an analysis problem in chemical sensing the number of input elements is equal to the dimension of the response vector and the number of output elements is equal to the number of identified compounds.

A number of known input compounds is necessary to train the network in a calibration step with the aim of setting weighting factors for the several signal lines. After this teach-in process the network is able to determine these compounds.

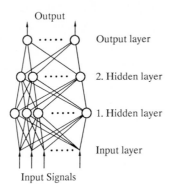

Output

Output layer

2. Hidden layer

1. Hidden layer

Input layer

Input Signals

Figure 6-5.
Model of a three-layer perceptron: neural net with two hidden units [40].

6.3   Practical Examples

In Table 1-3 typical sensor parameters x' have been considered which are obtained with individual sensors under equilibrium conditions or during rapid changes in the partial pressure of the environment that lead to characteristic initial slopes or relaxation times (see Figure 6-2). Parameters which describe the static sensor response and parameters describing the dynamic properties of the sensor response can both be used as measuring signals for a multi-component analysis.

A variety of modulation techniques are becoming increasingly important for producing different signals for analyzing gas mixtures with different compositions. In this context, two different types of experiments may be carried out:

1. The partial pressure of a specific gas may be varied periodically at a certain frequency. The sensor signal x' follows with a characteristic phase and amplitude change as a function of frequency (Figure 6-2). This modulation can be done by a so-called *pressure modulation* technique, which has been applied, eg, to characterize solid electrolyte surfaces [42]. Another method is the use of a catalytic filament in front of the sensing element which produces

temperature-dependent reaction products. As a result, the corresponding partial pressure of the gas may be modulated by triangular [10], sinusoidal, step-functional [2, 11] etc. power inputs to the filament, or by an additional modulation by switching stationary gas streams between catalytic and non-catalytic media [2, 3]. This method is called *indirect modulation*.

2. Sensor selectivities may be changed periodically at a constant gas composition. As examples sensor temperatures of pellistors, oxidizing or reducing potentials of electrochemical sensors [43], or voltages of electronic conductivity sensors may be varied periodically with characteristic frequency-dependent changes in amplitudes and phases of sensor responses. This method is called *direct modulation*.

In the following sections examples are discussed which present different strategies for solving the problem of single-component and multi-component analysis.

6.3.1 Examples of Single-Component Analysis

6.3.1.1 Artificial Noses

The stimulation to develop an artificial nose was based on the search for similarities between chemoreceptors in living organisms and artificial chemical sensors. These similarities concern the recognition procedure for obtaining selectivities and cross-sensitivities to several gas components, as mentioned in the introduction. As an example, when stimulating the olfactory sensors of rabbits with low concentrations of organic compounds (odors) such as anisole, cyclopentane, and octane, specific electrical patterns can be deduced from the different receptors by using implanted microelectrodes as shown in Figure 6-6 [44–45]. These signals make possible the detection of individual specific odors by subsequent pattern recognition in the brain and/or in the computer.

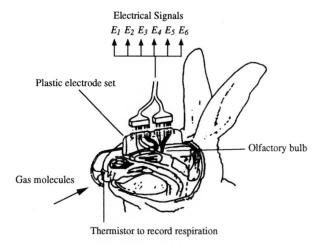

Figure 6-6. The olfactory sense of a rabbit is stimulated by different gases at low concentrations. Electrical patterns as deduced from implanted microelectrodes serve as patterns for the subsequent gas analysis in the brain or in the computer [45].

Analogous to this experimental approach to the evaluation of data from a "natural nose", Ikegami et al. [45] described an "artificial nose" consisting of a thick-film sensor array (Figure 6-7) kept at 400 °C. They used an 8-bit microcomputer to identify and determine the concentrations of single gas components. Since different semiconductor oxides and metal dopants were used for the individual conductivity sensors with different sensitivities to the different gases, specific patterns of conductance changes could be deduced on exposure to a single gas component. Using evaluation schemes described in more detail below, the microcomputer identifies the gas by calculating similarities with stored standard patterns. The latter were obtained in a calibration with different concentrations of the same gas. This also makes possible the quantitative determination of concentrations by using the sensor element with the highest sensitivity to this specific compound.

To improve the performance of the sensor array, drift-induced influences in the pattern recognition procedure must be avoided. Therefore, Ikegami et al. used the relative conductivity

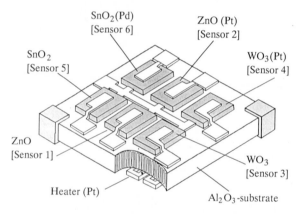

Figure 6-7. Scheme of a sensor array consisting of ZnO, Pt-doped ZnO, WO_3, Pt-doped WO_3, SnO_2, and Pd-doped SnO_2. These oxides were deposited by thick-film technology on alumina substrates with a platinum heater on the rear side [45].

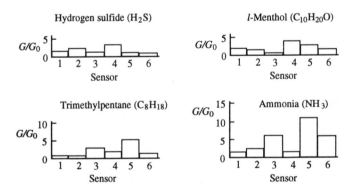

Figure 6-8. Specific patterns obtained from conductivity changes G/G_0 of a sensor array on exposure of different gases. G_0 is the conductivity in air; G denotes the conductivity with 10 ppm of gas [45].

changes of the six-sensor array as signal patterns. Typical results obtained after exposure to 10 ppm of anisole, cyclopentane, octane, ℓ-menthol, and ammonia in air are shown in Figure 6-8. They demonstrate the principal possibility of identifying different compounds with this setup.

In the evaluation of this array of electronic conductivity sensors, the partial pressure-dependent conductance changes (G/G_0) of the ith sensor on exposure to a single gas component k at constant oxygen partial pressure are to a first approximation given by

$$\left(\frac{G}{G_0}\right)_i = x_i \sim (1 + K_{ik} p_k)^{\beta_{ik}} . \tag{6-19}$$

where x_i is the sensor signal and K_{ik} and β_{ik} are parameters describing the non-linear sensor characteristics which can be determined from a previous calibration. To reduce the influence of p_k on the signal pattern, the sensor signals x_i have to be normalized. Neglecting differences of the parameter K_{ik} for different sensors i but the same gas k, the following parameter α_{ik} can be deduced from the logarithmic form of Equation (6-19):

$$\alpha_{ik} = \frac{\log x_i}{\sum\limits_{i=1}^{m} \log x_i} = \frac{\beta_{ik} \log (1 + K_k p_k)}{\sum\limits_{i=1}^{m} \beta_{ik} \log (1 + K_k p_k)} = \frac{\beta_{ik}}{\sum\limits_{i=1}^{m} \beta_{ik}} . \tag{6-20}$$

For $K_{ik} = K_k$, the response α_{ik} of the ith sensor on exposure to the gas k can be calculated from the conductivity changes, which are independent of the partial pressure. A normalized pattern of this sensor array is obtained for a single gas component. However, a slight partial pressure dependence of α_{ik} was observed experimentally as Equation (6-20) is only valid to a first approximation. Nevertheless, the following procedure for evaluating Equation (6-20) enhances the performance of the pattern recognition. The vectors $A_k = (\alpha_{ik})_{k = \text{const}}$ deduced from Equation (6-20) characterize the response of the sensor array upon gas exposure. The pattern recognition to identify unknown gas components starts with the calculation of the similarity S_k between the unknown normalized pattern $A = (\alpha_i^*)$ and the patterns $A_k = (\alpha_{ik})$ obtained by the preceding calibration with L_k different partial pressures of the same gas k:

$$S_k = \frac{1}{L_k} \sum\limits_{\ell=1}^{L_k} \sum\limits_{i=1}^{m} w_{ik} (\alpha_i^* - \alpha_{ik})^2 \tag{6-21}$$

where w_{ik} characterizes the dependence of $\alpha_{ik} = \alpha_{ik} (p_k)$ on the partial pressure for a specific sensor and a specific gas. This statistical weight w_{ik} is given by

$$w_{ik} = \frac{1}{\sigma_{ik}^2 \sum\limits_{i=1}^{m} \sigma_{ik}^{-2}} \tag{6-22}$$

and is low for a high partial pressure dependence and vice versa. The parameter σ_{ik} is deduced from the $m \times L_k$ matrix $[\alpha_{ikl} (p_k)]$ with each value $\alpha_{ikl} (p_k)$ obtained for the lth partial pressure of the same gas k according to

$$\sigma^2_{ik} = \frac{1}{L_k} \sum_{\ell=1}^{L_k} \alpha^2_{ik\ell} - \left(\frac{1}{L_k} \sum_{\ell=1}^{L_k} \alpha_{ik\ell} \right)^2 . \qquad (6\text{-}23)$$

Thus, considering also the partial pressure dependence of the normalized patterns $A_k = (\alpha_{ik})$, the best agreement between unknown and standard calibration patterns is determined by a minimum of the similarities S_k. In this case, the unknown single gas component is identified as shown quantitatively in Figure 6-9.

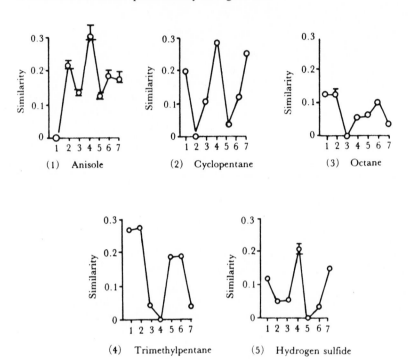

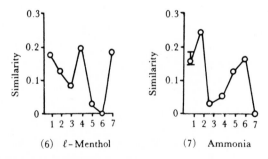

Figure 6-9. Similiarities S_k between the normalized pattern obtained from the six-sensor array on exposure to anisole (1), cyclopentane (2), octane (3), trimethylpentane (4), hydrogen sulfide (5), ℓ-menthol (6), or ammonia (7), and the stored standard patterns from preceding calibrations [45].

All gases investigated in this study were identified with good accuracy. As an example of the limitations of this approach, similarities and, as a result, possible interferences between these gases occur after exposure of the array to cyclopentane but not to hydrogen sulfide.

In all cases the partial pressure of the gases can be determined from the sensor element with the highest sensitivity. This is demonstrated in Figure 6-10.

It should be mentioned again here that the patterns obtained so far are only determined by a single component in air. Interference effects between different gases have not been studied. These would, however, lead to completely different patterns and thus require more sophisticated data evaluation.

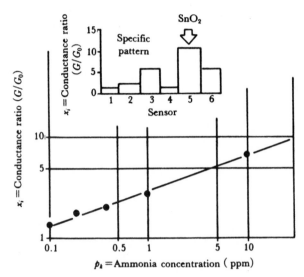

Figure 6-10.
Specific pattern for ammonia obtained from a six-sensor array. The partial pressure p_{NH_3} is determined from sensor 5 with the highest sensitivity to detect NH_3 [45].

6.3.1.2 *Zeolite Sensors*

In contrast to the one-dimensional sensor signals used for the artificial nose, this section deals with sensor arrays consisting of two-dimensional sensor signals, ie, each single sensor produces two linear independent signals. The signal x_1 is deduced from the first derivative of the sensor response $x_1 = \partial x'/\partial t \big|_{t_0}$ and the signal x_2 from the sensor response x' at a certain time t' after starting the gas exposure at $t = t_0$, ie, $x_2 = x'(t')$, holds. The value x_2 was used, eg, as the steady-state fullscale deflection (fsd) signal [9].

Müller and co-workers used MOS gas sensors with a zeolite filter covering the conventional Pd gate electrode as shown schematically in Figure 6-11 [9, 46]. Hydrogen, methane, and acetylene were used as test gases.

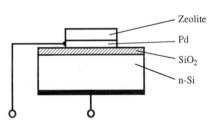

Figure 6-11.
MOS sensor element with zeolite coating [9].

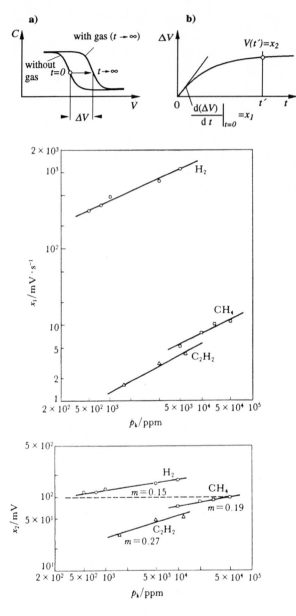

a)

b)

Figure 6-12.
Signals of the MOS gas sensor.
(a) Capacitance/voltage ($C(V)$) characteristics before and after gas exposure, ie, for $t = 0$ and $t \rightarrow \infty$. (b) Time-dependent voltage shift ΔV for a given capacitance C during gas exposure [9].

Figure 6-13.
Signals x_1 and x_2 deduced from MOS sensors [9].

Catalytic reactions within the zeolites and/or at the metal/insulator interface produce hydrogen atoms which pass through the palladium. As a result, the $C(V)$ characteristic of the MOS capacitors is shifted and partial pressure-dependent signals $x_1 = \partial \Delta V / \partial t \big|_{t_0}$ and $x_2 = \Delta V \big|_{t'}$ are monitored with typical results as shown in Figure 6-12.

Figure 6-13 indicates that both signals depend on the gas and its partial pressure. From simple model calculations for this type of chemical sensor, the signal x_1 is expected to be proportional to the partial pressure and the signal x_2 to follow a power law with $x_2 = p_j^n (n \neq 1)$. To a first approximation, Müller and co-workers determined a ratio $x_1/x_2^{2.4}$ which is independent of p for the same gas in certain partial pressures ranges. The invidual values of $x_1/x_2^{2.4}$, however, are characteristic for a specific gas, H_2, C_2H_2, or CH_4, as a single component:

$$x_1/x_2^{2.4} \approx 4 \cdot 10^{-4} \quad \text{for} \quad 500 < p_{H_2} < 10000 \text{ ppm}$$

$$3 \cdot 10^{-4} < x_1/x_2^{2.4} < 6 \cdot 10^{-4} \quad \text{for} \quad 2000 < p_{C_2H_2} < 20000 \text{ ppm}$$

$$x_1/x_2^{2.4} \approx 2 \cdot 10^{-3} \quad \text{for} \quad 10000 < p_{CH_4} < 50000 \text{ ppm} .$$

Two-dimensional plots of x_1 versus $x_1/x_2^{2.4}$ allow both the determination of the partial pressure p_j and the identification of the gas k thus denoted as representations of p_k versus k.

In this pattern recognition method, the gas flow with the unknown gas must be switched on and off at the MOS sensor. If heated MOS sensors reach their constant temperature faster than their steady-state response signals x_2 for a specific gas, a stepwise modulation between a low and a higher temperature can be used to switch between an insensitive and a sensitive state of the sensor. Two signals $x_1 = dx'/dt|_{t_0}$ and $x_2 = x'(t')$ are deduced from the sensor with the time t_0 denoting the time at which the temperature is changed and t' that at which a steady-state signal is reached.

The pattern recognition method was also applied to a four-sensor array with MOS structures without zeolites covering the Pd overlayer and with MOS structures using zeolites with different pore sizes, ie, 0.3, 0.4, and 0.9 nm. Typical values x_1 and x_2 for this array obtained after exposure to hydrogen, ammonia, methanol, acetone, benzene, and toluene are shown in Figure 6-14.

This setup allows the identification of unknown single gas components if standard spectra of all gases to be detected are taken and stored in the computer. This is tested for hydrogen, methane, and acetylene with four sensors. The pattern recognition is based on the calculation of the correlation coefficient ρ_k^* between the signal pattern obtained for an unknown gas (represented by the asterisk) and the standard pattern for the gas k. The parameter ρ_k^* is defined by

$$\rho_k^* = \frac{\sum\limits_{i=1}^{m} (x_i^* - \bar{x}^*)(x_{ik} - \bar{x}_k)}{\sqrt{\sum\limits_{i=1}^{m} (x_i^* - \bar{x}^*)^2 \sum\limits_{i=1}^{m} (x_{ik} - \bar{x}_k)^2}} \tag{6-24}$$

where x_i^* denotes the signal of sensor i for the unknown gas, and $\bar{x}^*$ the average signal of the four sensors. The values $\bar{x}_{ik}$ denote the average signal of the sensor i for the gas k determined in a calibration at different partial pressures with

$$\bar{x}_{ik} = \frac{1}{L_k} \sum\limits_{\ell=1}^{L_k} x_{ik\ell} \tag{6-25}$$

where L_k is the total number of partial pressures. The average value $\bar{x}_k$ for the sensor elements is

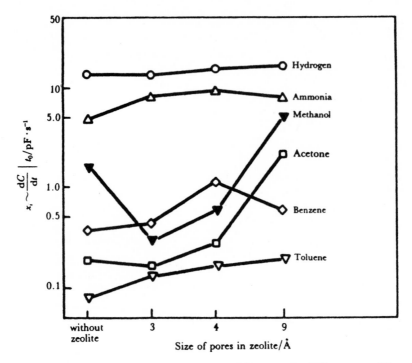

Figure 6-14. Signals x_1 obtained from different MOS sensors for different gases [9].

$$\bar{x}_k = \frac{1}{m} \sum_{i=1}^{m} \bar{x}_{ik} \, . \tag{6-26}$$

An unknown gas is identified if $0.8 < \rho_k^* < 1$ holds for the correlation coefficient as shown quantitatively in Figure 6-15. For $r < 0.8$, more than one or an unidentified gas component is present. For the determination of the partial pressure of the identified gas k, Müller and co-workers do not use the signal of the most sensitive sensor but take a parameter Z_k^* weighted with the respective sensitivity as a criterion for the "magnitude" of the spectrum:

$$Z_k^* = \sum_{i=1}^{m} x_i^* \bar{x}_{ik} \, . \tag{6-27}$$

The calibration values $Z_{k\ell}$ are obtained from the calibration signals $x_{ik\ell}$ determined for ℓ different partial pressures:

$$Z_{k\ell} = \sum_{i=1}^{m} x_{ik\ell} \bar{x}_{ik} \tag{6-28}$$

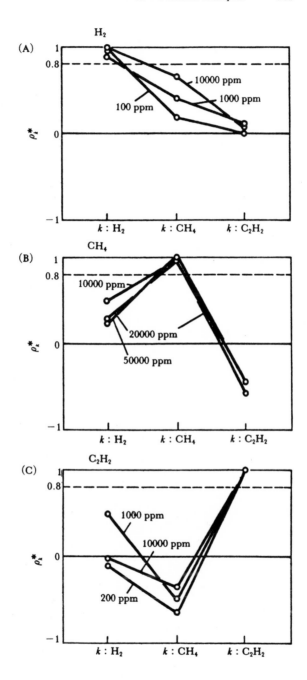

Figure 6-15.
Correlation coefficients ϱ_k^* for
(a) H_2, (b) CH_4, and (c) C_2H_2 as
functions of the partial pressures
p_{H_2}, p_{CH_4}, and $p_{C_2H_2}$ for the MOS
sensor covered with zeolite of 0.3
nm pore size [9].

with the average value x_{ik} corresponding to Equation (6-23). For $Z_{k\ell} < Z_k^* < Z_{k\ell+1}$, the
partial pressure p^* of the identified gas is within the interval $[p(\ell), p(\ell + 1)]$. As a typical
example, Figure 6-16 shows the calibration characteristics $Z_{k\ell}$ for hydrogen, methane, and
acetylene.

If the number of sensor signals exceeds the number of gases, this redundance can be used to detect a defective sensor element. Subsequently the signal of this sensor can be excluded from the evaluation and hence the measurement accuracy can be improved.

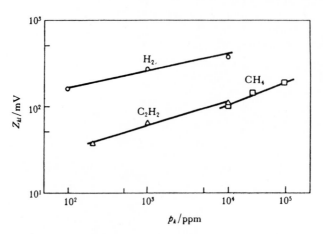

Figure 6-16. Partial pressure dependence of the parameter $Z_{k\ell}$ for H_2, CH_4, and C_2H_2 obtained by calibration [9].

6.3.1.3 Odor-Sensing Systems with Arrays of Mass-Sensitive Sensors

In this example presented by Nakamoto and co-workers [47, 48], six quartz-resonator sensors with different coating materials were used for odor sensing. The quartz resonators were AT-cut with a resonator frequency of 10 MHz. Odorants are adsorbed on the sensing membranes over quartz resonators, giving a decrease in the resonance frequency. The membrane materials were (1) epoxy resin, (2) triolein, (3) squalane, (4) acetylcellulose + triolein, (5)

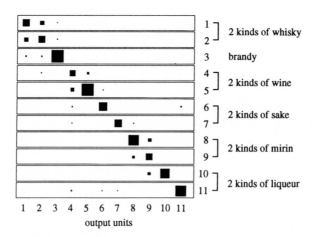

Figure 6-17.
Activity level of output units when input signals are used from which ethanol solution signals had been subtracted. Mirin is a Japanese cooking saké [48].

acetylcellulose + diethylene glycol, and (6) di-*n*-octyl phthalate. With these sensors odors of alcoholic drinks were tested. The frequency shifts varied from 100 Hz to 5 kHz, depending on the coating materials and the kind of odorant.

For data evaluation a three-layer neural network using the back-propagation algorithm was used. The number of input units was six (the same as the number of sensors) and the number of output units was equal to the number of the types of odors.

Experiments on odor recognition for alcoholic drinks were performed using the network with six intermediate units. Figure 6-17 shows the results for eleven kinds of drinks. The network was trained by three million sweeps and the probability of obtaining a correct answer was about 67%. The odor-recognition ability was enhanced to 81% by subtracting the data patterns of pure ethanol solution from those of the drinks. By increasing the number of intermediate units the probability was improved to 88%.

6.3.1.4 *Arrays of Electrochemical Sensors and Catalysts*

A portable instrument capable of detecting, identifying, and monitoring hazardous gases and vapors at and above the parts per million (ppm) level has been developed in a cooperative project by Argonne National Laboratories and Transducer Research [4]. The instrument is a small, microprocessor-controlled device, composed of a compact array of amperometric gas sensors, each of which is adjusted to respond differently to electrochemically active gases and vapors. Two noble-metal filaments precede the sensor array. They are used in different heating modes to oxidize electrochemically inactive compounds. The reaction products are detected by the sensors.

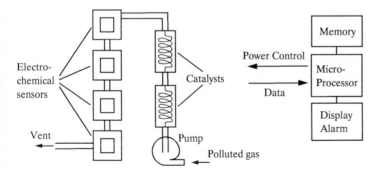

Figure 6-18. Experimental setup with sensor array, catalytic filaments, data acquisition, and control unit [49].

Figure 6-18 shows a schematic diagram of the experimental setup. Under microprocessor control, a constant concentration of the unknown vapor (in air) is passed through the filament chamber and the sensor array. The steady-state signal is monitored in four operating modes of the filaments as summarized in Table 6-1. This procedure generates 16 individual signals on exposure to specific gases, as shown in Figure 6-19, which yield a vector $X^* = (x_1^*, x_2^*, x_3^*,$

Table 6-1. Sensor channels 1–16 as determined by four operating modes and four sensors [4]

Operating mode		Type of sensor			
	Sensor:	1	2	3	4
	Working electrode:	Au	Au	Pt	Pt black
	Bias voltage:	-200 mV	$+300$ mV	$+200$ mV	O mV
No filament		1	5	9	13
Pt filament, 750 °C		2	6	10	14
Rh filament, 450 °C		3	7	11	15
Rh filament, 600 °C		4	8	12	16

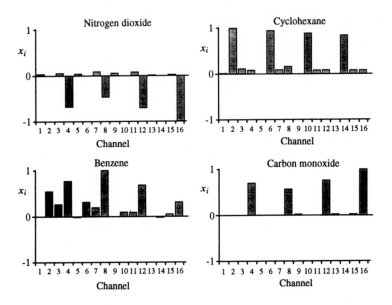

Figure 6-19 Typical normalized patterns obtained from the sensor array with sensors from Table 6-1 and corresponding sensor signals x_i after exposure to NO_2, cyclohexane, benzene, and CO in air [4].

..., x_{16}^*). As is well known for the amperometric detection principle used here, the sensor signals x_i ("channels") are proportional to the partial pressure of the gas k and $x_i \sim p_k$ holds.

The identification of an unknown gas starts with a zero calibration of the array leading to a vector X_0 which corresponds to the zero signals of the sensors. Exposure to an unknown gas yields the vector X^*. The elements of this vector are normalized with the largest signal set to $|x_i| = 1$. Since $x_i \sim p_k$ holds for the sensors chosen here, a normalized vector X^*_{norm} is obtained, independent of the partial pressure p_k:

$$X^* = \gamma^* X^*_{norm} \tag{6-29}$$

where γ^* is the sensitivity of the array to the unknown gas. The pattern recognition uses as the simplest algorithm the Euclidian distance in a 16-dimensional space. Neglecting drifts in the sensitivity, the Euclidian distance Δ_k^* between the vector X_{norm}^* and the standard vectors X_k obtained by a preceding calibration and storage in the computer memory is calculated by

$$\Delta_k^* = |(X^* - X_0)_{norm} - X_k| . \tag{6-30}$$

Even with the limited computer power of a portable instrument, about 50 compounds can be identified in a few seconds.

Two important features are necessary to produce reliable results with this simple algorithm:

1. the sensor signal should be linearly correlated with the gas concentration over the whole detection range; and

2. the sensors should have different sensitivities to different gases, to produce different patterns also for similar gases.

With the additional requirements of cheap and lightweight instrumentation, amperometric sensors fulfil these features well. The detected signal is the current between the working electrode and counter electrode. It is linear with the chemical reaction rate provided that the concentration of the gas is not too high. The sensitivity of the sensor is influenced by the material of the working electrode (gold and platinum with high and low surface areas) and by the redox potential of the working electrode. The adjustment may be chosen so as to obtain the optimum working conditions to generate patterns with a large Euclidian distance between each gas. The reliability for identifying a specific gas becomes high if the filament conditions are chosen well.

The instrument leads to reliable results if one gas or vapor component is present in the analyte sample with a predominant concentration. Analysis of gas mixtures is not possible in this approach because only recognition parameters, which depend on the gas concentration of one component, have been used. Based on this setup, which uses electrochemical sensors and catalytic filaments, methods for increasing the selectivity have been described. A modified setup with more sophisticated data evaluation is described in Section 6.3.2.6 in its application to a "real" multi-component analysis.

Another field of application of this instrument is the test of spoiled grain using pattern recognition methods based on neural networks [49]. The unique properties of neural networks have significant potential for handling low-quality information. The analysis of a grain sample aims at a classification of grain of good quality or grain of bad quality, such as "sour", "insect", "musty" or "COFO" (commercially objectionable foreign odor).

For the experimental setup a configuration with four electrochemical sensors biased at different oxidizing potentials and a single rhodium filament were chosen. The filament was operated at 25, 450, 750 and 850 °C, which produced an array of sixteen data points per analysis. The odor samples were generated by heating the grain to 60 °C and flushing with a measured volume of air. The effluent air was passed through an ice trap to collect a "non-volatile" fraction and a liquid nitrogen trap to collect the "volatiles". The two fractions were run separately and in duplicate. Typical vectors of different classes of grain quality are shown in Figure 6-20. The rate of correct classification was found to be 65% (see Table 6-2). Because the test conditions had changed during the measurements, another element was added to the data vectors to differentiate the measurements made before and after a sensor was changed. The neural net was retrained with the new 17-element vector. The rate of correct classification was thereby increased to 83%.

Table 6-2. Summary of the accuracy of the neural network algorithm for identifying vapors drawn from wheat samples, left: training data; right: evaluated data [49].

No.	Data set for training	Accuracy of identification (%)	No.	Wheat samples data set	Accuracy of identification (%)
1	Original data	100	1	Total data	100
2	5% error added	100	2	Train on 55% of data set	65
3	10% error added	98	3	Add channel for test conditions	83
4	15% error added	92			

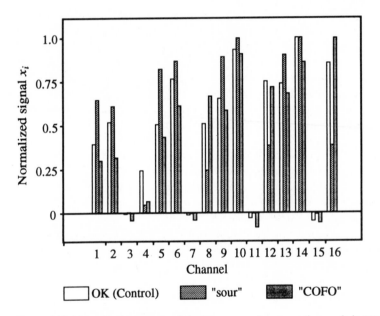

Figure 6-20. Normalized responses of the four sensor–four operating mode instrument to "good" (OK), "sour", and "COFO" grain [50]; "COFO" means "commercially objectionable foreign odor".

6.3.1.5   Selective Identification with Catalyst/Electrochemical Sensor Systems

This study also uses electrochemical sensors as detectors and a catalytic filament as modulator. With special operating conditions of the catalyst the selectivity of the system to single gases is increased [11]. Information from a modulated output signal is used for identifying and quantifying different individual components in a gas mixture. An unknown gas passes through a reaction chamber with a heated filament. The output products of the reaction chamber are detected by an electrochemical sensor. The concentration of the reaction products is modulated by varying periodically the temperature of the catalytic filament.

The electrochemical sensor produces a current $i(t)$ which is proportional to the chemical reaction rate at the sensor. If the number of moles of gas in the volume V_0 increases from 0 to n_c at $t = 0$, the resulting current response is

$$i(t) = m \cdot \frac{n_c}{V_0}\left(1 - \exp\left(-\frac{t}{\tau_S}\right)\right) \tag{6-31}$$

where m is the intrinsic cell sensitivity and τ_S the sensor time constant.

In the reaction chamber a chemical reaction takes place in which an active reactant C is formed from electrochemically inactive components A and B:

$$\text{A} + \text{B} \xrightarrow{k} \text{C} \tag{6-32}$$

where k is the reaction rate constant. The reaction rate constant is assumed to depend on the temperature only:

$$k = A\,\exp\left(-\Delta E/R_G\,T\right) \tag{6-33}$$

where A is a pre-exponential factor for the particular oxidation, ΔE is the activation energy for the conversion of A to C, and R_G is the gas constant per mole.

For a constant filament temperature, with a step-functional change in the number of moles of component A in the reaction chamber at $t = 0$, the following equation is obtained:

$$i(t) = m\left(\frac{\tau_{res}}{\tau_{res} + \tau_a}\right)\left(\frac{n_a}{V_0}\right)\left(1 + \frac{\tau_S}{\tau^* - \tau_S}\exp\left(-\frac{t}{\tau_S}\right) - \frac{\tau^*}{\tau^* - \tau_S}\exp\left(-\frac{t}{\tau^*}\right)\right). \tag{6-34}$$

The first term in parentheses on the right-hand side represents the fraction of A converted to C, the second term represents the initial concentration of A in the reaction chamber, and the last term yields the time dependence. The factor τ_{res} is the residence time in the reaction chamber, and τ_a is the characteristic chemical reaction time. This leads to $(\tau^*)^{-1} = (\tau_{res})^{-1} + (\tau_a)^{-1}$. At the steady state ($t \to \infty$) the last term of Equation (6-34) approaches unity.

A periodic variation in the filament temperature, such as a sinusoidal variation, can yield information which can assist in differentiating between different species that produce detectable reaction products (C) from reactions with different rates (eg, different rate constants and activation energies). For simplicity, in the analysis it is assumed that the variations ΔT in temperature about the mean value T_0, which may be given by

$$T = \Delta T \sin(\omega t + \varphi) + T_0, \tag{6-35}$$

are sufficiently slow that the solution for the steady-state current

$$i(\infty) = mC_a R \qquad \text{with} \qquad C_a = n_a/V_0 \tag{6-36}$$

is valid for all times during the temperature cycle. This condition will be met if the period of the modulation is much greater than τ^*.

Gas mixtures can be analyzed by this detector system if there are independent non-interfering species present and the system response to each is linear. It is more likely that such approximations will be valid if low concentrations and low conversion ratios are maintained.

For a constant filament temperature a sensitivity S_{ij} (μA ppm^{-1}) for the species i at temperature T_j and flow-rate u_j can be defined:

$$S_{ij} = i_i(\infty)/C_i|_{T_i, u_j} \tag{6-37}$$

where $C_i = n_i/V_0$ is the concentration of the species i. The matrix S_{ij} has to be calibrated for each species i, $i = 1, \ldots, n$ and n different sets of conditions $j = 1, \ldots, n$. S_{ij} is a square matrix which has to be inverted to determine C_i.

For periodic variations of the filament temperature, an expression for the total AC response at conditions (u_j, T_j) is obtained:

$$i_{ACj}(\infty) = \Delta T \sum_i C_i s_{ij} \tag{6-38}$$

where

$$s_{ij} = m_i \frac{\Delta E}{R_G T^2} \frac{u/k_i n_b}{(u/k_i n_b + 1)^2}\bigg|_{T=T_j, u=u_j}. \tag{6-39}$$

The factor s_{ij} depends on the temperature, the flow-rate u and the activation energy ΔE_i. In order to ensure that the system of equations can be solved, s_{ij} must obey the same restrictions as S_{ij}.

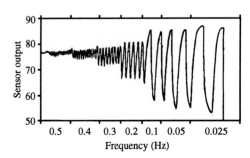

Figure 6-21.
Electrochemical cell response to 511 ppm of benzene in air at different filament frequencies with square power input, temperature variation between 650 and 870 °C; the flow-rate was $u = 120$ cm^3 min^{-1} and the reaction volume was $V_0 = 1$ cm^3 [11].

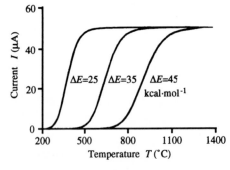

Figure 6-22.
Calculated current versus temperature curves for three different activation energies, ΔE; the flow-rate was $u = 1$ cm^3 min^{-1} [11].

Figure 6-21 shows the response of an electrochemical CO sensor for different frequencies of the filament temperature cycle when a benzene-air mixture is introduced. The amplitude of the temperature cycle is between 650 and 870 °C. For high modulation frequencies the oscillation is damped by two effects, slow sensor response and slow filament response. At low frequencies the steady-state response $i(\infty)$ is nearly achieved at each temperature extreme.

Simulations of the steady-state sensor response to changes in the raction rate due to changes in the activation energy (ie, species) and the filament temperature (ie, rate constant) are shown in Figure 6-22 for different activation energies ($\Delta E = 25, 35, 45$ kcal mol^{-1}). A response of 50 µA represents 100% conversion of A to C. The position of the i versus T curves is seen to depend on the activation energy. Figure 6-23 shows a simulated mixture with the activation

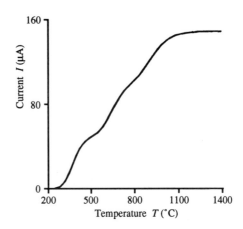

Figure 6-23.
Calculated steady-state sensor response to a mixture of species with activation energies of 25, 35, and 45 kcal mol^{-1} [11].

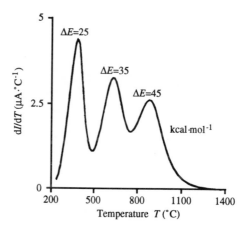

Figure 6-24.
Derivative of response curve in Figure 6-23 with respect to temperature T. The derivative is proportional to the magnitude of the AC sensor response to small sinusoidal variations of the filament temperature [11].

energies from Figure 6-22. When the derivative with respect to temperature is taken of the mixture curve, a "spectrum" is obtained with peaks corresponding to different activation energies (Figure 6-24). The peak heights correspond to the magnitude of the AC sensor response, and the location of the peak identifies the operating point P_0 and the temperature T_0 at which the sensor exhibits the greatest temperature dependence for the species with the particular activation energy ΔE_0. This kind of spectrum could be generated experimentally by modulating the temperature in a sine wave while also ramping the temperature from low to high.

Modulation of the filament about T_0 should not produce a modulated sensor signal for species with activation energy significantly different from ΔE_0. This is illustrated in Figure 6-25. The activation energy of this species was $\Delta E_0 = 33.7$ kcal mol^{-1}. The absolute current scale depends on the sensitivity of the sensor.

With a similar experimental setup, Otagawa and Stetter [10] investigated a method for the selective detection of airborne chemicals. They used a rhodium filament connected to an electrochemical CO sensor. Figure 6-26 shows the steady-state signals of the sensor plotted versus the temperature of the rhodium filament. From the pseudo-activation energy for the formation of electroactive products the reactions and chemicals can be divided into three general categories: those having low (hydrogen cyanide, ammonia and acrylonitrile), medium

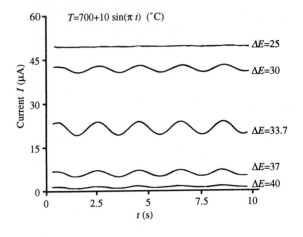

Figure 6-25.
Calculated response of a sensor for activation energies from 25 to 40 kcal mol^{-1}. The filament temperature varies sinusoidally from 690 to 710 °C, with a period of 2 s [11].

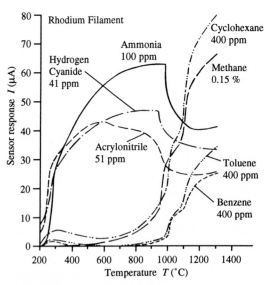

Figure 6-26.
Steady-state signal of an electrochemical CO sensor at various temperatures of a rhodium filament [10].

(cyclohexane and methane) and high (benzene and toluene) pseudo-activation energies for the formation of electrochemically active products detected by the CO sensor. In the temperature range 600–1200 °C vastly different responses from chemicals are obtained. Figure 6-27 shows the exciting modulation of the filament temperature in the important temperature range between 600 and 1000 °C and the corresponding sensor response of a component with medium activation energy. Figure 6-28 shows three types of modulated sensor response. For classification the ratio b/a as defined in Figure 6-27 is used and tabulated under the traces. The parameters a and b were observed to be proportional to the concentration of the chemical vapor in the range between 0 and 400 ppm at a temperature cycle time of 20 or 40 s.

This method allows the determination of a kinetic parameter (eg, pseudo-activation energy) that provides the information required to make the single-sensor system selective for particular chemicals.

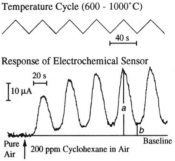

Figure 6-27.
Response of the electrochemical sensor to 200 ppm of cyclohexane in air with the filament cycled between 600 and 1000 °C. The ratio b/a represents an important parameter for the classification of different gases [10].

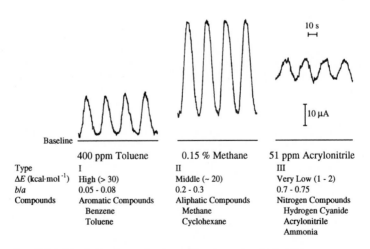

Type	I	II	III
ΔE (kcal·mol^{-1})	High (> 30)	Middle (~ 20)	Very Low (1 - 2)
b/a	0.05 - 0.08	0.2 - 0.3	0.7 - 0.75
Compounds	Aromatic Compounds	Aliphatic Compounds	Nitrogen Compounds
	Benzene	Methane	Hydrogen Cyanide
	Toluene	Cyclohexane	Acrylonitrile
			Ammonia

Above the columns: 400 ppm Toluene | 0.15 % Methane | 51 ppm Acrylonitrile

Figure 6-28. Three types of modulated patterns observed using a filament cycle of 20 s [10].

6.3.2   Examples of Multi-Component Analysis

6.3.2.1   Pellistor Arrays

The pattern recognition method described in this section for analyzing single and binary mixtures of reducing gases is based on their catalytic oxidation at differently doped (Pt, Pd, Rh, Ir) ceramic $BaTiO_3$ pellistors (Figure 6-29) [51].

The enthalpy changes ΔH (heats of combustion) during these reactions lead to a change in the electrical power ΔP needed in an automatic control system to operate these devices isothermally. The value of ΔP is given by the reaction rate dr/dt (moles of the reducing gas oxidized per second), by the catalytically active area A_{eff} of the pellistor, and by ΔH [52]:

$$\Delta P \sim A_{eff} \cdot \frac{dr}{dt} \cdot \Delta H . \tag{6-40}$$

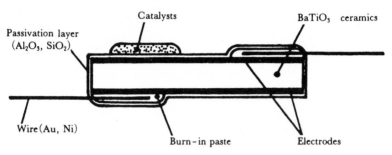

Figure 6-29. Design of microcalorimeters based on $BaTiO_3$ PTC ceramics [51].

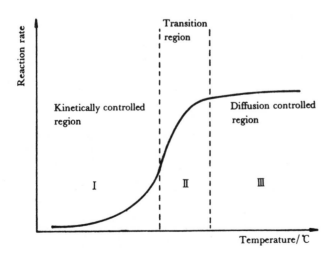

Figure 6-30. Reaction rate of the catalytic oxidation of reducing gases such as H_2 and CO versus pellistor temperature. Typical temperature regions are marked for different operation modes of the pellistors [51].

The reaction rate dr/dt shows a strong temperature dependence. At low temperatures the reaction coefficient, which is characteristic for a specific gas, controls the total reaction rate (kinetically controlled region I) as shown in Figure 6-30. Because of the low heat of combustion ΔH and, hence, the low electrical power change ΔP in this region, the accuracy of the detection of a reducing gas is poor and is therefore not used for its identification. In region III, all reducible species are oxidized nearly independently of the temperature. The reaction rates are completely determined by the diffusion of molecules to the pellistor surface. Determined by the diffusion laws, a linear dependence between ΔP and the partial pressure p_k of the gas is usually observed in this region, which allows the determination of p_k. In the transition region II the temperature dependence of the reaction rate is characteristic for a specific gas and also for a specific catalyst with high reaction rates and therefore large ΔP values. The latter can be measured accurately with isothermally operated ceramic $BaTiO_3$ used as microcalorimeters (Figure 6-31).

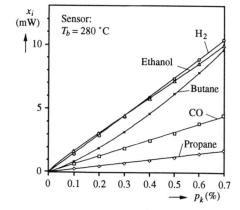

Figure 6-31.
Typical partial pressure (p_k) dependences of the pellistor signals for hydrogen, ethanol, butane, carbon monoxide, and propane [51].

As mentioned above, in many cases the measured signals x_i show linear dependences on the partial pressures of single gas components, eg, for H_2, C_2H_5OH, CO, or propane even for temperatures within region II, ie,

$$x_i = \gamma_{ik} p_k \qquad (6\text{-}41)$$

holds with a constant sensitivity γ_{ik} over the whole range of p_k (Figure 6-32). For H_2, for example, with a detection limit of about 40 ppm, this range extends to values of up to 7000 ppm. As an exception, also shown in Figure 6-32, the sensitivity towards butane, $\gamma_{C_4H_{10}}$, is a weak function only of its partial pressure, $p_{C_4H_{10}}$. This is a result of a low reaction order, $n < 1$.

Riegel and Härdtl [51] operated their gas sensors at different temperatures to detect single gas components in air by the specific signal pattern as shown in Figure 6-32. If the catalytic oxidation rates of the individual components in binary mixtures in air are superimposed on each other without perturbation, a simple linear approach can be applied successfully to perform quantitative measurements of both components in these mixtures. In Figure 6-33a the calculated overall heat generation of the parallel and undisturbed oxidation of two combustible gases at the same catalyst is shown. Figure 6-33b shows the measured data for pure butane, hydrogen, and a mixture of the two. Using two microcalorimetic sensors operated at 220 and 320 °C, the maximum error for quantitative analysis of a butane-H_2 mixture in a concentration range between 0 and 6000 ppm for each gas was found to be 30%. Using an additional sensor at 280 °C, the error caused by the nonlinear response of these sensors to butane (see Figure 6-32) was reduced to 20%. From the signals from three sensors a quadratic expression for the response to butane was calculated.

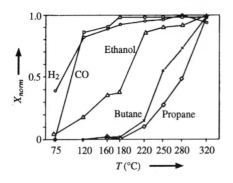

Figure 6-32.
Different patterns obtained for different gases at pellistors operating at temperatures T between 75 and 320 °C. For normalized signals $x_{norm} \approx 1$ diffusion-controlled oxidation (region III in Figure 6-30) occurs [51].

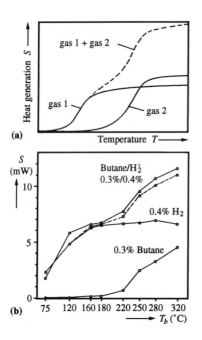

Figure 6-33.
(a) Expected overall heat generations of the parallel and undisturbed oxidation reactions of two combustible gases with the same solid-state catalyst [51].
(b) Measured overall heat generation of the simultaneous combustion of 0.3% of butane and 0.4% of hydrogen in air at a porous platinum catalyst. The dashed line is the calculated superposition of the individual reactions [51].

6.3.2.2 MOSFET Gas Sensor Arrays

A gas sensor array with three pairs of Pd-gate MOSFETs and Pt-gate MOSFETs has been studied during exposure to gas mixtures containing hydrogen, ammonia, ethylene, and ethanol [53]. Each pair is operated at a different temperature, 150, 200 and 250 °C. The signals of the six sensors are analyzed by three different PLS (partial least-squares) models, linear (ℓPLS), modified linear (mPLS) and a nonlinear (nPLS) method. For hydrogen a modified linear PLS model gives the best prediction, whereas for ammonia a linear PLS model gives the best results. The results indicate that for determining different components in a gas mixture, different pattern routines should be employed. The methods chosen in this study could be applied to the determination of individual components in a gas mixture. A very important problem is to consider matrix effects caused by the interaction of the gases from the mixture in the gas phase or at the hot catalytic metal surface of the MOSFET. For example, chemisorbed nucleophilic ammonia may react with partly oxidized ethylene and/or ethanol, thus changing the selectivity and sensitivity of the sensor array.

6.3.2.3 SnO₂ Sensor Arrays

With the aim of describing the basic principle of multi-component analysis with SnO_2 sensor elements, we describe the determination of two partial pressures with a sensor array consisting of three different chemically modified SnO_2-based conductivity sensors [5, 54]. In this characteristic example we determine the partial pressures p_{CO} and p_{CH_4} in air by starting

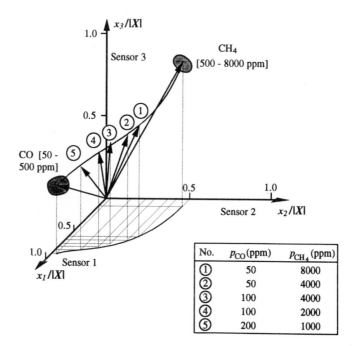

No.	p_{CO} (ppm)	p_{CH_4} (ppm)
①	50	8000
②	50	4000
③	100	4000
④	100	2000
⑤	200	1000

Figure 6-34. Schematic presentation of results from the multi-component analysis to determine CO and CH₄ in humid air (50% r.h.) with three different SnO₂-based sensors. Each combination of partial pressures p_{CO} and p_{CH_4} leads to a normalized vector X (x_1, x_2, x_3) with a specific orientation. Variations of pure CO and CH₄ in the ranges 50–100 and 1000–8000 ppm, respectively, lead to changes in the vectors indicated by the circles. Further explanations are given in the text [54].

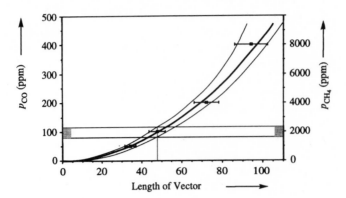

Figure 6-35. Determination of the absolute values of the partial pressures from the lengths of the vectors $X(x_1, x_2, x_3)$ for a specific ratio $p_{CH_4}/p_{CO} = 20$ (point 4, Figure 6-34) [54].

with the responses x_1, x_2, and x_3 of three different sensors. Here, the response x_i is defined as the relative conductance $G_i/G_{i,0}$ of the sensor i where $G_{i,0}$ is the conductance in air. The vectors $X(x_1, x_2, x_3)$, which depend on p_{CO} and p_{CH_4}, are divided by their total lengths $\sqrt{x_1^2 + x_2^2 + x_3^2}$ to obtain the plot of normalized vectors $X_{norm}(x_1, x_2, x_3)$ (Figure 6-34). Each vector $X_{norm}(x_1, x_2, x_3)$ corresponds to a specific ratio of partial pressures p_{CO}/p_{CH_4} from which the absolute values p_{CO} and p_{CH_4} can be deduced in the ranges 0–500 ppm for p_{CO} and 0–8000 ppm for p_{CH_4} (Figure 6-35) by taking into consideration the total length of X.

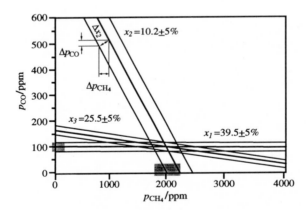

Figure 6-36. Graphical solution of a multi-component analysis for the particular example of the $CO-CH_4$ mixture corresponding to point 4 in Figure 6-34. The measuring signals $x_i = G_i/G_{0,i}$ indicated are identical for each thick line. They were obtained by measuring conductances at different combinations of CO and CH$_4$ partial pressures with the different sensors i and make possible the unequivocal determination of individual gas concentrations in mixtures. The uncertainty in the determination of partial pressures by assuming a variation of ±5% with respect to x_i (thin lines) characterizes the error in determining p_{CO} and p_{CH_4} by this analysis. Further explanations are given in the text [54].

To illustrate the formal concept of multi-component analysis applied to this simple example, we finally calculate the parameters of analytical chemistry as introduced in Section 6.2.3. We apply the definitions given there to describe the pattern recognition procedure with three different SnO$_2$-based sensors (samples 1, 2, and 3).

For the example shown in Figure 6-36, we determine the accuracy in the quantitative analysis of CO and CH$_4$ partial pressures by measuring their conductances after CO and CH$_4$ exposure. The partial specificities Ψ_i can be calculated from the partial sensitivities γ_{ik} according to Equation (6-13) by taking into account only the sensor responses x_1 and x_2. The minimum of Ψ_i, ie, the selectivity of the total analysis procedure, is determined to be -0.6. The CO specificity is relatively high (160 $\leq$ Ψ_{CO} $\leq$ 27) if compared with the CH$_4$ specificity (-0.99 $\leq$ Ψ_{CH_4} $\leq$ -0.96). The latter data characterize statistical errors in the determination of CO and CH$_4$ partial pressures which are observed in statistical variations of the measuring parameters, x_1, x_2, and x_3. Assuming a variation of 5% with respect to x_1, x_2, and x_3, we find the determination of the CO partial pressure to be far more accurate in comparison with the determination of the CH$_4$ partial pressure. The accuracies are influenced directly by the partial sensitivities γ_{ik} of the sensors with quantitative results indicated graphically in Figure 6-36. The sensor response x_3 appears to be redundant and can thus be neglected in the calculations of specificities and selectivities from the calibration matrix.

6.3.2.4 *SnO$_2$ Sensor Parameter Arrays*

In this example, different sensor signals of the same SnO$_2$-based sensor are used for pattern recognition. These parameters are the DC conductance G, change of work function $\Delta\Phi$, and catalytic activity r, ie, reaction rate of CO and CH$_4$ oxidation [5, 54, 55].

Figure 6-37 shows the schematic setup for measuring the sensor properties during exposure of reducing gases in air. In the following, we determine relative changes of different sensor properties x_i of SnO$_2$-based sensors (Figaro type 812, operating temperature 630 K).

a. For relative conductance changes G/G_0, Equation (6-42) was found to hold to a good approximation:

$$x_1 = \frac{G}{G_0} = \left(\frac{p_i}{p_{i,0}}\right)^{n_{i,G}} \tag{6-42}$$

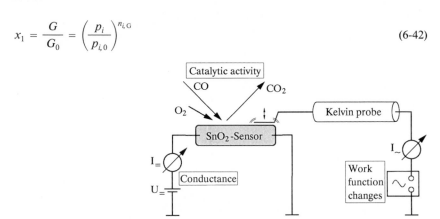

Figure 6-37. Schematic setup for measurements of conductances, changes of work functions, and catalytic activities (reaction rates) of SnO$_2$-based sensors [5].

where $p_{i,0}$ is a given constant gas partial pressure (reference) and the exponent $n_{i,G}$ is a gas-specific characteristic parameter.

b. In polycrystalline SnO_2 sensors, the changes in conductivity during exposure of gases are coupled with variations of the barrier height, $e\Delta V_s$, at the grain boundaries. This effect and additional variations in dipole moments lead to strong changes of the work function $\Delta\Phi$ during gas exposure [56]. The dependence of the relative changes $\Delta\Phi/\Delta\Phi_0$ on the partial pressures of reducing gases may to a good approximation be described by an equation similiar to Equation (6-42):

$$x_2 = 10^{\frac{\Delta\Phi}{\Delta\Phi_0}} = \left(\frac{p_i}{p_{i,0}}\right)^{n_{i,\Delta\Phi}}. \tag{6-43}$$

Here, the gas specific characteristic sensor parameter is the exponent $n_{i,\Delta\Phi}$.

c. The catalytic reactions of reducing gases, such as CO, occuring at the SnO_2 surfaces lead to detectable concentrations of oxidation products in the gas phase. For the particular example of CO in air, the catalytic rates of CO_2 may to a good approximation be described by

$$x_3 = r_{CO_2} = k_0 \exp\left(-\Delta E/kT\right)(p_{CO})^{n_{CO,cat}}(p_{H_2O})^{n_{H_2O,cat}} \sim p_{CO_2} \tag{6-44}$$

where the exponents $n_{CO,cat}$ and $n_{H_2O,cat}$ are reaction orders referring to CO and H_2O, the absolute values of $n_{CO,cat}$ and $n_{H_2O,cat}$ being 1 and 0, respectively, k_0 is a pre-exponential constant (entropy factor) and ΔE is the activation energy of CO oxidation [56–58].

Following a vectorial representation, the different sensor properties relative conductance, relative work function, and catalytic activity are used with typical results as shown in Figure 6-38. As indicated by the distinct normalized signals, a determination of CO and H_2O partial pressures in air may thus be performed with only one SnO_2-based sensor. In this particular case, measuring of the third parameter, r_{CO_2}, ie, the catalytic activity, is not necessary but does enhance the accuracy.

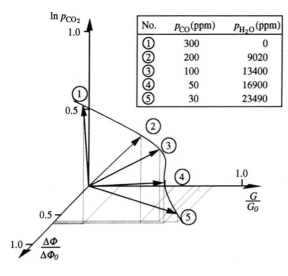

No.	p_{CO} (ppm)	p_{H_2O} (ppm)
①	300	0
②	200	9020
③	100	13400
④	50	16900
⑤	30	23490

Figure 6-38.
Vectorial representation of normalized signals G/G_0, $\Delta\Phi/\Delta\Phi_0$, and catalytic activities p_{CO_2} obtained by the same sensor for different partial pressures of CO and H_2O [5].

6.3.2.5   *Gas Analyses by Partial Model Building (PMB)*

Preclassification of classes of possible gas mixtures was a method introduced into practical use by Horner and Hierold as Partial Model Building [59]. The idea is to evaluate the responses from non-selective sensors with pattern recognition. Most pattern recognition algorithms, such as principal component model building (PCMB) [60] or partial least-squares (PLS) methods [61], use linear regression setups to find estimators for the parameter set describing the sensor behavior. These estimators can be used to predict several dependent variables, eg, the concentrations of n different gases p_i, from independent variables, such as m sensor signals x_i ($n \leq m$), by solving a system of linear equations. As most chemosensors have non-linear characteristics, such a linear model can only be valid within a small concentration range ΔP. To overcome this problem, a non-linear pattern recognition model has been developed, which combines a non-linear transformation of the data set with a linear regression model [62].

This "transformed least-squares (TLS) model" has been applied to a calibration set of measurements with sintered SnO_2-based gas sensors and shows good results for the prediction of two-component gas mixtures over a wide concentration range.

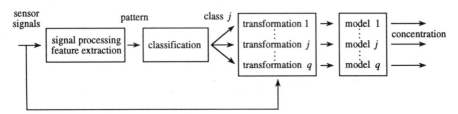

Figure 6-39. Flow chart of a partial model-building algorithm [6].

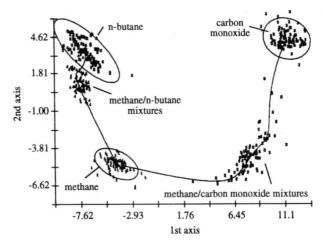

Figure 6-40. Distribution of the feature vectors on the discriminant plane (five classes; sample size for each class: $n = 100$) [6].

Introducing several classes of binary gas mixtures and single gases, the predicted error of an overall model increases owing to certain restrictions on the transformation functions. To overcome this problem, a partial model building procedure is used.

Figure 6-39 shows a flow chart of the algorithm. For the classification of the different gas mixtures and single gases, the pattern vectors of the five sub-sets of the calibration set were projected onto a discriminant plane. The position of the discriminant plane in feature space can be calculated by a generalized Karhunen-Loeve expansion [63]. This feature selection technique transforms the original pattern vectors into a new coordinate system, where all the discriminatory information is compressed in the first few axes. For q classes all the information can be compressed in a $(q - 1)$-dimensional hyperplane. Figure 6-40 shows the distribution of the five classes of the calibration set on the first two axes of such a hyperplane. The pattern vectors representing single gases can be described by ellipses, whereas the pattern vectors representing gas mixtures are distributed between those ellipses.

6.3.2.6 Response Functions of Catalyst/Electrochemical Sensor Systems

With the system described in Section 6.3.1.4, an operation and evaluation method was developed for the quantitative analysis of gas mixtures. One electrochemical sensor and one catalytic platinum filament operated at different temperatures were used. The sensor is exposed to a step-functional change of the gas concentration realized by switching stationary gas streams with solenoid valves. For the evaluation, the response characteristics of the sensor at a given filament temperature are used. An electrochemical sensor produces a current $i(t)$ (measured signal x_i') as a response to the change in the reaction rate (analytical information). The sensor current i and reaction rate r are correlated linearly as long as the system is in an equilibrium state. The dynamic properties of the sensor response, eg, to a step-functional change in the gas concentration, can be described by the response of an equivalent circuit for the electrochemical cell, represented by a capacitor C and a resistor R with time constant $\tau_S = RC$ as an exponential function:

$$i(t) = mC_G(1 - e^{-t/\tau_S}) . \tag{6-45}$$

The apparent time constant τ_S is specific for an interaction of a specific gas with the sensor surface. Different reaction mechanisms of different gases result in typical response times $\tau_{S,i}$ on the sensor. Thus, the time dependence of the sensor current $i(t)$ caused by a pure gas in air can be described formally by two separate contributions: a sum of exponential functions $R_G(t)$, which are time dependent but concentration independent, and a time independent but gas concentration dependent term $y_G C_G$. The intrinsic cell sensitivity m given in Equation (6-45) is a constant of the apparatus.

$$i_G(t) = mR_G(t)y_G C_G \tag{6-46}$$

with

$$R_G(t) = \sum_i A_i(1 - \exp(-(t - t_0)/\tau_i)) \tag{6-47}$$

where A_i is the part of the amplitude which corresponds to the time constant τ_i in the exponential function and t_0 is the time at which the sensor starts to respond. The sum of the A_i is calibrated to 1.0. In general, for a good approximation of the response function more than one function is required. As an example, for perchloroethylene four or five functions are required when the reaction products from an catalytic filament at 900 °C are analyzed.

For CO, the sensitivity $y_{CO} C_{CO}$ is a linear function of the concentration. For the CO sensor the cell sensitivity m is calibrated with a value of $y_{CO} = 1$. The calibrations for the other gases are related to the CO value by y_G values different from 1 or by functions of the concentration $y_G (C_G)$.

For the treatment of the cross-sensitivity effects in pairs of gases, Equation (6-43) has to be written as vector equation with y as matrix:

$$i_{total}(t) = m\mathbf{R}(t)(\mathbf{Y}C) . \tag{6-48}$$

The diagonal elements of the matrix $\mathbf{Y}$ are the calibration factors for the pure gases $y_G (C_G)$, the nondiagonal elements are the cross-sensitivities: $y_{i,j} = y_{i,j}(C_i, C_j)$. This matrix $\mathbf{Y}$ has to be calibrated for each pair of possible gases at different ratios and different concentrations. This method is suitable for mixtures of two gases. For three or more gases terms of higher order have to be taken into account: $y_{i,j,...,n} = y_{i,j,...,n}(C_i, C_j, ..., C_n)$. For practical use in an instrument, calibration of more than three gases at different ratios and concentrations should be avoided.

For an analysis of a mixture of different gases the measured sensor response $i_{total}(t)$ can be fitted by a linear combination of the gas-specific but concentration-independent functions $R_i(t)$ with a resulting $\mathbf{B}$ vector:

$$i_{total}(t) = m(R_1(t) B_1 + R_2(t) B_2 + ... + R_n(t) B_n) \tag{6-49}$$

with

$$B_i = B_i(C_1, C_2, ..., C_n) . \tag{6-50}$$

The subscript n is the number of gases for which the system is calibrated. The real concentrations are calculated from these factors by applying the inverse matrix $\mathbf{Y}^{-1}$ to $\mathbf{B}$:

$$C = \mathbf{Y}^{-1} \mathbf{B} . \tag{6-51}$$

Up to now this method has been discussed only for the sensor without consideration of the catalytic filament. The filament temperature can be used as further dimension and the current $i(t)$ then becomes a vector with elements $i_{T_k}(t)$ with k different filament temperatures T. The catalytic effect of the filament transforms electrochemically inactive compounds, such as perchloroethylene on a platinum electrode sensor, to detectable reaction products. This is shown in Figure 6-41. Whereas CO and formaldehyde give strong responses with a cold filament, for benzene a maximum sensor response is observed at a filament temperature of 700 °C. Perchloroethylene shows a typical response above 600 °C which is characterized by superimposed oxidizing and reducing reactions. These reactions have different time constants which gives the typical response shown in Figure 6-41 for perchloroethylene. Figure 6-42 shows an example of the quantitative analysis of a mixture of CO and formaldehyde with a Nafion-

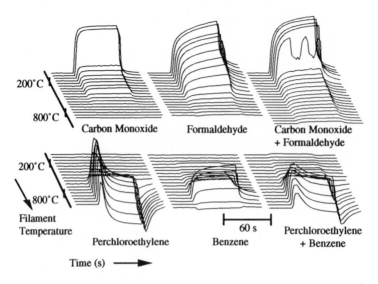

Figure 6-41. Typical response characteristic of the electrochemical CO sensor to four different pure gases and two two-component mixtures. The concentrations are carbon monoxide 100, formaldehyde 200, carbon monoxide/formaldehyde 200/50, perchloroethylene 660, benzene 500, and perchloroethylene/benzene 500/500 ppm. The filament temperature is the parameter of the curves. The monitoring time for one curve was 120 s. Exposure to the test gas occurred between 30 and 90 s after the start.

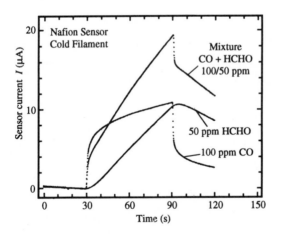

Figure 6-42.
Typical response characteristics of the Nafion sensor for carbon monoxide, formaldehyde, and a mixture. All curves are monitored with a cold filament. Each curve consists of 1024 data points. The gas exposure occurred between 30 and 90 s [3].

based electrochemical sensor. Evaluating the response functions of the characteristics to the pure components and the mixture, it was found that the CO signal was decreased to 62% in the presence of formaldehyde compared with the pure CO sample whereas the formaldehyde signal was increased to 112% in the presence of CO. This is considered in the model by non-diagonal matrix elements in the **Y** matrix, which are different from zero. With a cold filament the CO and formaldehyde responses are not influenced by benzene or perchloroethylene. Therefore, for an analysis of a four-component mixture with these gases the problem can be divided into two temperature ranges, as shown in Figure 6-43. In the high-temperature range the response characteristics are also influenced by CO or formaldehyde but with the knowledge from the low-temperature measurements of these components they can be taken into account. Figure 6-44 shows the typical response of a commercial electrochemical sensor to pure perchloroethylene, a mixture of perchloroethylene and benzene and a four-component mixture. The influence of humidity can be neglected because of the aqueous electrolyte of the sensor.

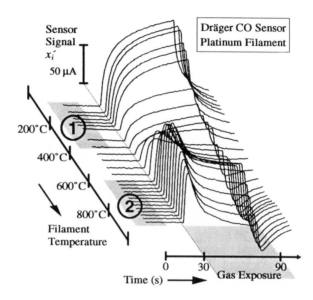

Figure 6-43.
Time dependence of the sensor signal with filament temperature as parameter. The test gas was a mixture of 500 ppm of perchloroethylene, 500 ppm of benzene, 75 ppm of formaldehyde, and 100 ppm of carbon monoxide, humidified to 30% r.h. For details, see text.

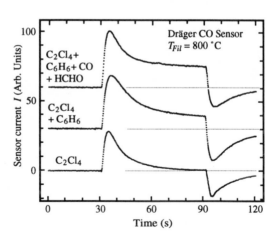

Figure 6-44.
Response of an electrochemical CO sensor to pure perchloroethylene (330 ppm), a mixture of perchloroethylene (500 ppm) and benzene (500 ppm), and a mixture of CO (100 ppm), formaldehyde (50 ppm), benzene (500 ppm), and perchloroethylene (500 ppm). All curves were taken with a filament temperature of 800 °C [3].

This method is very versatile because two-dimensional information is generated by a suitable data preprocessing method. Although, the separation of the sensor response into a part that contains the time constants and a part that contains the concentration dependence may be not valid in a strict sense, it works for practical applications and can be used with low expense in the hardware. Cross-sensitivity effects are considered in a two-by-two gas interaction matrix element. Although this empirical approach leads to specific parameters for each type of sensor, preliminary results indicate that the cross-sensitivity matrix is unique for the same type of sensor, and individual sensors of the same type require only calibrations for pure gases to be performed.

6.4 Outlook

If the sensor properties can be treated as state functions in the thermodynamic sense, subsequent pattern recognition techniques are extremely powerful for the quantitative determination of components in mixtures. Under these conditions, the concept of multi-component analytical chemistry is useful in describing quantitatively the accuracy of the determination of partial pressures or concentrations in pattern recognition techniques. This concept is simple for linear response functions in which independent matrix components of the calibration matrix are observed. In more general cases, local linearization relating concentration variations with sensor responses is necessary and can be used to generalize the simple concept.

Most of the experimental work done so far has dealt with the identification of one component only. A few more recent approaches aimed at the quantitative determination of different concentrations in mixtures have been reported.

The examples presented in this chapter indicate that it is useful to define and apply the terms sensitivity, selectivity, and partial specificity to chemical sensor arrays in exactly the same way as is already done in analytical chemistry.

Future developments in this rapidly growing field are aimed at designing reliable long-term stable sensors with controlled interface properties, at new modulation techniques and different types of sensor arrays, and at developing more sophisticated pattern recognition techniques based on theoretical concepts of information theory, ie, cluster analysis, classification methods, and factor analysis. In addition to these standard methods of chemometrics, new concepts are becoming increasingly important in chemical sensing: neural networks are applied for pattern recognition or for interpreting low-level information. Another concept is the fuzzy logic approach, which may be used for sensor systems with different types of measurement principles [64].

6.5 References

[1] Zaromb, S., Stetter, J. R., "Theoretical Basis for Identification and Measurements of Air Contaminants Using an Array of Sensors Having Partly Overlapping Selectivities", *Sens. Actuators* **6** (1984) 225–243.
[2] Stetter, J., Findlay, M., Jr., Maclay, G. J., Zhang, J., Vaihinger, S., Göpel, W., "Sensor Array and Catalytic Filament for Chemical Analysis of Vapors and Mixtures" in: *Proceedings of 5th International Conference on Solid-State Sensors and Actuators (Transducers '89) and Eurosensors III, Montreux, Switzerland, 1989; Lausanne, CH: COMST; and Sens. Actuators* **B,1** (1990) 43–47.

[3] Vaihinger, S., Göpel. W., Stetter, J. R., "Detection of Halogenated and Other Hydrocarbons in Air: Response Functions of Catalysts/Electrochemical Sensor Systems", presented at Eurosensors IV, Karlsruhe, FRG (1990), and *Sens. Actuators* **B,4** (1991) 337–343.

[4] Stetter, J. R., Jurs, P. C., Rose, S. L., "Detection of Hazardous Gases and Vapors: Pattern Recognition Analysis of Data from an Electrochemical Sensor Array", *Anal. Chem.* **58** (1986) 860–866.

[5] Weimar, U., Kowalkowski, R., Schierbaum, K. D., Göpel, W., "Pattern Recognition Methods for Gas Mixture Analysis: Application to Sensor Arrays Based upon SnO_2" in: *Proceedings of 5th International Conference on Solid-State Sensors and Actuators (Transducers '89) and Eurosensors III, Montreux, Switzerland, 1989; Lausanne, CH: COMST;* and *Sens. Actuators* **B1** (1990) 93–96.

[6] Horner, G., Hierold, C., "Gas Analysis by Partial Model Building", *Sens. Actuators* **B,2** (1990) 173–184.

[7] Ikegami, A., Olshi, T., Kaneyasu, M., "Discrimination of Chemical Compounds and Functional Groups by Pattern Recognition", *Hybrid Circuits* **16** No. 5 (1988) 19–22.

[8] Shurmer, H. V., Gardener, J. W., "Intelligent Vapor Discrimination Using a Composite 12-Element Sensor Array", *Sens. Actuators* **B1** (1990) 256–260.

[9] Müller, R., Horner, G., "Chemosensors with Pattern Recognition", *Siemens Forsch.- Entwickl.-Ber.* **15** (1986) 95–100.

[10] Otawaga, T., Stetter, J. R., "A Chemical Concentration Modulation Sensor for Selective Detection of Airborne Chemicals", *Sens. Actuators* **11** (1987) 251–264.

[11] Maclay, G. J., Stetter, J. R., Christesen, S., "Use of Time-dependent Chemical Sensor Signals for Selective Identification", *Sens. Actuators* **20** (1989) 277–285.

[12] Kaiser, H., "Zur Definition von Selektivität, Spezifität and Empfindlichkeit von Analyseverfahren", *Fresenius' Z. Anal. Chem.* **260** (1972) 252–260.

[13] Carey, W. P., Beebe, K. R., Sanchez, E., Geladi, P., Kowalski, B. R., "Chemometric Analysis of Multisensor Arrays", *Sens. Actuators* **9** (1986) 223–234.

[14] Wienke, D., Danzer, K., "Evaluation of Pattern Recognition Methods by Criteria based on Information Theory and Euclidean Geometry", *Anal. Chim. Acta* **184** (1986) 107–116.

[15] Varmuza, K., *Pattern Recognition in Chemistry;* Heidelberg: Springer, 1980.

[16] Jurs, P. C., Isenhour, T. L., *Chemical Applications of Pattern Recognition;* New York: Wiley-Interscience 1975.

[17] Kowalski, B. R., Bender, C. F., "The K-Nearest Neighbor Classification Rule (Pattern Recognition) Applied to Nuclear Magnetic Resonance Spectral Interpretation", *Anal. Chem.* **44** (1972) 1405–1411.

[18] Danzer, K., Singer, R., Mäurer, F., Flórián, K., Zimmer, K., "Mehrdimensionale Varianz- und Diskriminanzanalyse spektrographischer Daten von Glasperlenfunden", Fresenius' *Z. Anal. Chem.* **318** (1984) 517–521.

[19] Kowalski, B. R., "Measurement Analysis by Pattern Recognition", *Anal. Chem.* **47** (1975) 1152A–1162A.

[20] Belchamber, R. M., Betteridge, D., Chow, Y. T., Sly, T. J., Wade, A. P., "The Application of Computers in Chemometrics and Analytical Chemistry", *Anal. Chim. Acta* **150** (1983) 115–127.

[21] Kowalski, B. R., "Chemometrics", *Anal. Chem.* **52** (1980) 112R–122R.

[22] Frank, E., Kowalski, B. R., "Chemometrics", *Anal. Chem.* **54** (1982) 232R–243R.

[23] Varmuza, K., "Chemometrie" in: *Computer in der Chemie,* Ziegler, E., (ed.); Heidelberg: Springer, 1984, pp 131–153.

[24] Beebe, K. R., Kowalski, B. R., "An Introduction to Multivariate Calibration and Analysis", *Anal. Chem.* **59** (1987) 1007A–1017A.

[25] Wold, S., et al., "Multivariate Data Analysis in Chemistry" in: *Chemometrics: Mathematics and Statistics in Chemistry;* Dordrecht: Reidel, 1984, pp. 17–95.

[26] Osten, D. W., Kowalski, B. R., "Background Detection and Correction in Multicomponent Analysis", *Anal. Chem.* **57** (1985) 908–917.

[27] Draper, N., Smith, H., *Applied Regression Analysis;* New York: Wiley, 1981, p. 327.

[28] Wold, H., *Festschrift Jerzy Neyman;* New York: Wiley, 1966, pp. 411–444.

[29] Jöreskog, K., Wold, H. (eds.) *Systems under Indirect Observation: Causality-Structure-Prediction, Part II;* Amsterdam: North-Holland, 1982.

[30] Wold, S., et al., *Food Research and Data Analysis;* Barking: Applied Science, 1983, pp. 147–188.

[31] Wold, S., Martens, H., Wold, H., *Matrix Pencils, Lecture Notes in Mathematics;* Heidelberg: Springer, 1983, pp. 286–293.

[32] Geladi, P., Kowalski, B. R., "Partial Least Squares Regression (PLS): A Tutorial", *Anal. Chim. Acta* **185** (1986) 1-17.

[33] Ho, C.-N., Christian, G. D., Davidson, E. R., "Application of the Method of Rank Annihilation to Quantitative Analyses of Multicomponent Fluorescence Data from the Video Fluorimeter", *Anal. Chem.* **50** (1978) 1108-1113.

[34] Ho, C.-N., Christian, G. D., Davidson, E. R., "Application of the Method of Rank Annihilation to Fluorescent Multicomponent Mixtures of Polynuclear Aromatic Hydrocarbons", *Anal. Chem.* **52** (1980) 1071-1079.

[35] Ho, C.-N., Christian, G. D., Davidson, E. R., "Simultaneous Multicomponent Rank Annihilation and Applications to Multicomponent Fluorescent Data Acquired by the Video Fluorimeter", *Anal. Chem.* **53** (1981) 92-98.

[36] McCue, M., Malinowski, E. R., Rank Annihilation Factor Analysis of Unresolved LC Peaks", *J. Chromatogr. Sci.* **21** (1983) 229-234.

[37] Lorber, A., "Quantifying Chemical Composition from Two-Dimensional Data Arrays", *Anal. Chim. Acta* **164** (1984) 293-297.

[38] Sánchez, E., Kowalski, B. R., Generalized Rank Annihilation Factor Analysis", *Anal. Chem.* **58** (1986) 496-499.

[39] Pau, L. F., Johansen, F. S., "Neural Network Signal Understanding for Instrumentation, *IEEE Trans. on Instrum. Meas.* **39** (1990) 558-564.

[40] Wolf, T., "Neuronen im Computer", *microcomputer,* April (1990) 92-108.

[41] Rummelhart, D. E., Hinton, G. E., Williams, R. J., "Learning Internal Representation by Error Propagation" in: *Parallel Distributed Processing: Exploration in the Microstructure of Cognition, Vol. 1: Foundations,* Rummelhart, D. E., and McClelland, J. L. (eds.); MIT Press, 1986.

[42] Dubbe, A., Wiemhöfer, H. D., Schierbaum, K. D., Göpel, W., "Kinetics of Oxygen Interaction with Pt/CeO₂ Sensors: Application of a New Pressure Modulation Spectroscopy", *Sens. Actuators,* **B, 4** (1991) 23-28.

[43] Ege, G., Kohler, H., Göpel, W., "Selectivity Enhancement of Amperometric Gas Detection Cells: Linear Potential Sweep Voltammetry (LPSV) After Fixed Potential Operation", in Proceedings of Eurosensors IV, Karlsruhe, FRG, 1990, and *Sens. Actuators* **B, 4** (1991) 519-524.

[44] Moulton, D. G., *Olfactory and Taste;* Oxford: Pergamon, 1963, pp. 71-84.

[45] Ikegami, A., Kaneyasu, M., "Olfactory Detection Using Integrated Sensor", in: *Proceedings of 3rd International Conference on Solid-State Sens. and Actuators (Transducers '85), Philadelphia, PA,* IEEE Catalog No. 85 CH 2127-9, 1985, pp. 136-139.

[46] Müller, R., and Lange, E., "Multidimensional Sens. for Gas Analysis", *Sens. Actuators* **9** (1986) 39-48.

[47] Nakamoto, T., Fukunishi, K., Moriizumi, T., "Identification Capability of Odor Sensing Using Quartz Resonator Array and neural Network Pattern Recognition", *Sens. Actuators* **B1** (1989) 473-476.

[48] Ema, K., Yokoyama, M., Nakamoto, T., Moriizumi, T., "Odour-Sensing System Using a Quartz-Resonator Sensor Array and Neural-Network Pattern Recognition", *Sens. Actuators* **18** (1989) 291-296.

[49] Stull, J. O., Stetter, J. R., Penrose, W. R., Zaromb, S., "Development and Evaluation of a Portable Personal Monitor for Detection and Identification of Hazardous Chemical Vapors" in: *Proceedings of Hazardous Materials Conference, Long Beach, CA, USA, December 3-5 , 1985;* Wheaton, IL: Tower Conference Mgt. Co.

[50] Penrose, W. R., Stetter, J. R., Findlay, M. W., Buttner, W. J., Cao, Z., "Arrays of Sensors and Microsensors for Field Screening of Unknown Chemical Wastes" *in: Proceedings of 2nd International Symposium on Screening Methods for Hazardous Wastes and Toxic Chemicals, Las Vegas, NV, February 12-14, 1991.*

[51] Riegel, J., Härdtl, K. H., "Analysis of Combustible Gases in Air with Calorimetric Gas Sensor Based on Semiconducting BaTiO₃ Ceramics" in: *Proceedings of 5th International Conference on Solid-State Sensors and Actuators (Transducers '89) and Eurosensors III, Montreux, Switzerland, 1989; Lausanne, CH: COMST;* and *Sens. Actuators* **B1** (1990) 54-57.

[52] Riegel, J., "Untersuchungen zur Gasanalyse mit Mikrokalorimetern auf der Basis keramischer PTC-Widerstände", *Fortschr. Ber. VDI* Reihe 8, No. 194, Düsseldorf, 1989.

[53] Sundgren, H., Lundström, I., Winquist, F., Lukkari, I., Carlsson, R., Wold, S., "Evaluation of a Multiple Gas Mixture with a Simple MOSFET Gas Sensor Array and Pattern Recognition", *Sens. Actuactors* **B2** (1990) 115–123.

[54] Schierbaum, K. D., Weimar, U., Göpel, W., "Multicomponent Gas Analysis: an Analytical Chemistry Approach Applied to Modified SnO_2-Sensors", *Sens. Actuators* **B2** (1990) 71–78.

[55] Schierbaum, K. D., Weimar, U., Kowalkowski, R., Göpel, W., "Conductivity, Work Function, and Catalytic Activity of SnO_2 Based Gas Sensors", *Sens. Actuators* in press.

[56] Kowalkowski, R., Schwarz, E., Göpel, W., "Gas-Sensing and Catalytic Activity of Metal Oxide Sensors: Conductivity, Work Function and Mass Spectrometric Studies" in Aucouturier, J.-L., et al. (eds): *Proceedings of 2nd International Meeting on Chemical Sensors, Bordeaux, France 1986,* p. 191.

[57] Clifford, P. K., "Microcomputational Selectivity Enhancement of Semiconductor Gas Sensors" in Seiyama, T., et. al. (eds): *Proceedings of 1st International Meeting on Chemical Sensors, Fukuoka, Japan, 1983; Tokyo: Kodansha,* pp. 153–158.

[58] Clifford, P. K., Tuma, D. T., "Characteristics of Semiconductor Gas Sensors: I. Steady State Gas Response", *Sens. Actuators* **3** (1982/83) 233–254.

[59] Horner, G., Hierold, C., "Gas Analysis by Partial Model Building", *Sens. Actuators* **B2** (1990) 173–184.

[60] Gerhard, A., "Modellierung eines Verknüpfungsarrays zur Erkennung von Gasgemischen", *Report: Sonderforschungseinheit "Sensorik mit Mustererkennung",* Feb. 3, 1987, Siemens AG Munich.

[61] Wold, S., Martens, H., Wold, H., *The Multivariate Calibration Problem in Chemistry Solved by the PLS Method, Lecture Notes in Mathematics;* Heidelberg: Springer, 1983, pp. 286–293.

[62] Hierold, C., Müller, R., "Quantitative Analysis of Gas Mixtures with Non-Selective Gas Sensors, *Sens. Actuators* **17** (1989) 587–592.

[63] Kittler, J., Young, P. C., "A New Approach to Feature Selection Based on Karhunen-Loeve Expansion, *Pattern Recognition,* **5** (1973) 335–352.

[64] Altrock, C. V., "Über den Daumen gepeilt", *Computer & Technik* **3** (1991) 78–92.

7 Liquid Electrolyte Sensors: Potentiometry, Amperometry, and Conductometry

Friedrich Oehme, Görwihl, FRG

Contents

7.1 Potentiometry

7.1.1 Introduction

Potentiometry is one of the most widely used electroanalytical methods and its history dates back about 100 years. For a long time, potentiometric measurements were restricted to the determination of pH values and, to a lesser extent, to oxidation-reduction potentials. End-point determinations in volumetric analysis, however, soon became of interest. This situation changed rapidly when, in 1965, ion-selective electrodes became available and, as a consequence, potentiometry experienced an unexpected renaissance.

All measuring devices in potentiometry are called electrodes. The term sensor is used for the active part of the electrode only (see Chapter 1).

7.1.2 Fundamentals

The term potentiometry is related to the measuring technique used, ie, the measurement of potentials (voltages, electromotive forces (e. m. f. s)). The voltage source is a form of galvanic cell which consists of a measuring electrode and a reference electrode.

Since many measuring electrodes have high internal resistances (up to $10^9\ \Omega$) and are also liable to polarize, potential measurements have to be carried out under zero-current conditions. This calls for input stages in the amplifier with impedances of the order of $> 10^{12}\ \Omega$ and off-set currents of $> 10^{-12}$ A. The design of circuits with these specifications is no longer a problem. In most cases the preamplifiers utilize field effect transistors which also act as impedance transformers.

The main task of a potentiometric measuring electrode is to respond quickly and with high selectivity to changes in the concentration of the sensed ion in a sample.

7.1.2.1 The Nernst Equation and Its Modifications

The relationship between the potential E of a measuring electrode and the activity a_i of a ionic species i to which it responds is governed by the Nernst equation:

$$E = E_0 \pm (RT/nF)\ln a_i \tag{7-1}$$

where E_0 is the standard potential for $a_i = 1$ mol/L, R is the gas constant ($8.3144\ \mathrm{J\,K^{-1}mol^{-1}}$), F is the Faraday constant ($9.6485 \cdot 10^4\ \mathrm{C\,mol^{-1}}$), T is temperature [K], n is the number of elementary charges on the ion i to be sensed, the $+$ sign represents cations and the $-$ signs anions.

In practice, the Napierian logarithm in Equation (7-1) will be replaced by the Briggs' (decadic) logarithm:

$$E = E_0 \pm 2.303\,(RT/nF)\log a_i \ . \tag{7-2}$$

Apart from the number of elementary charges n, the term in parentheses in Equation (7-2) depends only on the temperature T. The term 2.303 (RT/F) is called the slope factor, slope, or (in German standards) Nernst voltage, and expresses the theoretical change of the potential E of a measuring electrode for a tenfold change in the activity a_i of the sensed ion.

Table 7-1 lists some values of the slope factor for temperatures between 0 and 100 °C.

Table 7-1. Values of the slope factor 2.303 (RT/F) at a number of selected temperatures.

Temperature (°C)	Slope factor (mV)
0	54.20
5	55.19
10	56.18
15	57.17
20	58.17
25	59.16
30	60.15
40	62.13
50	64.12
60	66.10
70	68.09
80	70.07
90	72.05
100	74.04

From Equation (7-2), it follows that there is a temperature dependence of the potential E of 0.1984 mV/K for $n = 1$. This is of importance when automatically compensating for the effects of temperature on the potential E will be applied. To do this, another sensor measures the temperature of the sample, and the signal from that sensor is fed to a temperature-compensating circuit.

Equation (7-2) is also helpful in estimating the error due to uncertainties in the measured potential E. For a temperature of 25 °C an uncertainty of ± 1 mV results in an error in a_i of $4n$ %, where n is the number of elementary charges on the sensed ion.

Until now it has been assumed that the electrode under consideration responds only to the ionic species i. Such ideal behavior of an ion-selective electrode will be found only rarely, and in most cases cross-sensitivities to other ions are observed. For the simple case of only one other ionic species j with a number of elementary charges m being present, an extended form of Equation (7-2) can be used:

$$E = E_0 \pm 2.303\,(RT/nF)\log\,(a_i + k_{ij}^{m/n} a_j)\,. \tag{7-3}$$

The newly introduced parameter k_{ij} governs the selectivity of the electrode and/or the degree of cross-sensitivity. Only for $k_{ij} = 0$ does it make sense to call an electrode specific. It will be shown later that k_{ij} may vary from 10^{-13} to 10^2 (!). Equation (7-3) is usually called the Nikolskii equation.

All theoretical and thermodynamic considerations in potentiometry make use of the activity a_i of ions. Sometimes activity values are required, eg, in medical applications. It is a great advantage of potentiometry that the activity of free ions can be determined, such as Ca^{2+} activities in body fluids. The analytical chemist, also however, needs to know the concentration c_i of the ions. The two are related by a simple equation:

$$a_i = c_i \cdot f_i \tag{7-4}$$

where f_i is the individual activity coefficient of the ionic species i, which depends on the ionic strength I, which will be influenced by the concentration c and the number of elementary charges n of all ions present in the sample:

$$I = (1/2) \sum c_i n_i^2 . \tag{7-5}$$

For dilute solutions with $c < 10^{-2}$ mol/L, f_i and I calculations can be based on the Debye-Hückel equation, provided that the approximate ionic composition of the sample is known. In more concentrated solutions strong interionic forces come into play which invalidate all "ideal" laws.

To overcome this dilemma special methods of a sample preparation can be used. One of the objectives is to keep the ionic strength I and consequently f_i constant. Hence the system of interest can be calibrated with solutions of known concentrations of the ion i of interest provided that they are treated in the same way. Further details of these techniques are presented in Section 7.1.5.1.

Such requirements are meaningless in measuring pH values which are based on thermodynamic considerations of the activity a_{H^+} of hydrogen ions. All considerations so far have related the electrode potential E to the activity a_i of an ionic species on a logarithmic scale. An entirely different situation will be found, however, for measuring electrodes which respond to the oxidation reduction potential (ORP) of a (reversible) redox system.

In the case of a redox system which consists of the two corresponding species a_{ox} and a_{red} the potential E can be described using the Peters equation:

$$E = E_0 + 2.303 \, (RT/nF) \log (a_{ox}/a_{red}) \tag{7-6}$$

where the parameters are as defined in Equation (7-1), but now n is the number of electrons which participate in the redox equilibrium in accordance with

$$a_{ox} + ne^- \rightleftharpoons a_{red} . \tag{7-7}$$

An example of such a reaction is

$$Fe^{3+} + e^- \rightleftharpoons Fe^{2+} . \tag{7-8}$$

According to Equation (7-6), the potential E depends on the ratio of two ionic activities.

The situation will be more complicated when in addition to the two participating species in the redox couple other ions and/or molecules are involved in the redox reaction. Such a situation can be found for the system

$$MnO_4^- + 5e^- + 8H^+ \rightleftharpoons Mn^{2+} + 4H_2O . \tag{7-9}$$

The Peters equation can be adapted to such systems according to the basic scheme

$$\text{ox} + ne^- + mH^+ \rightleftharpoons \text{red} \tag{7-10}$$

with

$$E = E_0 + 2.303\,(RT/nF)\log\,(a_{\text{ox}}/a_{\text{red}}) + 2.303\,(RT/nF)(m/n)\log a_{H^+}\,. \tag{7-11}$$

Since a_{H^+} according to an earlier definition of the pH value can be expressed as

$$\text{pH} = -\log a_{H^+} \tag{7-12}$$

the oxidation-reduction potentials of such systems may depend strongly on the pH value. In Equations (7-6) and (7-11) E_0 is again the standard potential of the redox system. The E_0 values of different redox systems form an electrochemical series which expresses the power of oxidation or reduction, as will be discussed in Sections 7.1.4.1 and 7.1.4.3.

7.1.2.2 Measuring Chains (Potentiometric Cells)

Measuring Electrodes

Almost all of the commercially used measuring electrodes are membrane electrodes. Figure 7-1 illustrates a typical design of such electrodes.

Two different principles can be used to make contact to the membrane within the body of the electrode: either with the use of internal filling solutions or a solid-state contact with an electronic conductor connected to the cable from the amplifier.

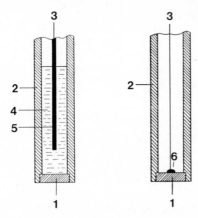

Figure 7-1.
Basic construction of potentiometric membrane electrodes. Left: internal connection using a reference electrolyte. Right: solid-state connection. 1, Membrane (sensor); 2, electrode body; 3, cable connection to amplifier; 4, reference electrolyte; 5, reference system; 6, silver-filled epoxy glue.

In the first instance an internal reference system is established, which in most cases is based on the 1st or 2nd kind electrodes (see Section 7.1.2.3). The main advantage of making connections in these ways is that by selecting the activity a_{me}, where "me" is the ion which is sensed by the membrane, and a_{re}, where "re" is the ion which is controlling the internal reference

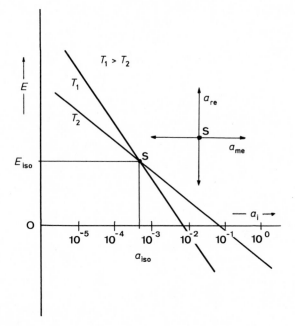

Figure 7-2.
Plot of two isotherms for an ion-selective electrode. For an ideal ion-selective cell including the external reference electrode the parameters should be $E_{iso} = 0$ mV, $a_{iso} =$ middle of the range. The inserted cross indicates which way the two parameters can be shifted. "re" represents the ion to which the internal reference system responds and "me" the ion which is sensed by the internal membrane surface.

system, both of the parameters at the isotherm intersection point, S, can be influenced. Figure 7-2 gives some more information relating to this effect.

Usually a_{me} is selected in such a way that it is in the middle of the dynamic measuring range of the electrode. This results in minimization of the effect of a changing sample temperature in those instances where no automatic temperature compensation is used. In addition, the potentials formed are due to reversible electrode reactions resulting in non-drifting interface potentials.

Compared with pH glass electrodes (see Section 7.1.3.4), few papers have dealt with the determination of the parameters at the isotherm intersection point [1,2]. This technique can also be applied to glass electrodes [3].

The alternative method making solid-state contact to the internal membrane surface with an expoxy glue filled with silver powder, is restricted to solid-state membranes with ionic conduction involving Ag^+ ions, such as Ag_2S, AgX (X = halide), or mixtures of Ag_2S with +CuS, CdS, or PbS. Only in such cases can reversible and zero-drift interface reactions be observed. Figure 7-3 illustrates this kind of interface reaction.

Fjeldly and Nagy have shown that impedance spectroscopy can be used to test the function of reversible or blocked interfaces [4].

The "coated-wire electrodes" sometimes used in experimental studies consist of a metal wire (Pt, Ag) which is covered with a gel matrix membrane. The metal/membrane interface is blocked in each case. Not much is known about the electrode reactions. A comment from Meyerhoff and Fraticelli is of interest: "The question as to what is actually occurring at the interface between the metal wire and the polymer film in conventional coated-wire electrodes continues to be a mystery" [5]. Hulanicki and Trojanowicz claim that at the surface of the wire an oxygen half-cell reaction will take place [6].

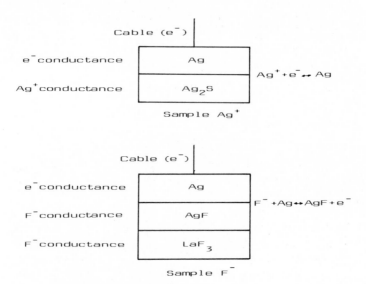

Figure 7-3. Solid-state connection of ion-selective membranes. The kind of conductivity is indicated on the left and the reaction at the critical interface on the right. "Ag" indicates a silver-filled epoxy glue which is mainly used for making the contact. An L(Cap.)aF$_3$ sensor (lower) needs an additional layer of AgF to provide reversible interfaces.

Table 7-2. Examples of membranes as sensors for potentiometric measuring electrodes.

Type of membrane	Composition	Ion which can be sensed
Solid state membranes		
Crystal cuts	LaF$_3$	F$^-$
Pressed pellets	Ag$_2$S	Ag$^+$, S^{2-}
	Ag$_2$S + AgX (X = Cl, Br, J)	X$^-$
	Ag$_2$S + MeS (Me = Cu, Cd, Pb)	Me^{2+}
Blown glass membranes		
	Li$_2$O − BaO − SiO$_2$	H$^+$
membranes	Na$_2$O − Al$_2$O$_3$ − SiO$_2$	Na$^+$
Liquid-/Gel matrix membranes *)		
Charged carriers	Organo-phosphoric esters	Ca^{2+}
	o-Phenanthrolin-nickel	NO$_3^-$
Neutral carriers	Valinomycin	K$^+$
	Nonactin	NH$_4^+$

*) The best description of the membrane structure is expressed by the lengthy term "plasticized membrane electrodes". Other terms such as "plastic membranes" and/or "polymer membranes" are less explicit. In this Chapter the term "gel membranes" will be used.

Questions relating to solid-state contacts to ion-sensing membranes are of great interest since in the case of ion-selective field effect transistors (ISFETs), solid contacts with electron conductors are the only available option [7, 8].

The physical and chemical properties of the sensing membranes in membrane electrodes are wide ranging. Table 7-2 lists a number of selected examples.

Solid-State Membranes

The first intensively studied solid-state membranes were pH-sensitive glasses. Glasses in general form an extended three-dimensional matrix of ions. The difference from crystalline solids lies in the absence of symmetry and periodicity [9, 10]. According to MacInnes and Dole [11], the matrix consists of SiO_4 tetrahedra linked together by oxygen atoms. The cations are distributed within the interstitial spaces of the lattice (see Figure 7-4).

The sensitivity and selectivity of the glasses to monovalent cations can be changed by altering the nature of the cations and by using lattice modifiers. It can be shown by experiment that glasses which respond to cations are ionic conductors in which the charge cariers are usually sodium ions. The H^+ response can be understood by assuming an ion-exchange equilibrium in a leached ("swollen") layer at the surface of the glass membrane. This is why the water content of the membrane is important. Hubbard et al. [12] have shown that there is a correlation between the H^+ response and the hygroscopic properties of a glass and it is well known that proper electrode functioning can be achieved only by the „formation" of a pH glass electrode for several hours in water.

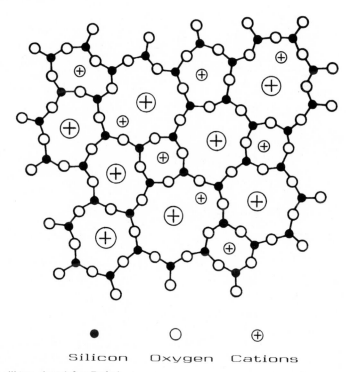

Silicon   Oxygen   Cations

Figure 7-4. The matrix of a silicate glass (after Perley).

The understanding of glass electrodes was, for a long time, based mainly on thermodynamic theoretical considerations and less on the applied electrochemistry of glass surfaces [13]. Baucke introduced the beam-etching technique together with secondary ion mass spectrometry (SIMS) for studying leached glass layers [14]. From the information obtained by this technique a dissociation mechanism was developed for the formation of functional groups at the glass surface which departs completely from the classical model [15].

Glass electrodes were for many years fabricated as thin-walled bulbs of Corning 015 glass. A serious disadvantage was the occurrence of a sodium-induced error at pH values greater than 10 caused by the presence of sodium ions. Intensive studies by Eisenman [16], Nikolskii et al. [17] and Wegman and Simon [18] overcame this problem by changing the basic composition of the glass from $Na_2O - CaO - SiO_2$ to $Li_2O - BaO - SiO_2$. In addition, lattice modifiers have been found which significantly lower the internal resistance of glass membranes. The advantage is that thicker and mechanically more stable glass membranes can now be used. The fabrication of membranes by a glass-blowing technique, however, is still the same.

Solid-State Ionic Conductors

Silver sulfide (Ag_2S) in its different forms is an excellent conductor of silver ions. Its extremely low solubility product of $K_L = 10^{-50}$ allows the measurement of silver ion activities down to 10^{-20} mol/L, provided that silver can be extracted from a mixture of complexed silver compounds.

The Nernst equation is valid for silver-sensing electrodes based on Ag_2S:

$$E = E_0 + 2.303\,(RT/F)\log a_{Ag+} \, . \tag{7-13}$$

In a suspension of silver sulfide in pure water, because of the low solubility product of Ag_2S, an ionic equilibrium will be reached in accordance with

$$K_L = a_{Ag+}^2 \, a_{S2-} \tag{7-14}$$

and

$$a_{Ag+} = \sqrt{K_L/a_{S2-}} \, . \tag{7-15}$$

Inserting Equation (7-15) in Equation (7-13) results in

$$E = E_0^* - 2.303\,(RT/2\,F)\log a_{S2-} \, . \tag{7-16}$$

This is the equation for a sulfide-sensing electrode and is based on the fact that any addition of sulfide to the solution in which the equlibrium is described by Equations (7-14) and (7-15) will increase a_{S2-} and at the same time lower a_{Ag+}.

Similar two-fold solubility equilibria have allowed the development of two different types of ion-selective electrodes for anions and cations. In the first, mixtures of Ag_2S with AgX (where X = Cl, Br, I) pressed into pellets have led to halide-selective electrodes. Secondly, mixtures of Ag_2S with CuS, CdS, and/or PbS and pressed pellets made thereof provide electrodes which are sensitive to heavy metal ions such as Cu^{2+}, Cd^{2+}, and Pb^{2+}.

Silver halides (AgCl, AgBr, and AgI) are also good ionic conductors for silver ions. Their solubility product is much higher than that of Ag_2S, but the pure silver halides are difficult to handle and are light sensitive, which is manifested as a potential shift.

Lanthanum Fluoride

In 1966, Frant and Riseman published a paper which sought to prove that lanthanum fluoride is an excellent sensor for fluoride ions [19]. It was later shown that LaF_3 is a pure ionic conductor of fluoride ions. Its solubility product is 10^{-29}, which is low enough to allow sensitivity to fluoride ion activities down to 10^{-7} mol/L.

The advantage of ion-selective electrodes based on these principles was one of the main reasons for the renaissance of potentiometry. For most of these solid-state sensors, the ionic conduction controls the selectivity whereas the low solubility fixes the detection limit.

Liquid and Gel Matrix Membranes

There is no chemical difference between these two types of ion-selective electrodes but mechanically there are significant differences. The matrix membranes contain the same ion-active substances as the liquid membranes. The organic solvents act as a plasticizer for polymers such as polyvinyl chloride (PVC) or polyvinyl acetate (PVAc). For this reason in this section the term liquid membrane will be used generally.

The ion-active substance is an organic molecule — an "ion carrier" — which together with a suitable solvent forms the ion-active phase. The term ion carrier implies that the ion to be sensed and the organic phase form a kind of adduct. The organic solvent strongly influences the properties of the membrane. The ion carrier calcium di(n-decyl) phosphate with n-decanol as solvent, for instance, results in a membrane which cannot distinguish between Ca^{2+} and Mg^{2+} ions. An electrode based on such a combination can, however, be used as a total water hardness-sensing device. Replacing the n-decanol with the strongly polar solvent dioctyl phosphonate changes the properties of the membrane markedly; selectivity to Ca^{2+} is obtained. The selectivity in relation to Mg^{2+} is expressed by the constant $k_{Ca,Mg} = 5.10^{-2}$ (for more details see Section 7.1.5.3).

The ion carriers can be divided into two different groups. The first consists of charged ion carriers which can form ionic species and will undergo ion-exchange processes with the ion to be sensed. The example of organophosphoric acid has been given already. Others are Ni-o-phenanthroline to sense nitrates and quaternary ammonium derivatives for a number of anions, eg, chloride.

The second group of ion carriers are called "ionophores". The compounds of interest are neutral ion carriers. Inclusion and/or chelate complexes can be formed with the ion of interest. The group of nonactins [20, 21] for monovalent cations and valinomycin [22, 23] for potassium are typical and analytically important examples.

Table 7-3 gives more information relating to liquid membranes.

Theoretical considerations of electrode mechanisms date back to the time of initial electrode development [24, 25]. Simon, Morf, and co-workers [26, 27] improved the theory and synthesized a large number of substances to be used as liquid membranes (Figure 7-5).

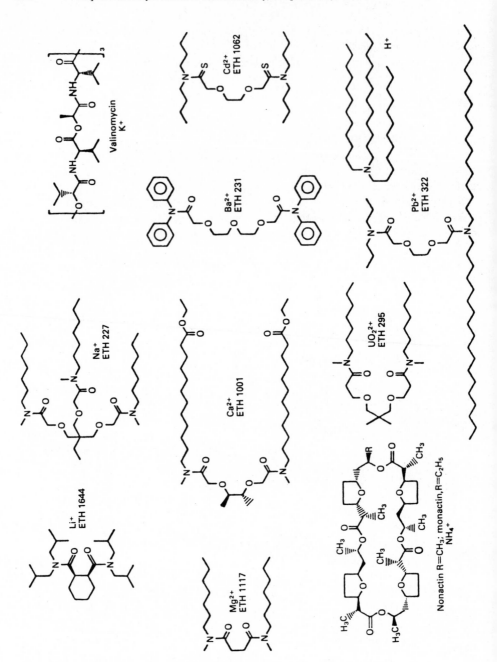

Figure 7-5. A number of newly synthesized ionophores [28, 29]. The preferentially sensed ions are indicated.

The development of liquid membranes extended the field of direct potentiometry and in doing so created a significant market relating to many different analytical applications (see Section 7.1.5.4).

Table 7-3. Parameters which influence the function of liquid membranes.

Properties of the ion-active substances
- Charged carrier molecules capable of ion-exchange,
- Neutral carriers capable of complex and/or inclusion compound formation.

Properties of the organic solvents
- Polarity, dielectric constant,
- Partition coefficient of the ion/molecule adduct between water and the organic solvent.

Thermodynamic aspects of the electrode reaction
- Reaction enthalpy,
- Dehydration enthalpy of the hydrated ions to be sensed

7.1.2.3 Reference Systems

In potentiometry reference systems [30] are needed for two different reasons, first to make contact with the internal surface of membrane electrodes (see Figure 7-1) and second to form external reference electrodes which, together with the measuring electrode, complete the measuring chain. Good reference electrodes, unlike measuring electrodes, should show a potential that does not depend on the properties of different samples. This is achieved in the case of external reference electrodes as shown in Figure 7-6, by positioning the reference system in contact with a solution of constant composition. The electrical contact to the sample is made using a diaphragm.

Most of the reference systems used are based on electrodes of the 2nd kind following the scheme $M/MX/X^-$, where M is a metal (Ag, Hg in the monovalent state, Tl) that forms sparingly soluble salts MX. By selecting the activity (or concentration) a_{X^-} the potential of the system is fixed. The mechanism for such electrodes is that the electrode M detects the activity of M^{n^+} which arises from the solubility of MX in relation to the solubility product of MX. This again is ruled by the a_{X^-} activity, as has been shown earlier (see Equations (7-13) and (7-16)). Most reference systems in practice make use of the combination $Ag/AgCl/Cl^-$ with potassium chloride as the reference electrolyte. Table 7-4 lists some of the most important requirements for reference electrodes.

Constant and reproducible potentials can be achieved with electrode systems with a high exchange current density of the reaction M/M^{n^+} [31, 32]. The temperature stability depends mainly on the chemistry of the system. For instance, calomel (Hg_2Cl_2) disproportionates to form Hg and $HgCl_2$ at temperatures above 60°C. The widely used calomel electrodes based on the system $Hg/Hg_2Cl_2/Cl^-$ will for this reason be used preferentially in laboratories for sample temperatures close to room temperature. Changing the chemistry to $Hg,Tl/TlCl/Cl^-$, ie, replacing mercury with an amalgam of thallium and calomel with thallium(I) chloride provides a system which is stable up to 130°C [33].

Table 7-4. Requirements for reference systems.

1. Stability of the potential (better than ± 0.1 mV)
2. Repeatability of the potential (± 0.5 mV)
3. Temperature stability (up to 130 °C)
4. Lack of hysteresis of the potential when subjected to sudden temperature changes
5. Constant liquid junction potential at the diaphragm
6. Free flowing liquid junction
7. Minimized risk of poisoning of the reference system by the samples
8. Easy maintenance

Another important factor in making temperature-stable reference electrodes is the amount of MX available. The solubility of AgCl in KCl increases with increasing temperature and a $[AgCl_2]^-$-complex is formed. Using a simple chloridized silver wire is not therefore a good way of realizing such an electrode, since insufficient AgCl is available. Cartridges made of a mixture of Ag with AgCl or of partially reduced AgCl give much better results (see Figure 7-6).

The most frequently used reference electrodes are listed in Table 7-5.

An important component in establishing the optimum working conditions for a reference electrode is the diaphragm. This guarantees that the reference electrolyte and/or the reference system are in electrical contact with the sample solution. A number of different diaphragms are in practical use (Figure 7-7).

The presence of a structure such as the diaphragm which allows free flow results in a permanent leakage of the reference electrolyte. The leakage rate is of the order of 0.5 mL/d under normal working conditions. The electrolyte must therefore be replenished and the body of a reference electrode is equipped with a refilling aperture (see Figure 7-6). In order to minimize maintenance in industrial applications, a reservoir of the electrolyte is usually connected to the reference electrode.

All the reference systems considered are electrodes of the 2nd kind and make use of solubility equilibria. This may lead to hysteresis when the temperature changes suddenly. Redox systems are free from such complications and can also be used as reference systems. These "Ross electrodes" are of special interest in industrial pH measurement [34]. A disadvantage is the shift of the parameter pH_{iso} in an isopotential diagram to pH 1 (see Section 7.1.3.3.). To avoid errors due to temperature fluctuations, high-precision automatic temperature-compensating circuits must be used.

A crucial factor with the reference is the liquid junction potential generated at the diaphragm. Depending on the concentration and the mobilities of the ions on each side of the diaphragm, values up to ± 15 mV can be observed [35]. KCl as reference electrolyte is used in high concentrations for most applications, a good compromise owing to the same mobility of the K^+ and Cl^- ions ("equitransference"). Applying ion-selective electrodes to the analyses of dilute sample solutions was the objective of the development of "equitransference reference electrolytes" [36].

The diffusion of ions from the sample through the diaphragm into the reference electrolyte can be slowed down by the addition of gelling agents (eg, polyacrylamide) to the electrolyte.

Undesirable reactions of the reference electrolyte with constituents of samples can be avoided by using double-junction reference electrodes working with two electrolytes. The internal electrolyte is the reference electrolyte adapted to the reference system used. The outer electrolyte can be selected in such a way that chemical interferences will no longer exist.

Figure 7-6.
Construction of a silver/chloride reference electrode. (Courtesy of Ingold).
1, Diaphragm; 2, reference electrolyte; 3, reference system (cartridge of Ag/AgCl); 4, internal connection; 5, refilling aperture; 6, electrode head; 7, cable.

Table 7-5. Specifications of the most frequently used reference electrodes.

Model/name	Chemistry of the reference system	Standard potential at 25 °C a) b)	Application/comments
Calomel electrode	$Hg/Hg_2Cl_2/sat.KCl$	$+244$ mV	$t_{max.} + 60$ °C
Silver chloride electrode	$Ag/AgCl/sat.KCl$	$+198$ mV	$t_{max.} + 130$ °C
Thalamid electrode	$Hg,Tl/TlCl/sat.KCl$	-577 mV	$t_{max.} + 130$ °C
Mercury sulfate electrode	$Hg/Hg_2SO_4/K_2SO_4$ (1 Mol/L)	$+641$ mV	Chloride-free reference electrolyte, $t_{max.} + 60$ °C

a) Term misleading, values are for the concentrations of the reference electrolytes specified and not for 1 mol/L,
b) Potential of the cell $Pt-H_2$/reference electrode

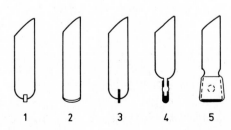

Figure 7-7.
Different types of diaphragms of reference electrodes. 1, Porous plug; 2, porous disc (lower resistance, increased leak rate); 3, twisted Pt wires; 4, exchangeable porous plug; 5, sleeve-type diaphragm (demountable for easy cleaning, low resistance, high leak rate).

Reference electrolytes in which the KCl concentrations are high will crystallize at lower temperatures. The precipitated crystals may block the diaphragm, leading to an unacceptable increase in its internal resistance. In most cases $5 \cdot 10^3 \, \Omega$ is the upper tolerated limit, except when amplifiers with two high-impedance inputs are used.

The use of low-temperature reference electrolytes will overcome these problems. They are aqueous mixtures with glycol or glycerol [DIS 1].

7.1.2.4 Half-Cells and Measuring Chains

All types of measuring electrodes and/or reference electrodes form a half-cell, the potential of which cannot be measured in isolation. A combination of two half-cells is always required for practical use. Such a combination is called a measuring chain or a potentiometric cell. According to the German Standard DIN 19260 [37], the term galvanic cell is applicable. The overall potential E_{cell} is the sum of a number of potentials:

$$E_{cell} = E_{er} - E_{ir} + E_{me} - E_{mi} \pm E_d \tag{7-17}$$

where E_{er} is the potential of the external reference electrode, E_{ir} the potential of the internal reference electrode, E_{me} the potential of the membrane (the sensor) of the measuring electrode in contact with the sample, E_{mi} the potential of the membrane in contact with the internal reference electrolyte, and E_d the potential of the liquid junction to be found at the diaphragm of the external reference electrode. The potentials E_{er}, E_{ir} and E_{mi} are constant, which means that E_{cell} depends only on E_{me} and E_d. The latter is either negligibly small or can be made constant.

7.1.3   pH Measurements

7.1.3.1   Definitions of pH

In 1909 the Danish chemist Sörensen proposed that the negative Briggs' logarithm of the hydrogen ion concentration c_{H^+} be called "pH":

$$pH = -\log c_{H^+} . \tag{7-18}$$

Such a definition has no underlying thermodynamic meaning and is usually called pcH.

More extensive studies to find a more fundamental definition of pH based on the activity a_{H^+} of the hydrogen ion resulted in the relationship

$$pH = -\log a_{H^+} . \tag{7-19}$$

It is called paH. All these attempts failed, however, for different reasons to establish a fundamental definition [13].

Today in all existing National Standards the definition of pH is an operational one based on the voltages of two cells with the following configurations.

Cell 1:

Reference electrode/sample solution X/Pt-H$_2$, where Pt-H$_2$ represents a standard hydrogen electrode (see Section 7.1.3.2). The voltage E_x will be measured.

Cell 2:

Reference electrode/buffer solution/Pt-H$_2$. The voltage E_b will be measured. The pH$_x$ of the sample solution is then related to the pH$_b$ of the buffer solution by

$$pH_x = pH_b + (E_b - E_x)/S \qquad (7\text{-}20)$$

where S is the theoretical slope valid for the working temperatures (see Table 7-1).

To complete the definition, it is necessary to determine the formulations of a number of buffer solutions of precisely known pH. Their pH values have been determined with a measuring chain built up of a hydrogen electrode and a reference electrode without transference (that is without a diaphragm). A number of theoretical considerations arrived at values which are close to the paH value mentioned earlier [13].

Table 7-6. pH values of five standard buffer solutions [38].

t (°C)	A	B	C	D	E
0		4.003	6.984	7.534	9.464
5		3.999	6.951	7.500	9.395
10		3.998	6.923	7.472	9.332
		3.999	6.900	7.448	9.276
20		4.002	6.881	7.429	9.225
25	3.557	4.008	6.865	7.413	9.180
30	3.552	4.015	6.853	7.400	9.139
35	3.549	4.024	6.844	7.389	9.102
40	3.547	4.035	6.838	7.380	9.068
45	3.547	4.047	6.834	7.373	9.038
50	3.549	4.060	6.833	7.367	9.011
55	3.554	4.075	6.834		8.985
60	3.560	4.091	6.836		8.962
70	3.580	4.126	6.845		8.921
80	3.609	4.164	6.859		8.885
90	3.650	4.205	6.887		8.850
95	3.674	4.227	6.886		8.833

The compositions of the solutions are: *)
A KH tartrate (saturated at 25 °C)
B KH phthalate (0.05 mol/kg)
C KH$_2$PO$_4$ (0.025 mol/kg)
   + Na$_2$HPO$_4$ (0.025 mol/kg)
D KH$_2$PO$_4$ (0.008695 mol/kg)
   + Na$_2$HPO$_4$ (0.03043 mol/kg)
E Na$_2$B$_4$O$_7$ (0.01 mol/kg)

*) As can be seen the concentration is based on a molality (mol/kg) concept.

In practice, the hydrogen electrode $Pt-H_2$ within a pH range from 2 to 12 can be replaced with a glass electrode. Equation (7-20) remains valid.

Table 7-6 lists the pH values of five standard solutions [38].

Equation (7-20) should be restricted to conditions where the expected pH_x of the sample solution and pH_b of the buffer solution are similar, otherwise slight differences in pH_x can be observed. One of the reasons is the variation of the liquid junction potential at the diaphragm of the reference electrode. The errors in pH_x introduced by these and other effects are normally small, of the order of ± 0.05 pH. There is an alternative, however, which may be helpful in more critical cases.

Two standard solutions, of pH_b' and pH_b'', will be used, the pH values of which are on each side of the expected value pH_x of the sample. Again, the potentials E are measured in the solution x, b' and b''. The value of pH_x then can be calculated from

$$(pH_x - pH_b')/(pH_b'' - pH_b') = (E_x - E')/(E_b'' - E_b') . \tag{7-21}$$

In addition to the IUPAC Recommendations there are many National Standards. Of special interest are the two German Standards DIN 19266 [39] and DIN 19267 [40]. DIN 19267 describes preferred buffer solutions for industrial pH-measuring devices for use at high concentrations.

As will be shown, pH glass electrodes at pH values greater than 10–12 in the presence of high concentrations of sodium ions are subject to errors which are called sodium errors (alkalinity error). Sodium-free buffer solutions are available to check pH glass electrodes for this error. Detailed information relating to the preparation of such solutions was given by Filomena et al. [41]. The pH of these solutions range from 7.9 to 12.5 (at 25 °C).

Efforts have also been made to establish an operational pH scale in organic solvents. For both theoretical and practical reasons success may only be possible for amphiprotic solvents which will mix with water. A suitable procedure has recently been endorsed by IUPAC [42, 43] for pH standards based on potassium hydrogenphthalate in solvent-water mixtures. In such cases pH glass, stored in water when not in use, shows potentially good stability.

Measurements in aprotic solvents are more difficult. They are of interest in relation to the definition of an acidity scale in solvents [44, 45]. Another important application is the volumetric analysis of very weak acids and bases in organic solvents. pH-indicating electrodes allow the end-point detection of so-called potentiometric titrations [46, 47].

7.1.3.2 The Hydrogen Electrode

The hydrogen electrode [48, 49] is formed by a small wire or a small piece of platinized platinum foil over the surface of which pure hydrogen gas is bubbled. "Platinized platinum" is made by galvanic deposition of finely divided "black" platinum [13]. Figure 7-8 shows an example of a practical hydrogen electrode.

The half-cell potential of a hydrogen electrode is

$$E_h = E_h^0 + 2.303 \ (RT/F) \ \log \ (a_{H^+}/\sqrt{p_{H_2}}) . \tag{7-22}$$

Thus the potential E_h depends on the hydrogen pressure and the observed potential must be corrected to the reference pressure of 1013 hPa.

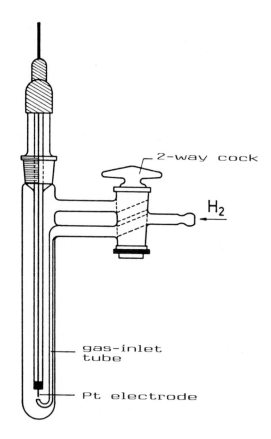

2-way cock

H₂

gas-inlet tube

Pt electrode

Figure 7-8.
Hydrogen electrode. The reference electrode to complete the cell and/or another half-cell is not shown. In the indicated position of the two-way tap hydrogen bubbles over, the platinum. When a stable potential has been reached the tap is turned to the other position.

The standard potential E_h^0 is by convention set to zero at all temperatures. Equation (7-22) then simplifies to

$$E_h = 2.303 \ (RT/F) \log a_{H^+} \ . \tag{7-23}$$

The hydrogen electrode has the ideal slope S values as listed in Table 7-1. For this reason it is used mainly for measurements which are theoretically directed. Its disadvantage, apart from being difficult to use, is the influence that redox systems have on E_h.

The definition of half-cell potentials is always based on the hydrogen electrode. It is defined as the voltage of a potentiometric cell in which the electrode on the left is a standard hydrogen electrode and that on the right the electrode under consideration. An example may be given of an electrode of the 1st kind realized by the arrangement Ag/Ag_+. The cell in question then has the form

$$Pt\text{-}H_2/H^+//Ag^+/Ag \tag{7-24}$$

where the double solidus // indicates that a diaphragm or an "electrolytic key" must be used to separate the two half-cells because of the redox sensitivity of the hydrogen electrode.

Under standard conditions in which all ion activities are 1 mol/L and $p_{H_2} = 1013$ hPa, a cell voltage of $+800$ mV will be generated; this is the standard potential of the silver electrode.

7.1.3.3   The Quinhydrone Electrode

Quinone and hydroquinone easily form a 1 : 1 molecular adduct called quinhydrone [50]. It is only slightly dissociated in aqueous solution.

$$HO\text{-}C_6H_6\text{-}OH \rightleftharpoons O = C_6H_4 = O + H_2 \qquad (7\text{-}25)$$
$$\text{Hydroquinone} \qquad \text{Quinone}$$

A certain partial pressure of H_2 will therefore build up in the solution.

A quinhydrone electrode consists of a piece of smooth Pt wire dipped in a solution saturated with quinhydrone. The properties of such a half-cell can be related to that of a hydrogen electrode. The Nernst equation is valid. The use of a hydroquinone electrode is limited, however, to pH values below 6. At higher pH values the dissolved oxygen in the sample solution will damage the hydroquinone component of the quinhydrone adduct.

Quinhydrone electrodes were at one time important pH-measuring electrodes. In special cases they are still used instead of a hydrogen electrode as a 2nd kind of absolute reference electrode because the potential can be calculated precisely from

$$E_{qu} = E_{qu}^0 - S \cdot pH \qquad (7\text{-}26)$$

where S is the theoretical slope (see Table 7-1). Standard potential values E_{qu}^0 can be found in monographs dealing with pH measurement [13].

7.1.3.4   Glass Electrodes

The history of pH glass electrodes began with the observations of Cremer in 1906. He studied the electromotive properties of organic tissues and tried to simulate their properties using thin glass membranes. In 1909 Haber and Klemensiewicz fabricated a practicable measuring chain using a soft glass as a membrane. They pointed out that it was easier to use than the hydrogen electrode. Twenty years later MacInnes and Dole published the precise composition of a pH-sensing glass which was made commercially and has been used as Corning 015 for many years [51, 52].

The development of pH glass electrodes correlates closely with the development of techniques for measuring low voltages from high internal resistance sources. The need to measure under zero current conditions in earlier times could only be effected using a compensating device which had been developed by Poggendorf [DIS 2, 53]. By 1930 so-called electrometer tubes were available for direct-reading pH meters. They were replaced by field effect transistors about 35 years later.

This short historical survey shows how closely the development of electrodes is linked with that of the measuring devices. The latest step has been the introduction of microprocessors, resulting in pH meters which are very easy to use [54].

Today pH glass electrodes have reached very high standards. Because of their good performance characteristics they are the most widely used chemical sensors. Table 7-7 details the main advantages of glass electrodes.

One of the disadvantages is the presence of the sodium error. This is observed in solutions with pH values above 12 in the presence of sodium ions. The selectivity constant $k_{H,Na} = 10^{-13}$ for modern pH glasses means, however, that the error in most cases is negligible. It is important that the sodium error increases strongly with increasing temperature [DIS 3]. Figure 7-9 gives an indication of the effect of the sodium error in two different pH glasses.

For well understood reasons, glass electrodes cannot be used in acidic fluoride-containing solutions because of the etching of glasses by hydrofluoric acid. Other limitations on the use of glass electrodes are listed in Table 7-7.

It has already been mentioned that glass membranes are made by glass blowing. The shape of the membrane can therefore be varied as shown in Figure 7-10. The legend of Figure 7-10 suggests some typical applications of glass membranes depending on their shape.

The shape of membranes is of special interest when automatic cleaning of the glass electrodes is required. It is necessary to do this since only a clean surface of any sensor guarantees a fast response. Another consideration is the maintenance-free in situ mounting of electrodes. The techniques used in automatic electrode cleaning systems are shown in Figure 7-11 [DIS 4].

Table 7-7.   Properties of pH glass electrodes.

1. Advantages
   - the Nernst equation is valid over a pH range of at least 2 to 12,
   - the stability of the electrode potential is of the order of ±1 mV/week,
   - due to the use of liquid internal reference systems it is possible to optimize the parameters E_{iso} (zero potential) and pH_{iso} (isopotential) of the point of isotherm intersection,
   - redox systems have no effect on the electrode potential,
   - the sodium error is in most cases negligible,
   - due to the availability of pH amplifiers with high impedance inputs of $>10^{12}$ Ω the internal resistance of glass membranes is no longer a problem, thick-walled and mechanical robust membranes can be used,
   - glass electrodes are the subjects of National and/or International Standards, compatibility of electrodes from differents makers has been reached by these efforts,
   - glass electrodes, together with apropriate reference electrodes can stand temperatures up to +130 °C,
   - shape of the membrane in glass electrodes can be adapted for many different uses,
   - the life time of glass electrodes under normal working conditions is at least 1 year.

2. Disadvantages
   - glass electrodes only can be made by manual or semi-automatic procedures piece by piece,
   - each electrode shows individual parameters which must be taken into account with calibration steps (see [54]),
   - periodic re-calibration is necessary,
   - glass electrodes will be destroyed by acidic fluoride solutions,
   - highly hygroscopic solutions "de-leach" the hydrated layer of the glass membrane and the electrode will cease to function,
   - some surfactants interact with the hydrated layer of the glass membrane causing a malfunction.

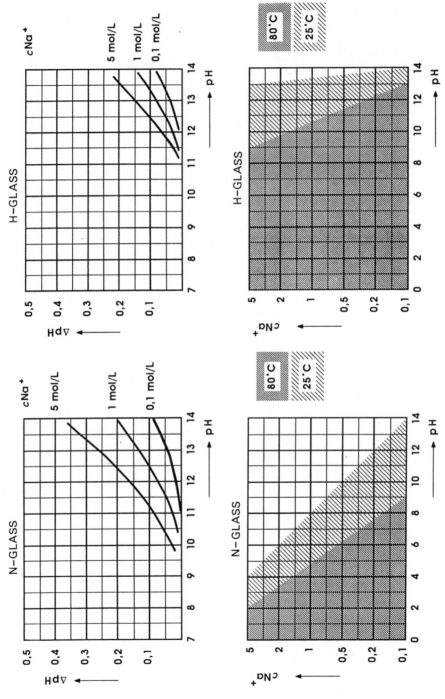

Figure 7-9. Sodium error of two different pH glasses. Top: pH error and sodium ion concentration. Bottom: usable pH ranges showing dependence on sodium concentration and on temperature. (Courtesy of Schott-Geräte GmbH [DIS 3]).

Figure 7-10.
Differently shaped glass membranes. 1, Classical bulb;
2, cylinder, preferred for mechanical cleaning (see
Figure 7-11); 3, cone, preferred for chemical cleaning (see
Figure 7-11); 4, head; 5, flat membrane for surface
measurements; 6, needle, eg, for insertion measurement in
meat and cheese.

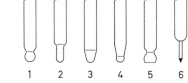

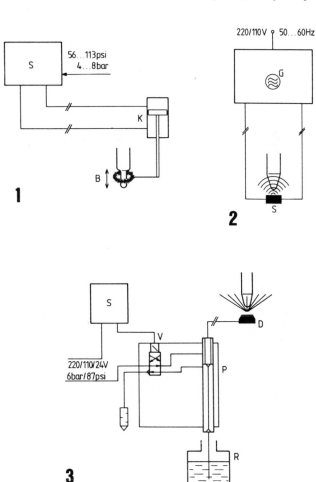

Figure 7-11. Different systems for automatic electrode cleaning. (Courtesy of Polymetron). 1, Mechanical
cleaning by brush wiping; 2, applying an ultrasonic field; 3, chemical cleaning with a solution that is delivered to the electrode by the nozzle D.

It has been pointed out earlier (Section 7.1.2.3) that in order to complete the potentiometric
cell a reference electrode must be added to the glass electrodes. This should be chosen so that
its reference systems is identical with that within the glass electrode. Cells of this kind are
called symmetrical. Adjusting the pH value of the internal filling solution of the glass elec-

trode to 7 results in a cell which in accordance with Figure 7-12 is characterized by the parameters $pH_{iso} = 7$ and $E_{iso} = 0$ mV. Such a cell, even without automatic temperature compensation, shows very little temperature sensitivity. Slight deviations from the parameters specified can be tolerated. In all other cases a pH amplifier must be used which allows compensation for larger deviations of pH_{iso} and E_{iso}; this is done by shifting voltages as shown in Figure 7-13. The requirements for such amplifiers are specified in the German Standard DIN 19265 [55].

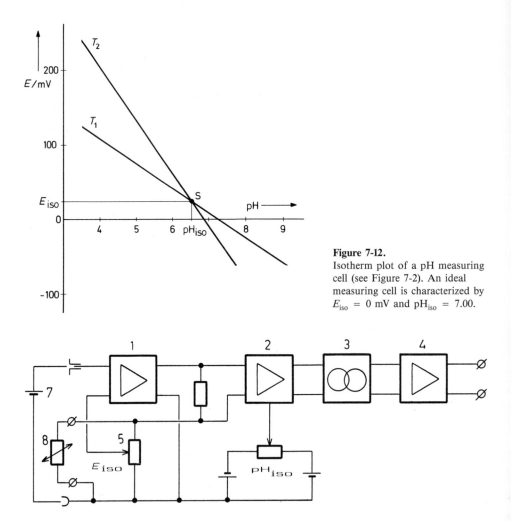

Figure 7-12.
Isotherm plot of a pH measuring cell (see Figure 7-2). An ideal measuring cell is characterized by $E_{iso} = 0$ mV and $pH_{iso} = 7.00$.

Figure 7-13. Circuit of a pH-measuring adaptation according to DIN 19265, permitting the adaptation of the parameters of isotherm plots (see Figure 7-12). 1, High-impedance input stage; 2, temperature compensation circuit; 3, decoupling stage, to prevent error signal of following instrumentation (eg, controller) via common ground loops; 4, power amplifier; 5 and 6, potentiometer/shifting to handle the parameters E_{iso} and pH_{iso} (c_{iso} for ion meters); 7, cell voltage; 8, resistance thermometer.

For a number of industrial applications automatic temperature compensation is important, because the calibration of the measuring chain is carried out at room temperature whereas the sample temperatures may be very different. A typical example is the control of pH in boiler feed water, which was treated in detail by Midgley and Torrance [56, 57]. Tauber has reported on the behaviour of glass electrodes at elevated temperatures [58].

Mechanically the combination of a glass electrode with a reference electrode can be realized in two ways, either by separately mounting the two individual electrodes in a holder and/or probe, or by using a combination electrode which contains both electrodes within a common electrode body. Details of the construction are shown in Figure 7-14.

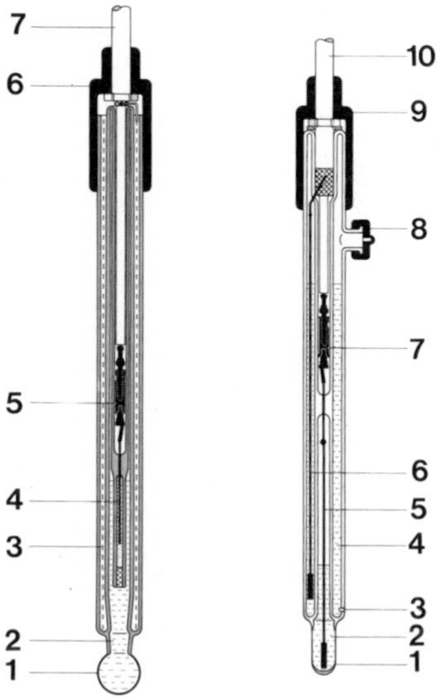

Figure 7-14.
pH glass electrodes. (Courtesy of Ingold). Left: single (half-cell) class electrode. 1, Glass membrane; 2, internal reference electrolyte (buffer); 3, metal screen; 4, internal reference system; 5, internal connection; 6, electrode head; 7 cable. Right: combination electrode with an integrated "external" reference electrode (= complete pH measuring cell). 1, Glass membrane; 2, internal reference electrolyte (buffer); 3, diaphragm; 4, external reference electrolyte; 5, internal reference system; 6, external reference system; 7, internal connection; 8, refilling opening; 9, electrode head; 10, cable.

Despite the high degree of mechanical stability of modern electrode systems, for most industrial applications they need additional protection. The combined electrodes and protection are usually called probes. The design of such probes includes means of adapting them to suit specific applications. Immersion probes, flow-through probes, demountable probe, and probes which can be equipped with automatic cleaning systems are distinct types. It is outside the scope of this book to deal in detail with such systems. The best source of information is the documentation provided by instrument makers [DIS 5, DIS 6].

7.1.3.5 Enamel Electrodes

There are a number of applications of pH-measuring equipment in process control where the sample may cause severe mechanical stress on the probes. Glass electrode-based probes cannot withstand such forces.

Pfaudler Werke [DIS 7] have developed a pH probe which is based on a very sturdy steel tube, the surface of which is covered with a special type of enamel. The formulation of the enamel, silicate, is such that it responds to the pH of aqueous samples. Each probe has an individual slope which is of the order of 55 mV/pH. The working range is from pH 0 to 10 with an accuracy of 0.1 pH. The sodium error increases strongly with increase in temperature, which may limit the application of the probe to pH values below 6–8 at temperatures above 140 °C. The measuring chain has a high degree of asymmetry. pH_{iso} is between 1 and 3.

In spite of these limitations, there are a number of heavy-duty applications in which the probe works satisfactorily. A practical realization of such an enamel pH probe is shown in Figure 7-15 [DIS 7].

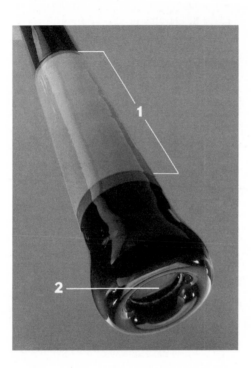

Figure 7-15.
Lower end of an enamel pH measuring probe for heavy-duty industrial applications (Courtesy of Pfandler). 1, pH-sensitive enamel zone; 2, flat ground steel diaphragm connected with the remote reference electrode. Typical dimensions (diameter x length, mm): 38 × 670, 83 × 1500, 127 × 2500, 180 × 3300.

7.1.3.6 Antimony Electrodes

It is known from the early work of Kolthoff and Hartong in 1925 [59] that metallic antimony in contact with antimony (III) oxide responds to pH changes in aqueous solutions. The electrode process could be formulated as follows:

$$Sb_2O_3 + 6H^+ + 6e^- \rightleftharpoons 2Sb + 3H_2O \; . \tag{7-27}$$

The oxide required will be formed at the surface of the metal by hydrolysis, which is one of the reasons why antimony electrodes need some time to become activated. An immediate electrode response can be achieved by adding antimony (III) oxide to antimony in the melting pot when pouring the electrodes. Mechanical cleaning of the electrode surface will not cause a loss of electrode function. Figure 7-16 illustrates how such a cleaning system works [60].

Figure 7-16.
Mechanical cleaning system for ring-shaped metal (antimony) electrodes. (Courtesy of Polymetron). 1, Ring-shaped metal electrode; 2, holder for two alumina rods with edges continuously rotating (ca. 6 rotations/min) – the driving gear and motor are placed in the head of the probe; 3, replaceable porous-plate diaphragm (made of alumina), connected with a tube filled with reference electrolyte to make contact with the remote reference electrode.

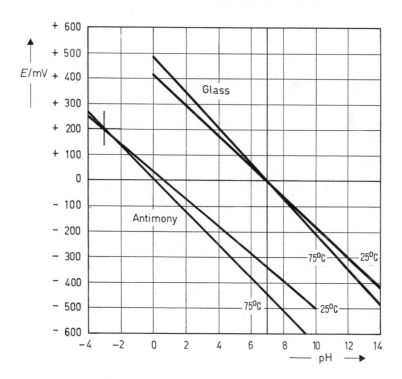

Figure 7-17. Isotherm plots for a pH measurement using a glass electrode and/or antimony electrode. The antimony cell is highly asymmetric. Automatic temperature compensation can only be achieved by utilizing an amplifier as shown in Figure 7-13.

Antimony electrodes have a very low internal resistance and for this reason they were used in early direct-reading pH measuring instruments. Today their use is restricted to applications in which glass electrodes are not suitable. The absence of a sodium error is also an advantage, eg, in controlling strong caustic solutions in the scrubbing of acid gases.

The slope of antimony electrodes is between 52 and 57 mV/pH at 25 °C, depending on its configuration. The accuracy is influenced by a number of parameters and is typically ±0.15 pH. Chlorides introduce a potential shift but do not influence the pH response. Sulfides, have a deleterious effect, even at low concentrations. This is also true of those substances, such as citrates, which form complexes with Sb^{3+} ions. Buffer solutions for calibrating antimony-based measuring cells should be based on phthalates, phosphates, and carbonates.

In comparison with glass electrode cells, those with antimony electrodes show a high degree of asymmetry, as can be seen in Figure 7-17 [61].

The use of pH amplifiers which conform to the specification in DIN 19265 allows the adjustment of antimony cells.

Oehme and Ertl studied the properties and possible applications of antimony in depth [61]. One of the most important industrial uses is in the neutralization of waste water which contains hydrofluoric acid using a slurry of lime. The electrode cleaning mechanism (Figure 7-16) keeps the electrode surface clean from the CaF_2 which is formed. Other features of antimony electrodes have been discussed elsewhere [62].

The antimony electrode is the only commercially used oxide electrode.

7.1.3.7   Other Oxide Electrodes

A number of other oxide electrodes have been studied. None of them, however, has achieved a standard of performance high enough for it to be used in significant numbers. One of the objectives in studying oxide electrodes is to have mirco-electrodes available for in vivo monitoring.

De Rooij and Bergveld [63], following the work of Perley and Godschalk [64], investigated iridium electrodes. In air-saturated solutions, iridium is covered by an adsorbed monolayer of oxygen atoms which take part in a redox equilibrium reaction which involves hydrogen ions. Improved properties were found with iridium that had been oxidized anodically or by heating it in an oxygen atmosphere to produce a surface layer of iridium oxide, IrO_2. The best behaviour, however, was achieved through the electrochemical formation of IrO_2. Cyclic voltammetry has been used to study the formation and dissociation of the oxide layer. The useful range according to Perley and Godschalk is from pH 0 to 14 with an almost ideal slope.

De Rooij and Bergveld reported that over the pH range 6.5–8.0 a random drift of ±2 mV per 100 h occurred. Kinoshita et al. [65] studied the system Pd/PdO and obtained data similar to those of de Rooij and Bergveld. In this system thermal oxidation resulted in the optimum behavior. The useful pH range was 2.5–8.0. The approach of Fogg and Buck [66] was different. They used a number of semiconducting oxides attached to a PTFE-bound graphite rod, known as the Růžička electrode [DIS 8]. The base metal is missing in each case. The work included TiO_2, Ta_2O_5, RuO_2, OsO_2, IrO_2, and PtO_2. The electrode systems were checked for response, accuracy, hysteresis, redox, and other (cation) interferences.

For most of the oxides the useful pH range is 2–11, with the exception of PtO_2 (5–10). The response varies over a wide range from under-Nernstian to over-Nernstian values. The accuracy is ± 2 mV for RuO_2 and OsO_2, but as much as ± 30 mV for Ta_2O_5. All the oxides tested showed a pronounced hysteresis in the sequence pH 2–pH 12–pH 2. Redox systems cause errors which are due to the electronic conduction of the oxides. It seems doubtful that a Růžička electrode is adequate for testing oxides.

Ta_2O_5 is of special interest since it is the pH-sensitive gate material of pH-ISFETs developed by Klein and Kuisl [67]. The authors reported different and much better results for this material than were obtained for the other oxides.

The application of yttria-stabilized zirconia (ZrO_2) as a pH-sensing material dating back to studies of Niedrach [68] should also be mentioned. ZrO_2 conducts by using O^{2-} ions as charge carriers. The pH response is possibly dependent on the transport of OH^- ions through the membrane. The main target of the work was the provision of robust and temperature-resistanced sensors for pH monitoring in boiler feed water. A serious drawback of ZrO_2 is its high internal resistance, which is of the order of $3 \cdot 10^9$–$6 \cdot 10^9$ Ω at 25 °C and $3 \cdot 10^8$ Ω at 95 °C. This is one of the reasons why all the studies were carried out at elevated temperatures close to the boiling point of water. The last report relating to ZrO_2 was from Light and Fletcher [69], who described the test procedure in some detail. No such sensors are currently commercially available.

7.1.3.8 Gel Membranes

Gel membranes based on polymer matrices are of prime importance in fabricating ion-selective electrodes. Simon and co-workers [28], in addition to investigating sensors for a number of ions, also studied H^+ ions closely. They synthesized a number of new pH-sensing substances and reported in detail on their behavior [29]. Ionophore 7 (N-octadecylmorpholine) has been widely utilized. Its dynamic pH range is from 2 to 12, with some interference from sodium and potassium ions.

Practical use of similar substances was made by Orion [DIS 9]. The sensor covers the pH range 0–4. It is considered to be suitable for pH sensing in solutions that contain hydrofluoric acid.

7.1.3.9 Application of pH Measurements

Because of their wide-ranging applications, the requirements for pH-measuring instrumentation can be very different, as Table 7-8 shows.

The kind of data required from measurements is also very different, as evidenced in the information in Table 7-9.

The manifold information that is available from pH measurements is the reason why this technique is one of the most widely applied in all fields of analytical chemistry and process control.

Table 7-8. Requirements on pH measuring instrumentation.

Special demand	Typical application
Small sample volume	Childrens hospital
High precision	Blood pH ±0.001
Extreme pH values	Caustic in scrubbers for acid gases, pH 15, Detoxification of chromates, pH 2
High temperature stability	Sterilization of fermentation broth, +117 °C
Saturated salt brines with temperatures up to 130 °C	Fertilizer production
Automatic electrode cleaning by ultra sonic fields	To prevent electrode fouling by oil films
Knowledge of the parameters of the intersection point of the isotherms	Automatic temperature compensation of highly unsymmetric cells, special amplifier

Table 7-9. Fields of application of pH measurements.

1. pH values as quality parameter
    - Clinical analysis of blood and body fluids,
    - control of milk on its arrival at dairies,
    - freshness of meat,
    - neutrality of treated industrial wastes,
    - acidity of rain,

2. pH values as process control parameters
    - pH optima of microorganism during fermentation,
    - alkalinization of boiler feed water with NH_3,
    - precipitation of heavy metal ions in industrial waste water (see Figure 7-18), [70]
    - adaption of redox potentials by pH control (see Figure 7-19)
    - exhaustion of caustic gas scrubber solution,
    - optimization of the dissociation of weak electrolytes (see Figure 7-25),
    - pH adaptation of detoxification reactions of industrial waste water (eg, cyanides pH 12, nitrites pH 4, chromates pH 2)

3. Endpoint detection of "potentiometric titrations"
    - acidimetry: volumetric analysis of acids by titration with bases of known concentration,
    - alkalimetry: volumetric analysis of bases by titration with acids of known concentration,
    - titration of very weak acids and bases in organic solvents.

7.1.4   Measurements of Oxidation-Reduction Potentials

7.1.4.1   Redox Electrodes

The more theoretical aspects of redox potential measurement have been treated earlier (see Section 7.1.2 and Equations (7-6) to (7-12)). The electrodes required to measure redox potentials should ideally be pure electronic conductors which respond to the "electron pressure" of a redox system without being influenced by other parameters. The closest approximation to such ideal electrodes is the use of noble metals, platinum and gold, as electrode material.

Surprisingly, both metals react readily with oxygen and are normally covered with thin layers of oxides. This can easily be demonstrated by cyclic voltammetry [71]. Platinum also has a tendency to form thin layers of hydrides at its surface when in contact with strongly reducing agents. It is clear from this behavior that the previous history of an electrode both in use and in terms of pretreatment has a strong influence on its properties. For the same reason, in weak solutions and in the presence of dissolved oxygen stable electrode potentials will be reached only slowly. The repeatability under such conditions is poor, and uncertainities of ±25 mV have to be accepted. Storing the electrode in a buffer solution of pH 4 saturated with quinhydrone guarantees a chemically clean and fast-responding electrode [72]; this pretreatment procedure is realizable in laboratories only, of course.

Another disadvantage of gold electrodes is that they behave as electrodes of the 2nd kind in the presence of chlorides and/or cyanides. Silver as electrode material shows this behavior much more strongly, which is one of the reasons why silver is not used as the electrode material in redox electrodes (see Figure 7-22). The only outcome of this situation is that electrodes must be selected on the basis of trial and error, despite the wealth of published material available [73].

It has been pointed out earlier that any redox electrode is only a half-cell which must be completed with an appropriate reference electrode to form a potentiometric cell (see Section 7.1.2.3).

The industrial applications of redox electrodes include processes during the course of which precipitates will be formed, eg, the neutralization of industrial wastes which contain heavy metal ions (see Figure 7-18). Here again, automatic cleaning procedures of the electrode surface are of prime importance. Figure 7-16 gives an example of how this can be done [74].

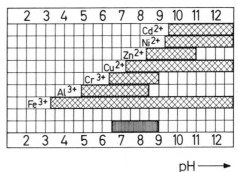

Figure 7-18.
pH ranges for the precipitation of metal hydroxides during the neutralization of industrial waste water. The neutrality of water by convention is from pH 6.5 to 9.0. On request the range can be extended to pH 9.5 (10) to handle water which contains Ni^{2+} and/or Cd^{2+} ions.

7.1.4.2 Redox Potentials and Influence Parameters

Equation (7-9) to (7-11) indicate that some of the redox systems include hydrogen ions in the reaction. In such cases the redox potential depends strongly on pH. A typical example of such a system is shown in Figure 7-19. The redox potential of a hypochlorite solution is influenced by the pH value while at the same time the likelihood of chemical reactions is pH dependent.

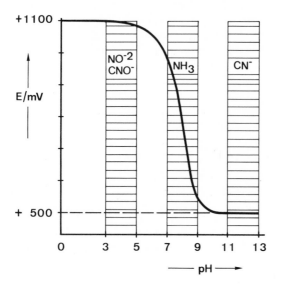

Figure 7-19.
pH dependence of the redox potential E of a hypochlorite solution. Only within the marked pH ranges can a number of important toxic substances be detoxified.

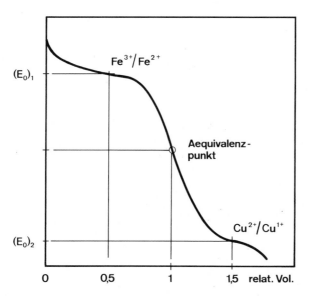

Figure 7-20.
Record of the redox potential E during the course of a titration of an Fe^{3+}-containing solution with a reagent consisting of Cu^+ in strong HCl. The E_0 values correspond to the standard potentials of the two redox systems involved in the reaction.

Most of the redox reactions take place between two different redox systems, both of which can be characterized by their standard potential E_0. This is done in the following example:

Redox system 1:

$$Fe^{3+} + e^- \rightleftharpoons Fe^{2+} \quad E_0 = +770 \text{ mV}$$

Redox system 2:

$$Cu^{2+} + e^- \rightleftharpoons Cu^+ \quad E_0 = +167 \text{ mV}$$

The question of whether two redox systems can react with each other will be answered by the "300 mV rule": "Two redox systems can react when their standard potentials E_0 show a difference of at least 300 mV" [74]. This statement, however, does not have a bearing on the speed of the reaction [75]. Applying the rule to the two redox systems given above leads to the possible reaction

$$Fe^{3+} + Cu^+ \rightleftharpoons Fe^{2+} + Cu^{2+} . \qquad (7\text{-}28)$$

In fact, iron (III) solutions can be titrated with copper (I) solutions. The reagent is prepared by dissolving $CuCl$ in 10% HCl. The end-point determination is carried out by sensing the redox potential during the course of the titration. Figure 7-20 illustrates the redox potential-volume record obtained during such a titration.

Jolly gave an excellent overview of the use of redox potentials in inorganic chemistry [75]. Bühler and Galster [73] dealt with the principles and problems of measuring redox potentials.

7.1.4.3 Applications of Redox Potential Measurements

Unlike the pH value, the redox potential can very rarely be used quantitatively. It is usually a process parameter providing information relating to the state of equilibrium of redox systems. Table 7-10 lists a selection of typical applications.

Ring-shaped redox electrodes can be cleaned automatically by the same methods as have been shown to be usable for antimony electrodes (see Figure 7-16). Also of practical importance is the control of the function of redox electrodes. Similarly to the procedure for pH electrodes, a number of calibration solutions, so-called "redox buffers", are available. They consist of solutions of redox systems which are highly reversible and strongly poised at the same time. The term "poising" originates in a proposal by Clark [80]: a solution is poised when it opposes a change in its redox potential on addition of oxidants and/or reductants. Since many redox systems have a pH-dependent redox potential, the buffers are also pH buffers. A number of typical redox buffers are listed in Table 7-11.

Table 7-10. Fields of application of redox potential measurements.

1. Redox potentials as quality parameters
   - chlorination of drinking water to guarantee a germicidal condition
   - to make sure that there is sufficient "hydro" (sodium dithionite) in a bath of vat dyestuff,
   - to monitor the redox potential as an auxiliary parameter during fermentation processes,
2. Redox potentials as process control parameters
   - detoxification of waste water from plating shops, eg, cyanides and chromates, the redox measuring instrumentation is part of a closed control-loop, see Figure 7-22 and 7-23 [70, 76]
   - automatic concentration control of peroxide bleach liquor [DIS 10],
3. Endpoint detection of "potentiometric titrations"
   - a large variety of inorganic and organic substances can be titrated in which case the reagent will be selected by its standard potential (see Figure 7-21, [77, 78], automatic analyzers play an important role in this field of analytical chemistry [79].

Table 7-11. Composition and properties of redox buffers.

Composition	pH	E_h (mV) b)
1. Potassium tetroxalate 0.05 mol/L a)	1.68	+600
2. Potassium hydrogenphthalate, 0.05 mol/L a)	4.00	+462
3. Potassium dichromate, 5%, pH adjustment with NaOH	7.80	+345
4. Sodium hydrogensulfite, pH adjustment with NaOH, with HCl	8.50	+ 35
	5.90	+175

a) Commercially available buffers can be used, must be saturated with quinhydrone.
b) Potential referred to the hydrogen electrode. It is not of interest in the routine checking of redox electrodes but for calibration of cells for rH-measurements.
   The rH concept should not be used, [80].

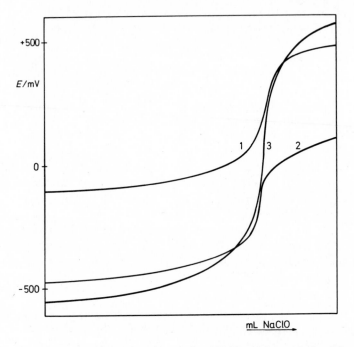

Figure 7-21. Titration of a cyanide solution (pH 12) with hypochlorite while recording of the redox potential R. Different metal electrodes were used to follow the reaction: 1, platinum; 2, silver; 3, gold. Gold gives double the potential change and is the electrode usually utilized in the detoxification of cyanide.

To check a redox measuring cell, two redox buffers are needed. The only parameter of importance is the slope S of the cells, which can be estimated in the simplest way using buffers 1 and 2 in Table 7-11:

$$S = \Delta E/\Delta S . \tag{7-29}$$

The slope in each case should be close to the theoretical value (see Table 7-1). The tolerance of E is of the order of $\pm 5-10$ mV.

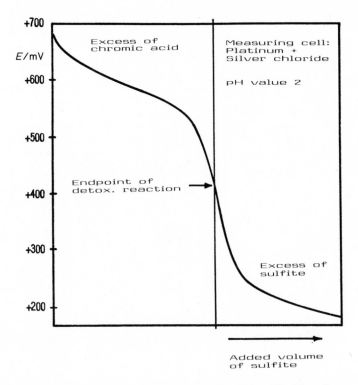

Figure 7-22. Record of the redox potential E during the detoxification of chromic acid with sodium hydrogensulfite ($NaHSO_3$).

7.1.5   Concentration Measurements with Ion-Selective Electrodes

7.1.5.1   Principles of Direct Potentiometry

The determination of concentrations by potentiometric measuring techniques was for a long time related to "potentiometric titrations" only. In a titration a reagent solution is added to a measured amount of the sample until the end-point is reached. The volume of the reagent solution allows the calculation of the sample concentration. This procedure follows the rules of volumetric analysis [79, 81].

With the development of ion-selective electrodes another principle could be used, based on the correlation of an electrode potential with the concentration to be measured. This direct-reading method is called direct potentiometry. The pH measurements according to the older definitions of pH could be termed the first direct-reading methods.

In all potentiometric methods the cell potential E is influenced by the activity a_i of an ionic species. The analyst, however, requires information relating to the concentration c_i. The key to converting one into the other is the activity coefficient f_i, which has been discussed (see Section 7.1.2 and Equations (7-4) and (7-5)). The dependence of f_i on the ionic strength I is illustrated in Figure 7-23.

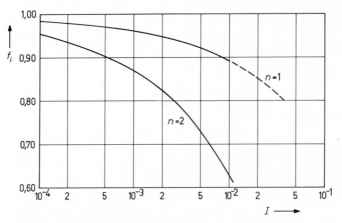

Figure 7-23. Dependence of the activity coefficient f_i on the ionic strength I for ionic species of charge $n = 1$ and 2.

As has been pointed out earlier, the value of I is usually unknown. The only way to overcome this problem is to prepare the sample in such a way that an ionic strength adjuster can be added to it. The latter consists of a solution of a similar electrolyte (containing ions to which the measuring electrode will not respond) in such a concentration that any fluctuation of I of the sample will be overcompensated. The ionic strength I is still unknown but it is at least kept constant. This leads to a simple analytical procedure in accordance with Figure 7-24. In this schematic illustration the adjustment of pH is included. The reason is that the degree of dissociation of weak electrolytes is strongly influenced by the pH value. The dissociation equilibrium is shifted in relation to the pK value:

$$pK = -\log K \tag{7-30}$$

where K is the dissociation constant of the weak electrolyte of interest. A distinction can be made between K_s values (for acids) and K_b values (for bases). At room temperature they are linked by the equation $pK_s + pK_b = 14$.

To reach complete dissociation according to Figure 7-25, the pH of the sample has to be adjusted to a value of at lest $pH = pK_s + 2$ and/or $pH = pK_b - 2$.

Table 7-12 gives some practical examples of the need for pH adjustment.

Table 7-12. Typical examples for pH adjustment of weak electrolytes.

Weak electrolyte	pK	Ion to be sensed	pH	
Hydrofluorid acid	$pK_s = 3.1$	F^-	5.2	
Hydrogen sulfide	$pK_s = 13$ a)	S^{2-}	14	b)
Ammonium hydroxide	$pK_b = 9.2$	NH_4^+	7.2	

a) 2nd step of dissociation,
b) Due to the high pK_s value only a compromise can be reached with a (constant) degree of dissociation of 90%.

Table 7-13. Possibilities and methods of sample preparation [82].

Desired result	Realized by
– Constant ionic strength	Addition of ion-strength adjustors as indifferent electrolytes at rather high concentrations.
– Complete dissociation of weak electrolytes	Addition of buffers with a pH which shows a difference of at least 2 units to the pK value, eg pH 5 for F^--selective electrodes
– Improvement of selectivity in case of pH interference	Addition of buffers to raise or lower the pH depending on the interference
– To prevent electrode fouling by precipitates	Addition of complexing agents, eg, EDTA in case of carbonates
– To prevent electrode poisoning	Addition of a reagent which destroys the poisioning substance, eg, H_2O_2 in case of sulfides
– Decomplexing so that the total ion concentration is available	Addition of stronger complexing agents for the cation, eg, NTA in the case of the $[AlF_6]^{3-}$ complex
– Chemical auxiliary reactions for cations for which no electrode is available	Addition of a defined amount of a ion which can be sensed and which complexes the cation, eg, F^- to sense Al^{3+} and measuring the drop of F^--concentration
– To protect low sulfide concentrations from air oxidation	Addition of anti-oxidants, eg, ascorbic acid

By using different sample preparation techniques, a number of other measurements can be achieved, as is shown in Table 7-13.

The crucial aspect of utilizing ion-selective electrodes is their selectivity. From Equation 7-3, a selectivity constant k_{ij} has been introduced. This has to be determined experimentally and the method by which this is done influences the value obtained. For this reason all the infor-

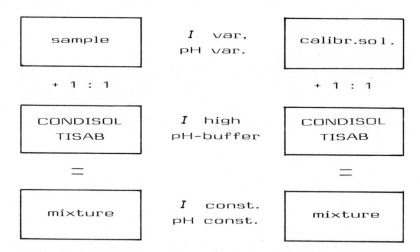

Figure 7-24. Principle of sample preparation to optimize the working conditions in direct potentiometry. CONDISOL (Polymetron) is "conditioning solution" and TISAB (Orion) is "total ionic strength adjustment buffer". The composition of such solutions is explained in Table 7-13 and in [98].

mation given in the documentation of electrode manufacturers (see Section 7.1.5.3) should be considered as rough estimates only. Figure 7-24 provides a guide for determining k_{ij} values.

A reliable method is to plot E against $f(c)$ for the ion i to be sensed and also for the interfering ion j. For a selected constant concentration c_k the accompanying cell potentials E_i and E_j will be obtained. The selectivity constant k_{ij} follows from

$$k_{ij} = (E_i - E_j)/S \qquad (7-31)$$

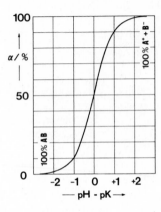

Figure 7-25.
Influence of pH on the degree of dissociation of weak electrolytes. pK $= -\log K$ is the dissociation constant of the electrolyte under consideration (see Table 7-12).

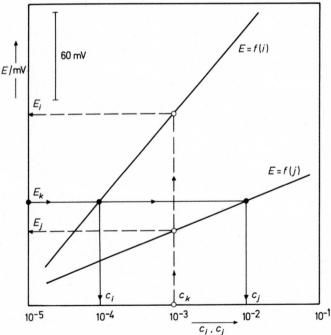

Figure 7-26. $E = f(c)$ plots for an ion-selective electrode, one with the ion i to be sensed and the other with an interfering ion j. From such plots the selectivity constant k_{ij} can be calculated. For more details, see text.

where S is the theoretical slope corresponding to the temperature of the solutions (see Table 7-1). the valency state n of the ions must be taken into account, however, by using a value of S/n.

From Figure 7-26 a second method follows. For a fixed value of E_k, the corresponding values c_i and c_j have to be found. Then,

$$k_{ij} = c_i/c_j . \tag{7-32}$$

The best method for checking the errors introduced by a limited selectivity, however, is to work with the plots provided in Figure 7-27. It is clear from this figure how the response of a calcium-selective electrode is influenced by the presence of magnesium. In practice, as the basis for the preparation of test solutions, the real sample solution should be used and not ion-free pure water.

Direct potentiometry is a powerful method in analytical chemistry but it is by no means *as simple as a pH measurement* to use a slogan used in advertisements!

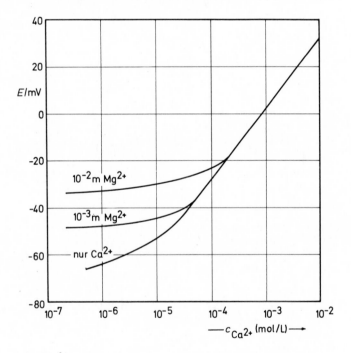

Figure 7-27. Interference of increasing Mg^{2+} concentrations on the response of a calcium-selective electrode.

7.1.5.2 Methods of Evaluation

The frequently used simple means of gaining analytical information is to use $E = f(c)$ plots with semi-logarithmic graph paper, the concentration axis being the logarithmic one, as shown in Figure 7-28. Even simpler is a technique based on the use of analog-reading ion meters.

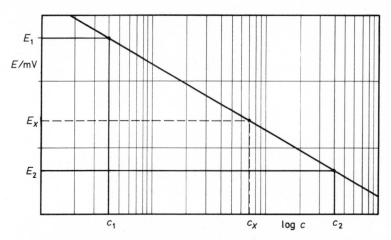

Figure 7-28. Calibration graph for an ion-selective measuring cell. Semi-logarithmic graph paper is used, the log axis of which is used for plotting the concentration. Two calibration solutions of known concentrations c_1 and c_2 lead to the corresponding cell potentials E_1 and E_2. From the potential E_x of the sample the concentration c_x can easily be determined.

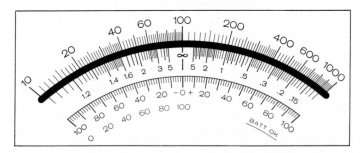

Figure 7-29. Scale of an ion meter (Orion Model 407) with a logarithmic divided scale for concentration measurements without using calibration graphs. The incremental scale will be used for known addition and known subtraction work.

The scale of the indicating meter is as shown in Figure 7-29. This principle is valid only as long as a linear portion of the concentration plot can be assumed.

As will be shown later, some of the commercially available ion-selective electrodes have an unsatisfactory detection limit. This important parameter can be determined following Figure 7-30.

In the case of a chloride-selective electrode, using a solid-state sensor based on a mixed pellet of $Ag_2S + AgCl$, the loss of linearity in a semi-logarithmic plot (see Figure 7-26) starts at a concentration of 10^{-4} mol/L Cl^-. By having more calibration points in the curved region of the graphs the detection limit can be lowered, at least when working with model solutions. It should be remembered, however, that the selectivity problems in such a region are almost unknown, despite a number of publications relating to this aspect [83, 84].

A different evaluation technique is the standard addition method, in which a known amount of the sensed ion is added to a known sample volume. Using an ion meter with a

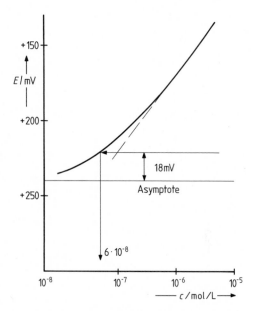

Figure 7-30.
Determination of the detection limit of an ion-selective electrode. The potential E of the asymptote is obtained by immersing the electrodes in CONDISOL/TISAB (see Figure 7-24) which is free from the ion to be sensed. At a distance of 18 mV a parallel is drawn to its intersection with the calibration graph. By dropping a perpendicular to the concentration axis the detection limit is obtained [97].

logarithmic scale the procedure is very simple, as Figure 7-31 illustrates. In the standard addition method, the added volume contains an ion to which the measuring electrode responds.

The principle of the standard subtraction technique is different. The added volume causes a reaction with the sensed ion. An example is the addition of a cadmium solution to a sample in which a measure of the concentration of sulfide is required, and a sulfide-selective electrode is immersed.

The meter scale shown in Figure 7-29 has an "increment range" facility, which is useful for the addition and/or subtraction technique. Detailed information is available [85, 86].

There are a number of simple equations which are the basis of such methods. Greater precision, however, can be achieved where the data handling is done with ion meters which include a microprocessor. One of the best known ion analyzers is made by Orion [DIS 11]. A detailed description was given by Avdeef and Comer [87].

Microprocessor-based ion meters can make use of similar evaluation methods known as multiple standard addition and/or Gran plot titration [88]. Such systems have become very easy to use. Cammann, however, pointed out that there is always a risk of making large errors when handling unknown systems and by mismatching the chemistry with the sample properties [89].

Nevertheless, the simultaneous development of ion-selective electrodes and modern data handling instrumentation has created a new and wide field of potentiometric application which is called in general direct potentiometry.

A. Calibrate unknown.

figure 1

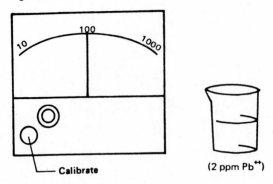

Calibrate   (2 ppm Pb^{++})

B. Add known amount of ion being measured.

figure 2

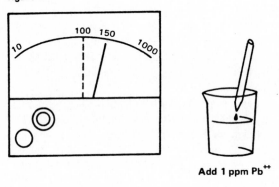

Add 1 ppm Pb^{++}

Figure 7-31.
Concentration measurement with an ion meter which is equipped with a logarithmic scale (see Figure 7-29). The electrodes immersed in the beaker have been omitted.
a) Calibrate unknown; b) add known amount of ion being measured; c) calculated original concentration of the unknown log-scale reading. (Courtesy of Orion).

C. Calculate original total concentration of the unknown from log-scale reading.

7.1.5.3   Commercially Available Ion-Selective Electrodes

All available ion-selective electrodes are membrane electrodes (see Figure 7-1 and Table 7-2).
Details of the construction of glass membrane electrodes have been discussed together with the consideration of pH glass electrodes (see Figure 7-13). Modifications to the construction of other membrane types are described here.

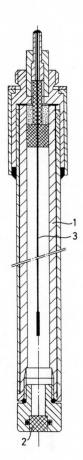

Figure 7-32.
Practical realization of an ion-selective electrode with a solid-state membrane as sensor [109]. 1, Electrode body (13 mm diameter); 2, membrane (sensor, diameter 5 mm, thickness 5 mm); 3, internal reference system.

Typical of ion-selective electrodes using solid-state membranes is the design shown in Figure 7-32. An electrode kit which can be used to realize electrodes with solid-state and/or gel membranes is illustrated in Figure 7-33. Another example using cartridges for gel membranes can be seen in Figure 7-34.

The chlorine-sensing electrodes, which consist of a combination of a redox electrode and a iodide-selective electrode, are of novel design. The electrode is a combination electrode and does not need a reference electrode to complete the cell. When using such an electrode, a defined amount of an acidic iodide reagent is added to the sample. Free chlorine reacts with the iodide and converts an equivalent part of it to iodine. The redox electrode then senses the ratio of free iodine to the residual concentration of iodide whilst the iodide-selective electrode responds to the iodide concentration [90, DIS 12].

Film electrodes, used to sense blood electrolyes such as Na^+, K^+, and Ca^{2+} [DIS 13], are different in construction. All the interfaces of such an electrode, as shown in Figure 7-35 are reversible.

All of the analytical information of interest is given in Table 7-14.

In spite of the many ion-selective electrodes available, it is clear that only a limited number of electrodes are widely used. In addition to pH glass electrodes this is valid for electrodes to sense K^+, Na^+, Ca^{2+} cations and for F^-, Cl^-, NO_3^- and S^{2-} anions.

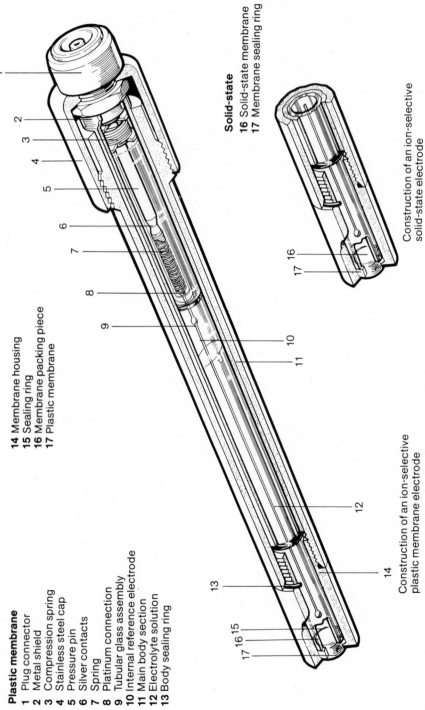

Plastic membrane

1 Plug connector
2 Metal shield
3 Compression spring
4 Stainless steel cap
5 Pressure pin
6 Silver contacts
7 Spring
8 Platinum connection
9 Tubular glass assembly
10 Internal reference electrode
11 Main body section
12 Electrolyte solution
13 Body sealing ring

14 Membrane housing
15 Sealing ring
16 Membrane packing piece
17 Plastic membrane

Solid-state

16 Solid-state membrane
17 Membrane sealing ring

Construction of an ion-selective
solid-state electrode

Construction of an ion-selective
plastic membrane electrode

Figure 7-33. Electrode kit for ion-selective electrodes. All parts, the membrane included, can be demounted and replaced on request. (Courtesy of Philips [91]).

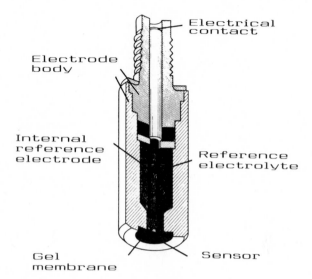

Figure 7-34.
Membrane holder ("cartridge") of an ion-selective electrode with a gel membrane. (Courtesy of Orion). The cartridge fits on to the electrode body. Easy electrode maintenance has been achieved by this construction.

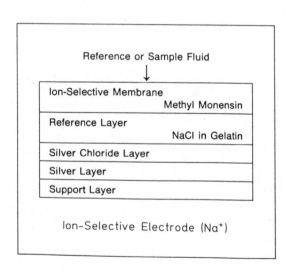

Figure 7-35.
Layers of a "film electrode" following the principles of gel membranes. Each interface of the package is reversible. A complete electrode chip has two such units which form the half-cells of the system. Dimensions 25×25 mm, 1.2 mm thick. (Courtesy of Kodak [DIS 13]).

Table 7-14. Specifications of commercially-made ion-selective electrodes

Ion to sense	Membrane type	Concentration range mol/L a)	pH range	Interferences b)	Made by c)
Amonium	Gel	10^{-6} – 10^{-1}	4 – 7	(K^+) $5\cdot10^{-2}$	P
Bromide	Solid-state	$5\cdot10^{-6}$ – 10^{-1}	1 – 11	(CN^-) 25, (I^-) 20	O,P
Cadmium	Solid-state	10^{-7} – 10^{-1}	2 – 8	Ag^+, Cu^{2+}, Pb^{2+}, Fe^{2+} interfere	O,P
Calcium	Gel	$5\cdot10^{-7}$ – 10^{-1}	3 – 12	(Mg^{2+}) 10^{-2}, (Na^+) 10^{-4}	O,P
Chloride	Solid-state	$5\cdot10^{-5}$ – 10^{-1}	1 – 10	(CN^-) 400, (I^-) 20, (Br^-) 2	O,P
Chloride	Gel	$5\cdot10^{-6}$ – 10^{-1}	2 – 8	no interference by reducdants	O
(Chlorine)	Solid-state d)	10^{-7} – 10^{-4}	2 – 4	strong oxidants interfere	O
Cupric	Solid-state	10^{-8} – 10^{-1}	1 – 5	Ag^+, Cd^{2+}, Pb^{2+}, Fe^{2+} interfere	O,P
Cyanide	Solid-state	10^{-6} – 10^{-3}	11 – 13	(I^-) 3	O,P
Fluoride	Solid-state	10^{-6} – 10^{-1}	4 – 8	(OH^-) 10^{-1}	O
Fluoborate	Gel	$7\cdot10^{-6}$ – 10^{-1}	2 – 8	Acetate, Phosphates interfere	O
Hydrogen-Ion	Glass e)	10^{-14} – 10^0	(1 – 14)	(Na^+) 10^{-13}	O,P,I
Iodide	Solid-state	$5\cdot10^{-8}$ – 10^{-1}	2 – 12	(CN^-) $5\cdot10^{-2}$, S^{2-} interferes	O,P
Lead	Solid-state	10^{-6} – 10^{-1}	2 – 8	Ag^{2+}, Cd^{2+}, Cu^{2+}, Fe^{2+} interfere	O
Nitrate	Gel	10^{-5} – 10^{-1}	3 – 12	(Cl^-) 10^{-2}, (HCO_3^-) $5\cdot10^{-3}$	O
Perchlorate	Gel	$7\cdot10^{-6}$ – 10^{-1}	3 – 12	(NO_3^-) $5\cdot10^{-2}$	O
Potassium	Gel	10^{-6} – 10^{-1}	2 – 12	(NH_4^+) 10^{-2}, (Na^+) 10^{-5}	O,P
Silver	Solid-state	10^{-7} – 10^{-1}	1 – 7	Hg^{2+}, S^{2-} interfere	O,P
Sodium	Glass	10^{-7} – 10^{-1}	7 – 11	(H^+) $5\cdot10^{-1}$, (K^+) 10^{-2}	
Sulfide	Solid-state	10^{-7} – 10^{-1}	12 – 14	Hg^{2+}, Ag^+ interfere	O,P
Thiocyanate	Solid-state	$5\cdot10^{-6}$ – 10^{-1}	4 – 7	S^{2-}, OH^-, Cl^- interfere	O
Water hardness	Gel	$5\cdot10^{-6}$ – 10^{-3}	5 – 8	two-valent cations interfere	O

a) Most of the electrode manufactures quote the upper detection limit as 10^0 mol/L which is not meaningful, however, since no ionic-strength adjustmend can be made at such high concentrations,

b) Selectivity constants k_{ij} for the listed up ions, detailed information in [91],

c) O = Orion [DIS 9], P = Philips [91], I = Ingold [DIS 5],

d) Combination electrode "iodide + platinum", special sample preparation (see text),

e) pH Glass electrodes.

7.1.5.4 Application of Ion-Selective Electrodes

The main fields of application are listed in Table 7-15.

In determining which ion-selective electrodes are suitable for particular applications, a very realistic approach is taken by a number of the electrode manufacturers [97–100]. Since the initial development of ion-selective electrodes, several thousand papers have been published, of which only a small fraction have led to practical use of the electrodes.

Table 7-15. Selected fields of application of ion-selective electrodes [92]

Water analysis [93]

Drinking water for Cl_2, NO_3^-, NO_2^-, F^- (SiF_6^{2-}) a)
Process water for hardness (Ca^{2+} + Mg^{2+}), Cl_2 [122] [123]
Waste water for S^{2-}, CN^-, NH_4^+ (NH_3), NO_3^-, NO_2^-
Boiler feed water for Na^+ [94]

Food and beverage analysis [95]

Dairy products for Na^+ (salt)
Meat products for Ca^{2+}
Snack foods for Na^+ (salt)

Emission monitoring [DIS 14]

Stack gases for HCl and HF after phase change
from gas to liquid, automatic analyzers

Clinical analysis

Blood and body fluids for Na^+, K^+, Ca^{2+} and Cl^-,
automatic analyzers for large sample numbers [94, DIS 15]

End-point determination [79]

Cd-selective electrode for complexometric titrations, Pb-selective electrodes for titration of sulfate [96]

a) LaF_3-based fluoride-selective electrodes sense all the F^- ions in the silicofluoride complex as "free" ions

7.1.6 Gas-Sensing Electrodes

7.1.6.1 Principle of Sensing Dissolved Gases

The first membrane-covered gas-sensing electrode was developed by Clark, who used an amperometric sensing technique to detect dissolved oxygen (see Section 7.2 and [101]). Subsequently Severinghouse and Bradley used the same principle for potentiometric measuring electrodes [102]. Ross et al. widely investigated other possible applications of this type of gas-sensing electrode [103].

In all these electrodes the dissolved gas diffuses across a gas-permeable membrane into a thin layer of a suitable electrolyte located between the membrane and the surface of the potentiometric measuring electrode. Both acidic and/or alkaline gases can be sensed. CO_2, NO_x (NO_2), and SO_2 are examples of acidic gases and NH_3 is an example of an alkaline gas. The electrolyte has to be modified to match the properties of the gas of interest. Typical is an electrolyte solution the pH of which is changed by the gas diffusing across the gas-permeable membrane. Under stationary conditions and making use of the Henry-Dalton law, a simple electrode reaction can be written:

$$E = E_0 \pm 2.303 \, (RT/F) \log p_{gas} \, . \tag{7-33}$$

This is a modification of the Nernst equation in which the $+$ sign represents acidic gases and the $-$ sign alkaline gases.

In the development of these electrodes, one of the main problems is to find a membrane which has a diffusion coefficient for a given gas which is great enough to secure a practically useful electrode response. Ross et al. [103] studied a number of materials. Most of the gas-permeable membranes are micro porous.

The construction of such a gas-sensing electrode is shown in Figure 7-36.

As for ion-selective electrodes, sample preparation is necessary so that ions will be converted in to gases. Table 7-16 includes the pH ranges to be controlled.

Unlike ion-selective electrodes, the temperature dependence of gas-sensing electrodes is governed by the diffusion across the membrane and by the partition coefficient of the gas between the sample solution and the electrolyte behind the membrane. As a consequence, automatic temperature compensation cannot be used. Samples and calibration solutions have to be kept at constant temperature.

Another important point is that the partial pressure of water in the samples to be measured must be close to that of the electrolyte film, otherwise, because of water vapor transport across the membrane, the concentration of the electrolyte will vary, which will result in a drift of the cell voltage.

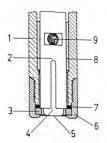

Figure 7-36.
Construction of a gas-sensing electrode. (Courtesy of Orion). 1, Electrode body; 2, body of a pH glass electrode with a flat membrane (4); 5, gas-permeable membrane; 8, electrolyte which forms a thin film between 4 and 5; 9, reference system. The electrode is of the combination type and does not need an external reference electrode.

Table 7-16. Specifications of gas-sensing electrodes

Gas	Concentration range, mol/L	pH range	Interferences, comments
CO_2	10^{-4} — 10^{-2}	1 — 4	NO_2 (NO_x), to destroy by the addition of amidosulfonic acid, SO_2, to oxidies with permanganate
NO_2	$5 \cdot 10^{-6}$ — $5 \cdot 10^{-3}$	2 — 4	In NO_x only the NO_2 component reacts. SO_2, to destroy with permanganate
SO_2	10^{-6} — 10^{-2}	1 — 3	NO_2, to destroy with amidosulfonic acid
NH_3	$5 \cdot 10^{-6}$ — 10^{-1}	11 — 14	volatile amines

Electrode makers: Orion [DIS 9], Kent-EIL [DIS 16]

7.1.6.2 Application of Gas-Sensing Electrodes

There are three main fields of application: the sensing of dissolved CO_2 and of NH_3 in blood and other body fluids, measurement of NH_3 in waste water control, and measurement of CO_2 for process control in fermenters.

In clinical analysis, microprocessor-based instruments which are called blood-gas analyzers are used [DIS 17]. Havas reported in detail on these and other biological applications [104].

Waste water control is concerned mainly with the treatment of sewage. Depending on the loading of a treatment plant, NH_3 (overload) and/or nitrite and nitrate (normal situation) will be found. Passing the water sample over a reducing column filled with amalgamated zinc converts the nitrites and nitrates to NH_3 [105].

Process control for CO_2 in fermenters is carried out with a special industrial probe made by Ingold [106, DIS 18].

The extent to which the ammonia titration in the Kjeldahl procedure for total nitrogen in organic matter can be replaced by ammonia-sensing electrode technology remains to be seen [107]. The only electrodes sold in substantial numbers are for NH_3 and CO_2. Attempts to analyze free gases with these types of electrodes have failed because of the unacceptably long recovery time following a decrease in concentration [108].

7.2 Amperometry

7.2.1 Introduction

Compared with potentiometry and conductometry, amperometric methods have reached their maturity for a wide range of applications rather late. This is surprising, because as early as 1913 the first practical amperometric sensor unit was developed by Rideal and Evans [110]. It was a two-electrode cell using platinum as the working electrode and copper as the counter electrode. This working arrangement, although it has several associated problems, is still used to measure dissolved chlorine in water [DIS 19].

Much more interest has been focused on a voltammetric method which originated in 1922. Its main feature is the use of liquid mercury as the working electrode. This method was termed polarography and over many years its inventor, Heyrovský, and others have laid a sound theoretical basis to the technique [111, 112].

In the late 1950s, Clark [113] and Tödt [114] independently returned to the conventional method of amperometry working with solid-state electrodes. They developed two different approaches for measuring dissolved oxygen. Schwabe [115] in 1965 published a paper which described for the first time the many possible applications of amperometry in liquid analysis. In 1970 Energetic Science [116] filed a patent application for gas analysis using potentiostatic three-electrode amperometry. With the foundation of City Technology [DIS 20] in 1977, the development of amperometric gas sensors began on a broad base and a number of sensors for toxic and noxious gases came on the market. In 1979 the first enzymatic biosensor using amperometry to detect hydrogen peroxide was made by Yellow Springs Instruments [DIS 21]. The latest in the family of amperometric sensors is a detector for organic components previously separated by high-performance liquid chromatography and/or ion chromatography [117, DIS 22].

Today amperometry is growing quickly and, together with potentiometry, conductometry, and polarography, it has become one of the most widely used electroanalytical methods.

7.2.2 Definitions

The classification and nomenclature of electroanalytical techniques is an urgent need since too many arbitrary names are used which are either incorrect or misleading. Although a number of recommendations are now available [118–121], the situation remains rather confused, particularly with respect to amperometry.

The classification in the IUPAC Compendium [120] is based on the nature of electrode reactions in electroanalytical systems. Such a procedure is not very helpful for analysts without an electrochemical understanding. From this point of view, the earlier approach of Kolthoff and Elving [119] makes more sense: the electrode reactions are used as the guideline in this approach also, but the interpretation is easier to understand.

Table 7-17, based on the comprehensive terminology given by Delehay et al. [118], summarizes a selection of terms which are mainly concerned with voltammetry and amperometry.

In an even more general way, Meites [111] defines voltammetry as an electroanalytical method that deals with the effect of the potential of an electrode on the current that flows through it. Once a voltammogram (see Figure 7-37) is available, the potential of the electrode can be fixed within the plateau of the limiting current.

The difference between voltammetry and polarography is that solid-state working electrodes are used, in the former.

An external voltage source is unnecessary when the two electrodes of an amperometric cell generate a voltage of their own which fits the limiting current plateau. It makes no sense to call such a cell a galvanic cell [122]. The term is misleading, since Hersch has used it for a related but different technique [123]. There is also a German Standard which deals with the fundamentals of potentiometric measuring cells which are called "galvanic cells" [124].

The frequently used term "fuel cell" for self-supporting amperometric cells is also unacceptable. Fuel cells are not operated in the limiting current modes [125]. Also, to call for example, the metal lead, which is the counter electrode of a Mackereth Oxygen Cell, "fuel" has little meaning.

Table 7-17. Electroanalytical methods with steady state electrode processes [118]

Relationship	Quantity being controlled	Quantity being measured	Method/Terminology
$f(E, I, c)$	c	$I = f(E)$	Voltammetry, with solid state electrodes a)
	c	$I = f(E)$	Polarography, with dropping mercury electrodes
	E	$I = f(c)$	Amperometry, with solid state electrodes
	E	$I = f(c)$	Endpoint determination by amperometric titration, solid state electrodes

a) Voltammetry has precedence over amperometry. Without knowing the $I = f(E)$ function, the voltammogram, the working point of an amperometric measuring electrode cannot be fixed

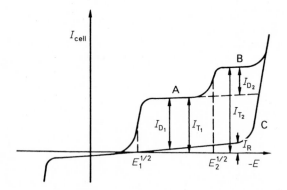

Figure 7-37. Voltammogram of a solution with two electrode-active substances A and B. In the case of the "empty" supporting electrolyte the working electrode of the cell is polarized and only a small residual current I_R will flow. The final current increase of C is due to the electrolysis of the electrolyte. After the addition of A and B with increasing voltage $-E$ a first current increase is caused by the more noble A. A plateau will be reached within which the current I_{D1} is diffusion controlled. The same is valid for B and I_{D2}. In both cases A and B act as depolarizers of the working electrode of the cell. The diffusion current I_D depends linearly on the concentration of a depolarizer. The half-wave potentials $E^{1/2}$ are related to the position of a depolarizer within the electrochemical series. The total currents I_T are the sum of I_D and I_R. Only in the case of trace analysis has I_R to be taken into account.

Another conflicting terminology is the term "polarographic cells". Polarography is an electrochemical technique which makes use of liquid mercury as the working electrode [118]. The frequently used term "electrochemical cell" is the least troublesome. It does not, however, give any methodological information since such a term embraces potentiometry, conductometry, and similar technologies.

In summary, it can be stated that amperometry is concerned with two- or three-electrode measuring cells with solid-state working electrodes. These are polarized in such a way that the limiting current will be reached whenever an electrode-active substance, a so-called depolarizer, is present. The limiting current has a linear dependence on the concentration of depolarizer.

7.2.3 Fundamentals

A simplified circuit for amperometric measuring cells is illustrated in Figure 7-38. The effective voltage E_{WE} at the working electrode depends on a number of other parameters, according to

$$E_{WE} = E_{POL} - E_{CE} - IR . \tag{7-34}$$

It can be seen that the stability of the voltage E_{CE} of the counter electrode is as important as the voltage drop IR across the cell.

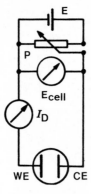

Figure 7-38.
Measuring circuit for two-electrode amperometric cells. E, Direct current source; P, variable resistor to select the cell voltage; I_D, current indicator; WE working electrode; CE counter electrode (reference electrode).

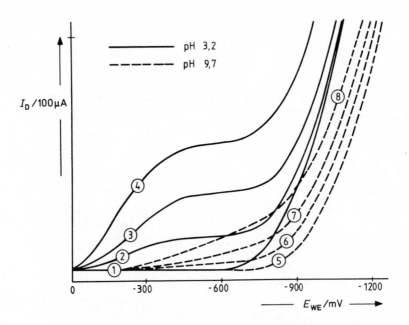

Figure 7-39. Voltammograms of hypochlorite solutions of increasing concentration for two different pH values. Concentrations: 1, 5:0 mg/L, 2, 6: 3 mg/L, 3, 7:6 mg/L, 4, 8:10 mg/L ClO^-, Only at lower pH values can an established plateau of the diffusion current I_D be observed [126].

The counter electrode of two-electrode cells acts as a kind of reference electrode, a premise which can only hold when its potential E_{CE} is independent of the flowing current I. Electrodes which fulfil this requirement are called unpolarizable electrodes.

The unaviodable voltage drop IR can be accommodated when the limiting current is well established. In contrast to Figure 7-37, which shows an idealized voltammogram, a more realistic example is given in Figure 7-39.

In the case of a sample pH lower than 3.5, the plateau is of a span which can stand a fairly large shift of E_{WE} caused by the voltage drop IR. At elevated pH values, however, no plateau at all will be formed and only a slight bend of the current curve can be observed. It approaches the final current increase quickly.

In such a situation, a simple two-electrode cell cannot be used. Another principle has to be used, viz, a three-electrode cell with an additional reference electrode. The reference electrode should be positioned as close as possible to the working electrode. In scientific studies

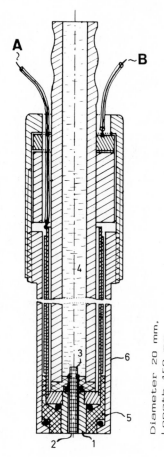

Figure 7-40.
Amperometric working electrode with a modified Luggin capillary [126]. 1, Working electrode; 2, 3, glass-fiber bundle in a coaxial drilling in the working electrode; 4, KCl solution of a remote reference electrode B; 5, insulator; 6, metallic electrode body which acts as counter electrode; A and B electrode leads. The system is a complete three-electrode measuring cell.

Figure 7-41.
Measuring circuit for three-electrode amperometric cells. E, direct current source; P. variable resistor to preselect the voltage of the working electrode WE; RE, reference electrode which controls the preselected voltage of the working electrode WE with the help of the potentiostat; CE, counter electrode; Pt 100, temperature sensor for automatic temperature compensation; I_D, current indicator.

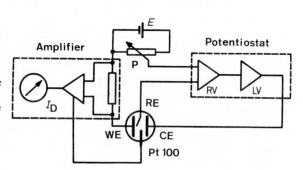

this can be achieved by the use of so-called Luggin capillaries [127], but this is not a practical method for analytical applications. For these a much better construction is shown in Figure 7-40 [126].

The measuring circuit also has to be modified. Instead of the simple d.c. voltage source in Figure 7-38, an electronic potentiostat will be used. This compares a preselected voltage difference between E_{WE} and E_{RE} and compensates any deviation automatically. Figure 7-41 shows the principle of such an arrangement [128, 129].

Another important feature of potentiostatic amperometry with three-electrode cells is that the voltage drop IR no longer contributes to E_{WE} (see Equation (7-34)).

For all amperometric cells, the limiting current I_D is linked with a number of other parameters in accordance with the equation

$$I_D = (n \cdot F \cdot A \cdot D \cdot c)/\delta \tag{7-35}$$

where n is the number of electrons participating in the electrode reaction, A is the area of the working electrode, F is the Faraday constant ($9.6485 \cdot 10^4$ C/mol), D is the diffusion coefficient of the electrode-active species in a sample, c is the concentration of the species (mol/L), and δ is the thickness of the diffusion layer at the surface of the working electrode. The thickness δ depends strongly on the rate of stirring and/or the flow-rate of the sample. The relative movement of the sample superimposes forced convection on the transport of the electrode-active species to the surface of the working electrode.

Equation (7-35) is only valid when there is no effect due to migration of the electrode-active species in the electric field applied across the electrodes, an effect which, in the case of charged ionic species (cations, anions), will easily take place. To prevent the contribution of a migration current to the diffusion current, I_D, an appropriate supporting electrolyte has to be added to the sample. Supporting electrolytes are strong and/or weak and contain cations and anions with such high standard potentials that the final current increase of a voltammogram will not be influenced by the discharge of these ions. The frequently observed influence of pH on electrode reactions can be taken into account by formulations of supporting electrolytes which also behave as pH buffers.

7.2.3.1   Usable Potential Ranges

The working electrode material, the properties of the solvent, and the pH value strongly influence the potential range which is available for analytical tasks. For aqueous solutions Figure 7-42 illustrates this for some frequently used electrode materials [126].

The final current increase is caused by the electrolysis of the solvent water, which means the generation of hydrogen at negatively polarized electrodes and of oxygen at positively polarized electrodes. The foregoing electrode reaction involves the discharge of H^+ and/or OH^- ions and it is therefore understandable that the pH of the solution will influence the usable working range.

According to Meites [111], this dependence can be expressed by a number of simple equations for the most negative potential E^- and the most positive potential E^+ in mV in relation to a saturated calomel electrode:

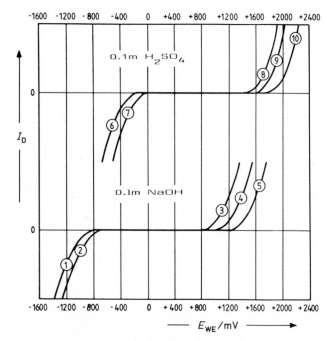

Figure 7-42. Usable potential ranges of working electrode materials [126]. 1, Boron carbide, gold; 2, platinum; 3, boron carbide; 4, platinum; 5, gold; 6, boron carbide; 7, platinum, gold; 8, boron carbide; 9, platinum; 10, gold. The final current increases in the + voltage ranges are due to oxygen generation whereas in the − ranges hydrogen is being formed.

Platinum electrodes:

$$E^- = -60 \cdot pH \tag{7-36}$$

$$E^+ = 1200 - 60 \cdot pH \tag{7-37}$$

Gold electrodes:

$$E^- = -600 - 60 \cdot pH \tag{7-38}$$

$$E^+ = 1450 - 60 \cdot pH \tag{7-39}$$

The correlation with our results is poor. Nevertheless, similar equations can be derived for specified experimental conditions and they are helpful for an initial assessment of the possibility of using the electrode arrangement for amperometric analysis.

The selection of the solvent and of the supporting electrolyte has a much greater influence on the usable potential range. Organic solvents, however, for practical reasons can only be applied for amperometric gas sensors in which, because of the greater solubility of many electrode-active gases, an increase in sensitivity is obtained. The main advantage of organic solvents is the extended potential range, as shown in Figure 7-43.

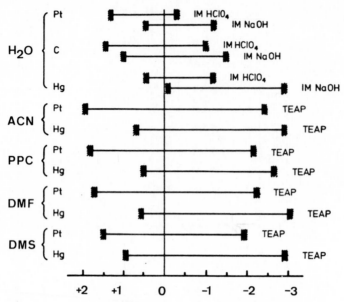

Figure 7-43. Usable potential ranges of working electrode materials in organic solvents and tetraethylammonium perchlorate (TEAP) as supporting electrolyte [130]. ACN, Acetonitrile; PPC, propylene carbonate; DMF, dimethylformamide; DMS, dimethyl sulfoxide.

The extension of the potential range is of practical interest insofar that in aqueous solutions many chemical substances of interest cannot be determined amperometrically since their working potential ranges are hidden behind the final current increase caused by water electrolysis. The results shown in Figure 7-43 are related to the use of TEAP as a supporting electrolyte. The latest development in this area comes from the use of tetrabutylammonium hexafluorophosphate as the supporting electrolyte [DIS 23]. This extends the working range in acetonitrile to more than +3000 mV. In this potential range, Fleischmann and Pletcher effected the oxidation of several aliphatic hydrocarbons [131].

7.2.3.2 Cyclic Voltammetry for Studying Electrode Reactions

To optimize the potential of an amperometric working electrode, the half-wave potential $E_{1/2}$ should be known (see Figure 7-37). The best way of obtaining such information is to use cyclic voltammograms. This technique utilizes a triangular sweep generator [132] which changes the potential of the working electrode at a rapid but constant rate to a preselected value, followed by a potential decrease at the same rate back to the starting value. An example of such a voltammogram is shown in Figure 7-44.

The interpretation of this voltammogram is that at the start the platinum is covered with residual adsorbed hydrogen from earlier cycles. With increasing positive polarization this is removed oxidatively until a dead current zone is reached which is characteristic of a clean metal surface. In the next step OH^- will be adsorbed and platinum oxides may form. Finally, oxygen is generated. Reversing the polarity of the voltage sweep leads to the delayed

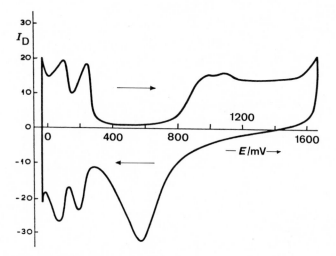

Figure 7-44. Cyclic voltammogram of a platinum electrode in 0.5 M sulfuric acid. Electrode area, $1.25 \cdot 10^{-3}$ cm^2; scanning speed, 30 V/s. The solution has been deaerated (after [133]).

reduction of the oxides and finally to adsorption of H$^+$ ions and the generation of hydrogen.

It is clear that in solutions of sulfuric acid platinum can be used as the electrode metal from 0.4 to 0.8 V versus to a standard hydrogen electrode.

Studies of this kind should be carried out with rotating disc electrodes to guarantee an even thickness of the diffusion layer δ (see Equation (7-35)) [134, 135]. The usefulness of cyclic voltammetry has been described elsewhere [136, 137]. It should be pointed out that when using membrane-covered working electrodes (see later) the electrode reaction has to be tested without the membrane.

7.2.3.3 Standard Potentials and Half-Wave Potentials

Because of a lack of the necessary test equipment, it is frequently attempted to correlate the half-wave potential $E_{1/2}$ with the standard potential E_0 of the redox reaction responsible for the electrode reaction. Such a procedure is unacceptable for a number of reasons.

One of the main objections is that the redox systems under consideration are frequently highly irreversible, in which case other parameters contribute to the value of E_0 (eg, how E_0 has been determined (information which is often unavailable), the influence of poising of the system, and which pH dependence is observed). In addition, it is known that many electrode-active substances may form redox systems with entirely different ways of reacting. It is not usually known which reaction is actually taking place. For example, several different reactions by which nitrogen oxide (NO) can be reduced can be cited:

		E_0 (mV)
$2\,NO + 6\,H^+ + 6\,e^- \rightleftharpoons 2\,NH_2OH$		$+380$
$2\,NO + 10\,H^+ + 10\,e^- \rightleftharpoons 2\,NH_3 + 2\,H_2O$		$+730$
$2\,NO + 2\,H^+ + 2\,e^- \rightleftharpoons N_2O + H_2O$		$+1590$
$2\,NO + 4\,H^+ + 4\,e^- \rightleftharpoons N_2 + 2\,H_2O$		$+1680$

It is even more doubtful as to what extent tabulated half-wave potentials can be used. Most of them are related to polarographic techniques using liquid mercury as the electrode material where the electrode reaction is different from that at solid-state electrodes.

Tables providing data for solid-state working electrodes [134] are based on unstirred solutions, which is again different from the approach taken in amperometry. It might be useful, however, to utilize the well developed theoretical considerations on polarographic half-wave potentials [111].

7.2.3.4 Selectivity in Amperometry

In considering the electrochemical analysis of systems with more than one electrode-active species, amperometry in general is not a very selective technique. Only that component associated with the first wave (see Figure 7-37) can be determined without an error being introduced by other species which are indicated at more negative and/or more positive half-wave potentials. A good example is the amperometric determination of dissolved chlorine in drinking water in the presence of dissolved oxygen. In this case no interference by oxygen will be observed. The reverse of this analytical task, ie, the determination of dissolved oxygen in the presence of chlorine, is far more difficult. However, membrane-covered working electrodes can be used to improve the selectivity, eg, the speed of diffusion of oxygen through silicone rubber is about 1000 times faster than that of chlorine [138].

In environmental air analysis, the monitoring of the constituents CO, NO, NO_2 and SO_2 is much more difficult. Table 7-18 shows the magnitude of cross-sensitivities which must be taken into account.

Table 7-18. Cross sensitivity of a three-electrode carbon monoxide cell to other gases [DIS 20]. The values shown are valid for the Type 3 F sensor

Test gas used	concentration	approx. CO equivalent
Hydrogen	100 ppm	40 ppm
Sulfur dioxide	100 ppm	65 ppm
Carbon dioxide	25% vol.	0
Methane	5% vol.	0
Ethylene	100 ppm	80 ppm
Hydrogen sulfide	10 ppm	35 ppm
Chlorine	10 ppm	-2 ppm
Nitrogen oxide	100 ppm	20 ppm
Nitrogen dioxide	10 ppm	-6 ppm

Table 7-19. Chemical scrubbers to remove the interference of other gases [DIS 24]

Scrubber Model	Used with this analyzer	Removes
MS-150	H_2S, HCN	SO_2, NO_2, Cl_2, HCl
MS-50	H_2S, Cl_2, NO_2	SO_2
MS-166	HCN	SO_2, NO_2, Cl_2, HCl
MS-325	HCN	H_2S
MS-170	NO_2, Cl_2	SO_2, H_2S, HCN, HCl
FB-166	SO_2	Cl_2, NO_2, HCl
FB-170	SO_2	H_2S
56	NO	SO_2, NO_2, H_2S, HCN
158	CO	NO_2, SO_2, H_2S, HCN
MS-100	CO	alcohols, aldehydes

Improved selectivity is frequently achieved by sampling the gas through a scrubber. Table 7-19 lists the types of scrubbers which are avaiable to reduce interferences in multi-component gas mixtures [DIS 24].

Another approach that has been utilized to minimize cross-sensitivities is the use of multi-sensor systems [139]. It is debatable, however, how far the data processing of the signals from different sensors really can provide better selectivity. There are too many dependent variables both known and unknown. The pattern-recognition requirements will quickly reach the capacity of even a medium-sized computer. An associated but different problem is that test gas mixtures will not always be stable and chemical reference methods are generally too slow to determine the actual composition of a gas mixture. Nevertheless, a number of computerized gas analysis systems using amperometric cells as sensors are available [140, DIS 25].

7.2.3.5 Detection Limits (Sensitivity)

The parameters of an amperometric cell which control the sensitivity can be seen clearly from Equation (7-34).

The simplest way to enhance the sensitivity is to increase the area A of the working electrode, a procedure which is being used successfully with membraneless working electrodes [DIS 26, DIS 27].

In the case of membrane-covered working electrodes, there are still problems to overcome. The first is that the electric field generated by the voltage applied across the electrodes cannot reach a significant distance from the edge of the working electrode to its center in the thin film of electrolyte behind the membrane. The field will diminish gradually, which makes a greater cross section of the electrode meaningless. The situation can be improved by inserting a porous spacer (eg, lens paper) between the membrane and the working electrode (see Figure 7-49). The increased response time resulting from such a procedure can be tolerated as long as it does not exceed 60 s of the 90 % values. Another improvement is based on making fine radial engravings in the surface of the working electrode [DIS 27]. Both measures also improve the rate of transport of the reaction products from the surface of the working electrode.

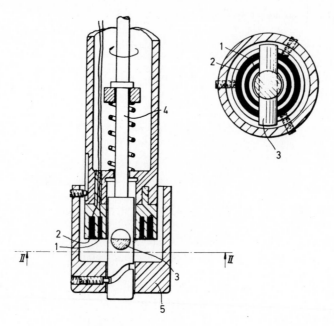

Figure 7-45. Membraneless amperometric oxygen sensor with automatic electrode cleaning. 1, Working electrode; 2, counter electrode; 3, carrier for two alumina grinding stones; 4, axial drive for 3 and for the "sampling beaker" 5.

For membraneless working electrodes, a faster sample flow causes a decrease in the thickness layer δ in Equation (7-34). The resulting current increase improves the signal-to-noise ratio of the cell, which in the final analysis corresponds to the lower detection limit.

Using these techniques, detection limits as low as 1 ppb of dissolved oxygen [DIS 28] or 10 ppb of dissolved chlorine [DIS 19] can be reached.

7.2.3.6   Temperature Coefficients

The diffusion coefficient D of the electrode-active species depends on the temperature, and consequently this also has an effect on the limiting current I_D. A different temperature dependence is observed with membrane-covered working electrodes, in which the permeation of the gas to be detected follows a different mechanism [141, 142]. Finally, in dissolved oxygen measurements the degree of oxygen saturation of a solution is frequently of interest. It changes with the temperature (and with the barometric pressure). For a pressure of 1013 hPa the saturation value at 20 °C is 8.84 mg/L O_2, and at 30 °C it is 7.35 mg/L O_2 [143].

It is clear that these different relationships are not always amenable to temperature compensation with a single temperature sensor and/or temperature-compensating circuit in the amplifier of the oxygen meter. Optimized temperature compensation is achieved using microprocessor-based signal- and data-handling techniques [DIS 29].

7.2.3.7 Liftetime of Amperometric Cells

The lifetime of a sensor is an important factor not only economically; frequent maintenance is also a cost factor. In membrane-covered amperometric cells the limiting factor of the cell lifetime is the volume of the electrolyte (see Figure 7-49). As the electrode reaction takes place the electrolyte undergoes a change in its concentration and/or composition. For example, a Clark cell can be considered. At the counter electrode, which is made of silver, an electrode reaction takes place which consumes the bromide (chloride) ion from the electrolyte converting Ag into AgBr:

Cathode reaction:

$$O_2 + 2\,H_2O + 4\,e^- \rightleftharpoons 4\,OH^-$$

Anode reaction:

$$4\,Ag + 4\,X^- \rightleftharpoons 4\,AgX + 4\,e^-$$

where X represents Cl^- or Br^-.

Once the halide concentration has been depleted to a critically low value, silver oxide (Ag_2O) will be formed in preference to AgX.

This results in a change of the electrode potential E_{CE} (see Equation (7-34)) of about 350 mV, which is enough to shift the potential E_{WE} of the working electrode out of the range of the plateau of the limiting current I_D (see Figure 3-37). As a consequence, the amperometric oxygen cell fails to work properly.

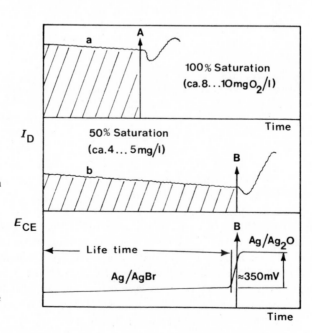

Figure 7-46.
Lifetime parameters of an amperometric oxygen cell [144]. Upper part: change of the diffusion current I_D with time for two different values of oxygen saturation of an aqueous solution. The cell fails to work properly at A and/or B where B is 2A under the specified conditions. Lower part: change of the potential E_{CE} of the counter electrode under B conditions. At the end of the lifetime the electrode reaction is changing.

Figure 7-46 illustrates this situation and shows that the product of current and time can be made to control the change in the anode reaction [144].

A different situation pertains in amperometric cells used in gas analysis which make use of an "air electrode" as counter/reference electrode. Figure 7-47 summarizes the principle and shows that the reaction at the working electrode is fully compensated by that at the counter electrode [145].

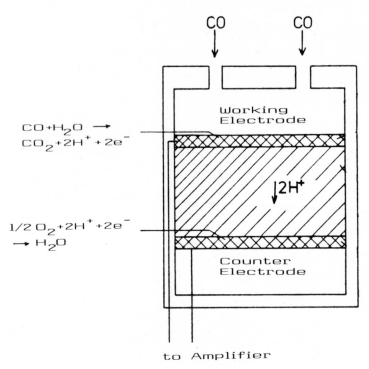

Figure 7-47. Electrode reactions of an amperometric CO gas sensor [145]. The sensor is working with an air electrode as counter electrode. In respect of the composition of the sulfuric acid used as electrolyte an unlimited sensor life could be predicted owing to the counter-balanced electrode reactions.

Such a cell might be expected to have an indefinite working life since, irrespective of the amount of current flowing through it, there will be no change in the composition of the electrolyte H_2SO_4. However, because both of the electrodes are catalytically activated using finely divided layers of noble metals, there is in practice a gradual loss of activity. The lifetime is therefore estimated empirically; it is usually 1–2 years [DIS 20].

All those amperometric cells which work without membrane-covered working electrodes are free from these life-limiting chemical effects (see Figures 7-48 and 7-49). Since such electrodes are usually equipped with automatic electrode-cleaning systems, it is the mechanical effects that will limit the lifetime. The abrasion of the electrodes in a "Züllig Oxygen Probe" [146] is about 0.02 mm/d, which corresponds to a lifetime of about 2 years.

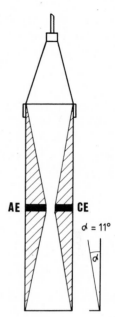

Figure 7-48.
Amperometric oxygen depth-profile sensor [114]. Immersing the sensor at constant speed in a lake, for instance, guarantees a constant flow of the water along the membraneless electrodes AE and CE.

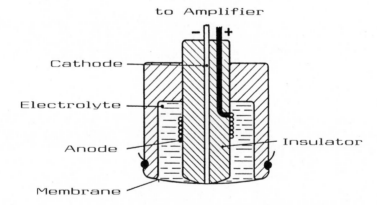

Figure 7-49. Membrane-covered amperometric sensor based on the Clark principle. The surface of the anode formed by an AgCl-covered silver wire should be at least ten times that of the cathode. To stabilize the thin electrolyte film between cathode and membrane a thin porous structure (eg, lens paper) can be inserted.

7.2.3.8 *Classification of Amperometric Cells Based on Details of Construction*

Three different types of cell of different construction can be distinguished, each of which has its own specifications.

Membraneless Working Electrodes

Cells of this design were the first type of amperometric sensors, as has been pointed out (Section 7.2.1 [110]) and modified versions are still in use [DIS 19]. Their main disadvantage is the badly defined potential of the counter electrode. Voltammograms of such systems have never been recorded and all of the development is based on empirical procedures.

These limitations also apply to the oxygen sensors developed by Tödt [114]. An example of such a cell construction is shown in Figure 7-48.

Similar systems have been studied in some detail by Züllig [146]. He added a mechanical automatic cleaning system for both the electrodes (see Figure 7-45) and investigated in detail means of influencing the operating parameters of the sample, particularly the change in the electrolytic conductivity, the pH value, and the presence of redox systems. Industrial oxygen probes have been used successfully to control the aeration of the biomass in sewage plants, for which a high precision is not required.

Membrane-Covered Working Electrodes

The development of these types of oxygen sensors derives from Clark's work [113]. Figure 7-49 illustrates the cell design.

The working electrode is covered by a polymer membrane which is permeable to oxygen. The main advantage is that the electrode reaction takes place in a thin film of an electrolyte which is optimized to the oxygen reaction. Depending on its pH value, two different reactions can take place:

Acidic electrolyte:

$$O_2 + 4\,H^+ + 4\,e^- \rightleftharpoons 2\,H_2O$$

Alkaline electrolyte:

$$O_2 + 2\,H_2O + 4\,e^- \rightleftharpoons 4\,OH^-$$

The second reaction is used in a modified Clark sensor which is usually called the Mackereth sensor [142, 147]. The electrolyte selected is a solution of KOH. Pairing silver as the working electrode with lead as the counter electrode results in an effective voltage of the working electrode which is within the width of the plateau of the limiting current of oxygen reduction. No external polarizing voltage is needed, but there is no reason to call such a cell a "fuel cell", as was pointed out in Section 7.2.2.

Kane and Young extended the application of amperometric cells with membrane-covered working electrodes to the determination of dissolved chlorine [148].

Cells with Gel Electrolyte

This third type of cell is characterized by the use of electrolytes which are solidified by the addition of hydrophilic organic polymers or by the addition of inorganic adsorbents such as silica gel [145, 149]. The use of such cells is restricted to gas analysis.

Since the working electrode is embedded in an electrolyte gel, no membrane will be needed. Cells of this design respond quickly to the gaseous species which can be sensed. When its concentration decreases, however, long recovery times can be observed. This is a consequence of the slow diffusion of the products of the electrode reaction through the highly viscous electrolyte.

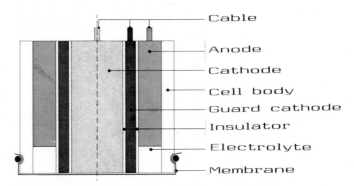

Figure 7-50. Membrane-covered amperometric oxygen sensor. (Courtesy of Orbisphere [151, DIS 32]. An additional guard-ring cathode lowers the residual current (see Figure 7-37) to negligible values so that the lower detection limit of oxygen traces has been reached.

7.2.4 Specifications of Commercially Available Amperometric Cells

7.2.4.1 *Liquid Analysis*

Dissolved oxygen

Monitoring dissolved oxygen is one of the domains of amperometry. Owing to its importance, an IEC document has been prepared which specifies membrane-covered amperometric cells in detail [122]. Importantly, the details of the calibration techniques are included. Calibration is based on the saturation of temperature-stabilized water with oxygen-nitrogen mixtures of known oxygen partial pressure. Even more convenient is the "partial pressure calibration" introduced by Schuler [144, 150], which can be carried out either in water saturated with air or in ambient air saturated with water vapor.

Table 7-20 lists a number of oxygen analyzers for which the technical specifications are provided. It is logical to consider the cells together with the "meter" as a complete sensing unit.

In addition to oxygen, the measurement of dissolved chlorine is another important application of amperometric liquid analysis, but for a number of reasons this application is more complex. By definition, a distinction can be made between free chlorine, which includes Cl_2, hypochlorous acid (HOCl) and the hypochloride anion (ClO$^-$), passive chlorine in the chloramines group and total chlorine, the sum of both. In most cases the concentration of free chlorine is required. It should be realized that there is a pH-dependent equilibrium of HOCl and its anion ClO$^-$ with a pK value of 7.5 (see Figure 7-25). The two components have different diffusion coefficients D in Equation (7-35) and different half-wave potentials. Figure 7-39 illustrates this conflicting situation.

Table 7-20. Specifications of commercially available oxygen analyzers based on amperometry

Analyzer Model	Cell specification a)	Measuring range of O_2	Other comments	References
DIGOX K 301	Membrane-less 3-electrode cell, WE: Ag, CE: SS, RE: Ag/AgCl, ext. polar. voltage	0 – 100 µg/L 0 – 1000 µg/L	The complete sensor unit is made up of a kit of discrete sub-units, eg, a flow controller and a "salinator" for water of low conductivity. High pressure sampling systems are available for boiler feed water analysis	[DIS 30]
Self-cleaning oxygen probe	Membrane-less 2-electrode cell, WE: amalgamated Ag, CE: Zn or Fe, ext. polar. voltage	0 – 5 mg/L 0 – 15 mg/L	Both of the ring-shaped electrodes are being cleaned continuously by the rotation of 2 alumina grinding stones (see Figure 7-49)	[146, DIS 31]
Model 2110 Sensor	Membrane-covered cell, 2-electrode cell with an additional guardring cathode	0 – 20 ppb 18 – 199 ppb, 0.18 – 1.99 ppm, 1.80 – 19.9 ppm	The guard-ring cathode prevents any oxygen diffusion from the electrolyte reservoir resulting in a negligibly low residual current, (see Figure 7-50)	[151, DIS 32]
TRIOX EO 200	Membrane-covered 3-electrode cell	0 – 80 mg/L 0 – 700% sat., 0 – 1500 hPa pO$_2$	The reference electrode controls the potential of the counter electrode by which principle an electrolyte exhaust will be detected early	[DIS 29]
O$_2$ Measuring System	Membrane-covered 2-electrode cell, WE: Pt, CE: Ag/AgCl, PTFE/ Silicon membrane, reinforced by a steel mesh, ext. pol. voltage	0 – 135 0 – 270 0 – 1065 hPa pO$_2$	Steam sterizable probe, up to 130 °C, pressure max. 6 bar, control of oxygen demand/saturation in bioreactor (fermentors)	[DIS 33]
BOD-M3 Analyzer b)	Membrane-covered 2-electrode cells, extern. pol. voltage	0 – 10000 mg/L BOD	Short response-time monitor for Biological Oxygen Demand (BOD) using two O$_2$-cells at the inlet/outlet of an integrated mini-bioreactor. The oxygen demand is proportional to the organic load of the continuously flowing water sample	[152, DIS 34]

Table 7-20. continued

Analyzer Model	Cell specification a)	Measuring range of O_2	Other comments	References
TOXALARM b)	Membrane-covered 2-electrode cell	Sensitivity: 0.25 mg/L Ag^+, 10 mg/L Cu^{2+}, 100 mg/L 2,4-D, 700 mg/L Kresole	The continuous flowing water sample is mixed with a microbial broth and saturated with air, after passage over an integrated mini-bioreactor the oxygen concentration is measured again. Toxic substances block the metabolism and the O_2 concentration increases from zero	[DIS 35]

a) WE working electrode, CE counter electrode, SS stain-less steel, RE reference electrode
b) amperometric oxygen measurements are being used to monitor indirectly the BOD and/or toxicity of water samples

Table 7-21. Specifications of commercially available chlorine sensors based on amperometry

Analyzer Model	Cell specification a)	Measuring range of chlorine	Other comments	References
DEPOLOX 3	Membrane-less 2-electrode cell, WE: Pt, CE: Cu	0-0.5-1.0-2.5-20 mg/L suitable also for ClO_2 and O_3	Continuous electrode cleaning by whirling sand particles, flow controller. pH value for chlorine 4-8, for ClO_2 below pH 4.	[DIS 19]
DIGOX OZ-201	Membrane-less 3-electrode cell, WE: Au, CE: SS, RE: Thalamid b)	0-0.1-0.3-4.0-9.9 mg/L, same range for O_3, too	Potentiostatic 3-electrode amperometry, switch-over ranges for Cl_2 and O_3, flow controller	[DIS 30]
CHLOROMAT	Membrane-less 3-electrode cell, WE: Au, CE: SS, RE: Ag/AgCl	0-200 μg/L, 0-200 mg/L	Continuous cleaning of the large-area working electrode by whirling glass balls, flow controller	[DIS 36]
DULCOMETER CHLORINE CELL CLE II	Membrane-covered 2-electrode cell, WE: Au, CE: Ag/AgCl	Cl_2/HOCL range 0-2, 0-15 mg/L	pH range 3.5-8, but constant, change of sample flow without influence whenever >0.5 L/min	[DIS 37]

a) WE working electrode, CE counter electrode, SS stain-less steel, RE reference electrode
b) Thalamid reference system Hg,Tl/TlCl/Cl^-, Section 7.1 Potentiometry, Table 7-5

For membrane-covered amperometric chlorine cells, both Cl_2 and $HOCl$ will pass through the membrane whereas ClO^- shows no permeation. Finally, there are no simple calibration procedures similar to those for oxygen measurements. The use of colorimetric test kits as advised by some manufacturers is not recommended. Only with photometric equipment can a reliable chlorine calibration be carried out.

These considerations clearly show that the amperometric determination of dissolved chlorine is not a precise technique, despite the „precise specifications" in advertisements and data sheets. In relation to the considerations of commercially available oxygen-measuring instruments, Table 7-21 describes in some detail cells and meters for dissolved chlorine.

Several other compounds are detected with amperometric cells. However, only hydrazine analyzers will be considered briefly. Hydrazine is used to remove dissolved oxygen from boiler feed water to prevent oxygen corrosion. Hydrazine N_2H_4 and the corresponding cation $N_2H_5^+$ are partners of a pH-dependent equilibrium, the pK value of which is 8.2 (see Figure 7-25). At this pH only 50% of free hydrazine is available for the anodic oxidation in an amperometric cell. For different reasons boiler feed water is usually conditioned by ammonia to maintain a pH value of 9.2–9.5, which corresponds to about 90% of free hydrazine. The usual concentration range of hydrazine meters is 0–500 ppb N_2H_4, which can easily be covered by two-electrode cells without any additional pH correction [DIS 38–DIS 40].

7.2.4.2 Gas Analysis

Basic Consideration

It has been pointed out earlier that the construction of amperometric cells for gas analysis in many cases is similar to that of cells for liquid analysis. Some important differences have to be considered, however.

Gases such as carbon monoxide or nitrogen oxide can be determined only with catalytically activated working electrodes. The fabrication of such electrodes is fairly simple. Porous polymer membranes can be metallized with finely divided metals such as platinum or rhodium by sputtering and/or chemical vapor deposition. In some instances screen printing is an even simpler alternative. The membrane and the working electrode form a unit of the cell design.

Another approach which broadens the field of application is based on auxiliary chemical reactions. Two typical examples are as follows. Acid gases which would not be detectable with an amperometric cell will react with an electrolyte behind a membrane which is an iodate-iodide mixture:

$$IO_3^- + 5 I^- + 6 H^+ \rightarrow 3 I_2 + 3 H_2O$$

The two components will only react with one another in the presence of protons, which are produced by the dissolution of acidic gases in the aqueous electrolyte mixture. The free iodine formed can be detected easily by a reducing polarized working electrode similar to the mechanism for determining dissolved chlorine.

Ammonia and organic amines will react with an anodically polarized working electrode made of copper:

$$4 NH_3 + Cu \rightarrow [Cu (NH_3)_4]^{2+} + 2 e^-$$

The use of membraneless cells, unlike liquid analysis, is no longer limited to samples whose conditions match the electrode reaction of interest. Figure 7-51 illustrates a cell in which a pH-modified electrolyte optimizes the determination of chlorine [DIS 41]. A similar device utilizes a wick to deliver the electrolyte to the electrodes [DIS 42].

Another important point in the development of gas-sensing units is in the field of personnel protection, where the availability of portable monitors which are light-weight, small, and do not need much power is essential. The simple function control of such monitors is of importance because catalytically activated electrodes, for instance, can easily be poisoned by various chemical species. To facilitate performance testing, test gas units have been developed. These make use of the technique of breaking an ampoule in a chamber of defined volume. Others depend on electrolytic test gas generation [DIS 43–DIS 45].

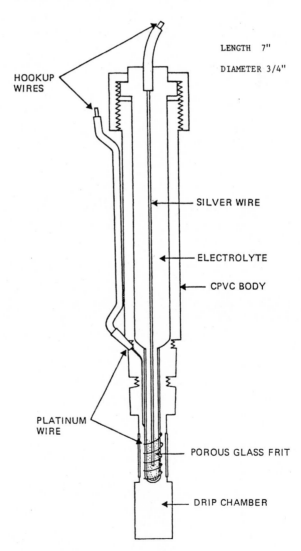

LENGTH 7"

DIAMETER 3/4"

HOOKUP WIRES

SILVER WIRE

ELECTROLYTE

CPVC BODY

PLATINUM WIRE

POROUS GLASS FRIT

DRIP CHAMBER

Figure 7-51.
Realization of a membraneless amperometric gas sensor for chlorine. (Courtesy of MDA [DIS 41]). The electrolyte penetrates from its internal reservoir a porous glass frit around the outside of which is wound a platinum wire. This wire is the working electrode (cathode) whereas the counter electrode is positioned in the electrolyte reservoir. Owing to the permanent flow of electrolyte it has to be replenished periodically.

Cells for Gas Analysis

Oxygen deficiency:

In coal mines, underground sewage channels, and sewage treatment plants there is a permanent danger for personnel of the sudden and uncontrolled escape of large volumes of carbon dioxide, which cannot be detected by portable gas-sensing equipment. Such an event, however, can drastically reduce the oxygen concentration in the air. It is therefore sensible to monitor the concentration of available oxygen and to set an audio and/or visible alarm to the critical level of oxygen deficiency. The amperometric cells for such oxygen meters are normally two-electrode cells based on the Mackereth principle [147], which means that they do not need an external polarizing voltage. One of the most widely used sensors has no membrane but uses a diffusion barrier to restrict oxygen diffusion to the cathode to a defined but low value. The lifetime of such a sensors is about 100000% h, which means 5000 h of use at an oxygen concentrations of 20% [DIS 46].

Toxic gases:

These are gases which either show a high degree of toxicity to living individuals or which are noxious in the sense of air pollution. There is therefore a large group of gases and/or volatile chemical compounds which are of interest. Based on information from two instrument makers, Table 7-21 lists the type of compounds, the detection ranges, and some additional information.

Amperometry is not a very selective or precise method in such applications. Nevertheless, it is of great practical importance because of its simplicity and cheapness.

Flue gas analysis:

A special field of application of amperometry is in the analysis of flue gases in both power stations and households using fossil fuels. Controlling the concentration of carbon monoxide enables the authorities to check the economy of an oven or burner [154, DIS 49]. Another aim is to check that concentration limits are not being exceeded. Special flue gas analyzers have been developed which make use of the data available from a number of different gas sensors to improve the selectivity [139, 140, DIS 25] (see Section 7.2.3.4).

7.2.4.3 Chromatographic Detectors

High-performance liquid chromatography (HPLC) has become a very powerful analytical tool during the last two decades. In parallel with the development of the separation technique, the need arose for detectors which would provide a signal proportional to the concentration for each of the zones separated by the chromatographic column (see Chapter 10). Many of the detectors used are electrochemical types, which, apart from conductivity monitoring in ion chromatography, means amperometry for HPLC in separating organic mixtures. The relevant features of such amperometric detectors are listed in Table 7-23. The specifications were collected from the information sheets of different instrument makers [DIS 50–DIS 52].

Table 7-22. Toxic gases to be detected by amperometry and measuring ranges

Gas to be detected	Composition	Measuring range COMPUR [DIS 47]	BIONICS [DIS 48]	
Ammonia	NH_3	–	0–75	ppm
Arsine	AsH_3	–	0– 0.1	ppm
Boron trichloride	BCl_3	–	0–15	ppm
Bromine	Br_2	–	0–3	ppm
Carbon monoxide	CO	0–650 mg/m^3	–	
Chlorine	Cl_2	0– 15 mg/m^3	0–3	ppm
Diborane	B_2H_6	–	0–3	ppm
Dichlorsilane	SiH_2Cl_2	–	0–15	ppm
Hydrazine	N_2H_4	0–1.5 mg/m^3 b)	0–10	ppm
Hydrogen cyanide	HCN	0–100 mg/m^3	0–30	ppm
Hydrogen selenide	H_2Se	–	0–1	ppm
Hydrogen sulfide	H_2S	0–150 mg/m^3	0–10	ppm
Nitrogen dioxide	NO_2	0–200 mg/m^3	0–15	ppm
Phosgene	$COCl_2$	0– 5 mg/m^3	0–5	ppm
Phosphine	PH_3	–	0–0.5	ppm
Silane	SiH_4	–	0–60	ppm
Sulfur dioxide	SO_2	–	0–15	ppm

a) COMPUR is citing normally the ten-fold of the MAK value of the gas. They have been recalculated with the help of a MAK table [153]. MAK = maximal tolerable concentration in the ambient air which for a daily exposition of 8 h is without any influence on the health of a person.
b) TRK values of carcinogeric substances.

Table 7-23. Important features of amperometric HPLC detectors

1. High sensitivity,
2. selection of the sensitivity, a)
3. wide dynamic range,
4. fast response,
5. no hysteresis,
6. extremely small detector volume,
7. optimal signal to noise ratio,
8. shielding against static electricity
9. auto zeroing, a)
10. offset current compensation, a)
11. selection of the polarization voltage, a)
12. supression of temperature effects.

a) Inclusive of amplifier properties

Of methodological interest is the fact that glassy carbon and wax-impregnated graphite are being used as materials for the working electrodes. Cells with two or three electrodes can be used. The working electrodes in most cases are polarized positively, which corresponds to the anodic oxidation of organic compounds. Table 7-24 compiles some of the most interesting compounds together with the values of the polarization voltages.

Table 7-24. Organic substances to be detected amperometrically in HPLC [155]

Substance	Polarization voltage
1. Aromatic hydroxy compounds	
1.1 Phenoles	+1200 mV
1.2 Halogenated phenols	+1200 mV
1.3 Hydroxy biphenyls	+ 800 mV
1.4 Catecholes	+ 800 mV
1.5 Methoxy phenols	+ 800 mV
1.6 Hydroxy cumarines	+1000 mV
1.7 Flavones	+1000 mV
1.8 Oestrogens	+1000 mV
1.9 Tocopherols	+ 800 mV
1.10 Anti oxidants	+ 800 ... +1200 mV
2. Aromatic Amines	
2.1 Aniline	+1000 mV
2.2 Benzidine	+ 600 mV
2.3 Sulfonamides	+1200 mV
3. Indoles	
3.1 Indolyl-3-compounds	+1000 mV
3.2 5-hydroxo indoles	+ 800 mV
4. Phenothiazine	+1000 mV
5. Mercaptanes	+ 800 mV
6. Others	
6.1 Ascorbic acid	+ 800 mV
6.2 Vitamine A	+1000 mV
6.3 Carotine	+ 800 mV
6.4 Purines	+ 800 ... +1000 mV

7.2.4.4 Biochemical Sensors

Biochemical sensors will be treated in detail elsewhere (Chapter 12). A brief outline, together with an evaluation of the measuring techniques is, however, appropriate at this juncture.

A typical biochemical sensor is a combination of a receptor and a transducer. The receptor reacts specifically with just one component of a mixture. The products of such biochemical reaction are frequently simple inorganic ions or molecules such as H^+, NH_4^+, CN^- and/or NH_3, O_2, H_2O_2, and CO_2. It is up to the transducer to sense the concentration of these reaction products and to transduce the indication to an electrical signal [156, 157].

Most of the practically used biochemical sensors contain enzymes as biochemical receptors. An example is the determination of glucose:

$$\text{Glucose} + O_2 + H_2O \xrightarrow{\text{GOD}} \text{gluconic acid} + H_2O_2$$

where GOD is the enzyme glucose oxidase. The dissociation of the weak acid gluconic acid leads to hydrogen ions. The overall reaction products are H^+ and H_2O_2, and the concentration of both is a function of the glucose concentration in the substrate. In terms of the use of the transducer, it could be a pH-sensitive device or a sensor to detect H_2O_2.

A number of different transducer techniques are being used, a critical comparison of which is given in Table 7-25. It is apparent that photometry and amperometry are the most widely used measuring principles.

The first enzymatic sensor using amperometry as the transducer technique was proposed by Clark [161] and has been realized commercially by Yellow Springs Instruments [DIS 21]. The construction of such an amperometric "glucose electrode" is illustrated in Figure 7-52.

Table 7-25. Comparison of the most widely used transducer techniques for biochemical sensors

Transducer technique	Comments
Potentiometry (see Section 7.1.3)	Detection of H^+, NH_3, CO_2. Due to the buffer action of the substrate a number of difficult to handle problems arise [158, 159].
Conductometry (see Section 7.3.5)	The background conductivity of the substrates strongly influences the sensitivity and the calibration. Only a limited number of applications are practicable [160, DIS 53].
Photometry (see Chapter 10)	For a wide variety of chemical substances photometric detection principles have been developed. It is one of the most used transducer techniques [DIS 54, DIS 55].
Amperometry	The technique is limited to the detection of O_2 and H_2O_2. Nevertheless the field of application is a very broad one. Apart from photometry it is the most important transducer technique [154, 161].

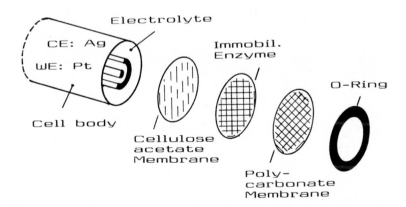

Figure 7-52. Construction of a multi-membrane amperometric glucose sensor. (Courtesy of YSI [DIS 21]). The polycarbonate membrane acts as dialyzer which holds back molecules with molecular weights over 1500. The cellulose acetate membrane has the task of delivering the pH-optimized electrolyte from the reservoir in the electrode body to the enzyme layer. At the same time the reaction product hydrogen peroxide will reach the electrode with uniform distribution via this membrane.

In such an electrode, the hydrogen peroxide formed will be reoxidized to oxygen:

$$H_2O_2 \rightarrow 2\,H^+ + O_2 + 2\,e^-\ .$$

This principle is applicable to a number of other enzymatic reactions, eg, for the determination of lactate [DIS 21].

Biochemical sensors have generally found a wide field of application in human and veterinary health care, in the control of fermentation broths and in foodstuff analysis [162, 163].

7.2.5 Applications of Amperometry

Amperometric cells are much more suitable for practical applications than other electrochemical sensors. In the preceding sections it was therefore inevitable that typical applications should be mentioned when details of cell construction were considered. For this reason, the brief summary in Table 7-26 will be sufficient to summarize the more important applications of amperometry.

Finally, another amperometric technique which has not been discussed and which has not been included in Table 7-26 should be mentioned, namely amperometric titration, which deals with the end-point detection of redox systems [164–166].

One of the best known examples is the classical dead-stop titration used in iodimetry. The determination of free chlorine, for instance, can be carried out in such a way that potassium iodide is added to the sample, which has been made strongly acidic. Chlorine releases an equivalent amount of iodine from the iodide. The resulting iodine-iodide redox system is highly reversible and the iodine acts as the depolarizer for an amperometric working electrode. Platinum is used as the electrode metal in most instances. A current will flow, the intensity of which depends linearly on the iodine concentration. In iodimetry thiosulfate is usually used as the reagent for titrating free iodine. During the course of the titration the current decreases as the iodine concentration falls. At the equivalence point the current will reach zero, which in older terminology is the "dead-stop" point.

The titration diagram itself consists of two straight lines which intersect at the equivalence point. The use of such dead-stop titrations is implemented in the Karl Fischer titration to determine water in liquids [135, 167].

For other amperometric titrations comprehensive studies to develop mathematical forms of the different titration curves have been carried out [165]. This has been done for amperometric cells with one and/or two polarizable electrodes; the latter technique is called biamperometry. In spite of all these efforts, with a few exceptions, modern automatic titrators rarely make use of such cell systems. The Karl Fischer titration is such an exception. In all other instances the difficult to handle shape of the titration curves together and their poor repeatability are the main reasons why the redox systems of interest will usually be analyzed by well known redox titrations (see Section 7.1.4.1).

Table 7-26. Application of amperometry

Application	Analytical task and other comments
Water treatment	
Drinking water hygiene	Controlled addition of chlorine, chlorine dioxide and/or ozone as disinfectant a)
Cooling water treatment	Periodical chlorination of the cooling water in power stations to prevent the growth of microorganism which would reduce the efficiency of the system a)
Boiler feed water control	Monitoring of dissolved oxygen (ppb level) and of hydrazine b)
Sewage plant optimization	Controlled aeration of the bio mass via in-line oxygen measurements
Water quality parameters	
Dissolved oxygen D. O.	The D.O. is one of the most significant quality parameter of surface waters (rivers, lakes, ...). Of additional interest is its change during the course of a year. b)
Biological oxygen demand B. O. D.	The loading of waters (surface waters, sewage plant effluents) with organic matter is characterized by the B. O. D. On-line analyzers monitor short time B. O. D. values via the breathing rate of microorganism. b)
Toxicity of waste water	Protection of the bio mass of biological sewage plants against toxic substances in the incoming wastes (eg, cyanides, copper ion, phenoles). Toximeters make use of the breathing rate of microorganism in an integrated mini bio-reactor. It is monitored by two D. O. cells the signals of which are fed to an evaluating amplifier. b)
Personell Protection	
Coal Mining	Monitoring of oxygen deficiency and of carbon monoxide with portable instruments c)
Inspection of underground sewage channels	Monitoring of oxygen deficiency and of hydrogen sulfide with portable instruments c)
Chemical plants	Portable monitors for toxic gases in the ambient air, eg, Cl_2, $COCl_2$, HCl, HF, HCN, H_2S c)
Rocket silo control	Fixed installations and/or portable monitors to monitor NO_2 and hydrazines c)
Chromatographic detectors	Detection of zones of organic substances separated by HPLC columns d)
Biochemical sensors	
Enzymatic bio sensors	Combination of amperometric hydrogen peroxide electrodes with layers of immobilized glucose oxidase. Monitoring of glucose and lactate with analyzer with a high sample through-put in clinical analysis e)

a) see Table 7-21   b) see Table 7-20   c) see Table 7-22 and Section 7.2.4.2
d) see Table 7-23 and 7-24 and Section 7.2.4.3   e) see Section 7.2.4.4

7.3 Conductometry

7.3.1 Introduction

Unlike most other electrical methods of chemical analysis, conductivity measurements are non-selective. All ions in a solution, depending on their mobility, charge, and concentration, will be detected. The sensors are passive, not active as in potentiometry. They also need an external voltage source to generate the signal, which is subsequently fed to an amplifier. Despite this apparent disadvantage, conductivity measurements play an important role, mainly in chemical water monitoring and related disciplines. With the exception of pH measurements, it is the most often used electrochemical technique.

The development of conductivity measurements can be traced back to Friedrich Kohlrausch (1840–1910), who with, from the modern point of view, very simple instrumentation but with a high degree of experimental skill, studied the behavior of a wide variety of electrolytes and derived a number of laws with which his name is now associated. He also specified calibration solutions of such accuracy that 100 years later corrections of only 0.1% were required.

7.3.2 Definitions and Units

The electrolytic conductivity κ is defined as the conductance G of a two-electrode cell the electrodes of which have an area A of 1 cm^2 and a separation distance d of 1 cm. Such cells can be characterized by the cell constant $k = d/A$ in accordance with Figure 7-53. As can be seen, the electric field between the electrodes is not homogeneous over the cross section of the cell. For this reason, calibration solutions of known conductivity must be used to determine the value of k. Generally the conductivity then follows from

$$\kappa = G\,k\,.$$

(7-40)

The unit of electrolytic conductivity is Siemens/cm. This has recently been replaced with (S/m) [168, 169] a decision which may cause much confusion since all the S/cm values used previously must be multiplied by a factor of 100. The instrument-making industry and the users of conductivity measuring methods have not yet accepted this decision and continue to use S/cm.

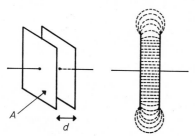

Figure 7-53.
Illustration of the principle of a two-electrode cell. The electric field lines do not end at the edge of the electrodes (right). The cell constant $k = d/A$ therefore cannot be calculated from the geometric dimensions, but has to be determined with calibration solutions.

The range of conductivities which are observed cover about 10 decades, beginning with the conductivity of ultra-pure water $(5 \cdot 10^{-8}$ S/cm) and ending with values close to 1 S/cm for strong inorganic acids. For practical reasons a number of sub-units are used, eg, 1 mS/cm = 10^{-3} S/cm and 1 µS/cm = 10^{-6} S/cm.

In the English literature until now the "mho" (reciprocal ohm, ohm^{-1}, Ω^{-1}) has been used. It is another term for the conductance G, which is the reciprocal of the ohmic resistance R.

The mho/cm concept is treated in a similar way in terms of the above-mentioned sub-units. Finally, in the water-treatment industry a third way of expressing conductivities is found. Instead of the conductance G, the resistance R of a cell and the cell constant k are used to give $\Omega \cdot$ cm values. A "specific resistance" of 10 M$\Omega \cdot$ cm corresponds to a conductivity of 0.1 µS/cm.

7.3.3   Principles of Conductivity

Compounds with strong polar bonding dissociate in solvents with a high dielectric constant resulting in positively charged anions and negatively charged cations. The insertion of two similar metal electrodes in such a solution with a d.c. voltage applied across them causes migration of the ions to the oppositely charged electrodes, at which the ions will be discharged, a process which is called electrolysis. For this reason such solutions and/or the dissolved substances are called electrolytes.

It is possible to distinguish between strong and weak electrolytes. In strong electrolytes dissociation is always complete, whereas weak electrolytes are only partially dissociated, which results in a far lower contribution to the conductivity.

Ions migrating in an electrical field of constant strength have to overcome the friction caused by the viscosity of the solvent. The accelerating force K_i and the friction R_i counterbalance each other in such a way that the ions migrate with a constant speed w_i:

$$w_i = K_i/R_i = (z_i\,e_o E)/(300 \cdot 300 \cdot \eta \cdot r_i) \tag{7-41}$$

where z_i is the charge number of the ions (eg, z_i = 2 for Ca^{2+}), e_o is the elementary charge $(1.60219 \times 10^{-19}$ C), E is the strength of the applied electrical field (V/m), η is the viscosity of the solvent (N s/m^2), and r_i is the radius of the migrating ion.

Dividing w_i by E leads to the velocity of migration u_i in a field of 1 V/m. Multiplying u_i by the Faraday constant $(F = 9.64846 \times 10^4$ C/ mol) a parameter λ_i is obtained which is called mobility of the ion i. It expresses the total amount of electrical charge transported by 1 mol of ions in a standard electric field.

To maintain electrical neutrality, each dissolved electrolyte contains the equivalent number of anions and cations. The total of both mobilities λ_i^- and λ_i^+ is called molar conductivity Λ. Table 7-27 lists selected ionic mobilities.

It is apparent that the mobilities for most ions are similar despite their different radii. Surprisingly, the smallest ion, Li$^+$, has the lowest mobility. The reason for this anomaly lies in the hydration of ions in aqueous solutions, which means that each individual ion is surrounded by water dipoles. These are orientated in such a way that the opposite charges of the ion and the end of the dipole attract each other. The ions in the surrounding water layer forms

Table 7-27. Ionic mobilities λ_i for some cations and anions

Cations	0 °C	λ_i (Ω^{-1} cm^2) a) 18 °C	25 °C	100 °C
H$^+$	225	315	350	637
Li$^+$	19	33	39	120
Na$^+$	26	43	50	150
K$^+$	40	64	74	200
Ag$^+$	33	53	62	180
1/2 Ca^{2+}	31	50	60	187
1/3 La^{3+}	35	60	70	220
Anions				
OH$^-$	105	174	200	446
Cl$^-$	41	66	76	207
J$^-$	42	67	77	–
NO$_3^-$	40	63	71	189
MnO$_4^-$	36	53	63	–
1/2 SO$_4^{2-}$	41	69	80	256

a) After Lange and Göhr [170]. All the values found in the literature up to the present do not conform with the ISO recommendations.
All the values have been extrapolated to $c = 0$.

a cloud with diminishing dipole orientation as the distance increases. The effective radius of an ion is therefore increased and the hydrated ion as a whole migrates in the electrical field.

In Table 7-27 the mobilities of H$^+$ and OH$^-$ ions are seen to be exceptions. For both of these ions a different charge transport mechanism has to be assumed, which is based on the formation of hydrogen bridges between the ions and the water dipoles.

The values of λ_i given in Table 7-27 have been extrapolated to a concentration $c_i = 0$, because with increasing concentration (c) interionic forces which reduce the mobility of the ions will increase. The decreased Λ_c values can be calculated with the help of the Debye-Hückel relation making use of $\Lambda = \lambda_i^- + \lambda_i^+$:

$$\Lambda_c = \Lambda_0 - (A\Lambda_0 + B \sqrt{c}) . \tag{7-42}$$

The constants A and B depend on the properties of the solvent and on temperature. Their values can be found in data tables [171–173].

For aqueous solutions and 1,1 electrolytes (for which $z_i^+ + = z_i^-$ holds) and a temperature of 25 °C, Equation (7-42) simplifies to

$$\Lambda_c = \Lambda_0 - (0.229 \, \Lambda_0 + 60.3 \sqrt{c}) . \tag{7-43}$$

When Λ_c is known, the corresponding electrolytic conductivity κ can be calculated:

$$\kappa = \Lambda_c \cdot c \cdot 10^{-3} . \tag{7-44}$$

The temperature dependance of κ follows from that of the ionic mobilities (see Table 7-27). The temperature coefficient $(d\kappa/dt) \cdot (1/\kappa)$ is positive and of the order of $1-3\%/\,°C$. Knowledge of this makes it possible to calculate the conductivity from listed values at other temperatures.

7.3.4 Conductivity Measurements

7.3.4.1 Calibration Solutions

A conductivity meter will only indicate the conductance G. To correlate it with the conductivity κ the cell constant k must be known.

In the case of two-electrode cells with parallel and identical electrodes, the relationship between the area A and the distance d leading to $k = d/A$ will hold only within a fairly large error band. The inhomogeneous electrical field at the edge of the electrodes cannot be taken into account and the active area obtained by platinizing the electrodes cannot be calculated, hence calibration solutions of accurately known conductivities must be available. There is an urgent need for other cell constructions which depart markedly from the two-electrode principle (see Section 7.3.4.3). In such cells no "mechanical approach" to establish k is meaningful.

As has been indicated, Kohlrausch originally worked out the formulation of a group of KCl solutions for use as conductivity standards. All the solutions are based on the molarity concept, which makes it possible to use the principle of dilution. Table 7-28 gives more detailed information.

Jones and Bradshaw [174] and other workers checked the reliability of Kohlrausch's data and only found a correction of the order only of 0.1% to be necessary. The disadvantage of

Table 7-28. Conductivity of calibration solutions (S/cm)

Temperature (°C)		Concentration (mol/L)		
		1	0.1	0.01
15	$\kappa\ =$	0.09254	0.01048	0.001147
20		0.10209	0.01167	0.001278
25		0.11180	0.01288	0.001413

To make the solutions weigh 47.555 g of dry KCl and dissolve it in 1 litre of ion-free water. The 1 mol/L solution will be diluted 1 : 10 and/or 1 : 100 to obtain the lower molarities.

Table 7-29. Calibration solutions of high conductivity (S/cm)

Solution/concentration		100 (%/grad)
NaCl, sat. at 25 °C	0.251	2.04[a]
NaOH, 15.0% wt./wt.	0.406	2.49
KOH, 27.5% wt./wt.	0.626	2.15
H_2SO_4, 367 g/L b)	0.826	1.47

a) The temperature coefficient $(d\kappa/dt)\,(1/\kappa)$ is valid for a reference temperature of 25 °C.
b) The concentration of commercially available H_2SO_4 has to be determined by titration.

these solutions is that the upper limit of conductivity is about 100 mS/cm. In practice, conductivities can be observed which are as much as ten times higher. In this regard a table that provides very precise data regarding NaCl solutions up to the point of saturation may be helpful [175].

Much higher conductivities can be realized with solutions of strong inorganic acids and bases. As will be shown later, the $\kappa = f(c)$ plot for such solutions passes through a maximum at which a slight change in concentration will not influence the conductivity. The precision of the data in Table 7-29 is however, poor [176].

7.3.4.2   Cells with Two and More Electrodes

Generally, for all two-electrode cells it is postulated that their resistance R_x in contact with a sample should be kept within the limits 100 $\Omega < R_x < 100$ kΩ. This can be achieved only with a range of cell constants from 10^{-2} to 10 cm^{-1}. Such cells make use of a different construction to that of the two-electrode concept. A selection of commercially available cells is illustrated in Figures 7-54 to 7-56.

The construction shown in Figure 7-57 is unusual. The field lines of the electric field which penetrate the sample are entirely inhomogeneously distributed. As a result, the cell has a fixed constant only within a limited range of conductivities. Its advantage is, however, that it is insensitive to air and/or gas bubbles in the sample.

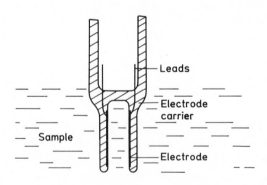

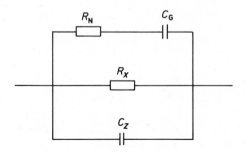

Figure 7-54.
Practical realization of a two-electrode cell. At higher frequencies shunt effects can be observed via the glassy electrode carrier (C_G) and the resistance R_N within the solution (for the simplified equivalent circuit of the cell, see Figure 7-58) [177].

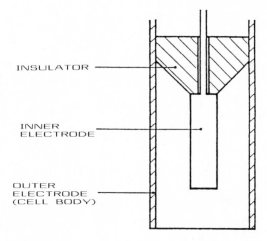

INSULATOR

INNER
ELECTRODE

OUTER
ELECTRODE
(CELL BODY)

Figure 7-55.
Two-electrode cell with concentric electrode arrangement. The cell is free of shunt effects (see Figure 7-58) [177].

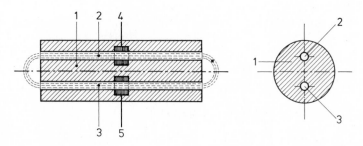

Figure 7-56. Two-electrode cell based on the principle of a "long current path" to achieve large values of the cell constant k (values up to $k = 100\ \text{cm}^{-1}$) [176]. 1, Cell body; 2, 3, channels filled with the sample; 4, 5, electrodes.

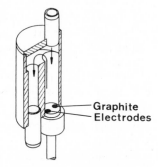

Graphite
Electrodes

Figure 7-57.
Two-electrode cell with a heterogeneous field distributions. The arrows indicate the sample flow. Cells of this type are insensitive to bubbles and particles in the sample. (Courtesy of WTW, see [178]).

The equivalent circuit of two-electrode cells is shown in Figure 7-58. The considerations are based on the resistance concept, which has been used earlier and facilitates the electrical treatment of the effects. R_x, the resistance of the cell in contact with a probe, is included in a network of other resistances and capacitances. A short discussion illustrates how carefully the entire concept of conductivity cells should be handled.

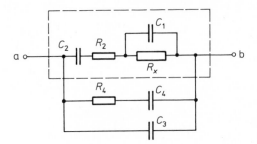

Figure 7-58.
Equivalent circuit of a two-electrode cell. The resistance of interest, R_x, is part of a multi-component network of other resistors and capacitors. The dashed-line rectangle includes all components deriving from electrode reactions; the others are caused by connections and cables (for more details, see text).

C_3 is the capacity of the cable from the cell to the amplifier. In industrial applications it is of the order of 100 pF/m. R_4 and C_4 are strongly dependent on the construction of the cell. Both components are part of a shunting circuit which will be observed, for instance, in cells made in accordance with Figure 7-54. The contribution of R_4 and C_4 is, however, negligible in concentric cell constructions following the principle shown in Figure 7-56. The other components within the rectangle surrounded by a dashed line are part of electrode reactions to be found at the interfaces of the electrodes with the samples. In the potentiometric sense (see Section 7.1.2) the interfaces are blocked and a meaningful analysis can only be achieved by working with a.c. sources of optimum frequencies. Because of the importance of such considerations, the next section is dedicated to the electrode reactions within conductivity cells.

Polarization and Frequency Selection

The term "polarization" is associated with the electroanalytical discipline called polarography, in which the electrodes must be polarizable. The presence of electrode depolarizing redox systems and/or other electrode-active substances causes a current to flow, the intensity of which is proportional to the concentration of these substances.

In conductometry the situation is different; polarization is a highly undesirable side effect of the contact of electrodes having electronic conduction with a solution showing ionic conduction. Even without applying an external potential the electrode/solution interface is blocked and covered with a layer of opposite charges which act as a capacitance shown as C_2 in Figure 7-55. In any case, a small leakage current will flow which is expressed by R_2 in the equivalent circuit of a cell. Both of these components are connected in line with the required resistance R_x.

To overcome the effect of the polarization, C_2 has to be made as high as possible whereas R_2 should be minimized. Another possible way of removing the influence is to select a sufficiently high measuring frequency to make the impedance $X_2 = 1/\omega C_2$ negligible compared with R_x.

Kohlrausch established empirically that the platinization of platinum electrodes with a layer of platinum black strongly reduces polarization effects. Also, the selection of a frequency of the order of 1–5 kHz even for a cell with a cell constant $k = 1$ cm^{-1} guarantees a wide dynamic range.

High measuring frequencies, however, may cause other problems because of the presence of the unavoidable capacity C_1, the impedance X_1 of which acts as shunt for R_x. The conflict can only be resolved by using low measuring frequencies. Much work has been done by different workers to determine the apparent cell resistance for a number of selected low frequencies and to extrapolate the "R_x" values to $\omega = 2 \cdot \pi \cdot f = \infty$ [179, 180]. Such a procedure depends on the availability of high-precision measuring equipment including a tunable frequency generator. It is practicable only in research laboratories and is not useful for field work.

Oehme [181] has shown that the determination of the cell constant k with a number of test solutions of known and increasing conductivity is a far simpler way of establishing the useful range of a conductivity measuring cell. From Figures 7-59 and 7-60 it can be seen that over a wide range k will remain constant but rises when the end of the dynamic range of the cell has been reached.

Testing different electrode materials and applying variable measuring frequencies is a quick and simple way of assessing the usefulness of different cell constructions.

Rommel carried out similar but more detailed work [182]. He made explicit quantitative statements with regard to C_2 and R_2 and worked out a number of rules to circumvent errors in conductivity measurement [177, 178].

When polarization effects are under control, other measures can be taken to overcome the different shunting effects. The rule is to keep all internal cell connections and/or the electrode itself away from any contact with the solution avoiding (thin) layers of dielectrics as insulators or supporting materials (see Figure 7-54). Parker reported such studies and developed a cell which is free from undesirable shunting effects [183].

Another study approached the problem from the point of view of amplifier technology. By equipping measuring instruments with phase-sensitive rectification the troublesome capacity C_1 can be eliminated [182].

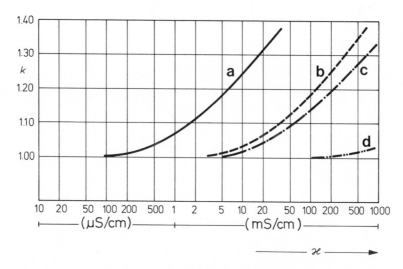

Figure 7-59. Dependence of the usable conductivitiy range of a two-electrode cell on the electrode material. The apparent increase in the constant k is caused by polarization effects. The cell can be used only within the range of a constant value of $k = 1.00$ cm^{-1} [13]. (a) Stainless steel 316 SS; (b) sintered carbon; (c) stainless steel covered with a sintered layer of titanium carbide TiC; (d) platinized platinum. Measuring frequency, 50 Hz. Platinized platinum is by far the best electrode material.

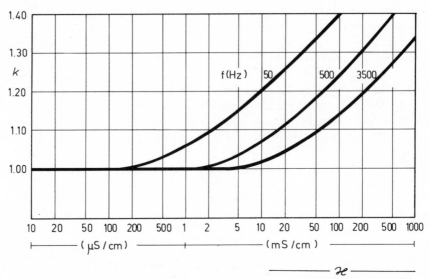

Figure 7-60. Dependence of the usable conductivity range of a two-electrode cell on the measuring frequency. Electrode material, stainless steel 316 SS. The apparent increase in the cell constant k is caused by polarization. The cell can be used only within the range of a constant value of $k = 1.00$ cm^{-1} [181].

It is therefore clear that even apparently simple two-electrode cells need careful development. They are the most widely used conductivity sensors because they are cheap and are possibly applicable even to difficult conditions, eg, temperatures up to 250 °C and pressures up to 40 bar [DIS 56].

Conductivity Meters for Two-Electrode Cells

Two-electrode cells can be used with many commercially available instruments. Therefore, it will be useful to mention briefly the different types of conductivity meters. The oldest principle used is the Wheatstone bridge, which is still found in a number of instruments for

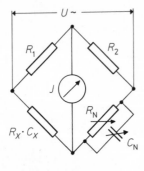

$$R_X = R_N \cdot R_1 / R_2$$
$$C_X = C_N \cdot R_1 / R_2$$

Figure 7-61.
Simplified circuit of a Wheatstone bridge to measure the resistance R_X of two-electrode conductivity cells. The cell has to be tuned manually with the help of R_N and C_N to the minimum of the meter J.

Figure 7-62.
Circuit of a direct-reading bridge
to measure the resistance R_X of
conductivity cells. Depending on
the ratio R_3/R_X the read-out of I
is nearly linear or logarithmic.

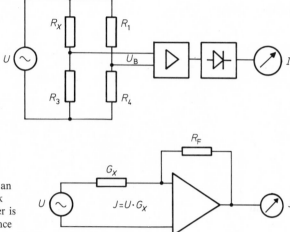

Figure 7-63.
Circuit of a conductivity meter using an
operational amplifier with a feed-back
resistor R_F. The reading I of the meter is
linearly proportional to the conductance
G_X.

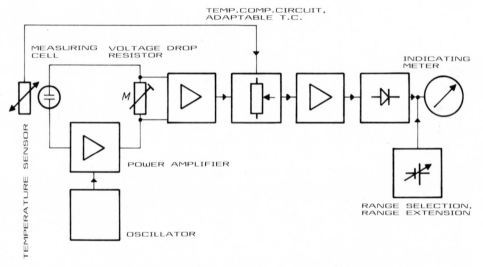

Figure 7-64. Detailed circuit of an industrial conductivity meter using the voltage-drop principle. The
current in the circuit containing the measuring cell causes a voltage drop across the standard
resistance M. By changing M the range can be selected. With the help of a temperature sensor automatic compensation of temperature effects can be achieved.

reasons of simplicity and universal usage. A disadvantage is, however, the need to adjust the
bridge manually. Figure 7-61 illustrates such a bridge.

Modified Wheatstone bridges are among the most precise instruments used for scientific
work. The bridge principle can be used for direct-reading instruments, as shown in Figure
7-62. The best means of obtaining a linear response is to use operational amplifiers in accordance with Figure 7-63. A box diagram of the circuit of a complete conductivity meter is
shown in Figure 7-64. Means for automatic temperature compensation are standard in all industrial applications.

Multi-Electrode Cells

The main disadvantage of two-electrode cells is that any physical and/or electrochemical change in the electrodes directly influences the result. This is true for all of the effects considered as polarization, for electrode corrosion by chemically aggressive solutions, and for the spoiling of the electrodes, eg, by oil films or precipitates. To overcome these disadvantages, cells have been developed which utilize at least four electrodes. In the simplest case two of the electrodes are connected to a constant-voltage source which results in the flow of a constant current I. Within the field generated by these "current electrodes", another two "voltage electrodes" are placed. They measure the voltage drop $U = I \cdot R_x$ by means of a high-impedance amplifier. An improved system is illustrated in Figure 7-65. Here an internal control loop maintains a constant voltage drop U across the voltage electrodes. To achieve this, the current has to be controlled so that it is proportional to the conductivity. A practical realization of this principle is shown in Figure 7-66 [DIS 57].

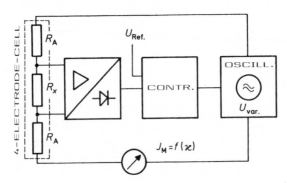

Figure 7-65.
Principle of a four-electrode conductivity meter. A controller circuit compares a preselected voltage U_{Ref} with the voltage drop across the resistance R_X. Any deviation will be compensated by a variation of the voltage U_{var} of the oscillator. R_X is free from polarization effects. The resistors R_A take over all of the difficult to control electrode reactions. The current I_M depends linearly on the conductivity of the sample.

Figure 7-66.
Practical realization of a multi-electrode conductivity cell, showing two current electrodes and four voltage electrodes which are positioned within equi-potential lines of the electrical field. (Courtesy of Polymetron [DIS 57]).

The planar electrode configuration considered until now can be replaced with ring-shaped electrodes [DIS 58]. In none of the cases polarization effects will be observed at the voltage electrodes. They can also be covered by non-conducting films without affecting the measured conductivity. Unlike the two-electrode systems, the cell and the amplifier are an integrated unit. Another point of practical interest is that with just one cell the entire range of conductivities can be covered, except for the lowest range below 1 μS/cm.

7.3.4.3 Contactless Cells

For electrochemical electrode reactions, cells with direct contact of the electrodes to the conducting solutions can only be a compromise. A different situation can be expected for contactless cells. Independently of each other, several groups developed contactless conductivity cells which use two entirely different concepts — inductive and capacitive. The following sections will illustrate the principles in some detail.

Inductive Conductivity Cells

In 1958 two papers were published in which Calvert et al. [184] and Salomon [185] illustrated the principle and the construction of inductive contactless cells. The concept is simple. Two coils of a transformer are magnetically shielded from each other; the only coupling loop is formed by a non-metallic and non-conducting tube which is filled with the sample. Figure 7-67 illustrates the principle.

It is seen that the voltage induced in the secondary coil depends only on the input voltage of the primary coil and the conductance G of the sample. Typical working frequencies are from 50 to 500 Hz. The entire range of conductivities of practical interest can be covered by one cell. Of special interest is that even the highest conductivities of the order of 1 S/cm can be measured. By selecting organic polymers as the cell body, a high degree of chemical resistance can be achieved. This, however, imposes a limitation on the admissible temperatures and pressures. A practical example of such a cell is shown in Figure 7-68 [DIS 59].

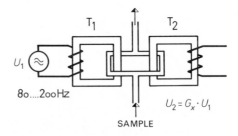

Figure 7-67. Principle of an inductive contactless conductivity cell. The two halves T_1 and T_2 of a transformer are magnetically shielded from each other. The only coupling between the two coils is achieved by a loop of a non-conducting material which is filled with the sample. The output voltage U_2 depends linearly on the conductance G_x.

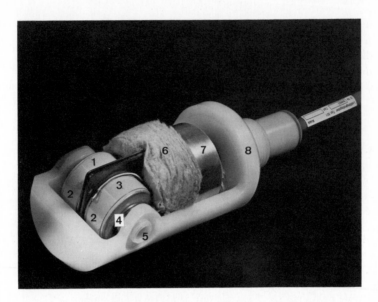

Figure 7-68. Section through an inductive contactless conductivity cell. (Courtesy of Knick [DIS 59]).
1,3, Two halves of a transformer with a magnetic shield between; 2,2, primary/secondary
coil (see Figure 7-67); 4, ring-shaped temperature sensor for automatic temperature compensation; 5, sample channel; 6, thermal insulation; 7, preamplifier; 8, cell body made of
Tefzel.

Capacitive Conductivity Cells

The principle here is entirely different from the foregoing. As shown in Figure 7-69, two
ring-shaped electrodes are placed on the outside of a non-conducting cell body. A high-frequency voltage of the order of 1–100 MHz is applied across these electrodes. A finite part of
the electrical field penetrates the internal cross section of the cell body, which is filled with
the conducting sample.

The equivalent circuit of such a cell is also shown in Figure 7-69. Depending on the frequency and/or the conductance, the cell can be treated as a conductance G_{res} or a capacity
C_{res}. Each of the equations contains the cell parameters C_s and C_p, and in the latter a contribution of the dielectric constant ε of the solution is included:

$$C_{res} = G_2 C_s + \omega^2 C_s C_p^2 + \omega^2 C_s^2 C_p / G^2 + 2(C_s + C_p)^2 \tag{7-45}$$

$$G_{res} = G\omega^2 C_s / G^2 + \omega^2 (C_s + C_p)^2 \tag{7-46}$$

where

$$C_p = \varepsilon C_0 \tag{7-47}$$

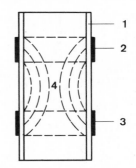

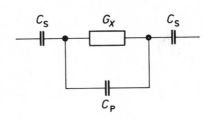

Figure 7-69.
Principle of a capacitive contactless conductivity cell.
1, Non-metallic and non-conducting cell body; 2,3, ring-shaped electrodes on the outside of the body; 4, sample volume which is penetrated by the electrical high-frequency field. The equivalent circuit of the cell shows two capacitances C_S which are caused by the conducting electrodes, the cell body (= dielectric) and the conducting sample. C_p contains a contribution of the dielectric constant of the sample solution, G_X is the conductance of the sample.

and C_0 is the vacuum capacity of the cell. As expected, the response characteristics derived using these equations are far from linear, as confirmed in Figures 7-70 and 7-71. The dynamic range of the C_{res} plot covers three decades of conductance whereas whilst the G_{res} values pass through a maximum.

Despite these serious disadvantages of capacitive cells, much experimental work has been carried out, which in most instance was related to the end-point detection of "high-frequency titrations". The measuring principle is fairly simple, as shown in Figure 7-72.

In C_{res} evaluations, the variable capacitor C has to be tuned to reach the maximum indication of the meter U. G_{res} measurements depend on this tuning, in which case the value of U

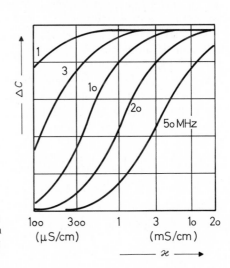

Figure 7-70.
Response characteristic of a capacitive contactless conductivity cell using the resulting capacity C_{res} (see Equation (7-45)). The range of optium sensitivity can be shifted by changing the working frequency and/or the cell configuration.

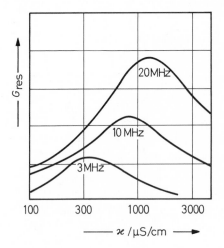

Figure 7-71.
Response characteristic of a capacitive contactless conductivity cell using the resulting conductance G_{res} (see Equation (7-46)). The mexima make the readings ambiguous. The principle therefore should be used only for samples the conductivity of which varies within known limits.

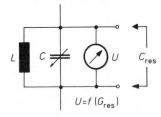

Figure 7-72. Tuned high-frequency circuit formed by a coil L and a variable capacity C to connect contactless capacitive conductivity cells. The circuit has to be tuned to resonance (= maximum reading of the meter U). From the capacity C the parameter C_{res} follows (see Equation (7-45) and Figure (7-70)). The value of the resonance voltage U allows the determination of G_{res} (see Equation (7-46) and Figure (7-71)).

is a measure of the conductance G of the solution. In both cases the measuring cell is connected in parallel to the resonance circuit consisting of an inductance L and a variable capacitance C.

By appropriate modifications, such measuring circuits can be made direct-reading. Such instruments were available [DIS 60–DIS 62] mainly for the detection of end-points during the course of titrations. However, the titration plots were difficult to interpret and the method is no longer used. In some instances such "titrimeters" have been used for conductivity monitoring, mainly in relation to the quality control of rivers [188, 189].

7.3.5 Application of Conductivity Measurements

Conductometric applications cover a very wide field in science and technology. Of special interest are studies dealing with the formation of inorganic and/or organic complexes. Table 7-30 gives some detailed information.

Another important field of basic electrochemistry is the determination of dissociation constants of weak inorganic or organic acids and bases.

Table 7-30. Fundamental chemical studies based on conductivity measurements

Chemical parameter	Procedure
Solubility of sparingly soluble salts [190]	To measure the suspension of the salt in ion-free water, to calculated the concentration from Equation (7-44) using tabulated λ_0 values,
Study of organic cis-/trans polyoles [191]	To measure the change (increase) of a saturated solution of boric acid due to the addition of the polyole,
Study of the structrue of ammino-cobaltiates and/or of nitrito-cobaltiates [192]	To compare the conductivity of 10^{-3} mol/L solutions and to calculate Λ_c values

Using Ostwald's dilution law, a number of simple measurements on solutions of known concentration allow the dissociation constant K to be calculated:

$$K = c \cdot \Lambda_c^2 / \Lambda_0 (\Lambda_0 - \Lambda_c) .\tag{7-48}$$

Most of the tabulated values of K in chemistry handbooks are based on such measurements. A check of the validity of the law can be carried out, following Kraus and Bray, in a simple and elegant way [193].

Microbial metabolism frequently includes the release and/or consumption of ionic components, resulting in a change of the conductivity of the system. Oehme studied the metabolism of carbohydrates by *Proteus mirabilis* [194]. Depending on the kind of carbohydrate, ammonia and/or pyruvic acid will be formed, changing the conductivity of the broth.

Recently the Malthus-AT Analyzer came on the market, which can monitor the sterility of foodstuffs and soft drinks. The samples are incubated in conductivity cells of a special design and the conductivity is recorded for several hours [DIS 53].

An unusual application is the monitoring of the content of sealed ampoules by means of contactless capacitive cells. The high-frequency field applied penetrates the glass wall of the ampoules easily and detects changes in conductivities due to abnormalities [195].

The possibility of "conductometric titrations" has been mentioned. This is the end-point detection of titrations with the help of conductivity measurements. The procedure is to plot or record κ as a function of the volume of reagent added to the sample. Apart from "potentiometric titrations" (see Section 7.1.3.9 and Table 7-9), the titration curves do not show a sigmoidal shape but rather a number of straight lines intersect each other at the equivalence point. This is shown in Figure 7-73 for a mixture of two weak organic acids titrated in an organic solvent. The different steps in the dissociation can be seen [196].

Despite the availability of a number of monographs [186, 187] this technique has been replaced to a large extent by automatic titrators using potentiometric principles. One of the main reasons is that in conductometric titrations the conductivity of an ionic "background" of the sample influences the shape of the titration diagram, although the mechanism whereby this occurs is not fully understood.

For many industrial applications the measurement of conductivities is of great importance. Two main fields of application can be identified: the use of conductivity to monitor concentration and the use of conductivity as a quality parameter.

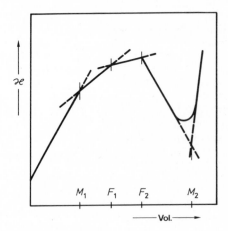

Figure 7-73.
Differentiating titration of a mixutre of fumaric acid (F) and maleic acid (m) in pyridine. The reagent for the titration is tetrabutylammonium hydroxide. By such a conductometric titration equivalence points can be determined which correspond to the intersection of the reaction lines. Both steps in the dissociation of the weak acids can be observed.

In the first case plots of the form shown in Figure 7-74 are of interest. The appearance of maxima can lead to ambiguity, but in many technological processes the concentration will vary within only a limited range. This is true, for instance, in monitoring the concentration of sulfuric acid made by the contact process. The range of interest is from 95.5 to 99.0%, which means that the descending part of the conductivity curve following the second maximum is used. Interestingly, the conductivity of concentrated sulfuric acid is low (of the order of 10 mS/cm) and such solutions behave equally well as an electrolyte and as a hydrogen-bonded liquid with "nearly organic" properties. The conductivity cell for such applications must be carefully designed, with a strong preference for contactless inductive cells.

Another typical application is the concentration monitoring of hydrogen peroxide bleach liquor for cotton fabrics. The process is run at high pH values where the peroxide is liable to deteriorate. The liquor therefore has to be stabilized by the addition of water-glass (sodium tetrasilicate), which itself is an electrolyte. Figure 7-75 illustrates the behavior of such a multi-component system.

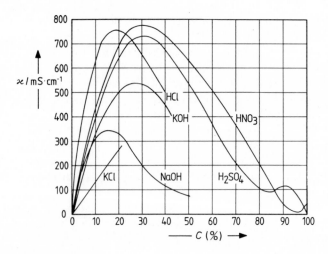

Figure 7-74.
Conductivity plots for strong electrolytes (values for a temperature of 25 °C). In most cases only the increasing part of the curves will be used for analytical applications. An exception is made in the case of sulfuric acid. The contact process leads to an acid of concentration 95.5–99.0%, which corresponds to the drop in conductivity following the second maximum.

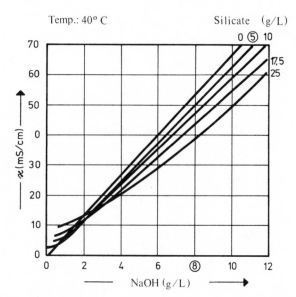

Temp.: 40° C

Silicate (g/L)

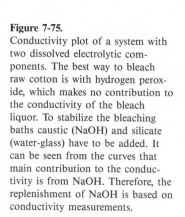

Figure 7-75.
Conductivity plot of a system with two dissolved electrolytic components. The best way to bleach raw cotton is with hydrogen peroxide, which makes no contribution to the conductivity of the bleach liquor. To stabilize the bleaching baths caustic (NaOH) and silicate (water-glass) have to be added. It can be seen from the curves that main contribution to the conductivity is from NaOH. Therefore, the replenishment of NaOH is based on conductivity measurements.

Many other industrial applications can be found in the literature [176] and in the documentation of instrument makers [197, 198]. One of the most important applications is in the quality control of water. Table 7-31 gives an overview.

In conductivity meters, microprocessor-based instruments not only facilitate usage but also in many instances allow the solution of difficult problems. A typical example is the temperature coefficient of the conductivity of ultra-pure water which is required to rinse semiconductor prints after etching. The temperature coefficient will change from 8%/ °C at 25 °C for water which is close to the theoretical conductivity of 0.04 µS/cm as to low as

Table 7-31. Conductivity of water as quality parameter and in water treatment

Parameter/Process	Procedure
Total dissolved salts of surface water	The conductivity (S/cm) leads to the concentration: $c = 6.4 \cdot 10^5 \cdot \kappa$ (mg/L) Böttger equation [199]
Chlorinity of sea water	The chlorinity Cl can be estimated out of the salinity S (both in ppt) with direct reading conductivity instruments [200]: $Cl = (S - 0.03)/1.805$
Purity of steam condensate using water-cooled condensors	Sampling through a cartridge of an ion-exchanger in the H^+-form, converting all cations into H^+-ions (increased sensitivity!)
Effluent monitoring of ion-exchangers for water-desaliniation	a) Continuous effluent monitoring indicates the break-through, b) rinsing of the regenerated exchangers by conductivity control
Rinsing water control	Fresh-water replenishment of rinsing baths in plating shops with "water savers" (closed control loop conductivity measurement) [201]

2.5%/ °C at 25 °C for water only slightly loaded after passing a "clean bench" and having a conductivity of about 1 µS/cm. Conventional means of temperature compensation (see Figure 7-64) cannot be used but the development of microprocessor based amplifiers has overcome the problem [202, DIS 63].

At the end of this short description of applications, another very important example should be mentioned, namely the development of conductivity cells to work as detectors in ion chromatography [117]. In analyzing mixtures of cations and anions, use is made of ion-exchange columns which are able to separate the different components by rinsing with, an eluent (a weak electrolyte of defined pH). The effluent from such a separation column is fed to a microvolume conductivity cell which detects the zones through their effect on the conductivity of the eluent [203, 204]. With a cell volume of the order of 10–25 µL a fast response and nearly ideal peaks can be realized. Macroporous electrodes have been developed which solve the problem of field-line distortion and which show a large linear conductivity range [205].

Conductometric Gas Analysis

An important aspect of conductometric application is the different methods of gas analysis. In each case the first step of the analysis is a phase change which transfers the gaseous component in a gas mixture and/or air into a solution in which the conductometric measurement is carried out. To achieve this, nebulizers, bubbler spirals, or impinger flasks are used. The aqueous solution can be water but in most instances solutions will be used which enable chemical auxiliary reactions to increase the conductometric response.

All the methods are concerned with acidic gases (eg, HCl, SO_2, CO_2) or alkaline gases (eg, NH_3). Opekar and Trojánek described a system in which SO_2 is preconcentrated into a polydispersed aerosol of water. The nebulized droplets condense on impact on the sample preparation vessel, the walls of which contain embedded electrodes for conductivity measurement [206]. The useful range is 0.02–2.2 mg SO_2/m^3 air with a detection limit of $2 \cdot 10^{-3}$ mg/m^3. So far the system has only been tested under laboratory conditions.

The Ultragas gas analyzer [DIS 64] uses, instead of ion-free water, a solution of hydrogen peroxide as the absorber. The peroxide oxidizes the SO_2 to SO_3, which together with water forms sulfuric acid instead of the weak sulfurous acid. By this chemical auxiliary reaction a much more pronounced change in conductivity is observed with a considerably extended linear response range from 0 to 5000 mg SO_2/m^3.

To measure CO_2, another auxiliary reaction employing NaOH solutions is used. The absorption of CO_2 converts the NaOH into Na_2CO_3, replacing the OH^- ion of high mobility by the CO_3^{2-} ion of much lower mobility [DIS 64].

Bartscher and Schmidts studied the change in conductivity of NaOH solutions of different concentrations caused by the absorption of CO_2 [207]. They showed that the addition of aminoethanol improves the efficiency of absorption of CO_2 by NaOH solutions. Another type of conductometric CO_2 analyzer was tested by Brukenstein and Symanski under laboratory conditions [208]. In this approach the gaseous and/or liquid sample is in contact with a hydrophilic, gas-permeable membrane behind which the absorber liquid flows. Liquid samples have to form acids to release CO_2 from carbonates.

A similar concept was used by Carlson [209] to detect CO_2 and/or NH_3. In the latter case the sample is made alkaline to release NH_3 from ammonium ions. The GAT N-360 ammonia analyzer makes use of this principle [210].

Organic halogens can be measured after their conversion to HCl (or HF). This can be achieved by pyrolysis [DIS 65] or by reaction with hydrogen [DIS 66]. High sensitivity can be achieved, eg, 100 ppb of Freon 12 [DIS 66].

7.4 References

7.4.1 Literature

[1] Light, T. S., in: *Ion-Selective Electrodes,* Durst R. A. (ed.); Washington, DC: NBS, Special Publ. 312, 1969, 349–373.

[2] Oehme, F., *Gewässerschutz–Wasser–Abwasser* **39** (1979) 111–128.

[3] Clerc, Z., Stefanac, R., Simon, W., *Helv. Chim. Acta* **48** (1965) 54–64.

[4] Fjeldly, T. A., Nagy, K., *J. Electrochem. Soc.* **127** (1980) 1299–1303.

[5] Meyerhoff, M. E., Fraticelli, Y. M., *Anal. Chem.* **54** (1982) 27 R–31 R.

[6] Hulanicki, A., Trojanowicz, M., in: *Ion-Selective Electrodes, 2nd Symposium at Matrafüred, Hungary, Oct. 18–21,* 1976, Pungor, E. (ed.); Budapest: Akadémia Kiadó, 1977, 139–150.

[7] Janata, J., Huber, R. J., *Solid State Chemical Sensors;* New York: Academic Press, 1985.

[8] Oehme, F., *GIT Fachz. Lab.* **30** (1986) 595–602.

[9] Zachariasen, W. H., *J. Am. Chem. Soc.* **54** (1932) 3841.

[10] Perley, G. A., *Anal. Chem.* **21** (1949) 394.

[11] MacInnes, D. A., Dole, M., *J. Am. Chem. Soc.* **52** (1930) 29–48.

[12] Hubbard, D., Hamilton, E. H., Finn, A. N., *J. Res. Nat. Bur. Standards* **22** (1939) 299–352.

[13] Bates, R. G., *Determination of pH,* 2nd Ed.; New York: J. Wiley & Sons, 1973.

[14] Baucke, F. G. K., *J. Non-Cryst. Solids* **14** (1974) 13–31.

[15] Baucke, F. G. K., *J. Non-Cryst. Solids* **73** (1985) 215–231.

[16] Eisenman, G., *Glass Electrodes for Hydrogen and Other Cations;* New York: Dekker, 1967.

[17] Nikolsky, B. P., Shultz, M. M., Pashekhovna, N. V., *Zh. Fiz. Khim.* **33** (1959) 1922–1945.

[18] Wegman, D., Simon, W., *Helv. Chim. Acta* **47** (1964) 1181–1193.

[19] Frant, M., Riseman, L., *Science* **154** (1966) 1553–1555.

[20] Simon, W., Pioda, L., *Chimia* **23** (1969) 72–90.

[21] Simon, W., *Angew. Chemie* **82** (1970) 433–470.

[22] Pioda, L., Stankova, V., Simon, W., Stefanac, R., *Mikrochem. J.* **12** (1967) 125.

[23] Frant, M., Ross, J. W., *Science* **167** (1970) 987–990.

[24] Eisenman, G., in: *Ion-selective Electrodes,* Durst, R. A. (ed.); Washington, DC: NBS, Special Publ. 312, 1969, 1–54.

[25] Ross, J. W., in: *Ion-selective electrodes,* Durst, R. A. (ed.); Washington, DC: NBS, Special Publ. 312, 1969, 57–88.

[26] Morf, W. E., Wuhrmann, P., Simon, W., *Anal. Chem.* **48** (1976) 1031–1037.

[27] Morf, W. E., Amman, D., Simon, W., *Chimia* **28** (1974) 65–67.

[28] Simon, W., Pioda, L., *Chimia* **23** (1974) 375.

[29] Oesch, U., Brozka, Z., Xu, A., Rusterholz, G., Amman, D., Pretsch, E., Simon, W., *Anal. Chem.* **58** (1986) 2285–2289.

[30] Ives, D. J. G., Janz, G. J., *Reference Electrodes;* New York: Academic Press, 1961.

[31] Delahay, P., *Double Layer Electrode Kinetics;* New York: J. Wiley, 1965.

[32] Marcus, R. A., *Ann. Rev. Phys. Chem.* **15** (1964) 155.

[33] Baucke, F. G. K., *Electroanal. Chem.* **33** (1971) 135–144.

[34] Comer, J., *Internat. Labmate* **8** (1984) No. 6.

[35] Conway, B., *Electrochemical Data;* Amsterdam: Elsevier, 1952.

[36] *Orion Newsletters* **1** (1969) pp. 21–23, Orion Research, Cambridge, MA 02139, U.S.A.

[37] *DIN 19260, pH Messung, allgemeine Begriffe;* Berlin: Beuth.

[38] IUPAC, *Manual of Symbols and Terminology for Physicochemical Quantities;* New York: Pergamon Press, 1979.
[39] *DIN 19266, Standardpufferlösungen;* Berlin: Beuth.
[40] *DIN 19267, Technische Pufferlösungen;* Berlin: Beuth.
[41] Filomena, M., Camoes, G. F. C., Covington, A. K., *Anal. Chem.* **46** (1974) 1547–1551.
[42] Covington, A. K., *Pure Appl. Chem.* **57** (1985) 531–534.
[43] Mussini, T., Covington, A. K., Longhi, P., *Pure Appl. Chem.* **57** (1985) 865–888.
[44] Hammett, L. P., *Chem. Rev.* **13** (1933) 61–75.
[45] Schwarzenbach, G., Sulzberger, R., *Helv. Chim. Acta* **27** (1944) 348–360.
[46] Fritz, S., *Acid-Base Titrations in Non-Aqueous Solvents;* Boston: Allyn and Bacon, 1973.
[47] Kucharsky, J., Safarik, S., *Titrations in Non-Aqueous Solvents;* Amsterdam: Elsevier, 1965.
[48] Hildebrand, J. H., *J. Am. Chem. Soc.* **35** (1913) 847.
[49] Perley, G. A., *Trans. Electrochem. Soc.* **92** (1947) 485–492.
[50] Biilmann, E., Jensen, A. L., *Bull. Soc. Chim.* **41** (1927) 151–160.
[51] MacInnes, D. A., Dole, M., *Ind. Eng. Chem., Anal. Ed.* **1** (1929) 57–59.
[52] MacInnes, D. A., Belcher, D., *J. Am. Chem. Soc.* **53** (1931) 3315–3324.
[53] *Ostwald-Luther's Hand- und Hilfsbuch physikochemischer Messungen,* Drucker, C. (ed.); New York: Dover Publications, 1943, p. 546.
[54] Busch, J., Graabeck, A. M., Halmvig, H., *Internat. Labmate,* May (1982) 92–96.
[55] *DIN 19265, Supplementary measuring apparatus* ; Berlin: Beuth.
[56] Midgley, D., *Analyst.* **112** (1987) 573–585.
[57] Midgley, D., Torrance, K., *Analyst* **107** (1982) 1297–1308.
[58] Tauber, G., *Process Automation* (1984) 51–58.
[59] Kolthoff, I. M., Hartong, B. D., *Rec. Trav. Chim.* **44** (1925) 113–120.
[60] Oehme, F., *Kommunalwirtschaft,* No. 9 (1964).
[61] Oehme, F., Ertl, S., *Chemie-Technik* **24** (1980) 447–450.
[62] Ives, D. J. G., Janz, G. J., *Reference Electrodes;* New York: Academic Press, 1961, Chapter 7.
[63] de Rooij, N. F., Bergveld, P., in: *Monitoring Vital Parameters During Extracorporeal Circulation,* Kimmich, H.P. (ed.); Basel: Karger, 1981.
[64] Perley, G. A., Godschalk, J. B., *US-Pat. 2 416 949.*
[65] Kinoshita et al., *Electrochim. Acta* **31** (1968) 29–35.
[66] Fogg, A., Buck, R. P., *Sensors and Actuators* **5** (1984) 137–146.
[67] Klein, M., Kuisl, M., *VDI-Ber.* Nr. 508 (1984) 275–279.
[68] Niedrach, L. W., *Science* **207** (1980) 1200–1204.
[69] Light, T. S., Fletcher, K. S., *Anal. Chim. Acta* **175** (1985) 117–126.
[70] Weiner, R., *Die Abwässer der Galvanotechnik;* Saulgau, FRG: Leuze Verlag.
[71] Burke, L. D., in: *Electrodes and Conductive Metal Growth,* Trasatti, S. (ed.); Amsterdam: Elsevier, 1980.
[72] Unpublished work, Oehme, F., Polymetron AG, CH-8617 Mönchaltdorf, Switzerland.
[73] Bühler, H., Galster, H., *Redoxmessung,* Dr. W. Ingold AG, CH-8902 Urdorf, Switzerland.
[74] Schenk, G. J., *J. Chem. Ed.* **41** (1964) 32–36.
[75] Jolly, W. L., *J. Chem. Ed.* **43** (1966) 198–201.
[76] Oehme, F., Rhyn, H., *Neue Zürcher Zeitung,* März, No. 875/876 (1965).
[77] Davis, G., in: *Comprehensive Analytical Chemistry,* Wilson, C. L., Wilson, D. W. (eds.); Amsterdam: Elsevier, 1964, 65–165.
[78] Berka, A., Vulterin, J., Zyka, J., *Massenanalytische Oxidations- u. Reduktionsmethoden;* Leipzig: Akad. Verlagsgesellschaft, 1964.
[79] Oehme, F., Richter, W., *Instrumental Titration Techniques;* Heidelberg: Hüthig Verlag, 1987.
[80] Clark, W., *Oxidation-Reduction Potentials of Organic Systems;* Baltimore: Williams and Wilkins, 1960.
[81] Schulze, G., Simon, J., *Maßanalyse;* Berlin: de Gruyter, 1986.
[82] Oehme, F., *Ionen-selective Elektroden;* Heidelberg: Hüthig 1986.
[83] Frazer, J. W., Balaban, D. J., Brand, H. R., Lanning, S. M., *Anal. Chem.* **55** (1983) 855–860.
[84] Jain, R., Schultz, J. S., *Anal. Chem.* **56** (1984) 141–146.
[85] *Orion Newsletters,* Feb., Jul.–Aug., Nov.–Dec., 1970; and Jan.–Feb., 1971, Orion Research, Cambridge, MA 02139, U.S.A.

[86] Midgley, D., Torrance, K., *Known-Addition and Known-Subtration Potentiometry;* Chichester: Wiley 1978.

[87] Avdeef, A., Comer, J., *Internat. Lab.,* May (1987) 54–63; June 44–52 and *Internat Lab.,* June (1987) 44–52.

[88] Gran, G., *Analyst* **77** (1952) 661–667.

[89] Cammann, K., *Das Arbeiten mit Ionen-selectiven Elektroden;* Berlin: Springer, 1977.

[90] Rigdon, L. P., Moody, G. J., Frazer, J. W., *Anal. Chem.* **50** (1978) 465.

[91] *Ion-Selective Measurements in the Laboratory,* Philips Science & Industry Office, Pye Unicam Ltd., Cambridge CB1 2PX, G. B.

[92] *Handbook of Elektrode Technology,* Orion Research, Cambridge, MA 02139, U.S.A.; Key word: electrode.

[93] *Guide to water and Wastewater Analysis,* Orion Research, Cambridge, MA 02139, U.S.A.

[94] Oehme, F., Bänninger, R., *Chem. Rundschau,* No. 19 (1983).

[95] *Guide to Food and Beverage Analysis,* Orion Research, Cambridge, MA 02139, U.S.A.

[96] *Orion Newsletters 1,* No. 6, 1969, p. 35–38, Orion Research, Cambridge, MA 02139, U.S.A.

[97] *Handbook of Electrode Technology,* Orion Research, Cambridge, MA 02139, U.S.A.

[98] *Guide to Ion Analysis,* Orion Research, Cambridge, MA 02139, U.S.A.

[99] *Bulletins der Deutschen Metrohm,* Deutsche Metrohm GmbH, D-7024 Filderstadt, FRG.

[100] *Ion-Selective Electrode Instruction Manuals,* Radiometer A/S, DK-2400 Capenhagen NV.

[101] Clark, L. C., *Trans. Am. Soc., Artif. Int. Organs* **2** (1956) 41.

[102] Severinghaus, W., Bradley, A. F., *J. Appl. Physiol.* **13** (1958) 515.

[103] Ross, J. W., Riseman, J. H., Krueger, J. A., *Pure Appl. Chem.* **36** (1973) 473–487.

[104] Havas, J., *Ion- and Molecular-Selective Electrodes in Biological Systems;* Berlin: Springer, 1985.

[105] *Handbook of Electrode Technology,* Key word: Nitrate in wastewater, Orion Research, Cambridge, MA 02139, U.S.A.

[106] Bühler, H., *Messen in der Biotechnologie;* Heidelberg: Hüthig, 1985.

[107] "Total Nitrogen in Foods", in: *Guide to Food and Beverage Analysis,* Orion Research, Cambridge, MA 02139, U.S.A.

[108] Scarano, E., Naggar, P., Belli, R., *Anal. Letters* **16** (A 10) (1983) 273–278.

[109] Kalman, L., Swiss Pat. 620, 298.

[110] Rideal, E. K., Evans, U. R., *J. Soc. Publ. Anal.,* No. 7 (1913).

[111] Meites, L., *Polarographique Techniques,* 2nd Ed.; New York: Interscience, 1965.

[112] Heyrovsky, J., Kuta, J., *Principles of Polarography;* New York: Academic Press, 1966.

[113] Clark, L. C., *US Patent 2 913 386,* 1959.

[114] Tödt, F., *Elektrochemische Sauerstoffmessung;* Berlin: de Gruyter, 1958.

[115] Schwabe, K., *Chemie-Ing.-Technik* **37** (1965) 483–492.

[116] Energetic Science Inc., New York, N. Y., *US Patent 3 776 832.*

[117] Fritz, J. S., Gjerde, D. G., Pohlandt, Ch., *Ion Chromatography,* Heidelberg: Hüthig, 1982.

[118] Delehay, P., Charlot, G., Laitinen, H. A., "Classification and Nomenclature of Electroanalytical Methods", *Anal. Chem.* **32,** No. 3 (1960) 103 A–108 A.

[119] Kolthoff, I. M., Elving, P. J., in: *Treatise on Electroanalytical Chemistry;* New York: Interscience Publishers, 1963, Chapter 43.

[120] *IUPAC Compendium of Analytical Nomenclature;* Oxford: Pergamon Press, 1978.

[121] "IUPAC Reports: Recommended Terms, Symbols and Definitions for Electroanalytical Chemistry", *Pure Appl. Chem.* **51** (1979) 1159–1174.

[122] *Draft Document 66D/25: Expression of Performance of Electrochemical Analyzers, Part IV: Dissolved Oxygen in Water Utilizing Membrane-Covered Amperometric Sensors;* International Electrotechnical Commission (IEC), Rue de Varembé, CH-1211 Genf 20, CH.

[123] Hersch, P., *Galvanic Analysis, Reprint 6213,* Beckman Instruments, Fullerton, CA, U.S.A.

[124] *DIN 19261: pH Messung, Begriffe für Meßverfahren mit galvanischen Zellen;* Berlin: Beuth.

[125] Stucki, S., *Chimia* **42,** No. 3 (1988) 94–99.

[126] Oehme, F., Ertl, S., *Chemie-Technik* **8** (1979) 95–100.

[127] Bockris, J. O'M., Azzam, A. M., *Trans. Farad. Soc.* **48** (1952) 145.

[128] MacDonald, D. D., *Transient Techniques in Electrochemistry;* New York: Plenum Press, 1977.

[129] Kissinger, P. T., in: *Laboratory Techniques in Electrochemistry;* New York: Marcel Dekker, 1984.

[130] Rach, P., Seiler, H., *Polarography and Voltammetry in Trace Analysis,* Heidelberg: Hüthig, 1987.

[131] Fleischmann, H., Pletcher, D., *Tetrahedron Letters* **60** (1968) 6255–6258.
[132] Albahadily, F. N., Mottola, H. A., *J. Chem. Ed.* **63** (1986) 271–274.
[133] Koryta, J., *Lehrbuch der Elektrochemie;* Vienna: Springer, 1975.
[134] Adams, R. N., *Electrochemistry at Solid Electrodes;* New York: Marcel Dekker, 1969.
[135] Eberius, E., *Wasserbestimmungen mit Karl-Lösungen,* Weinheim: Verlag Chemie, 1958.
[136] Mabbott, G. A., *J. Chem. Ed.* **60** (1983) 697–682.
[137] Kissinger, R. T., Heineman, W. R., *J. Chem. Ed.* **60** (1983) 701.
[138] Ross, J. W., Riseman, J. H., Krueger, J. A., *Pure Appl. Chem.* **36** (1973) 479, Table 3.
[139] Carey, W.-P. et al., *Sensors and Actuators* **9** (1986) 223–234.
[140] Fabian, L., *Chemie-Anlagen-Verfahren, Dec.* (1986) 95–102.
[141] Hitchman, M. L., *Measurement of Dissolved Oxygen,* New York: Wiley, 1978.
[142] Mancy, K. H., Okun, D. A., Reilley, C. N., *J. Electroanal. Chem.* **4** (1962) 65.
[143] *International Oceanographic Tables,* National Institute of Geography of Great Britain, Wormley, Surrey, G. B.
[144] Oehme, F., Schuler, P., *Gelöst-Sauerstoff-Messung;* Heidelberg: Hüthig, 1983.
[145] Kitzelman, D., Deprez, J., *German Patent Application DE 3033 796 A 1,* 1980.
[146] Züllig, H., *Gas-Wasser-Fach Wasser/Abwasser,* **118** (1977) 227–234.
[147] Mackereth, F. J. H., *J. Sci. Instr.* **41** (1964) 38.
[148] Kane, P. O., Young, J. M., *J. Electroanal. Chem.* **75** (1977) 255–267.
[149] *US Patent 4 477 403,* 1983, Teledyne Industries, Los Angeles, CA.
[150] Schuler, P., *GIT Fachz. Lab.* **24** (1980) 799.
[151] *Technical Notes: Revolutionary Guard Ring Electrode Makes Low Level Oxygen Detection a Technical Reality,* Orbisphere Laboratories, CH-1222 Vésenaz/Geneve.
[152] Riegler, G., *Korresp. Abwasser* **31** (1984) 369–377.
[153] *Maximale Arbeitsplatzkonzentrationen;* Weinheim: VCH Verlagsgesellschaft, 1985.
[154] Scheller, F. W., in: Biosensors, Turner, A. P. F., Karube, J., Wilson, G. S. (eds.); Oxford: Oxford University Press, 1987.
[155] *Metrohm Information 1980, 10, No. 2,* pp. 11–13, Metrohm AG, CH-9100 Herisau.
[156] Schmidt, H.-L., Kittstein-Eberle, R., *Naturwissensch.* **73** (1986) 314–321.
[157] Havas, J., *Ion- and Molecule-Selective Electrodes in Biological Systems;* Berlin: Springer Verlag, 1985.
[158] Chandler, G. K., Eddowes, M. J., *Sensors and Acutators* **13** (1988) 223–228.
[159] van der Schoot, B. H., Bergveld, P., *Anal. Chim Acta* **199** (1987) 157–160.
[160] Oehme, F., *Chemiker Ztg.* **85** (1961) 257–262.
[161] Clark, L. C., in: *Biosensors,* Turner, A. P. F., Karube, J., Wilson, G. S. (eds.); Oxford: Oxford University Press, 1987.
[162] McCann, J., in: *Biosensors,* Turner, A. P. F., Karube, I., Wilson, G. S. (eds.); Oxford: Oxford University Press, 1987.
[163] Home, P. D., Alberti, K. G. M., in: *Biosensors,* Turner, A. P. F., Karube, J., Wilson, G. S. (eds.); Oxford: Oxford University Press, 1987.
[164] Reilley, Ch. N., Murray, R. M., in: *Treatise on Analytical Chemistry,* Vol. 4, Kolthoff, I. M. (eds.); New York: Interscience Publishers, 1963, Chapter 43.
[165] Meites, L., "Amperometric Titrations", in: *Polarographique Techniques,* 2nd Ed.; New York: Interscience, 1965.
[166] Stock, J. T., *Amperometric Titrations;* New York: Krieger Publishing Co., 1975.
[167] Mitchell, J., Smith, D. M., *Aquametry,* Part III; New York: Wiley 1980.
[168] Draft ISO/TC 147 Electrolytic conductivity.
[169] *DIN 38404, Entwurf: Bestimmung der elektrischen Leitfähigkeit;* Berlin: Beuth.
[170] Lange, E., Göhr, H., *Thermodynamische Elektrochemie;* Heidelberg: Hüthig, 1962.
[171] *International Critical Tables of Numerical Data in Physics, Chemistry and Technology;* New York: McGraw-Hill, 1930.
[172] Conway, B. E., *Electrochemical Data;* Amsterdam: Elsevier, 1952.
[173] *Landoldt-Börnstein, Zahlenwerte und Funktionen aus Naturwissenschaften und Technik,* Vol. II, Part 6; Berlin: Springer, 1960.
[174] Jones, G., Bradshaw, B. C., *J. Am. Chem. Soc.* **55** (1933) 2245.

[175] Hewitt, G. F., *Tables of the Resistivity of Aqueous Solutions, Document AERE 3497;* Harwell, U. K.: Atomic Energy Research Establishment, 1960.
[176] Oehme, F., Bänninger, R., *ABC der Konduktometrie,* Polymetron AG, CH-8617 Mönchaltorf, CH.
[177] Rommel, K., *VGB Kraftwerk* **65** (1985) 417–421.
[178] Rommel, K., *Kleine Leitfähigkeitsfibel,* WTW, D-8120 Weilheim, FRG..
[179] Tomkins, R. P. T., Ganz, G. J., Anadalaft, E., *J. Electrochem. Soc.* **117** (1970) 906–907.
[180] Nichol, J. C., Fuoss, R. M., *J. Phys. Chem.* **58** (1954) 696.
[181] Oehme, F., *GIT Fachz. Lab.* **21** (1977) 15–21.
[182] Rommel, K., *LABO* **12**, No. 5 (1981).
[183] Parker, H. C., *J. Am. Chem. Soc.* **45** (1923) 1366, 2017.
[184] Calvert, R., Cornelius, I. A., Griffiths, V. S., Stock, D. I., *J. Phys. Chem.* **62** (1985) 47.
[185] Salomon, M., *Chem. Technik* **10** (1958) 207.
[186] Cruse, K., Huber, R., *Hochfrequenztitration;* Weinheim: Verlag Chemie, 1957.
[187] Pungor, E., *Oscillometry and Conductometry;* Oxford: Pergamon Press, 1965.
[188] Klutke, F., *Dechema Monographien* **17** (1951) 108–121.
[189] Klutke, F., Wuratz, H., *Zeitschr. Naturf.* **5** (1950) 441.
[190] Findlay, A., *Introduction to Physical Chemistry;* London: Longmans, Green & Co., 1933.
[191] Boeseken, J., *Rec. Trav. Chim. Pay-Bas* **58** (1939) 3; and **59** (1940) 745.
[192] Grinberg, A. A., *Einführung in die Chemie der Komplexverbindungen;* Berlin: Verlag Technik, 1955.
[193] Kraus, Ch. A., Bray, W., *J. Am. Chem. Soc.* **35** (1913) 1315.
[194] Oehme, F., *Chemiker Zeitung* **85** (1961) 257–262.
[195] Oehme, F., *Pharmazie* **10** (1955) 532–534.
[196] Kucharsky, J., Safarik, L., *Titrations in Nonaqueous Solvents;* Amsterdam: Elsevier, 1965.
[197] "Industrial Conductometry", *Data sheet TE 876,* Polymetron AG, CH-8617 Mönchaltorf, CH.
[198] *Produkte Handbuch,* Conducta GmbH, D-7016 Gerlingen, Germany.
[199] Howard, C. S., *Ind. Eng. Chem., Anal. Ed.* **5** (1933) 623.
[200] Hamon, B., *J. Sci. Instr.* **33** (1956) 329.
[201] Mohler, B., *Met. Finish.,* July (1977) 54.
[202] Schuler, P., Degner, R., *GIT Fachz. Lab.* **No. 6** (1985) 593–600.
[203] Johnson, E. L., *Internat. Lab.,* April (1982) 110.
[204] Franklin, G. O., *Internat. Lab.,* July/August (1985) 56–67.
[205] Jansen, K. H., Meier, K. D., *Labor Praxis,* May (1985) 438–453.
[206] Opekar, F., Trojánek, A., *Anal. Chim. Acta* **203** (1987) 1–10.
[207] Bartscher, W., Schmidts, W., *Z. Anal. Chem.* **203** (1964) 169–178.
[208] Bruckenstein, S., Symanski, J., *Anal. Chem.* **58** (1968) 1766–1780.
[209] Carlson, R. M., *Anal. Chem.* **50** (1978) 1528–1531; and **57** (1985) 1590–1591.
[210] Baumann, T., *Labor Praxis,* May (1988) 542–547.
Albery, W. J., Hitchman, M. L., *Ring-Disc Electrodes;* Oxford: Clarendon Press, 1971.
Böhm, H., *Techn. Messen* **50** (1983) 399–406.
Harman, J. N., "Electrochemical Generation of Pollutant Standards", in: *Calibration in Air Monitoring, ASTM STP* **598** (1976) 282–300.
Knox, J. H. (ed.), *High Performance Liquid Chromatography,* Edinburgh: University Press, 1978.
Heisz, O., *Hochleistungs-Flüssigkeits-Chromatographie,* Heidelberg: Hüthig 1987.
Anson, D., et al., *J. Inst. Fuel,* April (1971) 191–195.

7.4.2 Devices, Instruments and Suppliers

[DIS 1] Low Temperature Electrolyte FRISCOLYT, Ingold Electrodes Inc., Andover, MA 01810, U.S.A.
[DIS 2] SIEMENS AG, D-7500 Karlsruhe 21.
[DIS 3] Schott-Geräte GmbH, D-6238 Hofheim.
[DIS 4] Automatic cleaning systems for industrial probes, Polymetron AG, CH-8617 Mönchaltdorf.
[DIS 5] Ingold Electrodes Inc., Andover, MA 01810, U.S.A.

[DIS 6] Conducta GmbH & Co., D-7016 Gerlingen, Germany.
[DIS 7] Pfaudler Werke AG, D-6830 Schwetzingen, Germany.
[DIS 8] Electrode maker: Radiometer A/S, DK-2400 Copenhagen NV, Denmark.
[DIS 9] Electrode Model 93-01, Orion Research, Cambridge, MA 02139, U.S.A.
    Model 1520 Calcium Hardness Monitor.
    Model 1770 Chlorine Monitor.
[DIS 10] Automatic concentration controller ACC, Polymetron AG.
[DIS 11] Model 960 Autochemistry Systems, Orion.
[DIS 12] Model 97-70 Total Residual Chlorine Electrode.
[DIS 13] Ektachem DT 60 Analyzer and DTE Model, Eastman Kodak, Rochester, N. Y., U.S.A.
[DIS 14] On-line analyzer Sensimeter G-HCl, G-HF, Bran & Lübbe, D-2000 Norderstedt, Germany.
[DIS 15] Automated chemistry analyzers, NOVA analytical, Newton, MA 02164, U.S.A.
[DIS 16] Kent Industrial Measurements Ltd., Gloucestershire GL10 3TA, England.
[DIS 17] Model ABL 30 Acid-Base Analyser, Radiometer.
[DIS 18] Sterilizable carbon dioxide probe, Ingold [DIS 1, DIS 5].
[DIS 19] DEPOLOX 3 Amperometric Chlorine Monitor, Wallace & Tiernan, D-8870 Günzburg.
[DIS 20] City Technology Ltd., London EC1 OHE.
[DIS 21] Model 23 Glucose Analyzer, Yellow Springs Instruments, Yellow Springs, OH 45 387, U.S.A.
[DIS 22] Multiple Electrode Detectors, Bioanalytical Systems, Inc., West Lafayette, IN 47906, U.S.A.
[DIS 23] Merck AG, Art. No. 8173, D-6100 Darmstadt.
[DIS 24] Toxic Gas Monitoring Instrumentation, Interscan Corporation, Chatsworth, CA 91 311, U.S.A.
[DIS 25] Gasanalyse-Computer MSI 4000, Measuring Systems International, D-5840 Schwerte.
[DIS 26] Process Analyzer Chloromat, Polymetron AG, CH-8617 Mönchaltorf.
[DIS 27] Oxygen Analyzer, Cambridge Instruments Inc., Ossining, N.Y.
[DIS 28] Oxygen Monitor OXISTAT, Polymetron AG, CH-8617 Mönchaltorf.
[DIS 29] Microprocessor Oximeter OXI 2000, Wissenschaftlich-Technische Werkstätten, D-8120 Weil-
    heim.
[DIS 30] Dr. Thiedig & Co., D-1000 Berlin 36.
[DIS 31] Züllig, AG, CH-9424 Rheineck.
[DIS 32] Oxygen Measuring Systems, Orbisphere.
[DIS 33] Dr. W. Ingold AG, CH-8902 Urdorf and Ingold Electrodes Inc., Andover, MA, U.S.A.
[DIS 34] Siepmann und Teutscher GmbH, D-6107 Reinheim.
[DIS 35] LAR GmbH, D-1000 Berlin 61.
[DIS 36] Polymetron AG.
[DIS 37] Chemie und Filter GmbH, D-6900 Heidelberg.
[DIS 38] Hydrazine Analyzer HYDRAMAT, Polymetron AG.
[DIS 39] Oxyflux Hydrazine Analyzer, Hartmann & Braun AG, D-6000 Frankfurt.
[DIS 40] Use of the DIGOX K-301 Analyzer.
[DIS 41] MDA Scientific, Inc., Glenville, IL 60025, U.S.A.
[DIS 42] Chloralarm Chlorine Monitor, Draeger Safety Ltd., Chesham, Bucks. HP5 2AR, U. K.
[DIS 43] Portable Calibration Gas Chamber, General Monitors, Manchester M1 5AU, United Kingdom.
[DIS 44] Chlorine Gas Generator, Wallace & Thiernan.
[DIS 45] Test Gas Generator MONITOX 4100, Bayer Diagnostic and Electronic, D-8000 München 70.
[DIS 46] Standard Oxygen Sensor, Type C/S, City Technology Ltd.
[DIS 47] COMPUR Monitox 4100 Series.
[DIS 48] Bionics Instruments Co., Ltd., Higashiyamato, Tokyo, Japan and Bernt GmbH, D-4000
    Düsseldorf 1.
[DIS 49] Fuel Efficiency Monitor, Neotronics Ltd., Bishop's Stortford, Herts. CM22 6PU, U.K.
[DIS 50] ESA, Inc., Bedford, MA 01730, U.S.A.
[DIS 51] Bioanalytical Systems, Inc., West Lafayette, IN 47906, U.S.A.
[DIS 52] Spark Holland BV, NL-7800 AJ, Emmen, Netherlands.
[DIS 53] MALTHUS AT ANALYZER, Malthus Instruments Ltd., Stocke on Trent ST6 3AT, England.
[DIS 54] KODAK Ektachem DT 60 Analyzer, Eastman Kodak Co., Rochester, N. Y., U.S.A.
[DIS 55] Reflotron and Reflolux II Systems for clinical parameters, Boehringer Mannheim GmbH,
    D-6800 Mannheim 31.
[DIS 56] Industrial Probe XKU, Conducta GmbH, D-7016 Gerlingen, Germany.

[DIS 57] Multielektroden Geber, Polymetron AG, CH-8617 Mönchaltorf, Switzerland.

[DIS 58] Leitfähigkeitsaufnehmer M 54214, Siemens AG, D-7500 Karslruhe, Germany.

[DIS 59] Induktiver Leitfähigkeitsgeber Typ 501, Knick GmbH, D-1000 Berlin 37, Germany.

[DIS 60] Chemical Oscillometer, E. H. Sargent Comp., Chicago, U.S.A.

[DIS 61] HF Titrimeter nach Pungor, Metrimpex, Budapest, Hungary.

[DIS 62] HF-Titrimeter HFT 30C, WTW.

[DIS 63] SoluComp, Ultrapure water analyzer, Beckman Industrial, Cedar Grove, NJ 07009, U.S.A.

[DIS 64] Gas analyzer ULTRAGAS U3EK-Universal, Wösthoff GmbH, D-4630 Bochum 1, Germany.

[DIS 65] CWK-Monitor, Auer-MSA, Pittsburgh, PA 15235, U.S.A.

[DIS 66] Model 870 Total Volatile Halogen Analyzer, Tracor Atlas, Houston, TEX 77041, U.S.A.

8 Solid-State Electrochemical Sensors

Michel Kleitz, Elisabeth Siebert, Pierre Fabry, Jacques Fouletier,
Laboratoire d'Ionique et d'Electrochimie du Solide de Grenoble,
FRANCE

Contents

8.1   Introduction

8.1.1   Scope

Two popular sensors make use of solid electrolytes: the pH electrode based on a glass membrane and the oxygen sensor based on a ceramic tube. For historical reasons, the first has always been regarded as a sensor for liquid electrolyte analysis, as for ion-selective electrodes. It is described in detail in Chapter 7 "Liquid Electrolyte Sensors". The second is, however, a typical example of a solid-state electrochemical sensor. This chapter deals with the use of a solid electrolyte, in terms of basic principles, properties, cell assemblies and applications. These features will be described from a solid-state electrochemistry standpoint. Accordingly, the zirconia cell will frequently be referred to and used as an example. On the other hand, pH and ion-selective electrodes will be quoted only to illustrate particular points.

We will review existing devices and published experiments and also provide guidelines which could be useful to scientists and engineers in their search for new sensors. Many potentialities are still unexplored.

Other review papers [1–12, 246] have been published on solid-state electrochemical sensors.

The *solid electrolytes* which are the basis of the sensors concerned here are solids which exhibit a predominant ionic conductivity. There are many solid ionic conductors. A crystal of AgCl for instance, when examined under the usual conditions, exhibits electric conduction due to the mobility of silver ions Ag^+. In the case of the pH electrode mentioned above, the sensitive material is typically a Li^+ or Na^+ conductor, for the zirconia sensor, it is an O^{2-} conductor. These various solid ionic conductors will be categorized in the last part of this introduction, according to properties essential for their use in chemical sensors. Let us already mention that, in contrast to more conventional liquid electrolytes, generally only cations or anions are mobile in these materials.

In solid-state electrochemistry, such conductivities are usually described in terms of point defects (vacancies, interstitials, etc.). The utilization of such descriptions could introduce unnecessary difficulties for non-specialized readers. So, we will write all the reactions involved in the following derivations in terms of simple ionic species. Except for very specific problems (exchange of Ag^+ ions on the surface of AgCl, either with Ag_i^+ or V'_{Ag}, for example) this should not limit the information provided.

Other solids exhibiting a *mixed ionic-electronic conductivity* can be used as sensitive membranes or reference materials in chemical sensors. They will also be presented in this section. Their specific applications, not yet numerous, will be described in the relevant sections.

8.1.2   A Well-Known Example

The oxygen sensor previously mentioned can be fabricated in many different shapes and sizes. A common assembly is sketched in Figure 8-1. The essential element is a tube made of gas tight sintered stabilized zirconia (ZrO_2-Y_2O_3, 9 mol%). The size can greatly vary from 1 cm long and 2 mm in diameter up to several tens of cm in length and more than 1 cm in diameter. It is operated in a furnace which maintains its closed-end part at a uniform temperature, usually fixed between 600 and 800 °C. Both sides of the closed-end base are covered with

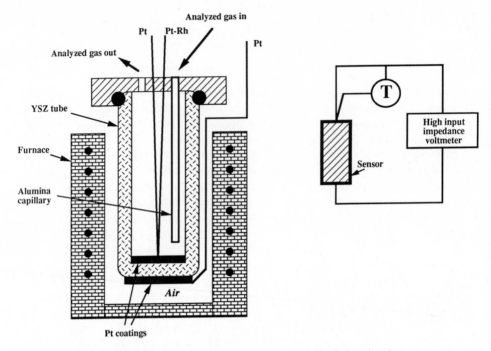

Figure 8-1. Typical potentiometric oxygen sensor based on a stabilized zirconia tube.

porous platinum coatings which form the *electrodes*. Two Pt leads connect them to the measuring instrumentation. An additional Pt-Rh wire forms a thermocouple with one of the Pt leads to measure the inner electrode temperature T. The analyzed gas usually flows over this electrode. The outside electrode is in contact with ambient air which is used as a reference oxygen pressure determining system. The corresponding pressure will be called P_R.

The voltage E of the inner *measuring electrode* with respect to this *reference electrode* is related to the measured oxygen partial pressure P, in the analyzed gas, by the thermodynamic Nernst equation:

$$E = (RT/4F) \ln (P/P_R) \tag{8-1}$$

where R is the gas constant and F the Faraday constant, ie,

$$2.303 \cdot R/4F = 0.4960 \text{ mV/K} \quad \text{for decimal logarithms,} \tag{8-2}$$

and at 750 °C: $2.303 \cdot RT/4F = 50.75$ mV $\tag{8-3}$

In this application, the zirconia cell obeys the theoretical law predicted for an electrochemical concentration cell.

8.1.3   Typical Features of the Potentiometric Sensors

Some of the qualities frequently encountered in solid electrolyte potentiometric sensors are typical features of this oxygen sensor.

The "heavy-duty" nature of the materials:
— All the components are solids and their sensitivity to mechanical parameters such as gravity, acceleration, and vibration is very small, if not negligible. They can also withstand relatively high pressures. Their surfaces can be cleaned by rubbing.
— Their chemical inertness can be very high, ensuring a long life-time and preventing any contamination of the analyzed systems.
— They are stable over wide temperature intervals. This makes them compatible with applications such as sterilization and in situ control of metallurgical processes. On the other hand, it must be stressed that many solid-electrolyte sensors can only be operated at high temperature, a serious drawback for some applications.

Signals that obey a theoretical equation:
— No calibration is usually required, good working conditions can be controlled easily, signal drift is reduced, etc.
— The logarithmic scale of the signal allows measurements of large variations with constant accuracy. With the oxygen sensor described above, variations from 1 atm down to 10^{-6} atm can easily be monitored with a few per cent relative accuracy. On the other hand, small variations, ie, from 0.20 to 0.21 atm, are not easily measured with a high accuracy on a logarithmic scale.

Suitable for miniaturization:
They can be miniaturized down to the microelectronic scale and can therefore form very light and low-energy consuming devices.

8.1.4   Various Electrochemical Types of Sensors Involving Solid Electrolytes

8.1.4.1   Potentiometric Cells

The oxygen sensor described above is a potentiometric device. Up to now this type of device has been by far the most widely developed. All conventional potentiometric sensors involve a *measuring electrode* and a *reference electrode* (cf. previous example). They are based on the Nernst law already mentioned above. It can be written as follows:

$$E = E° + kT \log a \qquad \text{with } k = R/nF \tag{8-4}$$

where E is the measured voltage between the measuring electrode and the reference electrode, $E°$ and k are constants ($E°$ depends on T), n is the number of electrons involved in the sensing electrode reaction, and a is the chemical activity of the analyzed species. Frequently the partial pressure P or concentration C of this species is equal or proportional to a (cf. Section 8.2.1). The generated voltage E is of the order of a few hundred mV and must be measured with a high input impedance apparatus, $E°$ is fixed by the chemical system used at the reference electrode. The temperature is an essential parameter that must be controlled and accurately measured as shown by Equation (8-4).

When dealing with technical cells, we can distinguish several cases. (The basic equations have been written for gas sensors. Generalization to other analyzed chemical species is straightforward).

— *case A :* A theoretical equation of type (8-1) is obeyed. The reference pressure P_R can be determined by a gas composition (air or pure oxygen, for example) or by an appropriate chemical system (a Ni/NiO mixture, for example).

— *case B :* The basic sensor equation is:

$$E = \beta + (RT/zF) \ln P \tag{8-5}$$

in which the constant β has to be determined by calibration. It is not simply related to thermodynamic parameters which could be measured independently. This is typically the case when the electrode reactions at the measuring and reference electrodes involve different ions.

— *case C :* The basic equation is:

$$E = \beta + \alpha \ln P . \tag{8-6}$$

As in case B, β is determined by calibration. Two sub-cases can be considered:

— *case C_I :* The constant α still is of the RT/zF form, but with an unusual integer value for z. This equation is likely to reveal a partial electrode reaction (Section 8.3.1.6).

— *case C_{II} :* The constant α also has to be determined by calibration.

— *case D :* The sensor voltage is not a linear function of $\ln P$:

$$E = f [\ln P] . \tag{8-7}$$

Such sensors can be used to analyze a wide variety of media: gases, liquids, metals, etc. As described in a previous paper [13] the basic diagrams of their utilization conditions vary significantly as a function of the electric conductivity of the analyzed system. The three typical cases are sketched in Figure 8-2.

(a) When the analyzed medium is an electronically conducting system, typically a metal (it could also be a semiconductor or a plasma) the solid-electrolyte sensing membrane can be directly put in contact with it. Then, the sensor signal is measured on this conductor with respect to the reference electrode. Thermo-emf should be appropriately corrected for, if necessary.

(b) When the analyzed species are ions dissolved in an ionically conducting solvent or a molten salt, the solid-electrolyte is also directly put in contact with the analyzed system, but then a second sensor must be used which, in some ways, is similar to the analyzing sensor. This is the case of the pH electrode (cf. example in Section 8.2.2.2).

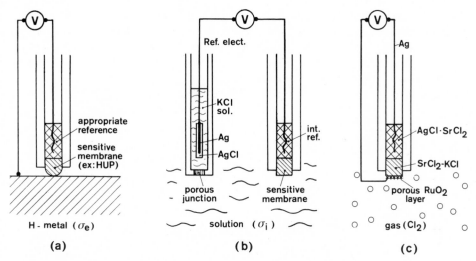

Figure 8-2. Various categories of potentiometric sensors [13] for: (a) electronically conducting media, (b) ionically conducting media, (c) electrically insulating media.

(c) When the analyzed medium is an insulator, typically a gas (it can also be a liquid or a solid) a metallic coating is provided on the surface of the sensitive solid-electrolyte membrane. The sensor signal is measured between the electrode formed in this way and the reference electrode. That is the case of the zirconia cell presented above.

The reasons for all these connections and more generally the principles of the electrochemical solid-state junctions which can be implemented in sensors will be presented in Sections 8.2.2 and 8.4.

In some cases, the sensor electrodes are connected to a resistive load and the resulting current is measured instead of the sensor voltage. Because of this operating mode, the device is then generally called a fuel cell sensor. Details will be given in Section 8.9.

8.1.4.2 Thermo-emf Cells

For special applications a cell rather similar to the potentiometric cells is fabricated so that the electrodes can be operated at different temperatures. A possible advantage of such a thermo-emf device is that it does not require a reference electrode. The whole cell is then immersed in the analyzed media. For the analysis of various melts, at high temperature, non-isothermal operation conditions are also sometimes preferred because they ensure a longer sensor life. Details will be given in Section 8.6.

8.1.4.3 Electrochemical Pumps and Amperometric Cells

Other sensors involving solid electrolytes are based on the Faraday law, the physics of which can be described as followed: because of the ionic nature of the electric conductivity in the solid electrolyte, any current passing through it will carry a corresponding flux of matter. For example, in the cell sketched in Figure 8-1, a current passing between the electrodes would in-ject/extract oxygen in/out of the analyzed gas. Because of the fixed number of electric charges carried by the mobile ions there is a fixed ratio between the current and the flux of matter being electrochemically pumped in this way. Therefore the measurement of the pumping cur-rent provides an easy and accurate determination of the quantity of matter being transferred from one electrode to the other. The corresponding Faraday law is written:

$$J = I/zF \qquad (8\text{-}8)$$

where J is the flux of pumped matter in mol/s, I, the pumping current in A, F, the Faraday constant, and z the number of electrons carried by the ions for one equivalent of a pumped species. For example, in the case of zirconia pumping oxygen, z is equal to 4 (1 mole of O_2 requires 2 O^{2-}, ie, 4 elementary charges).

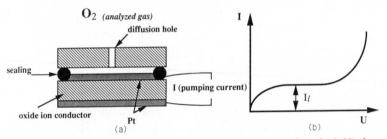

Figure 8-3. A simple amperometric oxygen sensor, (a) cross section, (b) $I(U)$ characteristic.

This property can be implemented in different direct or indirect ways. The most simple cell assembly designed for oxygen analysis is sketched in Figure 8-3 a. The pumping current I is passed in such a direction that oxygen is pumped always from the sensor chamber, up to the limit set by the inwards diffusion of oxygen from the analyzed gas through the diffusion hole. Then the diffusion is a simple function of the oxygen partial pressure to be determined and it is directly measured by the pumping current. The $I(U)$ characteristic of such a cell is shaped as shown in Figure 8-3 b. The diffusion-limited regime is evidenced by a plateau whose current I_l will constitute the measure of the analyzed gas pressure. Typical currents are in the range 1 μA to 1 mA. With respect to potentiometric devices, the main advantage of the ampero-metric sensors is their higher sensitivity to small variations of a gas pressure. They are also far less temperature dependent and do not require reference electrodes. On the other hand, geometrical parameters, such as the hole diameter in the sensor sketched above, become essen-tial and must be maintained strictly constant to avoid any drift of the performance characte-ristics.

In practice, a current is frequently measured in these devices instead of a potential difference in the potentiometric sensors. For this reason, they are called amperometric sensors (some-times pumping sensors).

In some cases, electrochemical pumping is used to modify, in a controlled way, the composi-tion of the analyzed system which is subsequently measured by a potentiometric device.

At present, all these sensing applications are subjects of very active research and develop-ment. They are all intended to analyze oxygen in gases. They will be described in detail in Sec-tion 8.10.

8.1.4.4   Conductivity Cells

Referring to a large set of data on the conductivities of solid composite electrolytes, Maier et al. [14] have assumed that mobile ions in solid electrolytes can be attracted by adsorbed nucleophilic species. A significant variation in the electric carrier concentration would be induced in this way in the solid electrolyte subsurface with a resulting variation in conductivity.

To evaluate the sensing properties which can be based on this type of chemical reaction, conductivity measurements were performed on various AgCl thin layers as a function of the NH_3 pressure in the surrounding gas (in the 5×10^3 to 10^5 Pa range). The observed variations in the sample resistances did confirm the assumption. These results are quite recent. They may open a new field of chemical sensing for the solid ionic conductors.

8.1.5   Brief History

As a preliminary remark, it must be stressed that the history of conventional solid-electrolyte gas sensors has been dominated by the development of the multiple versions of the oxygen sensor which is still the paradigm in this field. It has been commercially available since 1965.

Potentiometric measurements with solid-electrolyte cells showing well-defined variations of the measured voltage with either a gas pressure or a condensed-phase composition in contact with one electrode, can be traced back to the beginning of the century. For instance, Haber and co-workers [15, 16] observed that a cell based on a porcelain wall separating oxygen from an hydrogen-water vapor mixture generates a voltage which depends on the hydrogen pressure.

It is commonly considered that the real start was marked by the theoretical analysis of Carl Wagner [17] in 1957 and by the paper by Weissbart and Ruka [18] and the patent by Peters and Möbius [19] on the oxygen gauge in 1961.

The first important industrial application, launched in the early 70's, was steel processing control by disposable devices which analyze oxygen in molten iron [20–23]. Many countries such as Canada, Japan and Belgium consume more than 100 000 units a year.

The second important application dating from the late 70's is the λ-sensor for the air/fuel ratio control in internal combustion engine. It is now fabricated by millions and is in use in many cars of American, Japanese, German, and other makes. Among the most active groups in this field are Ford [25], Bosch [128–130], and Nissan [24].

Another application more difficult to evaluate quantitatively although quite successful, is atmosphere monitoring and control by home made sensors in research laboratories. Many publications mention solid-state electrochemical sensors.

The possible diversification towards analysis of gases other than oxygen has already been demonstrated with other ionic conductors. It can be as easy as oxygen analysis. Many papers have shown that SO_3 [26–27], H_2 [28–30], Cl_2 [31–34], and other species can be analyzed, with or without pretreatment, sometimes with a sensitivity well below 1 ppm and a high accuracy.

For ion analysis, besides the well-known pH glass electrode developed by Cremer [35] at the turn of the century, the LaF_3 single crystal electrodes, invented by Frant and Ross [36] and commercialized by Orion [37] for F^- ion analysis has been an unmatched success since 1966. The first AgCl electrode was introduced in 1937 [39]. Various glasses based on alumina and silica are traditionally used for sensing alkali ions on the basis of the work by Eisenman et al. in 1957 [38]. Numerous compounds of the Ag_2S family are also widely used for sensing anions and cations. Industrial development began in the early 60's [40]. New specific applications of scaffolded solid electrolytes have recently been suggested [13, 41, 42].

Also, since the early 60's [43, 44], the oxide ion conductors have been used for probing the O^{2-} concentration variation in molten salts.

The electrochemical pump which was specifically investigated for the first time in 1966 [45] had long been used mostly in laboratories. Philips [46] commercialized an oxygen demanding device for the characterization of natural water in 1974. Its analytical capabilities are now being systematically investigated mostly by Japanese groups [47, 48]. Here the main application envisaged is exhaust gas analysis of combustion engines operated in the lean regime.

Gas sensor miniaturization has been a direction of continuous efforts since the mid 70's especially with the utilization of thin-layer cells. A research team at Thomson CSF [49–51] has promoted several new ideas in this field. Obvious goals are the reduction of size and cost and also the lowering of the minimum operating temperature in the case of the zirconia sensor. The initial idea was that a temperature drop mostly increases the electrolyte resistance, which should be compensated for by a thinner electrolyte layer. Thin-layer potentiometric cells have unfortunately appeared very sensitive to short-circuits through pin-holes. Another now well-characterized difficulty encountered when lowering the temperature is the slowing of the electrode kinetics.

In Ion Sensitive Field Effect Transistors (ISFET's), the first attempts to use solid ionic conductors as sensitive membranes appeared in the late 70's [52, 53].

Developments in many of these new fields are rather slow, mostly because the markets are not yet well-defined and frequently not attractive enough for industrial production. This general situation contrasts with the simplicity of most of the sensor designs and with the facility, frequently stressed in the literature, to assemble new sensors for laboratory purposes.

8.1.6   Selection Criteria of Solid-Ionic Conductors

Many solids exhibit an ionic or a mixed ionic-electronic conductivity. Lists can be found in various books dedicated to this property [54–56]. For a continuous updating, the scientific review *Solid State Ionics* can also be referred to.

The first technical classification criteria is the nature of the phases they form. Single crystals, sintered ceramics, glasses, polymers, and even composites can be implemented for their ionic conductivity. The fabrication cost, the ease of shaping, the robustness, etc. markedly depend on these criteria. Special features of these phases themselves can also be of interest: a polymer, for instance, can also be used as a cement between sensor parts [197]. On the other hand, rigid materials, as the ceramics, will require compatible expansion coefficients with materials contacting them, when used or assembled at high temperature.

Another obvious criterion is the temperature range over which the ionic conductivity of the material is sufficient for the sensing purpose. It is commonly considered that 10^{-6} Ω^{-1} cm^{-1} is a comfortable lower limit for the conductivity of a sensing membrane, but far less conducting materials are frequently used. As mentioned above, solid-state conductivities generally decrease exponentially with decreasing temperature. Sometimes, a large conductivity drop makes this decrease even more dramatic. Stabilized zirconia which is an excellent ionic conductor at 1000 °C is a perfect insulator at room temperature and can no longer be used for sensing oxygen. To be more accurate on the utilization of this criteria, one can distinguish a high temperature domain above about 600 °C, a medium temperature domain around 300 °C, and a room temperature domain just above room temperature and possibly below. Polymers for instance are typical room temperature materials. β-Aluminas and Nasicons can be used from room temperature up to high temperature. $SrCl_2$ is a medium temperature material.

Another criterion is the nature of the mobile carrier in the solid electrolyte, ie, Na^+, H^+, O^{2-}, F^-. In the solid phases, except for polymers and a few other materials, only the cations or the anions are mobile. It is sometimes assumed that oxygen analysis requires an O^{2-} ion conductor, sodium analysis a Na^+ ion conductor and so on. As we will see in the next sections, such a selection generally insures excellent performance but is not at all a prerequisite. It has been clearly demonstrated that for instance a F^- or a Cl^- conductor can be used in an oxygen pressure sensing device [31, 58].

The last general criterion to be taken into account, especially for analysis at low temperature, is the existence of a network of covalent bonds in the material structure. In practice two different groups of solid ionic conductors can be distinguished whatever their crystallinity or amorphicity is: binary compounds and scaffolded materials.

Binary and pseudo-binary compounds such as $AgCl$, CaF_2, $PbCl_2$, K_2SO_4, stabilized ZrO_2, etc. are linked by rather ionic chemical bonds (except the oxides) and generally behave as predicted by the relevant thermodynamic laws when they are in contact with the analyte. When they are immersed in solutions, for instance, their behavior will be governed by their solubility products. Under high-temperature gas analysis conditions, they vaporize as predicted by thermodynamics. In terms of interference with the analyzed medium, one must also consider the risk of displacement reactions such as

$$CaF_2 + 1/2\ O_2 \longrightarrow CaO + F_2 \qquad \text{(8-9)}$$

or the possible dissolution of gas components in the material.

The other group of solid ionic conductors, scaffolded materials such as β-alumina, Nasicon, Hollandite, silicate glasses, polyethylene oxide salt polymeric complexes, etc. are frequently used under conditions where the covalent scaffolding can be viewed as a chemically inert supporting structure. Then, the only interaction with the analyzed media is the exchange of the mobile ions and possibly electrons. The only risk of interference is the dissolution of neutral species in the scaffolding (ie, solvent molecules) and the substitution of foreign ions for the ionic carriers. Both risks are markedly reduced with 3-dimensional rigid scaffoldings such as Nasicon, for instance. With 2-dimensional materials such as β-alumina, the risk is more serious because of the easy expansion of the network which can accommodate various chemical species. Such a risk is even greater with polymers.

Typical examples of solid ionic and mixed conductors are given in Table 8-1, with the values of all these criteria. This table is far from exhaustive. As sensors, all of the quoted cells exhibit reasonably good performance characteristics in terms of accuracy and response time.

Table 8-1. Typical examples of solid ionic conductors used in electrochemical sensors.

Materials	Anal. species	Phase	Mobile ion	Temp. range (K)	Cell	Performances	Ref.
ZrO_2 + (CaO, MgO, Y_2O_3)	O_2, $(O)_{melt}$ CO-CO_2 H_2-H_2O	fully or partially stabilized	O^{2-}	750–1100	ref./solid electrolyte/ME, O_2 ref.: air, Ni-NiO, Pd-PdO, Co-CoO, Cr-Cr_2O_3 ME: Pt, Ag	$10^{-5} - 10^{-22}$ Pa O_2 $\Delta P/P \approx$ 2–5% t_r:0.1 s to a few seconds	1–11, 20–25 51, 88, 98 104, 122–132 136, 137, 144, 148
	H^+			400–550	Cu-Cu_2O (or Hg-HgO)/Y.S.Z./ anal. solution	$2 <$ pH < 9	59–61
CeO_2 + (CaO, Y_2O_3)	O_2	sintered	O^{2-}	673–1273	air (or pure O_2), Pt/CeO_2-CaO (Y_2O_3)/Pt, O_2	$P(O_2) > 10^3$ Pa	62, 63
Bi_2O_3-MoO_3	O_2	sintered	O^{2-}	573–623	air, Ag/$(Bi_2O_3)_{0.84}(MoO_3)_{0.16}$/Ag, pure O_2	$t_r <$ 1 min at 623 K	64
SrCl$_2$-KCl-SrO	O_2	sintered	Cl^-	540–650	Ag/AgCl/SrCl$_2$-KCl-SrO/FePc,O_2	1–10^5 Pa	31
	Cl_2	sintered	Cl^-	350–500	Ag/AgCl/SrCl$_2$-KCl-SrO/RuO$_2$,Cl$_2$	$P(Cl_2)$:10^{-1}–10^5 Pa t_r:a few seconds	31, 32
	O_2	single cryst. or	F^-	300–373	Sn-SnF$_2$/LaF$_3$/Pt black, O_2	$P(O_2)$:10^4–10^5 Pa t_{90}:3–5 min	65, 163 164, 166
LaF$_3$	F^-	sputtered	F^-	273–353	ref. el./LaF$_3$/anal.solution/ ref.el.	10^{-1}–10^{-6} mol L^{-1} (0 < pH < 8)	37
	La^{3+}	film	F^-		Ag/AgF/LaF$_3$/anal. solution/ ref.el.	10^{-1}–10^{-6} mol L^{-1}	37
LaF$_3$ + EuF$_3$	O_2	sintered	F^-	320–420	Ag-AgF/LaF$_3$-EuF$_3$/Pt black, O_2	$P(O_2)$:4 10^3–10^5 Pa t_{90}: a few minutes to 20 min	66

Table 8-1. (continued)

Materials	Anal. species	Phase	Mobile ion	Temp. range (K)	Cell	Performances	Ref.
$PbSnF_4$	O_2	sintered	F^-	370–520	$Sn\text{-}SnF_2/PbSnF_4/Pc(Fe),O_2$	10^2–10^5 Pa	3, 160, 165
$\beta\text{-}PbF_2$	O_2, H_2, NH_3	thin film	F^-	400	$Bi/PbF_2/LaF_3/Pt, O_2$; $Bi\text{-}BiF_3/PbF_2/Pt, H_2 (NH_3)$		67, 68
AgX ($+ Ag_2S$) X: Cl, Br, I	Cl_2		Ag^+	420–500	$Pt, Ag/AgCl/Pt, Cl_2$	$P(Cl_2): 1$–10^4 Pa	33, 34
	X^-	sintered	Ag^+	273–353	ref. el./$AgX\text{-}Ag_2S$/anal. solution/ref.el.	$5\ 10^{-6}$–1 mol L^{-1} ($2 <$ pH < 12)	37
	Ag^+		Ag^+	273–353	ref.el.	10^{-7}–1 mol L^{-1} ($2 <$ pH < 12)	37
MS ($+ Ag_2S$) M: Cu, Pb	Cu^{2+} Pb^{2+}	sintered	Ag^+	273–353	ref.el./$MS\text{-}Ag_2S$/anal. solution/ref.el.	10^{-1}–10^{-6} mol L^{-1} ($2 <$ pH < 12)	37
KAg_4I_5	I_2	sintered	Ag^+	313	$Ag/KAg_4I_5/Pt, I_2$	$P(I_2) > 10^{-2}$ Pa t_r: a few seconds to a few minutes	69
$\beta\text{-}Al_2O_3$	Na (Na^+)	sintered pellet or thin	Na^+	473–633	$Pt/\alpha + \beta\text{-}Al_3O_3/\beta\text{-}Al_2O_3(Na)/Na$	$P(Na):10^{-5}$–1 Pa t_r: 10 min	70
	S_x (Ag^+)		Ag^+	363–1073	$Ag/\beta\text{-}Al_3O_3(Ag)/Ag_2S, [S]$		1, 71
$\beta'\text{-}Al_2O_3$	H_2 (Na^+)	film	Na^+	297	$Pd/\beta\text{-}\beta''\text{-}Al_2O_3(Na)/Pt, N_2 + H_2$	10–10^4 Pa	72
Sulfates $M_2SO_4\text{-}Ag_2SO_4$	$SO_2\text{-}SO_3$	two-phase system	M^+	500–1200	$O_2, Pt/Y.S.Z./K_2SO_4/Pt, SO_3, O_2$; $Ag/Ag_2SO_4\text{-}M_2SO_4/Pt$ (or V_2O_5), $SO_2 + SO_3$	> 1 Pa t_r: a few minutes	133 167–171
M: Na, K, Li				700–1100	$Pt, O_2 + SO_2 + SO_3/Na_2SO_4/\beta\text{-}Al_2O_3$ (or Nasicon)/$Na_2SO_4/Pt, SO_2 + SO_3 + O_2$	t_{99}: 1900 s at 750 K t_{99}: 60 s at 1100 K	75, 76

(continued on next page)

Table 8-1. (continued)

Materials	Anal. species	Phase	Mobile ion	Temp. range (K)	Cell	Performances	Ref.
Carbonates K_2CO_3	CO_2	sintered	K^+	575–1025	$Ag/K_2CO_3\text{-}Ag_2SO_4/Pt,CO_2,O_2$ $O_2,Au/ZrO_2\text{-}CaO/K_2CO_3/$ Au,O_2,CO_2	$P(CO_2) > 10$ Pa in air t_r: a few seconds to a few minutes	2, 73 77, 78, 85
Na_2CO_3			Na^+	600–1000	$O_2,CO_2,Au/Na_2CO_3/Nasicon/Au,O_2$	$P(CO_2)$: $1\text{–}10^5$ Pa t_r: a few seconds to a few minutes	78
Nasicon	$\dfrac{Na}{Na^+}$	sintered	Na^+	450–650	stainless steel/(Na)/Nasicon (or $\beta\text{-}Al_2O_3/\,Na_xCoO_2/$ stainless steel		79
				298	ref.el./Nasicon/anal. solution/ ref.el.	$10^{-4}\text{–}1$ mol L^{-1}	57
Silicate glasses	$\dfrac{H^+}{\text{alkali ions}}$	glass	alkali ions	293–400	Ag/AgCl/ref.solution/mem- brane/analyzed solution/ref.solution/Hg_2Cl_2/Hg	$2 < pH < 12$ $10^{-6}\text{–}1$ mol L^{-1}	37 37
Pyrex glass	$\dfrac{O_2,\ CO\text{-}CO_2}{H_2\text{-}H_2O}$	glass	Na^+	420–770	W,Na(liq.)/Pyrex glass/Pt, O_2	$1\text{–}10^5$ Pa (O_2) $10^{-10}\text{–}10^{-25}$ ($H_2\text{-}H_2O,CO\text{-}CO_2$)	80
	S_x			500–720	W,Na(liq.)/Pyrex glass/S_x	$1\text{–}10^3$ Pa	80
H.U.P.	H_2	sintered	H^+	293	Pd/PdH_x/H.U.P./Pt,H_2	$P(H_2) > 10^3$ Pa $\Rightarrow t_r$: 3–4 s	28–30 50,173–176
$SrCe_{0.95}Yb_{0.05}O_{3-\delta}$	H_2O	sintered	H^+,O^{2-}	873–1273	gas + H_2O,Pt/$SrCe_{0.95}Ce_{0.95}Y_{0.05}$ $O_{3-\delta}$/Pt,air + H_2O	$t_r \approx 15$ min	81, 82
$Zr(HPO_4)_2 \cdot n\ H_2O$	$\dfrac{H_2}{CO}$	sintered	H^+	298	air,Pt/H^+ cond./Pt black,gas	t_r: 10 s at $2\ 10^2$ Pa t_r: 2 min at 10 Pa in air	83

8.1.7 Warnings to the Uninformed Reader

Electrochemical terminology has been created in successive strata and the official regulations [84] are not strict enough to eliminate the inconsistencies of common usage. For the uninformed reader, some of these commonly accepted inconsistencies can be seriously confusing. Two, which can be irritating, will be encountered in this text. They are related to uses of the terms electrode and potential.

— *Electrode:* Properly speaking, for an electrochemist, an electrode is made of an electronic conductor in contact with an ionic conductor. The chemical species electrochemically reacting at this contact have also to be well-defined (see, for example, Reaction (8-15)).

In the sensor field, the term electrode frequently refers to a more complex system. For example, the *pH electrode* is made of a glass bubble, whose external surface is the sensing element. It contains a liquid in which a true electrode is immersed. In aqueous solution analysis, a *reference electrode* is another example, rather similar in its electrical complexity to the pH electrode.

In the following, an electrode will be a contact between an electronically and an ionically conducting material. Other acceptions will be indicated by enclosing the term in quotation marks.

— *Potential:* Expressions such as electrostatic potential, electrode potential, electrochemical potential will frequently be used.

In practice, without any other specific qualification, the term electrode potential will mean the electrostatic potential of the metal forming the electrode with respect to that of another piece of metal forming another electrode (sometimes it will be defined with respect to the electrostatic potential inside the ionic conductor, as it should be, in theoretical electrochemistry).

The electrochemical potential term may be more confusing, as we will see in the next section (cf. Equation (8-11)). It is related to the electrostatic potential, φ, by an equation of the form:

$$\bar{\mu} = C^{st} + zF\varphi \tag{8-10}$$

in which z is a charge number and F is the Faraday constant. Therefore, it does not have the same units as the electrostatic potential φ. In essence, $\bar{\mu}$ is an energy. The same remark applies to the chemical potential μ which has the same units as $\bar{\mu}$.

8.2 Basic Principles of Potentiometric Devices

8.2.1 Parameters and Basic Equations

In the derivations dealing with isothermal cells, the essential parameters are the partial free enthalpies of the electric carriers in the *solid ionic conductor* (s.i.c) and of the chemical species active at the *electrodes* (el). They are called their electrochemical potential, $\bar{\mu}$, when the electric energy is included, or their chemical potential μ otherwise. They are written in a way which shows the parameters regarded as variable in the derivations: the average local electric potential φ (the Galvani potential) and the local concentration C (or pressure P for a gas) of the corresponding species:

$$\tilde{\mu} = \mu + zF\varphi \tag{8-11}$$

where $\mu = \mu^\circ + RT \ln C.$ \hfill (8-12)

μ° is a constant, R and T have their usual meanings (cf. Section 8.1.2), and z is the number of elementary charges carried by the species (ie, -2 for O^{2-}). When C varies over a broad range and when the phase in which the species is dissolved is not ideal, the concentration term is written $RT \ln \gamma C$. The new parameter γ, called the *activity coefficient,* is introduced to account for possible deviation from the ideal Equation (8-12). It is a function of C, among other parameters. Sometimes, γC is simply written a, called the *chemical activity* of the species.

The essential arguments of any derivation are the following:
1) Electrons and electron holes are regarded as regular electrochemical species.
2) The driving force for the migration/diffusion of a species is the gradient of its electrochemical/chemical potential. The corresponding flux density J is typically written as:

$$J = -\tilde{u}C \text{ grad } \tilde{\mu} \tag{8-13}$$

where $\tilde{u}$ is the electrochemical mobility of the species (cf. Section 8.5.1.1).
If the species is electrically charged the corresponding partial electric current density, I, is

$$I = zJF. \tag{8-14}$$

3) Any equilibrated surface reaction can be translated into an equation relating the revelant electrochemical potentials. For instance, the oxygen electrode reaction occuring at the electrodes of the oxygen sensor previously described,

$$O_2 \text{ (gas)} + 4e^- \text{ (Pt)} \longleftrightarrow 2O^{2-} \text{ (YSZ)}, \tag{8-15}$$

will be translated into the following equation:

$$\mu_{O_2} + 4\tilde{\mu}_e = 2\tilde{\mu}_{O^{2-}}. \tag{8-16}$$

4) When two phases, called I and II, can exchange a species c, under equilibrium, the electrochemical potential values in both phases are equal for this species:

$$\tilde{\mu}_c^{II} = \tilde{\mu}_c^I. \tag{8-17}$$

5) The electric potential difference between two electric leads, called I and II, made of the same metal, is simply related to the corresponding difference between the electron electrochemical potentials $\tilde{\mu}_e$ in these leads. The electron concentrations being constant in the metals at a given temperature, one can write $\tilde{\mu}_e$ in the form:

$$\tilde{\mu}_e = \mu_e^\circ - F\varphi. \tag{8-18}$$

Therefore,

$$\varphi^{II} - \varphi^{I} = -\frac{1}{F} \, (\tilde{\mu}_e^{II} - \tilde{\mu}_e^{I}) + \frac{1}{F} \, (\mu_e^{\circ II} - \mu_e^{\circ I}) \, . \tag{8-19}$$

The metals of the leads being the same in nature, the second terms is nil and then:

$$\varphi^{II} - \varphi^{I} = -\frac{1}{F} \, (\tilde{\mu}_e^{II} - \tilde{\mu}_e^{I}) \, . \tag{8-20}$$

Referring to these simple arguments, the generation of a cell voltage can be described as follows. Let us take again, as an example, the zirconia oxygen sensor. At the measuring electrode, called II, Equation (8-16) can be established. At the reference electrode, called I, the same equation is also valid. One directly gets from these two equations:

$$-4 \, (\tilde{\mu}_e^{II} - \tilde{\mu}_e^{I}) = (\mu_{O_2}^{II} - \mu_{O_2}^{I}) - 2 \, (\tilde{\mu}_{O^{2-}}^{II} - \tilde{\mu}_{O^{2-}}^{I}) \tag{8-21}$$

which gives the cell voltage by applying Equation (8-20):

$$\varphi^{II} - \varphi^{I} = \frac{1}{4F} \, (\mu_{O_2}^{II} - \mu_{O_2}^{I}) - \frac{1}{2F} \, (\tilde{\mu}_{O^{2-}}^{II} - \tilde{\mu}_{O^{2-}}^{I}) \, . \tag{8-22}$$

This equation shows two contributions:

− a surface part $(\mu_{O_2}^{II} - \mu_{O_2}^{I})$ which can be correlated to the *surface specific sensitivity* of the electrodes
− a s.i.c. bulk part $(\tilde{\mu}_{O^{2-}}^{II} - \tilde{\mu}_{O^{2-}}^{I})$.

To define this second part we will say that the bulk of the s.i.c. functions as an *electrolytic transmission line*.

As shown by Equation (8-16), the surface sensitivity fixes the jump $(\tilde{\mu}_e - \tilde{\mu}_{O^{2-}})$. In other words, it fixes the level of the *significant* parameter $\tilde{\mu}_e$ with respect to the *reference* parameter $\tilde{\mu}_{O^{2-}}$. The bulk conditions fix the variations of this reference parameter.

To describe more clearly these functions and their variations we have introduced the so-called "double-band" diagrams [85–87]. In essence, they superimpose the description of the electronic distribution in the material to that of the mobile ionic carriers among the available crystallographic sites. To make such a superimposition meaningful, new ionic parameters have been introduced, called the *reduced electrochemical potentials W* (or the ionic Fermi levels). They are defined by the equation:

$$W_i = \frac{\tilde{\mu}_i}{-z_i} \tag{8-23}$$

where z_i is the ionic charge of the i ion. In contrast to the actual electrochemical potentials, they vary in parallel ways as functions of φ and the usual reasonings of semiconductor physics can be straightforwardly applied (for electrons: $W_e = \tilde{\mu}_e$).

In what follows we will describe the diagrams of the sensors in terms of variations of such reduced electrochemical potentials. The sensor voltage $(\varphi^{II} - \varphi^{I})$ is related to the electron reduced electrochemical potential in the electrodes by the same relation as Equation (8-20):

$$\varphi^{II} - \varphi^{I} = -\frac{1}{F}(W_e^{II} - W_e^{I}).$$

$$(8\text{-}24)$$

As an example, Figure 8-4 shows the diagram of a zirconia electrochemical cell. Under sensing conditions, the ion reduced electrochemical potential $W_{O^{2-}}$ is constant. This electrolytic transmission line maintains constant the level of the electrochemical reference.

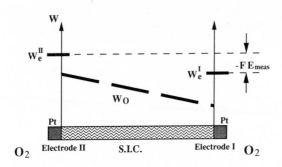

Figure 8-4.
Basic diagram of a solid-state electrochemical cell (in a potentiometric sensor, the reference level, W_O, is constant).

8.2.2   Description Mode

To illustrate this mode of description in greater details, let us examine two other examples.

8.2.2.1   First Example

Consider the following cell which has been assembled for analyzing Cl_2:

$$Cl_2: C^{II}/AgCl/PbCl_2/Pb/{}^{I}C .$$

$$(8\text{-}25)$$

(C has been added to the right hand side to eliminate the thermo-emf).
Under the measuring conditions AgCl is an Ag^+ conductor and $PbCl_2$ predominantly a Cl^- conductor.
Potentiometric conditions mean:

$$I = 0 .$$

$$(8\text{-}26)$$

This implies that the partial currents carried by the predominant carriers are nil:

in AgCl: $I_{Ag^+} = 0$ and therefore $W_{Ag} = C^{st}$

$$(8\text{-}27)$$

in $PbCl_2$: $I_{Cl^-} = 0$ and $W_{Cl} = C^{st} .$

$$(8\text{-}28)$$

At the measuring electrode II the basic sensing reaction is:

$$1/2 \ Cl_2 \ (gas) + Ag^+ \ (AgCl) + e^- \ (C) \longleftrightarrow AgCl \ . \tag{8-29}$$

Therefore:

$$\frac{1}{2} \ \mu_{Cl_2} + \tilde{\mu}^{II}_{Ag^+} + \tilde{\mu}^{II}_{e^-} = \mu^{II}_{AgCl} \tag{8-30}$$

and

$$W^{II}_e - W^{II}_{Ag} = \mu^{II}_{AgCl} - \frac{1}{2} \ \mu^o_{Cl_2} - \frac{RT}{2} \ \ln P_{Cl_2} \ . \tag{8-31}$$

(The sign on the indices such as Ag are left out for the sake of simplicity).
At the reference electrode I:

$$Pb + 2Cl^- \ (PbCl_2) \longleftrightarrow PbCl_2 + 2e^- \ (C) \tag{8-32}$$

$$W^I_e - W^I_{Cl} = -\frac{1}{2} \ \mu^I_{PbCl_2} + \frac{1}{2} \ \mu^o_{Pb} \ . \tag{8-33}$$

At the AgCl/PbCl$_2$ contact:

$$Ag^+ \ (AgCl) + Cl^- \ (PbCl_2) \longleftrightarrow AgCl \tag{8-34}$$

$$W^{cont}_{Cl} - W^{cont}_{Ag} = \mu^{cont}_{AgCl} \ . \tag{8-35}$$

If the transmission along the electrochemical lines is performed correctly, in other words, if W_{Ag} and W_{Cl} are constant, as expressed by Equations (8-27) and (8-28), the overall sensor equation can be obtained by appropriately adding Equations (8-31), (8-33) and (8-35). This gives:

$$W^{II}_e - W^I_e = -\frac{RT}{2} \ \ln P_{Cl_2} - C^{st} \tag{8-36}$$

with:

$$C^{st} = (\mu^{cont}_{AgCl} - \mu^{II}_{AgCl}) - \frac{1}{2} \ (\mu^I_{PbCl_2} - \mu^o_{Pb} - \mu^o_{Cl_2}) \ . \tag{8-37}$$

In principle, the first term of C^{st} is nil, but it can be useful to leave it in the equation to show the nature of the deviation from ideal conditions which may result from a chemical reaction at the AgCl/PbCl$_2$ contact.

Equation (8-36) shows that the assembled sensor can indeed be operated as a chlorine sensor. The corresponding diagram is shown in Figure 8-5.

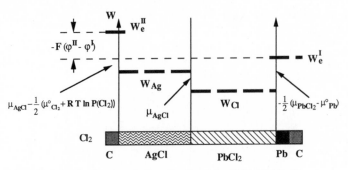

Figure 8-5. Basic diagram of the cell: Cl_2, $C/AgCl/PbCl_2/Pb/C$.

Here, the electrolytic transmission line which goes from one electrode to the other includes a *junction* between two ionic conductors. The junction between two different carrier materials such as AgCl and $PbCl_2$ will be called a *heterojunction*.

8.2.2.2   Second Example

The following electrochemical chain can be treated in the same way:

$$O_2, Pt^{II}/ZrO_2 - Y_2O_3/SiO_2 - Na_2O/\beta\text{-alumina}/^I Na/C/Pt \ . \tag{8-38}$$

Yttria-stabilized zirconia is an O^{2-} conductor while the silicate and β-alumina are Na^+ conductors.

The electrode reaction on the left hand side is the oxygen electrode reaction that we have already encountered Equation (8-15). On the right hand side, it is a sodium electrode reaction:

$$Na^+ \ (\beta\text{-}alumina) + e^- \ (C) \longleftrightarrow Na \tag{8-39}$$

which implies:

$$W_e^I - W_{Na} = \mu_{Na}^\circ \ . \tag{8-40}$$

At the *heterojunction* ZrO_2-Y_2O_3/SiO_2-Na_2O, the basic reaction is likely to be:

$$O^{2-} \ (YSZ) + 2Na^+ \ (silicate) \longleftrightarrow Na_2O \ (silicate) \tag{8-41}$$

$$W_O - W_{Na} = \frac{1}{2} \ \mu_{Na_2O}^{YSZ \, cont} \ . \tag{8-42}$$

At the $SiO_2 - Na_2O/\beta$-alumina contact, the basic reaction is a simple Na^+ ion exchange:

$$Na^+ \ (silicate) \longleftrightarrow Na^+ \ (\beta\text{-}alumina) \tag{8-43}$$

$$W_{Na}^{silicate} = W_{Na}^{\beta\text{-alumina}} \ . \tag{8-44}$$

The level of this electrochemical reference goes through the interface without any alteration. Such a contact between two materials conducting by the same ion will be called a *homojunction*.

The corresponding diagram is shown in Figure 8-6 and the sensor equation is:

$$W_e^{II} - W_e^{I} = \frac{1}{2} \mu_{Na_2O}^{YSZ \, cont} - \frac{1}{4} \mu_{O_2} - \mu_{Na}^{o} .$$ (8-45)

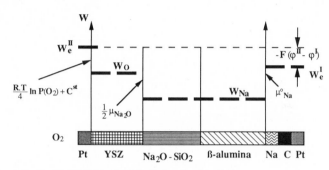

Figure 8-6.: Basic diagram of the cell: O_2, $Pt/YSZ/Na_2O-SiO_2/\beta$-alumina$/Na/C/Pt$.

If there are enough free oxide ions in the silicate, one can also assume that a simple oxide ion exchange prevails at the zirconia/silicate junction:

$$O^{2-} \, (YSZ) = O^{2-} \, (silicate) .$$ (8-46)

Then one can extend the W_O transmission line into the silicate. Inside this material, the distance $(W_O - W_{Na})$ is determined by the dissociation reaction of Na_2O:

$$O^{2-} \, (silicate) + 2Na^+ \, (silicate) = Na_2O \, (silicate)$$ (8-47)

which implies that:

$$W_O - W_{Na} = \frac{1}{2} \mu_{Na_2O}^{YSZ \, cont} .$$ (8-48)

Accordingly, to get the corresponding sensor equation, one just has to substitute $(W_O - W_{Na})$ for $\frac{1}{2} \mu_{Na_2O}^{YSZ \, cont}$ in Equation (8-45).

When the sodium activity in electrode I and the oxygen pressure in electrode II are maintained constant, this system can be used as a sensor for measuring the local activity of sodium oxide or the variation of the O^{2-} activity in silicates. More precisely, it performs a measurement in the vicinity of the contact with the zirconia tube. The overall device formed by the tube and the oxygen electrode inside it can be viewed as a Na_2O or O^{2-} "electrode". In its principle, it is similar to the pH "electrode".

These descriptions show the parts played by the different elements of a potentiometric sensor: the electrodes, the electrolytic transmission line, and the junctions which can be present along the line.

The essence of the guidelines we would like to give in the sections dealing with potentiometric sensors is the following: the two functions, surface sensitivity and electrolytic transmission, can be more or less, or even totally separated, in many different manners. Many new possibilities may result from this independence of the surface sensitivity with respect to the electrolytic transmission line. A well-known sensor implementing this principle is the glass pH electrode in which the transmission line is generally ensured by the Na^+ or Li^+ ions which are the main electric carriers of the glass forming the bulb. The surface sensitivity is due to other species such as OH surface groups attached to superficial Si atoms.

8.3   Surface Sensitivity

We have recently reviewed the surface reactions which can be presently implemented to sense a gaseous chemical species [88]. For a comprehensive discussion the reader is referred to this article. The various possibilities are only listed here. Examples of technical realizations and real-life characteristics will be given in Section 8.7.

8.3.1   Sensing of Neutral Species in Non-Conducting Media

This is typically the case of a gaseous species analysis in a gas phase (Figure 8-2c). The solid ionic conductor surface is coated with an electronic conductor which will "collect" the local value of the reduced electrochemical potential of the electrons (cf. Figure 8-4).

8.3.1.1   Sensing the Co-Species

O_2/YSZ is an example.

This is the most conventional situation. Examples are O^{2-} conductors used for sensing O_2 or Cl^- conductors for sensing Cl_2. The corresponding reactions have been written above.

It has been demonstrated that the overall Reaction (8-15) always proceeds via an intermediate adsorption or dissolution of the analyzed species at least on/in one of the solid phases in contact (for more details, see Reference [88]).

8.3.1.2   Sensing a Fixed-Component Parent

Cl_2/AgCl and O_2/β-alumina are examples.

The description is also straightforward in the case of a binary compound. For instance, AgCl, which is a Ag^+ conductor, can be used for sensing chlorine. Then, one writes:

$$Cl_2 \ (gas) + 2e^- \ (el) + 2Ag^+ \ (s.i.c.) \ \longleftrightarrow \ AgCl \tag{8-49}$$

and applies the reasoning leading to Equation (8-16).

When the s.i.c. is a more complex compound, any attempt to write a similar reaction shows that additional chemical conditions must be controlled at the surface. Let us examine for instance the oxygen electrode reaction at the surface of β-alumina which is a compound based on Na_2O and Al_2O_3. It exhibits a predominant Na^+ conductivity. The electrode reaction can be written either:

$$1/2 \ O_2 \ (gas) + 2e^- \ (el) + 2Na^+ \ (s.i.c.) \ \longleftrightarrow \ Na_2O \ (s.i.c.) \tag{8-50}$$

or

$$1/2 \ O_2 \ (gas) + 2e^- \ (el) + 2Na^+ \ (s.i.c.) + x \ Al_2O_3 \ \longleftrightarrow \ \text{β-alumina} . \tag{8-51}$$

Equation (8-50) shows that, for the sensor to be operated properly, the experimental conditions should maintain the concentration in Na_2O constant in the vicinity of the electrode surface.

Equation (8-51) shows a technical solution which consists in putting a small quantity of pure Al_2O_3 (in the α form, for instance) in local contact with the electrode.

8.3.1.3 *Implementing a Secondary Electrode Reaction with Gaseous Species*

The CO_2–CO/YSZ system is an example.

Any reaction between the main ionic carrier of the s.i.c. and a complex chemical system can, in principle, be implemented in a sensor. In essence, there is no difference with respect to the basic situation (Section 8.3.1.1). With the system quoted in the above title, we write:

$$O^{2-} \ (s.i.c.) + CO \ (gas) \ \longleftrightarrow \ CO_2 \ (gas) + 2e^- \ (el). \tag{8-52}$$

Two remarks about such a basic sensing reaction must be made:

– The sensor signal now measures only the pressure ratio of the gas components:

$$E = E° + \frac{RT}{2F} \ \ln \left(\frac{P_{CO_2}}{P_{CO}} \right) . \tag{8-53}$$

– In the solid electrochemistry literature, it is frequently assumed that the chemical system fixes the pressure of oxygen according to a reaction such as:

$$CO \ (gas) + 1/2 \ O_2 \ (gas) \ \longleftrightarrow \ CO_2 \ (gas) \tag{8-54}$$

and that the sensor essentially functions as described in Section 8.3.1.1 according to:

$$1/2 \ O_2 \ (gas) + 2e^- \ (el) \ \longleftrightarrow \ O^{2-} \ (s.i.c.) . \tag{8-55}$$

In the case of an electrode fully equilibrated with the surrounding gas, the two descriptions are obviously equivalent. In the case of a deviation from equilibrium, the equivalence is no longer valid. It is now well established that Reaction (8-52) must be regarded as a different electrode reaction with its own characteristics describing the real system. Any reference to the system of Equations (8-54) and (8-55) can then be seriously misleading. The response of the λ sensor analyzing non-equilibrated O_2, CO_2, CO mixtures gives a convincing illustration of this statement: under such conditions the sensor frequently appears more sensitive to the CO_2/CO system than to the oxygen species. Then, reaction (8-52) strongly interferes with reaction (8-55) and in this way undoubtedly proves its own existence.

8.3.1.4   Implementing a Secondary Electrode Reaction Involving an Additional Solid Phase

We will consider $Cl_2/AgCl/Ag$ β-alumina here as an example.

The basic sensing equation of the quoted example is:

$$1/2 \, Cl_2 \, (gas) + Ag^+ \, (s.\,i.\,c.) + e^- \, (el) \longleftrightarrow AgCl \, . \tag{8-56}$$

Such a reaction implies that, at the active "points" of the sensing surface:

- the analyzed gas and the Ag β-alumina surface are in intimate contact,
- the electrode material is present (and connected to the rest of the potential collector that it forms),
- the additional solid phase (AgCl) is also present, at a chemical activity of 1, which supposes the presence of a real crystal.

We are no longer dealing with a triple but with a quadruple electrode contact.

Cells of this type have been suggested to analyze Cl_2 [33], NO_2 [89], SO_2–SO_3 [90], and even AsH_3 [91].

In some devices, the technology is slightly different than expected from this description and the basic principle is, in fact, essentially different. The s.i.c. is covered by the additional phase and the electrode material is coated on top of it. One can then assume that the actual reaction occurs on the surface of the coated additional solid phase which acts as a second s.i.c., according to a reaction similar to Equation (8-49) and that the contact between the two s.i.c.'s (AgCl/Ag β-alumina in the example) functions as a homojunction permeable to common ions (Ag^+).

8.3.1.5   Diluted Co-Ion

The example given is the $O_2/O^{2-}/SrCl_2$ system.

None of the basic sensing Reactions (8-15), (8-49), (8-52), for instance, require the co-ion, O^{2-} or Ag^+, to be the main ionic carrier of the s.i.c.. In conventional aqueous electrochemistry, most frequently, the species involved in the investigated electrode reactions are in fact totally different from the ions of the supporting electrolyte. Such a situation can be extrapolated to the s.i.c.'s. This statement has now been confirmed by many experimental results. Some of the very first were obtained with a cell formed on $SrCl_2$, a Cl^- ion conductor, in which oxide ions were dissolved in the form of SrO [31]. The oxygen electrode reaction on this material was found to be rapid enough and its characteristics sufficiently reproducible to be exploited in a sensor.

Let us examine, in more details, the required conditions of this separation of the sensing reaction from the predominant ionic conduction process in the s.i.c., in other words, from the transmission line (cf. Section 8.2).

At the contact with the electrode quoted in the example, we can assume that the overall reaction still is Reaction (8-15). Referring to Equation (8-11) and Equation (8-16), we will then deduce that the electrode potential, φ^{II}, obeys the following equation:

$$\varphi^{II} = E^\circ + \frac{RT}{4F} \ln P_{O_2} - \frac{RT}{2F} \ln \gamma \, C_{O^{2-}} + \varphi^{I} \tag{8-57}$$

where E° is a constant and φ^I the average electrostatic potential in the s.i.c., at its surface.

For the sensor to work properly, in other words, for the electrode signal to depend only on the measured parameter P_{O_2}, all the other parameters $C_{O^{2-}}$ and φ^I should be kept constant. The concentration $C_{O^{2-}}$ can be easily kept constant if a rather large O^{2-} concentration has been initially added and if the O^{2-} ions mobility is rather small (working with a two phases s.i.c. guarantees a perfect constancy). The Galvani potential φ^I is, in principle, controlled by the predominant electric carriers as shown below.

Because of the potentiometric conditions one gets

$$I = 0 \tag{8-58}$$

and because of the predominance of the Cl^- ions as electric carriers:

$$I_{Cl^-} = 0 \tag{8-59}$$

$$\text{grad } \tilde{\mu}_{Cl^-} = 0 \tag{8-60}$$

$$\text{grad } (\ln \gamma C_{Cl^-}) - F \text{ grad } \varphi = 0 \tag{8-61}$$

If the local composition in the s.i.c. is maintained constant (essential condition also for the previous requisite) the first term of Equation (8-61) is nil at any point inside the s.i.c.. Therefore grad φ is nil and φ is kept constant.

In summary, if the composition of the s.i.c. is not altered by the operating conditions, the sensor we are examining will work properly as a sensor. Its electrode signal will obey the usual equation:

$$E = E^\circ + \frac{RT}{4F} \ln P_{O_2} \tag{8-62}$$

where E° is a constant (Equation of case B in Section 8.1.4.1).

This demonstrates that the surface sensitivity and the electrolytic transmission line can really function separately. This situation is described in Figure 8-7 (for the sake of simplification the γ coefficients have been omitted).

The W_O line now acts as a local intermediate reference. By maintaining the local composition constant in the s.i.c., its distance with respect to the main transmission line, W_{Cl}, is kept constant.

We will now list several possible extensions of this diagram.

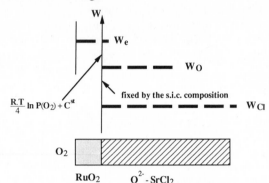

Figure 8-7.
Basic diagram of the electrode: O_2, $RuO_2/O^{2-}-SrCl_2$.

8.3.1.6   Partial Redox Reactions

Example O_2/O_2^{2-}

The implementation of a redox reaction in a sensor does not require the full oxidation or reduction of the electroactive species. A reaction such as:

$$O_2 \ (gas) + 2 \ e^- \ (el) \ \longleftrightarrow \ O_2^{2-} \tag{8-63}$$

can be treated as the more usual Equation (8-15). Experimental observations have been interpreted in this respect [92]. Referring to the previous derivation we can see that the only requirement is that the concentration in O_2^{2-} remains constant under the measurement conditions in the vicinity of the sensing surface.

A rather unexpected and interesting feature associated with the use of such partial redox reactions is the following. On lowering the measuring temperature, the transition which may occur from the full redox sensing reaction (Equation (8-15), for instance) to an incomplete reaction of type Equation (8-63) is generally accompanied by a marked improvement: a significant reduction of the sensor response time. This behavior can simply be analyzed from a description of the full reaction as a series combination of the partial reaction and an additional mechanism. For example, one could schematically write for the oxygen sensing reaction:

$$O_2 \ \xrightarrow{k_1} \ O_2^{2-} \ \xrightarrow{k_2} \ 2\,O^{2-} \ . \tag{8-64}$$

At high temperature, the k_2 rate constant is faster than k_1 and one only observes the overall reaction:

$$O_2 \ \longrightarrow \ 2\,O^{2-} \ . \tag{8-65}$$

On lowering the temperature, on the contrary, k_2 becomes slower than k_1. Then, the overall reaction rate is dominated by the second reaction and appears slower than the partial reaction which reaches equilibrium more rapidly. On further lowering the temperature, k_2 may become practically nil and the second step is no longer observed. Before that, this second step will manifest itself only as a drift disturbing the equilibrium of the partial reaction.

8.3.1.7 Surface and Adsorption Reactions

The reasoning developed in Section 8.3.1.5 can be extended to oxide ions only present on the s.i.c. surface:

$$1/2 \; O_2 \; (gas) + 2 \; e^- \; (el) \longleftrightarrow O^{2-} \; (s.i.c. \; surface) \; . \tag{8-66}$$

The same condition is required: as long as the concentration in O^{2-} and in the predominant ionic carriers are maintained constant up to the surface, the response of the sensor will be Nernstian.

Experimentally, this condition is likely to be more difficult to fulfill but intermediate situations, obtained, for instance, by ion implantation, may prove viable. The results obtained by Salardenne et al. [68] on $PbSnF_4$, O^{2-}-doped only near the surface, support this statement.

Another condition, implicit in the previous descriptions, which is likely to become more troublesome here, is related to the buffer capacity of the implemented redox reaction. To be efficient, a basic sensing reaction should not be impeded by any other redox reactions and especially by the electronic equilibration of the s.i.c. subsurface. This implies that the local electronic conductivity of the s.i.c. is very weak (as in many halides used under non-extreme conditions for instance). If, for any reason, the implemented redox reaction is not sufficiently predominant, deviations from the Nernst law may be observed and the sensor may require calibration.

Surface reactions directly suggest the extension of reaction Equation (8-66) to any chemical adsorption reactions which result in some reduction/oxidation of the adsorbate, such as:

$$O_2 \; (gas) + S \; (s.i.c. \; surface) + a \; e^- \; (el) \longleftrightarrow S\text{-}O_2^{a-} \tag{8-67}$$

where S is an adsorption site.

To be able to predict the behavior of a sensor based on such a sensing reaction, two subcases should be distinguished as above:
- *case I:* the concentration of adsorbed co-ions remains small as compared to the local predominant ionic concentration. Then, the derivation developed in Section 8.3.1.5 can be extrapolated rather easily. For instance, from Equation (8-67), one deduces that:

$$\mu_{O_2} + a\tilde{\mu}_{e-} = \tilde{\mu}_{S\text{-}O_2^{a-}} - \mu_S \; . \tag{8-68}$$

One can assume that the chemical component of the right hand side can be determined as a function of P_{O_2}:

$$\mu_{S\text{-}O_2^{a-}} - \mu_S = f(P_{O_2}) \; . \tag{8-69}$$

Then:

$$\mu_{O_2} + a\tilde{\mu}_{e^-} = f(P_{O_2}) - aF\varphi \qquad (8\text{-}70)$$

where φ is the electrostatic potential in the s.i.c. At the sensing surface we get:

$$W_e = -\frac{RT}{a} \ln P_{O_2} + \frac{1}{a} f(P_{O_2}) - F\varphi + C^{st} \qquad (8\text{-}71)$$

which can be converted into:

$$W_e - W_{main\ carrier} = -\frac{RT}{a} \ln P_{O_2} + \frac{1}{a} f(P_{O_2}) + C^{st}. \qquad (8\text{-}72)$$

That will be the basic sensing equation, which is likely to be of D type (Section 8.1.4.1).

– *case II:* When the concentration of adsorbed co-ions is comparable or higher than that of the main carriers, then, any change in their concentration is likely to also result in a variation of the local value of the electrostatic potential φ (because of a variation of the space charge). Then, the situation will be even more complicated in that there may not be a simple relationship between the adsorption isotherm and the basic equation of the sensing reaction.

8.3.1.8 Ionically Bonded Catalyst

The example shown is the O_2/Fe-Phthalocyanine/$PbSnF_4$-system.

To implement Reaction (8-67), the idea we have tested is the following: instead of using a rather inert "potential-collector" such as platinum as the electrode material, we used a catalyst known to adsorb the analyzed species. This material will provide the active adsorption sites S (Reaction (8-67)).

For the basic premises expressed by Equation (8-60) to be fulfilled, this catalyst must also exhibit some solubility for the main ionic carriers of the s.i.c., so that the transmission line maintains the electrolytic reference up to the catalytic surface. In practice, experiments were performed with iron phthalocyanine coated on a fluoride ion conductor.

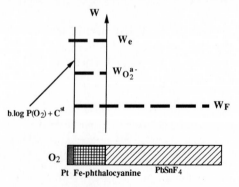

Figure 8-8.
Basic diagram of an ionically bonded catalyst: O_2, Fe-phthalocyanine/$PbSnF_4$.

This electrode was found to be sensitive to the oxygen pressure [58]. Figure 8-8 shows the diagram of this sensing electrode, under ideal conditions.

8.3.2 Sensing of Dissolved Species in Electronically Conducting Media

As mentioned in the introduction, any electronically conducting medium can be analyzed by simply pressing an appropriate s.i.c. against it (cf. Figure 8-2a). In practice, this has essentially concerned metals and the direct analysis of simple species such as O, S, H.

Theoretically, some of the complex situations encountered with insulating media (Section 8.3.1) can be envisaged here too. In terms of basic principles, the elementary sensing reaction of oxygen in metals, for instance, is very similar to Reaction (8-15):

$$O \ (metal) + 2\,e^- \ (metal) \ \longleftrightarrow \ O^{2-} \ (s.i.c.)\,. \tag{8-73}$$

Of course, in this case, one can also assume that oxygen is ionized in the metal and that the reaction is a simple ion exchange:

$$O^{2-} \ (metal) \ \longleftrightarrow \ O^{2-} \ (s.i.c.)\,. \tag{8-74}$$

Then, we will be dealing with a homojunction (see Section 8.4.1).

The equation corresponding to Reaction (8-73) becomes:

$$W_e - W_O = \frac{1}{2}\,\mu_O\,. \tag{8-75}$$

It can be exploited exactly as Equation (8-16) as long as the connection of the analyzed metal to an appropriate electric lead is isothermal. If not so, a thermo-emf correction has to be made (an easy way of determining unequivocally the sign of this correction consists in imagining the ideal circuit, without thermo-emf, connected in parallel with the real one).

The secondary electrode reaction

$$3\,O^{2-} \ (s.i.c.) + 2\,Cr \ (metal) \ \longleftrightarrow \ Cr_2O_3 \ (second \ phase) + 6\,e^- \ (metal) \tag{8-76}$$

has also been successfully implemented for analyzing chromium in molten steel [93]. Here again, the determining reaction can be viewed as a secondary chemical reaction according to:

$$O^{2-} \ (s.i.c.) \ \longleftrightarrow \ O \ (metal) + 2\,e^- \ (metal) \tag{8-77}$$

$$2\,Cr + 3\,O \ (metal) \ \longleftrightarrow \ Cr_2O_3 \ (second \ phase)\,. \tag{8-78}$$

Similarly, silicon was analyzed with $ZrSiO_4$ [94].

8.3.3   Sensing of Dissolved Ionic Species in Ionically Conducting Media

This is typically the case of electrolytic solution and molten salt analysis. An example is the fluoride ion "electrode" based on a LaF_3 single crystal. The sensing reaction is a simple ion exchange between the sensing material and the analyzed media:

$$F^- \ (s.\,i.\,c.) \longleftrightarrow F^- \ (anal.\ media) \ . \tag{8-79}$$

To be strictly correct, this description applies when the exchanged ions are mobile in both phases in contact, especially in the sensing membrane.

8.3.3.1   Preliminary Remark

To avoid any confusion, an essential difference to the previous situations must be made clear: here, the sensitive surface functions, properly speaking, as an ionic junction (as for the zirconia contact described in Section 8.2.2.2) and not as a true electrode. We have seen (Equation (8-16) and Figure 8-4) that a true electrode "transforms" the variations of the measured parameter (the oxygen pressure in the example) into variations of W_e, with respect to a reference W_O which is maintained constant. The variations of W_e can directly be probed and transmitted to a measuring millivoltmeter by a chain of electronic conductors.

The basic equation corresponding to Reaction (8-79) leads to a different sensing principle as shown below. This equation is:

$$\tilde{\mu}_{F^-} \ (s.\,i.\,c.) = \tilde{\mu}_{F^-} \ (anal.\ media) \tag{8-80}$$

from which one deduces that:

$$\varphi \ (s.\,i.\,c.) - \varphi \ (anal.\ media) = \frac{1}{F} \left[\mu_{F^-} \ (anal.\ media) \right] - \mu_{F^-} \ (s.\,i.\,c.) \ . \tag{8-81}$$

$\mu_{F^-} \ (s.\,i.\,c.)$ is assumed to remain constant, therefore:

$$\varphi \ (s.\,i.\,c.) - \varphi \ (anal.\ media) = (RT/F) \ln \left[C_{F^-} \ (anal.\ media) \right] - C^{st} \ . \tag{8-82}$$

Here, the sensing surface "transforms" the variations of the measured concentration into variations of the average electrostatic potential of the s.i.c. with respect to that of the analyzed media.

The essential differences with a true electrode are due to the following two points:

− To measure the variations of the s.i.c. electrostatic potential, one inevitably needs to contact an electronic conductor to this ionic conductor, in other words to form another electrode. Its characteristics must be maintained strictly constant, especially in terms of the $(W_e - W_{ion})$ distance, to avoid any interference with the measured variations.
− The electrostatic potential of the analyzed media ($\varphi \ (anal.\ media)$ in Equation (8-82)) must itself be connected to the other elements of the measuring electric chain by appropriate

junctions so that it can be treated as a reliable reference. Here again another electrode must be introduced in the chain and its characteristics must also be maintained strictly constant.

For these reasons, the measuring electric chain is generally more complicated for the analysis of ions in ionically conducting media than in the previous cases. Among other peculiarities, a second "electrode" called the "reference electrode" has to be immersed in the analyzed media (cf. Figure 8-2b) to "connect" the electrostatic potential of this media to the other elements of the measuring electric chain. In the example given in Section 8.2.2.2, the β-alumina and its sodium electrode will be viewed as the "reference electrode" in such a description.

Concerning terminology, it must be mentioned that in this type of application, the true electrode which is on the other side of the sensitive membrane is called the *internal reference electrode*. In the example given in Section 8.2.2.2, the oxygen electrode on YSZ is the internal reference electrode.

8.3.3.2 *Typical Sensing Reactions*

In addition to homojunctions (Equation (8-79)), other reactions similar to the ones described for the neutral species are also exploited. Sensing a *fixed component ion* (or parent) is the most common case. AgCl, for instance, is used for sensing Cl^- ions in solutions:

$$Ag^+ \ (s.i.c.) + Cl^- \ (sol) \ \longleftrightarrow \ AgCl \ . \tag{8-83}$$

Similarly Ag_3SI, which is an Ag^+ conductor can be used for sensing sulfur compounds, with restrictions made in Section 8.3.1.2.

Among the other analogs of the various sensing reactions described for the neutral species, one deserves special attention: the exchange reaction limited to the surface (cf. Section 8.3.1.7). Such a reaction occurs for instance at the surface of the pH glass "electrode". At this surface, the "Si-O" network ends are partly reduced into "Si-OH" in contact with water. Other H_2O molecules and OH^- ions certainly also become bonded to the surface. In this way, the surface transforms itself into a kind of acid which can exchange protons with the analyzed solution:

$$Surface \ (s.i.c.) + H^+ \ (sol) \ \longleftrightarrow \ Surface - H^+ \ . \tag{8-84}$$

This overall sensing reaction has been successfully used for years despite the complexity of, and the uncertainty in its microscopic description (cf. Chapter 7). Surprisingly enough, it is even one of the most accurate electrodes.

Concerning the basic principles of ion analysis in ionically conducting media, another specific possibility can be mentioned: the utilization of *mixed conductors as sensitive membranes*. On this principle, IrO_2, for instance, is used for sensing H^+ (cf. Chapter 7). The derivations made in Section 8.3.3 do not require that the so-called s.i.c.'s be pure ionic conductors. Equations (8-79) through (8-84) are valid, even in the presence of an additional electronic conductivity of the sensing membrane (because no electrons are involved in these reactions). A significant advantage of using such a mixed conductor is the following: no electrode, properly speaking, is needed to connect a mixed conductor to the measuring instrumentation. In principle, a simple electronic contact can provide a good anchoring of the "potential transmission" through an exchange of electrons:

$$e^- \; (s.\,i.\,c.) \quad \longleftrightarrow \quad e^- \; (contact.\; metal) \;.$$
(8-85)

The corresponding diagram is depicted in Figure 8-9. As clearly shown in this diagram, such an electrode will work properly if the distance $(W_e - W_H)$ is maintained constant in the sensing material. This implies among other things, that the composition of the sensing material is not modified by any reaction with the analyzed media and especially by the sensing reaction itself (in some respects, this type of cell is similar to the ionic bridges described in Section 8.5.2; compare Figures 8-9 and 8-13).

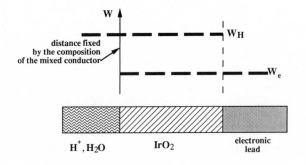

Figure 8-9.
Basic diagram of a mixed conductor used as a H^+ sensitive membrane.

Another more stringent condition is the absence of any interfering reaction between the electrons of the sensing membrane and the species of the analyzed media, in other words of any redox reactions. The utilization of such mixed conductors is, in fact, not recommended when redox reactions are likely to be active under the analysis conditions. This risk of interference is, obviously, a function of the additional electronic conductivity magnitude in the sensing membrane.

8.4   Ionic Junctions

Junctions are contacts between two ionic conductors. As we have seen in the examples of Section 8.2, a junction can be a simple element of the transmission line or the actual sensing component of the sensor. This is the case with the YSZ/silicate contact in the electrochemical cell in Figure 8-6. This also is the case with the glass surface in the pH "electrode".

8.4.1   Simple Ion Exchange and Homojunctions

The exchange of ions of the same nature between two solid ionic conductors has already been encountered in the previous sections (cf. Section 8.3.1.4 for instance).

In principle, and especially under potentiometric conditions ($I = 0$), the basic equation is likely to be very simple. In the case of AgCl in contact with Ag-β alumina, for instance, the exchange of silver ions is simply described by the reaction:

$$Ag^+ \ (AgCl) \longleftrightarrow Ag^+ \ (\beta\text{-alumina}) \tag{8-86}$$

which is governed by the equation:

$$\tilde{\mu}_{Ag^+} \ (AgCl) = \tilde{\mu}_{Ag^+} \ (\beta\text{-alumina}) \ . \tag{8-87}$$

When the Ag^+ ions ensure electrochemical transmission, Equation (8-87) means that the crossing of the contact interface does not alter the reference level.

Few experimental data are known on the characteristics of such simple exchange reactions. Investigation of contacts between identical ceria samples [95] and zirconia samples [96] has shown that the exchange is a straightforward reaction characterized by an exchange reaction resistance and a contact capacitive effect.

In the sensor field such simple ionic junctions have been implemented with four objectives:
— To provide a sensitive surface (cf. Section 8.3.3.1).
— To extend the range of application of a sensor by "connecting" the applicability ranges of two materials. An example is the oxygen sensor based on the following electrochemical chain [97, 98]:

$$O_2 \text{ dissolved in Metal/Thoria/YS Zirconia/O}_2, \text{ Pt} \ . \tag{8-88}$$

By intercalating thoria, which also is an oxide ion conductor, between stabilized zirconia and the analyzed metal, much lower oxygen activities can be measured in the metal. In thoria, adequate characteristics are maintained down to highly reducing conditions. On the other hand, the presence of zirconia allows us to use air as a reference system. This would not be possible with thoria alone. This material exhibits too high an electron hole conductivity under such oxidizing conditions.
— To implement a type of "secondary electrode". The example of AgCl coated on Ag-β alumina provides an illustration:

$$Cl_2, \text{ Pt / AgCl /Ag-}\beta \text{ alumina / Ag} \ . \tag{8-89}$$

In essence this coupling combines the surface sensitivity of one material (sensitivity of AgCl to Cl_2 in the example) and the high ionic conductivity of the other (β-alumina). AgCl, which is a rather poor conductor under the measuring conditions, is "supported" by β-alumina and can, therefore, be used as a very thin layer.
— To eliminate the detrimental effect of the electrochemical semipermeability in special cell assemblies (cf. Section 8.7.1.5).

8.4.2 Heterojunctions

Heterojunctions are contacts between two solids in which ions of different nature are mobile. An example which has been quoted several times in the literature is the following (cf. Section 8.7.2):

$$K_2SO_4 / ZrO_2 - Y_2O_3 . \tag{8-90}$$

The predominant carriers are K^+ ions in the sulfate and O^{2-} ions in the stabilized zirconia.

Referring to the fact that electrodes (and contacts) of potentiometric sensors are fully equilibrated, it is frequently assumed that all consistent sets of chemical reactions are equivalent and appropriate to describe the system. In Section 8.3.1.3 we have already stressed a misinterpretation resulting from an abuse of this broad degree of freedom. In the case of heterojunctions, the risk of misinterpretation is even more serious. In the absence of any well established model, we suggest the following conservative approach which has been applied in the previous sections. To describe a heterojunction, the basic reaction should be written with:

- only the mobile ions as ionic species,
- the chemical species present at the contact. This may reveal that additional species have to be considered. For instance, the basic reaction of contact (Equation (8-90)) will be written:

$$2 K^+ \textit{ (sulfate)} + O^{2-} \textit{ (YSZ)} + SO_3 \textit{ (gas)} \longleftrightarrow K_2SO_4 \tag{8-91}$$

from which one deduces that:

$$2 \tilde{\mu}_{K^+} + \tilde{\mu}_{O^{2-}} + \mu_{SO_3} = \mu_{K_2SO_4} \tag{8-92}$$

$$-W_K + W_O = -RT \ln P_{SO_3} + C^{st} . \tag{8-93}$$

This last equation shows that the distance between the two reference lines W_O and W_K in the zirconia and in the sulfate depends on P_{SO_3}, the pressure around the contact. This has been confirmed experimentally [26, 85, 99] and can even be implemented in a sensor. A similar conclusion was drawn on the calcia stabilized zirconia/potassium carbonate junction which can be used as an interface sensitive to CO_2 [100].

8.4.3 Chemical Stability

From this description, and those given in Section 8.2.2.2, two useful conclusions must be stressed:

- The homojunction is likely to be rather stable. As shown by Equation (8-87), the transmission line will not be disturbed as long as the common ions remain the main electric carriers in the materials in contact (and as long as they do not carry any current).

– On the other hand, any chemical reaction occurring at a heterojunction is very likely to alter the step of the reference level along the transmission line because it depends on the local chemical activity of a determining component.

8.5 Electrochemical Reference Transmission

In principle, this function should work straightforwardly as depicted in the examples of Section 8.2. The main possible sources of deviation from ideal behavior are the other electric carriers which can be present in the s.i.c., especially the electronic carriers. The resulting effect, called *electrochemical semipermeability,* has been abundantly investigated [101–108] and will be described in Section 8.5.1.

The requisites for a perfect transmission of the reference level are:

– strictly constant levels of the relevant reduced electrochemical potentials W through all bulky parts of the transmission line or at least steady state profiles,
– extension of this property up to the very surface where the sensing reaction occurs,
– fixed steps at the junctions (constant distances between the relevant reduced electrochemical potentials W).

The first (and the two other) requisite implies that no variable current crosses the system along the direction of the line. Among other things this requires an impedance of the measuring device much higher than that of the sensor. For an accurate evaluation of the detrimental effect on the reference level of any current, one can refer to the basic equation relating the current density per unit surface to the local variation of the carrier electrochemical potential:

$$I_i = -C_i \tilde{u}_i F z_i \text{ grad } \tilde{\mu}_i \tag{8-94}$$

where $\tilde{u}_i$ is the electrochemical mobility of the i carriers (cf. Section 8.5.1.1).

With a one-dimensional distribution of the parameters and constant C_i and $\tilde{u}_i$, this equation can be integrated from one electrode to the other into:

$$I_i \frac{l}{\sigma_i} = \frac{\Delta \tilde{\mu}_i}{F z_i} \tag{8-95}$$

where σ_i is the partial conductivity of the i carriers and l the length of the s.i.c.. This equation can also be written (after multiplication by the cross sectional area):

$$\Delta W_i / F = R_i I_i \tag{8-96}$$

where I_i is the total current carried by the i carriers over the entire surface of the material and R_i the relevant partial resistance. When these carriers are predominant, R_i is approximately equal to the overall resistance of the material and I_i equal to the overall current crossing it.

This equation shows that, in essence, the deviation ΔW_i of the reference level is a simple ohmic drop. As in a regular electric circuit, the utilization of a better conductor and a shorter length will reduce ΔW_i.

For similar reasons a high resistance makes the sensor more sensitive to electric disturbances.

It must also be mentioned that a rather resistant transmission line has also been observed to induce long response times after a variation of the measured parameter. Several interpretations have been put forward in terms of bulk chemical reequilibration of the material [88] and possible partial blocking at the grain boundaries [109] when sintered materials are used.

An overall resistance of 10^5 Ω is regarded as a comfortable upper limit for an element of a transmission line.

To avoid any misinterpretation, it must be stressed that a current crossing a potentiometric sensor also polarizes the electrodes (cf. Section 8.5.1). The ΔW_i deviation described above is not the only detrimental effect.

The second requisite is automatically fulfilled when the predominant electric carriers also are the co-ions electroactive in the basic sensing reaction. On the other hand, in the case of total decoupling between the electrochemical transmission and the sensing reaction, this requisite may appear more difficult to ensure. Deviations from its fulfillment are likely to result in a departure from the predicted Nernst law but could, however, give reproducible results (see Section 8.3.1.7).

Concerning the third requisite, as shown by Equations (8-86) and (8-87), for example, an homojunction is likely to work without any major problem (if the ion exchange rate is fast enough and not blocked by hydrated surfaces or other major alterations of the surfaces). On the other hand, a heterojunction will be markedly dependent on the local chemical reactions. One possible way to stabilize it, is a curing of the contact at a temperature higher than the measuring temperature.

8.5.1   Influence of the Minority Carriers – Electrochemical Semipermeability

To understand the effect of minority carriers on the alteration of the electrochemical reference line, let us refer to a diagram similar to that of Figure 8-4 and examine the behavior of *electronic minority carriers*. As shown in Figure 8-10 the electron reduced electrochemical potential is usually not constant throughout the material (when the sensor voltage is not nil).

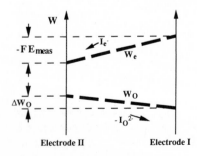

Figure 8-10.
Basic diagram of an oxygen sensor showing the counter currents of the electrochemical semipermeability (electrode polarizations can be viewed on such a diagram as alterations of the $[W_e - W_O]$ distances at the electrode surfaces).

If electronic carriers (e^- or h^+) are actually present and mobile in the s.i.c., they will, therefore, move along the W_e gradient in an effort to reequilibrate their distribution. Under steady state conditions, this results in an electric current from one surface of the material to the other. Under potentiometric measurement conditions, this current is not evacuated by the external circuit (because $I = 0$). It is compensated by a counter current carried by the predominant carriers, so that no net current crosses the materials.

We have seen in the introduction of this section that any current carried by the predominant carriers results in a variation ΔW_i of the reference level (Equation (8-96)). That will be the case here, too (cf. Figure 8-10). This variation obviously induces a corresponding effect on the measured voltage ($\Delta E_{meas} = -\Delta W_O / F$).

Let us stress that this current also results in a transport of matter from one electrode to the other (cf. Section 8.1.4.3), which may contaminate the analyzed system. It may also polarize the electrodes, ie, alter the distances ($W_e - W_O$) at the electrodes (cf. [104] and Section 8.7.1.5).

More generally, when the local reactions on the surfaces maintain different values of any minority carrier reduced electrochemical potential, then, a similar process is likely to occur. Its magnitude will increase with the partial conductivity (the concentration and mobility) of these minority carriers.

8.5.1.1 Analytical Description

An analytical description of the electrochemical semipermeability is, in principle, possible within the framework of the "generalized" Wagner theory [110, 111]. The essential steps of the derivation are the following:

- To work under appropriate conditions, a s.i.c. used in a potentiometric sensor should be a divariant thermodynamical system: all its compositions in point defects (or ions and free electrons) should depend only on one chemical parameter, besides temperature. Among other factors, the ($W_e - W_i$) distance should depend only on the chemical activity of the analyzed species. The selected determining chemical parameter will be called O (which may stand for atomic or molecular oxygen in the case of an oxide).
  Under these conditions, the electrochemical potentials of all the ionic carriers K^{z+} are linked to O by equations of the type:

$$\tilde{\mu}_{K^{z+}} = -z\,\tilde{\mu}_{e^-} + \alpha_K + \beta_K\,\mu_O \tag{8-97}$$

in which α_K and β_K are constant.
From this equation, one deduces that:

$$\text{grad }\tilde{\mu}_{K^{z+}} = -z\,\text{grad }\tilde{\mu}_{e^-} + \beta_K\,\text{grad }\mu_O . \tag{8-98}$$

- The flux densities J_k of the mobile species are governed by the general laws of irreversible processes. We will assume, that the interactions between the fluxes are negligible (although experimental observations have shown that a high electronic conductivity may influence the ionic conduction parameters [112[) and that, under isothermal conditions and in the absence of any forces other than the electric and chemical forces, they are in the form:

$$J_k = -L_k \text{ grad } \tilde{\mu}_k .$$ (8-99)

A common assumption is also that fluxes are proportional to the local concentrations in mobile species:

$$L_k = -C_k \tilde{u}_k .$$ (8-100)

The proportionality coefficient, $\tilde{u}_k$, is called the *electrochemical mobility*.
When the species K are charged, the J_k flux carries a current

$$I_k = -C_k \tilde{u}_k z_k F \text{ grad } \tilde{\mu}_k$$ (8-101)

where z_k is the number of electric charges of K^{z+}; $\tilde{u}_k z_k F$ is sometimes called the *electrical mobility*.
From the usual decomposition of $\tilde{\mu}$ one gets:

$$I_k = -C_k \tilde{u}_k z_k F \text{ grad } \mu_k - C_k \tilde{u}_k z_k^2 F^2 \text{ grad } \varphi .$$ (8-102)

$C_k \tilde{u}_k z_k^2 F^2$ is the *partial conductivity* σ_k of the K^{z+} species. Let us also define the *transference number* of the K^{z+} carriers

$$t_k = \frac{\sigma_k}{\sigma} \text{ with } \sigma = \sum_i \sigma_i$$ (8-103)

where the sum is extended to all i carriers, including the electronic ones.
— The total current density crossing the s.i.c. is simply equal to the algebraic sum (vectorial in the more general case) of all the partial currents:

$$I = \sum_k I_k .$$ (8-104)

This sum is also extended to the electronic carriers, the corresponding currents being simply expressed by equations of type Equation (8-101):

$$I = -\sum_k C_k \tilde{u}_k z_k F \text{ grad } \tilde{\mu}_k .$$ (8-105)

The basic idea of the "generalized" Wagner theory is to combine Equations (8-98) and (8-105). This gives:

$$I = \frac{\sum_k \sigma_k}{F} \text{ grad } \tilde{\mu}_e - \frac{\sum_k \beta_k \sigma_k}{z_k F} \text{ grad } \tilde{\mu}_O$$ (8-106)

where β_k are constant coefficients and

$$\text{grad } \tilde{\mu}_e = \frac{F\boldsymbol{I}}{\sigma} + \frac{\sum\limits_{k} \beta_k t_k}{z_k} \text{ grad } \mu_O \, .$$ (8-107)

By integration from one electrode to the other, one gets:

$$W_e^{II} - W_e^{I} = F \int\limits_{I}^{II} \frac{d\boldsymbol{I}}{\sigma} + \int\limits_{\mu_O^{I}}^{\mu_O^{II}} \frac{\sum\limits_{k} \beta_k t_k}{z_k} \, d\mu_O \, .$$ (8-108)

Under potentiometric conditions, the first term is nil and the only variable parameters are the transference number t_k. In principle they can be measured experimentally or deduced from adequate models. Therefore, Equation (8-108) allows us to calculate the ΔW_e variation (and the cell voltage according to Equation (8-24)) which results from the transport properties of the s.i.c. whatever they are.

– The basic Equation (8-105) can be exploited in another way. Under potentiometric condition ($I = 0$), it can be written, by grouping together the chemical and electrostatic components of the $\tilde{\mu}_k$'s:

$$0 = -\left(\sum\limits_{k} \frac{\sigma_k}{F z_k} \text{ grad } \mu_k \right) - \sigma \text{ grad } \varphi \, .$$ (8-109)

One deduces from Equation (8-109) that:

$$\varphi^b - \varphi^a = -\frac{1}{F} \sum\limits_{k} \int\limits_{\mu_k^a}^{\mu_k^b} \frac{t_k}{z_k} \, d\mu_k \, .$$ (8-110)

This equation, which is usually applied to membranes, gives the internal electrostatic potential variation *between two points a and b of the s.i.c.*

To clarify the difference between Equations (8-108) and (8-110) we have represented in Figure 8-11 the value of the parameters given by these equations.

Figure 8-11.
Electronic band diagram of a s.i.c. with minority electronic carriers. Variations of the electrostatic potential inside the material and of the electronic Fermi level W_e.

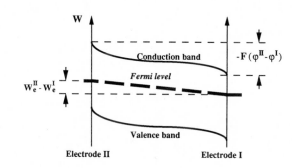

8.5.1.2   Approximations and Iterative Approach

A diagram which is frequently used to memorize the main effects of the electrochemical semipermeability is sketched in Figure 8-12.

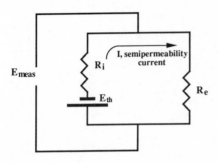

Figure 8-12.
Mnemonic circuit describing the "ohmic drop" effect of the electrochemical semipermeability.

E_{meas} is the measured sensor voltage, E_{th} the theoretical value. R_i and R_e are the average ionic and electronic resistances of the material. From this simple circuit one can easily calculate that the internal short-circuit current I is:

$$I \quad = E_{th}/(R_i + R_e) \tag{8-111}$$

and that:

$$E_{meas} = E_{th}\, R_e/(R_i + R_e) \tag{8-112}$$

or

$$E_{meas} = E_{th}\,(1 - \overline{t_e}) \tag{8-113}$$

where $\overline{t_e}$ is the average electronic transference number defined by:

$$\overline{t_e} = R_i/(R_e + R_i) \,. \tag{8-114}$$

Remember that, in the case of an oxygen sensor:

$$E_{th} = -\Delta\mu_{O_2}/4\,F \,. \tag{8-115}$$

This derivation can then be used as a mnemonic for the Wagner equation:

$$E_{meas} = -\frac{1}{4\,F} \int_{l_1}^{l_2} (1 - t_e)\, d\mu_{O_2} \,. \tag{8-116}$$

To get a quantitative evaluation of the effects, the following iterative calculation can sometimes be applied. The first order approximation is calculated on the assumption that the alteration of the reference level is negligible ($\Delta W_O = 0$ in Figure 8-10). The minority-carrier current is calculated under this assumption. Then, the alteration of the reference level is determined using the equation:

$$I_{\text{majority carriers}} = -I_{\text{minority carriers}} \tag{8-117}$$

and Equation (8-96).

If necessary, a second order approximation can be calculated from this approximated alteration of the reference level by following the same procedure, and so on.

This approach is especially appropriate when the contribution of the minority carriers is small, typically with minority-carrier transference numbers of the order of 1%. Then, one iteration is sufficient.

As an example, let us apply this first approximation procedure to the oxygen sensor based on stabilized zirconia (cf. Figure 8-10). (For the sake of simplification, the calculations are carried out with a unit surface for a unidimensional system).

− The first order assumption implies that:

$$W_{O^{2-}} = C^{\text{st}} \tag{8-118}$$

$$\text{grad } \tilde{\mu}_{O^{2-}} = 0 \tag{8-119}$$

$$\text{grad } \mu_{O^{2-}} - 2F \text{ grad } \varphi = 0 . \tag{8-120}$$

In YSZ the O^{2-} concentration is constant, therefore:

$$\text{grad } \mu_{O^{2-}} = 0 \tag{8-121}$$

and then:

$$\text{grad } \varphi = 0 . \tag{8-122}$$

− Under rather high oxygen pressure, the only minority carriers which can contribute a significant effect are the electron holes. Their current is given by (cf. Equation (8-101)):

$$I_h = -C_h \tilde{u}_h F \text{ grad } \tilde{\mu}_h \tag{8-123}$$

with:

$$\tilde{\mu}_h = \mu° + RT \ln C_h + F\varphi . \tag{8-124}$$

By combining with Equation (8-122) one obtains:

$$I_h = -FRT \tilde{u}_h \text{ grad } C_h . \tag{8-125}$$

− Steady-state conditions imply that at any point of the material:

$$I_h = I_{O^{2-}} \tag{8-126}$$

and

$$C_{O^{2-}} = C^{st} .$$

(8-127)

Equation (8-127) implies that:

$$dI_{O^{2-}} / dx = 0$$

(8-128)

and therefore that:

$$dI_h / dx = 0$$

(8-129)

which means:

$$grad\ C_h = C^{st} .$$

(8-130)

In a one-dimensional system the following equation holds:

$$grad\ C_h = \left(\frac{C_h^{II} - C_h^{I}}{l} \right)$$

(8-131)

where C_h^{II} and C_h^{I} are the electron hole concentrations at electrodes II and I and l the distance between these electrodes.

− At the surface electrodes we find:

$$1/2\ O_2 \longleftrightarrow O^{2-} + 2h^+ \quad \text{with}$$

(8-132)

$$\frac{C_h^2\ C_{O^{2-}}}{P^{1/2}} = K$$

(8-133)

where P is the oxygen pressure, K a constant, and

$$C_h = aP^{1/4} \quad \text{with } a = C^{st} .$$

(8-134)

− With Equations (8-125), (8-131) and (8-134) we obtain

$$I_h = -[(P^{II})^{1/4} - (P^{I})^{1/4}]\ FRTa\tilde{u}_h / l$$

(8-135)

or:

$$I_h = -\frac{RT(\sigma_h^{II} - \sigma_h^{I})}{Fl}$$

(8-136)

where σ_h^{II}, for example, is the electronic conductivity of YSZ in equilibrium with the oxygen pressure P^{II} prevailing at electrode II.

− Then, one calculates the alteration ΔW_O of the electrochemical reference from the Equation (8-96):

$$\Delta W_O = R_{O^{2-}} I_{O^{2-}} F \tag{8-137}$$

with:

$$I_{O^{2-}} = -I_h \tag{8-138}$$

and:

$$R_{O^{2-}} = l / \sigma_{O^{2-}} \tag{8-139}$$

$$\Delta W_O = \frac{RT(\sigma_h^{II} - \sigma_h^{I})}{\sigma_{O^{2-}}} . \tag{8-140}$$

- When the pressures P^{II} and P^{I} are sufficiently different, σ_h^{I}, for instance, will be negligible with respect to σ_h^{II} and Equations (8-136) and (8-140) can be simplified into:

$$I_h = -\frac{RT\sigma_h^{II}}{Fl} \tag{8-141}$$

$$\Delta W_O = \frac{RT\sigma_h^{II}}{\sigma_{O^{2-}}} . \tag{8-142}$$

- These expressions are especially useful to quantitatively describe the complex situation where electron holes are the main minority carriers at one electrode and electrons at the other [107, 113]. Then we refer to the steady-state condition:

$$I_e = I_h \tag{8-143}$$

and to an equation similar to Equation (8-141) for the free electron contribution. This gives:

$$\frac{RT\sigma_h^{II}}{Fl_1} = \frac{RT\sigma_e^{I}}{F(l - l_1)} \tag{8-144}$$

where electrons and electron holes are assumed to recombine at a distance l_1 from electrode II. Under these conditions, the internal short-circuit current is:

$$I_{O^{2-}} = \frac{RT(\sigma_h^{II} + \sigma_e^{I})}{Fl} . \tag{8-145}$$

- At any stage of the iteration, the contamination of the analyzed gas by the oxygen semipermeating through the YSZ can be calculated from the equation:

$$J_{O_2} = \frac{I_{O^{2-}}}{4F} . \tag{8-146}$$

- The polarization of the sensor electrodes, resulting from the electrochemical semipermeability, can be evaluated by referring to their steady state $U(I)$ characteristic, using the internal short circuit current $I_{O^{2-}}$ as a value of I.

8.5.2   Ionic Bridges

As we have seen in the description of the heterojunctions, the contact between two solids conducting by different ions may be chemically unstable in terms of transmission of an electrochemical reference. To solve this difficulty we have recently introduced the ionic bridge concept [13]. This component of an electrochemical transmission line is based on a material in which *two ions are mobile*. Let us imagine that we have to contact two conductors, one by Na^+ and the other by Ag^+ and let us insert between them a solid in which both Na^+ and Ag^+ ions are mobile. If the materials are appropriately selected, we will get a predominant exchange of Na^+ at one contact and Ag^+ at the other. Both contacts will work as homojunctions. The corresponding diagram is shown in Figure 8-13. As obvious in this diagram, the bridge will function correctly if the distance between the W's remains constant with time at any place on the bridge. That requires that the homojunctions work selectively, for instance, that the material on the right hand side does not dissolve Ag^+ from the bridge in an exchange reaction:

$$Ag^+ \ (bridge) + Na^+ \ (I) \longrightarrow Na^+ \ (bridge) + Ag^+ \ (I) \ . \tag{8-147}$$

Experiments [57] were performed with the following electrochemical lines:

$$Nasicon / AgCl-NaCl / Ag$$
$$Nasicon / AgI-NaI / Ag \ .$$

The bridges were found to work satisfactorily (cf. Section 8.8.2.4).

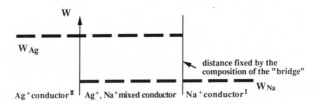

Figure 8-13. Basic diagram of an ionic bridge.

8.6   Thermo-emf Cells

The thermo-emf effects are:
− a source of error when the isothermal functioning requirement is not fulfilled,
− the very principle of a specific type of sensor,
− an inherent contribution to the sensor voltage in devices which are intentionally operated under non-isothermal conditions.

The general formula of a non isothermal concentration cell voltage is rather complex and includes the thermo-emf due to the electric leads, the partial entropies of all the electroactive species and the heat of transport of the ionic species. Most of these terms cannot be directly measured and are unknown. For a more detailed examination of these contributions, one can, for instance, refer to the paper by Jacob and Ramasesha [114].

In practice, all the experiments reported on potentiometric cells in which the measuring and the reference electrodes are operated at temperatures not too different, have shown that the cell voltage can simply be decomposed into three terms:

– two electrode potentials which appear not to be significantly altered by the temperature gradient through the s.i.c.,
– a thermo-emf effect.

In the case of oxygen electrodes, called I and II, this can be summarized as:

$$E = \alpha \, (T^{II} - T^{I}) + \frac{R}{4F} \, (T^{II} \ln P^{II} - T^{I} \ln P^{I}) . \tag{8-148}$$

Referring to the data by Goto [115], Fouletier et al. [116] have calculated for α a value of $0.514 \, \text{mV} \cdot \text{K}^{-1}$ for calcia stabilized zirconia. With $(ZrO_2)_{0.955} \, (Y_2O_3)_{0.045}$, Baucke [117] gives $0.474 \, \text{mV} \cdot \text{K}^{-1}$. Fabry [118] proposes a value which slightly depends on T. When the oxygen pressures are equal to 1 (second term nil), the positive electrode is the cold one.

In some respects, the temperature gradient plays a role similar to that of the minority carriers described in Section 8.5.1.

In terms of applications, let us first examine two conclusions which can be drawn from Equation (8-148).
– As a function of the measuring electrode parameters, it can be written as:

$$E = C^{st} + \frac{RT^{II}}{4F} \ln P^{II} . \tag{8-149}$$

This expression shows that the sensor still perfectly obeys the Nernst law in terms of the P variations, under the essential condition that the sensor temperature is measured at the measuring electrode.

A practical conclusion is: when temperature gradients may pose a problem and if one wants the sensor to at least obey an equation of type B (cf. Section 8.1.4.1), then the temperature used in the basic equation has to be determined at the measuring electrode. If it is measured at the reference electrode, the sensor basic equation will be of type C_{II} (Section 8.1.4.1) with a β coefficient to be determined by calibration. An error will result if a regular Nernst formula is applied.
– When both pressures P^{II} and P^{I} are equal, Equation (8-148) can also be written as:

$$E = C^{st} + \frac{R}{4F} \, (T^{II} - T^{I}) \ln P \quad \text{or as} \tag{8-150}$$

$$E = \beta + \gamma \ln P \tag{8-151}$$

where β and γ are constant for given temperatures of the electrodes. This last equation shows that the thermocell depicted in Figure 8-14 which is assembled on a simple bar of an ionic conductor can be used as a sensor. Several attempts have been made to develop such a sensor [119, 120] the essential quality of which is that it does not require an isolated reference electrode.

– Non-isothermal cells also are used for the analysis of very high temperature metal and glass melts (cf. Chapter 26). For various reasons, it is sometimes preferable to only immerse in the melt a rod of stabilized zirconia and to maintain the reference electrode in contact with it, above the melt level (cf. Figure 8-15). Under these conditions, the zirconia

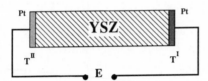

Figure 8-14.
Oxygen thermocell without any reference electrode.

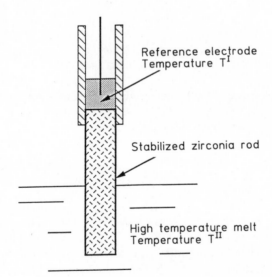

Figure 8-15.
Diagram of a non-isothermal sensor used in high temperature glass and metal melts.

surface where the measurement is performed and the reference electrode are usually at different temperatures. In glass melt analysis, this sensor assembly is used because stabilized zirconia is slightly soluble in the glass and is gradually consumed.

To simplify the thermo-emf correction of the measurements, Jacob [114] has developed thermo-compensated reference electrodes. By appropriately selecting the alloy/oxide system generating the reference oxygen fugacity, he has been able to obtain a zero overall thermo-emf. The $Ni/Ni_{0.65}Mg_{0.35}O$ mixture, for instance, has been demonstrated to be a correctly compensating reference electrode for liquid copper analysis.

8.7 Practical Examples of Potentiometric Sensors

The following sections will describe a few typical examples of sensors which are available on the market or which are known to work correctly in their laboratory-made versions. Because of very specific technological problems, the sensors used at very high temperature, for metal and glass melt analysis will be described in detail in a separate chapter (Chapter 26).

8.7.1 Oxygen Sensors

Depending on the application, an oxygen sensor can encounter severe operating conditions, eg, high temperatures, large temperature fluctuations and gradients, high gas velocities, aggressive dusts and vapors and mechanical shocks. These factors may lead to a premature degradation of the ceramic or the electrodes and should be taken into account.

In the industrial environment, gases are frequently far from equilibrium and free oxygen can coexist with reducing gaseous compounds.

When the sensor is incorporated in a feed-back control unit, its response time is one of the main parameters. Then, the electrode material and its microstructure are key features. Electrodes may also degrade for a number of reasons to be taken into account, eg, high temperature sintering, poisoning, etc.

The conventional zirconia-based oxygen sensor functions acceptably in the 600–1000 °C temperature range. Much effort has been devoted to improving its performance at very high temperatures, eg, above 1200 °C. Intensive research has also been carried out to reduce the working temperature for possible new applications in biotechnology and in the food industry.

Recently, various adaptations of microelectronic technologies have been attempted. The goals are: reduced cost and size and improved reproducibility.

8.7.1.1 Laboratory Design and Characteristic Performances

Zirconia sensors (cf. Figure 8-16) can be used for very accurate monitoring of oxygen pressure higher than 10^{-2} Pa in inert gases, in the 600–800 °C temperature range. Deviations from the straight lines predicted by the Nernst law (Equation (8-1)) are observed (see Figure 8-17) at low temperature because of an excessive impedance of the cell and at high temperature because of the polarizing oxygen semipermeability flux.

Calibration of an oxygen sensor is always done by referring to the Nernst law. It can be performed with a flowing gas either by varying the sensor temperature T with a constant measured pressure P or by varying P at constant T. In the first case, it is essential to connect a second sensor immediately after the calibrated one in the gas line to make sure that P is actually maintained constant. In the second case, the utilization of an oxygen pump (cf. Section 8-10) to vary P has been found to be a very convenient and accurate technique [121–125]. In both cases, the results can be presented in terms of a *domain of ideal response*. Figure 8-17 shows typical results of a sensor fabricated on an industrial stabilized zirconia tube [123, 124]. The test was performed at constant oxygen pressure. The domain of ideal response corresponding to an error smaller than 10% ($\Delta P_{O_2}/P_{O_2} < 0.1$) ranges from 600 to 800 °C at 10^{-2} Pa and

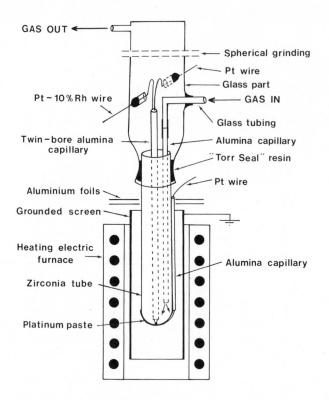

Figure 8-16. Laboratory oxygen zirconia sensor for gas analysis.

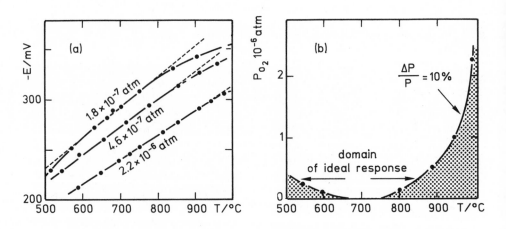

Figure 8-17. (a) Verification of the Nernst law as a function of temperature for a zirconia oxygen sensor.
(b) Determination of the domain of "ideal response" at different oxygen pressures [123, 124].

from 500 to 950 °C at 10^{-1} Pa. In the domain of ideal response, the accuracy of the determination ($\Delta P/P$) can be as good as 2%. The response time at 95% is a few seconds or less
at oxygen pressures higher than 10 Pa and about one minute in the $1-10^{-1}$ Pa range. It also
markedly depends on temperature.

The same device can be used to analyze gaseous mixtures such as CO_2-CO or H_2O-H_2 in
the oxygen pressure range $10^{-3}-10^{-22}$ Pa (Figure 8-18).

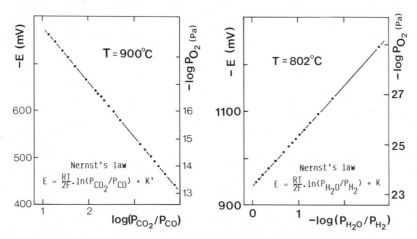

Figure 8-18. Verification of the Nernst law as functions of gas composition with (left) CO_2–CO and
(right) H_2O–H_2 mixtures [125].

For industrial applications, fully stabilized zirconia which is sensitive to thermal shocks is
often replaced by partially stabilized zirconia (ZrO_2-MgO 3–4 wt.%) or by a zirconia-yttria-
alumina two phase electrolyte.

8.7.1.2 Miniaturized Sensor

A miniaturized sensor (Figure 8-19) with an enclosed metal-metal oxide reference system has
found several applications [126]. The Pd-PdO reference mixture has an equilibrium oxygen

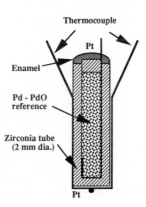

Figure 8-19. Oxygen minisensor [126].

pressure of the order of 10 Pa at a working temperature around 650 °C. This is a good compromise for an oxygen sensor. This sensor presents the following advantages: good stability, good heat shock resistance, and good resistance to high pressure (up to 10^7 Pa). It can be maintained at the working temperature with only 10 watts and it can be brought from room temperature to the working temperature in a few minutes. In situ measurements are possible without circulation of the analyzed gas. On the other hand, the temperature dependence of the equilibrium partial pressure of the reference system makes the sensor more sensitive to temperature fluctuations than the air-reference device. Moreover, the Pd/PdO system must periodically be regenerated by coulometry to eliminate oxygen which has permeated into the reference.

8.7.1.3 Pellet Oxygen Sensor

The SIRO$_2$ commercially available sensor (Figure 8-20) [127] is formed on a zirconia-based pellet or thin disc welded to the end of an alumina tube by heating at about 1900 °C until a film of the eutectic liquid is formed at the interface between the tube and the pellet or disc. The zirconia electrolyte expansion coefficient is reduced by using partially stabilized electrolyte or by addition (50 wt.%) of alumina to stabilized zirconia. In these ways, expansion coefficients close to that of alumina are obtained. This sensor is suitable for applications under severe industrial conditions: analysis of combustion exhaust gases, carburizing gas mixtures, molten copper, etc.

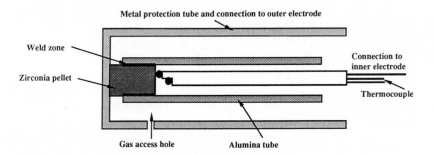

Figure 8-20. SIRO$_2$ pellet oxygen sensor [127].

8.7.1.4 Lambda Sensors

Considerable efforts have been made to develop closed-loop systems for controlling internal combustion engine emissions. The present system (Figure 8-21) combines a three-way catalyst and a zirconia sensor, called a *lambda sensor* [128–131]. A high efficiency with respect to the oxidation of CO and unburnt HC gases and to the reduction of NO$_x$ is achieved only over a very narrow air/fuel ratio range, near the stoichiometry value (ratio close to 15 : 1).

The response of the sensor shows an abrupt voltage step (more than 500 mV) at the stoichiometry point which can thus be easily detected. The sensor is generally shaped like a sparking plug. The external electrode, in contact with the exhaust gas, is made of a platinum layer which also acts as an oxidation catalyst. This electrode is covered with a spinel-type coating which protects it from the corrosive action of the gas and reduces the penetration of undesirable poisoning components containing lead, sulfur or phosphorus. The reference system generally is air. The device is inserted in the exhaust manifold and is heated by the exhaust gases. The response time is lower than 200 ms above 400 °C. The operating life time is of the order of 25 000 km.

Lambda conductivity sensors based on non-stoichiometric oxides such as TiO_2 or Nb_2O_5 now are serious competitors.

A major effort has recently been directed towards sensors for lean-burn engines, ie, with air/fuel ratios higher than 15 : 1. Conventional zirconia or semiconductor sensors have been found relatively insensitive to the composition of the corresponding exhaust gas. For this application, the amperometric devices (see Section 8-10) may soon appear to be more appropriate.

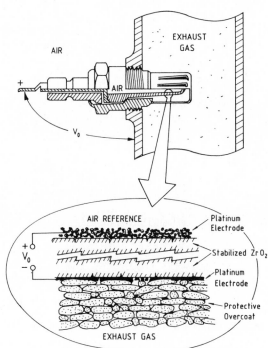

Figure 8-21.
λ-zirconia sensor for automotive exhaust gas control [131].

8.7.1.5 High-Temperature Sensors

The main problem encountered with the use of zirconia sensors at high temperatures are:

— the poor thermal shock resistance of fully stabilized zirconia tubes,
— the shrinkage of the reference metal-metal-oxide reference mixture,
— the existence of a rather high electronic conductivity inducing an oxygen semipermeability flux through the electrolyte.

Since 1970, MgO stabilized zirconia also is widely used in commercial oxygen sensors. The thermal shock resistance of this electrolyte is better but, after long term use at temperatures higher then 1200 °C, a degradation of its mechanical and electrical properties is observed. Destabilization of stabilized zirconia has been invoked as a possible cause for that.

A needle sensor (Figure 8-22), initially proposed by Janke [132], is now commercially available for steel melts, at 1600 °C. This low cost design exhibits a better heat shock resistance and a shorter response time after immersion than conventional tube sensors. The core part is a metallic conductive needle (Mo). It bears a thin coating of the reference Cr-Cr_2O_3 mixture and the electrolyte is coated by spraying.

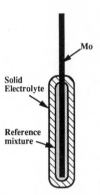

Figure 8-22. Zirconia probe for oxygen analysis in molten iron [132].

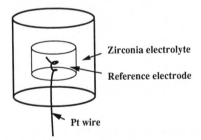

Figure 8-23.
Very high temperature oxygen sensor: isostatically pressed electrolyte body [133].

At very high temperature, a shrinkage of the reference mixture can destroy the intimate electrical contact between the electrode and the electrolyte. Recently Liu et al. [133] designed a compact sensor in which the reference pellet is completely encased by an isostatically pressed electrolyte body (Figure 8-23). Improved stability over current commercial oxygen sensors has been obtained. It is claimed that the sensor can last over 10 hours in steel melts, at 1600 °C.

Various electrochemical oxygen sensors for the liquid sodium used in fast breeder reactors have also been developed [134, 135]. In addition to thermal and mechanical shock resistance, such cells have to be insensitive to radiation and sodium corrosion. For example, Jakes [134] has proposed a cell based on a thoria crucible with a sealed-in Sn-SnO_2 or In-In_2O_3 reference system (Figure 8-24).

A minisensor has also been developed for the determination of thermodynamic properties of fuel rods and fission products. The minisensor is composed of a ThO_2-Y_2O_3 solid electrolyte tube (10 mm in length, 3 mm in diameter) containing a Fe-FeO reference (Figure 8-25). The sensor gives locally the oxygen activity on the surface of the oxide sample. Measurements of the oxygen potential in urania-based solid-solutions have been carried out with this device as a function of the O/M ratio between 600 and 1000 °C [137]. Such a cell has also been used for the continuous control of the oxygen redistribution in UO_{2+x} under a thermal gradient [136].

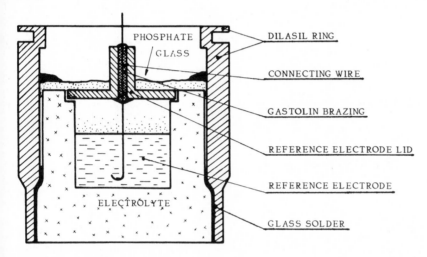

Figure 8-24. Oxygen sensor for liquid sodium analysis [134].

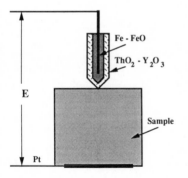

Figure 8-25.
Solid electrolyte miniprobe for measurement of the oxygen potential in solid oxides [136].

Reduction of the semipermeability flux by appropriate material selection:

As mentioned above, the main limitation in high temperature sensors is due to the electronic conductivity of the oxide electrolyte. The sources of error induced by this electronic conductivity have been reviewed in a variety of papers [1-6, 22, 101-108]. Many solutions have been promoted to reduce the resulting semi-permeability oxygen flux or its effects.

To visualize the determining parameters of this oxygen flux, under given experimental conditions, one can deduce its expression from the set of Equations (8-135) to (8-146). For instance, when the electron hole conductivity is predominant at both electrodes (under rather high oxygen pressures) it will be:

$$J_{O_2} = \frac{RT}{4 F^2 l} \; \sigma^\circ{}_h \, [(P^I_{O_2})^{\frac{1}{4}} - (P^{II}_{O_2})^{\frac{1}{4}}] \; . \tag{8-152}$$

The standard electron hole conductivitiy $\sigma^\circ{}_h$ is defined by the equation: $\sigma_h = \sigma^\circ{}_h \, P^{\frac{1}{4}}_{O_2}$.

Referring to equations of this type, we will conclude that the oxygen flux can be reduced by:

– using an electrolyte with a lower electronic conductivity (factor $\sigma^\circ{}_h$). In this respect, doped ceria was found to be even worse than zirconia [62].
– choosing a reference mixture with an oxygen chemical potential close to that of the analyzed system (factor $[(P^I_{O_2})^{\frac{1}{4}} - (P^{II}_{O_2})^{\frac{1}{4}}]$). Pd/PdO, at temperatures between 550 °C and 800 °C for inert gas-oxygen mixture analysis is an adequate reference system, for this reason. Similarly, in the case of a prevalent free electron conductivity at both electrodes, Cr/Cr_2O_3 is used for the measurement of oxygen dissolved in steel melts at 1600 °C,
– increasing of the electrolyte thickness (parameter l): plug-type sensors with an electrolyte thickness of 10–20 mm do prove to be more reliable for measurements in steel melts than tubular sensors (electrolyte thickness: 2–4 mm) [139, 140],
– using of special series connexions of solid electrolytes.
For the measurement of extremely low oxygen activities in Fe-O-C melts, Romero et al. [98] have proposed a bi-electrolyte plug (Figure 8-26) consisting of a $ThO_2\text{-}Y_2O_3$ pellet in contact with the melt and an intermediate $ZrO_2\text{-}CaO$ pellet in contact with the $Cr\text{-}Cr_2O_3$

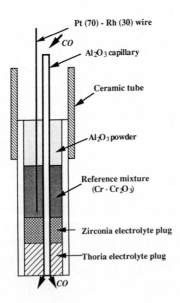

Figure 8-26.
Carbon-oxygen sensor for low oxygen contents in iron melts [98].

reference. An alumina capillary maintains a carbon monoxide pressure of 1 bar above the molten metal. Then the formula to apply would be:

$$J_{O_2} = \frac{RT}{4 F^2 l} \ [\sigma^\circ{}_e^{II} (P_{O_2}^{II})^{-\frac{1}{4}} - \sigma^\circ{}_e^{I} (P_{O_2}^{I})^{-\frac{1}{4}}] \tag{8-153}$$

in which $\sigma^\circ{}_e^{II}$ and $\sigma^\circ{}_e^{I}$ are the standard free electron conductivities in doped thoria and zirconia.

- using "unusual" solid electrolytes with low conductivities has also been found to be a possible solution. For sensors working at very high temperatures, calcium zirconate or mullite can be superior to stabilized zirconia [141].

Semipermeability and electrode buffer capacity:

The oxygen semipermeability flux also alters the electrode equilibria. The magnitude of the resulting error depends not only on the permeability flux but also on the buffer capacity of the analyzed and reference mixtures [104, 123, 248]. The buffer of a neutral gas-oxygen mixture has a high capacity for oxygen pressures greater than 1 Pa but a very low capacity below 10^{-1} Pa. CO-CO_2 mixtures exhibit good buffer capacities for P_{CO}/P_{CO_2} ratios in the $10^{-3} - 10^3$ interval.

The coexistence electrodes such as Cu-Cu_2O, Fe-FeO, Mo-MoO_2, Cr-Cr_2O_3, Pd-PdO, Ni-NiO, etc., often used as reference electrodes, can also be polarized by the oxygen flux permeating through the electrolyte. It has been demonstrated [142] that Fe-FeO, Cu-Cu_2O and Pd-PdO systems are less sensitive than Cr-Cr_2O_3, for instance. The Ni-NiO mixtures have been found to exhibit a poor buffer capacity and to be extremely polarizable.

Cell assemblies which deviate the semipermeability flux:

Special assemblies have been developed to prevent the oxygen semipermeability flux from reaching and disturbing the measuring electrode [1, 104, 116]. A simple design is the "solid electrolyte tip", depicted in Figure 8-27 [104]. A conical piece of stabilized zirconia is pressed in contact with the bottom of a closed-ended tube of the same composition. The oxygen semipermeability flux is dissipated at the tip end since this flux follows the path of lowest resistance. It does not reach and disturb the measuring electrode coated on the top part of this piece. The design is especially recommended for high temperature use [104] and oxygen analysis under vacuum [143].

Another assembly is formed by two contacting concentric tubes as shown in Figure 8-28 [116]. A gas of a composition close to that of the analyzed system flows in the guard space. The overpotential effect is completely eliminated if the two electrolyte surfaces in contact are not platinized. The only drawback of this assembly is a rather high contact resistance. This cell has been mainly used for the measurement of oxygen activities of geological samples and the characterization of complex phase systems [144-148]. Then, the analyzed materials are placed inside the smallest tube. The guard space between the two zirconia tubes is flushed by a gas mixture more reducing than the sample in order to avoid any stoichiometry alterations induced by the oxygen permeability flux. The outside of the external tube is in contact with air which is the reference system.

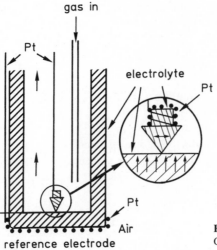

reference electrode

Figure 8-27.
Oxygen sensor with a solid electrolyte tip electrode [104].

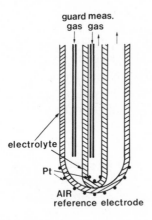

Figure 8-28. Differential oxygen sensor [116].

8.7.1.6   Low-Temperature Sensors

A great deal of research is currently aimed at developing oxygen sensors capable of operation at temperatures as low as possible for new applications. Several approaches have been proposed [149]:

New electrolytes

Oxide electrolytes exhibiting higher conductivities than zirconia at temperatures lower than 300 °C can, in principle, be used. CeO_2- or Bi_2O_3-based solid solutions are potential candidates for low-temperature sensors [62, 150, 247]. For example, $(Bi_2O_3)_{0.8} - (Er_2O_3)_{0.2}$ has a conductivity 50–100 times higher than stabilized zirconia [151–153]. However, these materials are easily reduced and the proposed applications are restricted to sensing rather high oxygen concentrations (1–100%). Moreover, the cells show rather long response times (about 3 minutes at 300 °C). Response times are acceptable only above 400 °C [153].

Thin layers

Badwal et al. [127] have proposed to replace the solid electrolyte pellet (see Section 8.7.1.3) by a thin disc sealed to the alumina tube by a butt weld to reduce the overall impedance of the cell.

By using hybrid device fabrication techniques such as silk-screen printing, chemical vapor deposition or reactive sputtering, thin-layer oxygen zirconia sensors have been fabricated. The planar structure appears to be the most reliable geometry [50]. A sensor is schematized in Figure 8-29. The reference electrode is composed of a thin layer of nickel or palladium, isolated from the ambient atmosphere by an enamel layer. The measuring electrode is a porous platinum layer. The electrolyte resistance is noticeably reduced because of the thinness of the layer, typically by a factor of 10^3. However, the response time becomes excessively long at temperatures lower than 400 °C.

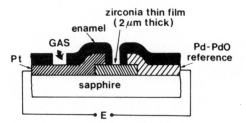

Figure 8-29.
Cross-section of a zirconia thin film sensor [50].

Catalytic electrode materials

Many systems have been tested:
- Oxide solid solutions of the type $(U, M)O_{2 \pm x}$, where M = Sc, Y, Pr, Dy, etc., have been systematically investigated [154–157]. These systems are much more effective than porous platinum down to 350 °C. Scandia doped urania mixed with platinum appears to give optimum performances.
- RuO_2 as an electrode material in an oxygen zirconia sensor responds down to a temperature of 225 °C with, however, a response time of several minutes [158].
- The perovskites such as $La_{1-x}Sr_xCoO_3$ also are effective electrodes [159].
- Recently, macrocyclic tetrapyrrolic complexes of cobalt and iron have been demonstrated to give exploitable responses down to 100 °C on LaF_3 and $PbSnF_4$ electrolytes [3, 160, 165].

Surface treatment

Surface treatments have also been proposed to improve sensor performances below 400 °C: HF treatment of stabilized zirconia [162] and pretreatment with water vapor of LaF_3-based oxygen sensors [163–164]. After treatment with water vapor at 90 °C, the response time of the

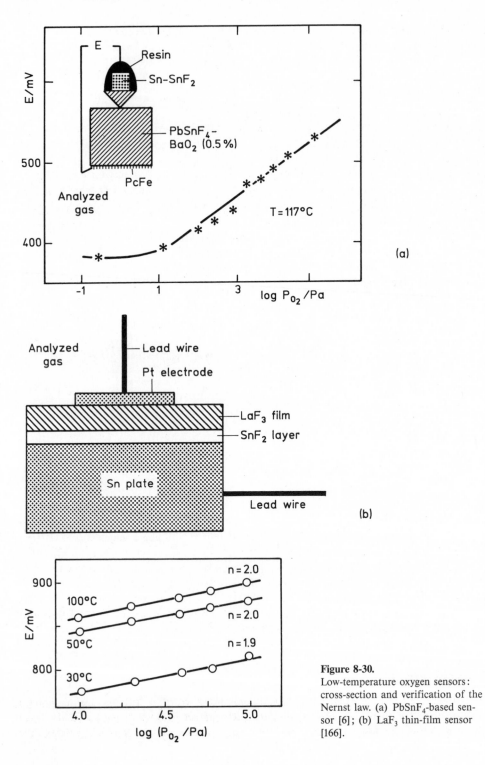

Figure 8-30.
Low-temperature oxygen sensors: cross-section and verification of the Nernst law. (a) PbSnF$_4$-based sensor [6]; (b) LaF$_3$ thin-film sensor [166].

LaF$_3$-based sensor was less than 1 min at 100 °C and 2 min at 25 °C, after changes in the oxygen partial pressure from 2×10^4 to 10^5 Pa.

Supporting electrolyte

Sensors using supporting electrolytes "saturated" with oxygen have been found to be promising (cf. Section 8.3.1.5). PbSnF$_4$ [3, 160, 165] and LaF$_3$ [65, 163, 164, 166] have been tested according to this principle down to 25 °C. At low temperature, a Nernst response corresponding to about 30 mV/dec. suggests a two-electron reaction at the sensing electrode (Reaction (8-63)). Peroxide ions have been dissolved in PbSnF$_4$ to favor such an exchange [3, 160]. Figure 8-30a shows the response of such a PbSnF$_4$-based sensor. A linear relationship is obtained in the $1-10^5$ Pa range of oxygen and the sensor can be operated down to 100 °C. The response time is of the order of one minute. In Figure 8-30b a corresponding LaF$_3$ thin-film sensor is shown.

8.7.2 Anhydride Sensors

The continuous monitoring of SO$_2$ and/or SO$_3$ concentrations in flue gases is part of an effort aimed at reducing atmospheric pollution (acid rain). The cells are generally composed of a sulfate-based electrolyte, a platinum measuring electrode and a reference electrode formed by a silver wire and dissolved silver sulfate in the electrolyte. Sometimes, the reference is formed by a sulfate mixture generating a well-defined SO$_3$, O$_2$ gas mixture or by a circulating gas (see Figure 8-31a and b). The use of a sulfate-based electrolyte for the detection and measurement of SO$_3$ and/or SO$_2$ was first suggested by Gauthier et al. [167, 168]. The recent development of these sensors has been reviewed by Worrell et al. [169–171].

In the measuring cell

$$SO_2 + SO_3 + O_2, Pt/M_2SO_4 - Ag_2SO_4/Ag \qquad (8\text{-}154)$$

the basic sensing reaction at the measuring electrode can be pagewritten as:

$$SO_3 + \frac{1}{2} O_2 + 2\,e^- \longleftrightarrow SO_4^{2-} \qquad (8\text{-}155)$$

corresponding to the Nernst equation:

$$E = C' + \frac{RT}{2F} \ln P_{SO_3} + \frac{RT}{4F} \ln P_{O_2}. \qquad (8\text{-}156)$$

As for any other anhydride sensors, the signal also depends on the oxygen pressure in the analyzed gas. This is not a problem for SO$_3$ analysis in air. On the other hand, in stack fumes, an oxygen sensor has to be used to determine this parameter. Devices based on heterojunctions similar to that described in Section 8.4.2 have been investigated to overcome this difficulty [99, 100]. In principle, their signals are independent of the oxygen pressure.

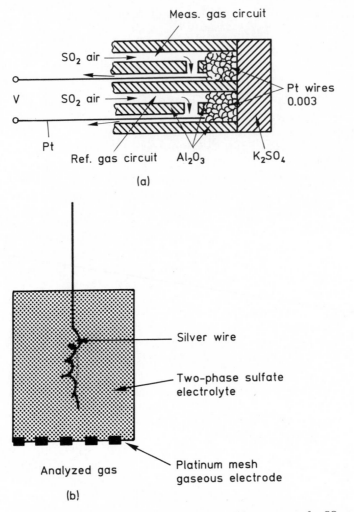

Figure 8-31. Cross-section of a potentiometric solid-state sensor for $SO_2 + SO_3$ analysis: (a) gaseous reference cell [26], (b) Ag/Ag^+ reference cell [133].

Assuming equilibrium conditions between SO_2 and SO_3,

$$SO_3 \longleftrightarrow SO_2 + \frac{1}{2} O_2 ,$$

(8-157)

and with

$$K = \frac{P_{SO_2} \cdot P_{O_2}^{1/2}}{P_{SO_3}}$$

(8-158)

the emf of the sensor can also be written as a function of P_{SO_2}:

$$E = C'' + \frac{RT}{2F} \ln P_{SO_2} + \frac{RT}{2F} \ln P_{O_2}.$$
(8-159)

Platinum is often used as a catalyst to accelerate the oxidation of SO_2 into SO_3. Vanadium pentoxide mixed in the electrolyte or platinum coating on a quartz fiber were also proposed to promote SO_2 oxidation [133].

As solid electrolytes, alkali metal sulfates such as K_2SO_4, Na_2SO_4, Li_2SO_4, Na_2SO_4 mixed with $Y_2(SO_4)_3$ and Al_2O_3 or SiO_2, have been investigated. A small drift has been observed due to solid solution formation between the electrolyte and Ag_2SO_4 when these two phases are distinct. Based on the principle described in Section 8.3.1.4, the utilization of β-Al_2O_3 and Nasicon has also been attempted [75, 76, 90].

Liu and Worrell [133, 169] have recommended the use of a two-phase mixture ($Ag_2SO_4 - Li_2SO_4$) between 510 °C and 560 °C (see Figure 8-31 b). In this temperature range, the boundary of the two-phase region of the phase diagram, between 21 and 35 mol % Ag_2SO_4 is essentially vertical, indicating that the Ag_2SO_4 activity is independent of temperature. The measured potentials are within ±3 mV of the calculated values in the range 3 to 10^4 vpm of SO_3. The response time, depending on the concentration range, is a few minutes. An emf stability of over six months has been obtained.

For a more exhaustive review of the sulfur-bearing species and anhydride analysis, the article by Jacob and Mathews [11] can also be referred to.

8.7.3 Chlorine Sensors

An electrochemical cell involving a $SrCl_2$-KCl (5 mol %) − AgCl (0.5 mol %) electrolyte can be used as a chlorine sensor (Figure 8-32a) [31, 32]. The reference electrode is formed by embedding a silver wire in the electrolyte (reference Ag/Ag^+). The measuring electrode is prepared by oxidation of ruthenium chloride in air (at 450 °C). Figure 8-32b presents the

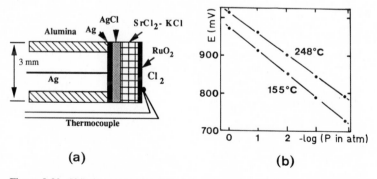

(a)   **(b)**

Figure 8-32. Chlorine sensor [31, 32]: (a) cross-section of the cell, (b) verification of the Nernst law at two temperatures.

variation of the emf of this cell as a function of the chlorine pressure in the $10^1 - 10^5$ Pa range. The response time is about 1 min at 120 °C. The presence of a large oxygen excess in the analyzed gas does not show any significant interference, at least for chlorine pressures above 10^2 Pa. A stability better than 0.5% of the emf was obtained over several months.

Hötzel et al. [33] and Mari et al. [34] have also suggested the following electrochemical chain for chlorine analysis:

$$Pt, Ag/silver~ion~conductor/AgCl/Cl_2, Pt \ . \tag{8-160}$$

The observed response times are rather long.

8.7.4   Hydrogen Sensors

The analysis of hydrogen has become important for the control and monitoring of various industrial processes: synthesis of ammonia and methanol, battery charging, measuring of hydrogen diffusivity in steel, semiconductor manufacturing, metallurgical processes, etc.

High temperature protonic conductors such as $SrCeO_3$-based solid electrolytes have been investigated [172, 245] but the majority of the tested sensors are based on room temperature protonic conductors. Protonic β''-alumina has been found to be a good H_3O^+ conductor, be-

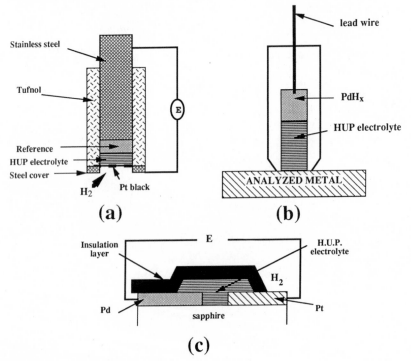

Figure 8-33. Hydrogen sensors: (a) sensor for gas analysis [30], (b) sensor for measurement of hydrogen in metals [174], (c) cross section of a thin-film hydrogen sensor [50].

tween 25 and 80 °C and is appropriate for this application. Sensors based on hydrogen uranyl phosphate tetrahydrate (HUO_2PO_4–$4H_2O$ or HUP) [28–30, 50, 173, 174] and other materials such as $Zr(HPO_4)_2$–nH_2O, Sb_2O_5–nH_2O, $H_3Mo_{12}PO_{40}$–nH_2O and the zeolites [175] have also been tested at room temperature. As solid state reference electrodes, Pd-PdH$_x$, H$_x$WO$_3$, H$_{0.34}$MoO$_3$, γ-MnO$_2$ and (α and β) PbO$_2$ [173, 176], associated with HUP, have been systematically essayed. The sensing electrode often is Pt black. As shown in Figure 8-33a, a typical cell is sandwich-shaped. Thin-film cells have also been fabricated (Figure 8-33c). When the hydrogen concentration is measured in a metal, this metal itself is used as an electrode (Figure 8-33b).

The response time varies between a few seconds to a few minutes in the 10 to 10^5 Pa range of hydrogen pressure.

Generally, a water vapor pressure has to be maintained around the electrolyte, to avoid dehydration.

8.8   Solid Ionic Conductors in pH and Ion-Selective Electrodes

Solid ionic conductors are commonly used in pH and "ion-selective electrodes":

— The first glasses for the pH measurement were based on sodium or lithium silicates. The present general formula is more complex. It can be written SiO_2, A_2O, MO_x, where A is an alkali metal (Na$^+$ and/or Li$^+$ and/or Cs$^+$) and M = Ca, Ba, Sr, La, Ce. To improve performances in very basic solutions and at high temperature, UO_2 or TiO_2 is partly substituted for SiO_2.
— AgCl is one of the basic components of many internal references (cf. Section 8.8.1).
— "Selective electrodes" for Na$^+$, K$^+$ and Ag$^+$ are based on glasses of the SiO_2-Al_2O_3-Na_2O system. The equivalent system with Li_2O is used for Li$^+$.
— Many mixtures based on Ag_2S are used for analyzing Cl$^-$, Br$^-$, I$^-$, S^{2-}, SCN$^-$, CN$^-$, SO_4^{2-} and Ag$^+$, Cu^{2+}, Hg^{2+}, Pb^{2+}, Cd^{2+}, Mn^{2+}, Zn^{2+}. This mixed-conductor has been selected for its very low solubility product.
— In the LaF$_3$ family, NdF$_3$ and SmF$_3$ have also shown good performances in F$^-$ "selective electrodes" but are too expensive. Doping with EuF$_3$ and CaF$_2$ has significantly improved the LaF$_3$ conductivity.

More information on these materials can be found in the basic books on pH and "ion-selective electrodes" [177–179].

Because of their solid nature, solid ionic conductors also are well adapted to the thin-layer technologies. In this respect, they are ad hoc sensing materials for ISFETs. For this purpose, AgCl and LaF$_3$ [180] have been the subject of successful investigations.

All these applications are described in detail in the relevant chapters (Chapter 7 and 10). Our purpose here is only to mention a few specific contributions coming from solid state electrochemistry.

8.8.1   Basic Electrode Reactions on the Ag/AgCl/Cl⁻ Electrode

This electrode mostly is an element of the internal reference electrode in "ion-selective electrodes" (Figure 8-34). It is commonly regarded as a secondary electrode sensitive to Cl^- ions. The basic electrode reaction is assumed to be a silver redox reaction:

$$Ag^+ \, (sol) + e^- \, (el) \longleftrightarrow Ag \, (el) \tag{8-161}$$

with an electrode potential given by:

$$E = E^\circ + (RT/F) \ln a_{Ag^+} \tag{8-162}$$

where a_{Ag^+} is the Ag^+ ions activity ($a_{Ag^+} = \gamma C_{Ag^+}$).

The associated chemical reaction is the AgCl solubility equilibrium which links the chloride and silver ion activities:

$$Ag^+ \, (sol) + Cl^- \, (sol) \longleftrightarrow AgCl \, (solid \; phase) \tag{8-163}$$

with:

$$(a_{Ag^+}) \, (a_{Cl^-}) = K_S \, (T) \; . \tag{8-164}$$

Combining Equations (8-162) and (8-164) gives:

$$E = E^\circ - (RT/F) \ln a_{Cl^-} \; . \tag{8-165}$$

However, as stressed by Buck [182], a more careful examination of the microscopic processes on the electrode surface requires an essential assumption: either the AgCl second phase fully covers the silver surface or not. (The argumentation is similar to that outlined in Section 8.3.1.4).

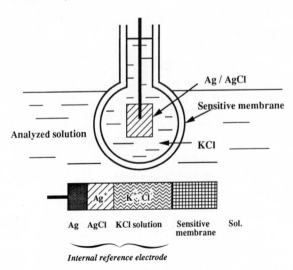

Figure 8-34.
Typical ion-selective electrode and corresponding electrochemical chain.

8.8.1.1 Non-Covering AgCl Coating

This assumption is consistent with the previous electrochemical description. A more detailed treatment of the corresponding reactions and equations, based on the decomposition of the electrochemical potentials into chemical and electrical contributions (Equation (8-11)), gives the conventional equation for the electrode potential:

$$\varphi_{metal} - \varphi_{sol} = E^\circ \, (Ag^+/Ag) + (RT/F) \ln K_s - (RT/F) \ln a_{Cl^-} . \tag{8-166}$$

This description shows that the electrode can work without any AgCl on its surface. It could be dispersed in the solution, for instance. However, when one examines the criteria for a rapid response to the Cl^- ions, it appears highly desirable to coat this second phase on the silver surface. (If the AgCl second phase were far away from the silver electrode, the whole solution would have to be equilibrated, according to Equation (8-163), before the electrode could respond to a change in Cl^- ion concentration. Primarily, it is only sensitive to Ag^+ ions).

In this case, the AgCl coating should be porous.

In this description, the electrons of the silver electrode are directly in contact with the analyzed solution. They can, therefore, participate in any possible redox reaction (for instance O_2/OH^-). If so, a mixed potential results with deviations from the ideal response. Induced errors of a few tenths of a millivolt cannot be neglected.

8.8.1.2 Covering AgCl Layer

Under this assumption, AgCl is regarded as an additional component of the electrochemical chain, inserted between the silver electrode and the solution [181]:

$$Ag/^{II} \, AgCl/^{I} \, Cl^- \, (sol) . \tag{8-167}$$

It is a silver ion conductor, because of mobile silver interstitial ions. Accordingly (cf. Section 8.4.2), the electrochemical reaction at interphase I will be written:

$$Cl^- \, (sol) + Ag^+ \, (AgCl) \longleftrightarrow AgCl \tag{8-168}$$

with:

$$\tilde{\mu}_{Cl^-} \, (sol) + \tilde{\mu}_{Ag^+} \, (sol) = \mu^\circ_{AgCl} \tag{8-169}$$

and:

$$W_{Cl}^{(sol)} - W_{Ag}^{I(AgCl)} = \mu^\circ_{AgCl} . \tag{8-170}$$

If no current crosses the AgCl layer, the predominant Ag^+ conduction implies:

$$W_{Ag}^{II} \, (AgCl) = W_{Ag}^{I} \, (AgCl) . \tag{8-171}$$

At interphase II the reaction is:

$$Ag \; (metal) \longleftrightarrow Ag^+ \; (AgCl) + e^- \; (metal) \tag{8-172}$$

with:

$$W_e^{II} - W_{Ag}^{II} \; (AgCl) = \mu_{Ag}^{\circ} \; . \tag{8-173}$$

By combining Equations (8-170), (8-171), and (8-173), we get:

$$W_e^{II} = \mu_{AgCl}^{\circ} - \mu_{Ag}^{\circ} + W_{Cl} \; (sol) \tag{8-174}$$

which is similar to Equation (8-166) (although written using a different terminology!).

As expected, under ideal conditions both descriptions lead to the same overall electrode potential. The details of these descriptions reveal the weak points of each situation. In the case of a non-covering layer it is the risk of interference with redox reactions. Here, it is the high impedance of the AgCl layer. AgCl is a rather poor conductor at room temperature ($\sigma \approx 10^{-8}$ S·cm^{-1}) and even a slight spurious electric signal will disturb the constancy of the electrochemical reference level W_{Ag} in it. According to Janz [183] AgCl ionic conductivity is higher on the grain surface than in the bulk. Small grain AgCl coatings would therefore give better electrode stability. Experimental observations on ionic conductivity of AgBr and AgI [185] support similar conclusions.

The processes conventionally used for fabricating the Ag/AgCl electrodes, which are based on anodic deposition or thermal preparation [183, 184] give porous layers. The first model would therefore more likely apply to existing electrodes, but one can also imagine that the pore surfaces in the AgCl coating partly contribute according to the covering-layer model.

Arevalo et al. [186] have observed that Ag/AgCl electrode characteristics do significantly depend on the porosity of the AgCl coating. Voltage stability, for instance, is improved if a thermal treatment is carried out below the melting point of AgCl, leading to a low porosity.

8.8.2   Solid-State Internal References

The internal reference systems of the "ion-selective electrodes" are generally based on an aqueous solution which exchanges ions with the internal surface of the sensitive membrane and with an appropriate electrode such as the Ag/AgCl system previously examined (Figure 8-34). The silver metal will ensure the connexion to the measuring instrumentation.

The presence of a liquid phase sets physical constraints which are not compatible with all desirable applications. For example, fermentor sterilization in the food industry requires high temperatures incompatible with the aqueous solution of the internal reference of the "pH electrode". Liquid reference systems are not easily miniaturizable. For these reasons and others, various attempts have been made to develop all solid-state internal reference systems. Nikolskii [187] reviewed them in 1985.

The internal references can be classified according to the conduction of the essential material forming them. This classification that we will use, also corresponds rather closely to the complexity of the systems.

8.8.2.1 Metallic Contacts

The most simple reversible electrochemical chain can be formed by coating a metal on the internal surface of the sensitive membrane (Figure 8-35 a). This metal should be selected so that it exchanges ions with the membrane. Metallic coatings of Na or Li can be used for this purpose inside a pH glass membrane conducting by either Na^+ or Li^+ ions. However, the device could be dangerous if broken, especially for in vivo applications. Shul'ts et al. [188] have proposed using an alloy containing a small concentration in alkali metal. This alloy could even be formed by in situ electrolysis. The drawback of this solution is the rather weak buffer capacity of such an internal reference. Silver on a sensitive membrane containing Ag^+ ions would certainly be appropriate. However, Ag (and Cu) coatings on the internal surface of the pH glass membrane have been found [187] to exhibit rather poor stability. This is probably due to too small a concentration in Ag^+ (or Cu^+) ions in the membrane, which makes the corresponding electrode potential unstable.

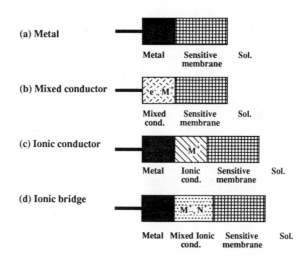

Figure 8-35.
Alternative solid-state internal references (for details see text).

8.8.2.2 Mixed Conductors

Another solution consists in coating a mixed ionic-electronic conductor (Figure 8-35 b). Various possibilities based on glassy materials and bronzes have been described by Nikolskii [187]. Positive electrode materials used in electrochemical batteries, such as $Li_x TiS_2$ and $Li_x V_2O_5$ have been suggested by Fog and Atlung [189]. The lithium content in these compounds can be optimized.

In this direction, Deportes et al. [190] have also investigated H_xMnO_2 in the pH electrode. In this case, the ionic exchange is limited to the membrane surface as it is on the outside surface, for the sensing purposes. This system has been patented [191] but is not used at present because of poor reproducibility after heat treatment above 100 °C.

8.8.2.3   Solid Ionic Conductors

A solid ionic conductor can also be inserted between a metal and the sensitive membrane (Figure 8-35c). The simplest application of this diagram has been developed by Lyalin et Turayeva [192] for the LaF_3 fluoride ion sensor. The electrochemical chain is the following:

$$M/^{II} MF_x/^I LaF_3/F^- \ (sol) \tag{8-175}$$

where M is a transition metal. In this case, the internal electrode reaction occurs at interphase II and a simple F^- ion exchange at interphase I. On a similar principle, Fjeldly and Nagy [193] have recommended Ag/AgF. Silver fluoride is attractive for another reason: it can easily be coated by melting. Ravaine et al. [195] used it as a powder in a resin composite, the electrode membrane was a F^- ion sensitive glass. AgF could be doped to improve its conductivity. Better ionic conductors such as PbF_2 or $PbSnF_4$ could also be used with Pb or Sn instead of Ag.

8.8.2.4   Ionic Bridges

With cationic sensitive membranes, the insertion of ionic bridges (cf. Section 8.5.2) will allow us to use more convenient metals than the alkali metals envisaged in Section 8.8.2.1 (cf. Figure 8-35d). Along this route, Fabry et al. [57] have investigated the following internal reference:

$$Ag/^{II} AgCl-NaCl/^I Nasicon/Na^+ \ (sol) \ . \tag{8-176}$$

The AgCl–NaCl solid solution is a mixed Na^+–Ag^+ conductor, and the Nasicon, a Na^+ conductor (cf. Section 8.8.3). A silver reaction occurs at interphase II and a Na^+ exchange at interphase I. Other systems, such as the AgI–NaI mixture, which exhibit a higher conductivity, were found to give shorter response time [197]. Unfortunately, the absence of an appropriate solid solution in this last system results in high instability. Other bridges based on PEO and PPO polymers were found more appropriate. These materials are known to dissolve a large variety of inorganic salts [196]. For an utilization with Nasicon, PEO was doped with AgI and NaI. In the solid solution, the silver ions are complexed into $AgI_x^{(x-1)}$. This reaction is favorable because it slows down a possible Ag^+ exchange with the Nasicon membrane. This system, which was patented [199] is not used industrially because of a hysteresis phenomenon due to polymer crystallization after heat treatment.

Finally, let us mention that Fjeldly and Nagy [198] have suggested using Ag/AgF in the pH electrode. To get an acceptable anchoring of the electrochemical transmission between the materials, they cured the AgF/glass contact at high temperature, up to the melting point of AgF. Under these conditions, a chemical reaction occurs which probably creates a sort of intermediate bridge according to the following diagram:

$$Ag/AgF // Ag^+, Li^+, Na^+ // Li^+, Na^+ // H^+ \ (sol) \tag{8-177}$$

intermediate   *sensing*

ionic bridge   *pH glass*

8.8.3 Three-Dimensional Scaffolded Fast Cation Conductors as Sensitive Membranes

The advantages of tri-dimensional framework fast ion conductors as sensitive membranes in ion "selective electrodes" have already been discussed [13]. This section will deal with a typical example, presently under pre-development investigation [13, 42]: Nasicon as a Na^+ sensitive membrane. The general formula of this compound is $Na_{1+x} Zr_2 Si_x P_{3-x} O_{12}$ [200]. It is a prototype example of fast cation transport in a tri-dimensional sub-array formed by a rigid oxide skeleton. The sensitivity of the Nasicon membranes is not very high. Nernstian behavior is observed only down to about 10^{-4} M in Na^+, but low impedance, rapid response and, above all, high selectivity make these membranes very attractive. (The observed low sensitivity is likely to be limited by a slight solubility of some elements of the compound).

Low impedance

As an example, the conductivity of the nominal composition $Na_3 Zr_2 Si_2 PO_{12}$ is of the order of 10^{-3} S cm^{-1} at room temperature. It is much higher than the conductivity of Na^+ glasses used in commercial Na^+ ion "selective electrodes". This makes the measurements significantly easier. Typically, the total impedance of a Nasicon based "electrode" involving a solid internal reference is lower than 10 kΩ. For a conventional "selective electrode" it is more than 1 MΩ.

Rapid response

After a change in the analyzed Na^+ concentration, the response of the Nasicon "electrode" is very fast, typically of the order of 10 ms. This has been confirmed by a detailed investigation by impedancemetry [201] of the Nasicon/Na^+ solution interphase. It has shown that in the 5×10^{-2} to 6.5×10^5 Hz frequency range, the ion exchange is simply characterized by a resistance (R_{ct}) in parallel with a capacitance (Figure 8-36). The corresponding exchange current densities ($i_0 = RT/FR_{ct}$) are high, typically higher than 0.4 mA cm^{-2} for a 0.1 mol l^{-1} NaCl solution. Taking into account the fact that the conductivity cavities in Nasicon are too small to allow water insertion, the ion exchange is described as:

$$(Na^+) \longleftrightarrow \langle Na^+ \rangle \tag{8-178}$$

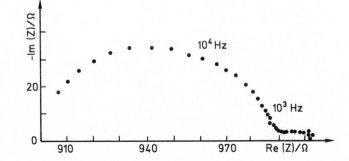

Figure 8-36.
Impedance response of the interface between Nasicon and a 0.1 mol L^{-1} Na^+ solution (area: 0.5 cm^2) [201].

where (Na^+) stands for hydrated sodium ions in the solution and $\langle Na^+ \rangle$ for sodium ions surrounded by fixed oxide ions in the Nasicon structure.

Selectivity

Generally, the relative sensitivity of an "electrode" to interfering ions is described in terms of Nikolskii [202] coefficients K_{XY}^{pot} characterizing the selectivity for an interfering ion Y relative to the primary ion X (the higher the K_{XY}^{pot}'s the worse the "electrode"). With solid state "selective electrodes", the interference may result from different mechanisms:

– The interfering ions Y may form a pure insoluble phase on the primary salt MX according to:

$$MX + Y \longrightarrow MY + X .\qquad(8\text{-}179)$$

This interference is frequently observed with silver halide membranes [178]. The selectivity coefficient can then be expressed as the ratio of the corresponding solubility products:

$$K_{xy}^{pot} = S_{MX}/S_{MY} .\qquad(8\text{-}180)$$

It is directly related to the equilibrium constant of Reaction (8-179).

– A mixed phase $MX_{1-\alpha} Y_\alpha$ can also be formed on the MX surface:

$$MX + \alpha Y = MX_{1-\alpha} Y_\alpha + \alpha X .\qquad(8\text{-}181)$$

A typical example is hydroxide ion interference in F^- analysis with LaF_3 [179]. In that case, a diffusion potential may generate into the corrosion layer and the ion transport properties have to be taken into account. The situation is somewhat similar to that of the glass electrodes. In the dry membrane, the Na^+ ions, for instance, are the predominant carriers throughout the whole material. When the membrane is in contact with the analyzed solution, some proton exchange occurs and, locally, the corresponding H^+ conductivity may become significant. It gradually vanishes deeper in the material. Eisenman [204] has given the following expression of the selectivity coefficient corresponding to this case:

$$K_{XY}^{pot} = \frac{\tilde{u}_Y}{\tilde{u}_X} K_{XY}\qquad(8\text{-}182)$$

where K_{XY} is the equilibrium constant of the surface ion exchange and $\tilde{u}_X$ and $\tilde{u}_Y$ are the ion mobilities in the membrane in the vicinity of the surface.

With Nasicon, the main interfering reactions to be feared are the ion exchanges, for instance:

$$\langle Na^+ \rangle + (K^+) \longrightarrow \langle K^+ \rangle + (Na^+)\qquad(8\text{-}183)$$

with a possible diffusion of the interfering ions inside the material. The rigid framework features of the material appear to limit these interfering reactions to small magnitudes. The conductivity cavities and bottlenecks between the cavities have been found to be very specifically suited to Na^+ ions, as if they were too small for big ions such as K^+ and too loose for small ions such as Li^+. The interfering coefficients of various cations and protons have been systematically determined [57] (Figure 8-37) and compared to those of a typical commercial Na^+ specific "electrode". The main results are shown in Table 8-2.

In the case of protons, for instance, the Nasicon selectivity coefficient is better than that of a commercial electrode by a factor of 5.

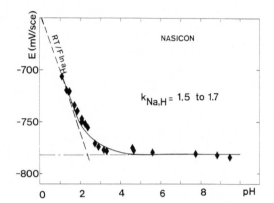

Figure 8-37.
Determination of the interference coefficient $K_{Na,H}$ of the Nasicon membrane [57].

Table 8-2. Selectivity coefficient of a Nasicon "electrode" for $[Na^+] = 10^{-2}$ mol L^{-1} [57].

Interfering ion	H_3O^+	K^+	Li^+	Ca^{2+}
Selectivity coefficient	1.5 to 1.7	$1.5 \cdot 10^{-2}$	$1.6 \cdot 10^{-2}$	$2.2 \cdot 10^{-2}$ to $2.7 \cdot 10^{-2}$

8.9   Fuel-Cell Type Sensors

A potentiometric sensor generates a voltage and in that sense can be viewed as an electrochemical battery generating an emf. After connection of its electrodes to an external resistive load, two types of response can be observed:

— either the battery is "strong" enough to sustain a steady-state current,
— or its electrodes are highly polarizable and the induced current falls to zero at a variable rate.

To our knowledge, the latter type of response has not been the object of any investigation for sensing purposes. On the other hand, the measurement of a current has been found attractive in some cases. When gases are analyzed in this manner and air used as a reference, the sensor works in fact, as a fuel cell. That is why this type of device is frequently called "fuel-cell" sensor.

Besides technical features which make a current measurement more convenient than a voltage measurement, the main advantage of this operating mode is its higher selectivity. Spurious electrode reactions which would interfere in potentiometric mode become negligible in fuel cell mode, because of their very low reaction rate (high polarizability). Furthermore, this advantage can easily be increased by using a catalyst active for the selected sensing reaction. By using different electrode materials, one can even favor different reactions at the two electrodes, and obtain a fuel-cell functioning with the same gas surrounding both electrodes. In this way, the sensor can work without any reference system.

Up to now, this fuel cell mode has been mostly developed with protonic conductors, and among them mostly with Nafion which is not regarded as a solid electrolyte in this chapter (details are given in Chapter 7). Miura and co-workers [206–208] have found that a sensor based on antimonic acid (Sb_2O_5–$2H_2O$) is sensitive to H_2, CO and NH_3. Pt black is used as the electrode material. The response time to 0.2% H_2 in air is about 10 s at room temperature (response at 90%). The response was found to be a linear function of the H_2 concentration in air from about 0.02 up to 1%. It also is a linear function of the CO-concentration from about 10 ppm up to 1000 ppm.

In the last version designed for H_2 and NH_3 analysis (Figure 8-38) two additional silver probes have been embedded in the vicinity of the platinum electrodes. A voltage is measured between these additional electrodes, giving a faster response.

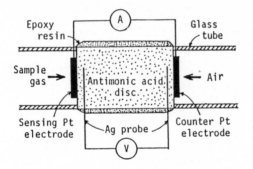

Figure 8-38.
Structure of the four-probe sensor using a proton conductor [207].

8.10   Amperometric Devices

As mentioned in the introduction, these devices are the object of very active investigations mostly spurred by the development of internal combustion engines operated in the lean regime. This regime which has been promoted by Toyota, offers high efficiency and low pollution but requires a strict control of the supplied air/fuel ratio which must be kept within a rather narrow interval. For this purpose a sensor capable of measuring small variations of the oxygen pressure is a key issue.

Since the first proposal by Heyne in 1973 [209, 210], many basic principles have been evaluated and results reported. Some are based on steady-state characteristics similar to those presented in Figure 8-3. Others imply more complex pumping modes. Takeuchi [47, 211] has recently reviewed them and proposed a classification based on various parameters:

- existence or not of a leak (diffusion) aperture;
- coupling or not with a regular potentiometric sensor;
- measurement of a limiting current or of a pumping time (or measurement of a voltage under constant current).

At present, the sensor diagram which appears to emerge is the diffusion-limited current device of the type described in the introduction (Figure 8-3). We will first concentrate on it and, in the subsequent section review the other cell assemblies and operating modes which have been demonstrated to work correctly as sensors.

All devices which will be presented here are oxygen sensors based on stabilized zirconia cells.

8.10.1 Diffusion-Limited Amperometric Sensors

This type of application is frequently compared to polarography with a limiting current measured to determine the concentration of the analyzed species. In fact, many ordinary electrodes exhibit limiting cathodic currents which are functions of the oxygen pressure [121, 225]. Ruka and Panson [214] even suggested to exploit this feature for gas analysis, but unfortunately stable performance characteristics could not be obtained. Only well defined diffusion barriers such as perforated plates or porous layers have been found capable of providing implementable responses. A typical geometry is shown with a perforated plate in Figure 8-39 and

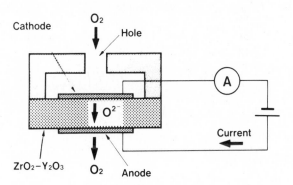

Figure 8-39.
Principle of a limiting current oxygen sensor with a leak aperture [226].

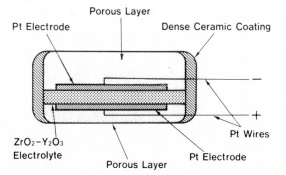

Figure 8-40.
Diagram of a sensing element of a limiting current oxygen sensor with a diffusion barrier [226].

with a porous diffusion barrier in Figure 8-40. The porous layer can be fabricated, for instance, by plasma spraying of a MgO–Al$_2$O$_3$ spinel. The physics of the corresponding diffusion mechanisms has been examined in detail by Dietz [212] in 1982. Recent results have been reviewed in 1988 by Takeuchi et al. [213] and by Usui et al. [215] in 1989.

Let us first recall two basic laws which are frequently referred to in the literature:

- All amperometric devices involve an electrochemical cell operated in pumping mode. This mode is governed by the Faraday law (Section 8.1.4.3) which correlates the number J_{O_2} of oxygen molecules being pumped per second to the pumping current I in Ampere:

$$J_{O_2} = \frac{I}{4F} \qquad (8\text{-}184)$$

where F is the Faraday constant.

- The flux of oxygen regularly diffusing through the leak aperture of an amperometric sensor (Figure 8-38) is given by:

$$J_{O_2} = -DS \frac{dC_{O_2}}{dz} + Jx_{O_2} . \qquad (8\text{-}185)$$

The first term describes oxygen diffusion in the gas. D is the corresponding diffusion coefficient, S the area of the diffusion hole, C_{O_2} the oxygen concentration and z the distance coordinate along the leak aperture. Table 8-3 gives typical values for D.

The second term describes the convection contribution which is, for example, predominant in the application described by Somov et al. (cf. Section 8.10.2). J is the convection flux and x_{O_2} the oxygen mole fraction.

Table 8-3. Normal oxygen diffusion coefficients in various gases at 400 °C and 1 atm [215].

Gas mixture	D (cm^2/s)
O$_2$–He	3.07
O$_2$–Ne	1.34
O$_2$–N$_2$	0.858
O$_2$–Ar	0.822

When one uses a sensor diagram with a perforated plate (cf. Figure 8-39), the operation procedure is the following: a voltage U is applied between the cell electrodes in such a way that oxygen is pumped out of the cell chamber (cathode potential smaller in algebraic values than its rest value). This voltage and the corresponding pumping current I are gradually increased until the chamber can be regarded as empty of oxygen. Under this condition, the oxygen flux through the leak aperture has reached its maximum value and the reaction rate of the oxygen electrode reaction:

$$O_2 \, (gas) + 4\,e^- \, (Pt) \longrightarrow 2O^{2-} \, (YSZ) \qquad (8\text{-}186)$$

which is equal to the inward diffusion flux of oxygen, cannot be further increased. The current I crossing the cell has reached its limit I_1. A plateau is observed on the $I(U)$ characteristic. A further increase of the voltage U activates new electrode reactions such as the reduction of CO_2 or the injection of electronic carriers in the solid electrolyte. The onset of a corresponding current increase is then observed (Figure 8-3b).

Diffusion along the aperture surface has been regarded as negligible [212]. Diffusion inside this hole depends on the relative values of the hole diameter and the mean free path of the oxygen molecules between two gas collisions. When the hole diameter is smaller than this parameter, a Knudsen mechanism is prevalent. Otherwise, it is a regular gas diffusion mechanism. With a pressure of the analyzed gas close to one atmosphere, this transition occurs with a hole diameter close to 10 μm [212]. Usui et al. [224] have shown that, by decreasing the overall pressure P_t, the regime gradually shifts from ordinary diffusion to Knudsen diffusion.

Under ordinary diffusion conditions, the limiting current I_1 varies as a function of the oxygen composition x_{O_2} as:

$$I_1 = -\frac{4 F S D_S T^{0.75}}{R l} \ln (1 - x_{O_2}) \tag{8-187}$$

in which F, R and T have their usual meanings (cf. Section 8.1.2), S and l are the hole area and length, D_S is the standard diffusion coefficient of oxygen in the analyzed gas. It is related to D (Equation (8-185)) by an equation of the type:

$$D = D_S \frac{1}{P_t} \left(\frac{T}{273} \right)^\alpha . \tag{8-188}$$

I_1 is independent of the overall pressure P_t of the analyzed gas and varies with the absolute temperature T as:

$$I_1 \sim T^{0.75} . \tag{8-189}$$

Under Knudsen diffusion conditions, I_1 is strictly proportional to the oxygen partial pressure $x_{O_2} P_t$:

$$I_1 = \frac{4 F d^3 \pi^{1/2}}{3 l \sqrt{MRT}} x_{O_2} P_t \tag{8-190}$$

where d and l are the hole diameter and length and M the oxygen molecular mass.

As shown by this equation, I_1 varies as function of the absolute temperature and overall pressure as:

$$I_1 \sim P_t T^{-0.5} . \tag{8-191}$$

All these laws have been verified experimentally with good accuracy [215, 216, 222, 227, 228] both with leak apertures and porous layers covering the measuring electrodes. The $I_1 = f(T)$

law for ordinary diffusion has been found, at times, to deviate from the predicted $T^{0.75}$ law [226].

It must be stressed that in the Knudsen regime, concentrations up to 100% of oxygen are rather easily analyzed. On the other hand, in the regular diffusion regime, the plateaus appear ill-defined at high oxygen concentration. The acceptable limit seems to depend on parameters which have not been fully explored yet. Some papers report a concentration limit of around 40% in O_2, others around 90%. (See also the operating mode investigated by Somov and Perfiliev, summarized at the end of the next section).

This operating mode based on the measurement of diffusion limited currents has been successfully extended to the analysis of other reducible gases such as H_2O and to gas mixtures such as N_2-O_2-CO_2 or N_2-O_2-H_2O [216, 223]. Then, two limiting currents are observed (Figure 8-41): the regular one corresponding to a diffusion limited inward flux of oxygen and

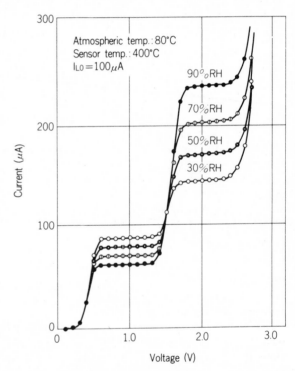

Figure 8-41.
Output (current-voltage) characteristics of a diffusion-limited amperometric sensor in wet air at 80 °C [217].

a second corresponding to the diffusion-limited electrochemical reduction of CO_2 or H_2O. The difference between the second and the first has been found to be a simple function of the CO_2 or H_2O concentration. Osanai et al. [217] have envisaged developing a humidity sensor on this principle. More complex responses have been found to result from the reaction between the retro-diffusing reduced species such as H_2 and the inward-diffusing oxygen molecules [223]. Combustible or unburned species analysis can also be attempted by reversing the pumping current. A reaction such as

$$C_3H_6 + \frac{9}{2} O_2 \longrightarrow 3\,CO_2 + 3\,H_2O \tag{8-192}$$

can also be limited, either by a slow inward flux of C_3H_6 or the outward flux of the combustion products.

The advantages of using porous layers over perforated plates are significant:

— The absence of a dead-volume chamber to be emptied before reaching the proper measuring conditions, markedly reduces the response time. Response times at 90% of the order of a few hundreds milliseconds have been reported [226] compared to several seconds obtained with perforated plates.

— The fabrication technology can be very simple with porous layers (cf. Figure 8-40). This geometry can easily be miniaturized. A sensor of only 1.7×1.75 mm^2 has been fabricated by Kondo et al. [227]. The device incorporates its own heating element (Figure 8-42) and consumes about 1 W under operating conditions.

— By appropriately selecting the pore distribution of the porous layer, an intermediate regime between ordinary and Knudsen diffusion can be obtained [213] with limiting currents independant of the temperature (but the cell voltage under which this current should be measured is likely to vary with temperature from less than 1 V up to 1.5 V).

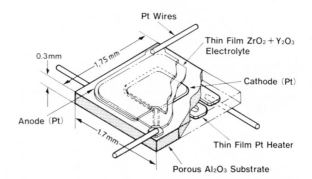

Figure 8-42.
Structure of a thin-film limiting current type oxygen sensor [227].

Measurements over long periods with a perforated plate [215, 220] have shown that the performance characteristics can be stable within a few per cent over at least 2 years. Long term extrapolations predict a 10 year life-time, mostly limited by the increase of the internal resistance of the pumping cell.

One special advantage of these amperometric sensors over zirconia potentiometric cells is their relatively low operating temperature. Temperatures as low as 400 °C can be envisaged. However, as the temperature is lowered, the plateaus become increasingly narrow (cf. Figure 8-43) and the selection of a measurement voltage becomes more delicate. This is mostly due to the ohmic drop in the solid electrolyte which becomes gradually predominant over the cell response. Using a thinner electrolyte may significantly improve the situation [213] and enable lower measuring temperatures. Recently, Liaw and Weppner [222] have shown that the utilization of a tetragonal zirconia electrolyte allows measurements down to 250 °C. The pumping capabilities demonstrated by Vinke et al. [221] for Bi_2O_3–Er_2O_3 solid solutions indicate that bismuth-based materials could also be good candidates for low temperature oxygen amperometric sensors.

Let us finally mention that Toyota [47] has developed an amperometric sensor on a technology similar to its conventional λ-sensor (Figure 8-44). The essential feature is the measuring electrode which is covered by an appropriate porous ceramic layer. The inner electrode is maintained in contact with ambient air. In this way, the sensor can be operated either as a regular potentiometric sensor or as an amperometric device. It can be used in both the rich and lean domains of the car engine.

More complex devices reported in the review paper by Takeuchi [47] consist of two electrochemical pumps with their own chambers. They can perform oxidizing and reducing analysis. One chamber can also be filled with pumped oxygen and function as a pseudo-air-reference for utilization in the rich regime.

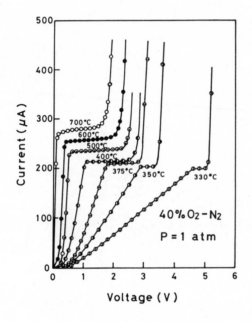

Figure 8-43.
Temperature dependence of the voltage-current characteristics of a perforated plate sensor in a 40% O_2–N_2 gas mixture [215].

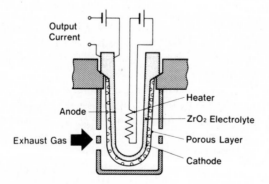

Figure 8-44.
Production type limiting current oxygen sensor [47].

8.10.2 Pump-Gauge Devices — Dynamic Pumping Modes

Initially, the pump-gauge device was developed for preparing and controlling inert gas-oxygen mixtures of well-defined compositions [110, 121, 230, 231]. It is simply made by coating 2 sets of platinum electrodes on a stabilized zirconia tube (Figure 8-45), one forming an electrochemical pump, the other a potentiometric sensor (gauge). With the pump, well-defined quantities of oxygen can be injected in (or extracted from) the gas circulating inside the zirconia tube. They are strictly proportional to the current passing through the pump according to the Faraday law (Equation (8-184)). With the sensor, the resulting gas composition can be checked. In passing, let us mention that such an electrochemical pump also is an essential equipment for calibrating the oxygen sensors intended to measure low oxygen pressures [123]. The puzzling variations of the zirconia sensor signals with analyzed gas flow rates, were definitely ascribed to the effect of the oxygen semipermeability through the zirconia wall thanks to its systematic utilization [104].

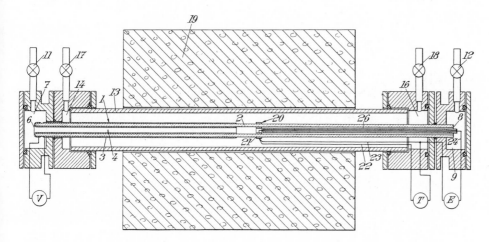

Figure 8-45. Original pump-gauge drawing [230]. (2) Stabilized zirconia tube, (19) electric furnace, (1, 3) platinum coatings forming the pump, (20, 21) platinum coatings forming the sensor, (11) gas inlet.

Quite obviously, this pump-gauge device can be used directly for combustible species titration in flowing inert gases [121, 232, 233]. For this purpose, the quantity of oxygen injected into the analyzed gas is gradually increased until the sensor indicates that all combustible species are being burnt and that an excess of oxygen prevails. Then, the pumping current provides an overall measure of the combustible species concentration by referring to the Faraday law. The titration point which corresponds to the transition from a reducing to an oxidizing atmosphere is marked by a sharp sensor voltage jump of about 500 mV.

In reported integrated devices [234, 235], the association of an additional electrochemical pump has also been used for "shifting the origin" of the measurements. For that, a constant current is passed through the pump to add or extract a fixed concentration of oxygen in the analyzed system. In this manner, Velasco [234] operated his λ-microsensor in exhaust gases containing oxygen excess (lean regime). Similarly, Ohsuga et al. [235] use a diffusion-limited current device to analyze reducing gases (rich regime) by adding a constant concentration of oxygen with an additional pumping cell (Figure 8-46).

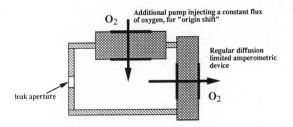

Figure 8-46.
Principle of the diffusion-limited amperometric device, investigated by Ohsuya et al. [235].

Behind the other cell assemblies and more sophisticated operating modes which have been reported in the literature, there are essentially two principles:

— The continuous coulometric measurement of oxygen being injected in or extracted from the sensor chamber. This gives more exploitable information than the simple steady-state limiting current;
— The addition of a second electrochemical cell operated as a potentiometric sensor which continuously measures the oxygen pressure inside the sensor chamber.

The original idea by Heyne [210] is a direct illustration of the first principle. The sensor diagram is similar to a diffusion-limited current sensor (Figure 8-39). Initially, no pumping current is passed through the sensor. The gas inside the chamber has the same composition as the analyzed gas, outside. Then, oxygen is rapidly pumped out. The end point is marked by a rather abrupt increase of the pumping voltage under constant current. By integrating the pumping current, one determines the amount of oxygen which was contained in the chamber. Its partial pressure P is calculated by:

$$P = n\,(RT/V) \tag{8-193}$$

where n is the number of oxygen molecules which had been pumped away. The cell coefficient RT/V, where V is the chamber volume, can be determined by calibration. The measurement is discontinuous.

The idea developed by Haaland [236] and improved by Franx [237] is a combination of the two principles. The cell (Figure 8-47) is a closed chamber with two electrochemical cells: a pump and a potentiometric sensor. It is operated as follows: initially, all oxygen is pumped away from the chamber (end point marked by a sensor signal of about 100 mV). Then it is pumped back in until the inside pressure is equal to that prevailing outside in the analyzed gas (end point: sensor signal = 0 mV). The integration of the pumping-in current gives the quantity of injected oxygen. Its partial pressure is also calculated by dividing it by the cell coefficient.

More sophisticated measurement by Hetrick et al. [238–240] are based on rather similar cells with one or two leak apertures (Figure 8-48). The authors determined the theoretical

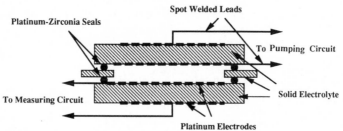

Figure 8-47. Haaland's cell [236].

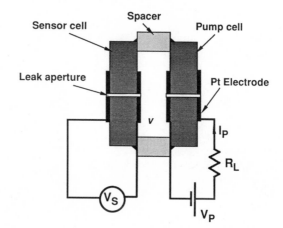

Figure 8-48. Hetrick's cell [238, 239].

equation relating, at any pumping time, the measured oxygen partial pressure P, the pumping current I and the potentiometric sensor voltage V:

$$I = 4F\sigma P \left[1 - \exp(-4FV/RT)\right] . \tag{8-194}$$

σ is, here, the diffusivity of the leak aperture, defined by:

$$J_{O_2} = \sigma \left(P_{\text{outside}} - P_{\text{inside}}\right) . \tag{8-195}$$

Plotting I versus $[1 - \exp(-4FV/RT)]$ gives P with a high accuracy.

The operating mode developed by Somov and Perfiliev [241, 242] is also based on correlations between pumping current and sensor voltage. The cell assembly is similar to that shown in Figure 8-39 with a leak aperture. Here, the aim mostly is to determine the impurity content in relatively pure oxygen. The idea is the following: as one continuously pumps oxygen out of the sensor chamber, the analyzed gas flows into it and impurities accumulate in it. The oxygen partial pressure which is measured by the potentiometric sensor, decreases accordingly and provides a measurement of this impurity enrichment. The rate at which this pressure decreases, compared to the magnitude of the pumping current, gives the impurity content in the analyzed gas.

Another idea by Maskell et al. [243, 244] also based on a relationship between pumping current and potentiometric sensor voltage, consists in applying an AC pumping current (frequency of the order of 0.1 Hz). The amplitude of the resulting voltage oscillation,

$$\Delta E = -\left(\frac{RT}{4F}\right)^2 \left(\frac{S}{V\omega P}\right) \cos \omega t , \qquad (8\text{-}196)$$

provides a direct measurement of the oxygen partial pressure P inside the cell chamber. For this application, a closed cell similar to that of Figure 8-47 is used and the average sensor voltage should be kept constant if one wants that the oxygen pressure inside the sensor chamber to be simply correlated to the measured pressure, outside.

8.11 References

[1] Sato, M., in: *Research Techniques for High Pressure and High Temperature:* Ulmer, G. C. (ed.); Berlin: Springer-Verlag, 1971, pp. 43–99.

[2] Gauthier, M., Belanger, A., Meas, Y., Kleitz, M., in: *Solid Electrolytes:* Hagenmuller, P., van Gool, W. (eds.); London: Academic Press, 1978, pp. 497–517.

[3] Kleitz, M., Siebert, E., Fouletier, J., in: *Chemical Sensors:* Seiyama, T., Fueki, K., Shiokawa, J., Suzuki, S. (eds.); Tokyo: Kodansha/Elsevier, 1983, pp. 262–272.

[4] Fouletier, J., *Sensors and Actuators* 3 (1982/83) 295–314.

[5] Maskell, W. C., Steele, B. C. H., *J. Applied Electrochem.* 16 (1986) 475–489.

[6] Fouletier, J., Siebert, E., *Ion-Selective Electrode Rev.* 8 (1986) 133–151.

[7] *Solid State Gas Sensors:* Moseley, P. T., Tofield, B. C. (eds.); Bristol: Adam Hilger, 1987.

[8] *Chemical Sensor Technology, Vol. 1:* Seiyama, T. (ed.); Tokyo: Kodansha/Elsevier, 1988.

[9] *Solid State Electrochemistry and its Applications to Sensors and Electronic Devices:* Goto, K. S. (ed.); Amsterdam: Elsevier, 1988.

[10] Subbarao, E. C., Maiti, H. C., in: *Advances in Ceramics, Vol. 24B, Science and Technology of Zirconia III:* Somiya, S., Yamamoto, N., Yanagida, H. (eds.); Westerville, Ohio: The Am. Ceram. Soc., 1988, pp. 731–747.

[11] Jacob, K. T., Mathews, T., in: *High Conductivity Solid Ionic Conductors:* Takahashi, T. (ed.); Singapore: World Scientific, 1989, pp. 513–563.

[12] *Chemical Sensor Technology, Vol. 2:* Seiyama, T. (ed.); Tokyo: Kodansha/Elsevier, 1990.

[13] Kleitz, M., Million-Brodaz, J. F., Fabry, P., *Solid State Ionics* 22 (1987) 295–303.

[14] Maier, J., Lauer, U., Göpel, W., *Solid State Ionics,* 40/41 (1990) 463–467.

[15] Haber, F., Fleischmann, F., *Z. Anorg. Chem.* 51 (1906) 245–288.

[16] Haber, F., Foster, G. W. A., *Z. Anorg. Chem.* 51 (1906) 289–314.

[17] Wagner, C., in: *Proc. Int. Comm. Electrochem. Thermo. and Kinetics (CITCE);* London: Butterworth Scientific Publ., 1957, pp. 361–377.

[18] Weissbart, J., Ruka, R., *Rev. Sci. Instrum.* 32 (1961) 593–595.

[19] Peters, H., Möbius, H. H., *Ger. (East) Pat. 21 673,* 1961.

[20] Rapp, R. A., *Techniques of Metal Research, Vol. 4, Part 2;* New York: Interscience Publ., 1970.

[21] Fischer, W. A., Janke, D., *Metallurgische Elektrochemie;* New York: Springer Verlag, 1975.

[22] Janke, D., in: *Applications of Solid Electrolytes:* Takahashi, T., Kozawa, A. (eds.); Cleveland: J.E.C. Press Inc., 1980, pp. 154–163.

[23] Nagata, K., Goto, K. S., *Solid State Ionics* 9/10 (1983) 1239–1256.

[24] Konno, K., *J. Soc. Automotive Eng. Jpn.* 31 (1977) 1182–1188.

[25] Logothetis, E. M., in: *Advances in Ceramics, Vol. 3, Science and Technology of Zirconia:* Heuer, A. H., Hobbs, L. W. (eds.); Columbus: Am. Ceram. Soc., 1981, pp. 388–405.

[26] Gauthier, M., Bellemare, R., Belanger, A., *J. Electrochem. Soc.* **128** (1981) 371–378.

[27] Worrell, W., Liu, Q. G., *Sensors and Actuators* **2** (1982) 385–386.

[28] Lundsgaard, J. S., Malling, J., Birchall, M. L. S., *Solid State Ionics* **7** (1982) 53–56.

[29] Schoonman, J., in: *Proc. Int. Sem. Solid State Ionic Devices:* Chowdari, B. V. R., Radhakrishna, S. (eds.), Singapore: World Scientific – Asian Society for Solid State Ionics, 1988, pp. 697–718.

[30] Kumar, R. V., Fray, D. J., *Sensors and Actuators* **15** (1988) 185–191.

[31] Pelloux, A., Quessada, J. P., Fouletier, J., Fabry, P., Kleitz, M., *Solid State Ionics* **1** (1980) 343–354.

[32] Pelloux, A., Fabry, P., Durante, P., *Sensors and Actuators* **7** (1985) 245–252.

[33] Hötzel, G., Weppner, W., *Solid State Ionics* **18/19** (1986) 1223–1227.

[34] Mari, C. M., Terzaghi, G., *Sensors and Actuators* **17** (1989) 569–574.

[35] Cremer, M., *Z. Biol. (Munich)* **47** (1906) 562.

[36] Frant, M. S., Ross, J. W., *Science* **154** (1966) 1553–1555.

[37] *Handbook of Electrode Technology:* Orion Research, Cambridge, MA 02139 (USA).

[38] Eisenman, G., Rudin, D. O., Casby, J. V., *Science* **126** (1957) 831–834.

[39] Kolthoff, I. M., Sanders, H. L., *J. Am. Chem. Soc.* **59** (1937) 416–420.

[40] Pungor, E., Toth, K., Havas, J., *Hung. Sci. Instrum.* **3** (1965) 2.

[41] Kleitz, M., Fabry, P., *Solid State Ionics for ISFET's,* Chemical Sensors Meeting, Rome (1984).

[42] Engell, J., Mortensen, S., *Int. Patent WO 84 01829,* 1984.

[43] Besson, J., Déportes, C., Darcy, M., *Compt. Rend. Acad. Sciences* **251** (1960) 1630–1632.

[44] Déportes, C., Darcy, M., *Silicates Industriels* **26** (1961) 499–504.

[45] Bullière, C., D.E.A., Grenoble, 1966.

[46] Butzelaar, P. F., Hoogeveen, L. P. J., *Philips Techn. Rev.* **34** (1974) 123–128.

[47] Takeuchi, T., *Sensors and Actuators* **14** (1988) 109–124.

[48] Asada, A., Yamamoto, H., Nakazawa, M., Osanai, H., *Sensors and Actuators* **B1** (1990) 312–318.

[49] Croset, M., Schnell, J. P., Siejka, J., *J. Vac. Sci. Technol.* **14** (1977) 777–781.

[50] Velasco, G., Schnell, J. P., Croset, M., *Sensors and Actuators* **2** (1982) 371–384.

[51] Pribat, D., Velasco, G., *Sensors and Actuators* **13** (1988) 173–194.

[52] Buck, R. P., Hackleman, E., *Anal. Chem.* **49** (1977) 2315–2321.

[53] Van der Spiegel, J., Lanks, I., Chan, P., Babic, D., *Sensors and Actuators* **4** (1983) 291–298.

[54] *Solid Electrolytes – Topic in Applied Physics:* Geller, S. (ed.); Berlin: Springer Verlag, 1977.

[55] *Solid Electrolytes:* Hagenmuller, P., van Gool, W. (eds.); New York: Academic Press, 1978.

[56] *Superionic Solids – Principles and Applications:* Chandra, S. (ed.); Amsterdam: North Holland Publ. Comp., 1981.

[57] Fabry, P., Gros, J. P., Million-Brodaz, J. F., Kleitz, M., *Sensors and Actuators* **15** (1988) 33–49.

[58] Siebert, E., *State Thesis,* Grenoble (1987).

[59] Neidrach, L. W., *J. Electrochem. Soc.* **127** (1980) 2122–2130.

[60] Neidrach, L. W., in: *Advances in Ceramics, Vol. 12, Science and Technology of Zirconia II:* Claussen, N., Rühle, M., Heuer, A. H. (eds.); Columbus: The Am. Ceram. Soc., 1984, pp. 672–684.

[61] Hettiarachchi, S., Kedzierzawski, P., Macdonald, D. D., *J. Electrochem. Soc.* **132** (1985) 1866–1870.

[62] Fouletier, J. Hénault, M., *Solid State Ionics* **9/10** (1983) 1277–1282.

[63] Dirstine, R. T., Gentry, W. O., Blumenthal, R. N., Hammetter, W., *Ceram. Bull.* **58** (1979) 778–783.

[64] Suzuki, T., Kaku, K., Ukawa, S., Dansui, Y., *Solid State Ionics* **13** (1984) 237–239.

[65] Kuwata, S., Miura, N., Yamazoe, N., Seiyama, T., *Chem. Letters* (1984) 981–982.

[66] Kuwata, S., Miura, N., Yamazoe, N., Seiyama, T., *Denki Kagaku* **51** (1983) 947–948.

[67] Birot, D., Couturier, G., Danto, Y., Portier, J., Salardenne, J., in: *Chemical Sensors:* Seiyama, T., Fueki, K., Shiokawa, J., Suzuki, S. (eds.); Tokyo: Kodansha/Elsevier, 1983, pp. 357–362.

[68] Salardenne, J., Labidi, F., Portier, J., Birot, D., in: *Proc. 2nd Int. Meet. on Chemical Sensors:* Aucouturier, J. L. et al. (eds.); Bordeaux: University of Bordeaux, 1986, pp. 323–326.

[69] Rolland, P., *Thesis,* University of Paris, 1974.

[70] Takikawa, O., Imai, A., Harata, M., *Solid State Ionics* **7** (1982) 101–107.

[71] Brisley, R. J., Fray, D. J., *Met. Trans.* **14B** (1983) 435–440.

[72] Velasco, in: *Chemical Sensors:* Seiyama, T., Fueki, K., Shiokawa, J., Suzuki, S. (eds.); Tokyo: Kodansha/Elsevier, 1983, pp. 239–244.

[73] Belanger, A., Gauthier, M., Fauteux, D., *J. Electrochem. Soc.* **131** (1984) 579–586.

[74] Côté, R., Bäle, C. W., Gauthier, M., *J. Electrochem. Soc.* **131** (1984) 63–67.

[75] Akila, R., Jacob, K. T., *Sensors and Actuators* **16** (1989) 311–323.

[76] Maruyama, T., Saito, Y., Matsumoto, Y., Yano, Y., *Solid State Ionics* **17** (1985) 281–286.
[77] Maruyama, T., Ye, X. Y. , Saito, Y., *Solid State Ionics* **23** (1987) 113–117.
[78] Maruyama, T., Sasaki, S., Saito, Y., *Solid State Ionics* **23** (1987) 107–112.
[79] Yao, P. C., Fray, D. J., *J. Applied Electrochem.* **15** (1985) 379–386.
[80] Yuan, D., Kröger, F. A., *J. Electrochem. Soc.* **118** (1971) 841–846.
[81] Fukatzu, N., Yamashita, K., Ohashi, T., Iwahara, H., *J. Jap. Inst. Metals* **51** (1987) 848–857.
[82] Nagata, K., Nishino, M., Goto, K. S., *J. Electrochem. Soc.* **134** (1987) 1850–1854.
[83] Miura, N., Yamazoe, N., in: *Chemical Sensor Technology, Vol. 1:* Seiyama, T. (ed.); Tokyo: Kodansha/Elsevier, 1988, pp. 123–139.
[84] MacGlashan, M. L., *Pur. Appl. Chem.* **21** (1970) 1–44.
[85] Kleitz, M., Pelloux, A., Gauthier, M., in: *Fast Ion Transport in Solids:* Vashishta, P., Mundy, J. N., Shenoy, G. K. (eds.); New York: Elsevier, 1979, pp. 69–73.
[86] Kleitz, M., *Solid State Ionics* **3/4** (1981) 513–523.
[87] Fernandes, R. Z. D., Aléonard, S., Ilali, J., Hammou, A., Kleitz, M., *Solid State Ionics* **34** (1989) 253–256.
[88] Kleitz, M., Siebert, E., in: *Chemical Sensor Technology, Vol. 2:* Seiyama, T. (ed.); Tokyo: Kodansha/Elsevier, 1989, pp. 151–157.
[89] Hötzel, G., Weppner, W., in: *Proc. 2nd Int. Meeting on Chemical Sensors:* Aucouturier, J. L., et al. (eds.); Bordeaux: University of Bordeaux, 1986, pp. 285–288.
[90] Saito, Y., Maruyama, T., Matsumoto, Y., Yano, Y., in: *Chemical Sensors:* Seiyama, T., Fueki, K., Shiokawa, J., Suzuki, S. (eds.); Tokyo: Kodansha/Elsevier, 1983, pp. 326–331.
[91] Kirchnerova, J., Bale, C. W., Skeaff, J. M., in: *Proc. 2nd Int. Meeting on Chemical Sensors:* Aucouturier, J. L. et al. (eds.); Bordeaux: University of Bordeaux, 1986, pp. 311–314.
[92] Siebert, E., Fouletier, J., Kleitz, M., *J. Electrochem. Soc.* **134** (1987) 1573–1578.
[93] Goto, K. S., Sasabe, M., Iguchi, Y., Iwase, M., Onou-ye, T., *Solid State Ionics* **40/41** (1990) 770–772..
[94] Gomyo, K., Sakaguchi, I., Shin-ya, Y., Iwase, M., *Solid State Ionics,* **40/41** (1990), 773–775.
[95] El Adham, K., Hammou, A., *Solid State Ionics* **9/10** (1983) 905–912.
[96] Fabry, P., Schouler, E., Kleitz, M., *Electrochim. Acta* **23** (1978) 539–544.
[97] Tretyakov, J. D., Muan, A., *J. Electrochem. Soc.* **116** (1969) 331–334.
[98] Romero, A. R., Härkki, J., Janke, D., *Steel Research* **57** (1986) 636–644.
[99] Gauthier, M., Bélanger, A., Fauteux, D., *Electrochem. Soc. Meeting,* Montreal (1982), abst. # 737.
[100] Gauthier, M., Bélanger, A., Fauteux, D., in: *Chemical Sensors:* Seiyama, T., Fueki, K., Shiokawa, J., Suzuki, S. (eds.); Tokyo: Kodansha/Elsevier, 1983, pp. 353–356.
[101] Smith, A. W., Meszaros, F. W., Amata, C. D., *J. Am. Ceram. Soc.* **49** (1966) 240–244.
[102] Patterson, J. W., Bogren, E. C., Rapp, R. A., *J. Electrochem. Soc.* **114** (1967) 752–758.
[103] Pal'guev, S. F., Gil'derman, V. K., Neuimin, A. D., *J. Electrochem. Soc.* **122** (1975) 745–748.
[104] Fouletier, J., Fabry, P., Kleitz, M., *J. Electrochem. Soc.* **123** (1976) 204–213.
[105] Iwase, M., Mori, T., *Met. Trans.* **9 B** (1978) 365–370.
[106] Iwase, M., Mori, T., *Met. Trans.* **9 B** (1978) 653–656.
[107] Kleitz, M., Fernandez, E., Fouletier, J., Kleitz, M., in: *Adv. in Ceramics, Vol. 3, Science and Technology of Zirconia:* Heuer, A. H., Hobbs, L. W. (eds.); Columbus: The Am. Ceram. Soc., 1981, pp. 349–363.
[108] Yamada, K., Murase, M., Iwase, M., *J. Applied Electrochem.* **16** (1986) 712–718.
[109] Heyne, L., den Engelsen, D., *J. Electrochem. Soc.* **124** (1977) 727–735.
[110] Kleitz, M., *State Thesis,* Grenoble, 1968.
[111] *Solid-State Electrochemistry,* D.E.A. Lectures, Grenoble.
[112] Levy, M., Fouletier, J., Kleitz, M., *J. Electrochem. Soc.* **135** (1988) 1584–1589.
[113] Fabry, P., *State Thesis,* University of Grenoble, 1976.
[114] Jacob, K. T., Ramasesha, S. K., *Solid State Ionics* **34** (1989) 161–166.
[115] Goto, K., Ito, T., Someno, M., *Trans. Met. Soc. AIME* **245** (1969) 1662–1663.
[116] Fouletier, J., Seinera, H., Kleitz, M., *J. Applied Electrochem.* **5** (1975) 177–185.
[117] Baucke, F. G. K., *Glastechn. Ber.* **56 K** (1983) 307–312.
[118] Fabry, P., *3rd Cycle Thesis,* Grenoble, 1970.
[119] Pizzini, S., Bianchi, G., *Chimica ed Industria* **55** (1973) 966–985.

[120] Briot, F., Vitter, G., *French Patent* 86 09778 (1986).

[121] Fouletier, J., Vitter, G., Kleitz, M., *J. of Applied Electrochem.* **5** (1975) 111–120.

[122] Kleitz, M., Fouletier, J., in: *Measurement of Oxygen:* Degn, H., Baslev, I., Brook, R. (eds.); Amsterdam: Elsevier Pub Co., 1976, pp. 103–122.

[123] Fouletier, J., Mantel, E., Kleitz, M., *Solid State Ionics* **6** (1982) 1–13.

[124] Fouletier, J., Siebert, E., Caneiro, A., in: *Advances in Ceramics, Vol. 12, Science and Technology of Zirconia II:* Claussen, N., Rühle, M., Heuer, A. H. (eds.); Columbus: The Am. Ceram. Soc., 1984, pp. 618–626.

[125] Caneiro, A., Fouletier, J., Bonnat, M., *J. Applied Electrochem.* **11** (1980) 83–90.

[126] Fouletier, J., Vitter, G., in: *Applications of Solid Electrolytes:* Takahashi, T., Kozawa, A. (eds.), Cleveland: JEC Press Inc., 1980, pp. 108–113.

[127] Badwal, S. P. S., Bannister, M. J., Garrett, W. G., *J. Phys. E: Sci. Instrum.* **20** (1987) 531–540.

[128] Dueker, H., Friese, K. H., Haeker, W. D., *SAE-Paper 75 0223, SAE-Automotive Engineering Congress,* Detroit, Feb. 1975, pp. 807–824.

[129] Hamann, E., Manger, H., Steinke, L., *SAE-Paper 77 0401, SAE-Automotive Engineering Congress,* Detroit, Feb. 1977, pp. 1729–1734.

[130] Gruber, H. U., Wiedenmann, H. M., *SAE-Paper 80 0017, SAE-Automotive Engineering Congress,* Detroit, Feb. 1980, pp. 1–12.

[131] Fleming, W. J., *SAE-Paper 80 0020, SAE-Automotive Engineering Congress,* Detroit, Feb. 1980.

[132] Janke, D., *Solid State Ionics* **3/4** (1981) 599–604.

[133] Liu, Q., in: *Proc. Int. Sem. Solid State Ionic Devices:* Chowdari, B. V. R., Radhakrishna, S. (eds.), Singapore: World Scientific – Asian Society for Solid State Ionics, 1988, pp. 191–204.

[134] Jakes, D., Kral, J., Burda, J., Fresl, M., *Solid State Ionics* **13** (1984) 165–173.

[135] Hobdell, M. R., Smith, C. A., *J. Nucl. Mat.* **110** (1982) 125–139.

[136] Ducroux, R., Fromont, M., Jean Baptiste, P., Pattoret, A., *J. Nucl. Mat.* **92** (1980) 325–333.

[137] Ducroux, R., Jean Baptiste, P., *J. Nucl. Mat.* **97** (1978) 333–336.

[138] Une, K., Oguma, M., *J. Nucl. Mat.* **110** (1982) 215–222.

[139] Janke, D., *Metallurgical Trans.* **13 B** (1982) 227–235.

[140] Janke, D., *Archiv. Eisenhüttenwes.* **54** (1983) 259–266.

[141] Goto, K. S., Pluschkell, W., in: *Physics of Electrolytes:* Hladik, J., (ed.); London: Academic Press, 1972, pp. 539–622.

[142] Worrell, W., Iskoe, J. L., in: *Fast Ion Transport in Solids:* van Gool, W. (ed.); Amsterdam: North Holland, 1973, pp. 513–521.

[143] Meas, Y., Fouletier, J., Passelaigue, D., Kleitz, M., *J. Chimie Phys.* **75** (1978) 826–834.

[144] Sato, M., Valenza, M., *Am. J. Sci.* **220 A** (1980) 134–158.

[145] Elliott, W. C., Grandstaff, D. E., Ulmer, G. C., Buntin, T., *Econ. Geol.* **77** (1982) 1493–1510.

[146] Hewins, R. H., Ulmer, G. C., *Geochim. Cosmochim. Acta* **48** (1984) 1555–1560.

[147] Arculus, R. J., Delano, J. W., *Geochim. Cosmochim. Acta* **45** (1981) 899–913.

[148] Amossé, J., Fouletier, J., Kleitz, M., *Bull. Minéral.* **105** (1982) 188–192.

[149] Badwall, S. P. S., Bannister, M. J., Garrett, W. G., in: *Advances in Ceramics, Vol. 12, Science and Technology of Zirconia II:* Claussen, N., Rühle, M., Heuer, A. H. (eds.); Columbus: The Am. Ceram. Soc., 1984, pp. 598–606.

[150] Arai, H., Eguchi, K., Yahiro, H., in: *Proc. 2nd Int. Meet. on Chemical Sensors:* Aucouturier, J. L. et al. (eds.); Bordeaux: University of Bordeaux, 1986, pp. 335–338.

[151] Takahashi, T., Iwahara, H., Esaka, J., *J. Electrochem. Soc.* **124** (1977) 1563–1569.

[152] Takahashi, T., Iwahara, H., *Mat. Res. Bull.* **13** (1978) 1447–1453.

[153] Verkerk, M. J., Burggraaf, J., *J. Electrochem. Soc.* **128** (1981) 75–82.

[154] Badwal, S. P. S., *J. Electroanal. Chem.* **146** (1983) 425–429.

[155] Badwal, S. P. S., Ciacchi, F. T., *J. Applied Electrochem.* **16** (1986) 28–40.

[156] Badwal, S. P. S., Ciacchi, F. T., Haylock, J. W., *J. Applied Electrochem.* **18** (1988) 232–239.

[157] Miura, N., Hisamoto, J., Kuwata, S., Yamazoe, N., *Chemistry Lett.* (1987) 1477–1480.

[158] Periaswami, G., Vana Varamban, S., Rajan Babu, S., Mathews, C. K., *Solid State Ionics* **26** (1988) 311–317.

[159] Arai, H., Eguchi, K., Inoue, T., *Shokubai* **29** (1987) 58–61.

[160] Siebert, E., Fouletier, J., Vilminot, S., *Solid State Ionics* **9/10** (1983) 1291–1294.

[161] Kuwata, S., Miura, N., Yamazoe, N., Seiyama, T., *Nippon Kagaku Kaishi* **8** (1984) 1232–1236.
[162] Obayashi, H., Okamoto, H., *Solid State Ionics* **3/4** (1981) 631–634.
[163] Yamazoe, N., Hisamoto, J., Miura, N., Kuwata, S., *Sensors and Actuators* **12** (1987) 415–423.
[164] Miura, N., Hisamoto, J., Kuwata, S., Yamazoe, N., *Chemistry Lett.* (1987) 1477–1480.
[165] Siebert, E., Fouletier, J., Le Moigne, J., in: *Proc. 2nd Int. Meeting on Chemical Sensors:* Aucouturier, J. L., et al. (eds.); Bordeaux: University of Bordeaux, 1986, pp. 281–284.
[166] Miura, N., Hisamoto, J., Yamazoe, N., Kuwata, S., Salardenne, J., *Sensors and Actuators* **16** (1989) 301–310.
[167] Gauthier, M., Chamberland, A., *J. Electrochem. Soc.* **124** (1977) 1579–1583.
[168] Gauthier, M., Chamberland, A., Bélanger, A., Poirier, M., *J. Electrochem. Soc.* **124** (1977) 1584–1587.
[169] Worrell, W. L., in: *Chemical Sensor Technology, Vol. 1:* Seiyama, T. (ed.); Tokyo: Kodansha/Elsevier, 1988, pp. 97–108.
[170] Worrell, W. L., *Solid State Ionics* **28/30** (1988) 1215–1220.
[171] Liu, Q. G., Worrell, W. L., in: *Phys. Chem. of Extractive Metallurgy:* Kuduk, V., Rao, Y. K. (eds.); Warrington, PA: The Met. Soc., 1985, pp. 387–396.
[172] Iwahara, H., Uchida, H., in: *Chemical Sensors:* Seiyama, T., Fueki, K., Shiokawa, J., Suzuki, S. (eds.); Tokyo: Kodansha/Elsevier, 1983, pp. 227–232.
[173] Schoonman, J., de Roo, J. L., de Kreuk, C. W., Mackor, A., in: *Proc. 2nd Int. Meet. on Chemical Sensors:* Aucouturier, J. L. et al. (eds.); Bordeaux: University of Bordeaux, 1986, pp. 319–322.
[174] Schoonman, J., Franceschetti, D. R., Hanneken, J. W., *Ber. Bunsenges. Phys. Chem.* **86** (1982) 701–703.
[175] Miura, N., Kato, H., Yamazoe, N., Seiyama, T., in: *Chemical Sensors:* Seiyama, T., Fueki, K., Shiokawa, J., Suzuki, S., (eds.); Tokyo: Kodansha/Elsevier, 1983; 233–238.
[176] Kahil, H., Forestier, M., Guitton, J., in: *Solid State Protonic Conductors III:* Goodenough, J. B. et al. (eds.); Odense: Odense Univ. Press, 1985, pp. 258–269.
[177] *Ion-Selective Electrode Methodology, Vol. 1:* Covington, A. K. (ed.); Boca Raton, Florida (USA): C.R.C. Press, 1979.
[178] *Studies in Analytical Chemistry 2 – The Principles of Ion-Selective Electrode and of Membrane Transport:* Morf, W. E. (ed.); New York: Elsevier Sci. Publ. Comp. 1981.
[179] *Ion Selective Electrodes, 2 nd ed.:* Koryta, J., Stulik, K. (eds.); Cambridge: Cambridge Univ. Press, 1983.
[180] Van der Spiegel, J., Lauks, I., Chan, P., Babic, D., *Sensors and Actuators* **4** (1983) 291–298.
[181] Buck, R. P., Mathis, D. E., Rhodes, R. K., *J. Electroanal. Chem.* **80** (1977) 245–257.
[182] Buck, R. P., *Analytical Chem.* **40** (1968) 1432–1439.
[183] Janz, G. J., in: *Reference Electrodes – Theory and Devices:* Ives, D. J. G., Janz, G. J. (eds); New York: Academic Press, 1961, pp. 179–230.
[184] Bousse, L. J., Bergveld, P., Geeraedts, H. J. M., *Sensors and Actuators* **9** (1986) 179–197.
[185] Young, V., *Solid State Ionics* **25** (1987) 9–19.
[186] Arevalo, A., Souto, R. M., Arevalo, M. C., *J. Applied Electrochem.* **15** (1985) 727–735.
[187] Nikolskii, B. P., Materova, E. A., *Ion Selective Electrode Rev.* **7** (1985) 3–39.
[188] Shul'ts, M. M., Ershov, O. S., Lepnev, G. P., Grekovich, T. M., Sergeev, A. S., *Zh. Prikl. Khim.* **52** (1979) 2487–2493.
[189] Fog, A., Atlung, S., *Ion Selective Electrode Device,* International Patent WO 83 03304.
[190] Déportes, C., Forestier, M., Kahil, H., in: *Proc. CIAME;* Paris: La Documentation Française, 1986, pp. 193–201.
[191] Déportes, C., Forestier, M., Kahil, H., *French Patent 84 19203,* 1984.
[192] Lyalin, O. O., Turayeva, M. S., *Zh. Anal. Khim.* **31** (1976) 1879–1885.
[193] Fjeldly, T. A., Nagy, K., *J. Electrochem. Soc.* **127** (1980) 1299–1303.
[194] Fjeldly, T. A., Nagy, K., Stark, B., *Sensors and Actuators* **3** (1982/1983) 111–118.
[195] Ravaine, D., Perera, G., Hanane, Z., in: *Chemical Sensors:* Seiyama, T., Fueki, K., Shiokawa, J., Suzuki, S. (eds.); Tokyo: Kodansha/Elsevier, 1983, pp. 521–526.
[196] Armand, M. B., *Ann. Rev. Mater. Sci.* **16** (1985) 245–261.
[197] Fabry, P., Montero-Ocampo, C., Armand, M., *Sensors and Actuators* **15** (1988) 1–9.
[198] Fjeldly, T. A., Nagy, K., Stark, B., *Sensors and Actuators* **8** (1985) 261–269.

[199] Hammou, A., Fabry, P., Bonnat, M., Armand, M., Montero-Ocampo, C., *French Patent 86 09765,* 1986.

[200] Hong, J. P., *Mat. Res. Bull.* **11** (1986) 173–182.

[201] Siebert, E., Caneiro, A., Fabry, P., *J. Electroanal. Chem.* (to be published 1990).

[202] Nikolskii, B. P., *Zh. Fiz. Khim.* **10** (1937) 495–503.

[203] Lewenstam, A., Hulanicki, A., Sokalski, T., *Anal. Chem.* **59** (1987) 1539–1544.

[204] Eisenman, G., *Anal. Chem.* **40** (1968) 310–320.

[205] Rudolf, P. R., Clearfield, A., Jorgensen, J. D., *J. Solid State Chem.* **72** (1988) 100–112.

[206] Miura, N., Kato, H., Yamazoe, N., Seiyama, T., *ACS Symposium Series,* Vol. 309; Washington: The American Ceramic Society, 1986, pp. 204–213.

[207] Miura, N., Kaneko, H., Yamazoe, N., *J. Electrochem. Soc.* **134** (1987) 1875–1876.

[208] Miura, N., Worrell, W., *Solid State Ionics* **27** (1988) 175–179.

[209] Heyne, L., Beekmans, N. M., Poolman, P. J., Eijnthoven, R. K., *Dutch Pat. Appl. 73 09537,* 1973.

[210] Heyne, L., in: *Measurement of Oxygen:* Degn, H., Baslev, I., Brook, R. (eds.); Amsterdam: Elsevier Pub. Co., 1976, pp. 65–88.

[211] Takeuchi, T., in: *Proc. 2nd Int. Meeting on Chemical Sensors:* Aucouturier, J. L. et al. (eds.); Bordeaux: University of Bordeaux, 1986, pp. 69–77.

[212] Dietz, H., *Solid State Ionics* **6** (1982) 175–183.

[213] Takeuchi, T., Igarashi, I., in: *Chemical Sensor Technology, Vol. 1:* Seiyama, T., (ed.); Tokyo: Kodanska/Elsevier, 1988, pp. 79–95.

[214] Ruka, R. J., Panson, A. J., *U.S. Patent 3 691 023,* 1972.

[215] Usui, T., Asada, A., Nakazawa, M., Osanai, H., *J. Electrochem. Soc.* **136** (1989) 534–542.

[216] Usui, T., Asada, A., Isono, Y., in: *Proc. of the 6th Sensor Symposium:* Tsukuba: IEE of Japan, 1986, pp. 279–283.

[217] Osanai, H., Nakazawa, M., Isono, Y., Asada, A., Usui, T., Kurumiya, Y., *Fujikura Technical Review* **17** (1988) 34–42.

[218] Asada, A., Usui, T., in: *Proc. of the 6th Sensor Symposium,* Tsukuba: IEE of Japan, 1986, 257–260.

[219] Usui, T., Asada, A., in: *Advances in Ceramics, Vol. 24B, Science and Technology of Zirconia III:* Somiya, S., Yamamoto, N., Yanagida, H. (eds.); Westerville, Ohio: The Am. Ceram. Soc., 1988, pp. 845–853.

[220] Asada, A., Yamamoto, H., Nakazawa, M., Osanai, H., in: *Chemical Sensors:* Turner, D. N. (ed.); Pennington: The Electrochem. Soc., Softbound Proceedings Series, 1987, pp. 132–141.

[221] Vinke, I. C., Seshan, K., Boukamp, B. A., de Vries, K. J., Burgraaf, A. J., *Solid State Ionics* **34** (1989) 235–242.

[222] Liaw, B. Y., Weppner, W., *Solid State Ionics* **40/41** (1990) 428–432.

[223] Takahashi, H., Saji, K., Kondo, H., Takeuchi, T., Igarashi, I., in: *Proc. of the 5th Sensor Symposium,* IEE of Japan, 1985, pp. 133–137.

[224] Usui, T., Nuri, K., Nakazawa, M., Osanai, H., *Jap. J. Applied Physics* **26** (1987) L2061–L2064.

[225] Kleitz, M., Besson, J., Déportes, C., in: *Proc. Journées Intern. Etude Piles à Combustibles,* Vol. III, Bruxelles: U.I.L. Bruxelles, 1966, pp. 35–41.

[226] Saji, K., Takahashi, H., Kondo, H., Takeuchi, T., Igarashi, I., in: *Proc. 4th Sensor Symposium:* IEE of Japan, 1984, pp. 147–151.

[227] Kondo, H., Takahashi, H., Saji, K., Takeuchi, T., Igarashi, I., in: *Proc. 6th Sensor Symposium:* IEE of Japan, 1986, pp. 251–256.

[228] Saji, K., *J. Electrochem. Soc.* **134** (1987) 2431–2435.

[229] Ruka, R. J., Weissbart, J., *US Pat. 3 400 054,* 1968.

[230] Besson, J., Déportes, C., Kleitz, M., *French Patent 1 580 819,* 1969.

[231] Yuan, D., Kröger, F. A., *J. Electrochem. Soc.* **116** (1969) 594–600.

[232] N. V. Philips Gloeilampenfabrieken, *French Patent 1 573 618,* 1969.

[233] Taimatsu, H., Kaneko, H., Kawagoe, M., *Solid State Ionics* **34** (1989) 25–33.

[234] Velasco, G., *Solid State Ionics* **9/10** (1983) 783–792.

[235] Ohsuga, M., Ohyama, Y., *Sensors and Actuators* **9** (1986) 287–300.

[236] Haaland, D. M., *Anal. Chem.* **49** (1977) 1813–1817.

[237] Franx, C., *Sensors and Actuators* **7** (1985) 263–270.

[238] Hetrick, R. E., Fate, W. A., Wassel, W. C., *Appl. Phys. Lett.* **38** (1981) 390–392.

[239] Hetrick, R. E., Fate, W. A., Wassel, W. C., *IEEE Trans. on Electron. Devices* **29** (1982) 129-132.

[240] Logothetis, E. M., Wassel, W. C., Hetrick, R. E., Kaiser, W. J., *Sensors and Actuators* **9** (1986) 363-372.

[241] Somov, S. I., Perfiliev, M. V., in: *Chemical Sensors:* Seiyama, T., Fueki, K., Shiokawa, J., Suzuki, S., (eds.); Tokyo: Kodansha/Elsevier, 1983, pp. 737.

[242] Somov, M. I., *Solid State Ionics* **36** (1989) 263-265.

[243] Maskell, W. C., Kaneko, H., Steele, B. C. H., in: *Proc. 2nd Int. Meeting on Chem. Sensors:* Aucouturier, J. L. et al. (eds.); Bordeaux: University of Bordeaux, 1986, pp. 302-306.

[244] Maskell, W. C., Kaneko, H., Steele, B. C. H., *J. Applied Electrochem.* **17** (1987) 489-494.

[245] Yajima, T., Koide, K., Yamamoto, K., Iwahara, H., *Denki Kagaku* **58** (1990) 547-550.

[246] Janke, D., *Solid State Ionics* **40/41** (1990) 764-769.

[247] Winnubst, A. J. A., Sharenborg, A. H. A., Burgraaf, A. J., *J. Applied Electrochemistry* **15** (1985) 139.

[248] Fabry, P., Kleitz, M., Deportes, C., *J. Solid State Chem.* **5** (1972) 1-10.

9 Electronic Conductance and Capacitance Sensors

Wolfgang Göpel and Klaus-Dieter Schierbaum, Universität Tübingen, FRG

Contents

9.1   Introduction

The chemically sensitive solid-state devices discussed in this chapter utilize changes in electronic or mixed electronic and ionic conductances ("conductance sensors"), and/or capacitances ("capacitance sensors") as electrical responses for the detection of molecules in the gas phase. Different denotations for these types of sensors are commonly in use such as homogeneous and heterogeneous semiconductor, Schottky diode, and dielectric sensors. They all fulfill the requirements of low-cost sensors. Table 9-1 shows typical compounds monitored by these types of sensors and typical detection ranges of these compounds in air.

Low-temperature conductance changes of inorganic single crystals upon changes in the gas components of the surrounding atmosphere were first described by Brattain and Bardeen for Ge (1953) [10] and by Heiland for ZnO (1954) [11]. Later, this principle was used for the first time in chemical sensors which were based upon SnO_2 polycrystalline materials to detect reducing gases in air [12, 13]. High-temperature conductance changes upon changes in the oxygen partial pressure are a well-known property of non-stoichiometric oxides since many years. Thermodynamic treatments decribe the underlying defect mechanisms (see, eg, Chapter 4.7).

More recently voltage-dependent conductance changes across interfaces in metal oxide/metal sandwich systems were used in diode sensors. The same holds for capacitance sensors with their earlier applications as humidity monitors.

In the following, we start with an overview on applications of commercial conductance and capacitance sensors in Section 9.2.1 and about "prototype" sensors which are currently developed in Section 9.2.2. Materials and fabrication technologies are treated in Section 9.3. The sensor response signals monitored by these types of sensors are explained in Section 9.4. Different types of conductance and capacitance sensors are described in detail in Section 9.5.

Table 9-1. Gaseous compounds to be monitored by conductance and capacitance sensors.

Compound	Detection range (pressure range)	Type of sensor	References
H_2	500–10000 ppm	Electronic conductance (SnO_2)	[1]
		Schottky diode (Pt/TiO_2)	[2]
H_2O		capacitance ($ZnCr_2O_4$–$LiZnVO_4$)	[3]
CH_4, C_3H_8, C_4H_{10}	500–10000 ppm	Electronic conductance (SnO_2)	[1]
CO	50–1000 ppm	Electronic conductance (SnO_2)	[1]
C_2H_5OH	1–300 ppm	Electronic conductance (SnO_2)	[1]
CH_3COOH	≥ 50 ppm	Electronic conductance (SnO_2)	[1]
NH_3	≥ 30 ppm	Electronic conductance (SnO_2)	[1]
NO_2	≥ 10 ppb	Electronic conductance (PbPc)	[4]
	≥ 100 ppb	Electronic conductance (SnO_2)	[5]
SO_2	≥ 10 ppm	Capacitance (polysiloxane)	[6]
AsH_3	≥ 0.6 ppm	Electronic conductance (SnO_2)	[7]
H_2S	5–100 pp,	Electronic conductance (SnO_2)	[1]
O_2	10^{-20}–10^5 Pa	Electronic conductance (TiO_2, $BaTiO_3$)	[9] [8]
	$10^{-7} - 10^5$ Pa	Schottky diode (Pt/TiO_2)	[8]

9.2 Specific Fields of Applications and Trends

9.2.1 Commercial Applications

9.2.2.1 Conductance Sensors

At present, most of the commercially applied conductance sensors are based upon SnO_2 and Fe_2O_3 bulk material. They are fabricated by companies like Figaro Inc. and Matsushita Electric Industrial Ltd. with a production output in the order of 10^6 sensors per year [14]. Typical polycrystalline ceramic SnO_2 sensors ("Tagushi sensors") are shown in Figure 9-1.

A low-voltage power supply is required for adjusting typical operation temperatures around 350 °C with the internal heating wire and a typical power consumption of 0.5–1 W. Usually, simple electronic circuits are sufficient for the signal preprocessing. Some typical application fields are summarized in Table 9-2.

Commercial thick-film conductance sensors based upon doped perovskites such as $BaTiO_3$ and $SrTiO_3$ are of increasing interest for the determination of oxygen partial pressures [9].

Figure 9-1.
Different Tagushi SnO_2-based conductance sensors (reproduction with kind permission of Figaro, Inc., Japan) with six different types TGS 816, TGS 203, TGS 501, TGS 109, TGS 8150, and TGS 813 in three rows starting from the upper left to the lower right side.

Table 9-2. Application fields of SnO_2-based commercial conductance sensors [15].

Industrial applications:	— Gas monitoring systems
	— Gas leak detector in factories
	— Analysis equipment
	— Fermentation control
	— Fire and toxic gas detector
Domestic applications:	— Gas leak alarm in houses, boats, etc.
	— Ventilation control in air conditioners, air cleaners
	— Automatic cooking control for microwave ovens
	— Breath alcohol detector
	— Humidity control in tape recorders, etc.

9.2.2.2   Capacitance Sensors

The most important commercial applications of capacitance sensors is the determination of the water partial pressures in air, ie, the humidity. A typical example of a thin-film humidity sensor is shown in Figure 9-2. Humidity sensors in general will be described in Chapter 20. That chapter will also include a comparison of capacitance sensors with other sensors operating with different detection principles.

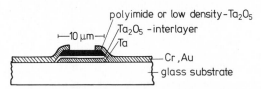

Figure 9-2. Cross-section of a capacitance sensor based upon Ta_2O_5 (or alternatively polyimide) layers to determine the relative humidity in air by monitoring changes in the capacitance upon formation of water dipoles [16]. The low density Ta_2O_5 layer is formed by anodic oxidation of sputtered Ta films. The polyimide layer is deposited by spin-on coating and subsequent polymerization.

9.2.2   Trends: "Prototype" Conductance and Capacitance Sensors

The cross-sensitivity of many conductance sensors, eg, to humidity, their low selectivity, and their limited long-term stability in aggressive environments often restrict practical applications. Therefore current research and development (R + D) efforts in many groups aim at an improvement of sensor performances.

— One approach is to replace the common ceramic fabrication technology to, eg, produce SnO_2-based devices by thick- or thin-film processes. A typical device based upon thin film silicon-technology is shown in Figure 9-3. A sensor array produced by thick-film technology is shown in Figure 6-7.

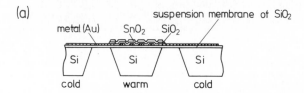

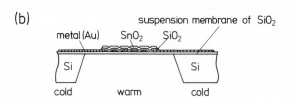

Figure 9-3.
Typical thin film sensor structures with integrated local heaters fabricated from silicon wafers [17]. The chemically sensitive SnO_2 layer is deposited by magnetron sputtering on a thin SiO_2-membrane with its low thermal conductivity. An Au layer is used for resistive heating. a) thermally isolated measuring part and b) measuring part with low heat capacity.

— An alternative approach to improve sensor performances is to use different ideal surfaces of single crystals or whiskers as chemically sensitive material [18, 19]. The samples are fixed and contacted on commercial thin- and thick-film heating elements with typical examples given in Figure 9-4.

— An improvement of sensor performances is generally achieved by

   — controlling the annealing (sintering) procedures of the oxide material and the dopants at higher temperatures,

   — using well-defined fabrication processes of the sensor base material including the proper choice of specific dopants,

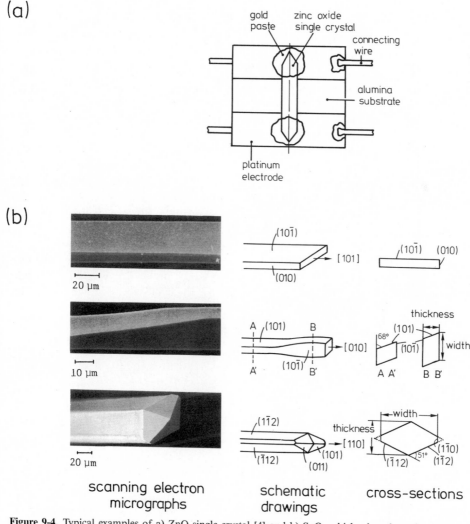

Figure 9-4. Typical examples of a) ZnO single crystal [4] and b) SnO_2 whisker-based conductance sensors [19] with scanning electron micrographs, schematic drawings and transverse cross-section of [101], [010], and [110] SnO_2 whiskers.

- carefully choosing operation temperatures for the specific detection of certain compounds,
- operation in ac-mode at different frequencies,
- operation with modulated temperatures and subsequent signal processing, and
- using filters and membranes as diffusion barriers for interfering molecules.

Details about the different technologies to produce the structures are discussed in the next chapter.

9.3 Survey on Materials and Fabrication Technologies

9.3.1 Materials

Most of the conductance sensors described in the literature are based upon inorganic semiconducting oxides with different metal and/or oxide additives and dopants. However, some organic semiconductors such as phthalocyanines (Pc) or a variety of polymers also show surprisingly high sensitivities to certain gaseous compounds. A survey on typical sensor materials for the detection of different gases in characteristic temperature ranges is given in Table 9-3.

Structures of some typical organic materials such as lead phthalocyanine (PbPc) and chemically modified polysiloxanes are shown in Figure 9-5.

Table 9-3. Chemically sensitive materials classified according to the detected gaseous compound. Additives, dopants, and operation temperatures are also indicated. Thick letters are used for Schottky-diodes; bold letters are used for capacitance sensors. All others are conductance sensors.

Gaseous Compound	Material (Operation Temperature in °C, range 100–700)	References
H_2	ceramic SnO_2 [SO_2] — H (~450 °C)	[20]
	thick film SnO_2 — H (~300 °C)	[21]
	single crystalline ZnO — H (~400 °C)	[22]
	thick film $Sb_{0.95}Sn_{0.02}O_2$ — H (~100 °C)	[23]
	SnO_2 whiskers — H (~600 °C)	[19]
	ZnO whiskers — (~300–500 °C)	[24]
	ZnO [Li] whiskers — (~500–700 °C)	[24]
	SnO_2 [Pd]	[25]
	SnO_2 [Sb_2O_3, Au] — (~100–600 °C) thin films	[26]
	(~100–300 °C) ceramic SnO_2 [Ag]	[27]
	ZnO [CuO]	[28]
	Pd/TiO₂ — evaporated Pd onto TiO_2 (001)	[2]
	Pd/CdS	[29]
PH_3	thick film Fe_2O_3 — H (~300 °C)	[24]

Table 9-3. continued

Gaseous Compound	Operation Temperature in °C	References
	100 200 300 400 500 600 700	
NH$_3$	⊢ ceramic Zn$_{1.55}$GeO$_{0.95}$N$_{1.73}$	[30]
	⊢ *NND polymer (N,N-dimethyl-3-aminopropyltrimethoxypolysiloxane)*	[6]
	⊢ *CPT polymer (3-cyanopropyltriethoxypolysiloxane)*	[6]
	SnO$_2$ [Pd, Bi, AlSiO$_3$]	[31]
	WO$_3$ [Pt]	[32]
	Cr$_2$O$_3$, Cr$_{1.8}$Ti$_{0.2}$O$_3$ ⊢————————————⊣	[33]
SO$_2$	⊢ ceramic SnO$_2$ [SO$_2$]	[20]
	⊢ *NND polymer (N,N-dimethyl-3-aminopropyltrimethoxypolysiloxane)*	[6]
CO	thick film SnO$_2$ ⊢	[21]
	⊢ thick film SnO$_2$ + ThO$_2$ + SiO$_2$	[34]
	single crystalline ZnO ⊢	[22]
	polycrystalline ZnO ⊢——⊣	[22]
	thick film α-Fe$_2$O$_3$ [Ti, Au] ⊢	[35]
	⊢ thin film LaCoO$_{3-x}$	[36]
	ZnO whiskers ⊢————————⊣	[24]
	ZnO [Li] whiskers ⊢————————⊣	[24]
	SnO$_2$ [Pd, ThO$_2$]	[37]
	⊢——⊣ ceramic SnO$_2$ [Pt, Sb]	[38]
	⊢————⊣ thick film SnO$_2$ [Pd, Cu]	[39]
	SnO$_2$ [PdO, MgO, ThO$_2$]	[40]
	TiO$_2$ [Pt]	[41]
	Co$_3$O$_4$	[42]
	⊢————⊣ thick film SnO$_2$ [Pt]	[43]
	PPA (polyphenylacetylene)	[44]
CH$_3$COOH	film ZnO ⊢—⊣	[21]
	⊢————⊣ Fe$_2$O$_3$ [Ti, Au]	[45]
	SnO$_2$ [Pd, Bi, AlSiO$_3$]	[31]
CH$_3$OH	⊢ N-methylpyrrol	[46]
C$_2$H$_5$OH	thick film SnO$_2$ ⊢	[21]
	ceramic ZnO [Ag, Pt, Au, Pd] ⊢——⊣	[47]
	WO$_3$	[48]
CH$_4$	thick film SnO$_2$ ⊢—⊣	[21]
	single crystalline ZnO ⊢—⊣	[22]
	polycrystalline ZnO ⊢———→	[22]
	ZnO [Li] whiskers ⊢————⊣	[24]
	PPA (polyphenylacetylene)	[44]
Carbon-hydrogen compounds	⊢————⊣ ceramic SnO$_2$ [SO$_2$ pretreatment]	[20]
	WO$_3$ [Rh, Pd, Pt, Ir]	[49]

Table 9-3. continued

Gaseous Compound	Sensor material (Operation Temperature in °C, scale 100–700)	References
Chlorinated carbon-hydrogen compounds	ZnO [V, Mo] ⊢⊣ (≈450)	[50]
	⊢⊣ (≈150) Pb/Pc/catalyst/sensor-combination	[51]
AsH_3	film SnO_2 ⊢⊣ (≈450)	[7]
H_2S	ceramic SnO_2 [SO_2] (≈100)	[20]
F_2	⊢⊣ (≈250) thin film PbPc	[52]
	⊢⊣ (≈225) thin film FePc	[52]
	⊢⊣ (≈275) thin film CoPc	[52]
	⊢⊣ (≈225) thin film CuPc	[52]
	⊢⊣ (≈225) thin film NiPc	[52]
CO_2	ceramic hydroxyapatite ⊢⊣ (≈375)	[53]
	PPA (polyphenylacetylene)	[44]
NO_2	⊢⊣ (≈250) thin film PbPc	[52]
	⊢⊣ (≈250) thin film CuPc	[52]
	⊢⊣ (≈250) thin film CoPc	[52]
	⊢→ (≈300) thin film ZnPc	[52]
	⊢⊣ (≈275) thin film NiPc	[52]
	⊢———⊣ (≈300–500) thin film SnO_2	[5]
	⊢⊣ (≈100) *NND polymer*	[6]
	NiO [Li, sulfanilic acid]	[54]
O_2	$BaTiO_3$ thick film ⊢→ (≈600)	[9]
	$SrTiO_3$ ceramic ⊢→ (≈600)	[55]
	Nb_2O_5 thin film	[56]
	TiO_2 [Pt] thick film ⊢→ (≈350)	[57]
	TiO_2 thin film ⊢→ (≈650)	[8]
	$LaCrO_3$ [SrO]	[58]
	$LaCoO_3$, $NdCoO_3$, $SmCoO_3$, $EuCoO_3$	[59]
	$FeTaO_4$, $CoTa_2O_6$, $NiTa_2O_6$ ⊢→ (≈550)	[60]
	$A(II)Fe_xB(III, IV)_{1-x}O_{3-z}$ ⊢→ (≈575)	[62]
	Pt/TiO₂ ⊢⊣ (≈300) evaporated Pt onto TiO_2(110)	[8]

a)   triclinic

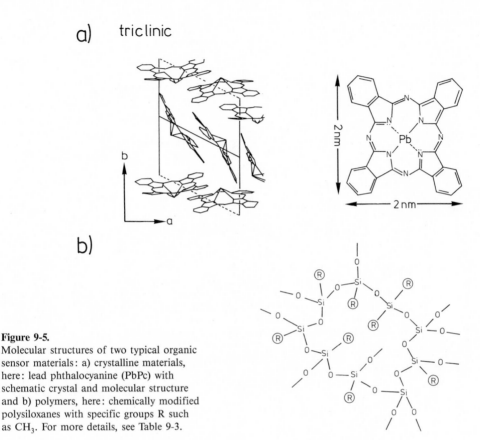

b)

Figure 9-5.
Molecular structures of two typical organic
sensor materials: a) crystalline materials,
here: lead phthalocyanine (PbPc) with
schematic crystal and molecular structure
and b) polymers, here: chemically modified
polysiloxanes with specific groups R such
as CH_3. For more details, see Table 9-3.

9.3.2 Fabrication Technologies

9.3.2.1 Ceramic Sensors

An example of ceramic sensor fabrication is shown schematically in Figure 9-6. It indicates
the different production steps of the practically most important conductance sensor based
upon SnO_2. For their adjustment of long-term stability, the sensors are finally sintered under
controlled conditions for several days.

9.3.2.2 Thick-Film Sensors

Structurized conductance and capacitance sensors may also be produced in thick-film
technology. As starting material for the chemically sensitive layers, pastes are prepared from
powders, inorganic additives, and organic binders. In contrast to pastes used for ceramic sensors, these pastes fulfill the specific requirements of thick-film technology (eg, viscosity). As
shown in Figure 9-7, they are usually printed onto electrode-covered ceramic Al_2O_3-substrates
with a heater and temperature sensor at the back- or frontside.

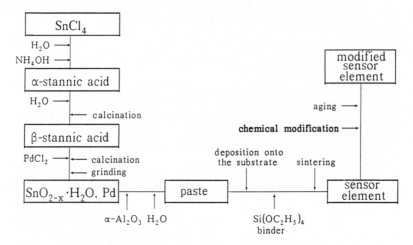

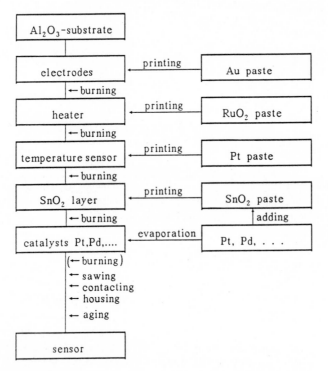

Figure 9-6. Schematic presentation of fabrication steps of polycrystalline SnO_2 ("Tagushi") sensors [62]. Often, the substrate is an Al_2O_3 ceramic tube with printed gold electrodes. The heater consists of a metal wire within the tube. The sensor element is bonded to metallic pins at a plastic or ceramic socket, and then encapsulated. Chemical modifications by adding metal salt solutions and subsequent sintering at temperatures $T \leq 600\,°C$ [63] lead to modified sensor elements with specialized sensor properties (see Section 9.5.1).

Figure 9-7. Schematic fabrication of thick film SnO_2 sensors [64].

9.3.2.3 Thin-Film Sensors

Compared with the processes described above, the fabrication of thin-film structures is based upon well-defined physical and chemical deposition techniques under clean conditions such as

- thermal evaporation,
- electron beam evaporation,
- sputtering (reactive sputtering with O_2 as sputter gas and magnetron sputtering),
- ion cluster beam deposition (ICB), or
- chemical vapor deposition (CVD).

The processes start with cleaning of the substrates. Commonly used are silicon wafers (usually covered with thermal oxides or nitrides), Al_2O_3 ceramic plates (with smaller roughness compared to thick film Al_2O_3 plates), sapphire, glass, or quartz. Adhesion layers (such as titanium or chromium thin layers) are often neccessary to prepare stable contacts. These layers are evaporated and subsequently covered with layers of noble metals (eg, gold or platinum) to manufacture electrodes, heating elements, and temperature sensors by means of common lift-off-processes. Electrochemical and ion etching-processes are currently developped for structurizing metal coatings. In a final step electrical connections are bonded or micro-welded onto contact pads of the sensor structures.

As a typical example of oxide-based thin-film sensor preparation, undoped polycrystalline SnO_2-films with different thicknesses have been prepared by thermal or by electron-beam evaporation of SnO_2 or by reactive sputtering of pure tin targets in oxygen.

- Bulk-doping with homogeneous dopant distribution may be achieved by starting from doped tin targets during the sensor preparation.
- Surface-doping is possible by subsequent evaporation of metals onto clean SnO_2 films. Specific dopant profiles are achieved by using ion implantation from solid and liquid sources with specific kinetic energies, ion current-densities, and times of the implanted dopants.
- Alternative cheaper techniques to dope involve evaporation and thermal treatment to oxidize dopant metals and to adjust specific dopant profiles as a function of annealing temperature (compare Figure 3-33) and time.

Epitaxial SnO_2-films with negligible grain boundaries could be prepared on Al_2O_3 single crystal substrates with a specific crystal orientation by using a gas-phase-transport reaction at high temperatures [65]. An alternative recent approach is atomic layer epitaxy (ALE) using organic tin-compounds as starting materials [66].

The fabrication of organic films with physical evaporation methods is only possible for thermally stable materials. Non-evaporable and polymeric layers may be deposited by

- dipping of the substrate into a solution and subsequent evaporation of the solvent,
- spin-on coating of monomers or oligomers and subsequent polymerization (thermally or by radiation),
- Langmuir-Blodget (LB) techniques,

- plasma polymerization,
- glow discharge polymerization, or
- evaporation of monomers and subsequent polymerization.

Some typical examples will be given in the Sections 9.5.3 and 9.5.6. For a comprehensive list of possible film preparation techniques, see Table 3-8.

9.4   Survey on Electrical Sensor Response Signals and Equivalent Circuits

9.4.1   Survey and Typical Sensor Designs

The electrical sensor response signals of conductance and capacitance sensors are based upon two different physical phenomena, ie,

- changes of the concentration of mobile charges in conducting materials and
- changes of the polarization of fixed charges in insulating material.

In real samples, both effects may occur simultaneously. For a homogeneous material, the conduction and polarisation in DC measurements, ie, in static electrical fields, are characterized by the specific conductivity σ and the relative permittivity ε. To describe the conductivity and permittivity in AC measurements, ie, in harmonic electrical fields of angular frequency $\omega = 2\pi v$, complex representations

$$\tilde{\sigma}(\omega) = \sigma + i\omega\varepsilon_0\tilde{\varepsilon}(\omega) \tag{9-1}$$
$$\tilde{\varepsilon}(\omega) = \varepsilon'(\omega) - i[\varepsilon''(\omega) + \sigma/\omega] \tag{9-2}$$

are commonly used with $i = \sqrt{-1}$, ε_0 as dielectric permittivity of the vacuum, ε' and ε'' as real and imaginary part of the permittivity $\tilde{\varepsilon}(\omega)$ (see Chapter 3.1 and Figure 3-5 at frequencies below 10^6 Hz). For a complicated system such as the equivalent circuit representing a general chemical sensor, $\tilde{\sigma}(\omega)$ and $\tilde{\varepsilon}(\omega)$ have different contributions from contacts, bulk, surface, and grain boundaries all of which which may contribute to the sensor response (Figures 9-8 and 9-9).

For a brief discussion of general response signals of conductance and capacitance sensors we will classify the different types of conductance and capacitance sensors according to their $G(C)$-V-characteristics, ie, the voltage-dependence of the sensor signal "conductance G" and "capacitance C". As illustrated in Figure 9-10 these parameters are monitored in DC or in AC measurements at frequencies $\omega = 0$ or $\omega > 0$ of the DC current I and voltage V or of the complex AC current $\tilde{I}(\omega)$ and voltage $\tilde{V}(\omega)$.

$$\tilde{I}(\omega) = I_0 e^{i\omega t + \varphi} \tag{9-3a}$$

and

$$\tilde{V}(\omega) = V_0 e^{i\omega t} \tag{9-3b}$$

and the complex admittance $\tilde{Y}(\omega)$, ie, the frequency-dependent conductance

$$\tilde{Y}(\omega) = \frac{\tilde{I}(\omega)}{\tilde{V}(\omega)} \tag{9-4a}$$

or the complex impedance

$$\tilde{Z} = \tilde{Y}^{-1} = \text{Re}(\tilde{Z}) + i \cdot \text{Im}(\tilde{Z}) . \tag{9-4b}$$

Here, I_0 and V_0 denote maximum amplitudes, t the time, and φ the phase angle between current and voltage. For resistances R the real part of the complex impedance is given by $\text{Re}(\tilde{Z}) = R$ and the imaginary part is zero ($\text{Im}(\tilde{Z}) = 0$). For capacitances C and inductances L the real part is zero and the imaginary part of the complex impedance is given by $i \cdot \text{Im}(\tilde{Z}) = (i\omega C)^{-1}$ and $i \cdot \text{Im}(\tilde{Z}) = i\omega L$, respectively.

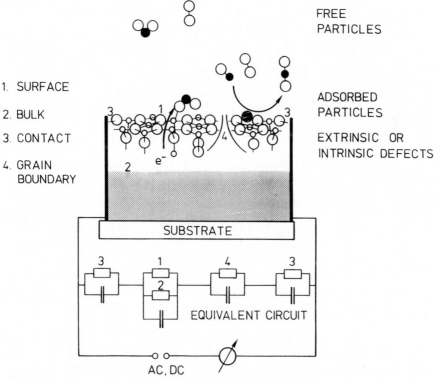

Figure 9-8. Schematic representation of elementary steps during the detection of molecules ("free particles") from the gas phase with changed concentrations of free electrons e^- of a conductance sensor based upon a metal oxide such as SnO_2. Adsorption, catalytic reactions involving intrinsic or extrinsic defects, or bulk diffusion lead to changes of the overall DC or AC conductance which can be separated into contributions of the surface (1), bulk (2), three phase boundaries or contacts (3), and grain boundaries (4), all of which may also contain capacitances. Also indicated is the equivalent circuit simulating the overall electrical response as a function of the frequency.

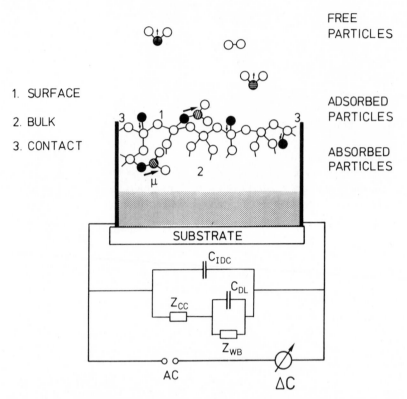

Figure 9-9. Schematic representation of elementary steps during the detection of molecules ("free particles") from the gas phase with dipoles of a capacitance sensor based upon a polymer such as polysiloxane. Adsorption of molecules at specific surface sites R (compare Fig. 9-5) may lead to the formation of localized dipoles μ (arrows). They induce a change in the relative permittivity ε due their orientation in an electrical field applied to the contacts. This can be monitored by a change ΔC in the overall capacitance. For details of the equivalent circuits to describe the overall response, see text.

We distinguish between

- conductance sensors with an ohmic behavior of the current-voltage curves (I-V), ie, with a conductance $G = dI/dV \neq f(V)$ independent of the voltage (Figure 9-10a) as for example homogeneous semiconductor sensors, and
- conductance sensors with non-linear Schottky-type I-V curves, ie, with a conductance $G = f(V)$ which depends on the applied voltage (Figure 9-10b) as for example heterogeneous semiconductor sensors such as Schottky diode sensors, and
- capacitance sensors with a voltage-independent capacitance and a complex admittance $\tilde{Y}(\omega)$ with $|\tilde{Y}(\omega)| = I_0/V_0 = \omega C \neq f(V_0)$ (Figure 9-10c) as for example dielectric sensors, and capacitance sensors with a voltage-dependent capacitance $C = f(V)$ and hence a non-linear admittance $\tilde{Y}(\omega) = f(V)$ as for example field-effect sensors based upon metal/oxide/semiconductor [MOS] sandwich structures which will be discussed in detail in Chapter 10.

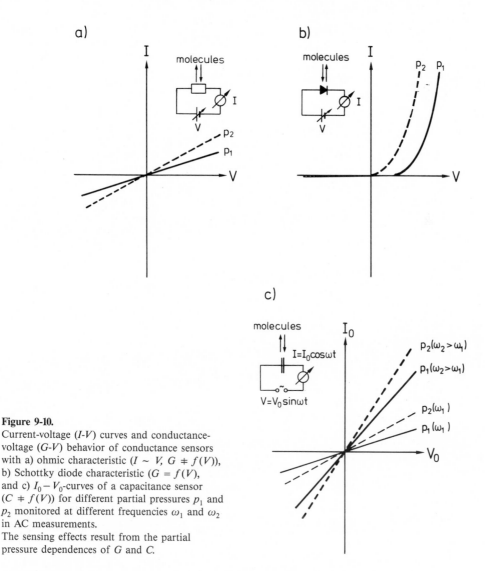

Figure 9-10.
Current-voltage (I-V) curves and conductance-voltage (G-V) behavior of conductance sensors with a) ohmic characteristic ($I \sim V$, $G \neq f(V)$), b) Schottky diode characteristic ($G = f(V)$, and c) $I_0 - V_0$-curves of a capacitance sensor ($C \neq f(V)$) for different partial pressures p_1 and p_2 monitored at different frequencies ω_1 and ω_2 in AC measurements.
The sensing effects result from the partial pressure dependences of G and C.

An alternative classification of conductance sensors is based upon the type of charge involved. If electrons or holes are the predominant charge carrier we use the term *electronic conductance sensor*. If electrons or holes as well as ions are involved, we use the term *mixed conductance sensors*. Sensors utilizing only ionic conductance are called *ion conductance* or *solid-state electrochemical sensors* (see Chapters 5 and 8).

Another alternative classification is based upon the type of the signal-determining elementary interaction process. We distinguish between *physisorption, chemisorption, catalytic reactions,* or *bulk defect equilibria* of the molecules to be detected by the chemically sensitive material. Models and physical-chemistry aspects of these elementary steps of detection have been treated in more detail in Chapter 4.

Typical experimental setups to determine conductances and capacitances of thin films are shown in Figure 9-11. Partial-pressure depending conductance changes parallel to the surface A may be monitored with the electrode arrangement shown in Figure 9-11 a. For quadratic shaped samples ($l^2 = A$), the conductance G is identical to the sheet conductance which is defined as

$$\sigma_\square = G \cdot \frac{l^2}{A} = R^{-1} \cdot \frac{l^2}{A} = \Delta\sigma + \sigma_b \cdot d \tag{9-5}$$

with $\Delta\sigma$ as surface conductivity in Ω^{-1}, σ_b as bulk conductivity in $\Omega^{-1}\mathrm{cm}^{-1}$, and d as thickness (see Chapter 4.5) [67].

Conductances of thin layers may be determined unequivocally by eliminating contact influences in a van-der-Pauw-arrangement (Figure 9-11 b) with point-like contacts [68]. In this case

$$G = I/V = (\pi/\ln 2) \cdot \sigma_\square \tag{9-6}$$

holds. For low-conductivity materials, structures with interdigital or comb electrodes (Figure 9-11 c) may be used in order to monitor conductances with low-cost instruments.

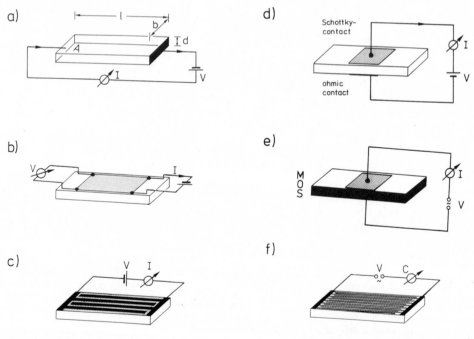

Figure 9-11. Survey on typical experimental setups for the determination of conductances and capacitances. a) plane-parallel contacts, b) van-der-Pauw arrangement, c) interdigital electrodes, d) Schottky diodes, e) metal-oxide-semiconductor MOS sandwich structures, and f) microstructurized interdigital condenser (third dimension of contacts more important if compared with c)).

Conductance changes perpendicular to the surface A are monitored in Schottky diode sensors with one Schottky- and one ohmic contact (Figure 9-11 d). To achieve large (partial) sensitivities (ie, large values $\gamma = \delta G / \delta p$, see Chapter 1.5), porous or permeable Schottky contacts are used with a large three phase boundary between the metal/semiconductor-interface and the gas phase. As shown in Figure 9-11 e, Schottky diode sensors are similiar to MOS sensors with a non-conducting oxide (O) as intermediate layer between the metal (M) and semiconductor (S) [69]. The latter use the change of the voltage-dependent capacitance C as sensor signal which may be determined in AC measurements of the admittance with different DC bias.

Capacitance changes of chemically sensitive samples may be monitored in an arrangement with plane-parallel contacts (compare Figure 9-11 a). Larger sensitivities are obtained with microstructurized interdigital structures (Figure 9-11 f) with optimized geometric dimensions of the combs, eg, small distances in the 1-10 µm-range and larger contact heights if compared with the structure Figure 9-11 c and large contact lengths in the 10 mm-range [6].

9.4.2 Equivalent Circuits

As already pointed out schematically in Figures 9-8 and 9-9 and discussed in more detail above, the different sensor types may be characterized by the frequency-dependent complex overall-admittance $\tilde{Y}_{tot}(\omega)$ or the reciprocal complex overall-impedance $\tilde{Z}_{tot}(\omega)$ (see Equation (9-4b)) as the electrical sensor response upon interaction with molecules. As shown in Figure 9-12, the frequency-dependent electronic signal measured as sensor response during gas exposure can be formally represented by different RC-equivalent circuits (Figures 9-8, and 9-9) with parallel and series arrangements of capacitances C, resistances $R = G^{-1}$, and Warburg impedances $\tilde{Z}_{WB}$ [72]. They may be attributed to frequency-dependent transport and relaxation processes of mobile and localized electrons and ions at contacts, at three phase boundaries, in the bulk, at the surfaces, and/or at grain boundaries. For a general improvement of sensor performances, one of these different frequency-dependent invidual contributions

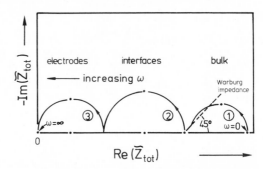

Figure 9-12. Complex impedance $\tilde{Z}_{tot} = \mathrm{Re}\,(\tilde{Z}_{tot}) + \mathrm{i} \cdot \mathrm{Im}\,(\tilde{Z}_{tot})$ of conductance and capacitance sensors with real (Re) and imaginary (Im) part for different frequencies. Different semicircles are determined by contributions from different charge transport and relaxation processes of dipoles, electrons, and/or ions. Warburg impedances (dotted line) result from diffusion processes and lead to straight lines with 45° slope [70]. Formally inductive effects with $\tilde{Z} = \mathrm{i}\omega L$ may also be observed in polycrystalline conductance sensors based upon SnO_2 in the presence of reducing gases [71]. The different contributions $\tilde{Z}_{tot}$ vary drastically for different sensors.

has to be utilized and optimized exclusively and all the others have to be suppressed or kept constant.

For conductance sensors with linear *I-V-*curves operated under DC conditions, ie, at a frequency $\omega = 0$, the ohmic part is measured only and $\tilde{Y}_{tot}$ is determined by the

- surface conductivity $\Delta\sigma$ in *chemisorption sensors* which utilize partial-pressure dependent changes of $\Delta\sigma = f(p)$ with $\sigma_b = $ const (compare Chapter 4.5)
- bulk conductivity σ_b in *bulk defect sensors* which utilize changes of $\sigma_b = f(p)$ with $\Delta\sigma = 0$ (compare Chapter 4.7), and
- grain boundary conductance in *microcrystalline* and *grain boundary sensors* (compare Chapter 4.8).

For conductance sensors with nonlinear *I-V-*curves operated under DC conditions, $\tilde{Y}_{tot}$ is given by the

- interface conductance in *interface* and *three phase boundary sensors* (compare Chapter 4.9).

For capacitance sensors utilizing interdigital electrode structures, the impedance $\tilde{Z}_{tot}$ is given by the geometric capacitance C_{IDC}, the Cole-Cole-impedance of the bulk ($\tilde{Z}_{CC}$), the capacitance C_{DL} due to a charge accumulation at the electrodes producing a dipole layer, and $\tilde{Z}_{WB}$ as Warburg impedance due to ion diffusion processes [70, 72].

9.5   Specific Data of Typical Conductance and Capacitance Sensors

9.5.1   SnO$_2$-based Ceramic Sensors

The most important type of sensors with respect to practical applications is the SnO$_2$-based Tagushi-sensor which utilizes the change of surface and/or grain boundary conductance for the detection of reducing gaseous compounds in air at operation temperatures around 350 °C. Since its first commercial introduction in 1968, different types have been developed and optimized empirically (Table 9-4) for different industrial and household applications.

Typical results of the partial-pressure dependence of the DC conductance for different compounds in air are given in Figure 9-13.

A serious problem is the cross-sensitivity particularly to H$_2$O which at low humidity values usually also affects the sensitivity to the compound to be detected. This is demonstrated in Figure 9-14 for the detection of CO in air. A strong reduction of CO sensitivity is observed at low partial pressures p_{H_2O}.

Their fast response is of great importance in warning systems. As a typical result, Figure 9-15 shows the time-dependence of the DC-conductance G of CO-sensitive sensors in synthetic humid air (50% relative humidity, r. h.) during repeated exposures to CO at different partial pressures $30 \le p_{CO} \le 300$ ppm. The response times are in the order of several minutes.

Table 9-4. The different types of Tagushi sensors and their specifications [1].

Type of sensor *former types*	Compound (decreasing sensitivity for identical partial pressures)	Detection range
TGS 109, TGS 109 M TGS 813, *TGS 813 C,*	$Iso - C_4H_{10} > C_3H_8 > CH_4 > CH_3 > OH > H_2 \gg CO$	500–10 000 ppm
TGS 816, *TGS 911*	$H_2 > iso - C_4H_{10} > CH_4 > C_2H_5OH \gg CO$	500–10 000 ppm
TGS 812, TGS 817	$C_2H_5OH > H_2 > iso - C_4H_8 > CO$	500–5000 ppm
TGS 711	$C_2H_5OH > H_2 > CO$	50–500 ppm
TGS 712 D	$C_2H_5OH > H_2 > CO$	20–200 ppm
TGS 100, TGS 800	$C_2H_5OH > H_2 > iso - C_4H_8 > CO$	1–300 ppm
TGS 203	$CO \gg H_2 \gg C_2H_5OH$	50–1000 ppm
TGS 822, TGS 823	$CH_3COCH_3 \approx C_2H_5OH \approx C_6H_6 \approx n - C_6H_{14} >$ $iso - C_4H_8 > CO \gg CH_4$	50–5000 ppm
TGS 814 D, TGS 824	NH_3	30–300 ppm
TGS 825	$H_2S > H_2$	5–100 ppm
TGS 880, TGS881	Volatile compounds during cooking of foods	
TGS 830	Freon 113	500–10 000 ppm
TGS 831	Freon 21, freon 22	500–10 000 ppm

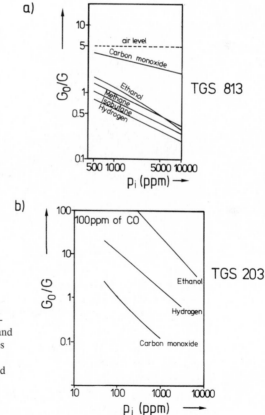

Figure 9-13.
Typical results of the inverse relative conductance G_0/G of Tagushi sensors TGS 813 a) and TGS 203 b) as a function of partial pressures of different gases in air [1]. A filter of the TGS 203 sensor consists of active carbon and reduces the cross-sensitivity to NO_2. Additionally, this type may be operated with modulated temperature.

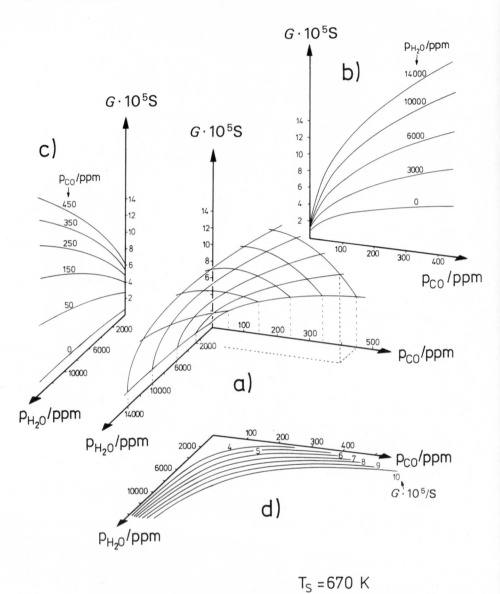

$T_S = 670$ K

Figure 9-14. Conductance G of a TGS 812 as a function of partial pressures p_{CO} and p_{H_2O} for a sensor temperature $T_S = 670$ K. a) Three dimensional plot of G depending on p_{CO} and p_{H_2O}. b) Projection on the $G - p_{CO}$ plane. c) Projection on the $G - p_{H_2O}$ plane. d) Projection on the $p_{CO} - p_{H_2O}$ plane to illustrate the effect of cross-sensitivity [73].

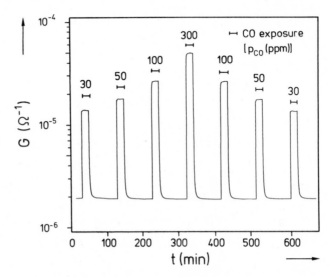

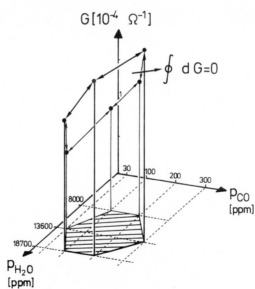

Figure 9-15. Time-dependent changes of the conductance G of a TGS 812 sensor in synthetic air (50% r.h.) and during exposure of CO with different partial pressures p_{CO} at constant temperature [73].

The conductance values G of different invidual sensors of the same type scatter because of the relatively simple and cheap manufacturing process. The conductance G of each invidual sensor, however, is a state function in the thermodynamic sense if variations of the partial pressures are kept within well-defined ranges at a constant temperature (see Chapter 4.2). A typical example is given in Figure 9-16 with $30 \leq p_{CO} \leq 400$ ppm and $8000 \leq p_{H_2O} \leq 18.700$ ppm. Large drifts of G and hence deviations from its state function behavior are observed in dry air and under strongly reducing conditions.

Figure 9-16.
The conductance G should be a state-function in the thermodynamic sense. This example shows the conductance G of a TGS 812 sensor as a function of partial pressures of CO and H_2O at a temperature $T = 760$ K [73]. Within the hatched area, $\oint dG = 0$ holds and the sensor operates completely reversible.

Within these limits of reversibility, several analytical expressions may be used to describe conductances as functions of partial pressures of different gas components. As an example, Clifford described the sensor response to the reducing gases CH_4 and CO in humid air by

$$G/G_0 = [p_{O_2}/p_{O_2}^0]^{-1} \cdot [1 + K_{CH_4} \cdot p_{CH_4} + K_{H_2O} \cdot p_{H_2O} + K_{CO} \cdot p_{H_2O} \cdot p_{CO} + K'_{CO} \cdot p_{H_2O} \cdot p_{CO}^2]^\beta$$

(9-7)

with G_0 as conductance in air, p_{O_2} as partial pressure of oxygen, $p_{O_2}^0$ as partial pressure of oxygen in an air reference, p_i as partial pressures of the different components i and k in the gas phase, K_{CH_4}, K_{H_2O}, K_{CO}, and K'_{CO} as characteristic parameters related with the rate constant for a particular reaction type [74]. Clifford found β to be temperature-dependent with typical values between 0.15 and 0.6.

For characteristic partial pressure ranges the Equation (9-7) may be simplified. This is particularly useful, if mixtures of only two or three gases are considered at a constant temperature for which

$$G/G_0 = \prod (p_i/p_{i,0})^{n_i}$$

(9-8)

often holds to a good approximation. Here, $p_{i,0}$ is a reference partial pressure of the gas component i with a value within the range of the state function characteristics of the conductance (see Figure 9-16). The value n_i is a characteristic parameter with typical values summarized in Tables 9-5 and 9-6. In CO/CH_4 mixtures, the n_{CH_4}-value depends also on the CO partial pressure which complicates the simple Equation (9-8) [75].

Important and also practically useful is the possibility to adjust an optimized operation temperature of SnO_2-based sensors for specific measuring conditions. Because of the strong temperature-dependence of equilibria and time constants of the different reactions involved in the detection of molecules, all sensor parameters such as their sensitivity, specificity, response and decay times depend significantly on the temperature. As an example, maxima of the sensitivity are observed for the detection of single components (Figure 9-17) which correspond to maxima in the catalytic turn-over rate of oxidation reactions of the detected molecules [77].

Table 9-5. Characteristic parameters of Equation (9-8) for differently doped sensors I and II [63].

Equation (9-8) is valid for 10 ppm $\leq p(CO)$, 100 ppm $\leq p(CH_4) \leq 20000$ ppm, and 7000 ppm $\leq p(H_2O) \leq 20000$ ppm.

	Gas	$G_0 \, [\Omega^{-1}]$	n_i		Remarks
Sample I (TGS 812)	CO	$8.04 \cdot 10^{-5}$	0.38		$p_{H_2O} = 13860$ ppm
	CO $\}$	$3.34 \cdot 10^{-4}$	$\{$ 0.47		$\}$ CO/H_2O-mixture
	H_2O		0.51		
Sample II (TGS 812)	CO $\}$	$4.54 \cdot 10^{-4}$	$\{$ 0.38	$\left(\dfrac{p_{CO}}{p_{CO,0}}\right)^{0.312}$	$\}$ CO/CH_4-mixture
	CH_4		0.028		$\}$ $p_{H_2O} = 13860$ ppm

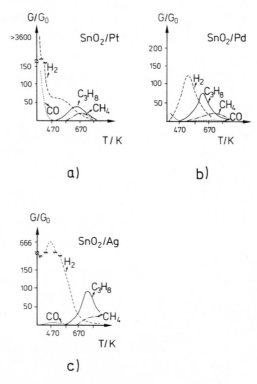

Figure 9-17.
Dependence of sensor signals (relative changes G/G_0) of polycrystalline SnO_2 sensors on the sensor temperature T for different noble metal dopants, a) Pt; b) Pd; c) Ag [76].

This temperature effect may therefore be used to improve the detection of a single compound. For different gases in mixtures their individual characteristic maxima are shifted or "smeared out" over a broader temperature range. This restricts the simple evaluation of signals from different sensors which are operated at different temperatures to obtain specific selectivity from a sensor array. It hence requires an additional pattern recognition evaluation for specific gas monitoring (see Chapter 6).

The sensitivity of SnO_2-sensors to detect specific molecules may also be improved by using metallic catalysts as dopants of SnO_2 (Figure 9-17). Only noble metals such as Pt and Au are thermodynamically stable under operation conditions at high oxygen partial pressures and temperatures between 300 and 400 °C. Two different models are commonly discussed to explain the role of surface dopants at supported catalysts to influence electronic charge transfer reactions: Metallic surface dopants may act as specific sites for the dissociation of O_2 and for the adsorption of molecules to be detected ("geometric factor") or they may pin the Fermi energy at the surface (Fermi energy control or "electronic factor") [78].

Most of the metals form oxides with the cations diffusing into the bulk of SnO_2. Usually, they form ternary compounds or, in the low concentration range act as additional donor or acceptor impurity states. This leads to a change in the conductance of the sensors if compared with the untreated ones and may be optimized with respect to specific gas sensitivities.

An overview about the influence of several metal salt dopants on the parameters n_i of Equation (9-8) of ceramic SnO_2 sensors is given in Table 9-6. Different models of the sensing mechanisms which must include the influence of percolation on the overall-conductance have been discussed for ceramic SnO_2 devices (see the schematic drawing of the percolation paths

Table 9-6. Characteristic parameters n_i (see Equation (9-8)) for chemically modified ceramic SnO_2 sensors. The values after the element symbols denote the amount of doping material in µg. For modified sensors denoted with X the Equation (9-8) is not valid [63].

Modified sensor	n_{CO}	n_{CH_4}	Modified sensor	n_{CO}	n_{CH_4}
Cr 5	0.38	0.32	Pd 1	0.51	0.36
Cr 10	0.34	0.28	Pd 5	0.52	0.40
Cr 20	0.43	0.32	Pd 10	0.20	0.15
Cr 50	0.23	X	Pd 20	0.18	0.15
Mn 10	0.58	0.36	Pt 5	0.57	0.48
Mn 20	0.49	0.36	Pt 10	0.72	0.61
Mn 50	0.59	0.40	Pt 20	0.57	0.43
Fe 10	0.45	0.32	Pt 40	0.11	0.15
Fe 20	0.46	0.32	Cu 1	0.41	0.28
Fe 50	0.40	0.32	Cu 5	0.46	0.32
Co 5	0.48	0.34	Cu 10	0.38	0.32
Co 10	0.48	0.34	Cu 20	0.36	0.30
Co 20	0.49	0.32	Cu 50	0.36	0.32
Co 50	0.45	0.40	Cu 100	0.39	0.31
Co 100	0.54	0.40	Ag 10	0.48	0.30
Co 200	0.11	0.11	Ag 20	0.46	0.30
Rh 5	0.50	0.32	Ag 50	0.53	0.34
Rh 10	0.46	0.35	Au 1	0.59	0.36
Rh 20	X	X	Au 5	X	0.45
Ni 10	0.48	0.30	Au 10	X	0.41
Ni 20	0.45	0.32	Au 20	0.11	0.06
Ni 50	0.41	0.32	SO_2	0.41	0.39
Ni 100	0.41	0.30			
			TGS 812	0.47	0.33
			TGS 813	0.33	0.40

in Figure 9-18). From X-ray diffraction [73] as well as from transmission electron microscopy (TEM) [79] investigations, small SnO_2-crystallites are known to be predominant and to lead to high specific surface areas with high sensor sensitivities (Table 9-7).

In spite of their complicated geometric and chemical structure, these sensors have surprisingly good and stable sensor properties for the detection of single components. This most probably results from the existence of small crystallites with sizes $\bar{l} < L_D/2$ (see Figure 9-16) and hence a homogeneous charging of crystallites upon gas adsorption. The role of the SiO_x-network on the electronic sensor properties and on the gas transport of detected molecules is not understood up to now.

There is an increasing interest to improve the device performances in general and to reduce the cross-sensitivity in particular by strategies discussed above in Section 9.2.2. Specific examples may illustrate recent success in this area:

— In contrast to the direct deposition of catalytically active material at the SnO_2 surface or in the bulk, the catalytically active part and the SnO_2 sensor may be separated as shown

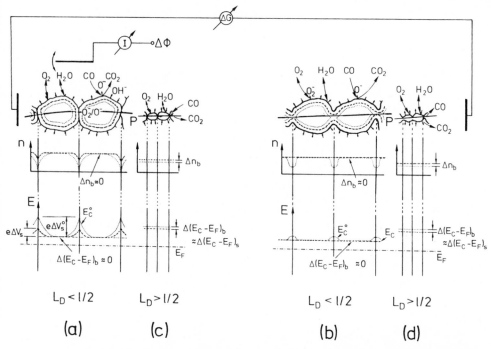

Figure 9-18. Grain-boundary conductivity sensor effect in SnO$_2$ ceramic samples and its atomistic understanding in a band scheme: The effect of O$_2$ (.....) and subsequent additional CO (---) exposure on grain boundary barriers is measured by changes in conductances ΔG and work functions $\Delta\Phi$. Also illustrated are corresponding changes in the electron concentration n and in the energy E of conduction band electrons over the conduction path "P" through the grains. E_C^0 denotes the conduction band edge in SnO$_2$, $e\Delta V_s^0$ the band bending at the grain boundaries in air, and E_F the Fermi energy.

After chemisorption of acceptor-type molecules, the following simplified cases are indicated: Case a) holds for a Schottky-barrier controlled conductance in non-sintered SnO$_2$: Here, changes of the band bending by a value $e\Delta V_s$ at the grain boundaries occur without variation of the bulk value of E_C, ie, $\Delta(E_C - E_F)_b \approx 0$ and $\Delta n_b \approx 0$. Case b) holds for an ohmic-contact-controlled conductance through sintered SnO$_2$ with grains larger than the Debye length of the electrons ($L_D < l/2$). Here, E_C^0 changes only at the necks and $\Delta(E_C - E_F)_b \approx 0$ holds. Cases c) and d) hold for grains smaller than the Debye length of the electrons ($L_D > l/2$) for non-sintered and sintered SnO$_2$-grains, resp. Here, no difference exists between the bulk and surface position of E_C, ie, $\Delta(E_C - E_F)_b \approx \Delta(E_C - E_F)_s$ holds. In these cases c) and d) the concentration of electrons n_b does not differ between pressed and sintered grains.

for the instrument in Figure 9-19. Here, organic molecules in the gas phase are decomposed at the catalytically active filament. The use of different filaments and different catalyst temperatures makes it possible to improve the sensitivity for specific components. Additionally, switching the gas stream between filament chamber and the bypass makes it possible to evaluate the response characteristics of the sensor. This setup had first been designed for the use of electrochemical sensors [82]. As an example, catalytically produced molecular fragments during the decomposition of chlorinated hydrocarbons are now successfully being monitored with SnO$_2$ sensors to detect selectively perchloroethylene [81].

Table 9-7. Thermal sintering properties and gas sensing properties of doped SnO_2; SA = specific area, $\bar{l}$ = mean diameter of grain [80]

Dopant 5%	600 °C		900 °C			Sensitivity (300 °C)	
	SA (m^2/g)	$\bar{l}$ Å	SA (m^2/g)	$\bar{l}$ Å	CO 800 ppm	H_2 800 ppm	i-C_4H_{10} 1000 ppm
None	24	138	6	272	10	21	9
Li	38	70	4	–	–	–	–
Na	44	100	5	–	43	79	40
K	54	92	5	330	42	38	20
Rb	55	78	–	–	–	–	–
Cs	62	90	12	262	–	–	–
Mg	52	52	22	125	10	120	30
Ca	40	54	20	110	66	246	110
Ba	54	57	33	86	44	162	58
Sr	58	51	23	99	35	151	71
V	51	67	6	382	5	6	32
Cr	41	71	–	99	1	1	1
Mn	37	61	24	140	1	2	3
Fe	39	72	20	182	1	1	1
Co	28	71	6	319	1	1	1
Ni	44	58	23	131	54	155	246
Cu	29	73	4	420	10	8	25
Zn	48	57	21	118	35	166	195
Ga	35	94	21	176	26	56	58
Nb	48	55	31	100	50	57	7
Mo	36	63	23	87	3	6	9
In	42	71	26	99	12	34	44
La	41	90	24	96	37	100	50
Ce	35	67	24	101	–	–	–
Pr	41	64	26	–	–	–	–
Nd	35	70	25	–	35	110	79
Sm	38	56	24	83	12	63	66
Gd	38	59	29	–	–	–	–
W	55	66	29	93	151	513	631
Tl	30	125	9	350	–	–	–
Pb	70	59	28	174	26	60	25
Bi	53	64	24	121	11	27	9
S	86	51	6	254	–	–	–
B	84	64	6	–	26	30	36
P	88	43	34	88	33	85	100
S-V	93	52	7	218	–	–	–
B-V	87	49	6	412	–	–	–
B-Ba	93	40	31	107	–	–	–
P-Ba	77	44	39	67	–	–	–
P-V	74	49	42	120	–	–	26

– An example for improving the sensor performances by cyclic variation of the operation temperature of the sensor itself is shown in Figure 9-20. Here, thin film sensors are used to detect selectively anesthetic agents such as ethylether, forane, halothane, and penthrane.

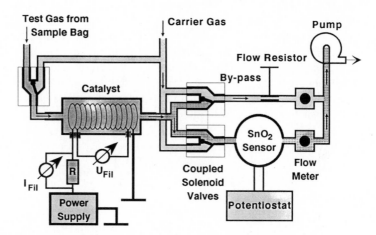

Figure 9-19. Experimental setup with a combination of different filaments (which at an optimized temperature act as catalysts to crack organic molecules) and an SnO₂-based sensor to, eg, monitor chlorinated hydrocarbons by their characteristic cracking patterns. The gas stream is switched between carrier and sample gas [81].

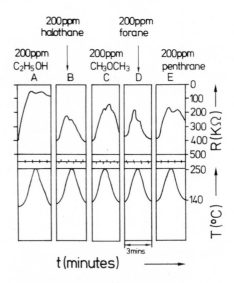

Figure 9-20. Detection of anaesthetic agents by cyclic changes of the sample resistance during variation of the sensor operation temperature ($140 \leq T \leq 250\,°C$) of SnO₂-based sensors. Characteristic resistance patterns are obtained for the different compounds [83].

9.5.2   SnO₂-Based Thin-Film Chemisorption Sensors

In this sensor specific chemisorption effects are monitored of those molecules from the gas phase which induce an electron transfer between the chemisorbed molecules and the sensor surface (see also Chapter 4.5). As a result, fast changes of the surface conductance $\Delta\sigma$ and of the work function $\Delta\Phi$ can be monitored with typical results shown in Figure 9-21 for a SnO₂ thin film structure to monitor NO₂ at temperatures $T \approx 200\,°C$.

The observed changes $\Delta\sigma_{eq}$ and $\Delta\Phi_{eq}$ are equilibrium values obtained after $t \gtrsim 100$ s by measuring the four-point conductance with point electrodes and the work function changes with an oscillating Kelvin probe. Both changes can be explained by a simple acceptor-type adsorption of NO₂ molecules at SnO₂ [85].

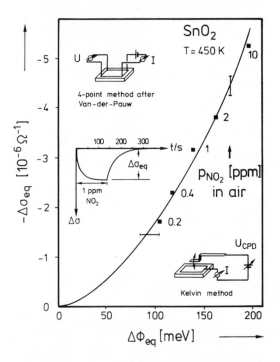

Figure 9-21.
Changes of the surface conductivity $\Delta\sigma_{eq}$ and of the work function $\Delta\Phi_{eq}$ during interaction of NO₂ with SnO₂ thin film sensors (thickness $d = 100$ nm) in air [84]. U$_{CPD}$ denotes the monitored contact potential voltage for $\Delta\Phi_{eq}$ measurements, I in the upper insert the monitored current for $\Delta\sigma_{eq}$ measurements.

Prototype devices which use only conductances as sensor signals lead to typical results as shown in Figure 9-22. A surprisingly low cross-interference to humidity and CO was found for these very thin film sensors of pure (ie, intentionally undoped) SnO₂.

In thin-film devices, concentrations and structures of grain sizes are mainly controlled by the evaporation rate and by the residual gas and substrate temperatures during and after film deposition. For thin film sensors in particular, the thickness of SnO₂ can be optimized towards a specific detection mechanism (Figure 9-23). As an example, low thicknesses should be used for NO₂ sensors based upon chemisorption effects.

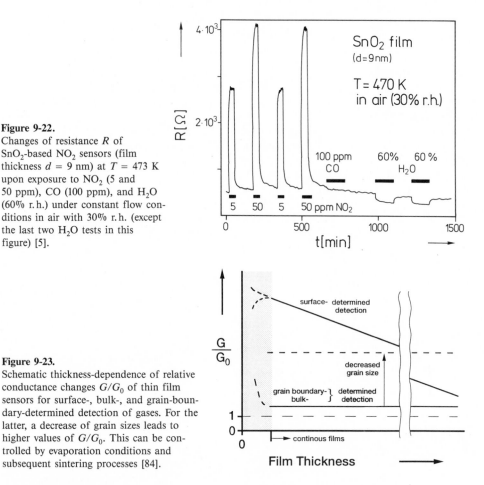

Figure 9-22.
Changes of resistance R of SnO$_2$-based NO$_2$ sensors (film thickness $d = 9$ nm) at $T = 473$ K upon exposure to NO$_2$ (5 and 50 ppm), CO (100 ppm), and H$_2$O (60% r.h.) under constant flow conditions in air with 30% r.h. (except the last two H$_2$O tests in this figure) [5].

Figure 9-23.
Schematic thickness-dependence of relative conductance changes G/G_0 of thin film sensors for surface-, bulk-, and grain-boundary-determined detection of gases. For the latter, a decrease of grain sizes leads to higher values of G/G_0. This can be controlled by evaporation conditions and subsequent sintering processes [84].

9.5.3   PbPc-Based Thin-Film Chemisorption Sensors

Extremely high sensitivities to detect NO$_2$ with $p_{NO_2} \geq 1$ ppb and to detect Cl$_2$ with $p_{Cl_2} \geq 200$ ppb with lead phthalocyanine (PbPc)-based conductance sensors have been reported since several years [4]. Small cross-sensitivities are reported for H$_2$S, NH$_3$, SO$_2$, H$_2$, CH$_4$, and CO. The PbPc is an organic p-type semiconductor with a molecular ring structure shown in Figure 9-5. In the absence of O$_2$, ie, under ultra high vacuum conditions (total pressure $p_{tot} \leq 10^{-10}$ mbar, $p_{O_2} \leq 10^{-14}$ mbar), the specific conductivity of PbPc is low ($\sigma \approx 2.8 \cdot 10^{-5} \ \Omega^{-1}$ at $T = 20\,°C$) but increases drastically in air as a result of O$_2$ bulk doping. The latter interaction type of O$_2$ molecules with PbPc has been identified from thickness-dependent measurements of the sheet conductance $\sigma_\square$ of thin PbPc films prepared by thermal evaporation onto SiO$_2$(011) single crystals (see Chapter 4.5 and Figure 4-25) [86].

Typical results of the time-dependent conductances G during exposure to NO$_2$ in air are shown in Figure 9-24. They indicate an acceptor-type of NO$_2$ chemisorption at the surface which increases the defect electron concentration in the bulk PbPc. The existence of a surface effect was deduced from the thickness-dependent chemisorption measurements with thin-film PbPc sensors (see Chapter 4.5 and Figure 4-25).

The detection mechanisms of NO_2 and O_2 by surface and bulk reactions, respectively, make it possible to optimize the sensors' sensitivity by choosing thick or thin layers. By choosing two sensors with different thicknesses, cross-sensitivities between these two molecules can be eliminated mathematically.

Lead phthalocyanine has also been applied in room-temperature NO_2 sensors. Their preparation requires a well-defined annealing procedure of thin PbPc films [87]. The latter leads to drastic variations of the surface composition (inorganic oxides are formed) and hence of the microstructure with a formation of very small and stable crystallites (compare Figure 9-18, for $L_D > 1/2$ in p-conducting PbPc if compared to n-conducting SnO_2 shown in this figure). Response and decay times are in the order of minutes.

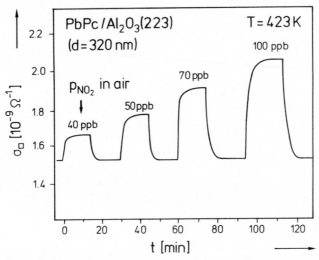

Figure 9-24. Time-dependent measurements of the sheet conductance of PbPc films during exposure to different partial pressures p_{NO_2} in air [86].

9.5.4   Sr TiO₃-, BaTiO₃-, and TiO₂-Based Bulk Defect Sensors

Polycrystalline bulk conductance sensors based upon binary and perovskite-type ternary metal oxides (compare Table 9-3 and Figure 4-19) can be used to determine the oxygen partial pressure p_{O_2} in the gas phase without a reference partial pressure. Typical devices are manufactured in thick or thin film technology with integrated heater and temperature sensor.

This type of conductance sensors is based upon a thermodynamically, ie, T- and p_{O_2}-controlled adjustment of the concentration of bulk point defects. The latter is coupled with the concentration of free electrons (see Chapter 4.7). Hence, the conductance G of non-stoichiometric oxides is given by

$$G \sim p_{O_2}^{m} \cdot \exp\left[-E_a/k_B T\right] . \tag{9-9}$$

Here, m characterizes ranges of p_{O_2} with one predominant type of defect out of the different possible types of charged point defects in the bulk. Usually, m varies from -0.25 to 0.25 (compare Table 4-2). The activation energy E_a summarizes the effects from activation energies of electron generation and corresponding mobilities. Typical results of $G(p_{O_2})$ are given in Figure 9-25 for BaTiO$_3$ ceramics at different temperatures.

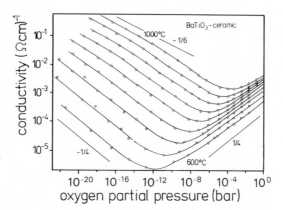

Figure 9-25.
Bulk conductivity σ_b of BaTiO$_3$-based sensors as a function of p_{O_2} at different temperatures [9] with different m-values between $-1/4$ and $+1/4$.

Depending on the sample preparation and hence the total concentration of acceptors, transitions from n-type to p-type semiconducting metal oxide are observed for oxygen partial pressures below around 10^2 bar. As a result, specific minima in the overall conductivities and hence nonequivocal correlations between p_{O_2} and G occur.

The response times of oxygen sensors are mainly determined by the bulk diffusion of both, electronic and ionic defects ("ambipolar diffusion") which is quantitatively given by the chemical diffusion coefficient $\tilde{D}$. The latter contains the self-diffusion coefficients of all mobile species. Typical response times for BaTiO$_3$ are $t_{90} \le 1$ s for $T \ge 873$ K. The ambipolar diffusion mechanism leads to a strong influence of the acceptor dopant concentration and in addition of the temperature- and partial-pressure-dependent concentration of intrinsic defects, ie, oxygen vacancies $V_O^{\cdot\cdot}$ on the absolute values of $\tilde{D}$ [88].

Due to the good stability of these oxides, bulk conductance sensors can be used to measure oxygen over a large range up to 25 decades of p_{O_2} even at high temperatures of $T \le 1473$ K with short response times and low cross sensitivities to other gases including H$_2$O.

The main disadvantages of bulk defect sensors are the strong influence of the temperature on the conductance and the nonequivocal sensor response around the conductance minimum. This minimum may be shifted by the temperature and may be adjusted by the extrinsic doping level. This makes possible the unequivocal application of these O$_2$ sensors in a specific range of p_{O_2}.

9.5.5   Pt/TiO$_2$-Based Schottky Diode Sensors

Schottky diode sensors utilize the partial-pressure-dependence of the (non-linear) interface conductance between semiconductors such as TiO$_2$ and CdS and noble metals such as Pd, Pt, and Au (compare Table 9-3).

As a specific example, Figure 9-26 shows the current-voltage-(I-V)-curves of a Schottky diode based upon Pt/TiO$_2$ for different O$_2$ partial pressures at constant temperatures. It consists of Schottky-type contacts (Pt/TiO$_2$) formed by low temperature evaporation of Pt onto rutile-phase TiO$_2$ [89]. The ohmic contacts consist of indium, zirconium, or platinum evaporated at high-temperatures.

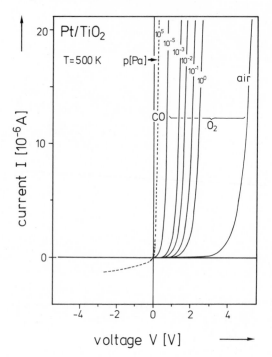

Figure 9-26.
Current-voltage-curves of Pt/TiO$_2$-junctions with ohmic Zr back contacts (the latter show Schottky behavior) for different oxygen partial pressures p_{O_2} at $T = 500$ K (solid curves) and for pure carbon monoxide (dashed curve).

Typical time-dependent measurements of the relative conductance I/I_0 at constant voltage V during the exposure to different CO-partial pressures are shown in Figure 9-27. Alternatively, the change of the voltage V may be monitored at a constant current I.

The voltage-dependent conductance of the Pt/TiO$_2$/Zr device (given by the slope dI/dV) is mainly determined by the geometric and electronic structure of the Pt/TiO$_2$ interface and of the TiO$_2$ (sub-)surface. A pinning occurs of the Fermi energy E_F relative to the conduction band E_C in an accumulation layer of TiO$_2$ at the ohmic back contact over the entire voltage range. This is essential because it leads to the required ohmic properties of the back contact.

As indicated in Figure 9-28, a negative voltage applied to the Pt contact which is equal to the contact potential difference V_{CPD} corresponds to the thermodynamic equilibrium under which no electrical current flows through the system. The higher work function Φ_{Pt} if compared with Φ_{TiO_2} leads to an intrinsic band bending $e\Delta V_s$ at the TiO$_2$ surface and hence to a Schottky barrier height in the absence of an applied voltage

$$\Phi_{SB} = e\Delta V_s + (E_C - E_F)_b = \Phi_{Pt} - \chi_{TiO_2} \tag{9-10}$$

between Pt and TiO$_2$. Here, χ_{TiO_2} denotes the electron affinity and $(E_C - E_F)_b$ the bulk position of the conduction band edge E_C relative to the Fermi energy E_F.

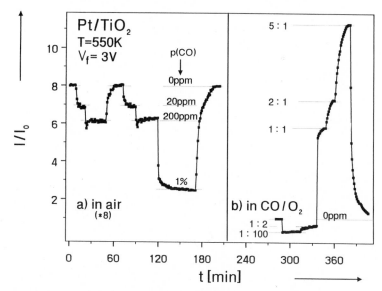

Figure 9-27. Typical response curves (relative current I/I_0 vs. time t with I_0 as conductance in air) monitored at a) an external voltage $V = 3$ V of Pt/TiO$_2$ Schottky-type sensors upon exposure to CO in humid synthetic air ($p_{H_2O} = 50\%$ r. h.) and b) for different CO/O$_2$-mixtures at a constant total pressure $p_{tot} = 10^5$ Pa [89].

As shown schematically in Figure 9-28 by dotted lines, flat-band conditions are adjusted by an external voltage which compensates for the work function difference $\Phi_{Pt} - \Phi_{TiO_2} \approx 0.9$ V. This value holds for reduced TiO$_2$ single crystals with $\Phi_{TiO_2} \approx 4.6$ eV. A constant position of the Fermi energy across the TiO$_2$ bulk even for externally applied voltages can be assumed due to the fact that the resistance R of the TiO$_2$ is by far lower than the total resistance of the Pt/TiO$_2$ structure. In the forward voltage regime, ie, with a positive voltage applied to the platinum, an accumulation layer is formed in TiO$_2$ (Figure 9-28, dashed lines) thereby leading to large interface conductances (compare Figure 9-26). In this region, the thermionic emission current is most probably enhanced by physical processes such as tunneling or recombination currents.

Chemisorption of gases such as O$_2$ and CO leads to interface states which change the work function at the three phase boundary. The latter can be monitored by a shift of the *I-V*-curves.

9.5.6 Polysiloxane-Based Capacitance Sensors

Chemically modified polysiloxanes which are optimized for the detection of gaseous compounds such as SO$_2$ and NO$_2$ may be utilized as sensitive layers for capacitance sensors [6]. Typical thin film devices are fabricated from quartz wafers with structurized interdigital electrodes and optimized geometric dimensions, ie, lengths and distances between both electrodes. After bonding the structure, the polysiloxane-layers with typical thicknesses of 1 μm are deposited by spin-on coating of different oligomeric co-polymers (see Table 9-3).

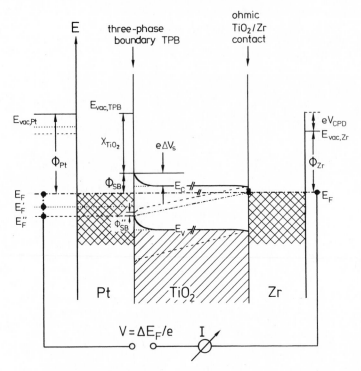

Figure 9-28. External voltage V as sensor effect and its atomistic understanding in a band scheme of the Pt/TiO$_2$/Zr structure with depletion in the TiO$_2$ subsurface layer (reverse voltage region $V = \Delta E_F \leq 0$) at the Pt/TiO$_2$ interface and with weak accumulation at the ohmic Zr/TiO$_2$ contact. $E_{vac,i}$ denotes the vacuum energy, E_F the Fermi energy, $\Phi_{Pt(TiO_2)}$ the work functions of the platinum (titanium dioxide), E_V and E_C the valence and conduction band edges of semiconducting TiO$_2$, respectively, and χ_{TiO_2} the electron affinity at the Pt/TiO$_2$ interface. Compensation of the band bending $e\Delta V_s$ leads to flat-band conditions (dotted lines) with a decreased Fermi energy E_F' of the Pt. In the forward voltage regime (dashed lines and E_F''), the Schottky barrier Φ_{SB} decreases and the voltage decay in TiO$_2$ becomes significant [89].

Typical results of time- and partial pressure-dependent measurements of the capacitance during exposure to different partial pressures p_{SO_2} of sulfur dioxide in nitrogen are shown in Figure 9-29.

The polysiloxane network may also be used as matrix for inorganic ions which form complexes with molecules from the gas phase. One example are Cu^{2+} ions which increase the sensitivity of the polysiloxane layer to NO$_2$ [6].

Chemically modified polysiloxanes with CN-groups make possible the detection of organic molecules such as n-hexane (C$_6$H$_{14}$) [90] with capacitance sensors.

These examples show that the application of organic compounds as materials for chemical sensors provide a lot of new possibilities to design sensors with specific adsorption sites for the molecules to be detected. Some strategies such as the use of macrocyclic compounds with specifically synthesized cages for neutral molecules have already been described in Chapter 4.10.

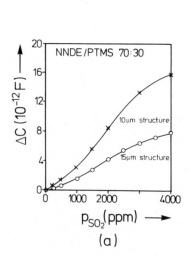
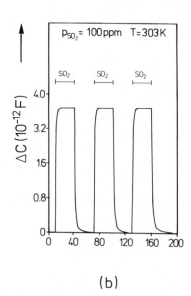

(a)   (b)

Figure 9-29. a) Capacitance changes ΔC of typical polysiloxane-based capacitance sensors as a function of the SO_2 partial pressure p_{SO_2} in N_2 [6]. b) Typical response curves during repeated exposures to SO_2.

9.6   References

[1] Figaro, Inc., Technical Report, 1990.

[2] Yamamoto, N., Tonomura, S., Matsuoka, T., Tsubomura, H., "A Study of a Palladium-Titanium Oxide Schottky Diode as a Detector for Gaseous Compounds", *Surf. Sci.* **92** (1980) 400–406.

[3] Yokomizo, Y., et al., "Microstructure and Humidity-sensitive Properties of $ZnCr_2O_4 - LiZnVO_4$ Ceramic Sensors" *Sens. Actuators* **4** (1983) 599.

[4] Bott, B., Jones, T. A., "A Highly Sensitive NO_2 Sensor Based on Electrical Conductivity Changes in Phthalocyanine Films", *Sens. Actuators* **5** (1984) 43–53.

[5] Schierbaum, K. D., et al., "Prototype Structure for Systematic Investigations of Thin-Film Gas Sensors", *Sens. Actuators* **B1** (1990) 171–75.

[6] Endres, H.-E., Kapazitive Gassensoren mit sensitiven Schichten aus Heteropolysiloxan, Thesis, 1989.

[7] Mokwa, W., Kohl, D., Heiland, G., "A SnO_2 Thin Film for Sensing Arsine", *Sens. Actuators* **8** (1985) 101–108.

[8] Kirner, U., et al., "Low and High Temperature TiO_2 Oxygen Sensors", *Sens. Actuators* **B1** (1990) 103–107.

[9] Schönauer, U., "Thick-Film Oxygen Gas Sensors Based on Ceramic Semiconductors", *Techn. Messen* **56** No. 6 (1989) 260.

[10] Brattain, W. H., Bardeen, J., *Bell Systems Tech. J.* **32** (1953) 1.

[11] Heiland, G., "Zum Einfluß von adsorbiertem Sauerstoff auf die elektrische Leitfähigkeit von Zinkoxydkristallen", *Z. Phys.* **138** (1954) 459.

[12] Seiyama, T., Kato, A., Fukiishi, K., Nagatini, M., "A New Detector for Gaseous Components Using Semiconductive Thin Films", *Anal. Chem.* **34** (1962) 1502–1503.

[13] Taguchi, N., *UK Patent Specification 1280809*, 1970.

[14] Oehme, F., *AMA-Tagung, Chemische und Biochemische Sensoren, Friedrichsdorf, FRG, 1987.*

[15] Takahata, K., "Tin Oxide Sensors – Development and Applications", *Chemical Sensor Technology* **1** (1988).

[16] Lüder, E., Kallfass, T., „Sensoren in Schichttechnik", *e & i* **107** (1990) 274.

[17] Demarne, V., Grisel, A., „An Integrated Low-Power Thin-Film CO Gas Sensor on Silicon", *Sens. Actuators* **13** (1988) 301–313.

[18] Bott, B., Jones, T. A., Mann, B., "The Detection and Measurement of CO Using ZnO Single Crystals", *Sens. Actuators* **5** (1984) 65–74.

[19] Egashira, M., Matsumoto, T., Shimizu, Y., Iwanaga, H., "Gas-Sensing Characteristics of Tin Oxide Whiskers with Different Morphologies", *Sens. Actuators* **14** (1988) 205–213.

[20] Lalauze, R., Bui, N., Pijolat, C., "Interpretation of the Electrical Properties of a SnO_2 Gas Sensor after Treatment with Sulfur Dioxide", *Sens. Actuators* **6** (1984) 119–125.

[21] Heiland, G., Kohl, D., "Problems and Possibilities of Oxidic and Organic Semiconductor Gas Sensors", *Sens. Actuators* **8** (1985) 227–233.

[22] Jones, A., Jones, T. A., Mann, B., Firth, J. G., "The Effect of the Physical Form of the Oxide on the Conductivity Changes Produced by CH_4, CO, and H_2O on ZnO", *Sens. Actuators* **5** (1984) 75–88.

[23] Yannopoulos, L. N., "Antimony-doped Stannic Oxide-based Thick-film Gas Sensors", *Sens. Actuators* **12** (1987) 77–89.

[24] Egashira, M., Kanehara, N., Shimizu, Y., Iwanaga, H., "Gas-Sensing Characteristics of Li^+-Doped and Undoped ZnO Whiskers", *Sens. Actuators* **18** (1989) 349–360.

[25] Oyabu, T., "Sensing Characteristics of SnO_2 Thin Film Gas Sensors", *J. Appl. Phys.* **53** (1982) 2785.

[26] Advani, G. N., Komem, Y., Hasenkopf, J., Jordan, A. G., "Improved Performance of SnO_2 Thin-Film Gas Sensors Due to Gold Diffusion", *Sens. Actuators* **2** (1981/82) 139–147.

[27] Yamazoe, N., Kurokawa, Y., "Catalytic Sensitization of SnO_2 Sensors", *Proceedings of the International Meeting on Chemical Sensors, Fukuoka, Japan* (1983) p. 35–40.

[28] Heiland, G., "Zur Theorie der Anreicherungsrandschicht an der Oberfläche von Halbleitern", *A. Phys.* **148** (1957) 15.

[29] Steele, M. C., MacIver, B. A., "Palladium/cadmium-sulfide Schottky Diodes for Hydrogen Detection", *Appl. Phys. Letters* **28** (1976) 687.

[30] Ghers, M., et al., "Oxygen Ionosorption on Compressed Semiconducting Powders of Zinc and Germanium Oxynitrides", *Sens. Actuators* **13** (1988) 263–273.

[31] Coles, G. S. V., Gallagher, K. J., Watson, J., "Fabrication and Preliminary Results on Tin(IV)oxide-based Gas Sensors", *Sens. Actuators* **7** (1985) 89.

[32] Willis, A. N., Silarajs, M., *U. S. Patent 4197089* (1980).

[33] Moseley, P. T., Williams, D. E., "A Selective Ammonia Sensor", *Sens. Actuators* **B1** (1990) 113–115.

[34] Prudenziati, M., Morten, B., "Thick-film Sensors: An Overview", *Sens. Actuators* **10** (1986) 65–82.

[35] Kobayashi, T., Haruta, M., Sano, H., Nakane, M., "A Selective CO Sensors Using Ti-doped α-Fe_2O_3 with Coprecipitated Ultrafine Particles of Gold", *Sens. Acutators* **13** (1988) 339–349.

[36] Arakawa, T., Tadaka, K., Tsunemine, Y., Shiokawa, J., "CO Gas Sensitivities of Reduced Perovskite Oxide $LaCoO_3$", *Sens. Actuators* **14** (1988) 215–221.

[37] Nitta, M., Haradome, M., "CO Gas Detection by ThO_2-doped SnO_2", *J. Electron. Mat.* **8** (1979) 571.

[38] Okayama et al., "Characteristics of CO Gas Sensor of Pt and Sb Dispersed SnO_2 Ceramics", *Proceedings of the International Meeting on Chemical Sensors, Fukuoka, Japan, 1983*.

[39] Lambrich, R., Hagen, W., Lagois, J., "Metal Oxide Films as Selective Gas Sensors", *Proceedings of the International Meeting on Chemical Sensors, Fukuoka, Japan, 1983*.

[40] Nitta, M., Haradome, M., "Oscillation Phenomenon in Thin-film CO Sensor", *IEEE Trans. Electron. Dev.* **26** (1979) 219.

[41] Esper, M. J. Logothetis, E. M., Chu, J. C., *SAE Automotive Eng. Congr. Series 790140, Detroit, Michigan, 1979*.

[42] Stetter, J. R., "A Surface Chemical View of Gas Detection", *J. Colloid. and Interface. Sci.* **65** (1978) 432.

[43] Lee, D.-D., Sohn, B.-K., "CO-Sensitive SnO_2/Pt Thick Film", in: *Proceedings of the 2nd International Meeting on Chemical Sensors, Bordeaux, France, 1986*.

[44] Hermans, E. C. M., "CO, CO_2, CH_4 and H_2O Sensing by Polymer Covered Interdigitated Electrode Structures", *Sens. Actuators* **5** (1984) 181–186.

[45] Haruta, M., Kobayashi, T., Sano, H., Nakone, M., "A Novel CO Sensing Semiconductor with Coprecipated Ultrafine Particles of Gold", in: *Proceedings of the 2nd International Meeting on Chemical Sensors, Bordeaux, France, 1986.*

[46] Bartlett, P. N., Ling-Chung, S. K., "Conducting Polymer Gas Sensors Part III: Results for Four Different Polymers and Five Different Vapours", *Sens. Actuators* **20** (1989) 287–292.

[47] Schultz, M., Bohn, E., Heiland, G., "Messung von Fremdgasen in der Luft mit Halbleitersensoren", *Tech. Mess.* **11** (1979) 450.

[48] Yamaguchi, S., *Mater. Chem.* **6** (1981) 505.

[49] Shaver, P. J., "Activated Tungsten Oxide Gas Detectors", *Appl. Phys. Lett.* **11** (1967) 255.

[50] Shiratori, M., Katsura, M., Tsuchiya, T., "Halogenated Hydrocarbon Gas Sensor", *Proceedings of the Intenational Meeting on Chemical Sensors, Fukuoka, Japan 1983.*

[51] Unwin, J., Walsh, P. T., "An Exposure Monitor for Chlorinated Hydrocarbons Based on Conductometry Using Lead Phthalocyanine Films", *Sens. Actuators* **18** (1989) 45–47.

[52] Jones, T. A., Bott, B., "Gas-induced Electrical Conductivity Changes in Metal Phthalocyanines", *Sens. Actuators* **9** (1986) 27–37.

[53] Nagai, M., Nishino, T., Saeki, T., "A New Type of CO_2 Gas Sensors Comprising Porous Hydroxyapatite Ceramics", *Sens. Actuators* **15** (1988) 145–151.

[54] S. R., Morrison, "Semiconductor Gas Sensors", *Sens. Actuators* **2** (1982) 329.

[55] Yu, Ch., Shimizu, Y., Arai, H., "Mg-Doped $SrTiO_3$ as a Lean-Burn Oxygen Sensor", *Sens. Actuators* **14** (1988) 309–318.

[56] Saji, K., Takahashi, T., Takeuchi, T., Igarashi, I., "Characteristics of TiO_2 Oxygen Sensor in Noneqilibrium Gas Mixtures", *Proc. of the Int. Meeting Chemical Sensors, Fukuoka, Japan, 1983* p. 185.

[57] Logothetis, E. M., Kaiser, W. J., "TiO_2 Film Oxygen Sensors Made by Chemical Vapour Deposition from Organometallics", *Sens. Actuators* **4** (1983) 333–340.

[58] Bethin, J., Chiang, C. J., Franklin, A. D., Snellgrove, R. A., *J. Appl. Phys.* **52** (1981) 4115.

[59] Arakawa, T., Takada, K., Tsunemine, Y., Shiokawa, J., "Characteristics of CO Detecting on The Reduced Perovskite Oxide $LaCoO_{3-x}$", *Proceedings of the 2nd International Meeting on Chemical Sensors, Bordeaux, France, 1986,* p. 115.

[60] Moseley, P. T., Williams, D. E., Norris, J. O. W., Tofield, B. C., "Electrical Conductivity and Gas Sensitivity of Some Transition Metal Tantalates", *Sens. Actuators* **14** (1988) 79–91.

[61] Williams, D. E., Moseley, P. T., "Dopant Effects on the Response of Gas-Sensitive Resistors Utilising Semiconducting Oxides", *J. Materials,* in press.

[62] Murakami, N., Yasunage, S., Ihokura, K.,"Sensitivity and Sintering Temperature of SnO_2 Gas Sensors, *Anal. Chem. Symp. Ser.* **17** (1983) 18.

[63] Weimar, U., Kowalkowski, R., Schierbaum, K. D., Göpel, W., "Pattern Recognition Methods for Gas Mixture Analysis: Application to Sensor Arrays Based upon SnO_2", *Sensors and Actuators* **B1** (1990) 93–96.

[64] Wiegleb, G., "Entwicklung eines Zinndioxid-Gassensors in Dickschicht-Technologie", Thesis, 1990, University of Witten/Herdecke, Witten, FRG.

[65] Schmitte, F.-J., "Eigenschaften von leitenden Zinndioxid-Schichten". Diploma Thesis, 1982, University of Aachen, FRG.

[66] Leskelä, M., Niinistö, L., "Chemical Aspects of the Atomic Layer Epitaxy (ALE) process, in: Suntola, T. and Simpson, M., Atomic Layer Epitaxy, Blackie, Glasgow and London, 1990.

[67] Göpel W., Lampe, U., "Influence of Defects on the Electronic Structure of Zinc Oxide Surfaces", *Phys. Rev.* **B22**, No. 12 (1980) 22.

[68] van der Pauw, L. J., *Philips Res. Rept.* **13** (1958) 1.

[69] Lundström, I., Söderberg, D., "Hydrogen Sensitive MOS-Structures Part 2: Characterization", *Sens. Actuators* **2** (1981/82) 105–138.

[70] Macdonald, J. R., *Impedance Spectroscopy;* Chichester: J. Wiley Sons, 1988.

[71] Auge, J., University of Magdeburg, Magdeburg FRG, Priv. Comm.

[72] Macdonald, J. R., "Simplified Impedance/Frequency Response Results for Intrinsically Conducting Liquids and Solids", *J. Chem. Phys.* **61** (1974) 3977–3996.

[73] Schierbaum, K. D., Weimar, U., Kowalkowski, R., Göpel, W., "Conductivity, Workfunction, and Catalytic Activity of SnO_2-Based Sensors", *Sens. Actuators,* **B3** (1991) 205–214.

[74] Clifford, P. K., "Microcomputational Selectivity Enhancement of Semiconductor Gas Sensors", in: *Proc. 1st Int. Meet. Chemical Sensors, Fukuoka, Japan, 1983*, p. 153.

[75] Schierbaum, K. D., Weimar, U., Göpel, W., "Multicomponent Gasanalysis: An Analytical Chemistry Approach Applied to Modified SnO$_2$-Sensors", *Sens. Actuators* **B2** (1990) 71–78.

[76] Yamazoe, N., Kurokawa, Y., Seiyama, T., "Effects of Additives on Semiconductor Gas Sensors", *Sens. Actuators* **4** (1983) 283–289.

[77] Göpel, W., "Chemisorption and Charge Transfer at Semiconductor Surfaces: Implications for Designing Gas Sensors", *Progr. Surface Sci.* **20** (1985), 9.

[78] Madou, M. J., Morrison, S. R., *Chemical Sensing with Solid State Devices;* San Diego: Academic Press, 1989.

[79] Yamazoe, N., "New Approaches for Improving Semiconductor Gas Sensors", in: *Proc. of the 3rd. Int. Meeting on Chemical Sensors, Cleveland, USA, 1990*.

[80] Xu, C., Tamaki, J., Miura, N., Yamazoe, N., "Influences of Additives on the Properties of SnO$_2$-Based Gas Sensors", *Intern. Symp. on Fine Ceramics, Arita, Japan, 1989*.

[81] Göpel, W., et al., "Sensor Array and Catalytic Filament of Chemical Analysis of Vapors and Mixtures", *Sens. Actuators* **B1** (1990) 43–47.

[82] Stetter, J. R., Jurs, P. C., Rose, S. L., "Detection of Hazardous Gases and Vapors: Pattern Recognition Analysis of Data from an Electrochemical Array", *Anal. Chem.* **58** (1986) 860–866.

[83] Wu, Q., Ko, W. H., "Micro-gas Sensor for Monitoring Anesthetic Agents", *Sens. Actuators* **B1** (1990) 183–187.

[84] Schierbaum, K. D., Weimar, U., Göpel, W., "Technologies of SnO$_2$-Based Chemical Sensors: Comparison Between Ceramic, Thick, and Thin Film Structures", *Conf. Proc. Sensor 91, Nürnberg, FRG, 1991*.

[85] Schierbaum, K. D., Wiemhöfer, H. D., Göpel, W., "Defect Structure and Sensing Mechanism of SnO$_2$ Gas Sensors: Comparative Electrical and Spectroscopic Results", *Solid State Ionics* **28–30** (1988) 1631–1636.

[86] Mockert, H., Schmeisser, D., Göpel, W., "Lead Phthalocyanine as Prototype Organic Material for Gas Sensors: Comparative Electrical and Spectroscopic Studies to Optimize O$_2$ and NO$_2$ Sensing", *Sens. Actuators* **19** (1989) 159.

[87] Sadaoka, Y., Jones, T. A., Revell, G. S., Göpel, W., "Effects of Morphology on NO$_2$ Detection in Air at Room Temperature with Phthalocyanine Thin Films", *J. Material Science* **25** (1990) 5257–5268.

[88] Müller, A., Härdtl, K. H., "Ambipolar Diffusion Phenomena in BaTiO$_3$ and SrTiO$_3$", *Appl. Phys.* **A49** (1989) 75–82.

[89] Schierbaum, K. D., Kirner, U., Geiger, J., Göpel, W., "Schottky-Barrier and Conductivity Gas Sensors based upon Pd/SnO$_2$ and Pt/TiO$_2$", *Sens. Actuators* **B4** (1991) 87–94.

[90] Haug, M., et al., "Controlled Selectivity of Polysiloxane Coating: Their Use in Capacitance Sensors", *Eurosensors* **V** Rome (1991).

10 Field Effect Chemical Sensors

Section 10.1 Device Principles
INGEMAR LUNDSTRÖM, Linköping Institute of Technology of Technology, Sweden

Section 10.2 Ion-sensitive Field Effect Transistors
ALBERT VAN DEN BERG, Centre Suisse d'Electronique et de Microtechnique SA, Neuchâtel, Switzerland,
BARTHOLOMEUS H. VAN DER SCHOOT, University of Neuchâtel, Switzerland,
HENDRIK H. VAN DEN VLEKKERT, PRIVA BV, De Lier, The Netherlands

Section 10.3 Field Effect Gas Sensors
MÅRTEN ARMGARTH, CLAES I. NYLANDER, Sensistor AB, Linköping, Sweden

Contents

10.1 Device Principles

Field effect devices are of two types: metal-insulator-semiconductor capacitors (MISCAPs) and transistors (MISFETs), as illustrated in Figure 10-1. The semiconductor is normally silicon and the insulator normally silicon dioxide. The structures are then called MOS devices. The interesting property of MIS devices is that the charge distribution at the insulator-semiconductor interface can be controlled by the potential on the metal.

For a p-type semiconductor as an example, a negative potential on the metal will accumulate (positive) holes at the semiconductor surface (Figure 10-2a). When the potential becomes positive, holes are pushed away from the semiconductor surface, and we are left with a depleted region at the surface (Figure 10-2b). If the positive potential is large enough, electrons start to accumulate at the semiconductor surface, forming a so-called inversion layer (Figure 10-2c). When inversion sets in, the depletion layer width remains constant with a further increase in the potential on the metal since the increase in the electric field is decaying over the insulator only. If we measure the (high-frequency) small signal capacitance of an MIS structure, we therefore obtain the result in Figure 10-1 b. For a MISFET, we obtain an I_D (V_G)

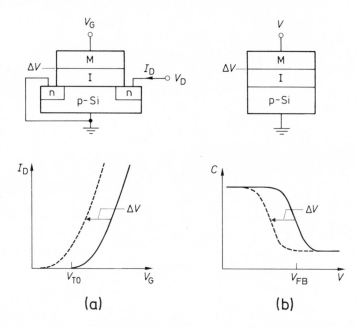

(a) (b)

Figure 10-1. Schematic illustration of metal-insulator-semiconductor structures and their electrical properties. The dashed lines indicate electrical characteristics with parallel shifted ΔV along the voltage axis. ΔV, V_{FB}, and V_T are discussed in the text. (a) MISFET and its I_D (V_G) curve (at a constant $V_D > 0$). The channel length L is the distance between the n regions. The channel width b is the dimension of the n region perpendicular to the plane of the paper. (b) MIS capacitor and its C (V) curve measured with a small AC voltage, v (at 1 MHz) superimposed on the DC potential, V.
C: capacitance, I_D: current in the channel (drain current), ΔV: voltage drop, V_{FB}: flatband voltage, V_G: gate voltage, V_D: drain voltage, V_{TO}: threshold voltage of the transistor.

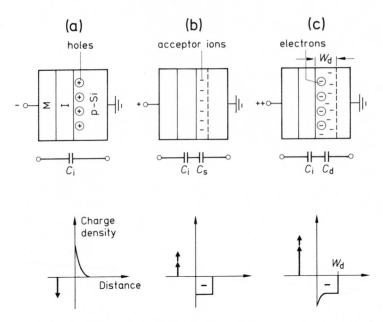

Figure 10-2. Schematic illustration of (a) accumulation of holes at the semiconductor surface. C_i: capacitance of the gate per unit area. (b) depletion of holes. C_i: capacitance of the insulator per unit area, C_s: capacitance of the depletion region. (c) accumulation of electrons (inversion). W_d: depletion layer width when inversion sets in, C_i: capacitance of the gate per unit area, C_d: capacitance corresponding to W_d. See the text for further description.

curve like that in Figure 10-1a. When inversion sets in, we obtain a conducting channel due to the electrons at the semiconductor surface. The principles of operation of field effect devices are therefore very simple and easy to understand. Fortunately, we do not have to elaborate on the mathematical details of the $C(V)$ and $I_D(V_G)$ curves to understand the chemical sensor uses of field effect devices. The physics of MIS structures are described in detail in [1] to which the interested reader is referred. We need, however, two key parameters, the so-called flatband voltage, V_{FB}, of an MIS structure and the so-called threshold voltage, V_T, of a MISFET.

The meaning of the flatband voltage is illustrated in Figures 10-3a and b for two cases: without and with charges in the insulator. The work functions of the metal, W_M, and the semiconductor, W_S, are defined in the diagrams. The flatband voltage, i.e., the potential difference between the metal and the semiconductor when the energy bands of the semiconductor are flat to the surface is

$$V_{FB} = (W_M - W_S)/q - \frac{Q_i}{\varepsilon_i} d_i .$$ (10-1)

Q_i is an equivalent insulator charge (per unit area) assumed to be located at the insulator-semiconductor interface, d_i the insulator thickness, and ε_i its dielectric permittivity. V_{FB} is often negative for a p-type Si MIS structure.

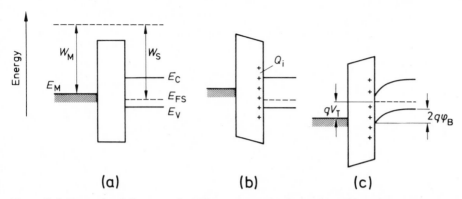

Figure 10-3. Energy band diagrams of a MIS structure under flatband conditions (for a p-type semiconductor). (a) Without charges in the insulator. W_M: work function of the metal, W_s: work
function of the semiconductor, E_M: Fermi energy of the metal, E_{FS}: Fermi energy of the
semiconductor, E_c: conduction band edge, E_v: valence band edge. (b) With charges in the
insulator. The charges are treated as an equivalent surface charge (Q_i) located at the insulator-semiconductor interface. (c) Energy band diagram when inversion starts. The electron concentration at the surface, n_s, is then equal to N_A. $2\varphi_B$ is the voltage drop across the
semiconductor surface. q: elementary charge, V_T: threshold voltage.

The threshold voltage of a MISFET is the gate voltage at which current starts to flow along
the semiconductor surface, which occurs when an inversion layer forms at the surface. This
is defined to happen when the concentration of electrons at the semiconductor surface, n_s,
is equal to the bulk concentration of holes ($= N_A^-$, the concentration of ionized acceptors).
We then have the situation in Figure 10-3 c, which means that to the flatband voltage we have
to add the voltage drop across the semiconductor, $2\varphi_B$, and the voltage drop across the insulator caused by the charge in the depletion region. This charge is $-qN_A^- W_d$, where W_d is
given by

$$W_d = \sqrt{\frac{\varepsilon_s 4\varphi_B}{qN_A^-}} \tag{10-2}$$

since the voltage drop across the depletion region is $2\varphi_B$. ε_s is the dielectric permittivity of
the semiconductor. Note that although the concentration of electrons at the semiconductor-
insulator interface is N_A^-, it drops away rapidly from the surface and does not contribute
significantly to the total charge in the semiconductor surface region. The extra voltage drop
across the insulator is $qN_A^- W_d \cdot d_i / \varepsilon_i$, and hence

$$V_T = \frac{W_M - W_S}{q} - \frac{Q_i}{\varepsilon_i} d_i + 2\varphi_B + \frac{d_i}{\varepsilon_i} \sqrt{4\varphi_B \varepsilon_s qN_A^-} \ . \tag{10-3}$$

V_T is therefore more positive than V_{FB} for an *n*-channel MISFET.
The definition of V_T implies that (for $V_D = 0$)

$$\sigma = \frac{\varepsilon_i}{d_i} (V_G - V_T) \tag{10-4a}$$

is the magnitude of the induced mobile charge (electrons) per unit area in the conducting channel at the semiconductor surface. If $V_D > 0$, the potential varies along the semiconductor surface, $V(x)$, where x is the distance from the source,

$$\sigma(x) = \frac{\varepsilon_i}{d_i}(V_G - V_T - V(x)).$$
(10-4b)

Further, the electric field along the conducting channel is $-dV/dx$. The current in the channel is

$$I_D = b \cdot \sigma(x) \mu \frac{dV}{dx}$$
(10-5)

where μ is the mobility of the electrons in the channel and b is the width of the channel. I_D must be constant along the channel and by integrating the equation above from $x = 0$ (the source) to $x = L$ (the drain), ie, from $V = 0$ to $V = V_D$, we find that

$$I_D = K_d\left((V_G - V_T)V_D - \frac{V_D^2}{2}\right)$$
(10-6)

where

$$K_d = \mu \frac{\varepsilon_i}{d_i}\frac{b}{L} \quad \text{or} \quad \mu C_i \frac{b}{L}.$$

$C_i \equiv \varepsilon_i/d_i$ is the capacitance of the insulator per unit area.

For $V_D = V_G - V_T$, the induced mobile charge $\sigma(x)$ becomes zero at the drain; the channel is said to be pinched off. If $V_D > V_G - V_T$, the pinch-off point (where $V(x) = V_G - V_T$) moves away from the drain and a depletion region occurs at the drain contact. Current continuity is sustained through a rapid transfer of the channel charge, which reaches the pinch-off point, across the depletion region. For $V_D > V_G - V_T$, the drain current remains constant and is given by

$$I_D = K_d\frac{(V_G - V_T)^2}{2}$$
(10-7)

which was obtained by replacing V_D with $V_G - V_T$ in Equation (10-6). The region $V_D > V_G - V_T$ is called the saturation region. A small increase in the current is observed when V_D is in the saturation region, because the effective channel length becomes smaller. A more thorough treatment of the transistor current can be found in [1]. Note that V_T is often negative for an n-channel FET, which means that a current flows also at $V_G = 0$.

FETs are sometimes used in a diode coupling where the gate and drain are connected. In this case, $V_D = V_G$ and for $V_T < V_D$ we find for the diode coupling

$$I_D = \frac{K_d}{2}V_D(V_D - 2V_T).$$
(10-8)

MIS capacitors (Figure 10-1b) are often used for exploratory work since they are easier to fabricate. It is also relatively easy to evaluate their electrical properties with the help of capacitance measurements as illustrated in Figure 10-1b.

The maximum capacitance (per unit area) is C_i. The capacitance at V_{FB}, the flatband capacitance per unit area, is given by [1]

$$\frac{1}{C_{FB}} = \frac{1}{C_i} + \frac{1}{C_S} \tag{10-9a}$$

$$C_S = \varepsilon_s \sqrt{\frac{q^2 N_A^-}{2 k_B T \varepsilon_s}} \tag{10-9b}$$

at flatband.

The minimum capacitance C_{min} (per unit area) is given by

$$\frac{1}{C_{min}} = \frac{1}{C_i} + \frac{1}{C_d} = \frac{d_i}{\varepsilon_i} + \frac{W_d}{\varepsilon_s} . \tag{10-10}$$

The description above makes it possible to understand the principles of chemical sensors based on field effect devices. Any chemically induced change, φ, ΔV, of V_T (or V_{FB}) shifts the electrical characteristics of the MIS structure along the voltage axis (Figure 10-1). Such changes can be due to changes in the charge content of the insulator (ΔQ_i), in the dielectric constant of the insulator ($\Delta \varepsilon_i$), in the thickness of the insulator (Δd_i), in the work function of the semiconductor due to polarization phenomena at the insulator-semiconductor interface (ΔW_s), or in the work function of the (metal) gate (ΔW_m), which can be due to a change in the bulk work function of the gate or polarization phenomena at the metal-insulator interface.

"Metal" above should be taken in a very general sense. Since the gate insulator normally has a very small electrical conductivity (ca. 10^{-14} (Ω cm)$^{-1}$), the gate "metal" can be a material with a conductivity much less than that of continuous metals. It can consist, eg, of discontinuous metals, (electrically conducting) polymers, semiconductors, and of an electrolyte as in the case of ion-sensitive field effect devices. It is observed that changes in ε_i and d_i also cause a change in the magnitude of the electrical characteristics.

Changes in Q_i may depend on changes in the number of so-called interface states, localized electronic sites, at the semiconductor insulator interface, or be due to charge injection or charge movements in the insulator.

Most of the successful chemical sensors based on the field effect depend on changes in the work function of the gate and especially polarization changes at the "metal"-insulator interface. This relates, eg, to gas-sensitive devices with gates of catalytic metals and actually also to the ion- sensitive devices where the gate insulator consists (partly) of an ion-selective layer and the gate "metal" of an electrolyte. These types of devices are in case of transistor structures called CHEMFETs, with the subclasses GASFETs and ISFETs; ISFETs and GASFETs are treated in detail in the following sections.

10.2   Ion-Sensitive Field Effect Transistors

10.2.1   Introduction

Since the first paper by Bergveld in 1970 [2], in which he introduced the ion-sensitive field effect transistor (ISFET), many different types of FET-based sensors have been presented [3]. After an explanation of some basic concepts, this chapter will give an overview of the enormous variety of sensors based on the FET principle. Further, some fundamental and practical problems that can limit the utilization of ISFETs for certain applications will be discussed.

10.2.1.1   Basic Concepts of Chemically Sensitive Field Effect Devices

The functioning of an ISFET can best be explained by comparing the sensor with a conventional MOSFET, a metal oxide semiconductor field effect transistor, as shown schematically in Figure 10-4. The ISFET is then a MOSFET in which the metal is replaced by a reference electrode and a solution with a certain pH (see Figure 10-5). The thermodynamics of the cell composed of reference electrode, solution, and MOSFET without a metal gate is not affected by the geometry or size of any of its components. The mechanism of operation can be described by the processes in each phase and at each interface. The arrangement of contacting phases, including the reference electrode, can be given as follows:

System 1: MOSFET:

$M'|Si|SiO_2|M''$

System 2: ISFET:

$M'|Si|SiO_2|$ solution pH $||$ saturated KCl $|$ AgCl $|$ Ag$'|M''$

Here M$'$ and M$''$ denote the respective metals contacting silicon and silicon dioxide or silver. When transforming the MOSFET into the ISFET, the interface $SiO_2|M''$ (system 1) has to be replaced with the corresponding interfaces of system 2. In this case, the term V_T which appears in Equation (10-6) that describes the drain current is no longer a constant parameter that depends only on solid-state properties. In this case it also contains all potential differences of the added interfaces:

$$V_T = E_{Ag/AgCl} + E_j - \Psi_0 + \chi + V_{ss} \qquad (10\text{-}11)$$

where V_{ss} includes the contributions of the potential contributions originating from the solid-state part of the system (ie, the same as Equation (10-3)):

$$V_{ss} = \frac{W_M - W_S}{q} - \frac{Q_i}{C_i} + 2\varphi_B + \frac{\sqrt{4\varphi_B \varepsilon_S q N_A}}{C_i}. \qquad (10\text{-}12)$$

$E_{Ag/AgCl}$ is the reference electrode potential, E_j is the liquid junction potential, for which frequently the Henderson approximation is used [5], Ψ_0 is the potential difference between the insulator surface and the bulk of the solution and χ is the surface dipole potential at the insulator-solution interface.

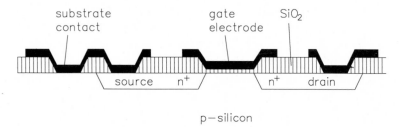

Figure 10-4. Cross section of a MOSFET.

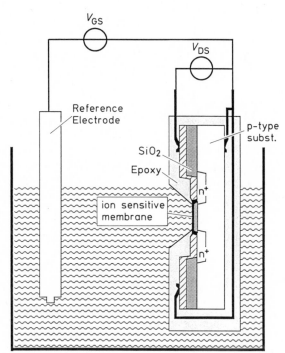

Figure 10-5.
Schematic diagram of an ISFET under operational conditions. V_{GS}: gate-source voltage, V_{DS}: drain-source voltage.

The potential difference Ψ_0, which is actually the pH-dependent term, is determined by the surface proton dissociation reactions as proposed by Levine and Smith [6], Davis et al. [7] and Smit and Holten [8]. Following the notation of James and Parks [9], the surface protonation reactions can be written as follows:

$$\text{SOH}_2^+ \rightleftharpoons \text{SOH} + \text{H}^+ \qquad K_{a1} = \frac{[\text{H}^+]\,[\text{SOH}]}{[\text{SOH}_2^+]} \exp\left(-\frac{q\,\Psi_0}{k_B T}\right) \qquad (10\text{-}13)$$

$$\text{SOH} \rightleftharpoons \text{SO}^- + \text{H}^+ \qquad K_{a2} = \frac{[\text{H}^+]\,[\text{SO}^-]}{[\text{SOH}]} \exp\left(-\frac{q\,\Psi_0}{k_B T}\right) \qquad (10\text{-}14)$$

where $[\text{H}^+]$ is the H^+ activity in the bulk solution, all other quantities in brackets are numbers of the differently occupied surface sites per unit area, and K_{a1}, and K_{a2} are chemical equilibrium constants. The activity coefficients for the surface sites are assumed to be constant and are incorporated in the equilibrium coefficients, as was discussed by Smit and Holten [8]. Healy and White [10] have given a derivation of these equations, showing the assumptions under which they are valid. The two reactions above give rise to a surface charge and a corresponding surface potential Ψ_0, which can be calculated to be given by [4,11]

$$2.303 \, (\mathrm{pH_{pzc}} - \mathrm{pH}) = \frac{q \, \Psi_0}{k_B \, T} + \sinh^{-1} \left(\frac{q \, \Psi_0}{\beta \, k_B \, T} \right) \tag{10-15}$$

where $\mathrm{pH_{pzc}} = (\mathrm{p}K_{a1} + \mathrm{p}K_{a2})/2$ is the pH at the point of zero charge, ie, when there is no net charge on the insulator surface and β is a dimensionless pH-sensitive parameter, given by

$$\beta = \frac{2q^2 N_s}{C_{dl} \, k_B \, T} \left(\frac{K_{a2}}{K_{a1}} \right)^{1/2} \tag{10-16}$$

where N_s is the surface site density ($= [\mathrm{SOH_2^+}] + [\mathrm{SO^-}] + [\mathrm{SOH}]$) and C_{dl} the linearized double layer capacitance per unit area at the insulator surface [11]. Equation (10-15) is an approximation valid for surfaces, such as oxides, where the site density is high enough to ensure that the surface charge is never close to its maximum value in the aqueous pH range. It is assumed that $\Delta \mathrm{p}K = \mathrm{p}K_{a2} - \mathrm{p}K_{a1}$ is large enough to ensure that $K_{a2}/K_{a1} \ll 1$. It has been shown that the formation of surface complexes between charged sites and counter ions has only a small effect on the Ψ_0/pH relationship, owing to the very large capacitances associated with these complexes [11, 12]. Around the point of zero charge, the relationship in Equation (10-15) can be linearized to yield

$$\Psi_0 = \frac{\beta}{\beta + 1} \, 2.303 \, \frac{k_B T}{q} \, (\mathrm{pH_{pzc}} - \mathrm{pH}) \, . \tag{10-17}$$

The preferred mode of operation of an ISFET is the constant-current mode with constant V_{DS}, using a feedback circuit which compensates for induced changes in I_{DS} by adjusting V_{GS}. This causes a parallel shift of the I_{DS}-V_{GS} characteristics, as already illustrated in Figure 10-1. For the linear range of I_{DS}, this results in:

$$V_{GS} = \frac{I_{DS}}{K_d \, V_{DS}} + E_{Ag/AgCl} - \Psi_0 + \chi + V_{ss} \, . \tag{10-18}$$

In the ideal situation Ψ_0 is the only varying term in Equation (10-18), which means that V_{GS} is, like Ψ_0, linearly related to the pH. However, in practice, V_{GS} depends not only on pH but also on temperature. Further, ISFETs often suffer from variation with time, ie, drift. These two problems will be discussed in more detail in Section 10.2.4.

Until now we only have treated the operation of the most simple pH-sensitive ISFET. In order to make such devices sensitive to, for example, potassium or sodium, ion-sensitive polymeric membranes can be deposited on top of the gate insulator. This results in system 3:

$$\mathrm{M'} \, | \, \mathrm{Si} \, | \, \mathrm{SiO_2} \, | \, \mathrm{ISM} \, | \, \mathrm{sol.} \, (a_i) \, | \, \mathrm{sat.} \, \mathrm{KCl} \, | \, \mathrm{AgCl} \, | \, \mathrm{Ag} \, | \, \mathrm{M''}$$

with ISM, the solvent polymeric membrane, sensitive to ion i and sol. (a_i) the solution with ion activity a_i. Similarly to the above-described approach, an expression can be found for V_T [13]:

$$V_T = E_{Ag/AgCl} + E_j - \frac{\mu_{i,\,sol.}}{z_i\,q} - \frac{\alpha_i^{ISM}}{z_i\,q} + V_{ss} \tag{10-19}$$

where μ_i is the standard chemical potential of ion i in the solution, z_i the charge number, and α_i^{ISM} the so-called real potential, consisting of the dipole potential at the ISM-insulator interface and the chemical potential of species i in the membrane. This equation is identical with that given by Buck and Hackleman [14] for a structure where SiO_2 is coated with AgBr.

The question of the thermodynamic stability of the term α_i^{ISM} is similar to the situation for coated wire electrodes (CWEs), first described by Hirata and Date [15], where the membrane-metal potential is also undefined. It was found that CWEs showed a potential dependence on oxygen partial pressure, especially for electrodes with platinum conductors [16, 17]. Therefore, the CWE function can be considered to be due to the formation of an oxygen half-cell on the membrane-metal interface due to water and oxygen penetration, which can be fast for solvent polymeric membranes [18]. A similar mechanism was proposed by Srianujata et al. [19], who used a CWE made of chlorinated silver wire, covered with a membrane (poly(vinyl chloride)) (PVC) and dioctyl phthalate (DOP), for metal ion determination. Prepared electrodes were soaked for 12 h in 0.1 M potassium chloride solution. According to these authors, the membrane contact with Ag/AgCl was through a thin film of aqueous salt (KCl) solution rather then directly. In both cases the electrode potential can be described using the standard expression for ion-selective electrodes with an internal filling solution. The stability of the devices now depends on the stability of the reference system and the concentration of the measured species i in the water layer between the membrane and reference system.

For ISFETs covered with polymeric membranes, and in particular potassium-sensitive membranes based on valinomycin, contradictory results regarding stability have been reported. A possible explanation of these contradictions could be the membrane-insulator interface, which may be stable under certain conditions as for CWEs. Fogt et al. [20] reported that pH-sensitive ISFETs, coated with a potassium-sensitive membrane, are subject to a number of unexpected interferences. Carbon dioxide, benzoic acid and acetic acid caused significant shifts in the output of these devices. Since conventional potassium membranes are normally not susceptible to these species [21], the mechanism involved must be related to the membrane-insulator interface. Supposing that there is a thin water layer at this interface, a change in the pH of this layer would be seen as a change in the output potential of the ISFET. Dissociation of species which diffuse through the membrane, such as CO_2 and benzoic acid, can easily induce a pH change of such a poorly buffered water layer. Similar results were reported by Harrison et al. [22] for MIS capacitors. Assuming that there is a hydrated layer between the membrane and insulator, the following system can be described for an ISFET covered with a potassium-sensitive membrane:

$$M' \,|\, Si \,|\, SiO_2 \,|\, sol.\ pH_1, \ a_{K_1^+} \,|\, ISM \,|\, sol.\ a_{K_1^+} \,|\, KCl\ sat. \,|\, AgCl \,|\, Ag \,|\, M''$$

which results in the following expression for V_T:

$$V_T = E_{Ag/AgCl} + E_j - \Psi_0 + x - \frac{k_B T}{q} \ln \frac{a_{K_2^+}}{a_{K_1^+}} + V_{ss}\,. \tag{10-20}$$

It can be seen from this equation that the stability of the system depends on the pH and the potassium activity of the hydrated layer. When these are stable, the system will react only to changes in the activity of potassium ions in the outside solution. The validity of this approach is discussed in [23] and [24].

So far we have discussed the basic concepts of the functioning of ISFETs together with the most commonly used modification method: ion-sensitive polymeric membranes. In the following we give a more detailed overview of modified ion-sensitive field effect transistors.

10.2.2   Directly Sensing Devices

10.2.2.1   Introduction

After the initial use of SiO_2 as a gate insulator for ISFETs, many different pH-sensitive gate materials have been investigated. Later membrane-covered ISFETs sensitive to a variety of (cat)ions were constructed, in addition to ISFETs that make use of an intermediate (reactive) membrane to measure uncharged chemical species. Thus, a wide variety of devices and possible ISFET-based sensors have been reported. In order to categorize this large diversity, we shall distinguish the devices according to their principles of operation.

As shown in the previous section, the ISFET combines two functions simultaneously: a selective detection of ions and a transformation of the sensed surface potential into a corresponding drain current variation. This variation results in a proportional output voltage variation when using an appropriate electronic control circuit. Two different cases can now be distinguished. In the first case, the measured potential variation corresponds directly to the chemical variable to be measured, as for instance in the case of pH ISFETs. These devices will be termed directly sensing devices (see Figure 10-6), and will be treated in this section. In the second case, an intermediate layer is used to transduce a variation of the (bio)chemical variable to be measured in a corresponding variation of a second chemical parameter (usually pH) which can be subsequently detected. These devices will correspondingly be termed in-

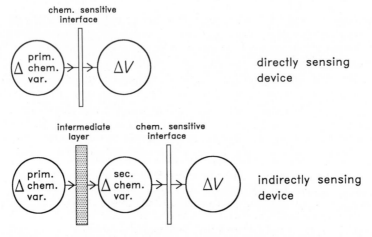

Figure 10-6. Schematic representation of directly and indirectly sensing ISFET devices. ΔV: voltage drop.

directly sensing devices, and will be discussed in Section 10.2.3. Examples of the second type of devices are, eg, CO_2 ISFETs and enzyme FETs. Membrane-covered ISFETs and chemically modified ISFETs will be considered as directly sensing devices, since no second chemical variable is used.

10.2.2.2 pH-Sensitive Devices

After the invention of the ISFET, initially the only gate material used was SiO_2 [2,25,26]. Although this material showed some pH sensitivity (20–40 mV/pH), the response appeared to be non-linear and lower than the expected Nernstian response. A Nernstian response is expected from Equation (10-15) when β is large. In addition, it was known that SiO_2 was not a very good ion-blocking material, a property that probably was responsible for the high drift rate and considerable sodium interference found with this material [4]. It was suggested by Bousse and Bergveld [27] that the existence of buried proton adsorption sites is responsible for the bad pH sensitivity, drift, and hysteresis.

The pH-sensing properties and drift behavior of the ISFET improved greatly with the introduction of other gate materials. Matsuo and Esashi [28] used a layer of Si_3N_4 on top of SiO_2, which showed a pH sensitivity of 45–50 mV/pH. The needle-like probe also made use of the excellent insulating properties of Si_3N_4, the device being surrounded completely by this material. Other layers were proposed which showed even higher pH sensitivities, such as Al_2O_3, with a reported sensitivity of 53–57 mV/pH [23,29], and Ta_2O_5, which shows about 55–59 mV/pH [29,30]. Both materials have the additional advantage of having a high dielectric constant, which keeps the effective gate insulator thickness small (see Equation (10-6). This provides a high electric sensitivity or transconductance. Also, the reported drift values of the latter two materials are acceptable, sometimes even down to 0.05 mV/h [23,29,30]. It has also been shown that the drift is decreased after the first immersion of the devices in solution (see Figure 10-7). In most of the work on pH ISFETs published so far these three gate materials, Si_3N_4, Al_2O_3, and Ta_2O_5, have been used. The materials each have their specific advantages and disadvantages.

As mentioned before, Si_3N_4, when deposited by low-pressure chemical vapor deposition (LPCVD), can serve as an excellent all-round probe-insulator, but shows a pH sensitivity

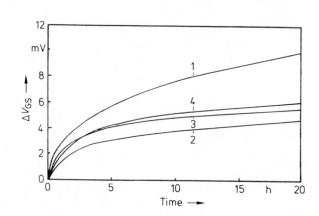

Figure 10-7.
Drift behavior of Al_2O_3 ISFETs at pH 7.60 (25 °C) (1) First immersion; (2) after dry storage for 1 week; (3) for 2 weeks; (4) for 18 weeks. ΔV_{GS}: gate-source voltage. (After [132])

which decreases with time. This effect is due to the slow but continuous transformation of Si_3N_4 into oxynitride which takes place in aqueous solutions. A brief dip in dilute HF removes the thin oxide layer and restores the sensitivity to >55 mV/pH, but the sensitivity decreases again thereafter. Recently it has also been found that the chemical response of Si_3N_4 can be very slow and that an almost Nernstian pH response can be measured, provided that enough time is allowed for the full response to develop [167].

Al_2O_3 shows a very low drift if fabricated under well defined conditions (see also Section 10.2.4.2.1), and has a satisfactory pH sensitivity. However, its deposition is not a standard IC technology process like the deposition of silicon nitride, and requires considerable process development.

Recently, Ta_2O_5 has received considerable attention, owing to its nearly Nernstian pH sensitivity, and relatively easy fabrication, although the first reports on the use of this gate material date from 1983 [30–34]. However, a non-negligible and partly irreversible light sensitivity prevents this material from being applied on a wider scale at present [33, 34]. More fundamental studies using physico-chemical characterization techniques such as those presented in [33] will be needed for a better understanding and control of the behavior of this pH-sensitive material. In Figure 10-8 typical pH sensitivities for ISFETs with SiO_2, Si_3N_4, Al_2O_3, or Ta_2O_5 as gate material are shown, as reconstructed from experimental results in Refs. [23, 28–30, 44, and 167]. Other more exotic insulator materials such as IrO_2 [35], ZrO_2 [36], borazon [37], Nb_2O_5, and Ti_2O_5 sometimes show nearly Nernstian sensitivities, but have only rarely been studied.

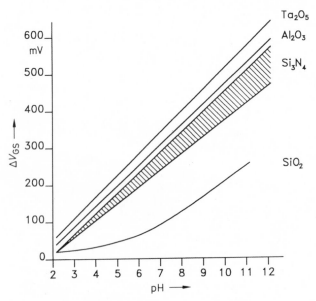

Figure 10-8. Typical pH-sensitivity curves for ISFETs with different gate insulator materials. ΔV_{GS}: gate-source voltage..

10.2.2.3   Devices Sensitive to Other Ions

Replacement or Bulk Modification of the pH-Sensitive Layer

Several methods have been used to make ISFETs sensitive to ions other than hydrogen (Figure 10-9). The first possibility is to replace or modify the original pH-sensitive gate material (Figure 10-9a). Pham and Hoffmann [38] demonstrated that the properties of the SiO_2 insulator could be modified by implantation of high doses of B, Al, Ga, In, and Tl, whereas implantation of Al and Li [32] or Al and Na [39] in Si_3N_4/SiO_2 resulted in sodium-sensitive devices. However, the sensitivities obtained were poor, especially in comparison with those obtained with membrane-covered ISFETs, and a significant pH interference was observed. Esashi and Matsuo [40] replaced the Si_3N_4 insulator with a layer of aluminosilicate or borosilicate, and achieved a Nernstian sodium sensitivity. Although this type of sensor has the advantage of possessing real solid-state properties, which ensures, for instance, a very long lifetime, a disadvantage is its considerable pH interference.

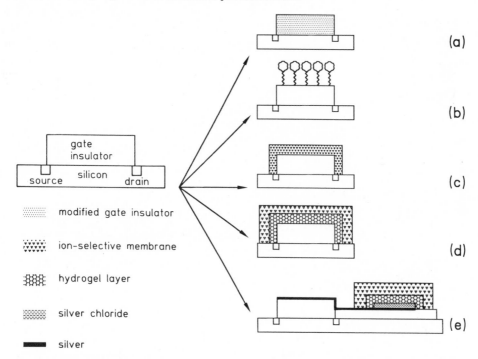

Figure 10-9. Schematic representation of pH-sensitive ISFET-devices: (a) bulk modification of gate insulator; (b) chemical surface modification; (c) membrane-covered ISFET; (d) membrane-covered ISFET with intermediate hydrogel; (e) extended gate FET.

Chemical Surface Modification

A second method consists in chemical surface modification of the original gate insulator. In this case the original chemical surface groups that provide the pH sensitivity are chemically coupled to a compound that is sensitive to other ions (Figure 10-9b). In this way, a silver-sensitive ISFET has been investigated [41]. A serious problem with this type of modified ISFETs is that even very small amounts of remaining pH-sensitive groups give rise to an important residual pH sensitivity [42]. This problem was circumvented by the structure proposed by Matsuo et al. [43], who first covered the pH ISFET with a pH-insensitive parylene layer, to which in a second stage potassium complexing molecules (crown ethers) were chemically bound to give a potassium-sensitive device. Although such a structure may seem promising at first sight, there are some practical problems. To obtain a high (Nernstian) potassium sensitivity, a sufficiently high crown ether density must be achieved. Further, the association constant of the crown ether with the potassium ion should be high enough to have a sufficiently high surface charge [44] (see Figure 10-10). The sub-Nernstian slope reported by Matsuo et al. seems to be due to these two effects. So far, efforts in this direction have remained mainly academic.

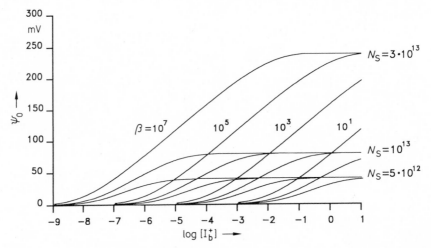

Figure 10-10. Calculated responses for ISFETs with chemically attached ionophores with different site densities N_s (cm^{-2}) and association constants β (dm^3/mol). Ψ_0: potential difference, $[I_b^+]$: bulk concentration of an ion I. (After [44])

Membrane-Covered ISFETs

Homogeneous Polymeric Membranes

The third way in which pH ISFETs can be made sensitive towards other ions is to cover them with an ion-sensitive membrane (Figure 10-9c). This is so far the easiest and most commonly applied method, although many practical problems still remain. The most frequently applied class of ion-sensitive membranes is polymeric membranes. The first report of ISFETs covered with such membranes appeared in 1975, when a potassium-sensitive ISFET was fabricated by Moss et al. [45], who covered the ISFET with a 300-μm thick plasticized PVC membrane. The composition of this membrane was closely comparable to that used in classical membrane electrodes. Several textbooks describing the theory and applications of such electrodes have appeared [46, 47].

Later, membrane-covered ISFETs of analogous construction sensitive to calcium [48, 49], sodium [50, 49], ammonium [50], and nitrate [51] were reported. Using the structures in an alternative way, such membrane-covered ISFETs have also been used to give a rapid and reliable determination of the selectivity of synthetic ionophores [52] (see Figure 10-11).

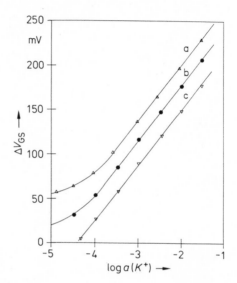

Figure 10-11.
Response curves of ISFETs covered with PVC membranes loaded with hemispherands (a and b) and valinomycin (c). ΔV_{GS}: gate-source voltage, a: activity coefficient of K^+. (After [52])

Although it appeared very easy to construct membrane-covered ISFETs with a specific sensitivity, the theoretical basis of its functioning remained troublesome. Janata and Huber [53] described schematically the thermodynamics of such structures, from which it appears that the membrane-insulator interface is badly defined (see also Section 10.2.1). This theoretical shortcoming was demonstrated experimentally in studies that showed a CO_2 interference for the membrane-covered ISFET [20] (see Figure 10-12). As mentioned in Section 10.2.1.2, a remedy for this effect was found in the use of an intermediate hydrogel containing both a pH buffer (pH < 4) and the ion to be sensed by the membrane [23, 24] (see Figure 10-9d).

Another theoretically unresolved point is whether it is possible to immobilize completely all membrane components in the membrane matrix. This issue is particularly interesting for the construction of ion sensors with relatively thin membranes. Recently, studies concerning the exact functioning of such membranes have been published that indicate more clearly the role of different membrane components and the electric properties of the membrane [54–57].

An important difficulty when using membrane-covered ISFETs is their limited lifetime. This problem is due mainly to two effects: leaching of membrane components and bad adhesion of the membrane to the solid support. The component that is especially subject to be washed out of the membrane is the plasticizer. To prevent its leaching, one approach is to synthesize plasticizers with a higher lipophilicity, since this property mainly determines the leaching rate, especially in thin membranes [58]. A different possibility is to use polymeric plasticizers [59], although with this solution a slightly lower sensitivity was experienced. The most promising approach in this respect, however, seems to be membranes that are intrinsically rubbery, eg, polysiloxanes. With these materials good sensitivities were obtained by Van der Wal et al. [60].

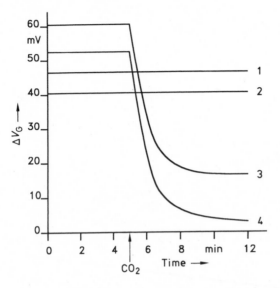

Figure 10-12.
CO_2 interference curves of ISFETs covered with PVC membranes with (1, 2) and without (3, 4) intermediate hydrogel membrane. After 5 min, CO_2 was bubbled through the solution. ΔV_G: gate voltage. (After [44])

ISFETs with photopolymerizable ion-sensitive membranes are fabricated by using methacrylate-type polymers [61]. Good calcium-sensitive devices for the use in flow-injection analysis (FIA) could be obtained with a moderate life time (<1 month) for the use. However, so far no on-wafer fabrication procedure to produce membrane-covered ISFETs has been reported.

To improve the attachment of the membrane to the substrate, several technological solutions have been proposed. Such solutions cannot be separated from the method of encapsulation of the whole device, especially since so far all ion-sensitive membranes used are applied to the gate after encapsulation of the ISFET. Therefore, membrane adhesion problems are closely related to encapsulation problems, unless an on-wafer fabricated membrane is used. Further consideration of this problem will be given in Section 10.2.4.1.

A suspended mesh [62] and also small cavities [63, 64] have been used to improve physically the adhesion of the membrane to the surface (see Figure 10-13). However, the adhesion problem seems to be rather of chemical origin, viz, the penetration of moisture between the membrane and the substrate, which provides an electric shunt to the formed membrane potential. Therefore, improving the attachment as a result of chemical modification of the membrane and/or the substrate probably provides a better solution [59–60, 61, 65, 66].

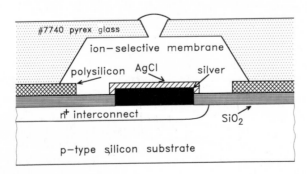

Figure 10-13.
Schematic representation of an ISFET structure with small holes to contain the ion-sensitive membrane. (After [64])

Finally, the centuries-old Japanese technology of fabricating Urushi lacquer membranes has been applied to produce durable sensors sensitive to calcium, sodium, potassium, and chloride ions [67, 68].

Other Membranes

Only a few membrane types other than those mentioned above have been used to render ISFETs sensitive to other ions. Langmuir-Blodgett films have been applied to obtain calcium-, potassium-, or pH-sensitive devices [69–73]. Such films are very delicate, however, and pinholes are easily formed, which restricts their practical applicability.

Heterogeneous membranes consisting of insoluble silver salts in silicone membranes are used to give chloride-, cyanide-, and iodide-sensitive ISFETs [74]. In this case the main function of the silicone membrane is to hold the finely divided grains of the salt together. Further, a fluoride-sensitive device was made by applying LaF_3 salt as the gate material [75].

Photoresist-type membranes have been used to make ion-controlled diodes (ICDs) and modified ISFETs with photolithographic methods [76, 77]. Unfortunately, the limited thickness and non-optimum composition of the membrane restrict their lifetime and ion-sensing properties. Nevertheless, this fabrication method opens the way to reproducible mass production.

10.2.2.4 Extended Gate Devices

The selective detecting and transducing functions that originally are both fulfilled by the gate insulator in the pH ISFET are clearly split in the later derived membrane-covered ISFET devices. Especially with the most elaborated structure, as shown in Figure 10-9e, a so-called extended gate FET (EGFET), in fact a solid-state, miniaturized version of a conventional ion-selective electrode is created with a direct membrane-potential amplification. The inner elec-

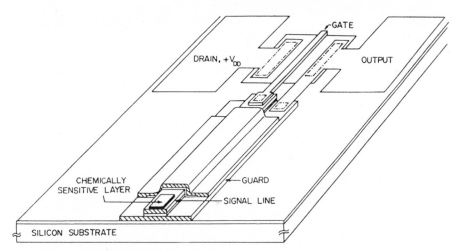

Figure 10-14. Schematic representation of an extended-gate structure with guarded signal lead. V_{DO}: drain-output voltage. (After [79])

trode here is formed by the Ag/AgCl-covered extended gate of a MOSFET. An early realization of such a structure was published in 1978 by Nagy et al. [78]. Here, a pH-sensing surface is no longer indispensable. The only requirement is to have a reversible electric contact with the hydrogel layer. Although extended gate structures may be considered to be nothing else than miniaturized solid-supported ion-selective membranes with an external FET-input amplifier, the similarity of the fabrication technique used for extended gate devices with that of membrane-covered ISFETs justifies their inclusion in this chapter.

An advantage of EG structures is that standard MOS devices can be used, which guarantees a very low (solid-state) drift of the amplifying part. The practical problem of the parasitic capacitance between the signal line and ground can be solved by using a bootstrap configuration as shown in the structure of Van der Spiegel et al. [79] (see Figure 10-14).

10.2.2.5   Practical Devices

Over the years, several practical ISFET devices have been proposed and realized. The simplest, used in laboratories, is the straightforward mounting of an ISFET on a printed circuit board and encapsulation with epoxy resin. A more specific device is the catheter-tip pH ISFET, where the ISFET is mounted in or at the tip of a catheter designed for in vivo biomedical purposes [80–83]. In this case, the small size of the ISFET plays an important role, since in principle multi-ISFET catheters can be constructed analogously, which is hardly realizable with miniature conventional glass electrodes.

Biomedical devices for extra-corporeal use often utilize flow-through structures [49, 90], where the small ISFET advantageously permits the construction of systems with a small dead volume. Such structures have the advantage that they can be recalibrated regularly, thus solving the drift problem.

The use of micro-fabrication techniques allows the construction of flow-through structures with a minimum internal volume and thus a small sample size and a fast response [84–86]. Various workers have applied ISFETs in flow-injection analyses (FIA) systems [87–89]. Here especially the fast response of the ISFET is an important property.

The needle-type ISFETs mentioned earlier are commercialized in the form of a pH pen (see Figure 10-15). The pH pen is one of the first commercially available products using an ISFET, but in its present version it should perhaps rather be considered as "electronic litmus paper" than a precise pH electrode, owing to its limited resolution of 0.1 pH. Another ISFET-based pH probe is marketed by IBM as part of their Personal Science Laboratory.

With the use of anisotropic etching, back-side contacted ISFETs have been realized [91–94]. These devices have the advantage of a relative simple encapsulation while they also possess a completely flat sensing surface. Here the problems are rather found in the technological complexity of the structure and the incompatibility with standard techniques.

Recently, SOS (silicon-on-sapphire) and SIS (silicon-insulator-silicon) structures have been published [95, 96]. Such structures have the advantage of an excellent electric insulation from the solution combined with relatively easy encapsulation. However, the devices become more expensive when using a sapphire substrate, whereas the silicon-silicon(oxide) bonding used in SIS structures is a complicated technique.

Finally, devices with integrated electronics have also been reported [97, 98]. Although with regard to discrete ISFET devices the thermal sensitivity may be reduced by differential

MOSFET/ISFET measurement, in principle it is only necessary to combine the MOSFET on the same chip. The rest of the electronics may well be supplied externally.

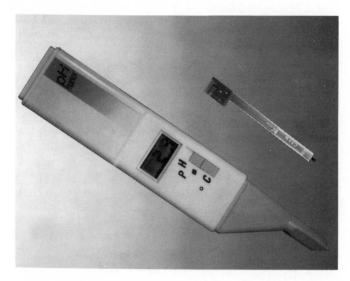

Figure 10-15. Photograph of a commercial pH pen with needle-type Si_3N_4 pH ISFET.

10.2.3 Indirectly Sensing Devices

10.2.3.1 Introduction

The direct measurement of ion concentrations with the use of ISFETs as described in the previous sections is based on the measurement of a potential difference. The voltage between an ion-sensitive layer and the solution develops as a result of the partition of ions across this interface and sensing is thus limited to species with an electric charge. As will be discussed below, only small ions that contribute to an actual transfer of charge between the solution and the membrane phase can be measured in this way. However, the practical application of ISFETs is not limited to the measurement only of ions. By the integration of ISFETs into more complex sensing systems, they can be used to detect changes in ion concentrations (usually pH changes) that develop as a result of a chemical reaction of an otherwise non-ionic species. On the other hand, because a field effect transistor is a voltage-sensing device, it can also be used to detect electrical phenomena that cannot be directly attributed to the activity of a single ion and that occur at the surface of a selective membrane

The use of an ISFET as the detecting element in a more complex chemical sensor can be attractive because of its relatively small size, as is important, eg, for the development of P_{CO_2} sensors for in vivo applications. The potential low cost of mass-produced ISFETs is also of interest with regard to the development of biosensors. Because of the nature of biologically active sensing layers, the lifetime of such sensors will be limited and there will be a need for cheap disposable products.

10.2.3.2 P_{CO_2}-Sensitive ISFETs

In 1958, Severinghaus and Bradley [99] described a sensor for the partial pressure of carbon dioxide (P_{CO_2}) based on a pH-sensitive glass electrode. The glass electrode measures the pH in a thin layer of a hydrogencarbonate solution that is separated from the analyte solution by a thin gas-permeable membrane. Carbon dioxide diffuses through the membrane until the inner and outer concentrations are equal and thus establishes the pH in the internal electrolyte solution.

Because the measurement of P_{CO_2} in blood is of great clinical importance, the development of ISFET-based Severinghaus electrodes for in vivo use seems of great interest. However, probably for reasons similar to those which limit the introduction of ISFETs for in vivo pH measurements, so far relatively few papers have appeared on the development of P_{CO_2}-sensitive FETs. Most of the work that has been published has been done by a group at Kuraray Co. in Japan, and their first report dates from 1980 [100]. The ISFET used is of a needle type, completely isolated by a layer of LPCVD silicon nitride as described in Section 10.2.2 [28]. An Ag/AgCl electrode is included on the chip to function as the internal reference electrode. After covering the electric bonding wires with epoxy, the chip is mounted in a nylon tube with the use of silicone rubber. The gate region of the ISFET is then covered with a hydrogel containing NaCl and $NaHCO_3$. After freeze drying of the hydrogel, the whole assembly is covered with a gas-permeable silicone-rubber membrane by dip coating. Figure 10-16 shows a cross section of the sensor mounted in a catheter only 0.8 mm in diameter. In later versions of the sensor [101], the hydrogel layer is replaced with a hollow fiber to contain the internal electrolyte. Apparently because of the practical problems that are involved with in vivo monitoring, a P_{CO_2} sensor of the same construction was later applied in a system for ex vivo use [102].

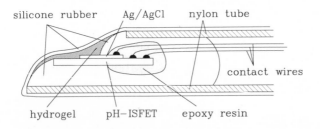

silicone rubber   Ag/AgCl   nylon tube

contact wires

hydrogel   pH–ISFET   epoxy resin

Figure 10-16.
Severinghaus electrode based on an ISFET. (After [100])

Because of the necessity for an internal electrolyte solution, the construction of Severinghaus-type electrodes based on ISFETs poses some practical problems regarding the electrical isolation of the transistor. One possible solution was demonstrated above with the use of an insulated needle-type ISFET. Another method for the isolation of the electrical connections in a P_{CO_2} sensor was described by Hu et al. [103]. Here an ISFET with back-side contacts is used [92] so that the solution and the electric contacts are well separated. Because the front side of the encapsulated chip is completely flat, the pH-sensitive gate of the ISFET can easily be brought into close proximity of the gas-permeable membrane and thus the reported 90% response time is less than 1 min.

Finally, a P_{CO_2} sensor based on a different principle to the conventional Severinghaus electrode was described by Van der Schoot and Bergveld [104]. The CO_2 concentration in the inner electrolyte layer of the sensor is determined by a coulometric acid-base titration while an ISFET is used as an indicator electrode to monitor the equivalence point. Because the ISFET is used only as an indicator during the titration, which lasts ca. 10 s, the device is insensitive to ISFET drift and shows excellent long-term stability.

10.2.3.3   ISFET-Based Enzyme Sensors

The application of enzymes as the selecting agent in ISFET-based sensing systems leads to the development of highly selective sensors. Such enzyme-modified ISFETs (EnFETs) can in principle be constructed with any enzyme that produces a change in pH on conversion of the concerning substrate. Although Janata and Moss had suggested in 1976 [105] that the construction of an EnFET should be possible, it was not until 1980 that the first practical results were published. Caras and Janata [106] described an EnFET sensitive to penicillin using a cross-linked penicillinase-albumin membrane. A second ISFET, covered with an unloaded albumin membrane, was used as a reference so that the differential output signal of the two FETs was insensitive to changes in the sample pH. Penicillinase catalyzes the hydrolysis of penicillin to penicilloic acid and hence the local pH in the membrane will be determined by the amount of penicillin in the sample. As such, this enzyme sensor was identical with similar sensors based on glass pH electrodes that had been published before. The main advantages of ISFETs are the minute amounts of enzyme needed to cover the pH-sensitive gate area and the simple application of a differential measurement set-up to compensate for the background pH of the sample. Figure 10-17 shows the measurement setup that is generally used for EnFETs. Because the two ISFETs are used in a differential mode, a simple metal wire can be used as a pseudo-reference electrode.

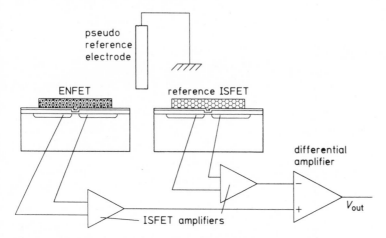

Figure 10-17. Differential measurement setup for EnFETs. V_{out}: output voltage. (After [107])

Later, ISFETs were used with the application of a wide variety of enzymes. The majority of the published papers concerns glucose oxidase (GOD) and urease for the measurement of glucose and urea, respectively. The application of EnFETs for the measurement of acetylcholine, ATP and neutral lipids has also been reported. The EnFET literature up to 1987 has been extensively reviewed [107].

The most frequently used enzymes for EnFETs are urease and GOD. Urease is an inexpensive, stable enzyme and is therefore excellently suited as a model enzyme for sensor studies. Urea is hydrolyzed by the enzyme and the reaction products will increase the pH in the membrane, provided that the original sample pH is below 9 [108]. Most results published on urea EnFETs show a rapid response and good sensitivity.

The popularity of glucose oxidase for use in EnFETs is less clear. The development of glucose sensors as a tool in diabetes control is, of course, very important. However, the sensitivity of GOD-modified ISFETs is generally poor, the change in output voltage being of the order of a few millivolts when the glucose concentration is changed from zero to a value in the physiological range. Glucose oxidase catalyzes the oxidation of glucose to gluconolactone, which is subseqently hydrolyzed to gluconic acid and thus gives a decrease in the membrane pH. The second, hydrolysis, step is relatively slow, however, and thus most of the gluconolactone is diffused out of the membrane before the acid is formed. Recently [109] the sensitivity of glucose EnFETs has been substantially improved by the addition of a second enzyme, gluconolactonase, to the membrane. Now the second reaction step proceeds much faster and thus the response is greatly enhanced. In spite of this improvement, it remains to be seen if glucose EnFETs can compete with amperometric glucose sensors.

The ISFETs used as the base element for the enzyme sensors can, of course, be mass-produced with IC technology. For the fabrication of truly inexpensive and thus disposable biosensors it is of great interest to apply the enzymatic membranes also in a batchwise process. Possible wafer-scale membrane deposition techniques are, for instance, the use of photo-cross-linkable membrane materials [110] or the lift-off technique for those materials which are not photo-sensitive themselves [111]. The latter method can even be used repeatedly on the same wafer for the production of multi-enzyme sensors [112], although not without the loss of some enzyme activity. An alternative for the production of multi-sensors is the use of an ink-jet nozzle, as commonly used in printers, for the localized deposition of the individual enzymatic membranes [113].

The operation of ISFET-based enzyme sensors is, of course, closely similar to that of their macroscopic counterparts based on glass electrodes and thus they exhibit the same practical limitations. Already from the first EnFET paper by Caras and Janata it was recognized that the response of the sensors depends strongly on the buffer capacity of the sample solution. In a strong buffer solution, the pH change by the enzymatically liberated protons or hydroxyl ions will be small and thus the sensitivity of the sensor is low, combined with a relatively large dynamic range. In a weak buffer the opposite is true. Moreover, both the enzyme activity and the buffer capacity of the sample solution are a function of pH. When the products of the enzymatic reaction are weak protolytes, even an additional pH-dependent factor will be included in the response of the devices. The complicated non-linear response of EnFETs has been theoretically investigated in a number of papers by Eddowes and co-workers [114–116] and Caras and co-workers [117–119]. The theoretically obtained responses were supported by experimental results using glucose oxidase- [115, 118] and penicillinase-modified FETs [119], respectively. Although these papers contributed to a more thorough understanding of the behavior of EnFETs, their main conclusion was that the buffer capacity plays a critical role in the response of these devices and thus has to be rigidly controlled in a practical application.

Recently, Van der Schoot and Bergveld described an ISFET-based enzyme sensor that overcomes the problems of the buffer dependence of the response. The pH-static enzyme sensor [120–122] performs a continuous coulometric titration of the enzymatic reaction products.

Figure 10-18 shows the measurement setup used. The titration is carried out with protons or hydroxyl ions electrogenerated at a platinum electrode inside the enzymatic membrane. As a result, the pH inside the membrane remains constant and the enzyme activity does not change. The generating current required to maintain this situation is now the sensor output signal, which under certain conditions is linearly dependent on the analyte concentration. Moreover, the response does not depend on the buffer capacity of the sample solution. In Figure 10-19 the response of this new enzyme sensor is compared with that of a conventional EnFET.

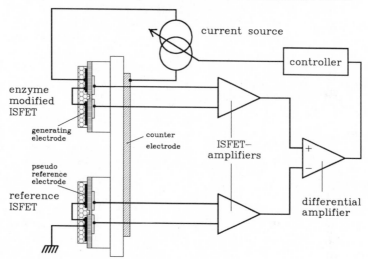

Figure 10-18. Control system for the pH-static enzyme sensor. (After [122])

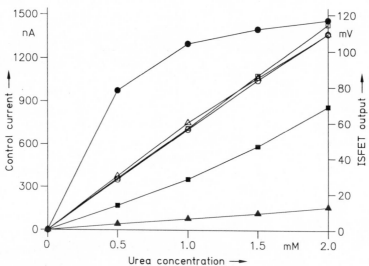

Figure 10-19. Comparison of the response of a conventional EnFET (solid symbols, right-hand axis) with that of a pH-static enzyme sensor (open symbols, left-hand axis) in 2 mM phosphate buffer of pH 7 (●,○), 10 mM phosphate buffer (■,□) and 50 mM phosphate buffer (▲,△). (After [122])

10.2.3.4   ISFET-Based Immunosensors

The development of immonosensors that directly react to the very specific binding action between antigens and antibodies is very challenging. The extremely wide range of pure monoclonal antibodies that can be produced makes it possible to envisage sensors for virtually any biological macromolecule. However, the principal problem is to transduce the molecular recognition action between antibody and antigen into a measurable signal.

In the age of solid-state chemical sensors it is of course natural to investigate also the possibilities of ISFETs as detecting elements in immunosensors. The first experiments in this direction were described by Janata and Huber [123], who adsorbed a layer of human albumin on the gate surface of an ISFET to study the antigenic reaction to rabbit antiserum. Since an immunological reaction as such does not generate ions, the ISFET should in this case be used for the measurement of changes in interfacial charge density when the reaction occurs. Although the preliminary experiments appeared to show some sensitivity and selectivity, the results were later interpreted as artifacts [124]. The main problem is that the charge in the interfacial protein layer should originate from the proteins only and not from the reversible double layer at the solid-state surface. In practical circumstances, this double layer will always be located much closer to the surface than the reacting charge sites in the protein film. It was shown by Nakajima et al. [125] that in solutions of very low ionic strength it is actually possible to measure the effects of protein adsorption on parylene-coated ISFETs. In more concentrated electrolyte solutions, however, the effects were totally absent. Thus, for a true immunosensing FET (ImFET) it would be necessary to create a truly capacitive interface at which the immunological binding sites can be immobilized. Until then, the concept of a directly immuno-sensing ISFET seems not to be feasible [126].

In spite of these practical difficulties, recently a directly sensing ImFET was described which appeared to be sensitive to human serum albumin [127]. The reported response was very low, however, and the antibodies were simply immobilized in a hydrophilic membrane. Therefore, until more experimental evidence is available, the true value of this ImFET concept remains to be established.

As described above, the direct detection of an immunological reaction with ISFETs remains difficult. However, the study of immobilized protein layers on the gate of an ISFET and the effects of changes in charge density as a result of an immunological reaction have led to the development of an indirectly operating ISFET-based immunosensor. With the ImmunoFET described by Schasfoort et al. [128], the protein layer on the gate is exposed to pulse-wise increases in electrolyte concentration. As a result, ions will diffuse into the protein layer and, because of a different mobility of anions and cations, transients in potential will occur at the protein-membrane|solution interface. The ISFET, being a voltage-sensitive device, is very suitable for the measurement of these transients. Since the mobility of ions is a function of the charge density in the protein membrane, changes in this charge density will influence the size and direction of the transients. By exposing the ImmunoFET to a pH gradient and a continuous series of ion concentration pulses, the isoelectric point of a protein layer can be detected and thus also the changes as the result of an immunological reaction.

10.2.4 Practical Limitations

10.2.4.1 Introduction

Although the ISFET concept has existed for over 20 years, practical applications still emerging only very slowly. In this section, we shall address some practical problems that have been, and partially still are, limiting factors in the commercial breakthrough of ion-sensitive field effect transistors. The problems can be divided into two categories, those which are inherent to the transistor itself, arising from limitations in the materials and technology used, and those problems common to the application of any solid-state ion sensor, viz, encapsulation and the need for a stable micro-fabricated reference electrode.

10.2.4.2 Inherent ISFET Limitations

Drift

In the past many different materials have been investigated as pH-sensitive layers. However, comparative data on baseline drift are scarce and difficult to compare owing to different fabrication methods, measuring techniques, and definition of terms.

Abe et al. [29] compared the properties of SiO_2, Si_3N_4, and Al_2O_3. They found that SiO_2 is not stable at all and that Al_2O_3 has a better stability than Si_3N_4. Si_3N_4 has a drift rate of about 0.6 mV/h and Al_2O_3 about 0.3 mV/h, after 10 h of stabilization. Akiyama et al. [129] reported a long-term drift rate of 8–20 mV/h for SiO_2 in an alkaline solution but they did not give data for their Al_2O_3 and Ta_2O_5 layers. Matsuo and Esashi [28] compared SiO_2, Si_3N_4, Al_2O_3, and Ta_2O_5 and found for Si_3N_4 a drift rate of 1 mV/h and for Al_2O_3 and Ta_2O_5 0.1–0.2 mV/h after 1000 min of operation at pH 7. The best results so far were published by Kuisl and Klein [30] for Ta_2O_5; they found drift rates of less than 0.1 mV/d. Although Al_2O_3 and Ta_2O_5 seem to be the most suitable materials for the fabrication of pH ISFETs, Si_3N_4 is the most widely used. Chauvet et al. [130] reported the influence of fabrication methods on the drift behavior of Si_3N_4. They found an improved drift behavior when the NH_3/SiH_2Cl_2 ratio was optimized during the Si_3N_4 deposition. The best value they gave is 0.05 pH/d (about 0.1 mV/h). The variation of the drift rates for Si_3N_4 reported in the literature can therefore probably be ascribed to the differences in fabrication. The data for Al_2O_3 seem to be the most independent of fabrication and measurement techniques.

The mechanism of the drift is not very clear and is probably different for each material. The drift behavior of SiO_2 can be explained by the theory of buried sites at the electrolyte gate material interface as far as chemical drift is concerned [27]. The drift of ISFETs with alumina was investigated by Ligtenberg [131] and Arnoux et al. [132], who independently measured the same effects. They claimed that the drift was caused by solid-state effects and not by an interface effect. Another explanation of this drift phenomenon is diffusion of ionic species at the grain boundaries of the alumina layer and at the same time diffusion of hydrogen traces, remaining from the chemical vapor deposition process, inside the grains. Optimization of the fabrication parameters, which influence both the grain size and hydrogen content, leads to drift rates for Al_2O_3 pH-sensitive ISFETs of about 50 μV/h or about 0.02 pH/d [133].

The drift of Ta_2O_5 was investigated by Gimmel et al. [33] and Voorthuyzen and Bergveld [34], who found that the drift of Ta_2O_5 layers is strongly influenced by light. Only under ideal circumstances and storage conditions can low drift rates, as mentioned by Kuisl and Klein [30], be achieved.

It can be concluded that much research is still needed to clarify and solve the drift problem of ISFETs. However, when drift is reproducible, as is the case with Al_2O_3, corrections can be made in order to compensate for the drift [134, 135].

Temperature Sensitivity

The temperature dependence of an ISFET is complicated but can be completely explained with standard MOSFET theory and the site-binding theory used in Section 10.2.1.2 [136]. The temperature dependence of the output signal of ISFETs can be divided into three parts:
– Temperature dependence of the solid-state part of the transistor.
– Temperature dependence of the interface potential, Ψ_0.
– Temperature dependence of the reference electrode and the pH of the solution.

By operating the ISFET in the constant-current mode the operating point determines the overall temperature dependence of the ISFET in the solution. By careful calibration the overall temperature dependence can be determined and compensated for when the temperature is measured simultaneously [134, 135]. In order to reduce the temperature sensitivity, one can measure differentially with a reference FET, a REFET, as is explained in the next section, or subtract from the ISFET the signal from an analogous MOSFET in order to eliminate at least the solid-state part of the temperature dependence [98].

To summarize, it can be said that the inherent problems of ISFETs are (partly) solved and understood. The accuracy of the measurements depends strongly on the ion-sensitive layer of the ISFETs and the measurement conditions. The most accurate measurements can be achieved when corrections are carried out (provided that the correction parameters are well defined and reproducible) for, eg, temperature and drift as is shown in Figure 10-20 for a pH-sensitive ISFET with alumina as the ion-sensitive layer.

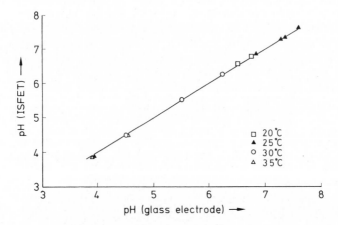

Figure 10-20. Comparison between a pH ISFET and a glass electrode. The ISFET output was automatically corrected for drift and temperature variations. (After [136])

10.2.4.3   External Limitations

Encapsulation

An important practical problem when working with chemical solid-state sensors, is the packaging of these sensors, especially with ISFETs. An additional problem is encountered when ion-selective membranes are deposited on top of the solid-state sensors, viz, the fixation of these membranes to the sensor and their compatibility with the packaging materials.

The encapsulation and packaging consist in principle of three steps:
− electronic isolation of active devices from the solution;
− lead attachment and encapsulation; and
− membrane attachment and fixation.

The final assembly of the encapsulated sensor and its carrier for custom use determines in principle which methods are used for each encapsulation step. Of course, depending on the techniques used, sometimes these steps can be carried out simultaneously. It is evident that the final application of the sensor determines the methods of packaging. Not only do the reliability and lifetime of the packaging play an important role in the decision regarding which methods will be used, but also the costs of the encapsulation and packaging have to be considered. In this respect a large difference exists, eg, between disposable sensors (one-time use during relatively short time periods, low price) and reusable sensors (repeatedly used, long time periods, higher prices). In this section a review of the techniques and materials for ISFET packaging will be presented.

Electronic Isolation of Active Devices from the Solution

In principle, all electrical components of a chemical sensor have to be isolated from their surroundings. Since the gate of an ISFET is exposed to the contacting solution, isolation from the solution is as important to the proper functioning as is the isolation of MOS components in circuits, from one another and from the environment.

The original ISFET, described by Bergveld [2], consisted of a MOS structure without gate metallization. However, the gate oxide, thermally grown silicon dioxide, loses its isolation property within a few hours of immersion in a solution. The mechanism of this breakdown appears to be consistent with the formation of micro-fissures in the top layer of the silicon dioxide which are filled with the solution [137] and not, as originally thought, with the total hydration of the silicon dioxide layer [105, 138, 139]. In order to isolate this gate oxide from the solution, another isolating layer, such as Si_3N_4, Al_2O_3, or Ta_2O_5, has to be placed on top of this gate oxide. However, after dicing the wafers into single chips, the substrate becomes exposed at the edges of the sensor. The simplest method of isolating these sides is encapsulation with epoxy-type resins. An excellent example of this method was given by Van der Starre et al. [81] and Ligtenberg et al. [82], where the epoxy was used not only to encapsulate the chip but also to encapsulate the bonding wires and to integrate the chip into the final package: a catheter. This catheter was used for in vivo pH measurements and contained a miniaturized reference electrode.

When using Si_3N_4 as insulator material [140, 141], another approach is possible. The sensors can be formed into needles by micromachining the wafer. These structures are then coated on all sides by LPCVD silicon nitride. However, the fabrication process of this device is not very compatible with standard IC technology.

Two other methods have been developed to isolate the border of the chip: application of diode isolation [142] or the silicon on sapphire (SOS) technique [143, 144]. The diode isolation technique consists of fabricating the n-channel ISFETs in p-type wells which are in turn implanted in an n-type substrate. Now, the reversely biased well-substrate diode ensures the electrical isolation of different ISFETs from each other and from the solution. The encapsulation of the edges of the chip is no longer critical. However, a light-induced photo-current in the diodes may give rise to some leakage current and hence the devices can be operated under limited illumination only.

The silicon on sapphire technique consists in manufacturing a FET on top of an insulator such as sapphire, thereby completely isolating the FET structure. These techniques are based on processes developed for integrated circuits and have the advantage that they can be carried out on the wafer level. Further, they do not interfere with subsequent packaging procedures.

Lead Attachment and Encapsulation

It is evident that for a proper functioning of ion microsensors the power-supplying leads and electrical signal leads have to be isolated from each other and from the solution. So far techniques such as wire bonding and tape automated bonding (TAB) [145, 146] have been used to attach these leads to the chip. For the packaging of normal integrated circuits these techniques present no problem, but the difficulty lies in the encapsulation of the bonds without covering the sensitive gate area. The currently applied method is to place the bonding pads on the sensor as far as possible away from the active gate region. With this configuration, the bonds can be isolated from the solution by covering them with epoxy resin without covering the gate. An improvement to these methods was developed by Ho et al. [145]. To prevent the contamination of the gate area with epoxy and to facilitate the encapsulation of the bonding wires, the gate areas are protected with photolithographically patterned material, such as Riston. After the bonding and encapsulation these materials are removed, leaving a well defined gate area.

The same well defined gate area can also be made by the field-assisted bonding of glass to silicon [147, 148], as shown in Figures 10-21 and 10-22. This electrostatic bonding of glass to silicon has the following advantages:
- excellent sealing properties;
- high resistance of glass to chemical attack;
- on-wafer processing;
- compatibility with membrane deposition; and
- the increased vertical dimension of the sensor allows wire bond encapsulation without difficulty.

Combined with the diode isolation technique, this method provides complete and easy encapsulation of the sensor. The glass encapsulation method has already been applied successfully to physical sensors such as silicon pressure transducers and solar cells. More recently this method was applied to the fabrication of planar silver/silver chloride electrodes [63, 149] and also the fabrication of ISFETs [136, 150]. The application of this technique to multi-ion sensor fabrication [85] is shown in Figure 10-23. It should be noted, however that, although the fabrication of these sensors can be made in a CMOS production line, the technology required to fabricate ISFETs suitable for anodic bonding is complicated.

A totally different approach is to place the contacts on the opposite face of the sensor, already suggested by Bergveld in 1972 [138] but without elaborating how this could be achieved. Over the years several techniques have been developed to create holes in a silicon wafer. The simplest method, which can be used on the wafer level, is the anisotropic etching

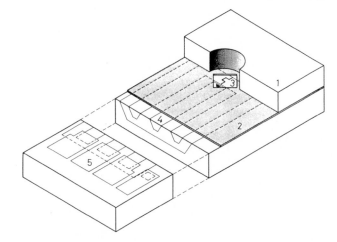

Figure 10-21.
Schematic diagram of a glass-encapsulated ISFET. (1) Glass plate; (2) polysilicon layer (completely covered by the glass); (3) gate area with protection diode; (4) p-well with drain and source; (5) contact pads.

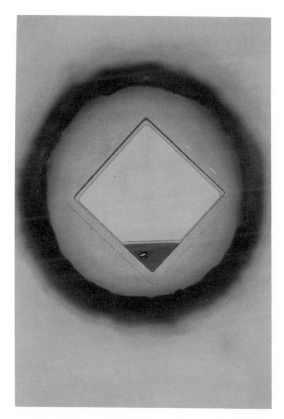

Figure 10-22.
Detail of gate area after glass encapsulation.

Figure 10-23.
Glass-encapsulated multi-sensor for flow-injection analysis (reprinted with permission from [85]).

of the silicon substrate with, eg, potassium hydroxide to form holes in the substrate, leaving a thin membrane with a controlled thickness [151]. Electrical contacts to the source and drain regions, formed on the front side of the chip, can be made by diffusion of the drain/source dopant through this thin silicon membrane. A cross section of an ISFET with back-side contacts is shown in Figure 10-24 [92]. Although this method has the disadvantage of requiring a relatively large space per contact, it has some attractive features, among which the flat surface of the encapsulated device is very important. This flat surface renders such an ISFET very suitable for, eg, the measurement of in vivo dental plaque pH for the investigation of the cariogenicity of food [92] or as part of carbon dioxide sensors with a fast response [103]. The same fabrication technique can also be applied for the integration of physical and chemical sensors. The simplest is the realization of a temperature sensor in combination with a pH sensor [152]. A more complicated design is a pH ISFET integrated with a pressure sensor [83, 92]. Extension of a back-side contacted sensor with a diode isolation is currently under investigation and increases the advantages of this method even more [93]. Another method which places the contacts on the other side of the active area is the silicon-insulator-silicon (SIS) three-layer structure [94, 96].

From the above it is evident that the methods of contacting the chemical sensor and encapsulation of the contacts depend strongly on the application. For multi-ion systems the method will probably be based on those utilized already for the encapsulation of integrated circuits

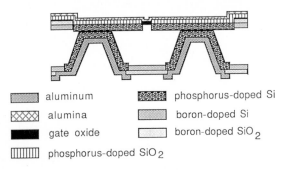

aluminum   phosphorus-doped Si

alumina   boron-doped Si

gate oxide   boron-doped SiO_2

phosphorus-doped SiO_2

Figure 10-24.
Cross section of an ISFET with back-side contacts.

such as tape automated bonding or anodic bonding to glass in combination with diode isolation. With respect to biomedical applications, especially for in vivo applications with the sensor mounted in a catheter, the wire bonding method followed by manual or automated encapsulation with epoxy resin will prevail and here the back-side contacted sensors with junction isolation can have great advantages.

Membrane Attachment and Fixation

As was discussed before, the sensitivity of ISFETs can be changed by depositing ion-selective materials on top of the gate insulator. Inorganic insulators and solid-state membrane materials such as Ag/AgCl [136, 150] are fabricated on-wafer and, apart from some adhesion problems, are compatible with the previously discussed packaging methods. However, these materials have no great variability in sensitivity, in contrast to solvent polymeric membranes where many ionophores exist to measure different ions. The fixation of these types of membranes on top of the gate is critical because any electrical shunt, either vertical or horizontal, through or around the membrane/solution potential generating interface will disturb the potential sensed by the FET.

To solve this problem, different solutions have been proposed. The simplest was the fabrication of epoxy wells in combination with silanization of the surface [154]. An improvement of this method was the photolithographically prepared well with Riston [145]. An elegant solution was also the fabrication of a suspended mesh of polyimide on top of the gate [62]. Further, the previously described method of encapsulation of anodically bonded glass provides a cavity into which these membranes can be solvent cast. Depending on the geometry of the glass, even liquid membrane materials (no support material) can be fixed into these cavities. Recently a system was described in which a micromachined silicon wafer was fixed with epoxy on top of another wafer containing the sensors. The silicon micromachined wafer had openings at the gate area into which the membranes can be solvent cast [155].

An additional problem arises when an intermediate hydrogel is necessary to improve the stability. A method to solve this problem of the fixation of a hydrogel and the subsequent deposition of the ion sensitive membrane is by the patterning of these materials with photolithographic techniques, as mentioned before. However, to achieve this the materials have to be modified to make them photosensitive, which is not a trivial task. Further, the encapsulation method still needs to be compatible with the materials used.

To summarize, it can be said that packaging of ISFETs can be performed using different approaches, depending on the final application. It can be expected that based on the various technologies outlined above fully integrated systems, such as flow cells, will be developed with inherent solutions for the packaging. For biomedical applications every use should be studied separately and the proper packaging method can then be determined with respect to the specific demands of the application. This also indicates that during the sensor design the packaging method should already be known in order to integrate the necessary provisions for a successful packaging of the chip.

Reference Electrodes

An important factor in the use of ISFETs as practical pH sensors is the reference electrode. To take full advantage of the small size of the ISFET, it is clear that there is a need for a miniaturized reference electrode. Principally two approaches have been followed to achieve this: reference FETs (REFETS), which are used in an ISFET/REFET/quasi-reference electrode setup, and miniaturized conventional reference electrodes.

In the first approach attempts have been made to cover the ISFET surface with a pH-insensitive layer, or to render the surface pH insensitive by chemical modification. Although the latter method appeared to be insufficient on practical and theoretical grounds [42, 44, 156], covering the pH-sensitive surface with a blocking polymer layer such as Teflon or parylene resulted at least in an acceptable pH insensitivity [157–159]. However, a non-negligible cation sensitivity was found instead, which appeared to be a function of the substrate bias voltage. In addition, REFETs using a completely unreactive and ion-blocking polymer on the gate suffer from a very low exchange current density, too low to define the interfacial potential. Therefore, they become very sensitive to all kinds of adsorption processes.

An alternative approach is to use a non-blocking polymer membrane on top of the gate, which exhibits a constant membrane potential as a function of pH and ionic composition [160, 161]. Such a behavior can be ensured by loading the membrane with equal amounts of membrane-confined anionic and cationic groups. In this way a stable potential is ensured, and a differential electronic setup using an arbitrary metal as a quasi-reference electrode gives a neat pH-sensitive output signal [160, 161] (see Figure 10-25).

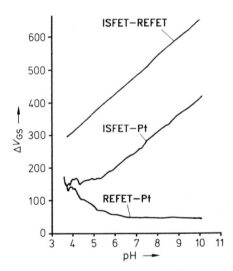

Figure 10-25.
Differential pH measurement with an Al_2O_3-ISFET/REFET/Pt. ΔV_{GS}: gate-source voltage. (After [44])

A slightly different approach is to use a quasi-REFET with a delayed pH response in a flow-through system [85]. In this case, the lowered mobility of ions in a p-HEMA hydrogel ensures a stable REFET potential during the time of measurement. A comparable hybrid solution using a small glass capillary to contact the external solution was reported in 1978 [162]. Finally, in enzymatic ISFET devices often an ISFET covered with a unloaded membrane is used as a reference to compensate for pH variations [107].

In the second approach, the structure of a conventional reference electrode (mostly of the Ag/AgCl type) is miniaturized partly [163] or completely [164–166] on a silicon wafer. In these cases the internal electrolyte volume is formed by an anisotropically etched cavity in the silicon wafer. A particularly important point in such miniature reference electrodes is the way in which the potential-determining liquid junction is defined. Whereas in conventional reference electrodes mostly a porous plug is used to control the inner electrolyte leakage, a porous glass plate [163], a porous silicon membrane [165, 166], or a very small opening size combined with reduced ion mobility through the use of a hydrogel have been tried with miniaturized versions. Although the porous silicon-based integrated reference electrode exhibits perhaps the best electrochemical performance, its wafer-scale fabrication appears to be very delicate. The above-mentioned devices have one problem in common: an acceptable lifetime coincides with a very high electrical resistance, which requires the use of a second (quasi-) reference electrode. This solution, however, has not been realized up to now.

10.2.5 Conclusions and Future Trends

Although the ISFET was introduced more than 20 years ago, it is only now, and in spite of some earlier attempts, that these devices have slowly started to emerge in the marketplace. Apparently, the initial claims for an easily mass-producible, inexpensive ion sensor were premature.

Principally there are two routes for the succesful commercialization of ISFETs. On the one hand there is a demand for specialized ISFET-based products such as pH catheters for biomedical use. Commercial introduction here is difficult because of the severe safety requirements placed on products for in vivo use. On the other, there is a possibility for a high-volume, low-price market, eg, for environmental control or in agriculture. In that case, not only is a low chip price important but even more so an inexpensive encapsulation method, as discussed in Section 10.2.4.3.

In addition to these commercial developments, scientific research has resulted in a better understanding of the functioning of ISFETs. Although basic studies on especially the pH-sensitive layer are still needed and are actually being carried out, most current investigations are rather directed towards devices sensitive to other ions instead of pH, and to indirectly sensing devices. In particular, the possibility of constructing miniaturized multi-sensors is an important incentive to continue research and development in this direction.

With respect to the design of ISFET-based sensors, the encapsulation is an often underestimated aspect. Especially with recently developed micromachining methods, new device realizations are conceivable which offer improved long-term performance of the sensor. Also, a miniaturized reference electrode, which is a necessary part of every electrochemical system, can be integrated with these techniques. Micromachining methods may also set the trend for future developments in miniaturized ion-sensor systems as, by using these techniques, complete chemical analyzer systems on a silicon chip can be envisaged.

10.3 Field Effect Gas Sensors

10.3.1 Introduction

Since the first paper by Lundström et al. [168] on the hydrogen gas sensitivity of palladium gate FETs appeared in 1975, a number of different FET-based sensors for various gases has been presented (see [169–171] and references cited therein). What at first seemed to be something of a curiosity now appears to be recognized as a whole new class of chemical sensors. This is partly due to the possibility for these sensors to be integrated with signal-processing electronic circuitry and the apparent scope for mass-fabricated single-chip multi-sensors.

Field effect devices have been used to detect gases such as hydrogen, ammonia, hydrogen sulfide, oxygen, unsaturated hydrocarbons and alcohols. Possible applications range from direct detection of hydrogen for various industrial purposes to indirect measurements for bioanalytical purposes.

This part of the chapter describes the principles of operation of the various types of field effect gas sensors, and their fabrication and measuring principles are also discussed briefly. The sensors are discussed in terms of characteristics such as sensitivity, selectivity, stability, and response kinetics. Also, various physical/chemical phenomena that have been studied in some detail are described, showing that the basic structures not only serve as a base for practical sensors but are also tools for research into areas such as surface physics and catalysis. In this context it is not possible or meaningful, to review all publications on gas-sensitive field effect devices and the selection of references is intended to be a guide to further reading. Nor is it practical to try to cover all applications that have been proposed. However, two applications, catalysis research under ultra-high-vacuum conditions and the use of the devices for bioanalysis, will be treated in some detail in order to give a flavor of the potential of these types of devices.

10.3.2   Principles of Operation

The basic mechanism behind the gas sensitivity of catalytic gate metal-oxide-semiconductor field effect devices is probably best understood by first studying the palladium gate field effect transistor (PdMOSFET) [168, 172, 173]. The detection mechanisms of palladium gate MOS capacitors (PdMOSCAPs) [174], metal insulator semiconductor (MIS), and MS Schottky diodes [175–179] and MIS switching devices [180] are very similar in principle. Other field effect devices such as the ultra-thin gate MOS (TMOS) ammonia sensor [181–183], the surface accessible FET [184], and the suspended gate FET [185] are based on different effects, as will be described later.

10.3.2.1   Basic Detection Principle of PdMOSFETs

It was initially observed that the electrical characteristics of Pd gate MOS transistors are sensitive to hydrogen gas in the ambient atmosphere [168–171]. Hydrogen shifts the drain current versus gate voltage curve towards lower voltages, as illustrated in Figure 10-26. Thin Pd foils are well known to be highly permeable to hydrogen and have therefore been used to separate hydrogen from other gases. When used as a gate metallization layer on an MOS transistor it specifically allows permeation of hydrogen to the metal/insulator interface.

Hydrogen gas molecules adsorb on the surface of palladium by dissociating into hydrogen atoms. These atoms diffuse rapidly through the metal and adsorb at the inner surface, where they become polarized. The resulting interface dipole layer is in equilibrium with the outer layer of chemisorbed hydrogen, and hence with the gas phase. The dipole layer gives rise to an abrupt step in the potential through the structure, which is often referred to as the voltage drop, ΔV. This voltage adds to the externally applied voltage and shows as a shift of the *I-V*

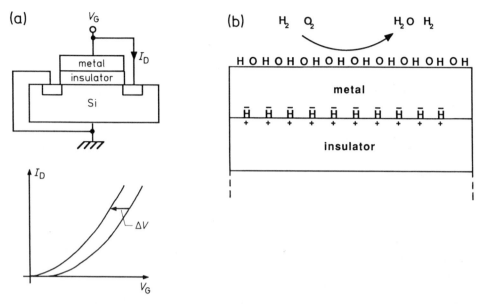

Figure 10-26. (a) Schematic illustration of a PdMOSFET structure. The device differs from an ordinary MOSFET in that the gate is metallized with palladium instead of aluminium or polysilicon. The current versus gate voltage characteristic is also the same, but shifts along the voltage axis when the device is exposed to hydrogen gas. This shift is always towards negative voltages irrespective of the device being a p-MOS or n-MOS FET. V_G: gate voltage, I_D: drain current, ΔV: voltage drop. (b) The hydrogen sensitivity of PdMOSFETs arises from catalytic dissociation of hydrogen molecules at the palladium surface. This process is reversible, but in the presence of oxygen most hydrogen atoms end up reacting with oxygen, producing water vapor. Hydrogen atoms penetrate the metal film and adsorb at the palladium/insulator interface where they form dipoles.

curve towards lower voltages. Taking V_{T0} to be the threshold voltage of the transistor in the absence of hydrogen, the apparent threshold voltage of the FET can be written as

$$V_T = V_{T0} - \Delta V \tag{10-21}$$

where

$$\Delta V = n_i \cdot p/\varepsilon_0 . \tag{10-22}$$

n_i is the number of hydrogen atoms (per unit area) at the inner interface and p is their dipole moment when adsorbed. Another way of looking at this phenomenon is to say that the voltage drop ΔV corresponds to a decrease in the effective work function of the metal at the metal/ insulator interface with the amount $e\Delta V$.

When hydrogen gas is removed from the ambient atmosphere, the hydrogen atoms at the outer surface recombine into hydrogen molecules, which desorb from the surface. In the presence of oxygen the recombination is dominated by water formation. Also, the metal/insulator interface, which is in equilibrium with the outer surface, is then emptied of hydrogen. Removal of hydrogen from the ambient atmosphere therefore restores the initial work function of the gate metal. The shift in the *I-V* curve is therefore completely reversible, even though full recovery may take a considerable time (hours) [186], as will be described in Section 10.3.5.2.

The chemical reactions on palladium take place even at room temperature, but a practical sensor should operate at an elevated temperature in order to react quickly enough to variations in gas concentration [168]. PdMOS devices are normally operated at 100–150 °C in order to be fast enough (see, eg, [170] and references cited therein).

10.3.2.2   Pd Gate MOS Capacitors

MOS capacitors (MOSCAPs), with their voltage-dependent capacitance characteristics, are very much easier to fabricate than MOS transistors. They are therefore often favored in the research laboratory. The basic mechanism behind the gas response is identical with that of the Pd gate MOSFETs but the dipole layer is instead detected as a shift in the capacitance-voltage (*C-V*) characteristics [169, 170, 174].

A typical *C-V* curve for a p-type silicon MOSCAP is shown in Figure 10-27. The flatband voltage (V_{FB}) of the capacitor (the voltage at which the electric field at the semiconductor surface is zero) is shifted by ΔV according to

$$V_{FB} = V_{FB0} - \Delta V \qquad (10\text{-}23)$$

where V_{FB0} denotes the flatband voltage with no hydrogen in the ambient atmosphere.

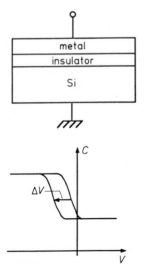

Figure 10-27.
A PdMOS capacitor is essentially a PdMOSFET structure with no source or drain contacts. The capacitance of a MOSCAP is voltage dependent, as illustrated. On exposure to hydrogen the *C-V* curve shifts exactly as does the *I-V* curve of a PdMOSFET. *C:* capacitance, ΔV: voltage drop, *V:* voltage.

10.3.2.3 MIS Schottky Diodes

Metal-thin oxide semiconductor Schottky barrier diodes [175, 176] and metal-semiconductor Schottky barrier diodes [177–179] are also gas sensitive, provided that the right gate metal is used (see Figure 10-28). Both the *C-V* and the *I-V* characteristics are influenced by gas exposure.

In the case of silicon substrates it is necessary to have a thin insulator (oxide) between the metal gate and the semiconductor [175, 176], otherwise electron states at the semiconductor surface will allow electrons to compensate for the voltage drop across the dipole layer [176]. This effect is known in semiconductor physics as Fermi level pinning by surface states. The possibility of palladium silicide formation has also been suggested [175]. Other semiconductors, such as CdS [177], ZnO [178], InP, and GaAs [179], have also been evaluated for use in gas-sensitive Schottky barrier devices.

Since the hydrogen dipoles affect the inversion layer in PdMOSFETs, they probably affect the height of the Schottky barrier in the Schottky diodes. Some evidence has been found, however, suggesting that hydrogen can penetrate the thin oxide and alter the electronic surface states in the semiconductor [175, 187]. There is still some confusion as to whether this may be the dominant effect in the Schottky barrier devices [171].

Thyristor-like switching devices with a palladium gate Schottky barrier have been shown to turn on or off on exposure to hydrogen [180]. By a suitable choice of the bias voltage, these devices can be made to switch from a conducting to a nonconducting state, or vice versa, at a predetermined hydrogen concentration.

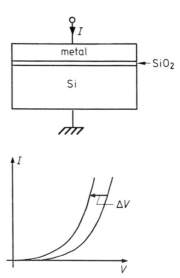

Figure 10-28.
A Pd Schottky diode is essentially a MOSCAP structure where the insulator is thin enough to allow a current to pass through. The thickness of the insulator is typically 0–2 nm. The capacitance-voltage curve shifts as for the MOSCAP; so does the *I-V* curve, as depicted. *I*: current, ΔV: voltage drop, *V*: voltage.

10.3.2.4 Thin Metal MOS (TMOS) Devices

In all the above-mentioned devices, the metal gate film is thick enough to be continuous, or at least dense enough to screen out any polarization of the outer surface. They are therefore sensitive only to hydrogen which is able to penetrate the metal and polarize the inner interface.

Other types of sensors based on ultra-thin (discontinuous) metal gates of Pt, Ir, etc., have been shown to be sensitive to other gases, such as ammonia [181–183] (see Figure 10-29). These devices are believed to operate with a different mechanism, especially since the ammonia sensitivity only occurs when the metal gate is thin enough to be discontinuous. This implies that the field effect arises from a capacitive coupling of a polarization of the outer metal surface to the semiconductor, through pores in the metal film [182, 183].

Both MOSFETs and MOSCAPs with ultra-thin metal gates have been shown to be gas sensitive. Again, the sensitivity shows as a shift of the electrical characteristics along the voltage axis, corresponding to an effective reduction of the threshold voltage or the flatband voltage [182, 183].

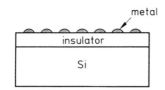

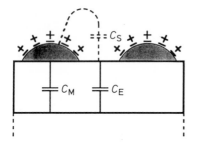

Figure 10-29.
Thin metal MOS devices differ from ordinary MOS devices in that the gate metallization layer is thin enough to be discontinuous. The space between metal grains allows the electrical polarization of the metal surface to affect the semiconductor. This capacitive coupling of the surface polarization makes it possible to detect gases that do not give rise to a detectable amount of hydrogen atoms at the metal/insulator interface when reacting on the metal surface. C_E, C_M: geometrical capacitances.

10.3.2.5   Related Devices

A type of device similar to the ultra-thin metal gate devices is that with holes in the gate metallization [188, 189]. These holes are defined by photolithography and are therefore much larger than the pores in the TMOS. However, the sensitivity mechanism is probably similar to that in TMOS devices, ie, outer surface polarization capacitively coupled through the openings in the metal. These devices have been reported to be sensitive to carbon monoxide [188, 189]. However, no direct comparison with the TMOS devices has been published.

Only FETs have been studied and it has been found that the slope of the *I-V* curve (the factor K_d in Equation (10-6)) also changes on gas exposure. This may be related to the fact that the threshold voltage is affected only at the holes and not between them. Such an inhomogeneous influence on the transistor channel must affect the overall characteristics of the transistor.

A different situation is found with the suspended gate field effect transistor, SGFET [185]. Here the gate consists of a metal mesh suspended about 100 nm above the insulator surface. Surface polarization of the metal, on both its outer and inner surfaces, which are now accessible from the gas phase, gives rise to surface potential changes of the semiconductor. Such a device is obviously sensitive to all gases capable of polarizing the metal or the insulator surfaces. The fact that it is not selective to one specific gas does not mean that it is not useful. It has been shown, for example, that coating the gate with polypyrrole increases the sensitivity to alcohols [185]. Modified versions of the device may, for example, be integrated in a sensor array.

Another means of obtaining a similar effect is to underetch the metal gate heavily, leaving an open space between the metal and the insulator at one end of the transistor channel. Such devices are recognized as surface-accessible FETs [184]. No further development of these has been reported.

10.3.3   Fabrication and Measuring Principles

10.3.3.1   Fabrication

The chemically sensitive MOSFETs can be fabricated by means of standard MOS technology, except for the gate metallization step. An MOS capacitor is, of course, the simplest device to make, although a slightly more sophisticated integrated circuit is preferred for a practical sensor. Most of the devices should be operated at an elevated temperature, and it is therefore worthwhile integrating a heating resistor, a temperature-measuring pn diode, and a MOSFET on the same chip.

A simple MOS fabrication process based on p-type silicon is as follows. Oxidize the silicon wafer in dry oxygen at 1200 °C to obtain a 100-nm thick insulator. Remove the oxide from the back of the wafer and deposit aluminium. Anneal this aluminium layer at 500 °C in forming gas in order to ensure a good electrical back-side contact. Vacuum deposit approximately 100 nm of palladium or other gate metal onto the wafers, patterning it to, eg, 1 mm^2 circular dots. Each of these dots constitutes an MOS capacitor which is individually contacted. In order to use this simple form of device, place the wafer on a temperature controlled heater which is in electrical contact with one terminal of the capacitance meter.

Much care has to be taken in obtaining a high-quality silicon dioxide, in order to minimize the normal MOS drift problems originating from alkali ion migration. This will otherwise turn out as a baseline drift, but will also affect the gas response characteristics, as will be discussed in Sections 10.3.5.1 and 10.3.5.2.

Since fairly inert metals are used, one has to address the problem of poor adhesion. A Pd film may actually peel off just as a result of the small mechanical stress induced by the absorption of hydrogen. A less drastic effect is blister formation in the film, which results in nonreversible changes in the device characteristics [190]. The quality of the gate metal layer is crucial for the overall performance of the sensor, especially in terms of reproducibility. This is even more important for the ammonia-sensitive ultra-thin metal gate devices, because their performance is highly dependent on the microscopic structure of the gate metal [182, 183].

It has been shown that the properties of the metal/insulator interface have a pronounced impact on the signal stability of the hydrogen-sensitive PdMOS devices [186]. Incorporation of a second insulating layer of, eg, alumina, tends to minimize the so-called hydrogen-induced drift (HID) (see Section 10.3.5.1). It is implied that this drift problem resembles that of ISFET devices made with silicon dioxide only.

10.3.3.2 Measuring Principles

The capacitance of an MOS capacitor is voltage dependent [191], as illustrated in Figure 10-27. The maximum capacitance in the accumulation region is simply that for a parallel-plate capacitor with an insulator thickness equal to the gate insulator thickness:

$$C_i = A_i \varepsilon_i / d \ . \tag{10-24}$$

In the voltage-dependent depletion region the capacitance follows approximately

$$C = C_i / \sqrt{1 + k(V - V_{FB})} \tag{10-25}$$

where k is a constant depending on the doping density in the semiconductor and the oxide thickness.

The flatband voltage is given by Equation (10-1) or, written in a slightly different way,

$$V_{FB} = \phi_{ms} - d_i Q_i / \varepsilon_i \tag{10-26}$$

where ϕ_{ms} is the difference between the effective work function of the metal (at the metal/insulator interface) and that of the semiconductor. The last term shows the effect of charges in the insulator and at the insulator/semiconductor interface and is important for the device stability. Q_i in Equation (10-26) is an effective oxide charge (see also Chapter 10.1).

The hydrogen dipoles affect ϕ_m and hence V_{FB}, which according to Equation (10-25) shifts the capacitance-voltage ($C(V)$) curve along the voltage axis. The best way to measure this is to monitor the voltage that must be applied in order to keep the capacitance constant. This does not need to be exactly where $V = V_{FB}$, but should be at a position where the slope of the $C(V)$ curve is significant. Going too far away from $C(V_{FB})$ may also cause changes in Q_i initiated by the electric field in the insulator.

The advantages of this technique over measuring the capacitance at a constant voltage are the linear relationship with the polarization and the fact that the electric field is kept constant, thereby minimizing changes in Q_i.

For a MOSFET, contacted according to Figure 10-26, the *I-V* relationship [191] is given by Equation (10-8) if $V_T < 0$ (note that $V_G \equiv V_D$ in Figure 10-26) and by

$$I_d = \frac{K_d}{2} \cdot (V_G - V_T)^2 \tag{10-27}$$

if $V_T > 0$ where V_T van be written as (see Equation (10-3)

$$V_T = V_{FB} + 2\phi_B + (d_i/\varepsilon_i) \sqrt{4q \varepsilon_i N \phi_B} \tag{10-28}$$

where ϕ_B is the difference between semiconductor mid-gap and the bulk Fermi level and N is the doping density in the semiconductor. As with the MOSCAP, it is preferable to monitor the voltage shift, here by maintaining a constant current. It is possible, of course, to measure the current at a constant voltage. However, nothing is gained in terms of higher sensitivity, since noise and drift are amplified as much as the signal is.

The same argument holds also for MIS and MS Schottky devices. Even though the *I-V* relationship for forward bias [191] is exponential:

$$I \sim \exp[q(V - \Psi)/k_B T] \tag{10-29}$$

where Ψ is the barrier height equal to the difference in effective metal work funktion and semiconductor electron affinity, nothing is gained in terms of sensitivity by operating the device at constant voltage.

10.3.4   Sensitivity and Selectivity

10.3.4.1   Hydrogen Sensitivity of PdMOS Devices

A simple theoretical model for the gas sensitivity of palladium gate MOS sensors was developed by Lundström et al. [168]. The basic assumptions in this model are as follows [168, 169, 171]:

A. Hydrogen molecules adsorb and dissociate on the metal surface in proportion to the partial pressure of hydrogen and the number of free adsorption sites.

B. Hydrogen desorbs from the metal surface by recombination into H_2 and by reacting with oxygen to form water:

$$H_2 \underset{d_1}{\overset{c_1}{\rightleftarrows}} 2\,H_a \quad \text{and} \quad O_2 \rightleftarrows 2\,O_a$$
$$4\,H_a + O_2 \rightarrow 2\,H_2O\,. \tag{10-30}$$

The water-producing reaction may consist of several different steps involving, eg, adsorbed OH groups. The reactions above balance each other, giving an equilibrium with a certain coverage of hydrogen atoms on the surface. Variations in the hydrogen and oxygen partial pressures shift this equilibrium.

C. Hydrogen atoms diffuse readily through the metal and adsorb at the metal/insulator interface. Owing to the very high diffusion constant there is always an equilibrium between the number of hydrogen atoms at the two surfaces:

$$H_a \underset{d}{\overset{c}{\rightleftarrows}} H_i\,. \tag{10-31}$$

D. Hydrogen atoms adsorbing at the metal/insulator interface become polarized, giving rise to a voltage shift, ΔV, which is proportional to the number of adsorbed hydrogen atoms at the interface, as given by Equation (10-22).

Assumptions *A* and *B* give the following relationship:

$$c_1 P_{H_2} F_e^2 = d_1 n_e^2 + f(P_{O_2})\, n_e^2 \tag{10-32}$$

where P_{H_2} and P_{O_2} are the partial pressures of H_2 and O_2, respectively, and n_e is the number of adsorbed hydrogen atoms H_a. F_e is the number of unoccupied adsorption sites on the surface, ie, sites with no hydrogen, oxygen, OH, H_2O, or contaminants. The left-hand side of Equation (10-32) is the probability of an H_2 molecule adsorbing and dissociating on two unoccupied sites, and the right-hand side is the probability of two H_as recombining into H_2 or H_2O. The function $f(P_{O_2})$ depends on what the actual reaction scheme for water formation is [169, 171, 192].

Assumption C gives, by analogy with the above,

$$c\, n_e\, F_i = d\, n_i\, F_e \tag{10-33}$$

where F_i is the number of unoccupied adsorption sites at the metal/insulator interface.

Assumption D can be expressed as

$$\frac{n_i}{F_i} = \frac{\Delta V}{\Delta V_{max} - \Delta V} \tag{10-34}$$

because sites at the interface can only be either unoccupied or occupied by hydrogen atoms.

Combining Equations (10-32), (10-33), and (10-34) yields

$$\frac{\Delta V}{\Delta V_{max} - \Delta V} = \frac{c}{d} \sqrt{\frac{c_1\, P_{H_2}}{f(P_{O_2}) + d_1}} \,. \tag{10-35}$$

Experimental results have been shown to fit this equation fairly well, giving a ΔV_{max} of about 0.5 V [168, 169, 171, 192]. Measurements of the oxygen dependence of the hydrogen sensitivity have been made in order to establish the function $f(P_{O_2})$. It turns out that it can be either $k_1 P_{O_2}$ or $k_2\sqrt{P_{O_2}}$ [192], depending on the device temperature [3, 26]. Armgarth et al. [193] showed that the first function was applicable at a temperature of 75 °C (see Figure 10-30 (top)). On the other hand, at 50 °C the results could only be fitted to the square-root dependence of $f(P_{O_2})$ (see Figure 10-30, bottom), indicating another dominating reaction path.

Equation (10-35) has a form like a Langmuir isotherm. This means that the response is nonlinear and saturates when the number of hydrogen atoms at the metal/insulator interface approaches a monolayer (see Figure 10-31). The differential sensitivity to hydrogen, in air, which is 25 mV/ppm around 1 ppm, falls to about 25 μV/ppm around 1000 ppm.

The very high differential sensitivity at low concentrations makes it very difficult to determine exactly the true zero point of the devices. Even trace amounts of hydrogen present in high-purity gases affect the baseline. This is even worse in the absence of oxygen because then the term $f(P_{O_2})$ in Equation (10-35) vanishes, resulting in an extremely high sensitivity to hydrogen. In ultra-high-vacuum experiments [194] where the partial pressure of hydrogen can be brought down to 10^{-10} Torr (equivalent to 10^{-7} ppm) or less, one still observes a hydrogen response which can be brought down by introducing oxygen into the vacuum chamber. An exact determination of the zero point at normal pressures is therefore impossible, and any estimation has to be made with a high-purity gas background containing as much oxygen as possible.

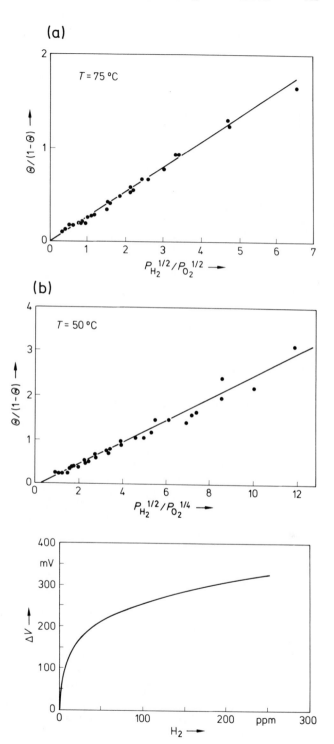

Figure 10-30.
Experimental results on a Pd-Al$_2$O$_3$-SiO$_2$-Si capacitor, fitted to Equation (10-35). Θ is the measured fraction of ΔV_{max}, which in this case was 570 mV. The top curve (a) shows the best fit at 75°C and the bottom curve (b) is the best fit at 50°C. The data had to be fitted to different functions of the oxygen pressure at the two temperatures in order to give straight lines. This indicates that different rate-limiting reactions between hydrogen and oxygen are dominating at different temperatures. P: partial pressure. (From [193])

Figure 10-31.
Response characteristics of a PdMOS hydrogen (H$_2$) sensor in air. Device temperature, 150°C. ΔV: voltage drop.

The applicability of the model above to inert gases (ie, with $f(P_{O_2}) = 0$) has been questioned [171, 194]. In inert atmospheres experimental sensitivity curves can be fitted to the Temkin isotherm, ie, a logarithmic dependence of the hydrogen concentration. Dannetun et al. [194] managed to do this over an eleven-decade range by combining the results from experiments in ultra-high vacuum, in high vacuum, and at atmospheric pressure.

Difficulty in comparing experiments made under different conditions is probably due to the hydrogen-induced drift (HID) discussed in Section 10.3.5.2. The possibility of HID has to be considered in careful analysis of experimental data.

10.3.4.2   Sensitivity to Hydrogen Sulfide

PdMOS devices are also sensitive to H_2S [195]. The H_2S sensitivity is nearly as high as the hydrogen sensitivity. The detection mechanism is believed to be similar to that for hydrogen, namely a dissociative adsorption of the molecule on the Pd surface. If the devices are exposed to H_2S in an inert atmosphere, they have been reported to be inactivated, possibly by sulfide deposition on the surface. However, exposure to air activates the sensor again. No permanent inactivation due to H_2S exposure in air was reported, although a long-term inactivation due to sulfide-containing air pollutants has been suggested [196]. The temperature dependence of the sensitivity is about the same as for hydrogen [195].

10.3.4.3   Sensitivity to Other Hydrogen-Containing Gases

By increasing the operating temperature, other hydrogen containing gases can also be detected [197–199]. The mechanism in this case is also dehydrogenation of the molecules, even though a higher temperature is needed to obtain a reasonable sensitivity. The atomic hydrogen formed is assumed to desorb via water formation. In order to avoid inactivation of the surface, the remainder of the dehydrogenated molecule has to be desorbed from the surface through one or several reaction steps.

Alcohols such as methanol, ethanol, propanol, and butanol have been detected at temperatures up to 240 °C [198, 200]. Figure 10-32 shows the response to ethanol as a function

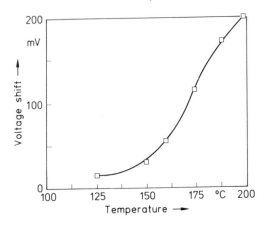

Figure 10-32.
Temperature dependence of the response to 0.8% of ethanol in air for a PdMOS device. The strong temperature dependence makes it possible to distinguish ethanol from hydrogen, which gives a large response at temperatures well below 150 °C. (After [198])

of temperature. The temperature dependence is much more pronounced for alcohols than for hydrogen.

Other hydrogen-containing molecules such as unsaturated hydrocarbons, eg, acetylene and ethylene, have been detected [197, 199, 201]. Pd gate devices operated at low temperatures have also been reported to be sensitive to saturated hydrocarbons at high concentrations [202]. It is, however, not likely that this is due to dehydrogenation of the hydrocarbon molecules. More probably, the sensor responded to trace amounts of hydrogen impurities in the gas [171].

10.3.4.4 Gas Sensitivity of TMOS Devices

MOS devices with ultra-thin gates of catalytic metals such as Pd, Pt, and Ir (TMOS) are reported to be very sensitive to ammonia [181–183] and to have a higher sensitivity to some hydrogen-containing gases than "thick" metal gate devices [199].

The ammonia sensitivity has been explained by the following model [182, 183]. On gas adsorption, the outer surface of the metal becomes polarized. This also happens on a thick metal but the polarization is then screened by the metal. Because a thin metal layer is discontinuous, the surface potential change can couple capacitively through the openings in the film. These openings are of the order of 1–10 nm, which is sufficient to allow the surface potential change to influence the electric field at the silicon surface. Simple arguments yield

$$\Delta V = \frac{\Delta \phi_S \, C_S \, C_E}{C_S \, C_E + C_M (C_E + C_S)} \tag{10-36}$$

where $\Delta \phi_S$ is the surface potential change due to adsorbed species and/or reaction intermediates. $\Delta \phi_S$ and C_S are complicated functions of the metal structure, whereas in a first approximation C_M and C_E can be regarded as simple geometric capacitances (see Figure 10-29 for a definition of the parameters in Equation (10-36)).

Figure 10-33 shows the response to different ammonia concentrations for a 3-nm Pt film. In this context it is worth pointing out that the same chemical reactions are known to occur

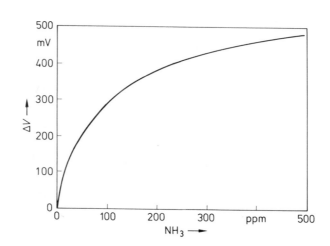

Figure 10-33.
Ammonia (NH$_3$) response characteristics of an Ir-TMOS device in air, where the discontinuous iridium gate metal is 3 nm thick. Device temperature, 150°C. ΔV: voltage drop. (Adapted from [181])

on the ordinary PdMOS device, but no liberated hydrogen atoms seem to be transferred to the interface. They probably desorb as water from the outer surface immediately. With a discontinuous film, however, reaction intermediates can give rise to measurable effects. The possibility of dehydrogenation of NH_3 molecules and detection of a hydrogen dipole layer has been excluded [183].

Only discontinuous gates from catalytic metals such as Pd, Pt, Ir, Ru, and Rh are sensitive to ammonia whereas discontinuous Fe, Al, and Ni gates are not [181–183]. This fact, together with the dependence on oxygen and temperature, strongly suggest that the adsorption process is chemisorption rather than simple physisorption. This conclusion is further supported by mass spectrometric experiments detecting reaction products such as H_2O, N_2, and N_2O from platinum and H_2O, N_2, and NO from iridium [203].

The capacitive coupling model can be described as an electrostatic edge effect around each metal grain. The model has been tested by fabricating gates where the grains were as large as 50 nm [182]. Exposure of such devices to ammonia results in a non-parallel shift of the *C-V* curve, indicating that a fraction of the surface area is not active. In the case of very small grains the shift is parallel, implying that the electric field from the edges overlap, resulting in an almost homogeneous field at the insulator/silicon interface.

The ammonia sensitivities of Pt and Ir gates have been measured from room temperature up to 250°C. Both metals exhibit a strong temperature dependence of approximately 1–2 mV/°C for NH_3 concentrations in the range 10–250 ppm. Iridium tends to be slightly more sensitive to ammonia, but the main difference between the two metals is that Pt gates generally respond faster.

Increasing the operating temperature yields sensitivity also to other gases. Figure 10-34 shows the sensitivity for three types of devices to five gases.

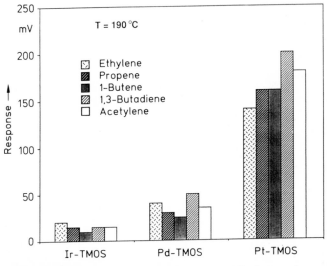

Figure 10-34. Response to 100 ppm (30-s pulses) of five hydrocarbon gases in air of three types of gas-sensitive FET sensors operated at 190°C. (Adapted from [199])

Several experimental observations support the idea of the ammonia sensitivity being due to a mechanism different to that of the normal hydrogen sensitivity of thick metal gates. In the case of hydrocarbon sensitivity for the thin metal gates, it is less evident that the capacitive coupling is solely responsible for the observed voltage shift. A voltage shift due to interfacial hydrogen may also occur.

10.3.4.5 Carbon Monoxide

Carbon monoxide sensitivity has been reported for devices with photolithographically made holes in the catalytic metal gates [188, 189]. The voltage shift, however, is in the opposite direction to the hydrogen-induced voltage shift [188]. A device with a PdO-Pd gate in which holes were etched has been reported to be sensitive to CO [189]. Those devices exhibited some cross-sensitivity to ethanol, butane, and methane. The maximum voltage shift due to CO exposure was about one order of magnitude lower than the ammonia-induced voltage shift for the ammonia-sensitive thin PtMOS devices. A capacitive coupling model has also been suggested to apply to the hole gate devices [189].

A 2.5-nm Pd gate MOS capacitor was reported to be sensitive to CO in the range 100–10000 ppm. With a 4-nm Pd gate no response was found [204].

10.3.4.6 Interferents

No comprehensive reports on different interferents for the different gas-sensitive catalytic gate MOS devices have been published. It is generally found, however, that oxidizing gases, such as oxygen, hydrogen peroxide, and chlorine, reduce the hydrogen sensitivity for PdMOS structures [167–171]. Sulfide-containing gases was discussed in Section 10.3.4.2.

The ultra-thin Pt and Ir gate structures are also sensitive to oxidizing gases. Water vapor can by itself give rise to a response in some cases, and has been reported to speed up the response to ammonia, still having little effect on the steady-state response [183].

The interference by oxygen on the hydrogen sensitivity for PdMOS devices is a minor problem in practical applications but has actually been utilized for the detection of small oxygen variations [205]. By flowing a hydrogen-containing carrier gas over the sensor, and mixing in consecutive pulses of an oxygen-containing reference gas and the test gas, small signal variations around a steady state were obtained. The maximum pulse response is limited by [205]

$$\Delta V_p = \Delta V_{max} \frac{1 - \sqrt[4]{P_{O_2}^{test}/P_{O_2}^{ref}}}{1 + \sqrt[4]{P_{O_2}^{test}/P_{O_2}^{ref}}} \tag{10-37}$$

assuming that $f(P_{O_2}) \approx P_{O_2}$ in Equation (10-35) (and neglecting d_1). This pulsed mode of operation gives a resolution of about 0.5% O_2, and minimizes influences from baseline drift. An example of results is given in Figure 10-35.

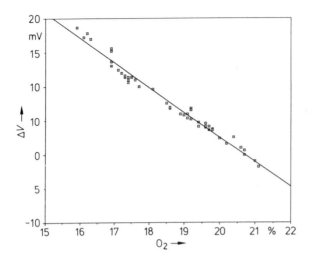

Figure 10-35.
Calibration graph for rapid oxygen determination by means of a PdMOS device kept in an atmosphere containing hydrogen. The response is the voltage shift that results from switching between a reference gas containing 20.7% O_2 and the test gas. Each pulse was 6 s long. This makes the system useful for, eg, clinical monitoring of exhaled oxygen, requiring time resolution of the order of seconds. ΔV: voltage drop. (From [205])

10.3.5   Stability and Kinetics

The phenomena that cause instability in catalytic gate MOS devices can be divided into three main categories: baseline drift, gas-induced drift and inactivation of the gas sensitivity.

Baseline drift is essentially dependent on process parameters as in conventional MOS technology, while the gas-induced drift is specific for these devices. Inactivation caused by chemisorbed contaminants is not as serious a problem as in industrial catalysis because the sensor response relies on an equilibrium rather than on a reaction rate. Inactivation in this context is therefore more a matter of slower responses than lower responses. Exceptions from this do exist, however, as further discussed at the end of Section 10.3.5.3. Generally the inactivation is a minor problem in practice, because continuous reactions, involving background hydrogen and oxygen, tends to clean the catalytic metal surface of contaminants.

10.3.5.1   Baseline drift

The baseline drift in gas-sensitive MOS devices results mainly from the same phenomena as does the threshold voltage drift in conventional MOS transistors. Sodium ions, arising from contamination of the insulator, drift across the structure under the influence of the electric field. The amount of sodium must therefore be kept low. Various methods are also known to prevent the unavoidable sodium ions from moving in the oxide. Another drift phenomenon is injection of electrons from the silicon into the oxide. The means of minimizing this is to grow the oxide under well controlled conditions and to anneal the oxide at a moderate temperature in, eg, forming gas.

Both the above-mentioned drift effects are affected by the electrical field through the oxide. It is therefore a good rule always to operate the devices so that the oxide field is kept as low as possible.

10.3.5.2   Hydrogen-Induced Drift

Even if the baseline drift can be brought to a minimum in the ordinary Pd- and PtMOS devices, they are still not stable enough for continuous monitoring of hydrogen. This limitation is due to a phenomenon referred to as hydrogen-induced drift (HID) [186]. As the name indicates, an otherwise stable PdMOS device starts to drift when it is exposed to hydrogen. Fortunately, the effect is reversible, but the time scale for recovery can be problematic. Figure 10-36 illustrates how the device gives a fast response followed by a slow drift. This additional shift of the flatband voltage cannot be described with a single time constant but has a wide range of time constants spanning from seconds to hours. It is even more predominant when the gas is removed because the recovery of the drift is much slower. The HID does, however, saturate in proportion to the hydrogen concentration. Devices suffering from HID can therefore be regarded as having an extremely long response time.

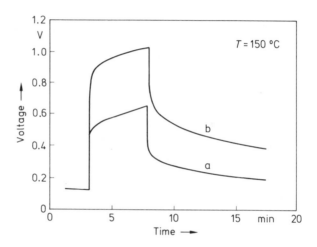

Figure 10-36.
PdMOS devices exhibit a hydrogen-induced drift (HID) which recovers very slowly after exposure to hydrogen. The drift is enhanced by sodium ion contaminations in the oxide. Curve a is recorded on a "clean" device and b is taken from a deliberately contaminated device. Both curves show the response to 500 ppm H_2 in an oxygen background. (From [186])

It was reported by Armgarth and Nylander that the HID does not occur on PdMAOS devices [206], ie, devices with a layer of alumina between the metal and the silicon dioxide. It was later reported that also other insulators such as Si_3N_4, Ta_2O_5 [207], and possibly BN [208] are similarly beneficial. These materials do, however, present a practical problem in that they substantially decrease the adhesion of the metal gate.

The fact that HID manifests itself mainly on SiO_2-based devices indicates that hydrogen not only polarizes the metal, but also interacts in some way with the insulator. Evidence for this was found by Nylander et al. [186], who made a series of experiments, indicating that protons probably interact with the first atomic layers of the oxide. It was found that this was related to the degree of sodium ion contamination of the oxide in such a way that HID can be increased by deliberately contaminating the oxide (see Figure 10-36), but could not be completely eliminated by careful fabrication. It is well known that sodium ions are present to some extent in even the most carefully fabricated MOS devices.

It was also found that, in the presence of oxygen, a significant part of the fast response to hydrogen is due to the onset of HID [186]. In the absence of oxygen, HID is virtually irreversible and is therefore observed only on the initial exposure to hydrogen (see Figure 1 in [186]).

The device behavior in argon was similar to that of a PdMAOS device. It is likely that the complexity of the observed response characteristics of PdMOS devices (see the end of Section 10.3.4.1) is partly due to HID. This is supported by the fact that the behavior depends on the choice of insulator.

Another type of gas-induced instability has been observed in ultra-thin PtMOS devices [182]. This effect has not yet been studied in detail, but it has been concluded that it is related to gas- and/or heat-induced structural changes in the metal films. Transmission electron microscopy has, for example, revealed grain fusion in discontinuous Pt film when exposed to hydrogen. This was correlated with a decrease in ammonia sensitivity.

10.3.5.3   Speed of Response

The HID effect makes it meaningless to define the response time in terms of time to obtain 90% of the final reading. Attempts to do so fail because of the dependence on the experimental conditions and device history. There are also considerable variations between different devices, even within fabrications batches.

It is, clear, however, that the initial derivative of the response is linearly proportional to the hydrogen concentration, as theoretically predicted [168, 171–173]. The initial derivative is also temperature dependent, higher temperatures giving faster responses. Hydrogen-sensitive devices can be operated at room temperature, but are then impractically slow. Operating temperatures above 100 °C are often recommended for practical purposes.

The linear relationship between initial derivative and gas concentration has been utilized with both hydrogen and ammonia sensors (see Section 10.3.6.2). The gas is then fed to the sensor as short pulses (typically 30 s) of constant duration. The heights of the response peaks become proportional to the initial derivative, which in turn is proportional to the gas concentration. Another advantage of working in a pulsed mode is obviously that baseline drift and hydrogen-induced drift are minimized. The reproducibility of measurements over a day becomes acceptable for many applications, but regular calibration is to be recommended.

If the carrier gas, as opposed to air, contains no hydrogen then the responses to the first pulses of test gas can be lower than for the following pulses of the same concentration. This effect has been described as oxygen inactivation [170, 209]. Ellipsometric measurements on PdMOS capacitors revealed [210] that inactivation is related to adsorption of molecules on the metal surface between hydrogen exposures. There is, however, no direct evidence that oxygen, rather than, eg, sulfur (cf, Section 10.3.4.2), is responsible for the inactivation.

10.3.5.4   Other Effects

Several other phenomena have been reported, mainly in PdMOS devices [171, 183, 209]. Some of these are relevant to the performance of the sensor, while some are interesting from a semiconductor physics point of view.

The catalytic metals used in catalytic gas sensors are all rather inert metals. They therefore adhere rather poorly to the insulator and care must be taken not to create mechanical stress in the films when depositing them. In the worst cases the films curl up when exposed to high concentrations of hydrogen. A less drastic effect, wich can cause irreversible changes of the baseline, is the formation of blisters in the metal film [190]. These have been studied and ways of avoiding them have been found [190, 211].

Catalyst poisoning is not a common problem when the gas sensors are being operated in normal atmospheres, but can be a serious problem in special applications. The effect of cadmium and silver has, for example, been studied in some detail by Petersson et al. [212, 213].

The gas sensitivity of TMOS devices has been shown to rely heavily on the microscopic structure of the metal film [182, 183]. Transmission electron microscopic studies show that the structure can be altered by high temperatures and by gases. Even ageing turns out to be sufficient to change the structure and the sensitivity.

The silicon/silicon dioxide interface presents few problems in gas-sensitive MOS transistors and capacitors [209]. MIS Schottky diodes, on the other hand, rely very much on careful fabrication of a high-quality thin oxide [187, 214]. Their strong dependence on oxide quality may be related to the fact that these devices are operated by running electrons through the oxide. It has been suggested that hydrogen penetrates these very thin oxides and interacts with electron energy states at the semiconductor surface. It has also been demonstrated how hydrogen can diffuse into the semiconductor to some extent and form complexes with boron doping atoms [215].

One phenomenon, which does not seem to be significant from a practical point of view, but may be scientifically interesting for the understanding of catalysts, is hydrogen spillover. Armgarth et al. [216] observed electrically that positive charges diffuse laterally out from palladium gates when exposed to hydrogen. Microscopic ellipsometry on capacitors supported the observation that p-type silicon outside the gate becomes electrically inverted during hydrogen exposures. Hydrogen spillover is known to occur, and be of practical significance, in catalysis and the PdMOS devices may therefore be useful for further studies of this effect.

10.3.6   Applications

Several applications of gas-sensitive field effect devices have been suggested and tested [169, 170, 217, 218]. The devices have been found to be especially useful in three different types of applications, namely for use in leak detectors and hydrogen alarms, in pulsed operation for quantitative measurements of hydrogen, ammonia, ethylene, etc., and oxygen (as described in Section 10.3.4.6), and in fundamental studies of heterogeneous catalytic reactions. The last two application areas are described in some detail below.

10.3.6.1   Catalysis Research

Palladium and platinum are both very important metals in the area of catalysis [219]. Chemical reactions on the surface of these metals are therefore subject to considerable research. Not only are practical catalysts being studied but also the fundamentals of the surface chemistry and the physics of noble metals. Much work is being done under ultrahigh-vacuum conditions in order to simplify the study of the complicated phenomena that occur. Various surface-sensitive analytical techniques are being applied, each of them generating clues to the understanding of catalytic reactions.

The gas-sensitive field effect devices have become a new tool in catalysis research in that they allow the measurement of the amount of hydrogen atoms in various reactions [171]. Kelvin probes are often applied to monitor changes in the surface polarization when chemical reactions take place on the metals. The MOS devices add a new dimension to these experiments by monitoring polarization of the inner interface. This is especially valuable because of the unique selectivity and sensitivity of these devices, not easily obtained with other techniques.

One excellent example of how the devices can be used for gaining a better understanding of catalytic reactions is the study by Dannetun et al. [201] of the dehydrogenation of unsaturated hydrocarbons on palladium surfaces at 200 °C. Palladium gate devices were used in conjunction with mass spectrometry, a Kelvin probe, Auger spectroscopy, ESCA, and electron energy loss spectroscopy. Experiments where made in which the C-V shift for acetylene was compared with that for hydrogen. It turned out that the same shift in the C-V curve is obtained if the acetylene pressure is adjusted to give the same flux of molecules to the palladium surface as a certain pressure of hydrogen does. It was concluded that, since acetylene contains the same number of hydrogen atoms as H_2, the acetylene molecules must be completely dehydrogenated by the palladium surface. The same result was obtained for other unsaturated hydrocarbons, indicating complete dehydrogenation of such molecules on a palladium surface at these temperatures, irrespective of their molecular bonding and geometries.

Similar experiments were made on saturated hydrocarbons, in which no dehydrogenation at all could be detected.

Dehydrogenation of unsaturated hydrocarbons was also studied on palladium surfaces covered with chemisorbed oxygen. It turned out that oxygen does not generally block the hydrocarbon dehydrogenation. Since no hydrogen was detected at the inner interface it must be evident that all hydrogen atoms are immediately consumed in the formation of water. One could therefore show that the water desorption rate is a good measure of the dehydrogenation rate on an oxygen-covered surface.

The high sensitivity to hydrogen in the absence of oxygen has been discussed in Section 10.3.4.1. Petersson et al. [220] also demonstrated how a palladium film covered with chemisorbed oxygen initially produces water rather than absorbing hydrogen. Water desorption on hydrogen exposure was monitored with a mass spectrometer, and when all the oxygen had been consumed (water desorption ceased) there was a considerable increase in the amount of hydrogen at the inner interface, detected as a shift of the C-V curve (see Figure 10-37).

10.3.6.2 Bioanalysis

Numerous papers have addressed the possibility of using the gas-sensitive FET devices for biosensing purposes [217, 218, 221–229]. Many biochemical reactions produce or consume gases such as hydrogen, ammonia, and ethylene. Therefore, MOS gas sensors are potentially very useful in biosensor technology, the advantages being that gaseous reactions products can be separated from the biological liquid, prior to detection.

In the bioanalytical application the gas sensor can be used either as a bioprobe or in a flow-through system. In the first case, enzymes are attached to a membrane separating the liquid phase from the gas phase. The disadvantage with this method is that only a limited amount

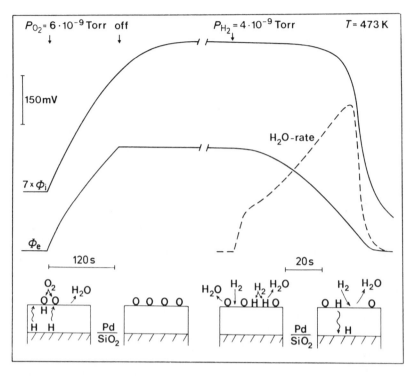

Figure 10-37. Simultaneous recordings of the inner interface potential Φ_i (*C-V* shift), the outer surface potential Φ_e (Kelvin probe shift) and the water production rate (mass spectrometer) on a PdMOS device. The experiment was made under UHV conditions. The device was first exposed to oxygen and thereafter to hydrogen. Oxygen consumes residual hydrogen, which is reflected by the potentials at both interfaces. When the oxygen-covered surface is exposed to hydrogen, water is produced, as reflected by the outer potential and the mass spectrometer. The inner interface potential remains constant until all oxygen has been consumed, whereafter it changes abruptly as the water production ceases. (From [201])

of enzymes is present and the probe is therefore susceptible to inhibition. In a flow-through system the enzymes are immobilized in a reaction column. The gas produced when the liquid sample passes through the column is separated from the liquid phase in a subsequent stage and passed across the sensor. Albeit a more complicated system with pumps, etc., the construction of a flow-through system is more straightforward than the bioprobe, and enzymes can be used in excess quantities. It is therefore less sensitive to inhibition. The possibility of operating the system in a pulsed mode makes it much more stable and reproducible than the bioprobe.

The biochemically important coenzymes NAD and NADH have [224], for example, been measured with a flow-through system including a PdMOS hydrogen sensor. Ammonia-sensitive TMOS devices have been used to measure ammonium ions in water and blood ([221] and references cited therein). By using different enzymes the same system could be used for the determination of urea, creatinine, and various amino acids. Very small sample volumes (<100 μl) were needed, and very high sensitivities were obtained (see Figure 10-38).

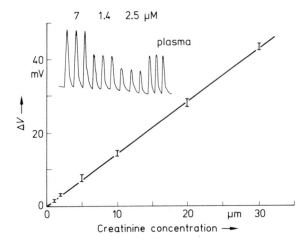

Figure 10-38.
Calibration graph for an enzymatic flow-through system for creatinine determination. The inset shows the actual response to 85-μl sample pulses injected every 4 min. The last three pulses are from three injections of a 25-fold diluted blood plasma sample. An iridium gate MOS device operated at 35°C was used as the ammonia sensor. ΔV: voltage drop (From [229])

In the above measurements there is normally an excess of enzymes, the ammonia concentration being a measure of the substrate concentration. If, on the other hand, there is a limited amount of enzymes the ammonia concentration becomes a measure of the enzyme activity. This was recently used for determination of mercury using mercury inhibition of the enzymes as the primary sensing mechanism [225].

Fermentation processes can be monitored with gas-sensitive FET devices by determining the hydrogen and ammonia gases produced in the process [218, 226]. One example of its usefulness is that local oxygen deficiency in a fermentor is readily detected by a hydrogen sensor in the air outlet, while an oxygen sensor in the liquid phase measures only the average oxygen level.

Gastrointestinal malfunctions have been diagnosed by simply breathing on an FET sensor. For example, lactose intolerance leads to an increased H_2 level in the blood, and hence in exhaled air [227]. Another biological application is the use of ethylene-sensitive TMOS devices to monitor ripening of fruits [228].

10.3.7   Future Trends

The discovery that catalytic gate MOS sensors can be made sensitive to gases other than hydrogen opened up a number of new opportunities. Apart from the obvious efforts to improve sensor parameters such as stability and durability, the work on finding new materials and structures for the detection of other gases will continue. For example, the detection of saturated hydrocarbons calls for higher operating temperatures. The FET devices cannot, however, be operated above 250°C, which is not high enough. An interesting development is therefore found in the work on so-called hot-spot sensors [230]. The catalytic part (hot spot) of the metal is laterally separated from the FET. The silicon is etched off underneath the hot spot by means of silicon micromachining, and a nickel foil heater is made. The idea is that hydrogen atoms obtained from dehydrogenation reactions on the hot spot will diffuse along a Pd film to the FET part of the chip, which is kept at a lower temperature. Preliminary experiments showed that the hot spot can be heated to up to 1000°C without affecting the temperature of the FET [231].

The fact that the selectivity and sensitivity of the gas-sensitive FET devices can be altered by a proper choice of material, metal film structure, and operating temperature opens up possibilities of constructing multi-sensor arrays. By means of modern pattern recognition methods, it should be possible to enhance the performance of this class of sensors considerably. One should be able to obtain a higher accuracy, and to reduce drift and noise, by simply using identical, redundant, multi-sensors. The selectivity can be improved to discriminate between different constituents in a gas mixture, or suppress the influence from interfering gases [232–233]. For example, it has been shown with an ordinary PdMOSFET that it is possible to determine simultaneously the hydrogen and ethanol concentrations in a gas mixture [33]. This was achieved by cycling the operating temperature up and down. The device is only sensitive to H_2 at low temperatures but becomes sensitive also to ethanol when the temperature is high enough.

Another approach is the development of practical measurement systems for new applications. Although a number of possible application areas have been reported in the literature, it is necessary to pursue the development of practical instruments and measurement methods in order to be able to utilize FET gas sensors fully.

10.4   References

[1] Sze, S. M., *Physics of Semiconductor Devices,* 2nd ed.; New York: Wiley, 1981, Chapter 8.

[2] Bergveld, P., *IEEE Trans. Biomed. Eng.* **BME-17** (1970) 70.

[3] Sibbald, A., *J. Mol. Electron.* **2** (1987) 51.

[4] Bousse, L. J., Thesis, 1982, University of Twente, The Netherlands.

[5] Henderson, P., *Z. Phys. Chem.* **59** (1907) 118 and **63** (1908) 325.

[6] Levine, S., Smith, A. L., *Disc. Faraday Soc.* **52** (1971) 290.

[7] Davis, J. A., James, R. O., Leckie, J. O., *J. Colloid Interface Sci.* **63** (1978) 480.

[8] Smit, W., Holten, C. L. M., *J. Colloid Interface Sci.* **7** (1980) 1.

[9] James, R. O., Parks, G. A., in: *Surface and Colloid Science* Vol. 12; New York: Plenum Press, 1982, p. 119.

[10] Healy, T. W., White, L. R., *Advan. Colloid Interface Sci.* **9** (1978) 303.

[11] Bousse, L., Meindl, J., in: *Geochemical Processes at Mineral Surfaces, ACS Symp. Series,* Vol. 323: Washington, DC: American Chemical Soc., 1986, p. 79.

[12] Bousse, L., de Rooij, N. F., Bergveld, P., *Surface Science* **135** (1983) 479.

[13] Bergveld, P., de Rooij, N. F., in: *Vaste Stoff Sensoren,* Deventer: Kluwer, 1980, p. 119.

[14] Buck, R. P., Hackleman, D. E., *Anal. Chem.* **49** (1977) 2315.

[15] Hirata, H., Date, K., *Talanta* **17** (1970) 883.

[16] May-Zurawska, M., Hulanicki, A., *Anal. Chim. Acta* **136** (1982) 395.

[17] Schindler, J. G., Stork, G., Struh, H. J., Schmid, W., Karaschinski, K. D., *Fresenius Z. Anal. Chem.* **295** (1979) 248.

[18] Oesch, U., Thesis, ETH, Zürich, Ch, 1979.

[19] Srianujata, S., White, W. R., Higuchi, T., Sternson, L. A., *Anal. Chem.* **50** (1978) 232.

[20] Fogt, E. J., Untereker, D. F., Norenberg, M. S., Meyerhoff, M. E., *Anal. Chem.* **57** (1985) 1998.

[21] Kobos, R. K., Parks, S. J., Meyerhoff, M. E., *Anal. Chem.* **54** (1982) 1976.

[22] Harrison, D. J., Verpoorte, E. M. J., Xizhong, L., *Proc. 4th International Conference on Solid-State Sensors and Actuators (Transducers '87), Tokyo, Japan, June 2–5, 1987:* p. 738.

[23] van den Vlekkert, H. H., Francis, C., Grisel, A., de Rooij, N. F., *Analyst* **113** (1988) 1029.

[24] Sudholter, E. J. R., van der Wal, P. D., Skowronska-Ptasinska, M., van den Berg, A., Bergveld, P., Reinhoudt, D. N., *Anal. Chim. Acta.* in press.

[25] Revesz, A. G., *Thin Solid Films* **41** (1977) 143.
[26] de Rooij, N. F., Bergveld, P., *Thin Solid Films* **71** (1980) 327.
[27] Bousse, L., Bergveld, P., *Sens. Actuators* **6** (1984) 65.
[28] Matsuo, T., Esashi, M., *Sens. Actuators* **1** (1981) 77.
[29] Abe, H., Esashi, M., Matsuo, T., *IEEE Trans. Electron Devices* **ED-26** (1979) 1939.
[30] Kuisl, M., Klein, M., *Messtechnik* **36** (1983) 48.
[31] Gimmel, P., Gompf, B., Schmeisser, D., Wiemhöfer, H. D., Göpel, W., Klein, M., *Sens. Actuators* **17** (1989) 195.
[32] Akiyama, T., *Pure and Appl. Chem.* **59** (1987) 535.
[33] Gimmel, P., Schierbaum, K. D., Göpel, W., van den Vlekkert H. H., *Sens. Actuators,* to be published.
[34] Voorthuyzen, J. A., Bergveld, P., *Sens. Actuators,* **B1** (1990) 350.
[35] Kreidler, K. G., Semancik, S., Erickson, J. W., *Proc. 4th International Conference on Solid-State Sensors and Actuators (Transducers '87), Tokyo, Japan, June 2-5, 1987:* p. 734.
[36] Sobzynska, D., Torbicz, W., *Sens. Actuators* **6** (1984) 93.
[37] Olszyna, A., Wlosinski, W., Sobczynska, D., Torbicz, W., *J. Cryst. Growth* **82** (1987) 757.
[38] Pham, M. T., Hoffmann, W., *Sens. Actuators* **5** (1984) 217.
[39] Ito, T., Inagaki, H., Igarashi, I., *Proc. 4th International Conference on Solid-State Sensors and Actuators (Transducers '87), Tokyo, Japan, June 2-5, 1987:* p. 707.
[40] Esashi, M., Matsuo, T., *IEEE Trans. Biomed. Eng,* **BME 25-2** (1978) 184.
[41] Bataillard, P., Clechet, P., Jaffrezic-Renault, N., Martelet, C., de Rooij, N., van den Vlekkert, H.-H., *Proc. 4th International Conference on Solid-State Sensors and Actuators (Transducers '87), Tokyo, Japan, June 2-5, 1987:* p. 772.
[42] van den Berg, A., Bergveld, P., Reinhoudt, D. N., Sudholter, E. J. R., *Sens. Actuators* **8** (1985) 129.
[43] Matsuo, T., Nakajima, H., Osa, T., Anzai, J., *Sens. Actuators* **9** (1986) 115.
[44] Van den Berg, A., Thesis, University of Twente, The Netherlands, (1988).
[45] Moss, S. D., Janata, J., Johnson, C. C., *Anal. Chem.* **47** (1975) 2238.
[46] Morf, W. E., *The Principles of Ion-Selective Electrodes and of Membrane Transport;* Amsterdam: Elsevier, 1981.
[47] Ammann, D., *Ion-Selective Microelectrodes:* Heidelberg: Springer, 1986.
[48] Moss, S. D., Johnson, C. C., Janata, J., *IEEE Trans. Biomed. Eng.* **25** (1978) 49.
[49] Sibbald, A., Whalley, P. D., Covington, A. K., *Anal. Chim. Acta* **159** (1984) 47.
[50] Oesch, U., Caras, S., Janata, J., *Anal. Chem.* **53** (1981) 1983.
[51] Teravaninthorn, U., Moriizumi, T., *Proc. 4th International Conference on Solid-State Sensors and Actuators (Transducers '87) Tokyo, Japan, June 2-5, 1987:* p. 764.
[52] Sudhölter, E. J. R., van der Wal, P. D., Skowronska-Ptasinska, M., van den Berg, A., Bergveld, P., D. N. Reinhoudt, *Rec. Trav. Chim. Pays-Bas* **109** (1990).
[53] Janata, J., Huber, R. J., *Ion-Selective Electrode Rev.* **1** (1979) 31.
[54] Tóth, K., Gráf, E., Horvai, G., Pungor, E., Buck, R. P., *Anal. Chem.* **58** (1986) 2741.
[55] Xie, S. L., Cammann, K., *J. Electroanal. Chem.* **229** (1987) 249.
[56] van den Berg, A., van der Wal, P. D., Skowronska-Ptasinska, M., Sudhölter, E. J. R., Reinhoudt, D. N., Bergveld, P., *Anal. Chem.* **59** (1987) 2827.
[57] van den Berg, A., van der Wal, P. D., Skowronska-Ptasinska, M., Sudhölter, E. J. R., Reinhoudt, D. N., Bergveld, P., *J. Electroanal. Chem.,* to be published.
[58] Oesch, U., Xu, A., Brzozka, Z., Suter, G., Simon, W., *Chimia* **40** (1986) 351.
[59] Harrisson, D. J., Teclemariam, A., Cunningham, L. L., *Proc. 4th International Conference on Solid-State Sensors and Actuators (Transducers '87), Tokyo, Japan, June 2-5, 1987:* p. 768.
[60] van der Wal, P. D., Skowronska-Ptasinska, M., van den Berg, A., Bergveld, P., Sudhölter, E. J. R., Reinhoudt, D. N., *Anal. Chim. Acta,* in press.
[61] Moody, G. J., Slater, J. M., Thomas, J. D. R., *Analyst* **113** (1988) 103.
[62] Blackburn, G. F., Janata, J., *J. Electrochem. Soc.* **129** (1982) 2580.
[63] Decroux, M., van den Vlekkert, H. H., de Rooij, N. F., *Proc. 2nd Int. Meeting on Chemical Sensors, Bordeaux, France,* **1986,** p. 406.
[64] Blennemann, H., Bousse, L., Bowman, L., Meindl, J. D., *Proc. 4th International Conference on Solid-State Sensors and Actuators (Transducers '87), Tokyo, Japan, June 2-5, 1987,* p. 723.

[65] Battilotti, M., Colilli, R., Giannini, I., Giongo, M., *Sens. Actuators* **17** (1989) 209.
[66] Satchwill, T., Harrison, D. J., *J. Electroanal. Chem.* **202** (1986) 75.
[67] Wakida, S., Yamane, M., Hiiro, K., *Sens. Mater.* **2** (1988) 107.
[68] Wakida, S., Yamane, M., Higashi, K., Hiiro, K., Ujihira, Y., *Sens. Actuators,* **B1** (1990) 412.
[69] Vogel, A., Hoffmann, B., Sauer, Th., Wegner, G., *Sens. Actuators,* **B1** (1990) 408.
[70] Brown, A. D., *Sens. Actuators* **6** (1984) 151.
[71] Miyahara, Y., Moriizumi, T., *Sens. Actuators* **7** (1985) 1.
[72] Takata, I., Moriizumi, T., *Sens. Actuators* **11** (1987) 309.
[73] Moriizumi, T., *Thin Solid Films* **160** (1987) 413.
[74] Shiramizu, B., Janata, J., Moss, S. D., *Anal. Chim. Acta* **108** (1979) 161.
[75] Fjeldly, T. A., Nagy, K., Stark, B., *Sens. Actuators* **3** (1982/1983) 111.
[76] Kawakami, S., Akiyama, T., Ujihira, Y., *Fresenius' Z. Anal. Chem.* **318** (1984) 349.
[77] Wen, C. C., Lauks, I., Zemel, J. N., *Thin Solid Films* **70** (1980) 333.
[78] Nagy, K., Fjeldy, T. A., Johannessen, J. S., *Proc. 153rd Ann. Meeting Electrochem. Soc.* **78** (1978) Abstract 108.
[79] van der Spiegel, J., Lauks, I., Chan, P., Babic, D., *Sens. Actuators* **4** (1983) 291.
[80] Thyboud, L., Depeursinge, C., Rouiller, D., Mondin, J.-P., Grisel, A., *Sens. Actuators,* **B1** (1990) 485.
[81] van der Starre, P. J. A., Harinck-de Weerd, J. E., Schepel S. J., Kootstra, G. J., *Crit. Care Med.* **14** (1986) 812.
[82] Ligtenberg, H. C. G., van den Vlekkert, H. H., Koning, G., *Proc. S. & A. Symposium of Twente University of Technology:* Deventer, NL: Kluwer, 1984, p. 123
[83] Kloeck, B., van den Vlekkert, H. H., de Rooij, N. F., Muntwyler, M., Anagnostopoulos, K., *Proc. Eurosensors II, Enschede, The Netherlands, 1988:* p. 541.
[84] Sibbald, A., Shaw, J. E. A., *Sens. Actuators* **12** (1987) 297.
[85] van den Vlekkert, H. H., de Rooij, N. F., van den Berg, A., Grisel, A., *Sens. Actuators,* **B1** (1990) 395.
[86] van der Schoot, B. H., Bergveld, P., *Sens. Actuators* **8** (1985) 11.
[87] Ramsing, A., Janata, J., Ruzicka, J., Levy, M., *Anal. Chim. Acta* **118** (1980) 45.
[88] Haemmerli, A., Janata, J., Brown, H. M., *Anal. Chim. Acta* **144** (1982) 115.
[89] Moody, G. J., Slater, J. M., Thomas, J. D. R., *Anal. Proc.* **23** (1986) 287.
[90] Gumbrecht, W., Schelter, W., Montag, B., Rasinski, M., Pfeiffer, U., *Sens. Actuators,* **B1** (1990) 477.
[91] Huang, J. C., Wise, K. D., *Int. Elec. Dev. Meeting, Tech. Digest* (1982) 316.
[92] van den Vlekkert, H.-H., Kloeck, B., Prongue, D., Berthoud J., Hu, B., de Rooij, N. F., Gilli, E., de Crousaz, Ph., *Sens. Actuators* **14** (1988) 165.
[93] Ewald, D., van den Berg, A., Grisel, A., *Sens. Actuators* **B1** (1990) 335.
[94] Sakai, T., Hiraki, H., Uno, S., *Proc. 4th International Conference on Solid-State Sensors and Actuators (Transducers '87), Tokyo, Japan, June 2–5, 1987:* p. 116.
[95] Partridge, S. L., *IEE Proc. I: Solid State Elec. Dev.* **133** (1986) 66.
[96] Sakai, T., Amemiya, I., Uno, S., Katsura, N., *Sens. Actuators,* **B1** (1990) 341.
[97] Lauks, I., van der Spiegel, J., Sansen, W., Steyaert, M., *Proc. 3rd International Conference on Solid-State Sensors and Actuators (Transducers '85) Philadelphia, PA, USA, June 11–14, 1985:* p. 122.
[98] Sibbald, A., *Sens. Actuators* **7** (1985) 23.
[99] Severinghaus, J. W., Bradley, A. F., *J. Appl. Physiol.* **13** (1958) 515.
[100] Shimada, K., Yano, M., Shibatani, K., Komoto, Y., Esashi, M., Matsuo, T., *Med. Biol., Eng. Comput.* **18** (1980) 741.
[101] Nakamura, M., Yano, M., Shibatani, K., Takakura, K., *New Mater. New Processes* **3** (1985) 367.
[102] Nakamura, M., Yano, M., Shibatani, K., Komoto, Y., *Med. Biol., Eng. Comput.* **25** (1987) 45.
[103] Hu, B., van den Vlekkert, H. H., de Rooij, N. F., *Sens. Actuators* **17** (1989) 275.
[104] van der Schoot, B. H., Bergveld, P., *Sens. Actuators* **13** (1988) 251.
[105] Janata, J., Moss, S. D., *Biomed. Eng.* July (1976) 241.
[106] Caras, S., Janata, J., *Anal. Chem.* **52** (1980) 1935.
[107] van der Schoot, B. H., Bergveld, P., *Biosensors* **3** (1987/88) 161.
[108] Adams, R. E., Carr, P. E., *Anal. Chem.* **50** (1978) 944.

[109] Hanazato, Y., Inatomi, K., Nakako, M., Shiono, S., Maeda, M., *Anal. Chim. Acta* **212** (1988) 49.
[110] Moriizumi, T., Miyahara, Y., *Proc. 3rd International Conference on Solid-State Sensors and Actuators (Transducers '85) Philadelphia, PA, USA, June 11–14, 1985;* p. 148.
[111] Kuriyama, T., Nakamoto, S., Kawana, Y., Kimura, J., *Proc. 2nd International Meeting on Chemical Sensors, Bordeaux, France, July 7–10, 1986;* p. 568.
[112] Murakami, T., Kimura, J., Kuriyama, T., *Proc. 4th International Conference on Solid-State Sensors and Actuators (Transducers '87), Tokyo, Japan, June 2–5, 1987;* p. 804.
[113] Kimura, J., Kawana, Y., Kuriyama, T., *Biosensors* **4** (1988) 41.
[114] Eddowes, M. J., *Sens. Actuators* **7** (1985) 97.
[115] Eddowes, M. J., Pedley, D. G., Webb, B. C., *Sens. Actuators* **7** (1985) 233.
[116] Eddowes, M. J., *Sens. Actuators* **11** (1987) 265.
[117] Caras, S. D., Janata, J., Saupe, D., Schmidt, K., *Anal. Chem.* **57** (1985) 1917.
[118] Caras, S. D., Petelenz, D., Janata, J., *Anal. Chem.* **57** (1985) 1920.
[119] Caras, S. D., Janata, J., *Anal. Chem.* **57** (1985) 1923.
[120] van der Schoot, B. H., Bergveld, P., *Anal. Chim. Acta* **199** (1987) 157.
[121] van der Schoot, B. H., Voorthuyzen, H., Bergveld, P., *Sens. Actuators* **B1** (1990) 546.
[122] van der Schoot, B. H., Bergveld, P., *Anal. Chim. Acta* submitted for publication.
[123] Janata, J., Huber, R., in: *Ion Selective Electrodes in Analytical Chemistry,* H. Freiser (ed.): New York: Plenum Press, 1980.
[124] Janata, J., Blackburn, G. F., *Ann. N. Y. Acad. Sci.* **428** (1984) 286.
[125] Nakajima, H., Esashi, M., Matsuo, T., *Nippon Kagaku Kaishi* **10** (1980) 1499.
[126] Janata, J., *Anal. Proc.* **24** (1987) 326.
[127] Gotoh, M., Tamiya, E., Karube, I., *J. Membr. Sci.* **41** (1989) 291.
[128] Schasfoort, R. B. M., Bergveld, P., Bomer, J., Kooyman, R. P. H., Greve, J., *Sens. Actuators* **17** (1989) 531.
[129] Akiyama, T., Ujihara, Y., Okabe, Y., Sugano, T., Niki, E., *IEEE Trans. Electron Devices* **ED-39** (1982) 1936.
[130] Chauvet, F., Amari, A., Martinez, A., *Sens. Actuators* **6** (1984) 255.
[131] Ligtenberg, H. C. G., Thesis, University of Twente, NL, 1987.
[132] Arnoux, C., Buser, R., Decroux, M., van den Vlekkert, H. H., de Rooij, N. F., *Proc. 4th International Conference on Solid-State Sensors and Actuators (Transducers '87), Tokyo, Japan, June 2–5, 1987;* p. 751.
[133] van den Vlekkert, H. H., unpublished results.
[134] van den Vlekkert, H. H., *Analysis* **16** (1988) 110.
[135] de Rooij, N. F., Haemmerli, A., *Proc. of the 6th general Conf. of the European Physical Society, Prague, 1984;* p. 597.
[136] van den Vlekkert, H. H., Thesis, University of Neuchatel, Neuchatel, CH, 1988.
[137] Cohen, R. M., Huber, R. J., Janata, J., Ure, R. W., Moss, S. D., *Thin Solid Films* **53** (1978) 69.
[138] Bergveld, P., *IEEE Trans. Biomed. Eng.* **19** (1972) 342.
[139] Moss, S. D., Janata, J., Johnson, C. C., *Anal. Chem.* **53** (1975) 2238.
[140] Wise, K. D., Weismann, R. H., *Med. Biol. Eng.* **9** (1971) 339.
[141] Matsuo, T., Wise, K. D., *IEEE Trans. Biomed. Eng.* **21** (1974) 485.
[142] Harame, D., Shott, J., Plummer, J., Meindl, J., *Int. Elec. Dev. Meeting. Tech. Digest.* (1981) 467.
[143] Akiyama, T., Komiya, K., Okabe, Y., Sugano, T., Niki, E., *Bunseki Kagaku,* **30** (1981) 754.
[144] Sanada, Y., Akiyama, T., Ujihara, Y., Niki, E., *Fresenius' Z. Anal. Chem.* **312** (1982) 526.
[145] Ho, N. J., Kratochvil, J., Blackburn, G. F., Janata, J., *Sens. Actuators* **4** (1983) 413.
[146] Cheung, P., Pace, S. J., Bigliano, R. P., *Proc. 4th International Conference on Solid-State Sensors and Actuators (Transducers '87), Tokyo, Japan, June 2–5, 1987;* p. 726.
[147] Pomerantz, D. I., *U. S. Pat. 3397278,* 1968.
[148] Ko, W. H., Suminto, J. T., Yeh, G. J., in: *Micromachining and Micropackaging of Transducers,* Fung, C. D. (ed.); Amsterdam: Elsevier, 1985; p. 41.
[149] Bousse, L. J., Schwager, F., Bowman, L., Meindl, J., *Proc. 2nd Int. Meeting on Chemical Sensors, Bordeaux, France, 1986;* p. 499.
[150] van den Vlekkert, H. H., Decroux, M., de Rooij, N. F., *Proc. 4th International Conference on Solid-State Sensors and Actuators (Transducers '87), Tokyo, Japan, June 2–5, 1987;* p. 730.
[151] Petersen, K. E., *Proc. IEEE* **70**, No. 5 (1982) 420.

[152] Hu, B., van den Vlekkert, H. H., Jeanneret, S., de Rooij, N. F., *Proc. Miconex '88, 3th Int. Instrumentation Conf. Peking, China, 1988.*

[153] Akiyama, T., Ujihara, Y., Okabe, Y., Sugano, T., Niki, E., *IEEE Trans. Electron Devices* **29** (1982) 1936.

[154] McBride, P. T., Janata, J., Comte, P. A., Moss, S. D., *Anal. Chim. Acta* **101** (1978) 239.

[155] Smith, R. L., Collins, S. D., *IEEE Trans. Electron Devices* **35** (1988) 787.

[156] Sudhölter, E. J. R., van den Berg, A., Skowronska-Ptasinska, M., van der Wal, P. D., Reinhoudt, D. N., Bergveld, P., *Sens. Actuators, Microtechnology for Transducers:* Deventer, NL: Kluwer, 1986, p. 97.

[157] Matsuo, T., Nakajima, H., *Sens. Actuators,* **5** (1984) 293.

[158] Fujihara, M., Fukai, M., Osa, T., *J. Electroanal. Chem.* **106** (1980) 413.

[159] Tahara, S., Yoshii, M., Oka, S., *Chem. Lett.* (1982) 307.

[160] Bergveld, P., van den Berg, A., van der Wal, P. D., Skowronska-Ptasinska, M., Sudhölter, E. J. R., Reinhoudt, D. N., *Sens. Actuators* **18** (1989) 309.

[161] Skowronska-Ptasinska, M., van der Wal, P. D., van den Berg, A., Bergveld, P., Sudhölter E. J. R., Reinhoudt, D. N., *Anal. Chim. Acta,* in press.

[162] Comte, P. A., Janata, J., *Anal. Chim. Acta* **101** (1978) 247.

[163] Yee, S., Jin, H., Lam, L. K. C., *Sens. Actuators* **15** (1987) 337.

[164] Prohaska, O., Goiser, P., Jackomowicz, A., Kohl, F., Olcaytug, F., *Proc. 2nd Int. Meeting Chem. Sensors, Bordeaux, 1986;* p. 652.

[165] Smith, R., Scott, D. C., *IEEE Trans. Biomed. Eng.* **BME-33** (1986) 83.

[166] van den Berg, A., van den Vlekkert, H. H., de Rooij, N. F., Grisel A., *Sens. Actuators* **B1** (1990) 425.

[167] Bousse, L., Hafeman, D., Tran, N., *Sens. Actuators,* **B1** (1990) 361.

[168] Lundström, I., Shivaraman, S., Svensson, C., Lundkvist, L., *Appl. Phys. Lett.* **26** (1975) 55–57.

[169] Lundström, I., Svensson, C., in: *Solid State Chemical Sensors,* Janata J., Huber, R. J., (Eds.); New York: Academic 1985, Chap. 1.

[170] Lundström, I., Armgarth, M., Spetz, A., Winquist F., *Sens. Actuators* **10** (1986) 399–421.

[171] Lundström, I., Armgarth, M., Petersson, L.-G., *CRC Crit. Rev. Solid State Mater Sci.* **15** (1989) 201–278.

[172] Lundström, I., Shivaraman, M. S., Svensson, C., *J. Appl. Phys.* **46** (1975) 3876–3881.

[173] Lundström, I., Shivaraman, M. S., Svensson, C., *Surf. Sci.* **64,** (1977) 497–519.

[174] Steele, M. C., Hile, J. W., Maclver, B. A., *J. Appl. Phys.* **47,** (1976) 2537–2538.

[175] Keramati, B., Zemel, J. N., *J. Appl. Phys.* **53** (1982) 1091–1099.

[176] Shivaraman, M. S., Lundström, I., Svensson, C., Hammarsten, H., *Electron. Lett.* **12** (1976) 483–484.

[177] Steele, M. C., Maclver, B. A., *Appl. Phys. Lett.* **28** (1976) 687–688.

[178] Ito, K., *Surf. Sci.* **86** (1979) 345–352.

[179] Poteat, T. L., Lalevic, B., Kuliyev, B., Yousuf, M., Chen, M., *J. Electr. Mater.* **12** (1983) 181–214.

[180] Ogita, M., Ye, D. B., Kawamura, K., Yamamoto, T., *Sens. Actuators* **9** (1986) 157–164.

[181] Winquist, F., Spetz, A., Armgarth, M., Nylander C., Lundström, I. *Appl. Phys. Lett.* **43** (1983) 839–841.

[182] Spetz, A., Helmersson, U., Enquist, F., Armgarth M., Lundström, I., *Thin Solid Films* **177** (1989) 77–93.

[183] Spetz, A., Dissertation No. 204, 1989, Linköping University, Sweden.

[184] Stenberg, M., Dahlenbäck, B. I., *Sens. Actuators* **4** (1983) 273–281.

[185] Blackburn, G. F., Levy, M., Janata, J., *Appl. Phys. Lett.* **43** (1983) 700–701; Josowicz, M., Janata, J., *Anal. Chem.* **58** (1986) 514–517.

[186] Nylander, C., Armgarth, M., Svensson, C., *J. Appl. Phys.* **56** (1984) 1177–1188.

[187] Fare, T. J., Zemel, J. N., *Sens. Actuators* **11** (1987) 101–133.

[188] Dobos, K., Strotman, R., Zimmer, G., *Sens. Actuators* **4** (1983) 593–598.

[189] Dobos, K., Zimmer, G., *IEEE Trans. Electr. Dev.* **ED-32** (1985) 1165–1169.

[190] Armgarth, M., Nylander, C., *IEEE Electr. Dev. Lett.* **EDL-3** (1982) 384–386.

[191] Sze, S. M., *Physics of Semiconductor Devices,* 2nd Ed.; New York: Wiley, 1981, Chapters 7 and 8.

[192] Lundström, I., *Sens. Actuators* **1** (1981) 403–426.

[193] Armgarth, M., Hua, T. H., Lundström, I., *Techn. Digest. Transducers '85, Philadelphia, PA, 1985;* pp. 235–238.

[194] Dannetun, H. M., Petersson, L.-G., Söderberg, D., Lundström, I., *Appl. Surf. Sci.* **17** (1984) 259–264.

[195] Shivaraman, M. S., *J. Appl. Phys.* **47** (1976) 3592–3593.

[196] Ruths, P. F., Ashok, S., Fonash, S. J., Ruths, J. M., *IEEE Trans. Electr. Dev.* **ED-28** (1981) 1003–1009.

[197] Ackelid, U., Winquist, F., Lundström, I., *Proc. 2nd Int. Meet. Chemical Sensors, Bordeaux, 1986;* pp. 395–398.

[198] Ackelid, U., Armgarth, M., Spetz, A., Lundström, I., *IEEE Electr. Dev. Lett.* **EDL-7** (1986) 353–355.

[199] Winquist, F., Lundström, I., *Sens. Actuators* **12** (1987) 255–261.

[200] Armgarth, M., Ackelid, U., Lundström, I., *Tech. Dig. Transducers '87, Tokyo, Japan, 1987;* pp. 640–643.

[201] Dannetun, H., Lundström, I., Petersson, L.-G., *Surf. Sci.* **173** (1986) 148–159.

[202] Poteat, T. L., Lalevic, B., *IEEE Trans. Electron Dev.* **ED-29** (1982) 123–129.

[203] Ackelid, U., Petersson, L.-G. to be published.

[204] Maclay, G. J., Jelley, K. W., Nowroozi-Esfahani, S., Formosa, M., *Sens. Actuators* **14** (1988) 331–348.

[205] Karlsson, J., Armgarth, M., Ödman, S., Lundström, I., *Anal. Chem.* **62** (1990) 542–544.

[206] Armgarth, M., Nylander, C., *Appl. Phys. Lett.* **39** (1981) 91–92.

[207] Dobos, K., Armgarth, M., Zimmer, G., Lundström, I., *IEEE Trans. Electr. Dev.* **ED-31** (1984) 508–510.

[208] Torbicz, W., Sobczynska, D., Olszyna, A., Fortunato, G., D'Amico, A., *Phys. Stat. Sol.* **86** (1984) 453–459.

[209] Armgarth, M., Dissertation No. 107, Linköping University, Sweden, 1983.

[210] Jansson, R., Arwin, H., Armgarth, M., Lundström, I., *Appl. Surf. Sci.* **37** (1989) 44–54.

[211] Choi, S.-Y., Takahashi, K., Matsuo, T., *IEEE Electr. Dev. Lett.* **EDL-5** (1984) 14–15.

[212] Petersson, L.-G., Dannetun, H. M., Karlsson, S.-E., Lundström, I., *Surf. Sci.* **117** (1982) 676–684.

[213] Petersson, L.-G., Dannetun, H. M., Lundström, I., *Phys. Rev. B* **30** (1984) 3055–3061.

[214] Fare, T., Spetz, A., Armgarth, M., Lundström I., *J. Appl. Phys.* **63** (1988) 5507–5513.

[215] Fare, T. L., Lundström, I., Zemel, J. N., Feygenson, A., *Appl. Phys. Lett.* **48** (1986) 632–634.

[216] Armgarth, M., Nylander, C., Svensson, C., Lundström I., *J. Appl. Phys.* **56** (1984) 2956–2963.

[217] Winquist, F., Lundström, I., Danielsson, B., in: *Applied Biosensors,* D. L. Wise (ed.); Boston: Butterworths, 1989, pp. 291–319.

[218] Hörnsten, G., Lundström, I., Elwing, H., in: *Bioinstrumentation: Research, Development and Applications,* D. L. Wise (ed.); Boston: Butterworths, 1989, pp. 47–91.

[219] Sinfelt, J. H., *J. Phys. Chem.* **90** (1986) 4711–4723.

[220] Petersson, L.-G., Dannetun, H. M., Lundström, I., *Surf. Sci.* **161** (1985) 77–100.

[221] Winquist, F., Spetz, A., Armgarth, M., Lundström, I., *Sens. Actuators* **8** (1986) 91–100.

[222] Winquist, F., Dissertation No. 158, Linköping University, Sweden, 1987.

[223] Hörnsten, E. G., Dissertation No. 197, Linköping University, Sweden, 1988.

[224] Winquist, F., Danielsson, B., Lundström, I., Mosbach, K., *Appl. Biochem. Biotechn.* **7** (1982) 135–139.

[225] Winquist, F., Lundström, I., Danielsson, B., *Anal. Lett.* **21** (1988) 1801–1816.

[226] Cleland, N., Dissertation, Royal Institute of Technology, Stockholm, Sweden, 1988.

[227] Berg, A., Eriksson, M., Barany, F., Einarsson, K., Sundgren, H., Nylander, C., Lundström, I., Blomstrand, R., *Scand. J. Gastroenterol.* **20** (1985) 814–822.

[228] Winquist, F., Lundström, I., Bergkvist, H., *Anal. Chim. Acta* **231** (1990) 93–100.

[229] Lundström, I., Spetz, A., Winquist, F., *Phil. Trans. R. Soc. London, Ser. B.* **316** (1987) 47–60.

[230] Enquist, F., Esashi, M., Armgarth, M., Lundström, I., Matsuo, T., *Tech. Dig. Transducers '87, Tokyo, Japan, 1987;* pp. 644–647.

[231] Enquist, F., personal communication.

[232] Müller, R., Lange, E., *Sens. Actuators* **9** (1986) 39–48.

[233] Sundgren, H., Lukkari, I., Lundström, I., Carlsson, R., Winquist, F., Wold, S., *Sens. Actuators* **B 2** (1990) 115–123.

11 Calorimetric Chemical Sensors

PETER T. WALSH, T. A. JONES †, Health and Safety Executive, Sheffield, UK

Contents

11.1 Introduction

The purpose of this chapter is to describe in detail the principles of operation, method of use, applications, limitations and usefulness of sensors based on calorimetry. In this type of sensor the presence or concentration of a chemical is determined by measurement of a temperature change produced by the chemical to be detected. The temperature change may be due to an exothermic or endothermic reaction or, where there is a temperature difference between transducer and the surrounding atmosphere, by a change in the thermal conductivity of the atmosphere. The liberation or abstraction of heat is conveniently measured as a change in temperature which can be easily transduced into an electrical signal.

Calorimetric sensors based on the measured of heat of reaction invariably utilise exothermic oxidation reactions because there is usually an adequate supply of oxygen in the ambient air to react with the detectable gas (fuel), and heats of oxidation are usually large thus ensuring adequate sensitivity. Furthermore, the sensor incorporates a catalyst in order to allow the oxidation reaction to proceed with a measurable rate at a convenient temperature of operation.

There are three types of calorimetric sensor; they differ in the way that the heat evolved is transduced. The catalytic sensor (this is the commonly used term for the first type of transducer, often known as the 'pellistor', although all three types utilise catalytic effects) employs platinum resistance thermometry, whereas the pyroelectric and Seebeck-effect sensors utilise these effects to measure the temperature change. The catalytic sensor has been investigated in greatest detail and widely exploited, finding widespread use in a range of instruments, throughout the world. The other two types have only been developed comparatively recently and have yet to be commercially exploited.

The sensors are primarily aimed at providing warning of a flammable gas hazard which can occur in many industrial, commercial and domestic environments. Methane is the most prevalent of flammable gases because of its use as a primary fuel in several countries and, since it is a product of decaying organic matter, it is also found in dangerous quantities in mines, sewers and waste tips. Hydrogen is still a major problem in those countries where it is used as a fuel and it also occurs as a product evolved from lead-acid accumulators. Liquid petroleum gas is used as a transportable fuel in a wide variety of applications ranging from heavy industry to leisure caravans and boats. Petrol is an obvious hazard in garages and enclosed car parks. Finally a wide range of hydrocarbons are commonly used in the chemical and petrochemical industries.

The hazard posed by a flammable gas is well defined. The concentration range over which the gas is flammable is defined by two limits, the lower explosive limit (LEL) and the upper explosive limit (UEL). There are extensive tables available detailing these limits [1]. It is common practice to endeavour to maintain the concentration levels below about 20%. LEL and the instrumental alarm or action level is usually set at or below this figure. The LEL for most gases lies between 1 and 5% V/V, therefore any sensor used for monitoring the hazard has to be capable of indicating concentrations in this range with a discrimination of better than 10% LEL. In principle this type of calorimeter can detect any flammable gas although they are mainly used for the commonly occurring species such as methane, ethane, propane, butane, hydrogen, and carbon monoxide.

Sensors based on thermal conductivity changes were among the earliest types of chemical sensors. They have been investigated thoroughly and reached a state of technological maturity.

The measurement method is a purely physical one, depending on the relation of the rate of heat loss from a hot body to the thermal conductivity of the surrounding gas. Their main areas of application are as low sensitivity, non-selective detectors in gas chromatographs and in process control systems. They are also used extensively to monitor high concentrations (1–100%) of certain gases in air or nitrogen i.e. those having thermal conductivities significantly different to that of the reference gas. Typical examples are hydrogen, methane and carbon dioxide.

Calorimetric sensors thus provide an essentially non-selective means of detecting comparatively high concentrations (0.1–100%) of gases and vapours. They have found extensive use in many situations e. g. measuring the flammability of an atmosphere. Indeed they are one of the commonest types of chemical sensor in industrial use. The following sections describe in detail the principles of operation of each of the four types of sensor outlined above, the current state of development is summarized and, finally, suggestions regarding possible future trends in the design and application of these sensors are made.

11.2   Catalytic Sensors

11.2.1   Overview

The development of the catalytic sensor derived from the need for a hand-held detector for methane to replace the flame safety lamp in coalmines. The use of the lamp required a degree of operator skill and provided a subjective method which could lead to inaccurate estimations of methane concentration. A convenient, low cost, hand held, reliable, accurate methane sensor was therefore required to detect methane over the range 0–5%. The most suitable device was found to be based on calorimetry and utilised the exothermicity of the combustion of methane to generate a signal. Since this work was initiated in the late 1950s the scope of the technique has broadened to include other industrial environments e. g. gas, petrochemical, etc, other flammable gases and different types of instrumentation including portable monitors/ alarms and fixed single and multiple sensor installations. The major uses are in the petrochemical industries and public utilities where the concentration of flammable gas needs to be monitored above ground, in underground tunnels and in pipes. For some applications the catalytic sensor is often used in conjunction with other gas sensors in order to provide warning of a combination of hazards, e.g. flammable, toxic and oxygen deficiency.

The accumulated knowledge of the catalytic sensor coupled with its acceptance and widespread use in industry resulted in British Standards [2] for the performance requirements of flammable gas detectors for industrial use being published in the early 1980s. The publication of the European Standard is expected in 1991.

11.2.2   Basic Principles of Operation

11.2.2.1   Sensor Form

The catalytic device measures the heat evolved during the controlled combustion of flammable gas in ambient air; the total oxidation of methane, for example, liberates 800 kJ/mol heat. A catalyst is usually required to sustain such a reaction at a reasonable temperature. If, under the conditions of measurement, the rate of reaction is dependent on the concentration of fuel then determination of the heat evolved provides a means of measuring gas concentration. Since the reaction occurs at the catalyst surface it is convenient to supply heat to the catalyst itself rather than the reacting gases and to measure the heat of reaction as a rise in temperature of the catalyst. Thus the basic constituents of a catalytic calorimetric gas sensor are a temperature sensor, a solid catalyst and a heater to maintain the catalyst at the operating temperature.

One of the most convenient temperature measuring devices is the resistance thermometer, since this allows a single coil of wire to act as both heater and temperature sensor. Platinum is the most suitable material because of the combination of a high temperature coefficient of resistivity, adequate ductility allowing coils to be fabricated easily, and inertness which enables it to resist oxidation at the elevated temperatures required to effect combustion of flammable gas.

11.2.2.2 Catalysis

The role of a catalyst is to increase the rate at which a thermodynamically feasible chemical reaction approaches equilibrium; the catalyst achieves this without itself becoming permanently altered by the reaction. Acceleration of the rate is accomplished by provision of alternative reaction paths, involving intermediates known as activated (transition) complexes, with lower activation energies than the uncatalysed mechanism. Catalytic sensors employ solid, heterogeneous catalysts i.e. having a different phase to the reactants and products. Here reaction involving the transition complex, an adsorbed species, occurs at the interface between solid and gas. Thus adsorption and desorption processes, the nature and concentration of the adsorbate, the strength of the adsorption bonds and the kinetics of the adsorption/desorption processes are critical to the performance of the catalyst and determine its role in gas sensing devices [3]. The energetics of an uncatalysed and catalysed reaction:

$$A + B \rightleftharpoons C + D$$

are shown in Figure 11-1. The general rate expression for this reaction is of the form:

$$r = k[A]^{\alpha}[B]^{\beta}[C]^{\gamma}[D]^{\delta} \qquad (11\text{-}1)$$

and

$$k = a \exp\left(-\frac{E}{RT}\right). \qquad (11\text{-}2)$$

Where a is the pre-exponential factor and E the activation energy for reaction. The uncatalysed, homogeneous reaction is characterised by an activation energy E_g. In the heterogeneously catalysed reaction either or both of the gaseous species A and B adsorb on the surface with a net heat of adsorption ΔH, which is always exothermic. The adsorbates, present in state I, react to form the transition complex (state *) which is the precursor to adsorbed products (state II). This process is characterised by an activation energy E_c, considerably lower than E_g. The adsorbed products then desorb having acquired the required activation energy.

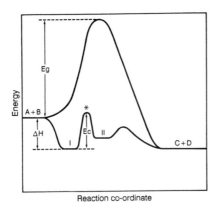

Figure 11-1.
Energy barriers for reaction A + B → C + D (g),
homogeneous gas phase reaction; (c), Heterogeneous
catalysed reaction.

It is apparent that if the heats of adsorption are too large, that is, if the species are too strongly adsorbed, the activation energy E_c may become too high for the reaction to be accelerated. However as the heat of adsorption, and hence E_c, decreases the surface coverage of reactants also falls as is shown later (see Equation (11-8)). Consequently, although the reaction becomes energetically more favourable, there are fewer species available for reaction. Thus the rate of reaction will be at a maximum when the strength of adsorption is as weak as possible, consistent with the adsorbate covering a large enough fraction of the surface.

Adsorption occurs because atoms or ions at the surface of the solid cannot fully satisfy their valency and coordination requirements. Thus there is always a net force acting inward; species are thus attracted from the gas phase in order to reduce the surface energy of the solid. If the interaction is weak then mediation is through van der Waals forces, similar to those involved in condensation. This process is known as physisorption and is characterised by low heats of adsorption essentially equivalent to heats of liquefaction (~ 10 kJ/mol). No chemical bonds are formed and therefore the electronic states of the catalyst and adsorbate are only weakly perturbed resulting in very low catalytic activity. In contrast if the solid possesses a high surface energy due to unsatisfied valency orbitals or surface ions with incomplete coordination shells, the gas may become adsorbed through an exchange of electrons with the surface, thus chemical bonding occurs, known as chemisorption. The role of physisorption is usually to provide a precursor for chemisorption. The nature of the bonds may be covalent, i.e. sharing of electrons with the surface, or ionic with donation of electrons to or abstraction from the solid. In both cases the heats of adsorption are large (>10 kJ/mol) and the electronic states of the adsorbate and adsorbant are measurably perturbed. This mechanism can lead to new reaction paths with different energetics from those of the gas phase reaction and forms the basis of heterogeneous catalysis.

A survey of the chemisorption properties of catalysts reveals that high chemisorption ability is largely confined to metals within the transition series. This is associated with the presence of partially filled d-bands, allowing interchange between these electrons and those of the adsorbate. Within the transition metal group the noble metals (Rh, Pd, Ir and Pt) are particularly active for oxidation reactions. First, the metals are efficient at dissociatively adsorbing gases such as hydrogen, oxygen and hydrocarbons. Gold, although classed as a noble metal, possesses a complete d-band, and interacts only very weakly with oxygen; it is therefore not an effective oxidation catalyst [4]. Second, the heat of adsorption (ΔH) of the reactants (i.e.

fuel and/or oxygen) is low enough to reduce the activation energy for oxidation (see Figure 11-1), yet ΔH is sufficiently high to ensure adequate surface coverage. Other transition metals adsorb oxygen too strongly which results in high activation energies for oxidation and, under certain conditions, they can form bulk oxides. This latter point further demonstrates the suitability of noble metals for oxidation reactions; even catalysed oxidation reactions sometimes require high temperatures at which oxidation of the metal to the metal oxide is rapid. Thus only noble metals, which are resistant to oxidation, can be used.

11.2.2.3   Kinetics

Any heterogeneous catalytic reaction can be broken down into a number of consecutive steps involving transport of the reactants and products to and from the surface, and the reaction itself on the catalyst surface. These steps are:

(1a)   reactants diffuse to the surface
(1b)   reactants diffuse through pore structure if the catalyst is porous
(2)   reactants adsorb on the surface
(3)   adsorbed species react on the surface to give products
(4)   products desorb from the surface
(5a)   products diffuse through pore structure if the catalyst is porous
(5b)   products diffuse away from the surface.

For a non-porous catalyst diffusion to and from the surface occurs by bulk gas diffusion through a 'depletion layer' enveloping the catalyst [5]. Additionally, in a porous catalyst diffusion of reactants and products through the catalyst pore structure must be considered. Steps (1a) and (5b) are dependent on the rate (r_d) of bulk diffusion to the catalyst and the rate may be represented thus,

$$r_d = D_{12} \, A \, \frac{dC}{dx} \; . \tag{11-3}$$

Where A is the area through which the gas diffuses, C the bulk gas concentration and x is the distance from the catalyst surface. The binary diffusion coefficient D_{12} may be estimated from the theoretical equation:

$$D_{12} = \frac{0.01858 \, T^{3/2} \, [(M_1 + M_2)/M_1 \, M_2]^{1/2}}{P \sigma^2 \, \Omega} \tag{11-4}$$

where T is the absolute temperature, M the molecular weight of the gases, P the total pressure (atm), Ω the collision integral, and σ a force constant [6]. Experimental data can invariably be fitted to a function of the form:

$$D_T = D_0 \left(\frac{T}{T_0} \right)^n \tag{11-5}$$

where D_T and D_0 are the diffusion coefficients at temperature T and T_0; the exponent n usually lies in the range 1-2 [7]. Thus it can be seen that the diffusion coefficient does not have a strong dependence on temperature, and, when the rate of diffusion is depicted on an Arrhenius plot, it can be considered as equivalent to a process having a very low activation energy ($\ll 1$ kJ/mol).

Steps (1b) to (5a) are activated rate processes and are represented by the usual Arrhenius expression (Equation (11-2)). For adsorption (step (2)) the pre-exponential term is dependent on the number of collisions of gas with the surface, a steric factor and the fractional coverage of the surface; the activation energy is that for chemisorption, denoted E_a. For desorption (step (4)) the pre-exponential term is the product of the specific rate constant for desorption and the fraction of sites available for desorption; the activation energy is that for desorption, denoted E_d [4]. The heat of adsorption Q is given by

$$Q = E_d - E_a . \tag{11-6}$$

For reaction on the surface (step (3)), the terms in the Arrhenius expression are analogous to those for adsorption. The pre-exponential factor, is governed by the collision number and steric factor and the activation energy (E_s) refers to reaction between adsorbed species. Where the gas must diffuse through narrow pores (Knudsen diffusion steps (1b) and (5a)) the measured activation energy (E_p) is half that of the intrinsic activation energy (E_s) where E_s refers to either a non-porous catalyst or one having very large pores [6]. Figure 11-2 shows an idealised Arrhenius plot for a porous catalyst over a wide temperature range.

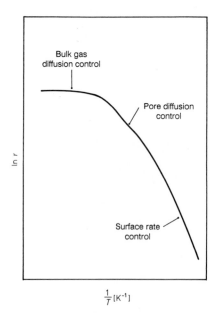

Figure 11-2.
Arrhenius plot showing rate (r) controlling processes for a porous element as a function of temperature T.

As described earlier, a convenient method of measuring the concentration of reactant (fuel) is to use the energy released in step (3). The overall rate of reaction and hence the rate of energy release (i.e. chemical power generated) will be controlled by whichever step is the slowest. The 'activation' energy terms in steps (1 a) and (5 b) are low, whilst those in steps (1 b), (2), (3), (4) and (5 a) are high (>20 kJ/mol). The rate and hence the signal will therefore be less temperature dependent if the rate is controlled by either steps (1 a) and (5 b), rather than steps (1 b) to (5 a). This also confers other benefits as is discussed in greater detail in Section 11.2.4.4.

The rate of reaction on the catalyst surface is governed by the normal rules for a heterogeneous reaction. For bimolecular reactions, between fuel and oxygen, there are two fundamental mechanisms: Langmuir-Hinshelwood, where the reaction proceeds through the adsorption of the two species on adjacent sites, and Eley-Rideal, where one species is chemisorbed and the other physisorbed. Thus if two gases A and B are adsorbed and react on the surface, the variation of the rate of reaction with pressure of gas A, at constant pressure of gas B, is of the form shown in Figure 11-3 a, if there is no competition between A and B for "sites" on the surface. This profile is obtained theoretically by adopting a rate expression

$$r = k\theta_A\,\theta_B \tag{11-7}$$

where θ is the coverage of species A or B. This equation represents the rate governed by the number of pairs of unlike species on adjacent sites. Expressions for θ are obtained from isotherms, the simplest being the Langmuir isotherm [4]. Thus, for no competition between sites

$$r = k\left[\frac{b_A\,P_A^{\frac{1}{m}}}{1 + b_A\,P_A^{\frac{1}{m}}}\right]\left[\frac{b_B\,P_B^{\frac{1}{n}}}{1 + b_B\,P_B^{\frac{1}{n}}}\right]. \tag{11-8}$$

Where b is the equilibrium constant for adsorption given by

$$b = a\,\exp\left(-\Delta H/RT\right). \tag{11-9}$$

The parameter P is the partial pressure of gas, and m and n are the number of fragments that the adsorbing molecule dissociates into. At constant P_B and for $m = 1$ Equation (11-8) reduces to

$$r = \frac{k^1\,b_A\,P_A}{(1 + b_A\,P_A)}. \tag{11-10}$$

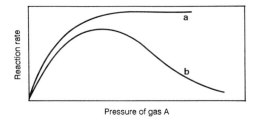

Figure caption along the axes: Reaction rate (vertical axis); Pressure of gas A (horizontal axis); curves labelled a and b.

Figure 11-3. Variation of reaction rate with gas pressure in a binary gas mixture (A + B); pressure of gas B constant. (a) No competition between A and B for sites, curve described by Equation (11-8); (b) Competition between A and B, curve described by Equation (11-11).

If A and B compete for sites on the surface, the curve takes the general form of Figure 11-3 b, and the rate equation is developed from Equation (11-7); assuming $m = n = 1$ gives

$$r = \frac{k\, b_A\, b_B\, P_A\, P_B}{(1 + b_A\, P_A + b_B\, P_B)^2}\, .$$

(11-11)

Depending on the relative values of b and P, different parts of the curve will be spanned as P is varied over the experimental range.

Most solid-state catalysts are not completely specific in their adsorption of gases and a catalyst for a given reaction may be capable of adsorbing other molecules besides the reactants. If these molecules are themselves strongly adsorbed on the catalyst, and decompose or react with one of the reactants to give a product which is itself strongly adsorbed, then the adsorption of reactants is inhibited and the catalysis of the reaction is impaired. This affects the rate equation (e. g. Equation (11-11)) by introducing an additional term $b_x\, P_x$ into the denominator such that

$$r = \frac{k\, b_A\, b_B\, P_A\, P_B}{(1 + b_A\, P_A + b_B\, P_B + b_x\, P_x)^2}\, .$$

(11-12)

If species x is strongly adsorbed compared to the reactants and $b_x\, P_x \gg 1$ then this reduces to:

$$r = \frac{k\, b_A\, b_B\, P_A\, P_B}{(b_x\, P_x)^2}\, .$$

(11-13)

These effects or "poisoning" are discussed in greater detail in Section 11.2.5.1, with particular emphasis on poisoning in porous catalysts.

11.2.2.4   Sintering

The reaction of the gas on the catalyst surface is controlled by chemical effects i. e. the kinetics of adsorption and reaction outlined above and the physical effect of the active surface area of the catalyst. In order to increase the rate of reaction, which is directly proportional to the surface area, it is necessary to maximise the surface area by employing polycrystalline catalysts having as small a crystallite size as possible. Such catalysts are generally, relatively unstable and will sinter with time. Moreover the rate of sintering of the catalyst is often increased when reaction is taking place on the surface. It is therefore common practice to disperse the catalyst crystallites over an inert, refractory material so that the total surface area of catalyst remains high; this is because the agglomeration of crystallites to their equilibrium surface area occurs to a much lesser extent than for an unsupported catalyst. It is also desirable to choose a catalyst which does not slowly react with one of the components of the gas mixture thus diminishing the active surface area of the catalyst. If sintering and catalyst degradation does occur it is advisable to ensure that equilibrium conditions are reached as quickly as possible. This can often be achieved by pretreatment of the catalyst prior to operation, generally at temperatures well above the operational level.

11.2.3 Types of Sensors

Having outlined the fundamental principles underlying the operation of catalytic sensors, it is appropriate to examine their construction and form in order to understand in greater detail how they operate and how ambient conditions can affect their performance. A chronological approach to the development of various types of catalytic sensor will be adopted since this best illustrates how developments in sensor fabrication took place in order to overcome operating difficulties as they arose. The greater part of this work was initially related to the detection of methane, particularly in mines. The applications subsequently spread to other working environments and to the detection of flammable gases and vapours other than methane. As stated earlier, the most convenient temperature measuring device is a platinum resistance thermometer since this allows a single platinum coil to act both as a heater and temperature sensor. Platinum is also an active catalyst for hydrocarbon oxidation, hence the simplest and oldest form of catalytic sensor is a platinum coil [8]. This form of sensor was used in many early methanometers and also for a number of fundamental studies of oxidation over platinum catalysts [9–11]. Although the platinum coil sensor is still used in some spot-reading instruments it is not well suited for use in continuous operation. Bulk metallic platinum is a relatively poor catalyst for the oxidation of methane compared to palladium and rhodium, hence the sensor must operate at high temperature (~1000 °C). At this temperature platinum evaporates at a significant rate, which is accelerated by the presence of oxygen and especially, combustible gas/air mixtures. This results in a reduction of the cross-sectional area of the wire, hence an increase in resistance, causing a drift in air ('zero-drift') and eventually leading to fusing of the coil. In order to overcome this deficiency of the simple platinum coil several means of increasing the lifetime of the devise were proposed [3], but the greatest improvement in sensor design has arisen from the deployment of more active catalysts that operate at significantly lower temperatures. The platinum wire coil in all cases is retained as the temperature sensing device but the catalyst is now separated from it.

This development led to an improvement in catalyst activity, allowing the temperature of operation to be lowered. Enhanced activity is achieved in two ways. First, catalysts with greater specific activity than platinum may be employed. Secondly, the catalyst may be prepared so as to expose a far greater surface area per unit mass, than for, say, a metallic coil, which has a geometric surface area of only around $0.1\ cm^2$ for typical coil dimensions. One example of this more active sensor is the "pellistor" first described by Baker [12] and shown in Figure 11-4. This sensor utilises palladium supported on thoria as the catalyst, which is deposited from aqueous solution onto the surface of a refractory bead (again deposited from aqueous solution) of ~1 mm diameter encapsulating the platinum coil. Palladium is a more active catalyst than platinum for the oxidation of certain hydrocarbons, including methane, which is oxidised at temperature around 500 °C. Encapsulation of the coil within a spherical bead in this fashion produces a device which is insensitive to orientation and also resistant to mechanical shock. This type of sensor is widely used in all types of flammable gas detection instruments. Other types of transducer having a similar form have been described in the literature [3].

The major limitation on the operation of catalytic gas sensing elements of the 'pellistor'-type is their loss of sensitivity on exposure to atmospheres containing catalyst poisons and inhibitors. The growth in demand for catalytic sensors to operate in hostile atmospheres where charcoal adsorbent protection is not suitable has stimulated the development of sensors which are resistant to poisons. These sensors generally take the form of a platinum wire coil embedded in a porous bead comprising of active catalyst either dispersed throughout a porous substrate or as discrete layers within the substrate. The major improvement lies in increasing the surface area of catalyst still further, compared to the essentially non-porous sensors described previously, by the use of high surface area, porous catalyst supports. The development of this type of sensor is discussed in more detail in Section 11.2.5.1.

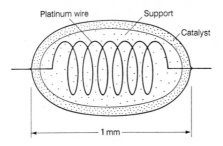

Figure 11-4.
A typical pellistor-type catalytic flammable gas sensing element.

11.2.4   Operating Conditions

In this section the typical operating conditions used to measure the response to flammable gases are described. The various aspects discussed here are obviously relevant to all types of catalytic sensor described above. They will, however, be illustrated with reference to the non-porous 'pellistor' type device [12], since this is the most common type of catalytic sensor in use.

All gas detection instruments consist of four basic components:

(1) A power supply

(2) A control and processing unit which regulates the power supplied to the sensor to maintain its operating temperature and process the signal for display

(3) The sensing unit which converts the gas concentration into an analogue signal

(4) A display unit which presents the signal in an appropriate form and/or takes appropriate action

First the power requirements of the sensor are discussed, then the various methods of obtaining an electrical signal from the sensor are described. Next, the sensing unit is considered; this comprises the sensor enclosure (determining sensor head geometry) which influences the rate of gas supply to the sensing element, and the sensing element itself. Finally, having characterized the electrical and geometric factors involved in the operation of the sensor, the physico-chemical factors affecting the usage of the device are addressed.

11.2.4.1   Power Requirements

For a given reaction and catalyst the total energy consumption of the sensor is controlled by its geometry. In all instruments, including continuously operating systems, hand-held and portable instruments it is desirable to minimise power drain. Heat losses from a body are reduced as its surface area is decreased, therefore the sensing element should ideally take the form of a small spherical bead. However the rate of reaction, and hence the rate of generation of heat and magnitude of the signal, is proportional to the catalyst surface area. Thus, in order to satisfy these conflicting requirements, the element should be fabricated as a small sphere of porous catalyst so that the reaction rate is governed by a large internal area, whilst heat losses and thus the power requirement are governed primarily by the external geometric area of the sphere. In practice the size of the signal is not a limiting factor (see Section 11.2.4.2) and the elements are commonly operated in a mass transport-controlled mode (see Section 11.2.4.4) and the reaction rate is independent of surface area. In such cases the principal advantage of a porous catalyst is enhanced resistance to poisons (see Section 11.2.5.1), not greater sensitivity.

Heat is lost from a sensor in the following ways:

(1) Conduction along the electrical connectors and support wires for the sensing element,
(2) Radiation from the surface,
(3) Conductance and convection through the surrounding medium.

The power lost by conduction along the support wires (P_{cw}) can be expressed thus:

$$P_{cw} = k_w \, a_w \, (1/L_1 + 1/L_2) \, (T_s - T_w) . \tag{11-14}$$

Where k_w is the thermal conductivity of platinum wire at an average temperature of $(T_s + T_w)/2$, a_w is the cross-sectional area of wire, L_1 and L_2 the lengths of the support wires, T_s the sensor temperature and T_w the temperature of the copper supports at the end of the platinum wire, assumed to be at room temperature. Typically in the pellistor described by Baker [12], approximately 90 mW are lost down the two supporting wires out of a total power consumption of 400 mW at an operating temperature of 550 °C, i.e. about 23% of the total. Thus it is important that the number and thickness of connections made to the sensing element should be minimised.

Heat losses by radiation (P_r) can also be calculated theoretically, assuming that the bead is spherical, the cavity is large compared to the dimension of the element and the surface of the element is at a uniform temperature T_s, then

$$P_r = 4 \, \pi \, r_s^2 \, \varepsilon \, \sigma \, (T_s^4 - T_a^4) \tag{11-15}$$

where r_s is the radius of the bead, ε is the emissivity of the sensing element (experiments in vacuum have established that, for sensors employing palladium-thoria as catalyst, the value of ε is approximately one [13]), and σ is Stefan's constant.

Heat losses by radiation and conduction can be calculated from the standard expressions given above but the treatment of heat loss by convection, coupled to conduction from a hot sphere is complex [14]. However, simple, empirical expressions operable over a narrow range of conditions, can be derived, one such expression is [13]:

$$P_{cd} + P_{cv} = (1 + \lambda r_s) \, P_{cd} . \tag{11-16}$$

Where P_{cd} and P_{cv} are the power drains due to conduction and convection in air respectively and λ is an empirically derived constant having a value of 17 ± 2 cm^{-1} for spheres where $0.02 < r_s < 0.05$ cm.

The rate of heat loss by conduction through gas in a spherically symmetrical system is given by:

$$P_{cd} = K_g \, 4 \pi \, r_s^2 \, \frac{dT}{dr} \tag{11-17}$$

where K_g is the thermal conductivity of the gas surrounding the element. K_g is temperature dependent, and can be represented as a series expansion:

$$K_g = K_0 + \alpha T + \beta T^2 \, ... \tag{11-18}$$

where K_0 is the thermal conductivity at temperature T_0.

The total power lost from a sphere (P_t) is thus given by

$$P_t = P_{cw} + P_r + (1 + \lambda r_s) \, P_{cd} \tag{11-19}$$

Figure 11-5 shows the measured power consumption at different temperatures for beads of different radius (with different diameters of support wires). The theoretically derived relationship (Equation 11-19) is also shown and agreement is good. In calculating the various contributions to the power drain it is assumed that the material from which the element is fabricated is a perfect conductor of heat, such that the coil and surface temperature are identical. This is a reasonable approximation for the essentially non-porous pellistor where a difference in temperature of approximately 40 °C was measured for a coil temperature of 500 °C, using optical pyrometry and pyroelectric temperature sensors [15]. However, for the porous, "poison resistant" type sensors having lower thermal conductivities, the differences in temperatures are typically around 80 °C [15]. This introduces greater error into the calculations and therefore should be considered when modelling porous catalytic sensors.

It is apparent that in order to reduce the power consumption of the sensor it is necessary to:

(1) lower the temperature, consistent with maintaining the required stability and mode of operation (see Section 11.2.4.4.), and

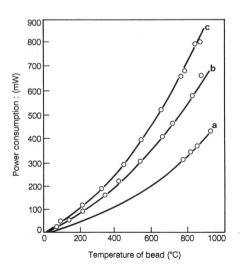

Figure 11-5.
Comparison of theoretical power consumption with experimental power consumption as a function of element temperature and bead radius. (a) bead radius 0.30 mm; (b) 0.45 mm; (c) 0.60 mm. Points are experimental values, curves are derived from Equation (11-19).

(2) reduce the size of the sensing element. The minimum size which can be achieved will depend upon the skill of the manufacturer. However reduction in size adversely affects the long term stability of the sensor making it more susceptible to drift and poisoning (see Section 11.2.5.1). Thus the environmental conditions will also determine the optimum size of the element.

11.2.4.2 Methods of Signal Measurement

Three methods of measuring the signal from the transducer can be identified. Two of these methods, the most widely used in commercial gas detection, are non-isothermal methods. The temperature of the sensing element is allowed to rise as a result of oxidation on the catalyst surface; the rate of reaction and hence concentration of flammable gas is derived from the increase in temperature. The circuit design is based on a Wheatstone bridge arrangement where the catalytically active sensing element and a similar but catalytically inert element form two arms in series on the bidge [16]. Power is supplied to the circuit to heat the elements to their operating temperature. When flammable gas is passed over the sensing element, reaction takes place one its surface increasing the power supplied to it, thus raising its temperature (resistance) until thermal equilibrium is achieved. This change is manifested as an out-of-balance voltage. The values of the fixed resistors in parallel with the elements are selected in order to balance the bridge in air. The two methods of measurement in use are distinguished by the mode of power supply, either constant voltage supply to the bridge, or constant current. The former is used in hand-held and portable instruments, whilst the latter is generally used in fixed monitoring installations.

Constant voltage

The equation describing small changes in the out of balance voltage (ΔV) across the Wheatstone bridge are:

$$\Delta V = R_d \Delta i + i \Delta R_d \tag{11-20}$$

$$-\Delta V = R_c \Delta i + i \Delta R_c \tag{11-21}$$

where R_d and R_c are the resistance of the detector and compensator, i is the current through the detector and Δi the change in current caused by the increase in resistance ΔR_d caused by the chemical reaction. Under normal operating conditions (eg, $\Delta V = 20$ mV; $R_d \simeq R_c = 3.5\ \Omega$; $i = 350$ mA) $\Delta R_c \simeq 0$; so that simplifying Equation (11-21) and combining with Equation (11-20) therefore gives

$$\Delta V = \frac{i \Delta R_d}{1 + R_d / R_c} . \tag{11-22}$$

It can be shown [16] that for small changes in ΔR_d

$$\Delta V = k r \Delta H \tag{11-23}$$

where r is the reaction rate, ΔH is the heat of combustion of the fuel and k is a constant for the system given by

$$k = \frac{iR_0 (D + 2BT)}{(1 + R_d/R_c) (dP/dT)} \tag{11-23 a}$$

where R_0 is the resistance of the detector at $0\,°C$, D and B are constants in the platinum resistance thermometer equation (with end wire correction), and P is the power supplied to the catalyst at temperature T [°C].

If the reaction rate is proportional to the fuel concentration then from Equation (11-23) the out-of-balance voltage of the bridge provides a measure of flammable gas concentration. The temperature of the element may also be affected by changes in the thermal conductivity of the ambient gas, particularly for those gases with thermal conductivities very different from that of air, e.g. hydrogen, methane. The incorporation of the non-active element compensates for such effects and for changes in ambient temperature by responding in a similar manner to the detecting element. It is therefore important to match closely the heat loss properties of the two elements, so that changes in the out of balance voltage will only be caused by a difference in reaction rate between detector and compensator.

Care must be exercised when detecting very reactive fuels such as hydrogen and methanol since oxidation may occur on the compensating element: this reduces the size of the signal. The problem may be overcome by operating at lower temperatures, alternatively the compensator can be isolated from the fuel in a sealed chamber or by employing a special type of constant resistance arrangement (see later).

Constant current

Again Equations (11-20) and (11-21) are appropriate, however for this method $\Delta i = 0$, thus

$$\Delta V = i\Delta R_d = -i\Delta R_c . \tag{11-24}$$

Here $\Delta R_d = -\Delta R_c$, thus ΔR_c is not negligible unlike in the constant voltage mode.

There are essentially two limitations in the use of the non-isothermal methods described above. Both are a consequence of allowing the temperature of the detector to rise during combustion. The constant k cannot be assumed to remain constant for large changes in temperature which are generated by high concentrations of flammable gas or reactions having very large heats of combustion. Moreover the temperature dependence of k is superimposed on that of the rate r (see Section 11.2.2.3) resulting in an observed decrease in ΔV with temperature at high temperature, even though the rate may be essentially independent of temperature. This is illustrated by comparison of the profiles from the non-isothermal and isothermal methods in Figure 11-6. Secondly, a finite time is required to re-establish thermal equilibrium at the new temperature following a change in the reaction rate. The delay is a function of the heat capacity of the sensing elements; for a typical pellistor-type element, this is of the order of a few seconds for a temperature rise of about 20 °C (as would be produced by an incremental change in methane concentration from 0 to 1% in air). The non-isothermal method does, however, have advantages which have ensured that it is the most widely used method. These are simplicity of circuitry (especially for the constant voltage mode), direct voltmeter display and compensation for variations in ambient conditions.

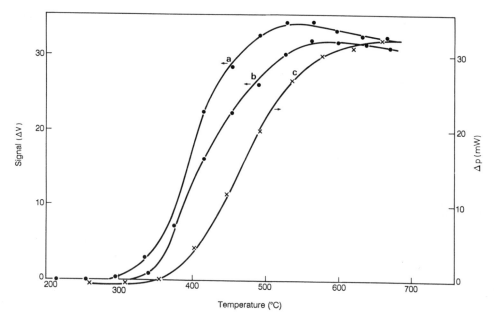

Figure 11-6. Response versus temperature for three methods of signal measurement. (a) constant voltage, (b) constant current and (c) constant resistance (temperature). Pellistor-type elements situated in a diffusion head; gas: 1% CH_4/air.

Constant resistance

This method is more widely known as the constant temperature or isothermal mode. It employs a Wheatstone bridge circuit as in the non-isothermal methods above; however the compensator is replaced by a fixed resistor and the out-of-balance voltage is maintained very close to zero by means of a feedback circuit which controls the power supplied to the bridge [17]. The heat generated during reaction, manifested through the resistance change, is sensed by the feedback circuit and the power is reduced to the bridge, thus ensuring that the element is maintained at constant temperature. The change in electrical power ΔP required to compensate for the chemical power is given by

$$\Delta P = -\Delta (i^2 R_d) = -2 i R_d \Delta i = -2 V\Delta V / R_d . \tag{11-25}$$

The parameters i and V refer to the current through and voltage across the detector. If P_a is the electrical power required to maintain the element at temperature T in the absence of reaction and P_f the electrical power to maintain the same temperature in the presence of a flammable gas, then

$$\Delta P = P_a - P_f = r \Delta H . \tag{11-26}$$

Hence the change in electrical power is a direct measure of the rate at which is heat liberated by the reaction [10] unlike the non-isothermal method. The response profile is shown in Figure 11-6. This relationship is particularly useful in fundamental studies of the kinetics of oxidation on catalytic elements which require constant temperature when determining reaction orders with respect to fuel and oxygen. Moreover, the time constant for the circuitry to maintain constant resistance is of the order of milliseconds, thus the speed of response of the isothermal method is faster than the other method. This is exploited in the development of a fast response flammable gas sensor [18], discussed further in Section 11.2.4.3.

The disadvantages of the isothermal method are that

(a) two measurements (current and voltage) are required to determine power, thus resulting in more complex circuitry. However, the change in bridge voltage across the detector and the fixed resistor in series with it can be measured [18], but this is not linearly related to the change in power dissipated, which is the true measure of reaction rate.

(b) the system cannot compensate for either changes in ambient temperature or the thermal conductivity of the gas; these have to be corrected for. This can be achieved electrically in the case of temperature by matching the temperature coefficient of the system with that of a thermistor or other temperature sensor and then subtracting the two signals [17]. Thermal conductivity effects can be compensated for, using linearising circuits, if the gas to be detected is known, although the effect is only significant for hydrogen and, to a lesser extent, methane.

High sensitivity operation

Finally, the conventional non-isothermal circuitry as described above can be slightly modified in order to measure low concentrations of flammable gas (0–1000 ppm) with a detection limit of about 20 ppm methane [19]. This is achieved by simply amplifying the out-of-balance voltage. Typically, an amplification factor of about 50 times is required; at higher factors the noise level becomes unacceptable. The major source of noise is thermal fluctuations arising from the hot element being situated in a cooler gas. Moreover, drift due to changes in ambient temperature is accentuated. It is therefore important to match the detector and compensator closely. This is difficult to achieve with the current fabrication methods used for pellistor-type elements. Commercial instruments for measuring low concentrations of flammable gas are available. They are used as 'sniffers' (i.e. semi-quantitative or qualitative) and, for the reasons cited above, do not have the accuracy of the LEL meters. Possible ways in which catalytic sensor design can be improved in order to measure low concentrations of flammable gas accurately are discussed in Section 11.2.6.

11.2.4.3   Sensor Enclosure

There are three basic designs of sensing head geometry; a typical example of each is shown in Figure 11-7. The most common enclosure is type (a) in which gas reaches the sensing element by diffusion through a metal sinter situated in the atmosphere under investigation. The response time of the sensor varies with the type of sinter and the separation between the sensing element and the sinter, however typical response times to 90% level (t_{90}) are in the range 10–20 s. In aspirated detectors (Figure 11-7b) gas is drawn along a tube by a manual or electric pump and then diffuses up a chimney to reach the sensor. This system is often used in portable and hand held instruments for sampling inaccessible regions e.g. down drains, in mine roof layers etc. The response time depends on the length of and flow rate along the aspiration tube, as well as the time taken to diffuse along the chimney. The third system (Figure 11-7c) is not commonly used in field instruments since the output of the device may be flow and orientation sensitive. Here the sensing element is positioned in the flow so that the gas is supplied directly to the sensing element and is not limited by diffusion across a sinter or orifice. The sensor is used to provide a fast response; e.g. $t_{90} \simeq 1$ s at a flow rate of 200 ml/min, an order of magnitude faster than the conventional diffusion-fed sensor heads [18].

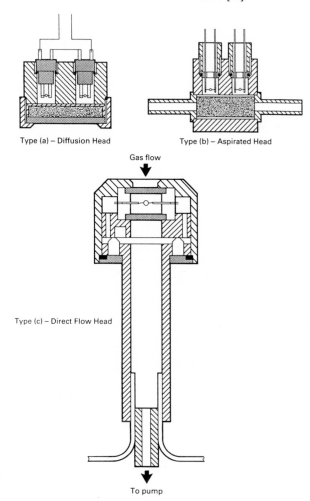

Type (a) – Diffusion Head   Type (b) – Aspirated Head

Gas flow

Type (c) – Direct Flow Head

To pump

Figure 11-7.
Sensor head configurations. (a) diffusion head, (b) aspirated head and (c) direct flow (fast response) head.

11.2.4.4 Physico-Chemical Factors

The signal from a catalytic sensor in a given concentration of flammable gas is governed by the rate of oxidation of that gas on the catalyst and the heat of oxidation. The rate of oxidation is in turn determined by the kinetics of the reaction as outlined previously in Section 11.2.2. A typical rate versus temperature plot obtained for either of the diffusion-fed systems outlined in Section 11.2.4.3 is shown in Figure 11-6c. At low temperatures the response is strongly dependent on temperature. This is characteristic of a surface controlled reaction. Under these conditions it is possible to determine the intrinsic kinetics of oxidation of fuels on various catalysts [9, 11, 20–23] and also to compare microcalorimetric kinetic data with conventional flow reactor data [10].

At higher temperatures the response becomes less temperature dependent, indicating that the reaction is diffusion controlled. For type (a) and (b) sensor heads (see Figure 11-7) the rate is controlled by diffusion through the sinter (or across an orifice) some distance away from the sensing element. The zone where diffusion takes place remains essentially at ambient temperature; it is not affected by the temperature of the element. Thus the rate (ΔP) is essentially independent of element temperature. In the flow-fed system (type (c) in Figure 11-7) no such plateau occurs, but the rate is only weakly dependent on temperature. This is characteristics of diffusion control close to the sensor surface; here the gas diffuses through a layer of fuel-depleted air surrounding the hot element and thus is influenced to a greater degree by the temperature of the element. Various models have been used to describe these conditions [5, 24]. In addition to predicting behaviour at high temperatures these models can allow more accurate determination of the kinetics of reaction by extrapolation to lower temperatures where the rate is neither completely surface reaction nor diffusion controlled [5, 11]. The catalytic sensor is not sensitive enough to allow measurement at very low temperatures where mass transport effects are minimal, nor can the temperature of onset of diffusion control be raised significantly (because of the geometry of the system) to extend the range of temperature where reaction kinetics can be determined in the absence of diffusion control.

In a practical instrument it is desirable to operate the element in the high temperature, diffusion controlled zone where the signal is directly dependent on flammable gas concentration and independent of both temperature and oxygen concentration. In this region the signal is proportional to the rate of mass transport (diffusion) of the fuel, essentially across a region remote from the sensor. The intrinsic activity of the catalyst is thus relatively unimportant; slight variations in catalyst surface area and intrinsic activity do not markedly affect the signal, only geometric factors govern its magnitude. Thus, the prime role of the catalyst is to permit operation in the diffusion controlled region, with its operational advantages, at as low a temperature as possible.

Diffusion controlled mode of operation

The signal generated by a flammable gas and air mixture under diffusion control has been considered by Firth et al. [25] who show that for any fuel, the response at the LEL is given by

$$\Delta V_{LEL} = k_1 \Delta H D_{12} \text{ [LEL: lower explosion limit]} \tag{11-27}$$

where k_1, is a constant. Equation (11-27) can be considered as a special case of Equation (11-3). It can also be demonstrated [25] that for many fuels there is an inverse relationship between the LEL concentration and the heat of combustion, i.e. the calorific value of any gas mixture at the LEL is approximately a constant. This relationship is closest for chemically similar fuels e.g. $C_1 - C_5$ alkanes. Thus Equation (11-27) can be approximated for many fuels to

$$\Delta V_{LEL} = K_2 D_{12} . \tag{11-28}$$

Moreover, Equation (11-28) may be further approximated for certain fuels to

$$\Delta V_{LEL} \approx k_3 \qquad\qquad (11\text{-}29)$$

because the diffusion coefficients for hydrocarbons in air vary by a factor of 4 over the range $C_1 - C_{10}$ and by only a factor of 2 over the range $C_3 - C_{10}$ [7, 25]. Thus, an important characteristic of catalytic sensors is their ability to provide an estimate of the explosiveness of any vapour or mixture of vapours based on Equation (11-29), when calibrated to read, say, 0–100% LEL in a standard fuel. If the composition of the vapour is known, simple correction factors based on the constants in Equations (11-27) and (11-28) may be applied to give more accurate readings. Diffusion and heat of combustion data for 93 flammable gases and vapours have been compiled in [25]. Table 11-1 shows an extract for a series of some commonly occurring fuels. In this case methane is adopted as the standard fuel and the concentrations of the other fuels estimated using Equations (11-27)–(11-29). For general flammable gas detection it is usually more accurate, and certainly safer, (i. e. over estimating concentration) to calibrate the instrument in a fuel other than methane; butane and n-pentane are ones commonly used.

Table 11-1. Comparison of the validity of Equations (11-27)–(11-29) in predicting the explosiveness of fuel/air mixtures.

100% LEL of gas	Signal (mV)	Concentration expressed as % (LEL) derived from		
		Equation 11-29	Equation 11-28	Equation 11-27
Methane	168.8	100	100	100
Ethane	129.0	76.4	114	113
Propane	106.8	63.3	121	115
n-Butane	98.5	58.4	118	100
n-Heptane	73.1	43.3	143	112
Methanol	154.6	91.7	124	107

Determination of calorific value

It is inherent in any method utilising rate of reaction under diffusion controlled conditions that errors are introduced when monitoring unknown mixtures of vapours having different diffusion coefficients. However, as stated above, there is a close correlation between LEL and ΔH. Thus an instrument which can measure the total heat of reaction rather than the rate of liberation of heat would provide a measure of explosiveness less dependent on the composition of the fuel, i. e. the dependence of the signal on D_{12} is removed. Such an instrument has been described [26] which operates by intermittently filling a reaction chamber with a known volume of the gas to be sampled. The signal is integrated following the complete oxidation of the fuel in the chamber. This integrated signal is a measure of the total heat of reaction and the explosiveness of the atmosphere and is relatively independent of the composition. The response (R) is directly proportional to the calorific value of the fuel (CV) and can be represented thus:

$$R = \gamma\,(CV) = \gamma\,c\,\Delta H = \gamma' \int_0^t \Delta V\,dt \qquad\qquad (11\text{-}30)$$

where γ and γ' are proportionality constants and c the fuel concentration. Expressed in terms of the LEL, then

$$R_{LEL} = \gamma_{LEL} \int_0^t \Delta V \, dt . \qquad (11\text{-}31)$$

11.2.5   Limitations on Sensor Performance

The catalytic sensor operates by oxidising flammable gas in air over a catalyst. Under diffusion control conditions the response is usually insensitive to variations in catalyst activity. However, there are occasions when gross changes in catalyst activity occur which markedly affect sensor behaviour; these are caused by poisoning and sintering. Moreover, there are problems associated with high fuel concentrations which can lead to coking of the catalyst, (a phenomenon related to poisoning) and the decrease in oxygen concentrations inevitably preventing complete oxidation of the flammable gas. These four factors — poisoning, coking, low oxygen concentrations and sintering, are discussed in turn.

11.2.5.1   Catalyst Poisoning

Outline of problem

The major limitation of catalytic sensors is their loss of sensitivity on exposure to atmospheres containing certain gases and vapours. This loss of sensitivity is particularly important since the device will give a falsely low estimation of the flammable gas concentration; the sensor will thus "fail to danger". Therefore it is essential that those species which can cause this effect be identified and to understand the way in which they influence the activity of the catalyst. It has become apparent that the principal contaminants which can impair catalytic sensor performance are silicone vapours, phosphate esters, halogen containing compounds, alkyl lead compounds and sulphur containing compounds. Exposure to these compounds is industry related e. g. in the mining industry the first three named vapours are usually encountered; whilst for the offshore petroleum industry hydrogen sulphide present in oil and gas, and silicones, present in fire retardants, are the usual sources of poisoning problems; alkyl lead compounds may influence determinations of petrol vapour concentrations at filling stations etc.

To overcome the problems of inaccuracy and failure due to poisoning, filters mounted in front of the detector element can be used. The most commonly used filter material is active charcoal. These filters adsorb and remove most common poisons including all those mentioned above. Granular charcoal is very effective against most forms of catalyst poison, however there are difficulties in fabricating such filters and the delay in response time which they introduce has led to the introduction of filters based on carbon cloth [27]. However the performance of such cloths can be very variable and some cloths themselves poison the sensing elements [27]. Table 11-2 shows data on the performance of some charcoal based protection methods. Charcoal, however, will also adsorb hydrocarbons in addition to the potentially harmful vapours. This limits the use of carbon-based filters to detectors for C_3 and lower-hydrocarbons, and gases such as hydrogen and carbon monoxide.

Table 11-2. Properties of some charcoal-based filters.

Sample No	Thickness (mm)	Description	Performance of experimental sensor head		
			Response time (s)	Breakthrough time for 500 ppm HMDS	Time to 30% loss of sensitivity in lab. air
1	1.0	Carbon blown into cotton or felt. Loading 5% W/W. Poor physical strength	–	2 s	–
2, 3	2.0	339-C carbon impregnated felt. Woven cotton backing. Not available commercially	25	10 h	–
4	0.5	Unwoven. About half area thermally compressed to improve rigidity	30	–	10 weeks
5	0.5	Stiff plastic foam, backed with cellulose mat	17	–	–
	2.0	Stiff plactic foam, no backing. Not available commercially		–	–
6, 7, 8	0.4	Wholly woven carbon cloth: derived from carbonization under CO_2 at 800 °C of Lewis acid treated cellulose or viscose	15	60 min / 120 min / 80–130 min	* / 15% loss after 5 years / 1 year
9	2	Impregnated foam plastic bonded to nylon backing	24	11 min	1 year
10	3.5	Two layers impregnated felt bonded together	12	35 min	18 days
11	0.7	Impregnated coarse buckram cotton with felt	17	2 s	15% loss after 2.5 years
12	1.0	Brushed felt: carbonized as samples 6–8	17	160 min	No loss after 1 year
13	2	207-C granular carbon (0.5–2 mm particles)	50	18–24 h	No loss after 4 years
14	–	Unprotected gas sensors	12	2 s	15% loss after 1–3 years

* Unwashed 115 days; washed 60 days.

The growth in demand for catalytic sensors to operate in hostile atmospheres and areas where routine calibration and maintenance are difficult and costly led to the development of sensing elements which are inherently resistant to poisons. These sensors are similar in form to the essentially non-porous pellistors. They consist of a platinum wire coil thermometer and heating element. The coil however is embedded in a porous bead comprising active catalyst dispersed throughout a porous substrate (see for example [28[) or as discrete layers within the substrate (see for example [29]). A typical example of the former type is shown in Figure 11-8. The use of a porous substrate invariably results in mechanically weaker devices.

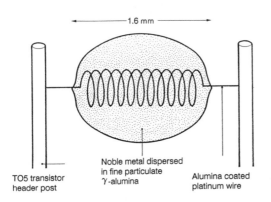

TO5 transistor
header post

Noble metal dispersed
in fine particulate
γ-alumina

Alumina coated
platinum wire

Figure 11-8.
A typical poison-resistant, porous
catalytic flammable gas sensing element.

Classification of contaminant gases

Contaminating gases can be classified into two groups on the basis of the irreversibility or reversibility of their effect on catalytic activity (sensor response). Thus a vapour (e. g. hexamethyldisiloxane (hmds), organo-lead vapor) is a poison if the response of the sensor remains impaired after the removal of offending species.

A vapour (e. g. certain halogenated hydrocarbons) is an inhibitor if the activity recovers on removal of the species. The difference in behaviour of the two groups derives from the manner in which they adsorb on the catalyst and the species which are subsequently produced. Both groups reduce the catalyst activity by adsorbing more strongly than one or other of the reacting species, thus reducing its surface coverage and hence the rate of reaction. An inhibitor adsorbs reversibly attaining an equilibrium coverage which is a function of temperature and gas phase concentration. Referring to Equation (11-12), once the inhibitor is removed from the gas phase ($P_x = 0$), the surface coverage of the reacting species A and B increase and the reaction rate recovers completely. A poison, however, is irreversibly adsorbed; coverage does not reach an equilibrium, but increases at a rate dependent on the gas phase concentration of the poison until coverage is complete. Thus the response of a catalytic sensor exposed to poison is reduced to zero irrespective of the concentration of the poison. However the time taken to deactivate the catalyst is dependent on the poison concentration.

Theory of catalyst poisoning

There is a wealth of knowledge relating to catalyst deactivation which can be tapped in order to understand catalytic sensor poisoning phenomena, particularly in porous catalysts.

This arises from similar problems encountered in the chemical and automotive industries to those in catalytic sensor technology.

The rate at which a catalyst is poisoned depends on the nature of the catalyst, the poison and the reaction under consideration. The catalyst may be porous or non-porous, the poison selective or non-selective and the reaction fast or slow compared with the rate of diffusion. The simplest case of a first order irreversible reaction is considered to illustrate the basic characteristics of catalyst poisoning. More general equations which could accommodate inhibition can be developed, but they are more difficult to solve, requiring various simplifications in order to solve them conveniently [30]. Characteristics for four different poisoning cases, taken from Hegedus and McCabe [30] are shown in Figure 11-9. Case (a) is when both

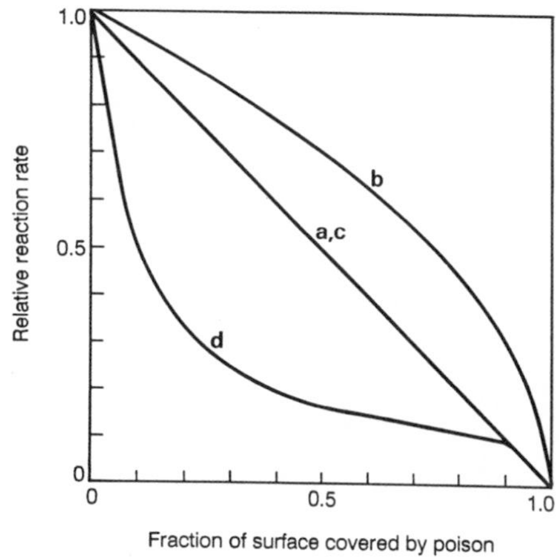

Figure 11-9.
Poisoning characteristics of a catalyst in the absence of external mass transfer control (after [30]). (a) reactant and poison uniformly distributed in the bead; (b) poison uniformly distributed, reactant not (diffusion effects); (c) reactant uniformly distributed, poison not; (d) selective poisoning or pore-mouth poisoning.

the reactant (fuel) and the poison are uniformly distributed over the catalyst bead, as would be obtained for a non-porous catalyst; case (c) occurs in a kinetically controlled reaction on a porous catalyst subjected to a non-selective poison. If however the poison is selective, i.e. the most active catalytic sites are preferentially poisoned, or the poison is adsorbed through multiple bonding, then curve (d) is obtained. In case (b) the poison is uniformly distributed throughout the porous bead but finite pore diffusion effects on the main reaction result in reactant concentration gradients. The form of the curve thus mimics antiselective poisoning behaviour in a non-porous catalyst. Finally, case (d) can also represent pore mouth poisoning, where the outer portion of porous catalyst has been poisoned by a sharply defined poison front. The abscissa in Figure 11-9 represents the fraction of surface poisoned, however it is more useful to be able to formulate the rate of poisoning with respect to exposure time. Also the effects of external mass transfer control (EMTC) in addition to internal mass transport (pore) effects described above, need to be considered for catalytic sensors. Such a model has been developed [31] which requires the solution of the following equations:

$$D_f \frac{\partial^2 c_f}{\partial r^2} - \frac{2}{r} \frac{\partial c_f}{\partial r} - k_f (a_0 - a_p)^m c_f^n = 0 \tag{11-32}$$

$$D_p \frac{\partial^2 c_p}{\partial r^2} - \frac{2}{r} \frac{\partial c_p}{\partial r} - k_p (a_0 - a_p)^a c_p^\beta = 0 \tag{11-33}$$

$$k_p (a_0 - a_p)^\alpha c_p^\beta = \frac{\partial c_w}{\partial t}.$$ (11-34)

The first two equations are the diffusion equations for a spherical catalyst where the concentration of the fuel and poison throughout the catalyst are c_f and c_p respectively. The third expression on the left hand side of Equation (11-32) relates to the rate of reaction on the pore walls, having an intrinsic rate constant of k_f, an initial surface area a_0 and a poisoned surface area, a_p. A similar expression in Equation (11-33) relates to the rate of poisoning having a rate constant k_p. The rate of accumulation of poison on the catalyst is given by Equation (11-34) where c_w is the concentration of adsorbed poison on the catalyst, having units of moles per volume of catalyst.

The effects of external mass transfer i.e. bulk diffusion control may be considered by introducing, in addition to the usual boundary conditions, the following conditions [31]:

$$D_f \frac{dc_f (t, r_e)}{dr} = K_f [c_f^b - c_f (t, r_e)]$$ (11-35)

$$D_p \frac{dc_p (t, r_e)}{dr} = K_p [c_p^b - c_p (t, r_e)]$$ (11-36)

where K is the mass transfer coefficient, c^b is the bulk gas concentration and $c (t, r_e)$ the concentration at the boundary of the spherical catalyst, radius r_e, at time t.

At this stage it is useful to introduce a parameter h, the Thiele modulus, in order to simplify solution. It is defined thus [31]:

$$h = r_e \left(\frac{k a_0^m [c^b]^{n-1}}{D} \right).$$ (11-37)

A high value for h implies that the reaction (or poisoning) rate is very fast compared with the rate of diffusion of reactant (or poison). A low modulus implies rapid diffusion, with reaction occuring throughout the whole of the catalyst. Another useful parameter is the Biot number (B), a dimensionless quantity related to the mass transfer coefficient [31] thus:

$$B = r_e K/D.$$ (11-38)

A low value of B ($\ll 1$) indicates strong external mass transfer control whilst $B \gg 1$ implies very little external mass transfer control, as is the case for the curves shown in Figure 11-9. The profiles obtained from solution of Equations (11-32)–(11-34) for various values of h_f and B, for $h_p = 10$ (i.e. a pore mouth poison) and $h_p = 0.1$ are shown in Figure 11-10. The ordinate represents the rate of reaction normalised against the initial rate and the abscissae represents a dimensionless time factor. It can be seen that increasing the external mass transfer control increases the lifetime of the catalyst and, under external mass transfer control, increasing the internal resistance (by increasing h_f) also prolongs catalyst life. The latter effect, however, does not exert as great an influence as the former.

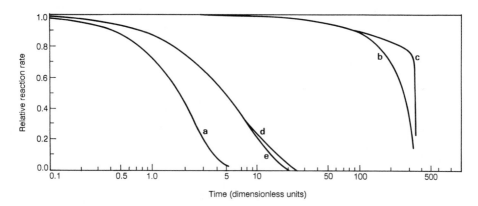

Figure 11-10. Time-dependent poisoning characteristics of spherical catalysts. Curves derived from Equations (11-32)–(11-38). (a) $B = 0.1$, $h_f = 1$, $h_p = 0.1$; (b) $B = 0.1$, $h_f = 1$, $h_p = 10$; (c) $B = 0.1$, $h_f = 10$, $h_p = 10$; (d) $B = 10$, $h_f = 1$, $h_p = 10$; (e) $B = 10$, $h_f = 10$, $h_p = 10$. (after [31]).

Similar conclusions were reached by Gentry and Walsh [32] who adopted a simpler approach to modelling the poisoning rate in porous catalysts with external mass transfer control. The method is based on a more empirical approach after the work of Frank-Kamenetskii [33]. The method essentially considers two consecutive first order reactions in overall steady state: the reaction rate on the bead in the absence of EMTC (R_e), and the rate of diffusion to the element (R_d). Since all the reactions are taken to be first order with respect to fuel the actual rate (R_a) can be expressed thus:

$$\frac{1}{R_a} = \frac{1}{R_e} + \frac{1}{R_d} .$$ (11-39)

The result of this approach when used with simplified expressions for the rates of reaction and poisoning, is shown in Table 11-3. The two values quoted for each parameter represent, typically, the limits that would be expected to occur in practice. Results for a typical rhodium poison resistant element [28] are shown for comparison. The two parameters which most strongly affect the poison resistance are the element radius (r_e) and the geometric factor g, related to the mass transfer coefficient K. Thus the elements most resistant to pore mouth poisons are those which are large and are strongly controlled by bulk gas diffusion i.e. external mass transport. Low values of g, which correspond to a high degree of EMTC, can be obtained by positioning the element behind fine sinters or small orifices. This will however, reduce the response of the sensor and increase the response time. Minor increases in poison resistance can be achieved by employing a catalyst with a large pore radius (r) and a high intrinsic rate constant (k_f).

Experimental studies of poisoning

The theory of poisoning as treated above is a much simplified representation of the actual reaction and poisoning mechanisms. Moreover, only irreversible poisoning has been con-

Table 11-3. Effect of parameters on the activity and poison resistance of a porous element.

Parameter	Value	Units	h_f	R_e (mol s^{-1})	α_{20}	t_{20} (h)
r_e	0.05	cm	5.9	0.31	0.16	0.7
	0.15		17.7	2.78	0.96	83
r_p	2.5×10^{-7}	cm	93	1.15	0.08	0.2
	2.55×10^{-5}		1.2	0.68	0.72	14
k_f	1×10^{-4}	cm/s	0.94	0.07	0.13	0.6
	0.1		29.7	2.80	0.58	11
g	0.25	cm	10.6	1.00	0.97	31
	1.0		10.6	1.00	0.29	2.8
Typical Rh element			10.6	1.00	0.58	11

h_f Thiele modulus for fuel oxidation
r_e element radius
r_p pore radius
k_f rate constant for fuel oxidation
g geometric factor related to the mass transfer coefficient
α_{20} fraction of surface covered by poison after 20% loss in observed response
t_{20} time taken to lose 20% of observed response

sidered. Despite these limitations the theory can be used to indicate the most significant factors in determining poison resistance. In general, major improvements in poison resistance may be obtained in three ways:

(a) the catalyst is chosen to have a high intrinsic resistance to poisoning, i. e. the value of k_p is small,

(b) the catalyst is operated under strong mass transport control, i. e. the intrinsic activity of the catalyst is high compared to the rates of bulk and pore diffusion, (i. e. $h_f > 1$, $B_f < 1$) and

(c) the pore structure of the support can be selected to restrict access of high molecular weight (large) poisons while allowing relatively free access of reactants to the catalyst surface, i. e. $h_p > h_f$.

A high intrinsic resistance of the catalyst to poisoning is dependent on the form of the interaction of the poison or inhibitor with the catalytic sites, adsorption effects, compound formation and poison induced surface reconstruction. Gentry and Jones [34] studied poisoning by hmds of several reactions on essentially, non-porous, pellistor type Pt/Al$_2$O$_3$ elements and concluded that at least three types of site exist. These sites could be distinguished by their ability to catalyse different oxidation reactions and the manner in which they interacted with hmds. The poison causes a slow linear reduction in rate for hydrogen oxidation (cf. Curve, a, Figure 11-9) while the rates of methane and propene oxidation are rapidly reduced (cf. Curve d, Figure 11-9). It must be emphasised, however, that these profiles may only be compared under the assumption that coverage increases linearly with time. Moreover the influence of hmds vapour on the reactions studied varies from total poisoning (methane oxidation), through inhibition (propene oxidation) to negligible effect (hydrogen oxidation). However, it seems likely that, in view of the essentially non-porous nature of the catalyst, hmds behaves as a selective poison, adsorbing irreversibly onto the most active sites which are essential for the oxidation of methane and responsible for the low temperature activity of propene and hydrogen oxidation. Irreversible adsorption of hmds leads to decomposition and formation of a silicious adlayer. Reversible adsorption occurs on the less active sites, inhibiting the high temperature oxidation of propene, but not leading to decomposition of the inhibitor and irreversible poisoning of these sites. Hydrogen oxidation appears to occur via an Eley-Rideal mechanism which does not require strong chemisorption of hydrogen. This reaction probably involves oxygen species adsorbed on sites which are virtually unaffected by hmds [34]. Similar patterns were observed by Cullis and Willatt [35] for the deactivation of methane and butane oxidation by hmds over Pd/ThO_2 pellistors. Butane behaved in a similar way to propene discussed above. This behaviour is also exhibited by lead compound poisons. The extent to which both silicon and carbon, formed by the decomposition of the organosiloxane compound, penetrate below the surface of the catalyst was found to be dependent on the nature of the precious metal and the catalyst support. Thus palladium which readily absorbs silicon into its bulk is more readily deactivated than platinum, and $Pd/ThO_2/Al_2O_3$, beads are less resistant than Pd/SnO_2 beads [35].

Halogenated hydrocarbons inhibit the oxidation of methane by adsorption on to sites on which normally the adsorption and activation of oxygen occurs, resulting in a deficiency of oxygen at the surface according to Cullis et al. [36]. An equilibrium is maintained between the inhibitor and reactants (fuel and oxygen); thus the extent of deactivation is decreased by raising the temperature of the catalyst. The influence of several poisons and inhibitors on non-porous pellistor and porous elements for the oxidation of methane and butane was investigated by Gentry and Walsh [37]. This work showed that butane oxidation is, in general, less inhibited by halogenated and sulphurated gases than methane oxidation, and both reactions are least inhibited on the least active metal, platinum. It was concluded that, in minimising the effects of halogenated and sulphurated gases, the choice of noble metal as well as the physical form of the catalyst plays an important role in determining poison resistance.

For hmds poisoning the choice of noble metal dispersed on a porous support is not as important. Increasing the activity of the catalyst by increasing the surface area is the prime factor, thus ensuring a high degree of mass transport control. This is illustrated in Figure 11-11 where the poisoning rates of various noble metals supported on high surface area (~ 100 m^2/g) γ-Al_2O_3 are compared to that of a typical Pd/ThO_2 pellistor operated in a diffusion head as shown in Figure 11-7a. The effect on the poisoning rate of operating the sensor in the fast response sensor enclosure (Figure 11-7c) is even more dramatic. Here the degree of EMTC is much less than in the diffusion-fed enclosure resulting in extremely low poison resistance (20% loss of signal in < 5 s, compared with ca. 100 s for the diffusion head). Although all the foregoing discussion has referred to porous elements having a fairly uniform distribution of noble metal throughout the support, there are elements which are composed of layers within the substrate [29]. Thus an outer non-active layer of support can act as a filter for poisons, although the activity of the element may be reduced because the fuel has to diffuse further into the element to react. The theory of poisoning where catalysts are distributed non-uniformly on supports, and in the absence of EMTC, is discussed in detail by Becker and Wei [38].

The pore structure of the catalyst influences the poison resistance of the sensing element indirectly by providing a large surface area which increases activity and, hence, poison resistance when the rate is mass transfer controlled. It can also directly affect poison resistance by restricting access of large, high molecular weight poisons whilst allowing reactants to diffuse more easily to the catalytic sites. This is best demonstrated by reference to zeolite catalysts used in flammable gas sensing elements [39, 40]. These materials possess a caged, micropore structure, rather than the meso- and macro-porous structures found in the standard poison resistant elements described previously. Palladium exchanged zeolites have been used to produce sensing elements with greatly increased resistance to poisons such as hmds. Improvements of the order of 1000 times greater than conventional pellistor types were obtained, compared to approximately 200 times greater for the conventional poison resistant type elements where the noble metal is dispersed throughout γ-Al$_2$O$_3$ [41]. In the zeolite elements, the pore structure is too fine for large poison molecules to penetrate significantly, but access to alkanes below about C$_6$ and oxygen is not significantly impaired. In terms of the theory developed earlier, this corresponds to $h_p > h_f$, as defined in Equation (11-37), and its effect on the poison resistance is shown in Figure 11-10. At the present time zeolite elements suffer from the disadvantages of having long response times ($t_{90} > 20$ s) and being unstable at normal operating temperatures (500 °C). Gradual thermal deactivation of the noble metal exchanged zeolite occurs, caused by the migration of the noble metal out of the cage structure; this is followed by sintering on the external surface of the alumino-silicate [41].

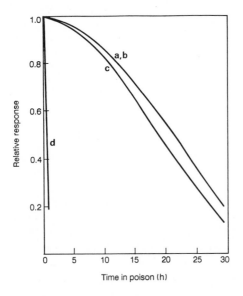

Figure 11-11.
Effect of time in poison (10 ppm hmds) on the relative response of catalytic elements. (a) Pd porous element; (b) Rh porous element; (c) Pt porous element; (d) Pd/ThO$_2$ pellistor (non-porous). Fuel, 1% CH$_4$/air; elements at 550 °C in a diffusion head.

Detection of inhibitors

The foregoing discussion has been concerned with minimising the deleterious effects of catalyst inhibitors and poisons. The phenomenon however can be turned to advantage by using the inhibiting effect on a fuel to detect and measure the concentration of the inhibitor [42]. A convenient fuel for this mode of operation is methanol because of the ease of supply of a constant concentration by means of a wick arrangement [42]. The kinetics of the oxidation reaction are relatively simple [5, 10] thus helping to ensure that the method is reliable.

The catalyst used in normal applications generally have a relatively high surface area so that under diffusion controlled conditions, fairly large concentrations of inhibitors must be present before they begin to affect the response of the sensing element. However by reducing the surface area of the catalyst the sensitivity towards inhibiting gases is markedly increased. Reducing its temperature also has the same effect. Firth et al. [42] describe such a sensor having potential to detect chlorine, sulphur dioxide and certain dihalomethanes. Sensing elements having very low surface areas are those based on platinum coils; they have been used to monitor oscillatory effects on the rate of carbon monoxide (the fuel) oxidation induced by isopropyl methylphosphonofluoridate (the inhibitor) at a concentration of 4 µg/l [43].

11.2.5.2 Effects of High Fuel Concentration

Catalyst coking

In low concentrations of hydrocarbon fuels in air (below the stoichiometric mixture concentration, approximately 9.5% mixture in air for methane) complete oxidation to carbon dioxide and water occurs when the sensor is operated under normal conditions. However in higher concentrations the hydrocarbon may be incompletely oxidised to carbon and carbon monoxide. This reaction probably proceeds via dehydrogenation [44] which occurs at an appreciable rate in an oxygen deficient atmosphere. Reduction of the surface oxygen coverage leads to a suppression of the main oxidation reaction and enhancement of the dehydrogenation reaction. The maximum deposition of carbon from methane on palladium catalysts at 600 °C occurs at a concentration of 40% CH_4/air [45].

Catalyst coking may have a marked effect on the operation of the sensing element because (a) surface carbon can poison the oxidation reaction; and (b) large deposits of carbon change the physical size, morphology and, to a lesser extent, the emissivity of the element, all of which alter the rate of dissipation of heat from the calorimeter and thus change the sensor output. Even when the carbon is subsequently burnt off, permanent damage to the sensor may have occurred because carbon growth in fissures in the catalyst increases the mechanical stress, eventually leading to breakdown of the catalyst support. It has been known [45] for many years that the incorporation of thoria with palladium in the pellistor type element greatly reduced the effects of coking arising from prolonged exposure to high concentrations of hydrocarbon fuels. The beneficial property of thoria is demonstrated in Table 11-4. Under normal operating conditions the catalyst mixture is essentially palladium oxide dispersed on thoria. At high hydrocarbon concentrations the palladium oxide is reduced to metal which, in the absence of thoria, sinters rapidly to produce more massive particles. It has been shown [44] that the role of thoria as a coke prevention agent is to decrease the palladium particle size. This reduce the rate of dehydrogenation which occurs preferentially on larger particles. Moreover dispersion of the noble metal by other means, e. g. porous elements as discussed in Section 11.2.5.1, also prevents excessive carbon deposition.

Another factor in determining the stability of elements in high methane concentrations is the mechanical strength of the alumina support [46]. This is demonstrated in Table 11-5 where stability is defined in terms of the time taken for the out of balance voltage in air to shift after repeated exposure to 40% CH_4/air mixture. It can be seen that stability is directly proportional to hardness or mechanical strength, defined in terms of a critical weight of crushing [46].

Table 11-4. Effect of 40% CH_4/air on zero level and sensitivity of various elements.

Element	Initial Sensitivity to 1% CH_4 + air (mV)	Zero Shift (mV)	Sensitivity Shift (mV)
28 Pd + ThO_2/α-$Al_2O_3^*$	27	3	2
65 Pd + ThO_2/α-Al_2O_3	30	12	-3
Pd/α-Al_2O_3	30	64	-12
13 Pd + Al_2O_3/α-Al_2O_3	20	3	1
Pd/glass	20	catalyst detached from support	
28 Pd + ThO_2/γ-Al_2O_3 (11 wt.% Pd loading)	25	2	1
Pd/γ-Al_2O_3 (11 wt.% Pd loading)	30	5	1

* (28 Pd + ThO_2) denotes mole fraction of Pd in Pd $-$ ThO_2 catalyst solution applied to bead is 28%.

Table 11-5. Variation of stability of aluminas in 40% CH_4/air with hardness of aluminas.

Hardness* (g)	Duration of stability (h)
500	1.0
1000	2.5
1800	4.2
2200	5.5
3000	7.5

* Hardness defined as the critical weight of crushing

Low oxygen concentration

Oxidation of the flammable gas in air is a prerequisite to obtaining a response from a catalytic sensor; thus, if the concentration of oxygen in the gas mixture is too low for complete oxidation of the flammable gas to occur, the response obtained will be low. As the concentration of flammable gas is increased in air from zero, the response of the sensor increases to a maximum at around the stochiometric ratio for total oxidation, and then decreases to zero at a concentration of 100% flammable gas. There is, therefore, a range of high concentrations of gas in air which will cause the sensor to indicate falsely that the gas concentration is below the LEL. For methane in air this concentration range is approximately 40–100% [45].

Various methods are employed to overcome this. The simplest method entails careful observation of the instrument reading as the gas enters the sampling chamber. As the concentration inside the cavity containing the sensor rises the signal will rapidly increase above 100% LEL before returning on scale. As the instrument is withdrawn from the gas to air the signal will again rise above 100% LEL before returning on scale. Alternatively an electronic 'latching' method may be employed, as specified, for example, in British Standard 6020 for flammable gas detectors [2]. This ensures that if the reading increases past 100% LEL, it remains there until manually reset.

Other instruments utilise thermal conductivity changes (see Section 11.5). In one example the sensor is operated at low temperature (ca. 200 °C) where the rate of catalytic oxidation is low compared to thermal conductivity effects. Having initially established whether the fuel concentration is high or low then, if the latter condition prevails, the sensor is operated at the normal temperature (ca. 500 °C) for determination of LEL.

Another example is where a separate transducer, e. g. thermal conductivity sensor (see Section 11.5) or oxygen sensor (see Chapter 3) is used in conjunction with the catalytic sensor to determine initially whether the concentration of flammable gas is below or considerably above the LEL.

11.2.5.3   Long Term Stability of Catalytic Sensors

The long term stability of a catalytic sensor in the absence of catalyst poisons and inhibitors is determined by the stability of the catalyst at its operating temperature (usually around 500 °C) in air. The two dominant factors governing catalyst stability are sintering of the catalyst and reactivity of the catalyst with oxygen. As discussed previously, dispersing the noble metal throughout a support helps to maintain a high surface area of catalyst. Moreover an initial treatment, at temperatures well above the operating temperature achieves an equilibrium sintered state before the sensor is commissioned. Nevertheless it is still advisable to operate sensors for several hours in air in order to stabilise their response, since the initial treatment tends to produce slightly more active catalysts. This enhanced response gradually decreases eventually attaining a stable level. This phenomenon may be related to the metal-metal oxide equilibrium on the catalyst surface. Palladium forms a stable surface oxide and the equilibrium ratio is dependent on the fuel oxygen ratio [47]. The high methane concentration (12% CH_4/air) used in the pretreatment reduces all the oxide to metal. Palladium metal is more active than palladium oxide for the oxidation of methane [47], thus an initially enhanced response is obtained. The Pd/PdO ratio then slowly re-equilibrates in air at the operating temperature, resulting in a decrease in the concentration of metal. This is the process which requires a 'running in' period to stabilise. Platinum does not form surface oxide layers of the type and extent formed on palladium, [48], however, it still undergoes sintering and thus requires pretreatment in a similar fashion to palladium.

11.2.6   Future Trends

The rate of development of catalytic sensors has slowed down following the introduction of sensing elements more resistant to poisons. Currently attention is focussed on the following areas, listed in approximate order of importance:

(a) increasing further the long term stability and poison resistance of the sensing element,
(b) reducing the power consumption of the sensor,
(c) providing advance warning of poisoning, and
(d) providing some degree of selectivity between flammable gases.

Increasing the general stability and lifetime of the sensor is a goal which continually occupies researchers in this field. It is undoubtedly a problem area, as discussed in Section 11.2.5, in which progress is slow and it is difficult to envisage dramatic improvements in this aspect of performance of catalytic sensors. There is little work in progress on the development of new catalysts specifically for gas detection, or on the design of novel catalytic sensor structures. Current work is concerned with marginal improvements in the production techniques, catalyst preparation and sensor operating conditions.

The power required by the sensor can be reduced (see Section 11.2.4) by miniaturisation, or operation at lower temperatures. Small catalytic sensors are available requiring around 200 mW of power to attain an operating temperature of ca. 500 °C. However, it is difficult to conceive of further reductions in the size of pellistor-type devices because of fabrication difficulties. Possibilities may be provided through the use of miniature ceramic tile substrates having trimmed thick film platinum resistance thermometers, as used in certain semiconductor sensors [49]. This method should give greater reproducibility of sensitivity between different sensors than the platinum coil method.

Any reduction in size or temperature decreases the poison resistance of the device. Thus a compromise must be reached between power consumption and poison resistance. It is anticipated that size reduction is only beneficial in 'clean' environments or where poison protection can be provided by external means, such as charcoal adsorbents; this restricts the use of the sensor to mainly CH_4, CO, and H_2 detection.

Greater scope for development of catalytic sensors lies in the utilisation of microprocessor technology for interrogation of an array of sensors composed of different catalyst and operated at different temperatures. It may be possible, therefore, to distinguish between various groups of flammable gases on the basis of their varying response-temperature profiles with various catalysts. The response profiles for various fuels on a conventional Pd/ThO_2 pellistor are shown in Figure 11-12. The signal at high temperature is an approximate measure of the toal flammability of the mixture where all gases contribute to the total signal; between

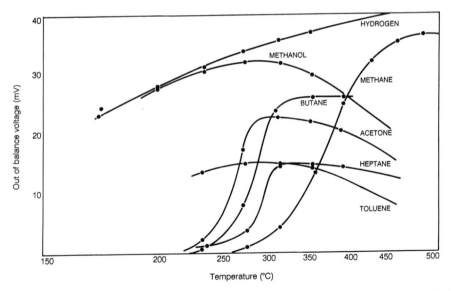

Figure 11-12. Response-temperature profiles of a Pd/ThO_2 pellistor element in various fuels. Fuel concentration, 20% LEL.

300–400 °C the signal is a measure of all gase except methane; whilst at 200 °C, the signal is dominated by hydrogen and toluene. A degree of discrimination is therefore possible using this technique.

Further selectivity may be achieved through the use of different catalysts which exhibit different activities to various fuels. Also, the application of pattern recognition techniques [50], in order to classify the data matrix composed of responses from multiple catalysts and temperatures, may be useful. Caution is advised for certain mixtures since there is evidence to suggest that certain flammable gases may modify the behaviour of others e. g. the detection of CO and H_2 mixtures in air [51].

The temperature interrogation method could also permit early indication of poisoning. Poisoning at lower temperatures usually occurs more rapidly than at elevated temperatures and can be detected before there is observable loss at high temperatures. Thus, by continually comparing the signals at the two temperatures it may be possible to detect the onset of poisoning before the 'normal' high temperature response becomes unreliable.

Improvements in this type of calorimetric sensor are likely to be marginal. However, they will, despite the numerous disadvantages listed here, continue to provide the main instrumental technique for quantitative in-situ measurement of explosive gases at concentrations up to the LEL.

11.3   Pyroelectric Sensors

11.3.1   Range of Use; Detection of Gases and Condensed Phases; Limitations

The sensitivity of the platinum coil thermometer employed in the catalytic sensor described previously is limited by the temperature coefficient of the resistivity of platinum. A more sensitive temperature sensor would, at similar noise levels, allow a lower detection limit to be achieved; alternatively the sensor may be operated at lower temperatures thereby providing substantial savings in power consumption. The pyroelectric sensor developed originally for the detection of infrared radiation [52] is such a temperature sensor; it is a very sensitive heat detector and has found application as a calorimetric sensor for chemical sensing [53].

Ferroelectric materials, i. e. those possessing a permanent electric dipole moment in the absence of an external field can form the basis of pyroelectric sensors. Usually the presence of this charge polarisation cannot be observed because of its neutralisation by stray charges trapped at the surface. However, by applying a large enough electric field to reverse the permanent moment, the surface charge can be detected. In some crystals, termed pyroelectrics, a change in temperature rather than applied field can change the spontaneous moment. This induces a change in the lattice spacings and thus produces a small change in the internal dipole moment. The temperature coefficient of the dipole moment is known as the pyroelectric coefficient (p) [52]. The device only responds to changes in temperature, therefore some form of temperature modulation is necessary in order to obtain a response.

The pyroelectric sensor behaves electrically as a current source and its equivalent circuit with amplifier is shown in Figure 11-13. The voltage across the pyroelectric is given by

$$V = i_p R = p A R \frac{dT}{dt}$$

(11-40)

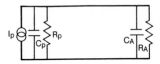

Figure 11-13.
Equivalent circuit for a pyroelectric device. p, pyroelectric, a, preamplifier (after [53]).

where A is the area and R is given by

$$R = \left(\frac{1}{R_p} + \frac{1}{R_A} \right)^{-1} .$$

(11-41)

Typically R_p is in the range $10^{10} - 10^{14}$ Ω and for lithium tantalate, a commonly used pryroelectric material, $p = 6 \times 10^{-9}$ C cm^{-2} K^{-1} [53]. Equation (11-40) holds providing the time constant of the circuit (RC, usually of the order of 10 ms) is much less than the thermal time constant.

The temperature change in the sensor arises from a net flux of heat into the pyroelectric, $\Delta P (t)$, expressed as power per unit area, thus Equation (11-40) may be written:

$$V = \frac{p R A \, \Delta P (t)}{C_v d}$$

(11-42)

where C_v is the heat capacity at constant volume and d is the effective thickness of the pyroelectric. The platinum resistor based catalytic sensor is capable of detecting a change in ΔR of approximately 100 μW; whereas the pyroelectric gas analyser is capable of a detection limit of about 2 μW. This latter figure is based on a theoretical calculation, assuming perfect matching of detector and reference electrodes [53]. By coating the crystals with a suitable material, e. g. adsorbent or catalyst, it is then possible to detect chemical species via their effect on the power drain of the sensor. In determining the power change induced by the interaction of chemical species on the surface, similar considerations to those outlined in Section 11.2.4.1 are required. First, there are heat losses caused by radiative transfer, thermal conduction and convection through the surrounding gas, and conduction through the end wires. Then there are the surface interactions: adsorption, desorption and reaction (e. g. oxidation) which will contribute to the power balance. If the pyroelectric is operated in differential mode where a chemically sensitive material is deposited on one electrode and the other electrode is a non-reacting reference, then the response of the sensor is solely dependent on surface interactions on the chemically sensitive material.

11.3.2 Device Fabrication

A schematic diagram of one type of pyroelectric gas sensor is shown in Figure 11-14, taken from [54]. Other electrode structures have been employed [53]; all however use fabrication techniques similar to those in silicon device lithography. In the device shown in Figure 11-14 two planar electrodes or interdigitated structures are deposited on one side of a 3 mm square, typically 230 μm thick, Czochralski grown, z-cut single crystal lithium tantalate wafer. A suitable adsorbent or catalyst is then deposited on one electrode, usually by thermal evaporation to a thickness of about 200 nm. A common electrode and heater film, capable of heating the sensor to a maximum of around 200 °C, are deposited, on the reverse side of the pyroelectric crystal.

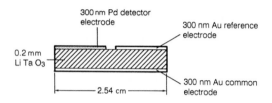

Figure 11-14.
Schematic diagram of a typical pyroelectric gas sensor based on a LiTaO$_3$ wafer (after [54]).

11.3.3 Applications of Pyroelectric Sensors

One of the first uses of a pyroelectric sensor for gas concentration measurement was for determination of exposure to carbon monoxide [55]. Polyvinylidene fluoride or lanthanum modified lead zirconate titanate were cited as suitable pyroelectric materials. The device was operated in differential mode with alternate electrodes coated with platinum black as the oxidation catalyst. The device was heated to 100 °C in order to obtain an adequate rate of reaction and thus effect a measurable change in the power drain. The sensor can be extended to determine other gases which oxidise under these low temperature conditions, for example, hydrogen and certain, reactive hydrocarbons such as methanol. The main applications, to date, for the sensor has been for the detection of hydrogen [54]. Sputtered palladium was employed as catalyst on the sensing electrode and sputtered gold as the inactive reference electrode. Signals arising from the reaction of 100 ppm hydrogen-in-nitrogen with oxygen on the catalyst surface were observed, corresponding to a temperature rise of 20–30 mK [54]. Further data on the H$_2$-O$_2$ interaction on evaporated palladium were obtained on a dual compensating sensor comprising a Au/LiTaO$_2$/Au reference sensor and a Pd/LiTaO$_3$/Au detecting sensor to reduce error signals caused by gas flux changes [56]. The two signals were passed into a differential amplifier whose output was then dependent only on processes occurring at the palladium surface. Exposure to 0.1% H$_2$/N$_2$ at room temperature produces a response due to the heat of adsorption/absorption of hydrogen on palladium. When the sensor is subsequently exposed to oxygen a large response is obtained. This arises from the net heat produced by the catalysed oxidation of adsorbed hydrogen (exothermic interaction) and the desorption of hydrogen as a result of competition with oxygen (endothermic interaction). The above studies demonstrate the sensitivity of the technique which allows measurement of heats of adsorption in addition to the conventional calorimetric determination of concentration through heats of oxidation.

11.3.4   Conclusions

The pyroelectric sensor has only recently been investigated as a calorimetric chemical sensor. It offers the advantage of greater sensitivity (to certain gases) together with the convenience of fabrication techniques used in silicon based devices. The temperature of operation is limited to around 100 °C, therefore it still may not be sensitive enough to detect methane. Its use will thus be restricted to detecting the more easily oxidised gases and vapours. It will also be subject to the usual drawbacks of catalytic devices: poisoning, ambiguity, etc., as described in Section 11.2.5. Thus, because of costs involved in pyroelectric device fabrication and the established position of catalytic sensors based on platinum resistance thermometry for the determination of concentrations around the LEL, it is envisaged that their market penetration would be low. However for certain applications where very low power consumption is imperative and where electrochemical cells may be inappropriate, then the pyroelectric gas sensor may be suitable. Its main application could well be as a research tool for exploring gaseous species and condensed phase interactions on surfaces.

11.4   Seebeck Effect Sensors

11.4.1   Range of Use; Limitations

Another method of temperature measurement, used in conjunction with a catalyst, for the determination of gaseous species by calorimetry is that based on the Seebeck effect. Here the establishment of a temperature differential between two metals or different regions of a semiconductor results in a potential difference, known as the Seebeck effect [57]. The potential is proportional to the temperature difference, thus the electric field E is given by

$$E = \frac{dV}{dz} = \alpha \frac{dT}{dz} \tag{11-43}$$

where V is the voltage, dT/dz the temperature gradient and α the Seebeck coefficient. Some values for α (units: $\mu V K^{-1}$) of some common semiconducting materials are as follows [57]: WO_3, 700; SnO_2, 930; MnO, 1700.

If, in the experimental arrangement, current is not allowed to flow through the semiconductor then the presence of a temperature gradient between the different regions of the semiconductor, caused by catalytic oxidation of a fuel, will result in a potential difference across the device. The device has potential as a very low-power sensor for the detection of certain, readily oxidisable fuels (e. g. H_2, CO) on a catalytically active surface deposited on a suitable metal oxide semiconductor. Data obtained from SnO_2 at room temperature using hydrogen as the test gas [57] show that the sensors developed to date can detect temperature differences of the order of 1 K (equivalent to approximately 100 ppm H_2). This is of the same order as that obtained from the platinum resistance thermometer employed in the catalytic sensor, operated at ca. 500 °C, discussed in Section 11.2.

11.4.2 Device Fabrication and Application

To date, Seebeck effect sensors have been fabricated from pressed pellets of two semiconductors: SnO_2 for hydrogen detection [57] and MoS_2 for H_2S detection [58]. A schematic diagram of the sensor is shown in Figure 11-15. Platinum and palladium catalysts at room temperature, either sputtered onto SnO_2 or by immersion of SnO_2 into an aqueous solution of the metal salt followed by thermal decomposition, were utilised to measure hydrogen concentrations ranging from 0.15–0.9% with a response time of the order of 1 min. For detection of hydrogen sulphide, platinum was employed as the catalyst. The response was reversible at room temperature, however the response was slower than for the hydrogen detection system, being of the order of several minutes.

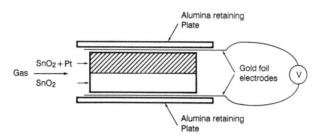

Figure 11-15.
Schematic diagram of a typical Seebeck gas sensor based on Pt/SnO$_2$ (after [58]).

11.4.3 Conclusions

The Seebeck effect chemical sensor is an even more recent development than the pyroelectric sensor and thus is only in the very early stages of development. Currently the sensitivity of the Seebeck effect sensor is comparable to the platinum resistor based calorimetric sensor, although this may be improved by optimisation of the physical structure of both the catalyst and semiconductor. Its attraction lies in the potential for very low power operation, i.e. operation at around room temperature and up to about 100 °C, for the detection of readily oxidisable gases such as hydrogen, and possibly carbon monoxide and the lighter oxygen containing hydrocarbons. In this respect it will also face competition from pyroelectric devices and commercially available sensors based on liquid electrolyte electrochemical cells (see Chapter 3).

11.5 Thermal Conductivity Sensors

11.5.1 Range of Use; Concentration and Types of Gases; Limitations

The three types of calorimetric sensor discussed above rely on the use of a catalyst in order to accelerate the rate of reaction which usually generates heat. They therefore depend upon chemical interaction to bring about a change in power consumption. Sensors based on the thermal conductivity or, more generally, heat transfer phenomena depend solely on a physical effect as a method of transduction. The transfer of heat from a hot body (the detector) to the surrounding gas is influenced by the thermal conductivity of that gas and this can provide a means of detecting and quantitatively measuring its concentration. This has formed the basis of one of the oldest types of gas detector, known by various names, viz. katharometer, hot wire or thermal conductivity detector. Its main application is in gas chromatography [59] as a reliable, non-selective and relatively low sensitivity detector (compared to flame ionisation detectors, for example). It is also used in process control and in field gas detection. This section will be principally concerned with the latter application where the detector is used on a routine basis in a range of different situations. The thermal conductivity detector is employed mainly to monitor high concentrations (1–100%) of flammable gases in air where it may be used in conjunction with the catalytic detector. Typical examples include monitoring natural gas leaks and landfill (waste tip) sites for high concentrations of CH_4. It also provides a safety check on the concentration of flammable gases in an inert environment, e.g. when flushing containers with nitrogen or in diesel engine exhausts. Other, less common, uses of the detector are for monitoring high concentrations of carbon dioxide and inert gases such as N_2, A and He.

11.5.2   Principles of Operation

At pressures above about 10^{-4} bar the thermal conductivity of a gas is independent of pressure and therefore, thermal conductivity devices are predominantly sensitive to composition change. In an ideal, binary gas mixture the net conductance K_m can be calculated from

$$K_m = x_1 K_1 + x_2 K_2 \tag{11-44}$$

where x is the mole fraction of component 1 or 2. Thus the conductivity change from one gas mixture to another (ΔK) is given by

$$\Delta K = K_1 - K_m = x_2 (K_1 - K_2) . \tag{11-45}$$

Very few gas mixtures obey this rule over the whole range of compositions and the thermal conductivity versus concentration plots may not only be curved but exhibit maxima or minima. At low concentrations of contaminant, however, the ΔK relationship is approximately linear.

Thermal conductivity is also dependent on temperature and is usually fitted to an expression similar in form to Equation (11-18). The optimum performance of a detector will be obtained when the two components (contaminant and air) have widely different values of K, thus minimising the temperature dependence of ΔK and improving the linearity of the detector.

The method most widely adopted to measure gas concentration is based on determination of heat loss from an electrically heated filament. It is apparent that many of the equations governing heat loss will be similar to those developed for the catalytic sensor in Section 11.2.4.1. For the special but common case of a metal filament coaxially situated within a cylinder then the power dissipated (P) is given by [59]:

$$P = \frac{2\pi K_m l (T_f - T_e)}{\ln (r_e/r_f)} + P_L \tag{11-46}$$

$$= \lambda K_m + P_L \tag{11-47}$$

where l is the filament length, T_f and r_f are the temperature and radius of the filament, T_e and r_e are the corresponding parameters for the cylindrical enclosure and λ is a constant. The quantity P_L represents the sum of heat losses by conduction through the end wires and electrodes, radiation and convection; these are described by the equations shown in Section 11.2.4.1.

An electrical signal is obtained by means of a Wheatstone bridge circuit. Here the detecting filament samples the atmosphere and the reference filament is housed in a sealed chamber to compensate for changes in ambient temperature. The equations governing the out of balance voltage developed in Section 11.2.4.2 are appropriate and for the constant voltage mode which is usually adopted for portable instruments, the change in bridge potential (ΔV) is

$$\Delta V = m \Delta R \tag{11-48}$$

where m is a constant, for small changes in the filament resistance (ΔR).

The change in the filament resistance is given by

$$\Delta R = \left(\frac{dR}{dT}\right) \Delta T \tag{11-49}$$

and

$$\Delta T = \frac{\Delta P}{\left(\dfrac{dP}{dT}\right)} . \tag{11-50}$$

Where $\dfrac{dP}{dT}$ is the temperature coefficient of the power dissipated by the filament at the temperature of operation. The change in power ΔP is obtained from Equation (11-47) thus:

$$\Delta P = \lambda \Delta K_m . \tag{11-51}$$

Hence, combining Equations (11-48)–(11-51) gives:

$$\Delta V = \lambda' \Delta K_m \tag{11-52}$$

where λ' is a constant.

It can be seen from the above equations that the value of ΔV is magnified for a given change in K_m by increasing the temperature difference between filament and enclosure. Sensitivity is thus proportional to the current flowing through the filament. Eventually, however, baseline stability (noise) at high temperatures limits the sensitivity.

11.5.3   Types of Sensors

As stated previously the commonest type of sensor is based on the metal filament. The metals invariably used are tungsten, platinum or nickel-iron alloy wires, the latter finds use because of its high coefficient of resistivity (approximately double that of platinum), thus improving sensitivity (see Equation (11-49)). Typical wire thicknesses and dimensions are similar to those used in catalytic sensors. The operating temperature of the filament is usually around 250 °C; this is high enough to ensure sufficient sensitivity, yet low enough to maintain baseline stability and to prevent catalytic effects masking thermal conductivity changes. Catalytic compensating elements ('pellistor blanks') composed of deactivated α-alumina on platinum resistance wire have also found use as thermal conductivity detectors. Catalytic effects can however be observed even at low operating temperatures (ca. 200 °C) with readily oxidisable fuels such as hydrogen and certain oxygenated hydrocarbons e. g. methanol, acetone. However one of the most frequent field uses of thermal conductivity detectors is for monitoring high concentrations of methane in air. Methane has a comparatively low rate of oxidation and therefore catalytic effects are not observed under normal operating conditions.

Usually the detecting filament is positioned in a side arm, diffusion chamber (see Figure 11-7a). This is to minimise the effects on the response of flow variations which influence the convective heat losses represented by P_L in Equation (11-40). This enclosure method results in response times of the order of 10 s, compared to around 1 s obtainable if the filament is positioned directly in the flow. In most cases however increased stability is preferable to a fast response time.

11.5.4   Summary

Calorimetric gas sensors based on changes in thermal conductivity are extremely useful devices for the detection of large concentrations of contaminants. They are non-selective and, unlike catalytic calorimeters, do not permit the monitoring of a hazard, i.e. the concentration of a flammable gas mixture, irrespective of the nature of the flammable gas components. However, provided that the application is clearly defined, i.e. it is known which gas is to be monitored and there are no interferents, then the thermal conductivity sensor is a reliable and extremely useful device. This is proven by its widespread use throughout industry. The sensor has reached a stage of technological maturity (the first device for gas detection was invented nearly 75 years ago) and little further development is envisaged.

11.6   References

[1] Zabetakis, M. G., *Bulletin 627,* US Bureau of Mines, Pittsburgh, Pa, 1965.
[2] *British Standard BS 6020:* Parts 1–5, 1981.
[3] Gentry, S. J., Jones, T. A., *Sens. Actuators* **10** (1986) 141–163.
[4] Bond, G. C., *Catalysis by Metals;* London: Academic Press, 1962.
[5] Scott, R. P., Watts, P., *J. Phys. E.* **14** (1981) 1009–1013.

[6] Satterfield, C. N., Sherwood, T. K., *The Role of Diffusion in Catalysis;* Reading, Ma: Addison-Wesley, 1963.
[7] Lugg, G. A., *Anal. Chem.* **40** (1968) 1072-1077.
[8] Baker, A. R., *The Mining Engineer (London)* (1969) 643-653.
[9] Gentry, S. J., Firth, J. G., Jones, A., *J. Chem. Soc. Faraday, Trans. I* **70** (1974) 600-604.
[10] Gentry, S. J., Jones, A., Walsh, P. T., *J. Chem. Soc. Faraday Trans. I* **76** (1980) 2084-2095.
[11] Cameron, P., Scott, R. P., Watts, P., *J. Chem. Soc. Faraday Trans. I* **82** (1986) 1389-1403.
[12] Baker, A. R., *UK Patent 892530,* 1962.
[13] Firth, J. G., *Int. Conf. of Safety in Mines Research, Tokyo, 1969.*
[14] Bird, R. B., Stewart, E. E., Lightfoot, E. N., *Transport Phenomena;* New York: J Wiley, 1960.
[15] Walsh, P. T., Unpublished work.
[16] Firth, J. G., *Trans. Faraday Soc.* **62** (1966) 2566-2576.
[17] Wilson, R., *UK Patent 1451231,* 1976.
[18] Dabill, D. W., Gentry, S. J., Walsh, P. T., *Sens. Actuators* **11** (1987) 135-143.
[19] Gentry, S. J., Unpublished work.
[20] Firth, J. G., Holland, H. B., *Trans. Faraday. Soc.* **65** (1969) 1121-1127.
[21] Firth, J. G., *Trans. Faraday. Soc.* **67** (1971) 212-215.
[22] Cullis, C. F., Nevell, T. G., Trimm, D. L., *J. Phys. E* **6** (1973) 384-388.
[23] Firth, J. G., Gentry, S. J., Jones, A., *J. Catalysis* **34** (1974) 159-161.
[24] Gentry, S. J., Walsh, P. T., *Analytical Proc.* **23** 1986 59-61.
[25] Firth, J. G., Jones, A., Jones, T. A., *Combustion and Flame* **21** (1973) 303-311.
[26] Firth, J. G., Hinsley, R. S., Jones, A., Jones, T. A., *British Patent 1 427515,* 1976.
[27] Gentry, S. J., Howarth, S. R., *Sens. Actuators* **5** (1984) 265-273.
[28] Dabill, D. W., Gentry, S. J., Hurst, N. W., Jones, A., Walsh, P. T., *UK Patent 2083630,* 1982.
[29] Edgington, B. R., Jones, E., *UK Patent 2121180,* 1983.
[30] Hegedus, L. L., McCabe, R. W., *Cat. Rev. Sci. Eng.* **23** (1981) 377-476.
[31] Hegedus, L. L., *Ind. Eng. Chem. Fundam.* **13** (1974) 190-196.
[32] Gentry, S. J., Walsh, P. T., in: *Solid State Gas Sensors,* Moseley, P. T., Tofield, B. C. (eds); Bristol, UK: Adam Hilger, 1987, pp. 32-50.
[33] Frank-Kamenetskii, D. A., *Diffusion and Heat Exchange in Chemical Kinetics* (Transl. by Thon, N); Princeton, N. J.: Princeton Univ. Press. 1955.
[34] Gentry, S. J., Jones, A., *J. Appl. Chem. Biotechnol.* **28** (1978) 727-732.
[35] Cullis, C. F., Willatt, B. A., *J. Catalysis* **86** (1984) 187-200.
[36] Cullis, C. F., Keene, D. E., Trimm, D. L., *J. Catalysis* **19** (1970) 378-385.
[37] Gentry, S. J., Walsh, P. T., *Sens. Actuators* **5** (1984) 239-251.
[38] Becker, E. R., Wei, J., *J. Catalysis* **46** (1977) 372-381.
[39] Firth, J. G., Holland, H. B., *Nature* **212** (1966) 1036-1037.
[40] Firth, J. G., Holland, H. B., *Trans. Faraday. Soc.* **65** (1969) 1891-1896.
[41] Gentry, S. J., Unpublished work.
[42] Firth, J. G., Jones, A., Jones, T. A., *Ann. Occup. Hyg.* **15** (1972) 321-326.
[43] Lamb, T., Scott, R. P., Watts, P., Holland, H. B., Gentry, S. J., Jones, A., *J. Chem. Soc. Chem. Comms.* (1977) 882-883.
[44] Gentry, S. J., Walsh, P. T., *Sens. Actuators* **5** (1984) 229-238.
[45] Baker, A. R., Firth, J. G., *The Mining Engineer (London)* (1969) 237-251.
[46] Firth, J. G., Holland, H. B., *J. Appl. Chem. Biotechnol.* **21** (1971) 139-140.
[47] Cullis, C. F., Keene, D. E., Trimm, D. L., *Trans. Faraday. Soc.* **67** (1971) 864-876.
[48] Gentry, S. J., Walsh, P. T., in: *Preparation of Catalysts III,* Poncelet, G., Grange, P., Jacobs, P. (eds.); Amsterdam: Elsevier, 1983, pp. 203-212.
[49] Bott, B., Jones, T. A., *Sens. Actuators* **5** (1984) 43-53.
[50] Carey, W. P., Beebe, K. R., Sanchez, E., Geladi, P., Kowalski, B. R., *Sens. Actuators* **9** (1986) 223-234.
[51] Dabill, D. W., Gentry, S. J., Holland, H. B., Jones, A., *J. Catalysis* **53** (1978) 164-167.
[52] Putley, E. H., in: *Semiconductors and Semimetals,* Willardson, R. K., Beer, A. C., (eds); New York: Academic Press, 1970, ch. 6.
[53] Zemel, J. N., in: *Solid State Chemical Sensors,* Janata, J., Huber, R. J. (eds.); Orlando, Fa.: Academic Press, 1985, ch. 4.

[54] D'Amico, A., Zemel, J. N., *J. Appl. Phys.* **57** (1985) 2460–2463.

[55] Taylor, A. L., *US Patent 3 861 879,* 1975.

[56] D'Amico, A., Fortunato, G., Ruihua, W., Zemel, J. N., in: *Transducers '85, 3rd Int. Conf. on Solid-State Sensors and Actuators, Philadelphia, June 1985;* IEEE Catalogue No. 85CH2127-9, p. 239–244.

[57] McAleer, J. F., Moseley, P. T., Bourke, P., Norris, J. O. W., Stephan, R., *Sens. Actuators* **8** (1985) 251–257.

[58] McAleer, J. F., Moseley, P. T., Norris, J. O. W., Scott, G. V., Tappin, G., Bourke, P., in: *Proc. 2nd Int. Meeting on Chemical Sensors, Bordeaux, July 1986,* pp. 201–204.

[59] Purnell, H., *Gas Chromatography;* New York: J. Wiley, 1962, pp. 276–294.

12 Optochemical Sensors

Section 1: Otto S. Wolfbeis, Joanneum Research, Graz, Austria
Section 2: Gilbert Boisdé, Commissariat à l'Energie Atomique, Saclay, France
Section 3: Günter Gauglitz, University of Tübingen, FRG

Contents

12.1   Optical and Fiber Optic Sensors

12.1.1   Introduction

The ideal sensor is a device that can be inserted into a sample and will display the result of a chemical analysis within a few seconds with sufficient precision and selectivity. No sampling, dilution (with its inherent drawbacks), or reagent addition is required, and the results may be displayed continuously and in real time. While such kinds of sensors are rare at present, there is growing interest by virtue of their real-time nature. Increasing concern about environmental quality and, in an increasingly cost-conscious world, considerable personnel savings in comparison with manual offline methods further contribute to the desirability of sensors. Hence, tremendous efforts have been devoted to the development of various sensing devices for use in analytical and clinical chemistry.

By definition, a sensor is a device that is able to indicate continuously and reversibly the concentration of an analyte or a physical parameter. Thus, a pH meter and a non-bleeding pH paper strip may be called true sensors, since they act continuously and fully reversibly. However, certain devices that are able to measure the concentration (or activity) of biomolecules have also been called sensors although they do not sense continuously but rather allow only a single determination. Some of these "sensors" have been included in this chapter, but they have usually been referred to as probes rather than as sensors.

Recent advances in optoelectronics and fiber-optic techniques have led to an exciting new technique called (fiber) optic sensing. Depending on the origin of the signal, the devices are classified as absorbance, reflectance, fluorescence, phosphorescence, Raman, light scattering, refractive index, or chemiluminescence sensors. These spectroscopic methods may be performed in both the conventional way and by the evanescent wave technique. Additionally, interferometry has also experienced some success in chemical sensing. Figure 12-1 shows the typical components of a fiber-optic chemical sensing device. It typically consists of a light source (except in chemiluminescence-based probes), a fiber coupler to funnel the light into the

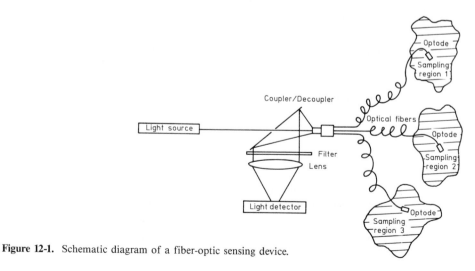

Figure 12-1. Schematic diagram of a fiber-optic sensing device.

fiber, the light guide, a decoupler where the returning light (the analytical information) is separated from the exciting light, and a light detection and amplification system which yields an electrical signal that can be converted, in a final step, into the desired analytical information.

This chapter is intended to give an introduction into this area, with sections covering both the theoretical and practical aspects of optical sensing. Emphasis is preferentially given to practical aspects rather than purely theoretical considerations. The literature surveyed here is considered to be representative up to the end of 1989. Several books [1–6] and reviews [7–16] cover the subject of optical sensing from various points of view. Some are confined to measurements of physical parameters [7], chemical [2, 3, 5, 9, 16], or clinical analytes [2, 3, 6, 10, 12], or remote sensing [13] and process control [4, 13, 14], whereas others discuss mainly optical problems [2, 11] or the large field of biosensors [3, 5, 6, 10, 12]. This chapter is intended to cover the various aspects of chemical sensing including optics, spectroscopy, analytical chemistry and biochemistry.

The dramatic progress that has been achieved in the past 10 years is a result of joint efforts by various specialists in their particular fields: optical spectroscopy, in particular fluorescence spectroscopy, has reached a very high level of sophistication and performance in terms of sensitivity, selectivity, and versatility. The communications industry has provided inexpensive optical fibers which allow the transmission of optical signals over large distances, even in the UV and NIR regions. Powerful lasers have become available as ideal light sources, although the price of lasers other than diode lasers and He–Ne lasers are still a limiting factor for their use in commercial instrumentation. Visible-light-emitting diodes (LEDs) and sufficiently sensitive photodetectors have become available at low prices which allow their use in simple and safe instrumentation. Finally, new methods in chemometrics together with powerful and small-sized calculators and microprocessors which allow data storage and rapid data processing, even in cases of complex signal-to-concentration relationships, have contributed significantly to the state of the art.

Doubtless, the strongest impact on fiber-optic sensing has come from optical fiber technology, an offshoot of the communications industry. The first sensors designed to collect information via fiber optics relied on the fact that alterations to a specific physical property of a medium being sensed would cause a predictable change in the light transmission characteristics of a fiber. Acoustic waves, acceleration, strain, position, and magnetic field are some of the physical properties measured with these initial sensors.

The field of application of fiber sensors in analytical chemistry greatly increased when classical indicator chemistry was coupled with the fiber-optic technique [6, 7]. As a result, sensing was no longer restricted to measuring physical properties that change the transmission of a fiber, but could be extended to numerous organic, inorganic, clinical, and biomedical analytes and parameters.

Various names have been given to optical sensors. In 1975, Lübbers and Opitz coined two words, first "optrode" (from optical electrode) [17], and later "optode" (from the Greek οπτικος οδος, "the optical way") [18] for their fluorescence-based device. Both expressions stress the fact that the signal is optical rather than electrical. Figure 12-2 shows schematic diagrams of the first optodes used for sensing oxygen in a gas flow-through cell. However, no fibers were used at that time.

Remote sensing with fiber optics was also named RFF (remote fiber fluorimetry). RFF emphasizes the fact that sensing can easily be performed over large distances. Other abbreviations that have been used include FOCS (fiber-optic chemical sensor), photode, photo-chemical sensors (a word that should be avoided since optical sensors do not rely on photochemical effects; in fact, photochemistry is a very undesirable phenomenon because it is the reason for the photobleaching of indicators). The term optode will be used throughout this chapter.

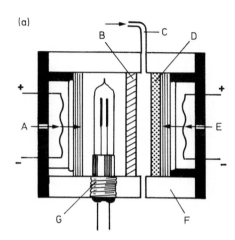

(a)

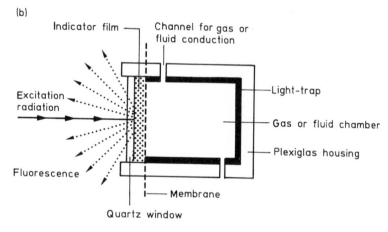

(b)

Figure 12-2. The first optodes. (a) The device of Bergman for continuous oxygen sensing (after Bergman, I., *Nature* **218** (1968) 396). A: reference photocell and filter; B: ultraviolet-transmitting filter; C: inlet for calibration gases and for gases sampled remotely; D: disk of porous vycor glass treated with fluoranthene; E: photocell with ultraviolet-absorbing filter; F: light-tight sintered metal cylinder for diffusion sampling; G: ultraviolet glow lamp. (b) The optode as designed in 1975 by Opitz and Lübbers (after Opitz and Lübbers, *Z. Naturforsch., Teil C* **30** (1975) 532).

12.1.2   Advantages of Optical Sensing

Depending on the field of application, optical fiber sensing can offer some of the following advantages over other sensor types:

(a) Optical sensors in principle do not require a reference element such as is required in all potentiometric methods where the difference between two absolute potentials is measured. In practice, however, all optodes work with an optical reference *signal*. The need for reference electrodes makes potentiometric instrumentation relatively costly. Moreover, the liquid–liquid junction between the two electrolytes is very sensitive to perturbations and can be considered to be the weakest "link" in all potentiometers.

(b) The ease of miniaturization allows the development of very small, light, and flexible fiber sensors. This is of great utility in cases of minute sample volumes and in designing small catheters for invasive sensing in clinical chemistry and medicine. Fluorescence optical sensor heads smaller than any electrochemical sensor (including field effect transistors) can now be manufactured.

(c) Low-loss optical fibers allow transmittance of optical signals over large distances, typically 10–1000 m, and even larger distances seem feasible when use is made of amplifiers currently used in optical telecommunications.

(d) Remote sensing makes it possible to perform analyses in ultra-clean rooms, when samples are hard to reach, dangerous, too hot or too cold, aggressive, located in harsh environments, or radioactive. Fibers appear to be resistant to radiation doses of the order of 10^7 rad or more.

(e) Because the primary signal is optical, it is not subject to electrical interferences by, eg, static electricity of the body or surface potentials of the sensor head. Fibers do not present a spark or fire hazard or risk to patients since there are no electrical connections to the body. Complete electrical immunity is also required when sensors are to be used in strong electric fields.

(f) Analyses can be performed in almost real time since no sampling with its inherent drawbacks is necessary.

(g) Since several fiber sensors placed at different sites can be coupled to one meter via a chopper or spatial multiplexers, the method allows multiple analyses with a single central instrument, thereby reducing the instrument costs. Siting of the spectrometer at a central location remote from the sensor head (which experiences varying experimental conditions) makes routine maintenance checking possible, assures that calibration will be preserved, and consequently renders the instrument more reliable.

(h) Coupling of small sensors for different analytes to produce a sensor bundle of small size allows the simultaneous monitoring of various analytes by hybrid sensors without cross talk of the single strands.

(i) Optodes are made from non-rusting materials and therefore are stable towards strong electrolytes. Since plastics are generally inexpensive, optical sensors are likely to possess great potential for use in disposable sensing.

(j) In many cases the sensor head does not consume the analyte at a measurable rate as, for instance, in the case of polarographic electrodes. This is of particular advantage in cases of extremely small sample volumes. Moreover, fiber-optic sensing is often a non-destructive analytical method.

(k) Optodes have been developed which respond to chemical analytes or physico-chemical parameters for which electrodes are not available.

(l) A fiber optic can transmit much more information than an electrical lead. High information density can be achieved since the optical signals can differ with respect to wavelength, phase, decay profile, polarization or intensity modulation. Thus, a single fiber may guide green and red light in one direction and blue and yellow light in the other. Therefore, a fiber can in principle guide a huge number of signals simultaneously. In practice, this may be exploited to assay several analytes at the same time because different analytes or indicators can respond to different excitation wavelengths. Moreover, the respective emissions are also spectrally different. Time resolution along with spectral selection offers a particularly fascinating new technique in fiber-optic sensing and can make superfluous the need for hybrid sensors.

(m) Since the signal obtained is optical, it is possible to couple the information directly into existing optical communication networks.

(n) Many sensors are simple in design and can easily be replaced by substitute parts, even when manufacturing the sensor head requires relatively complex chemistry. This opens the way for disposable sensing materials.

(o) Many fiber sensors can be employed over a wider temperature range than electrodes and are steam sterilizable.

(p) Optical sensors can offer cost advantages over electrodes, particularly when a single spectrometer is used in combination with several sensors. The raw material for glass fibers is sand, indeed an abundant source.

12.1.3   Disadvantages of Optical Sensing

Notwithstanding the many of advantages over other sensor types, optical sensors exhibit the following disadvantages:

(a) Ambient light interferes. Although this plays no role when the sensor is applied in a dark environment, it can result in a major limitation when sensing intrinsically colored or fluorescent substances such as blood. Therefore, they must either be used in dim or dark surroundings, or the optical signal must be encoded so that it can be resolved from ambient background light. It is advisable that the optical system of the sensor be isolated from the sample by a suitable optical filter.

(b) Optodes with indicator phases (Chapters 17 and 18) are likely to have limited long-term stability because of photobleaching or wash-out. Signal drifts can be compensated for by relating the signals obtained at two different excitation wavelengths or ratioing the signal to the intrinsic Raman scatter, or making use of lifetime measurements. Photobleaching efficiency increases with increasing light intensity. Consequently, a powerful laser should be used only when necessary, for instance, when long optical cables with their considerable attenuation makes lasers indispensable.

(c) Since in indicator phase sensors the analyte and indicator are in different phases, a mass transfer is necessary before a steady-state equilibrium and consequently a constant response are established. This, in turn, limits response times for analytes with small diffusion coefficients. This situation makes it desirable to keep the volume of the analyte-accessible indicator phase much smaller than that of the sample in order not to dilute the sample volume.

(d) Sensors with immobilized indicators have limited dynamic ranges compared with electrodes since the respective association equilibria obey the law of mass action. The corresponding plots of optical signal versus log (analyte concentration) are sigmoidal rather than linear as in the case of, eg, the Nernst relationship. Figures 12-3 and 12-4 show the relationship between analyte concentration and optical signal change in indicator-based sensors for the three most common spectroscopic techniques.

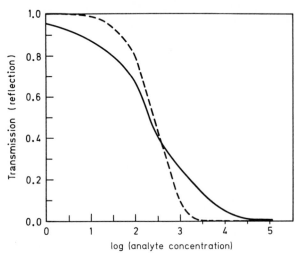

Figure 12-3.   The analytical range of indicator-based optical sensors. Solid line: change in absorption or fluorescence intensity with log (analyte concentration) when the indicator binds the analyte. Dashed line: corresponding plot when the signal is measured by reflectometry. Typical examples where such a signal-to-analyte ratio is obtained include pH sensors based on conventional pH indicators.

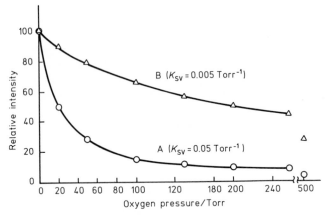

Figure 12-4.   Signal-to-analyte concentration plot of an oxygen sensor based on fluorescence quenching according to the Stern–Volmer equation for two different quenching constants (K_{sv}). Note that the analyte concentration is plotted linearly. Obviously, the sensitivity is highest at low analyte concentration.

(e) The fiber optics used at present have impurities of a spectral nature that can give a fluorescence and Raman background. Inexpensive (plastic) fibers are confined to wavelengths between ca 400 and 800 nm. Long-wavelength UV radiation is efficiently transmitted by glass fibers, but short-wavelength UV radiation is guided by expensive quartz fibers only. The intensity losses in very long fibers are further complicated by spectral attenuation and change in the apparent numerical aperture as a function of fiber length.

(f) Commercial accessories of the optical system have not yet been optimized. Stable and long-lived light sources, better connectors, multiplexers, terminations and optical fibers, and inexpensive blue lasers and LEDs are needed.

(g) More selective indicators have to be found for various important analytes and the immobilization chemistry has to be improved so as to achieve both better reproducibility, selectivity, and sensitivity.

(h) Many indicators suffer a reduction in sensitivity after immobilization or when dissolved in a polymer. In particular, the dynamic quenching efficiency is frequently drastically diminished. Consequently, the respective conventional fluorimetric method will be much more sensitive than the fluorosensor method.

12.1.4   Potential Fields of Application

Fiber-optic sensors can be expected to be of utility in analytical chemistry whenever the sample cannot be brought to the photometer. Hence, they may replace electrodes and even ion-selective electrodes in the remote determination of pH and ionic analytes, and of gases such as oxygen, carbon dioxide, ammonia, and methane. Moreover, optodes are tougher, cheaper, and smaller than the corresponding electrodes. Remote sensing may also be applied to various analytical problems such as in pollution and process control, jet and rocket machines undergoing testing, tracer studies in geology, clinical chemistry, and various biomedical applications. Their ruggedness makes possible their utilization in locations too inaccessible to either the instrument or the analyst.

12.1.4.1   Groundwater Monitoring

In the face of increasing public concern about the quality of drinking water, continuous monitoring of groundwater has become a major aspect of modern analytical chemistry. Rather than digging a well field with numerous boreholes large enough to admit sample collectors which are subsequently brought to a laboratory for analysis, it has been proposed [8] to introduce long-distance communication-grade fibers down to the groundwater level and to monitor pH, chloride, sulfate, uranium, organic pollutants, herbicides, and trace substances using the corresponding optodes. Several fibers may then be coupled to one spectrometer at a central location up to 1 km distant (Figure 12-1). The ability to make up to 50 unattended in situ measurements, using a reasonably priced centralized instrument, has been discussed [19] and should result in acceptable economy.

For nuclear waste repositories, environmental monitoring of nuclear installations, or the study of underground nuclear tests, the hazards associated with the samples can be avoided by leaving them safely underground. Even if the fiber is damaged by radiation it is easier and cheaper to replace than the whole monitor.

The principles of optical groundwater monitoring do, of course, also hold for geological tracer studies using highly fluorescent markers such as Rhodamine 6G, Eu(III), or Tb(III), and for pH gradient studies using pH indicators. They are detectable in boreholes in picomol amounts when lasers are used as excitation light sources.

12.1.4.2   Pollution Monitoring

Airborne laser sensors offer a rapid, real time, and efficient method for the continuous control of sea-water and air pollution. The possibility of remote airborne sensing of environmental parameters has been studied by several groups [19, 20]. Increasing efforts have been directed to the detection in the environment of polycyclic aromatic hydrocarbons (PAHs), which are by-products of many coal-processing factories including coal-liquefaction plants. Similarly, mineral oil spills released by ships can be distinguished from the sea-water background by this technique and can even be classified into sub-groups (see Chapter 18). Many of these methods may be transferred to fiber sensing schemes.

Air pollution sensing for industrial applications and environmental research can conveniently be performed with a network of specific optodes hooked up to a central measurement station by optical fibers. The sensors may be instantaneously concentration sensitive or be designed for cumulative measurements of the total integrated exposure. Working principles have been described [12] for formaldehyde, ammonia, nitrogen oxides, chloroform, hydrogen sulfide, and reactive hydrocarbons. Again, chemical analysis at multiple locations can be made from a central station using fibers that can readily multiplexed to the station.

12.1.4.3   Process Control

Production efficiency and product quality are critically dependent upon process control. More and more, online sensors are replacing classical sampling techniques, and analytical chemistry is emigrating from the laboratory to the factory. In view of the costs resulting from a failing chemical reaction and its environmental consequences, there are extreme reliability requirements on process control instrumentation.

On the other hand, most sensors for continuous process control are faced with a harsh environment. Typically, a reliability of better than 99% is required under extreme and rapidly varying temperatures, high noise and vibration, and substantial chemical exposure, but with a minimum of servicing and maintenance at long intervals. Further, the instrument is expected to remain accurate over a prolonged period, despite the absence of recalibration or even checks.

Clearly, process control is becoming a major field of application for sensing (see Chapter 17). The development of long-range, inexpensive, and high-performance fibers together with the design of optodes for the most important chemical and physical parameters have provided a new dimension of continuous monitoring, since the instrument can remain in the benign environment of the laboratory, where service and calibration are not longer a problem.

12.1.4.4   Remote Spectrometry

The Telephot remote photometer has been used successfully for several years in France for nuclear applications. Other instrumentation for remote spectrometry is commercially available from, eg, Oriel (Stamford, CT, USA), Guided Wave (Sweden), and Photonics Society (France). It is intended for use in radioactive areas, fermentation systems, high-voltage areas, explosive and dangerous areas, biological hazard stations, and in marine, river, and reservoir locations. Remote fluorimetry is not as sensitive as is conventional fluorimetry because of light attenuation by the fiber, and a considerably higher background in the case of single fibers.

Fibers have also been used to measure the intrinsic color or fluorescence of high-performance liquid chromatographic (HPLC) effluents. Remote fiber fluorimetry has been applied in combustion measurements and can provide a simple means for studying chemical processes in flames and combustion gases. Plain fibers have also found application for the measurement of methane levels in explosive environments via the NIR absorption of methane.

12.1.4.5   Biomedical Sensing

Possibly the greatest field of application is sensing clinically and biochemically important analytes such as blood gases, electrolytes, metabolites, enzymes, coenzymes, immunoproteins, and inhibitors. Sensors responding to these parameters are frequently called biosensors. Fluorosensors for blood analytes are being used and will be used in vivo as sensors for the continuous monitoring of the critically ill and as devices for testing blood samples in vitro. Continuous measurements of critical parameters which give warnings of life-threatening trends such as pH, oxygen, carbon dioxide, and blood pressure variations are well established in principle. One may expect from fiber optics to see continued improvements in biocompatibility, signal stability, ease of calibration, and sterilization.

Within the last decade, clinical practitioners have gradually moved from the diagnosis of established disease toward presymptomatic prognosis and preventive measures. Continuous tests for electrolytes, total protein, urea, glucose, creatinine, cholesterol, triglycerides, and other compounds in blood or urine are the subject of intense research and development to produce devices capable of automatically recording results by using computerized data logging and output systems. Existing sensors for oxygen and pH will be able to fulfil some of these needs. The corresponding sensors may be utilized to determine enzymatic activities which are accompanied by a change in pH or oxygen partial pressure.

Optical methods will also be applicable to immunosensors. Thus, the interaction of antibodies with antigens, a process known to be of outstanding selectivity, can be followed by various techniques (see Section 12.3.7 and Chapter 18). Evanescent wave sensors with immobilized antibodies on the waveguide surface, fluorescence polarization studies of labeled binding partners, fiber-optic decay measurements, or combinations thereof, offer numerous possibilities. Bilirubin, steroids, albumin, and enzymes have all been assayed by antibody-binding methods, both on planar waveguides and with fiber optics. Finally, the area of therapeutic drug monitoring is also of rapidly increasing significance, together with early detection of infectious diseases, and of the various forms of cancer.

12.1.4.6   Biotechnology

The latest field of application for optical sensors is in biotechnology, with its strict requirements for sterilization and long-term stability of sensors in bioreactor process control. Again, the excellent performance of optical gas sensors, in particular oxygen, suggests the construction of sensors with a design that specifically meets the requirements of biotechnologists. A major problem here is the coating of the sensor surface by protein films. They compromise response time and long-term stability.

In the following sections and in Chapters 17 and 18, an overview will be given of the present status of light guides as applied to chemical sensing problems, the scope of optical sensing techniques, and how they have been applied to perform chemical analyse in various representative fields.

12.2   Light Guides

A sensor, which is an instrument capable of reversibly measuring a physical or chemical quantity, requires a sensitive element and implies the conversion of the data from the magnitude to be measured to an electrical value that can be handled by the usual electronic techniques. For optical sensors, transduction into electrical signals is carried out by a light data element. Optical fibers are dielectric guides that conduct light over long distances. Fiber-optic sensors are devices with one or more fibers which detect and transport the optically encoded light data, which are representative of one or more parameters to be measured.

The interest in optical fibers for medical instrumentation [21] and in chemistry was demonstrated more than 30 years ago for the purpose of conveying images over short distances [22] (endoscopy), or taking remote chemical measurements by means of hypodermic needles [23]. Their application to "in vivo" oximetry, in which the oxygen saturation of the blood (ratio of oxyhemoglobin to hemoglobin) is determined by two-wavelength spectrophotometry, was investigated in 1964 [24]. The earliest fiber-optic colorimeters and pyrometers were produced shortly thereafter. Until 1973, the medical field was virtually the only one concerned with this technology [25]. With the development of plastic fibers made of poly(methyl-methacrylate) (PMMA) [26], followed by that of the silica fibers for the purpose of optical communications, new fiber-optic chemical sensors (FOCS) were developed.

The optode (see Section 12.1.1) is defined as "a device used at the interface between an optical fiber and a sample solution under study" [27]. This definition justified its extension to physical measurements [28]. However, the term is now specifically employed for chemical sensors, and usually applies to a physico-chemical reaction at the fiber end [29]. The optode can thus be defined as a fiber-optic probe for "observing" a chemical process at its end or (by extension) along its optical path. The optode, which is more characteristic of the type of measurement taken (absorption, luminescence, scattering, refractive index) than of the medium to be analyzed, can be applied to the determination of chemical species in liquids, gases, and solids. The fiber-optical chemical sensor (FOCS) thus requires a sensitive element (the optode), an optical transmitter (the fiber), a photo-receiver to convert the light into electric current, and, except for luminescence produced by the medium, a light source for illumination/or light excitation.

The recent advent of thin technology and of planar waveguides to make optical sensors further broadened the field of chemical application. They are considered as guided-wave chemical sensors (GWCS) rather than as optodes (hence as FOCS). They are also discussed here. Their operation is also based on interactions between a reaction phase and the optical properties of waveguides. The values measured (light intensity, phase shift, etc.) and the methods employed (absorption spectrometry, fluorescence, evanescent waves) are of the same type as those of the FOCS.

Chemists and biochemists not familiar with the concepts of guided optics are obliged to assimilate certain basic concepts if they wish to derive maximum benefit from optical fibers. These concepts are reviewed in this section, while the measurement methods (spectrometry, refractometry, interferometry) and recent significant results are discussed in Section 12.3.

12.2.1   Theory of Waveguides

The optical fiber has a physical structure that essentially tends to inhibit the natural divergence of light, while limiting the weakening of the power transmitted over long distances. In this case, the power received at the fiber end can be nearly invariant with its length. In practice, however, light exhibits a corpuscular aspect as well as a wave aspect.

12.2.1.1   Photons

The corpuscular aspect of light derives from its interaction with matter. The propagation of electromagnetic energy appears in the form of particles, photons, each of which is associated with an elementary energy (E) determined by the radiation frequency (v):

$$E = hv , \tag{12-1}$$

where h is Planck's constant ($6.6256 \cdot 10^{-34}$ J s).

Quantum physics assigns it a zero characteristic mass and a relativistic mass $m = E/c^2$. The photon travels in the vacuum at the speed $c = 299792.458$ km s^{-1} and in matter at a reduced speed $v = c/n$, where n is the refractive index of the medium. The wavelength is related to the frequency by $\lambda = v/v$, so that, in vacuum, $\lambda = c/v$ ($n = 1$). Two photons may be in the same state, determined by space and spin coordinates. Photons are hence superimposable. Thus, unlike electrons, two light beams can intersect without any problem.

A photon can propagate in a physical medium by exchanging energy with the medium. Absorption applies to processes of photon losses by photon–matter interaction. The absorption coefficient (a) (in cm^{-1}) obeys the empirical law of Urbach according to the equation

$$a = a_0 \exp (E/E_0) , \tag{12-2}$$

where E is the energy level of the photon at wavelength λ before (E_0) and after (E) absorption.

In simple linear scattering, the photon of frequency v deviates and leaves the light beam on interaction of the particles with the medium. The number of photons diffused per unit time is proportional to the average number of incident photons. If the wavelength of the light is much greater than the mean irregularity dimension of the material, scattering is of the Rayleigh type and proportional to K/λ^4. This factor, which is important at short wavelengths, limits the use of fibers in the UV range. Scattering with energy exchange designates processes of losses by interactions involving three particles. This factor is more pronounced at high light intensities. It produces stimulated Raman scattering and Brillouin backscattering. Raman scattering is observed when the wavelength of the diffused light is different from the exciting light by a change in polarization and vibrational energy of the molecules. Its intensity is directly proportional to the concentration of the species, and is characteristic of chemical bonds. Rayleigh scattering is of the elastic type and Raman scattering of the inelastic type.

12.2.1.2   Light Wave

Light is an electromagnetic wave with a transverse form. The wave can be represented by a surface that is deformed as a function of time, the propagation direction being normal to this surface. The concept of a plane wave serves to define a wide range of physical values:

- wavelength (λ) and wave number $v/v = 1/\lambda$ (the inverse of the wavelength);
- angular frequency $\omega = 2\pi v$;
- wave vector K of modulus $k = \omega/v = 2\pi/\lambda$;
- phase velocity $v = \omega/k$;
- refractive index $n = c/v$.

A wave group stretched slightly about its frequency v is manifested by a modulation that prpagates at a velocity u (group velocity) on the carrier of frequency v. According to De Broglie's equation, this velocity is related to the phase velocity by the equation

$$1/u = (1/v)\,(1 - (v/v)^2\,dv/dv). \tag{12-3}$$

A "non-dispersive" medium, in which the velocity is independent of the frequency, has no specific group velocity ($u = v$).

The resolution of the Maxwell equation relative to the magnetic field H (with a vector perpendicular to v) and to the electrical field (vector perpendicular to H) shows that the wave function ψ has a solution of the sinusoidal type:

$$\psi = A\,\sin(\omega t - kz) + B\,\cos(\omega t - kz). \tag{12-4}$$

The electromagnetic wave serves as a photon guide. The intensity of the light is proportional to the number of photons and the square of the wave amplitude.

12.2.1.3   Elements of Geometric Optics

Whereas light propagates in air in a straight line only, this does not apply to waveguides. In a waveguide, it can follow a curved path determined by the geometry and the various physical obstacles in the waveguide. However, if the medium is homogeneous, the light coming from a point source diverges in rectilinear rays. A number of optical concepts are hence conveniently used:

- The light flux passing per unit area (illumination) is inversely proportional to the square of the distance from the source.
- The Descartes–Snell laws on reflection and refraction indicate the direction of the light rays crossing two homogeneous and isotropic media. This direction given by the sine of the angles of incidence (θ_1) and of refraction (θ_2) depends on the refractive indices (n) of the two media:

$$n_1 \sin \theta_1 = n_2 \sin \theta_2 . \tag{12-5}$$

- Total reflection, which serves to propagate the light in a core with a refractive n_1 inserted in a cladding of refractive index n_2 is achieved for angles greater than the critical angle (θ_c) with

$$\theta_c = \sin^{-1} (n_2/n_1) . \tag{12-6}$$

- Fresnel reflections in the transfer of energy from one plane to another induce an optical power reflection coefficient in normal incidence on the plane diopter:

$$R = [(n_1 - n_2)/(n_1 + n_2)]^2 . \tag{12-7}$$

If reflection is not total, the electric field of the wave may be parallel $(R_\parallel)$ or perpendicular $(R_\perp)$ to the plane of incidence. The cancellation angle of $(R_\parallel)$ is the Brewster's angle of incidence.
- The conservation of the geometric extent indicates that the product of the emission surface area (S) and the solid angle (Ω) cannot be increased by optical means between the object and its image:

$$S_1 \Omega_1 = S_2 \Omega_2 . \tag{12-8}$$

A knowledge of the quantity $n^2 \, dS \, d\omega = \text{const.}$ is important in constructing optical links and optodes to achieve maximum efficiency in the light source–optical fiber coupling.

12.2.1.4 Light Across the Optical Fiber

Most recent work on fiber optics has been concerned with optical communications applications [30–32], including components and the associated electronics [33]. A review of the most significant patents concerning their fabrication, materials, components, and associated systems was also published recently [34]. Specific studies on fiber-optic sensors are beginning to appear at conferences and courses. The importance of chemical sensors has grown steadily since the first articles devoted to this subject [28, 29, 35] (see also Section 12.1.1). Recent synthesis are reference papers [3, 5, 9].

Figure 12-5 shows the simplest light guide (step index of the multimode type). The refractive index of the core (n_1) is higher than that of the cladding (n_2) to guarantee total reflection at the core/cladding interface. This property, applied by Van Heel as early as 1954, serves to isolate the fiber optically from the outer medium.

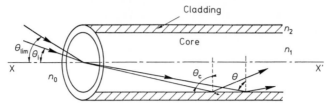

Figure 12-5. Light passage through step-index optical fiber. See text for details.

The guidance conditions are obtained when the angle of incidence (θ) of the light beam at the core/cladding interface is greater than the critical angle (θ_c). The limit angle (θ_{lim}) of acceptance of the beam in the waveguide defines the numerical aperture (NA):

$$NA = n_0 \sin \theta_{lim} = (n_1^2 - n_2^2)^{1/2} = n_1 \sqrt{2\Delta} \tag{12-9}$$

with

$$2\Delta = (n_1^2 - n_2^2)/n_1^2 . \tag{12-10}$$

Thus at an angle θ, smaller than θ_{lim} any ray is guided in the fiber. For a helical ray (which does not intersect the XX' axis), the guidance condition in total reflection ($\theta \geqslant \theta_c$) can also be applied. The solid angle formed by the acceptance cone of the fiber is

$$\Omega_{lim} = 2\pi (1 - \cos \theta_{lim}) . \tag{12-11}$$

The propagation of the light in the fiber is expressed by a dispersion equation with solutions each of which is characterized by two whole numbers, which define the structure of the wave function. These functions are propagation modes. Their number (whole, finite) is approximately $N = 1/2 \, V^2$, where V is the normalized frequency:

$$V^2 = a^2 (k_1^2 - k_2^2) , \tag{12-12}$$

where $2a$ is the waveguide diameter, k_1 is the wave number of the core, and k_2 is the wave number of the cladding. The propagation modes are designated by two letters (components of the field) and two whole numbers. The first is the azimuthal parameter and the second the radial parameter. The fundamental mode is denoted by HE_{11}.

The different modes propagated in the fiber are not attenuated in the same proportions. Above a certain distance (mode mixing), however, this attenuation per unit distance becomes constant. In terms of optical power, it is characterized by the attenuation A (in dB):

$$A = 10 \log (P_0/P) , \tag{12-13}$$

where P_0 and P represent the incident light and the transmitted light, respectively. This equation can be compared with the Beer–Lambert law (Equation 12-24):

$$\text{absorbance} = A/10 = \varepsilon \, cd \, . \tag{12-14}$$

The data rate of a fiber is limited by dispersion. Dissipation and intermodel coupling modify the form of the pulse received. The spread of this pulse (τ) in a fiber of length L is the modal dispersion:

$$\tau = t/L = (1/c)n_1 \sqrt{2\Delta} \tag{12-15}$$

a short distances, and

$$t = \tau \sqrt{L_c L} \tag{12-16}$$

at long distances for a coupling length L_c. The band width of an optical-fiber transmission link is thus affected during the propagation of a signal by a time dispersion (in pulse width) and spectral dispersion (in propagation time of the different modes). This band width is therefore a characteristic parameter of the fiber.

12.2.1.5 Modulations

Various possibilities of light modulation are available for making fiber-optic chemical sensors (FOCS):

- Amplitude modulation (or light intensity modulation), which is indicative of the variation in the number of the photons measured. It is evaluated in optical power and often given in relative intensity (fluorescence, Raman) or in attenuation units (absorbance or its inverse, transmittance). This modulation is applied to most optodes.
- Phase modulation, which is equivalent to a measurement of the time-of-flight of the photons. Using interferometric devices, the elongation of the optical path of a fiber with a single propagation mode (single-mode fiber) is determined. A relative effect (interference fringes) is determined between a reference and a sensing branch of coherent monochromatic light. It is also possible to observe the relative elongation of paths of several propagation modes of a multimode fiber illuminated by a laser (speckle pattern).
- Wavelength modulation (λ), which corresponds to the spectral variation in position (energy variation) or width (coherence length). Combined with intensity modulation, it provides the basis of spectrometry.
- Polarization modulation, characteristic of changes in birefringence, and hence of the variation in ellipticity (photon spin). It is often related to an amplitude modulation (polarimetry).

A fifth type, time modulation, is generally acknowledged, which is indicative of the variation in a signal with time (chronometry). In most fiber-optic sensors, it is used for signal processing (frequencies from a few Hz to a few MHz). It is essential in optical time domain reflectometry (OTDR), polarization OTDR, and frequency OTDR techniques.

12.2.2   Types of Fibers

Optical fibers can be examined according to various criteria: type of optical transmission, geometric configuration, core and cladding material, and hence spectral characteristics. Undesirable effects are also observed, with important repercussions in spectrometric techniques.

12.2.2.1   *Optical Transmission*

In the multimode step-index fiber, the refractive index (n_1) of the fiber core is uniform. The variation in refractive index from the core to the cladding (n_2) with ($n_2 < n_1$) is abrupt. An outer layer (with an intermediate refractive index or close to n_1) is often added for all-silica fibers (AS). Additional mechanical protection (jacket) is required. In plastic-clad silica fibers (PCS), the core may be wide (up to 1.5 mm) and the fiber has a large numerical aperture ($NA \approx 0.4$). Plastic and fluoride glass fibers may have even larger NA (0.6). "Telecommunications" fibers (50/125 or 100/140) of silica raise more difficult coupling problems in instrumentation than larger diameter AS fibers (200/280 or 400/440). Multimode fibers are commonly used for FOCS.

The multimode gradient-index fiber was designed to improve the band-width characteristics without reducing the numerical aperture (0.2 to 0.3). The optimum index profile of the core approaches a parabola. This type of fiber is used in time-of-flight spectrometry [36], in which the fiber is the dispersive element. In fact, the variation in photon velocity is exploited to characterize the spectral dispersion. This dispersion, relative to the refractive index (n) of the material, represents a relationship between the transit time (τ) of a monochromatic light pulse and the wavelength (λ) for a fiber of length L:

$$d\tau/d\lambda = (-\lambda L/c)\, d^2 n/d\lambda^2 . \tag{12-17}$$

The Selfoc fiber is a gradient-index fiber whose length is proportional to the wave phase. It usually serves as a lens to obtain couplings with high energy efficiency.

The single-mode fiber with a small numerical aperture (ca 0.1) has a very small diameter core (typically 2–5 μm), so that the light is guided nearly linearly. The thickness of the optical cladding must be at least ten times the core diameter. This fiber requires pointed sources (lasers, diode lasers) and raises difficult problems of connection. This type of fiber is used in interferometric techniques.

Planar waveguides are discussed in Section 12.2.3.

12.2.2.2   *Geometric Configuration*

Most of the configurations that have emerged in the area of fiber-optic sensors [37], such as multicore fibers and fibers with maintained circular or linear polarization, now have no chemical applications other than in polarimetry and interferometry.

Liquid fibers were recently used to improve transmission in the UV region or to enhance the sensitivity of the species to be measured. These are hollow silica fibers filled with a liquid of a suitable refractive index (such as CS_2 with a refractive index of 1.63 to measure I_2) [38].

Step-index fibers in incoherent bundles are easily divided to make bifurcated or multi-channel fibers. A specific ring-shaped fiber structure (light receiver) and a central fiber (emitter)
were recently recommended for Raman spectrometry [39].

12.2.2.3   Materials

The range of fibers usable for instrumentation has broadened considerably in the past
decade. The spectral domains concerned depend on the type of material involved. Figure 12-6
shows the relative spectral attenuations of the different fibers marketed today [14].

The most commonly used plastic fibers are made of PMMA. In irradiating environments,
polystyrene fibers may display better behavior. The minimum attenuations of these plastic
fibers are still high (>20 dB km^{-1}). Their main drawbacks are their sensitivity to elevated
temperature ($>85\,°C$) and to corrosive chemical environments, and their narrow spectral
zones ($0.5-0.8\,\mu m$). However, they are inexpensive, large-diameter fibers (up to 2 mm), making
them easy to cut and couple. The thinner cladding, which is easily removed, makes them attractive for surface chemical reactions. Their advantages for in vivo measurements were exploited in the first pH optodes with monofibers [40]. The addition of deuterium (which is
costly) helps to eliminate the CH absorption bands and extends the spectrum towards the
near-infrared region.

Silica fibers are the unanimous choice today for medium- and long-distance measurements,
and are the only single-mode products. The spectral limitations in the infrared are due to
SiO_2 vibration bands (GeO_2 bands for doped fibers) and, in the ultraviolet, to impurities and

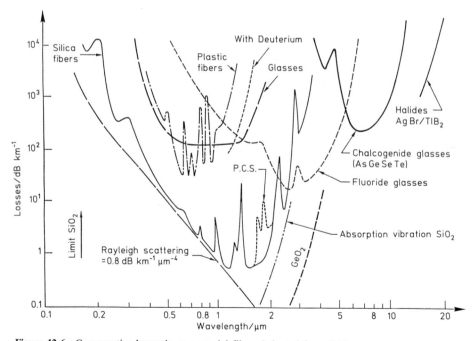

Figure 12-6. Comparative losses in commercial fibers (adapted from [14]).

to Rayleigh scattering. Figure 12-7 shows the present limits in this area [41]. The advantages offered by these fibers include their excellent transparency over a wide spectral range for long-distance measurements (typically several kilometers). These multimode and monomode fibers are benefiting from advances in optical communications (connectors, couplers, multiplexers). The easy removal of the PCS fiber cladding is an advantage for grafting chemical reactants to the silica or for direct measurement by surface reaction (refractive index, reactant cladding). In some industrial systems, they can be used at high temperatures (350 °C for AS polyimide-jacket fibers, and above 600 °C for thermo-coaxial jackets). Their very good resistance to ionizing radiation up to integrated doses of 10^{10} rad [42] makes them nearly indispensable in chemical nuclear environments [43]. However, although their purity and an OH doping are favorable in the ultraviolet region and under radiation, certain spectral properties (fiber drawing-peak, variation in OH vibration bands, formation of colored centers) restrict their use in absorption and fluorescence below 0.35 μm.

Fluoride glass fibers [44, 45], which are available in a complete range, are ideal for spectrometric measurements between 1.5 and 4.5 μm (5 μm for some compositions). Their absorption minimum lies at about 2.5–3.2 μm. Their value for long-distance optical communications (theoretical minimum attenuation about 10^{-3} dB km^{-1} at 3.2 μm and low dispersion) is one reason for the rapid growth of their performance (1000 dB km^{-1} in 1980, a few dB km^{-1} in

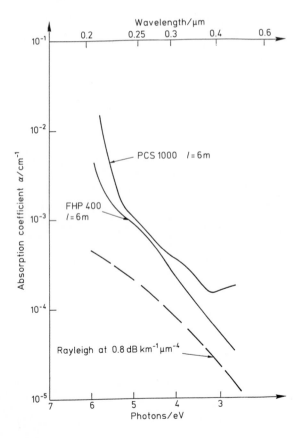

Figure 12-7.
Spectral evaluation of fibers in UV determination using quartz-silica fiber (France) (plastic clad silica PCS, 1000 μm, 300 ppm OH content) and fiber-high OH silica core–polyamide buffer type (FHP 400/440) from Polymicro Technologies (1200 ppm OH content) with a Hewlett-Packard 8450 spectrophotometer after elimination of coupler losses.

1988) [44]. They display good behavior in nuclear environments (low induced attenuation) and are easy to regenerate [46]. Their uses in IR and Fourier transform IR spectrometry are steadily growing (see Chapter 17).

Chalcogenide fibers form the basis of at least two elements of the family Ge, P, As, S, Se, Te [47]. With halogenides (CsBr, AgBr, AgCl, BrI, $ZnCl_2$), they are taking over in the infrared region (5–20 μm). Their attenuations and costs are still high. They are certainly useful for gas measurements (eg, CO_2 at 10.5 μm), which are now currently taken with poor sensitivity at wavelengths of second-, third- or fourth-order harmonics.

12.2.2.4   Undesirable Effects

Another limitation to the use of fibers stems from the undesirable optical effects they cause, particularly in fluorescence laser spectrometry and Raman spectrometry. The fluorescence spectrum of the fiber [48], owing to the impurities present, such as traces of molecular oxygen [49], usually requires an arrangement of the optodes with at least two fibers, one for light excitation and the other for reception.

Another undesirable effect occurs with optical claddings. The existence of evanescent waves at the core/cladding interface can induce absorption peaks or fluorescent effects due to the cladding. Thus, PCS fibers reveal the characteristic absorptions of the silicone cladding between 1.6 and 2 μm. Similar effects are observed with the increased attenuation induced under radiation in AS fibers due to the fluorine doping of the cladding [43].

12.2.3   Evanescent Waves

Light can remain in a waveguide (fiber) if the angle of incidence (θ) of the light ray reaching the waveguide/optical cladding interface (or liquid to be measured) is greater than the critical angle (θ_c). At the interface, however, light penetrates partially into the optical cladding, which may be the site of chemical reaction [50]. This penetration to a distance z outside the guide, related to a fraction of the wavelength (λ), depends on the bonding conditions at the interface [51]. The evanescent wave interactions at interfaces are the basis of internal reflection spectroscopy [52]. The electric field E of the evanescent wave along the propagation perpendicular direction z from the interface is given by

$$E_{(z)} = E_0 \exp\left(-z/d_p\right), \tag{12-18}$$

where d_p is the distance at which E_0, the electric field at the interface, decreases by a factor of 1/e of its value, and d_p is given by

$$d_p = \lambda/(2\pi n_1 (\sin^2\theta - \sin^2\theta_c)^{1/2}) \tag{12-19}$$

for wavelength λ, angle of incidence of ray θ and critical angle θ_c (Figure 12-8). The amplitude of the evanescent wave field can be analyzed by absorption and fluorescence methods. For a ray with angles relative to the fiber longitudinal direction of θ_y and to the transverse direction of θ_z [53]

$$d_p = \lambda/(2\pi n_1 (\sin^2 \theta_c - \sin^2 \theta_y \sin^2 \theta_z)^{1/2}) \,. \tag{12-20}$$

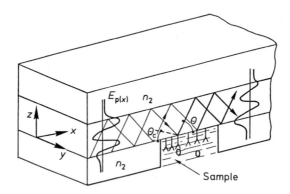

Figure 12-8.
Schematic configuration of a symmetric planar optical waveguide. See text for details.

The application of evanescence spectrometry to the production of chemical sensors has been widely investigated in the area of immunology (fluoroimmunoassays) [50, 51] and, in general, in biochemical fiber-optic sensors [54, 55]. This technique is used with optical fibers by coupling light from one fiber to another [56] or with planar waveguides. Figure 12-8 also shows the variation of an evanescent field of a planar waveguide by an antigen–antibody reaction. The "total internal reflection fluorescence" (TIRF) technique can be used at a fixed angle to monitor the introduction of biological molecules, or at a variable angle to determine simultaneously the concentration and thickness of the fluorescent layer adsorbed. The sensitivity obtained in TIRF is as high as 10^{-9} mol/L. In fact, if the excitation evanescent wave has the same electric field intensity as that of the induced fluorescence, the enhancement effect observed can be exploited in evanescence spectroscopy. The angles of incidence are close to the critical angle in order to maximize sensitivity.

The planar waveguide, which can be used with very fine polymer guide films, offers the advantage of increasing the reactive surface area, enabling mode and polarization selection, and allowing a layout adapted to the slit of the spectrometers [51]. The planar waveguide is a sort of slab with refractive index n_1 in which the light is guided in a sandwich between plates of refractive index n_2 ($n_2 < n_1$) (Figure 12-8). The guided modes are propagated in the x direction in accordance with a wave number and a structure of a standing wave that "fits" in the z direction [57]. The evanescent field produced at the surface of the waveguide is often analyzed in terms of leaky modes.

The intensity of the evanescent wave can be increased by the collective oscillation of a free electron plasma at the interface of a dielectric/metal layer. The photon momentum along the surface matches that of the electrons in the metal [58]. This effect, called the "surface plasmon resonance" or "surface plasmon polarization", occurs when the light is totally reflected in the dielectric, as when a thin layer of silver (thickness 50 nm) is coated on a flat glass or on a prism.

The optimum coupling angle is a few degrees wider than the critical angle. The resulting effect is a broad variation in the reflection coefficient at this "resonance angle". The modification of the properties of the sample, through the shift in this angle, serves to measure the sample–high-sensitivity waveguide interactions. The angular resolution of this type of measurement is better than $5 \cdot 10^{-2}$ degrees.

Planar multilayer structures are ideal for evanescence spectroscopy. Figure 12-9 shows two recently described examples [59]. In the first (surface reaction sensor), the variation in "quenching" is observed by transfer of the luminescence energy emitted by reactant groups incorporated in a layer of the guide structure. The waveguide is a Si_3N_4 layer deposited by CVD on a SiO_2 substrate. Dye molecules (Rhodamine) incorporated in polyurethane serve as donors, and an organic layer (thickness 10 nm) of chromoionophores (bromocresol purple pH indicator) serves as the acceptor.

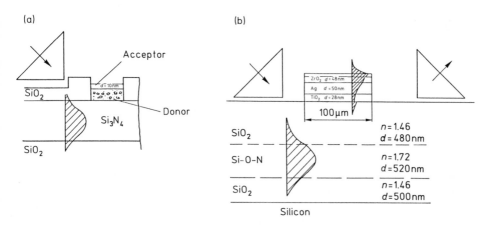

Figure 12-9. Examples of multilayer structures (adapted from [59]). (a) Total internal reflection fluorescence with energy transfer and quenching; (b) surface plasmon waveguide sensors. *d*: thickness of layer, *n*: refractive index.

In the second case, the structure is of the "surface plasmon waveguide sensor" (SPWS) type. It is composed of three layers, two of silica (SiO_2) and a silicon–oxygen–nitride core, modified over a distance of 100 µm by a multilayer structure, in which silver is the active surface material placed between a layer of ZrO_2 (excitation interface with Ag) and TiO_2 (to prevent Ag–SiO_2 coupling). This structure is coated with a thin organic layer capable of absorbing the chemical species to be measured.

12.2.4   Practical Aspects

12.2.4.1   *Optodes*

Figure 12-10 is a schematic representation of optodes. A light source (S) illuminates a fiber (or a bundle of fibers) towards one or more detectors (D) by means of one or more light-receiving fibers. The effect measured is generally observed at the end of the fiber(s) according to

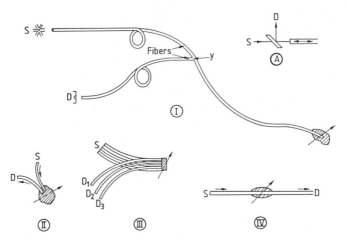

Figure 12-10. Schematic representation of optodes. A light source (S) illuminates a fiber towards one or more detectors (D). The effect is observed at the end of the fiber (monofiber I, multifibers II or incoherent bundles III) or the interface core/cladding (IV) (adapted from [14]).

diagrams I, II or III. Also, this may be caused by a variation in the properties of the optical cladding (variation in refractive index) (IV) or of the evanescent wave (by bringing two fiber cores close together). If the effect is analyzed in feedback on a single fiber, it is necessary to use an optical coupler in a Y shape (I) or with a beam splitter (A).

Three types of optodes can be distinguished [14]:

— "Passive" (extrinsic) optodes for which the optical fibers are merely light guides. The optical value measured results from a natural property of the medium (absorption, luminescence). These optodes, which are easier to build, have mainly been used for analyzing gases, monitoring, process control and Raman spectrometry. New types of fiber [39] and various arrangements have been proposed for long-distance spectrofluorimetry [28, 60] with intrinsic reference [61] and/or for on-line absorption measurements with high-sensitivity cells [62, 63].
— "Semi-active" or "extrinsic active" optodes, for which the physical value measured results from a chemical reaction located between the surrounding medium and a reagent placed at the end of the fiber, whose own optical properties nevertheless remain unchanged. This type of optode accounts for most of the products today. It is illustrated by Peterson's pH optode [40], various reflectance probes, and arrangements with a chemical jacket or sensitive film placed at the fiber end. Recently, some typical configurations have been analyzed [9].
— "Active" (intrinsic) optodes for which the surrounding medium modifies the natural optical properties of the fiber. Chemical reactions may alter the properties of the waveguide, either at the surface by a variation in the properties of the optical cladding (Figure 12-10, IV) or in the fiber core (scintillating fibers, liquid core fibers). By extension, optodes with surfaces specially treated to be reactive are also of the active type.

Many systems of "flow-through cells" (flow sensing) allow the detection of chemical species in a flowing reaction medium. These cells may be passive [60, 62, 63], semi-active [64] or active [65], especially if the monomode fibers are used in interferometry [66]. Further, the use of fibers in bundles split into two (bifurcated) or more channels makes multifiber optodes suitable for diversified arrangements. An interesting arrangement is obtained with central-fiber optodes (excitation) and peripheral fibers in a ring (detection) arranged linearly in front of the inlet slit of a spectrometer.

12.2.4.2 Sources and Detectors

Recent reviews have appeared concerning sources and detectors [67, 68]. Apart from conventional sources (xenon, deuterium, tungsten, halogen), the three major alternatives today in sources for FOCS are light-emitting diodes (LED), laser diodes and high-intensity diodes (spectrofluorimetry, Raman spectrometry). LEDs are applicable down to 450 nm only and laser diodes down to 630 nm only (except when frequency-doubled). Photographic flash bulbs offer the advantage of being cheap, intense and rich in UV radiation for special applications.

Detection by photomultipliers is gradually being replaced with the use of photodiodes (PIN, avalanche), which are inexpensive, and photodiode arrays (possibly intensified) for real-time spectrometry. Developments in semiconductors for optical communications [33] are essential for obtaining new components, directly coupled to optical fibers. Arrangements with inexpensive modulable LEDs, usable in a wide range of the visible spectrum, are promising for the detection and identification of many species [69], by using the resources of chemometric analysis [13].

12.2.4.3 Couplers, Multiplexers, and Networks

While the adaptation of fibers to conventional spectrometers [70] has become routine, it still requires a thorough knowledge of the dynamic range remaining in amplitude and wavelength [41], of the link budget (connectors, etc.), and excellent control of the source–fiber coupling under special conditions [71]. Couplers for fibers other than telecommunications products still need to be developed. Optical multiplexing techniques are frequency division multiplexing (FDM), polarization multiplexing (PM), time division multiplexing (TDM), space division multiplexing (SDM), wavelength division multiplexing (WDM), and their combinations. Recently, a WDM coupling with a photodiode array has been tested in fluorescence [72]. Space division multiplexing, shown in Figure 12-11, is essential for multipoint measurements in distributed sensors, especially if the measurement technique is expensive. The guided wave assembly (Figure 12-11) is applicable in spectrophotometry for some ten points. However, technical progress is still needed to make multiplexing operational, especially in spectrofluorimetry and Raman spectrometry, in order to offer the advantages of an optode structure in networks [28].

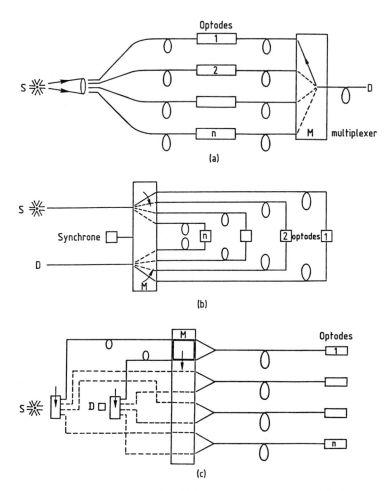

Figure 12-11. Fiber-optic spatial multiplexing with single source and single detector. (a) Multifiber from source; (b) monofiber from source and to detector; (c) monofiber optode.

12.3   Optical Sensing Principles

12.3.1   Principles of Spectroscopic Methods

12.3.1.1   Dispersion and Elastic Scattering

All spectroscopic detection methods depend on the interaction of electromagnetic radiation and matter [73]. Optical spectroscopy is a fast, relatively convenient, and non-destructive method for examining gases, liquids and solid samples, the last two even in thin films [74, 75].

In the process of optical sensing, especially electromagnetic radiation [76] in the following spectral ranges is used for quantitative analysis:

- ultraviolet (UV):   200   −380 nm, 50000−26000 cm^{-1}, 6.2−3.3   eV;
- visible (Vis):   380   −780 nm, 26000−13000 cm^{-1}, 3.3−1.6   eV;
- near-infrared (NIR): 780 nm−   3 μm, 13000− 3300 cm^{-1}, 1.6−0.4   eV;
- infrared (IR):   3   − 50 μm,   3300−   200 cm^{-1}, 0.4−0.025   eV.

During the interaction of the electromagnetic radiation with matter the frequency v (s^{-1}, Hz, hertz) of the interfering radiation does not change, but the wavelength λ (nm or μm) and the velocity v (m s^{-1}) of the propagating radiation are influenced according to the theory of refraction [77, 78]. Radiation can be considered either as a motion of particles (photons) or as a package of waves [77]. Both concepts (Newton, Huygens) are combined by the duality principle of de Broglie. They are necessary to explain different interactions of the radiation with matter: dispersion, absorption, diffraction, refraction, interference, and reflection [75, 79].

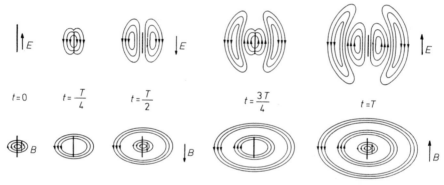

Figure 12-12. Electromagnetic radiation (at a far distance from the transmitter) is in phase for the electric *E* and magnetic *B* field vectors. The direction of propagation *x* is perpendicular to the two vectors. The wave is characterized by its wavelength λ, its frequency v and the velocity of propagation v.

The wave character of the radiation is shown in Figure 12-12. On the left an emitting dipole is symbolized by a metal bar which acts as a transmitter [77]. Moving electrons in the bar cause a change in charge distribution. This system represents a Hertz dipole emitting an alternating field perpendicular to the dipole's axis, containing an electric *E* and a magnetic *B* field vector. Both vectors are in-phase (have maxima and minima at the same time) according to electrodynamic theory [78] if the radiation is far from its source. They stand perpendicular to each other and to the direction of propagation of the wave. The change in the amplitudes of the vectors induces a change in the charge distribution of the molecules in matter as soon as the electromagnetic wave interacts with a sample [77, 80]. By this interaction the molecule is polarized [81]. This alternating field can induce in liquids

- either a rotation of the total molecule (orientation polarization), which takes place in the radiowave range;

– or a relative change in the charge distribution in the molecules, given by the relative positions of the nuclei and/or of the electron density (displacement polarization).Thereby either the nuclei are forced to vibrate relatively to each other (atomic polarization), and/or the electron density distribution varies relative to the position of the nuclei (electronic polarization).

The frequency of the radiation determines which type of polarization takes place [77, 82]. The refractive index is dependent on the frequency in those frequency regions where polarizability varies. This phenomenon is called *dispersion* (dn/dv or $dn/d\lambda$ [78, 82]). The amount of polarization is proportional to the dielectric constant ε [73, 81]. According to the electromagnetic theory of radiation (Maxwell), this dielectric constant is equal to the square of the refractive index at low frequencies (as long as the magnetic permeability of the medium approximates to 1). The refractive index does not vary in this frequency range. This is valid for water up to $v = 2 \cdot 10^9$ s^{-1} at room temperature. In the case of molecules which have no permanent dipoles, the dielectric loss even begins at higher frequencies (IR) [82].

Normally the refractive index increases with increasing frequency *(normal dispersion)* [77, 82]. This effect is responsible for the ability of prisms [75, 79] to separate different wavelengths. In these frequency regions the phase between the electric and the magnetic field vectors remains the same, the amplitude of the radiation is not reduced, and the radiation has the same intensity in front of and behind the sample (at a first approximation).

The polarized molecule acts as a new Hertz dipole. It sends out a secondary wave with a frequency identical with that of the radiation, which has induced the charge polarization (Huygen's principle [77]). This interaction can be treated according to oscillator theory [82]. The amount of polarization depends on the interference sphere (cross section of the particle), which is proportional to the square of the radius [83]. As mentioned above, induction of a Hertz dipole causes an emission of a secondary electromagnetic wave. In liquids and gases the molecules are statistically distributed in all directions in space. Therefore, non-polarized

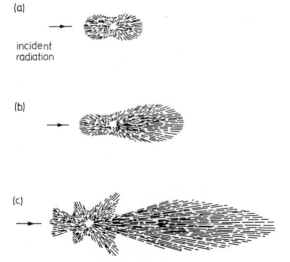

(a)

incident
radiation

(b)

(c)

Figure 12-13.
Scattering of radiation for different diameters of the molecule relative to the wavelength of radiation.
(a) Rayleigh scattering for diameters smaller than $\lambda/10$, (b) Rayleigh–Gans scattering for diameters approximately equal to λ; (c) Mie scattering for particles with diameters larger than λ. In all cases statistical orientation of the molecules and non-polarized radiation has been assumed.

electromagnetic radiation causes these molecules to send out electromagnetic waves in all directions (with a certain intensity distribution). However, small molecules (with a diameter smaller than 1/100 of the wavelength) do not show intense effects, because the polarization remains small (see above) [84]. For this reason the effects of scattering in all the directions in space are negligible and the reduction of the intensity of the radiation transmitted through matter is extremely small [75].

Larger particles have increased effective cross sections. Therefore, this elastic scattering has to be taken into account [85] also. It becomes measurable as soon as the diameter of the molecules comes close to $\lambda/10$ (in the visible region). The effect is called Rayleigh scattering [81]. The intensity of the elastic scattered light is the largest in foreward and backward directions (Figure 12-13) [86, 87]. This intensity distribution is only valid for statistically oriented particles and non-polarized radiation. In the case of linear polarization, in the plane of polarization almost no intensity can be observed (no emission of secondary waves except perpendicular to the induced dipole moment). Since dispersion increases with increasing frequency, Rayleigh scattering depends on the wavelength (proportional to λ^{-4}). This means that in the ultraviolet region the scattering is greater than that in the visible region [81].

As soon as the diameter of the molecules becomes close to the wavelength of the radiation, the intensity distribution acquires the shape shown in Figure 12-13b. This is called Rayleigh–Gans or Rayleigh–Debye scattering [86, 88]. If the diameter of the molecule becomes larger than the wavelength, it can have more than one center of scattering. The scattered waves interfere and the intensity distribution around the molecule changes its shape in comparison with the distribution discussed above. This type of scattering is described by Mie's theory [81, 86] (see Figure 12-13c).

12.3.1.2 Absorption

So far, the interaction of the electromagnetic radiation with matter did not cause any change in phase between the electric and the magnetic field vectors. For this reason the intensity of the radiation was not changed in principle. As soon as the dielectric matter absorbs energy from the electromagnetic radiation, the definition of the real dielectric constant can no longer be used. However, the definition of the refractive index remains valid. A phase difference occurs between the electric and magnetic vectors. Refractive index and dielectric constant become complex quantities. This effect can be used successfully in internal reflection spectroscopy (see Section 12.3.3).

In frequency regions of absorption, an energy transfer from radiation to the molecule takes place in addition to the above-mentioned polarizing effects. In the IR and UV-Vis regions the molecules have some eigenfrequencies (oscillator theory). They can be considered to be either resonant normal vibrations of the atoms relative to each other or changes in electron distribution relative to the position of the nuclei. Both cause a resonant uptake of energy from the radiation.

At these resonant frequencies the curve of dispersion has a different shape. The refractive index is drastically reduced with increasing frequency [77, 83]. This effect is called *anomalous dispersion*. In theory, the excitation of such normal vibrations should cause a so-called "resonance catastrophe", which means that dissociation of the bond would occur. However, the elongation of the atoms is damped by the force constant of the bond [82, 83]. This damping causes a phase difference between the electric and the magnetic field vectors. One observes that the maxima of the two no longer appear at the same time.

These areas of resonance and damping are shown in Figure 12-14 in the upper part of the dispersion curve as regions of extreme change in dielectric constant and refractive index, respectively. A phase difference causes a loss of energy of the radiation which results by transfer to the molecule which can no longer be polarized in-phase with the radiation. Either the nuclei and/or the electrons can no longer follow the alternating field because of their inertia. The curve of dispersion splits into two branches, one given by the imaginary and the other by the real part of the refractive index and the dielectric constant. The imaginary part of the dielectric constant (see lower part of Figure 12-14) corresponds to the absorptivity known in quantitative photometry (see Section 12.3.3.2).

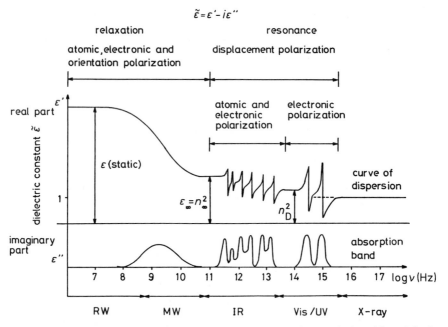

Figure 12-14. Dispersion curve. The dielectric constant ε is plotted versus the logarithm of the frequency. At low frequencies this curve represents n^2 according to Maxwell. In the case of absorption this dielectric constant becomes complex. In addition to the real part of the dispersion curve an imaginary part is found (lower curve: Gaussian-shaped absorption bands).

Vibrational excitation results in a new eigenstate of vibration with a new eigenfrequency, given by the motion of the atoms in the molecule relative to each other (normal coordinates). Each state corresponds to a certain energy and an equilibrium position of the vibration of the nuclei. The energy difference between two of these states is identical with the resonance energy, given by Einstein's equation:

$$\Delta E = h\nu .$$

(12-21)

Stable electron density distributions also correspond to energy levels and are called orbitals. In the case of electronic excitation the frequency of the radiation is in resonance with the energy difference between the energy levels of the electronic ground state and that of the electronically excited state. Such an orbital cannot contain more than two electrons, which have to differ in their spin according to Pauli's principle [73]. Such energy states with opposite electron spins are called singlets. Usually in organic molecules the ground state is a singlet state. The distribution of the electrons to the same or different orbitals results in a certain state with a defined energy and electron density distribution. Any excitation to the next (excited) electronic state has to be done by an electromagnetic radiation with a frequency ν, which is equivalent to that given by the energy difference between the two states involved.

12.3.1.3 Photophysical and Primary Photochemical Processes in the Molecule

In Figure 12-15 the different electronic states are indicated with a thick line to the left for the absorbing molecule [80]. Some smaller lines are also included, representing the different energy states of vibration (eigenresonances according to relative atomic motions in the molecule). The absorption is represented by a vertical straight line from the singlet ground state to the first excited state, which also has to be a singlet (no spin change is allowed [73]). In the latter case the opposite spins are no longer localized in one orbital, but in two different orbitals. This means that in the excited state one of the two electrons has been excited from a lower to a higher energy level. The result is a new electron density distribution. The different energy levels in the molecule (vibrational and electronic, no rotational) are shown schematically in Figure 12-15 in a simplified form. This representation of energy levels is called a Jablonski diagram. Absorption usually takes place to a state A*, which is both electronically and vibrationally excited.

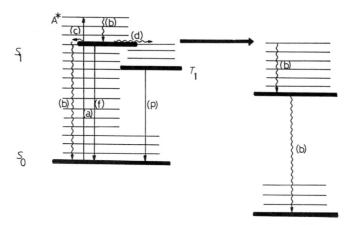

Figure 12-15. Jablonski diagram with electronic and vibrational energy levels, but without rotational levels. The diagram is given for the original excited molecule and (on the right-hand side) for a new product formed by a photochemical reaction. The different possible energy transitions are given by (a) absorption to A*, (b) thermal equilibration, (c) internal conversion (IC), (d) intersystem crossing (ISC), (f) fluorescence, and (p) phosphorescence. S and T designate singlet and triplet terms.

This excited state is unstable. The molecule tries to reach a more stable configuration as rapidly as possible. The energy dissipates to other vibrational, rotational or even translational energy states, either within the same molecule or by transition to another molecule that has come into the effective sphere, for example by impact [73, 89]. All these processes are called radiationless degradation, since the energy taken up by radiation is lost by radiationless processes. If the molecule remains in the same electronic state, this type of radiationless transition is called thermal equilibration [73, 80]. Usually the vibrational ground state in the first electronically excited singlet state lives relatively long (10^{-9}–10^{-6} s). Therefore, some competitive processes can take place starting at this energy level:

— Internal conversion (IC) in the same singlet state by isoenergetic transfer to the lower electronic state but at a very high vibrational energy level. As a consequence, thermal equilibration takes place in this new electronic state to the vibrational ground state.
— Return to the ground state (vibrational and electronic) by emission of radiation. This process is called fluorescence in the singlet term scheme [90, 91].
— Further, the excited singlet state can change its spin multiplicity, which means that one of the anti-parallel electronic spins in the different orbitals is turned around. The process is called intersystem crossing (ISC) to a triplet state with parallel spins (the spin multiplicity becomes 3). According to Pauli's principle, this state is only allowed if the two electrons are localized in different energy levels, one in the ground state and the other in the electronically excited state. The main process of energy dissipation from this type of excited state is a radiationless deactivation. However, under certain conditions one finds luminescence given by an emission of radiation from this triplet state. This so-called phosphorescence is an extremely forbidden process, since during this emission of radiation one of the spins has to be turned around. For this reason, this process only happens in competition with all other possible processes, if this excited state lives long enough. This means that all other possible processes (radiationless degradation, photochemistry) have to be less probable.
— Both the excited singlet and the excited triplet state can be the starting point for a photochemical reaction. Whereas in photophysical processes only the relative electron density distribution is changed in combination with some new interatomic distances [75], in the case of photochemistry the arrangement of the atoms in the molecules is changed. Examples are either photoisomerizations, which give a new conformation, or the breaking of bonds to form new molecules. These photochemical reaction pathways give a crossover to a new term scheme, shown in the right part of Figure 12-15.

It is a general rule that all the discussed processes are competitive. Which of them is favored depends on the properties of the molecules and on the surrounding medium. Therefore, the change in photophysical behavior is a sensitive tool for monitoring external effects (eg, fluorescence sensors). On the other hand, if the intensity of the excitation source is high, it may cause non-negligible photochemical reactions (see Section 12.3.5).

12.3.1.4   Nonelastic Raman Scattering

In Section 12.3.1.1, just the elastic scattering approach has been discussed. If the scattering is accompanied by an absorption process causing a change in vibrational and/or rotational energy, this scattering is called inelastic or Raman scattering [92]. Electronic excitation during the Raman process leads to a new, unstable electron density distribution, a so-called virtual excited energy state [75], which immediately deactivates back to equilibrium. Part of the energy is taken to induce vibrations and/or rotations in the molecule. The remaining energy is emitted as radiation at wavelengths longer than the Rayleigh wavelength [73, 92].

In addition to these emission lines at smaller energies, one finds some others (very weak) at higher frequencies. Usually some of the molecules in the electronic ground state are in an excited vibrational and/or rotational state at room temperature. They can also be excited to the virtual state mentioned above. This process is less probable because of the small number of such molecules. The resulting emission from the virtual state at shorter wavelength (higher frequency), which is called anti-Stokes, is very weak in comparison with the Stokes lines, which themselves are very weak in comparison with the elastic Rayleigh scattering [73, 93].

12.3.1.5 *Interactions Between Matter and Radiation as a Tool for Measurement*

The different types of interaction between matter and radiation are illustrated in Figure 12-16. The intensity of the incident radiation (Figure 12-16a) can be reduced by passing an absorbing medium. According to Figure 12-16b the excited state can emit in all directions in space (fluorescence, phosphorescence), since the distribution of molecules in space is statistical [73, 75]. Further, reflection can take place at an interface between media of different optical density. As long as the interface is structured (eg, a thin-layer chromatographic plate), refection is diffuse [94, 95]. The typical intensity distribution in the half-sphere according to Lambert's cosine law is given by Figure 12-16c. If the surface is polished one finds guided reflection (see Figure 12-16d) [52, 96].

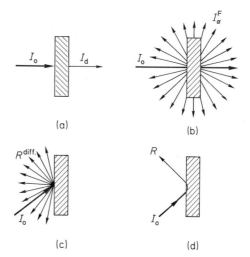

Figure 12-16.
Different interactions between radiation and matter: (a) transmission and absorption; (b) luminescence (fluorescence); (c) diffuse reflection; (d) guided reflection.

The relative amounts of reflected, absorbed, and transmitted intensity depend on the wavelength, the angle of incidence, and the ratio of refractive indices of the media [96]. Multilayers can cause multiple reflections and interferences (see Section 12.3.4). The measured intensity of the beams reflected at two interfaces depends on the thickness of the layer between these interfaces (see Section 12.3.4). As will be discussed later (see Section 12.3.3), total reflection can be observed under certain conditions for radiation intending to pass from a medium with a higher to a medium with a lower refractive index.

12.3.1.6   Comparison of the Different Spectroscopic Techniques with Respect to Potential Sensor Applications

Different photophysical and photochemical pathways have been discussed. Either absorption, fluorescence, and/or reflection, or even photochemical processes can be used in sensor techniques. Frequently some fluorescence techniques have been summarized as photochemical sensors [97]. According to the discussion given above, this term should be discarded. On the contrary, any competitive photochemical pathway reduces the quality of a fluorescence sensor (reduction in the concentration of the fluorescent indicator molecule). However, systems can be defined as photochemical sensors if the photoinduced modification of the sensor element can be correlated, for example, with the number of absorbed photons (see Section 12.3.6).

IR and UV-Vis spectroscopy are frequently used in analysis. Microwave spectroscopy has not yet achieved any significant analytical application except in gas analysis. IR spectroscopy is considered to be very efficient for the examination of structural changes in molecules. UV-Vis spectroscopy is preferably used for quantitative analysis. Because of the excitations of rotations and vibrations taking place in combination with any electronic transitions, the bands in the UV-Vis region are very broad, especially in solution. Only in very specific cases (perfluorinated solvents) is one able to resolve the vibrational fine structure.

Near-infrared (NIR) spectroscopy has become of increasing interest in recent years, because it allows the measurement of overtones of the normal vibrations [98, 99]. Since in the wavelength region between 1 and 2 μm higher overtones (second to fourth harmonics) are observed, the effects are small. Nevertheless, NIR spectroscopy can combine the advantages of IR and UV-Vis spectroscopy: structural information and photometry [98]. In IR spectroscopy the energy difference ($\Delta E = E_1 - E_0$) between the vibrational ground state (E_0) and the excited energy level (E_1) is very small. Therefore, at room temperature the excited state is also populated. The relative number of the molecules (N_0, N_1) in the two energy levels is, according to Boltzmann's law,

$$N_1 = N_0 \, e^{-\Delta E / k_B T}, \tag{12-22}$$

where $k_B = 1.38 \cdot 10^{-23}$ J K^{-1} (Boltzmann constant). Under these conditions, Lambert–Beer's law (see Section 12.3.2) is no longer valid. This cannot happen in NIR spectroscopy, which has been used in the past in food analysis [100] and has become more interesting in recent years [98].

12.3.2 Photometry

12.3.2.1 Quantitative Determination by Lambert–Beer's Law

UV-Vis spectroscopy has been stated to be quantitative because of the linear relationship between the concentration of the analyte and the measured absorbance. The incident light at the wavelength of measurement λ is absorbed. The decrease in its intensity in an infinitely small volume element (at position x) is proportional to the path length dx in this volume element, the concentration of analyte c_i, and the intensity $I_\lambda(x)$ at this element itself [80, 101]:

$$-dI_\lambda(x) \sim c_i I_\lambda(x)\,dx \,. \tag{12-23}$$

The proportionality constant is specific for the analyte's molecule i and depends on λ; it is called the molar absorption coefficient or absorptivity (ε_λ) [102]. Using this constant and integrating the equation above for the path length d, one obtains a linear relationship between the logarithm of the ratio of incident ($I(0)$) to transmitted ($I(d)$) intensity and the concentration:

$$A_\lambda = \log \frac{I(0)}{I(d)} = \varepsilon_\lambda\, c_i\, d \,. \tag{12-24}$$

A_λ is called the absorbance; the term optical density should no longer be used.

Two facts have to be realized when applying Lambert–Beer's law in practice:

- the absorbance is the sum of all the absorptions of different components in the analyte, and
- under certain conditions deviations of this law are observed.

Both chemical and physical deviations exist. Chemical deviations are caused by association or dissociation of particles. They can be avoided by working in very dilute solutions, in which no association appears. The maximum concentration depends on the analyte, since some of the substances associate more easily than others (eg, aromatic compounds or heterocyclics). Physical deviations are caused by a lack of monochromaticity of the radiation. A ratio of the natural band width of the absorption band to the spectral band width of the radiation of less than $10:1$ causes an error of more than 2% in quantitative applications [75, 103]. The error becomes large in the measurement of gases. Some part of the spectrum of the incident radiation is not absorbed by the sample. Therefore, this deviation has to be avoided either by use of line sources or good monochromators with appropriate small spectral band widths.

12.3.2.2 Components of an Optical Sensor System

According to Equation (12-24), the signal obtained for the analyte has to be compared with that of a reference [75, 103]. Either single- or double-beam set-ups are conventionally used. Their advantages are discussed in the literature [80, 104]. With a single-beam set-up the wavelength of interest or the total spectrum is measured one after the other sequentially for the reference and the analyte. In contrast, split-beam techniques in the double-beam arrangement allow the "parallel" measurement of both the sample and the reference beam. Therefore, the quality of the measurement in single-beam arrangements depends on the stability of the light source and the photomultiplier and on the time between the measurements of the signal and the reference [104].

The light source offers either a continuous spectrum or single lines. The latter type can be a mercury arc lamp (Figure 12-17) [105], which can be doped with certain other elements to increase the available number of wavelengths. Lasers show even better monochromaticity [106]. Nowadays gas lasers (He–Ne) are very cheap and are frequently used in optical sensors if their emission wavelength fits the application. Dye lasers allow a wide variation of wavelength [107], but the set-ups are expensive and the dyes used undergo some photodegradation [108].

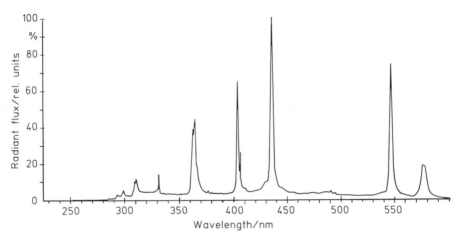

Figure 12-17. Radiant flux for a high-pressure mercury arc (HBO 500, Osram) versus wavelength. The spectral band width of the spectrometer is chosen to be very small.

Recently diode lasers and LEDs have been developed both consisting of semiconductor material (GaAs, GaP) [109]. In LEDs the transition of electrons between the conductivity and valence bands causes an emission of radiation by recombination in the p–n junction. The wavelength depends on the type of injection (Figure 12-18a). Diode lasers generally contain the same semiconductor material, but their mode of operation is different (Figure 12-18b) and their structures are more sophisticated [110]. A low junction current results in incoherent radiation (as in LEDs), which becomes coherent above a certain threshold current (usually pulsed operation) in diode lasers. Recent developments have made them better and cheaper [110, 111]. The band width of LEDs of about 40 nm is reduced to below 1 nm. Both types of light sources are compared in the literature [112, 113] for single-mode fiber systems.

On the other hand, continuous spectral sources (hydrogen/deuterium, xenon or halogen–tungsten lamps) allow the selection of wavelengths from a broad spectrum either by filters (broad band) or monochromators (small band). At wavelengths higher than 400 nm the relative intensity of the H_2/D_2 source decreases to one tenth of its ultraviolet intensity. This behavior is demonstrated in Figure 12-19. The tungsten lamp is usable in the visible region

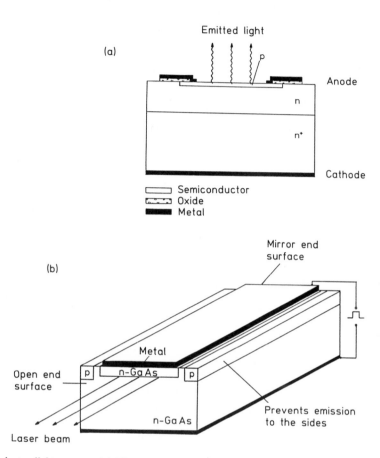

(a)

Emitted light

p

Anode

n

n⁺

Cathode

☐ Semiconductor
▱ Oxide
▰ Metal

(b)

Mirror end
surface

Metal

Open end
surface

p   n-GaAs   p

Prevents emission
to the sides

n-GaAs

Laser beam

Figure 12-18. Semiconductor light sources: (a) LED (light-emitting diode); (b) diode laser.

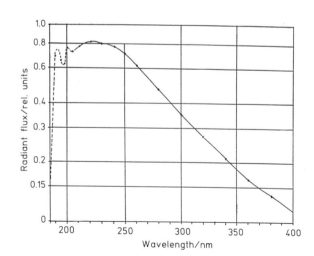

Figure 12-19:
Radiant flux for a deuterium lamp
(Quarzlampengesellschaft Hanau)
versus wavelength.

(300–900 nm). A xenon lamp has a spectrum between 200 nm and 1 μm. However, in the UV region beyond 300 nm xenon lamps emit relatively poorly, as can be seen in Figure 12-20. In addition, sharp emission lines (fingerprint region) in the spectrum (400–500 and 800–1000 nm) may cause compensation problems.

By use of colored glases or dielectric layers (interference filters), relatively broad wavelength ranges are selected. They should be applied only if a line source is used or low emission signals require high excitation intensities.

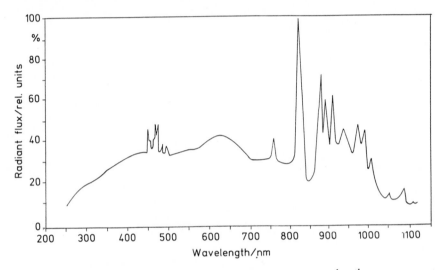

Figure 12-20. Radiant flux of a xenon lamp (XBO, Osram) versus wavelength.

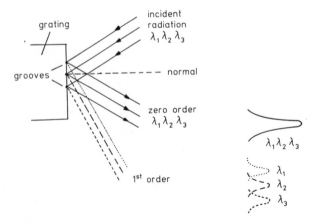

Figure 12-21. Principle of a grating. Interference of reflected waves causes an increase in intensity at different geometries for different wavelengths. These can be resolved by use of an exit slit behind the grating. The resolution depends on the number of grooves in the grating.

Nowadays most monochromators use gratings for spectral dispersion, even though prisms in the far-UV region show better properties. The advantage of a grating is the nearly linear dispersion. Separation of the different wavelengths occurs by interference of the light beams, reflected by the different grooves of the grating (Figure 12-21) [77].

Photomultipliers (head-on, side-on) [114] are used as detectors in high-quality sequential spectrometers. They are very sensitive in the UV and visible regions. Even though nowadays, they have been replaced with photodiodes in many applications, high-performance equipment and single photon counting devices (see Section 12.3.6) cannot manage without them. Incident photons force electrons to leave a photocathode and are focused to an anode. Some further electrodes (called dynodes) between the cathode and the anode cause an amplification of the signal, but also of the dark current. The measured current is proportional to the intensity.

The operation of photodiodes is based on the inner photoelectric effect occuring in the p–n or p–i–n junctions (a layer of intrinsic semiconductor material is sandwiched between n and p layers). In the photoconductivity mode the p–n junction is reverse biased (removal of opposite charges from the junction). Radiation is absorbed in this depleted region. Charge car-

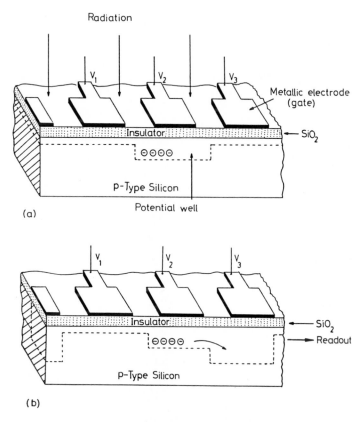

Figure 12-22. Charge-coupled device (CCD). (a) Free charges are formed during irradiation, which are trapped in wells caused by biased voltage; (b) synchronized charge transfer in a shift register for read-out.

riers are formed (electron–hole pairs), which are rapidly removed in the p and n regions by an applied electric field. The current is proportional to the number of photons absorbed. This type shows a very fast response (avalanche photodiode) but a relatively high noise level. The latter can be drastically reduced for photodiodes working in the photovoltaic mode, but these are slower and their linear response is limited by the photo-current [115]. Silicon diodes offer good performance and they are cheap. They are sensitive between 350 and 1100 nm (enhanced types start at 200 nm). Expensive germanium diodes work at longer wavelengths.

In Section 12.1 internal reference by multi-wavelength detection was mentioned as one of the advantages of optical sensors. Therefore, in the future such arrangements will replace the present single-wavelength detection. A possibility is given by use of an array of photodiodes, called a linear diode array, solid-state image sensor, or optical multi-channel analyzer [80, 116]. It has become commonplace to call these devices "CCDs" (charge-coupled devices), even though a large variety of photoelectric devices and methods for their read-out exist [117]. CCDs consist of an array of closely spaced metal–insulator–semiconductor capacitors (photodiodes). A positive external voltage causes a depletion region ("potential well"), in which the photogenerated electrons are temporarily stored (both signal and dark current, Figure 12-22a). After the chosen integration time the voltage is changed at the elements and the charges are shifted sequentially from diode to diode (transfer mode, Figure 12-22b). In this type of read-out a snychronized clock correlates the video signal with the illuminated element. One element after another is read to the video line. By multiplexing, selective access to storage areas parallel to the active sites is possible. This type allows the random read-out of selected diodes [75, 117]. Yearly up-dated buyers' guides review the state of the art of optoelectronic devices [118, 119].

Figure 12-23 shows a combination of a monochromator (without exit slit) and a diode array as an example of a modern optical sensor. The white light is dispersed in the monochromator and falls onto the diode array, each wavelength at another element. Therefore, a spectrum can

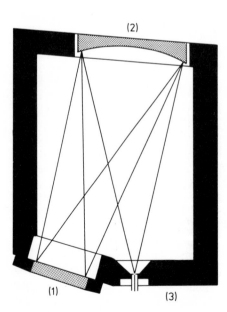

(2)

(1)

(3)

Figure 12-23.
Combination of a grating monochromator with a diode array. Zeiss MCS 23-UV/Vis diode array spectrometer. (1) Photodiode array with 512 photodiodes; (2) concave holographic grating; (3) entrance slit, 50 μm × 250 μm. The chassis and the holders for all optical parts are built from ceramic material. Dispersion of the white light at different wavelengths can be correlated with different diodes in the array. Electronics allow a separate read-out for each diode. The spectrum is recorded simultanuously.

be recorded simultanously for all the wavelengths with such a set-up. Even though the sensitivity of photodiodes cannot attain that of expensive photomultipliers, their signal-to-noise ratio is so good that they can be used in spectroscopic applications. Their quantum efficiency drastically decreases in the wavelength region below 250 nm, as Figure 12-24 demonstrates.

Usually a single-beam arrangement is used. The detection time of a total spectrum is in the region of 5–20 ms. Therefore, both the reference and the sample can be recorded within a short time. Nowadays the repetition rate of spectra acquisition is less limited by the time of recording (if high-intensity light sources are used), but more by the time to read out the information from the diode array, that means by the time for the analogue-to-digital conversion, and for data storage in the memory of the computer.

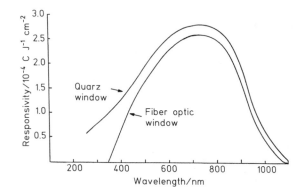

Figure 12-24.
Quantum efficiency of a usual silicon photodiode as a function of wavelength.

In diode arrays the optical arrangement is different to that in conventional photometers or spectrometers. Normally in UV-Vis spectrometry the sample is positioned between the monochromator and the detector. Therefore, only light of one wavelength at a time falls onto the sample. In diode arrays the sample is placed between the light source and the monochromator (as in an IR arrangement) and all wavelengths fall onto the sample at the same time. This ar-

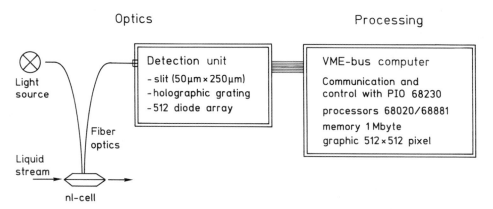

Figure 12-25. Spectral optical sensor system: modular arrangement of light-source, fiber optics, dispersion element, diode array, and process control equipment.

rangement for diode arrays is chosen because the different wavelengths do not fall onto different volume elements of the sample (in the case of sample positioning behind the monochromator). Such an optical arrangement is shown in Figure 12-25 (reflection arrangement in the chosen example). It uses a bifurcated fiber optic. All components can be used as modules. The disadvantage of this set-up is that the higher overall intensity might cause a photochemical reaction of photo-unstable samples. Therefore, some of the commercially available instruments have a photo-shutter in their light beam to avoid unnecessary irradiation of the sample by the light source.

12.3.3   Refractive Index

12.3.3.1   Nonabsorbing Medium

Previously it had been explained, that radiation penetrating matter changes the wavelength λ and the propagation velocity v crossing the interface of the two (transparent) media with different refractive indices. Light striking this interface from the direction of lower refractive index is partially reflected and partially transmitted. As shown in Figure 12-26, the reflected beam (I_r) has the same angle ($\alpha_i = \alpha_r$) with respect to the optical normal as the incident one (I_0). The transmitted beam is refracted according to Snell's law [77]:

$$n_1 \sin \alpha_i = n_2 \sin \beta . \tag{12-25}$$

In non-absorbing media the sum of the light intensities I_r (reflected) and I_t (transmitted) gives I_0. In this case the reflected relative to transmitted intensity can be calculated for any

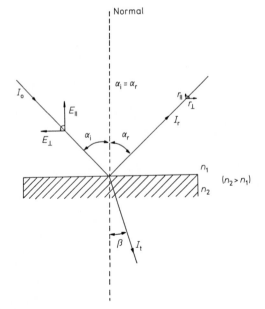

Figure 12-26.
External reflection for radiation I_0, incident from the optically rarer to the optically denser medium ($n_2 > n_1$). Incident and reflected beams have the same angle ($\alpha_i = \alpha_r$) with respect to the optical normal. The beam penetrating the interface into the denser medium is refracted with an angle β ($\beta < \alpha_i$).

angle of incidence α_i by use of the Fresnel relationships [77, 78]. In general, the amplitudes r of the reflected waves depend on their polarization. According to the direction of the vibration of the electric field vector, one distinguishes between a plane perpendicular ($r_\perp$) to the plane of incidence or parallel ($r_\parallel$) to it. Perpendicular polarization is known as transverse electric (TE) and parallel polarization is given by a transverse magnetic wave (TM) [78, 79]. The Fresnel equations give the reflection coefficients as ratios of the electric field vectors E of the incident and the reflected beam (which corresponds to the amplitudes of the wave) $r_\perp$ and $r_\parallel$ according to

$$r_\perp = -\frac{\sin(\alpha_i - \alpha_r)}{\sin(\alpha_i + \alpha_r)} \tag{12-26a}$$

$$r_\parallel = \frac{\tan(\alpha_i - \alpha_r)}{\tan(\alpha_i + \alpha_r)} . \tag{12-26b}$$

The geometries of the parallel and perpendicular components of E and r are plotted in Figure 12-26. At approximately normal incidence ($\alpha_i \rightarrow 0$) the ratio of reflected to incident radiant power (Φ_r, Φ_i) [77] is called the reflectivity (or reflectance) R. It is equivalent to the above-mentioned reflection coefficient squared:

$$R = \Phi_r/\Phi_i = r^2 . \tag{12-27}$$

Under these conditions the reflectivity is identical for both polarized beams with amplitudes $r_\perp$ and $r_\parallel$. In addition, there is no difference whether the light strikes the interface from the denser or rarer medium. This normal incidence can be approximated by

$$n_1 \alpha_i \approx n_2 \alpha_r; \quad \alpha_r \approx n_{12} \alpha_i , \tag{12-28}$$

using the definition

$$n_{12} = \frac{n_1}{n_2} . \tag{12-29}$$

Under these conditions ($\alpha_i \rightarrow 0$)

$$R = \frac{(n_1 - n_2)^2}{(n_1 + n_2)^2} = \frac{(n_{12} - 1)^2}{(n_{12} + 1)^2} . \tag{12-30}$$

If light strikes the interface from the rarer medium the effect is called *external reflection* (Figure 12-26). The values for the two components of the reflectivity (perpendicular ($\perp$) and parallel ($\parallel$) polarization) are plotted as the solid curve in Figure 12-27 for different angles of incidence, marked as external reflection [52]. Whereas the perpendicularly polarized reflectivity increases monotonically with the angle of incidence, that for parallel polarization passes through a minimum. It becomes zero at the Brewster's angle, α_B, given by

$$\alpha_B (\parallel) = \tan^{-1} n_{12} . \tag{12-31}$$

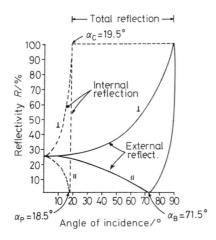

Figure 12-27.
Reflectivity R (%) versus the angle of incidence [52]. Both external and internal reflection are given (solid and dashed lines, respectively) for parallel ($\parallel$) and perpendicular ($\perp$) polarized radiation. Refractive indices are $n_1 = 1.33$ and $n_2 = 4$ for external reflection and $n_1 = 4$ and $n_2 = 1.33$ for internal reflection. Brewster's angle α_B is 71.5° for parallel-polarized external reflection. In the case of internal reflection the reflectivity is 100% (total reflection) for perpendicular- and parallel-polarized radiation between the critical angle $\alpha_c = 19.5°$ and 90°.

All parallel polarized radiation penetrates the denser medium. $r_\parallel$ becomes zero according to Equation (12-26b), since

$$\tan(\alpha_B + \alpha_B') = \tan 90° \to \infty \qquad (\alpha_B \text{ and } \alpha_B' \text{ are complementary})$$

The incident radiation is totally polarized, and therefore this angle is called a polarizing angle. In addition, one finds a phase shift of $\lambda/2$ at this interface. At grazing incidence ($\alpha_i = 90°$) the reflectivity becomes 100% for both the polarized beams.

If the light beam passes the interface from the denser to the rarer medium, the process is called *internal reflection*. The reflectivities can again be calculated by use of Equations (12-26a) and (12-26b). The result is given in Figure 12-27 as the dashed curve. Both reflectivities become 100% at a critical angle α_c, according to

$$\alpha_c = \sin^{-1} n_{21}, \tag{12-32}$$

where $n_{21} = n_2/n_1$. According to Figure 12-27, one defines a principle angle α_P, which is analogous to the polarizing angle α_B (Brewster's angle). Another name of this angle is *internal polarizing angle* α_P. It is the complement to α_B and is given by

$$\alpha_P = \tan^{-1} n_{21}. \tag{12-33}$$

At this angle the reflectivity of the parallel polarized beam is zero.

External reflection (rarer to denser medium) causes the transmitted beam to be refracted in the direction to the normal. In contrast, with internal reflection (denser to rarer medium) the transmitted beam is refracted away from the normal, as shown in Figure 12-28a. As long as the angle α_i is smaller than the critical angle α_c, the relative intensities of the reflected and refracted light beam are given according to Equations (12-26a) and (12-26b). At the critical angle the light no longer passes through the interface, as shown in Figure 12-28b. At the interface to the rarer medium all the light is grazing along the interface.

For angles larger than the critical angle all the light is reflected, ie, $I_0 = I_r$, and we have *total reflection*. The interface acts as a perfect mirror. This total reflection occurs for both the polarized beams at angles $\alpha_i > \alpha_c$. No phase shift occurs.

It should be mentioned that this totally reflected light has to be characterized by an elliptical polarization. This means that both the electric field vectors $r_\perp$ and $r_\parallel$ have a phase difference between 0 and π, which is constant with time [77].

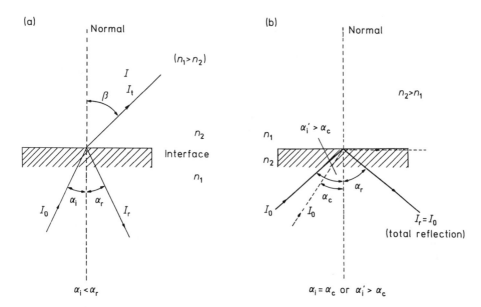

Figure 12-28. (a) Internal reflection. Incident radiation with an angle $\alpha_i < \alpha_c$ (critical angle), passing the interface from the denser to the rarer medium ($n_1 > n_2$) is partially reflected ($\alpha_i = \alpha_r$) and partially refracted into the rarer medium, whereby the angle in the rarer medium is larger than that in the denser medium ($\beta > \alpha_i$). (b) Incident radiation at an angle α_i' equal to a larger than the critical angle α_c. At the critical angle α_c the reflected beam is grazing the plane of the interface. At an angle $\alpha_i' > \alpha_c$ light no longer passes through the interface. The result is total reflection ($I_r = I_0$). Incident and reflected radiation have the same angle with respect to the normal.

12.3.3.2 Absorbing Medium

So far nonabsorbing media have been discussed. With an absorbing medium, the refractive indices have to be replaced by complex values, (explained in Section 12.3.1) according to

$$\tilde{n} = n(1 - i\kappa) . \tag{12-34}$$

The attenuation index κ is related to the absorption coefficient a according to

$$a = 4\pi n\kappa/\lambda . \tag{12-35}$$

The dimension of a is the reciprocal of length (m^{-1}). It characterizes the absorption per unit path length of a medium. It is correlated to the analytical absorptivity ε of a liquid by

$$a = \varepsilon c_i \,, \tag{12-36}$$

where c_i is the concentration.

Reflectivities for absorbing and nonabsorbing media do not differ substantially for external reflection, except in the case of metals (strongly absorbing) [52]. However, internal reflection is influenced considerably, especially at angles close to the critical angle α_c. Thus measurements at this angle offer an extremely sensitive tool for monitoring small changes in absorption coefficients.

In the case of external reflection it can be shown by use of Maxwell's equation [78] that standing waves are established in general at a total reflecting surface, because of the superposition of incident and reflected waves [120, 121]. In the case of metals this standing wave pattern shows a specific behavior. The electric field vector becomes zero at the interface to the metal. One obtains a node close to the surface (Figure 12-29), since a high conductivity exists in metals. The radiation penetrates the metal less than 10 nm.

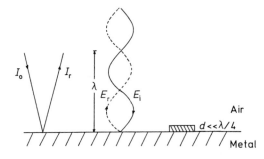

Figure 12-29.
External reflection at a metal surface. Incident and reflected waves of the radiation interfere and form a standing wave pattern, given by the electric field vector. Any absorbing or non-absorbing film covering the metal surface is just not detectable as long as the thickness d is much smaller than $\lambda/4$. In this case the small electric field vector is not influenced by the film [52].

Thin films of absorbing material with a thickness d of less than $\lambda/4$ are not observable. The reason is the node ($E = 0$) at the metal surface. No significant field amplitude can interact with these types of film. Accordingly, any absorbance measurements are difficult under these conditions [122, 123]. The properties of the standing waves are not affected by the denser medium.

An elliptical polarization mentioned above is also found at such reflecting metals (mirrors). The so-called ellipsometry [96, 124] can be used to observe monolayers of adsorbed molecules [125], which disturbs the phase of the differently polarized reflected beams.

12.3.3.3   Total Internal Reflection

For internal reflection the superposition of incident and reflected waves of nonabsorbing media differs from the behavior of metals. There is still a sinusoidal variation of the electric field vector with distance from the surface in the denser medium. The superposition depends on the angle of incidence. In Figure 12-30 the standing wave amplitudes are plotted near a totally reflecting interface. In the denser medium (n_i) one finds a standing wave, which decreases exponentially in the rarer medium (n_2) [52, 120]. It is called an evanescent wave with wavelength λ_e. In contrast to metals, the electric field vector may have large values at the surface. Each reflection at the interface can gather some information about the rarer medium. Therefore, multiple reflections allow amplification of the signal.

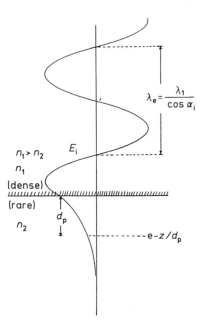

Figure 12-30.
Internal reflection from the denser to the rarer medium, with no metal surface involved. A standing wave pattern is formed. Its wavelength λ depends on the angle of incidence α_i and the wavelength of radiation in the rarer medium. The electric field vector is not zero at the interface. An evanescent wave penetrates the rarer medium to some extent, being damped with penetration depth d_p. This evanescent wave can couple to the rarer medium by two mechanisms: in the case of frustrated total reflection (FTR) the medium does not absorb (no energy loss), whereas in the case of attenuated total reflection (ATR) the medium absorbs. By coupling energy is transferred from the medium to the evanescent wave and the standing wave in the denser medium [52].

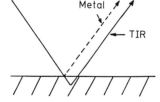

Figure 12-31.
Total reflection at a metal surface does not cause any displacement of the point of reflection with respect to the incident wave. In contrast, a non-metal surface causes a displacement described by Goos and Hänchen [127], since the incident wave penetrates into the other medium to a certain extent.

12.3.3.4   Penetration Depth

In Figure 12-30 it was shown that the electric field amplitude decreases exponentially with the distance from the surface. The depth of penetration d_p is given by

$$d_p = (\lambda_1/2\pi)\,(\sin^2\alpha - n_{21}^2)^{-1/2}\,. \tag{12-37}$$

This distance d_p is determined by the electric field amplitude which decreases to $1/e$ of its value at the surface [78, 126]. By definition, $\lambda_1 = \lambda/n_1$ is the wavelength in the denser medium and $n_{21} = n_2/n_1$ is the ratio of the refractive indices of the rarer to the denser medium. It can be shown that this penetration depth depends on the angle of incidence α_1 and becomes about one tenth of the wavelength in the denser medium if the angle is close to grazing incidence ($\alpha \approx 90°$). If the angle approaches the critical angle α_c, the penetration depth becomes indefinitely large. In addition, d_p is also proportional to the wavelength. The penetration depth becomes greater at longer wavelengths according to Equation (12-37).

Total internal reflection shows another peculiarity. The reflected radiation is somewhat displaced from the point of incidence (Goos–Hänchen shift) [127], which is demonstrated in Figure 12-31, where the reflection at a metal surface and total internal reflection (TIR) are compared.

12.3.3.5 Evanescent Waves

In the case of internal reflection, the evanescent wave couples to the rarer medium. One can distinguish two mechanisms by which this wave can couple to this medium. If this medium does not absorb, no energy will be transferred from the evanescent wave to it (frustrated total reflection, FTR), even though the reflected beam contains information about the optical properties (refractive index) of the rarer medium. It can cause a change in the amount of total internal reflection. The second mechanism takes place if the rarer medium absorbs. The absorption coefficient influences the reflectivity, which can deviate from 100%. Energy is extracted from the evanescent wave and transferred into the medium (attenuated total reflection, ATR).

Frustrated Total Reflection (FTR)

There are different ways to make the total reflection less than 100%, even though the conditions for total internal reflection apply. This can be achieved by a coupling mechanism of the evanescent wave in such a way, that the exponentially decreasing wave in the rarer medium penetrates the interface without loss and propagates further. Three methods are known for adjusting reflectivity continuously between 0 and 100% ($R + T = 100\%$ remains valid):

a) Another suitable optical transparent medium is placed close to the interface. Depending on the distance d, the reflectivity R varies between 0 and 100% and the remainder of the radiation is transmitted (T). An example is given by two prisms in close contact ($d = 0$, $R \rightarrow 100\%$; $d \gg \lambda$, $R \rightarrow 0\%$ and $T \rightarrow 100\%$) [128, 129]. In this case the same refractive indices have been assumed for both prisms.

b) Selection of a suitable angle of incidence in combination with the change in refractive index of one of the media [130]. Close to the critical angle any change in n_{21} causes a drastic change in reflectivity. This change can be achieved by electric fields, for example (Kerr effect) [77].

c) The dispersion can be used to vary the reflectivity drastically, since the refractive index depends strongly on the wavelength [78]. Any wavelength divergence from the chosen value influences the reflectivity.

All these effects result a strong dependence of the reflectivity. Correspondingly, FTR can be used to monitor even extremely small variations of the refractive indices in the rarer medium in contact with the interface. It is observable by a variation in the measured amount of reflectance with respect to transmission.

Attenuated Total Reflectance (ATR)

This type of total reflection occurs if an absorbing medium is placed in contact with the interface. The absorbing rarer medium causes an energy loss and the reflectivity is reduced from 100% to somewhat more than 0% (in theory 0% is impossible [52]). The remaining radiant power is not transmitted but converted into heat, for example [130]. The variation of reflectivity via the absorption coefficient of the medium in contact is called *attenuation*. A typical way to attenuate reflection is achieved by fixing the angle of incidence just above the critical angle and sweeping the wavelength through an absorption band. This attenuation can be modulated in the case of semiconductors by an electric field (Franz–Keldys effect) [131, 132]. Any change in the distance between the two media also influences the reflectivity.

Both FTR and ATR methods are used in multiple reflection methods. FTR is restricted to the change in the refractive index. Since no absorption takes place and the reflection is frustrated, a large number of multiple reflections are necessary to obtain large signals. In contrast, by use of ATR just a few reflections show a noticeable decrease in intensity because of the absorption of the reflected beam. ATR is influenced by a change in the absorption coefficient, which in general is accompanied by a change in the refractive index. Both methods are applied to obtain information about bulk materials [133] and thin films of polymers [134, 135], and in analyses of liquids [136].

12.3.4   Interferometry

12.3.4.1   Multiple Reflections

Two waves in phase can interfere. The interfaces between layers cause multiple reflections (Figure 12-32). Assuming monochromatic radiation, two beams a and b interfere in such a way that either the measured I_r of the reflected intensity is small ("extinction") or large ("amplification"). The phase difference depends on the ratio of the refractive indices (n_{12}), the physical path length d, and the angle α_i of the incident light to the normal [74, 78].

If the light source is polychromatic, the measured reflectivity depends at each wavelength on the phase difference mentioned above. A so-called interference pattern is obtained,

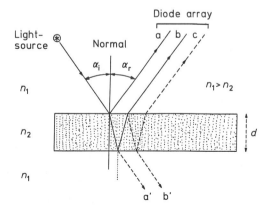

Figure 12-32.
Multiple reflections at a layer with two interfaces. The refractive index of this layer is higher than that of the surrounding ($n_2 > n_1$). Monochromatic light causes an amplification or extinction of the interfering beams a and b depending on the physical thickness of this layer d, the ratio of the refractive indices n_1/n_2 and the angle of incidence α_i. Using polychromatic light the reflected intensity depends on the observation wavelength. An interference pattern is obtained.

characterized by a wavelength-dependent modulation of the reflectivity [52, 78]. It can be measured by a diode array which simultaneously records the intensities at all wavelengths (see Figures 12-23, 12-25, and 12-32). The interference pattern is very sensitive to even small changes in refractive index and path length if α comes close to α_c.

The resulting interference pattern is used as a new detection or sensor principle [136]. Because of the sensitivity of the measurement to small changes in the optical properties, the effect was used in interferometric systems long ago. By dividing a light beam into two different paths and combining them later, any small change in the properties in one of the pathways can be detected by the modulation of the interference of the two waves. This principle is used, for example, in Fourier transform analysis by application of a Michelson interferometer.

12.3.4.2   Interferometers

The principle of a Michelson interferometer [77] is shown in Figure 12-33. Polychromatic radiation of the light source passing a cell is subdivided into two beams. One is reflected by a fixed mirror and the other by a mirror, which can be moved within a certain distance. Recombination of the two beams results an interference pattern caused by the polychromatic light with different intensities at each wavelength depending on the relative distance of the two mirrors from the beam splitter.

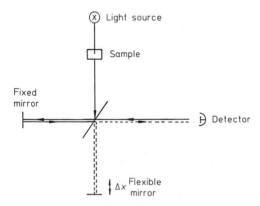

Figure 12-33.
Michelson interferometer. Polychromatic light is used and a half-transparent mirror separates the radiation into two beams. One beam is reflected by a fixed mirror and the other one by a flexible mirror. Radiation is modulated by the frequency of motion of the second mirror. The interference pattern obtained depends on the optical properties of the sample placed between the source of radiation and the mirror arrangement.

An oscillation of the second mirror superimposes a changing phase connection between the two beams. Since the polychromatic light is influenced differently at each wavelength by the sample, this modulation differs with wavelength [137, 138]. Mathematical treatment of the resulting interferogram transforms the dependence on position (cm) to a reciprocal dependence (wavenumber, cm^{-1}). Therefore, any information in the interferogram about the position of this mirror can be found in the mathematically transformed spectrum as a dependence on wavenumbers also. In this type of interferometer the sample is placed at a position where the beams are not split. The interferogram is produced by artificially changing the relative path lengths between the mirrors.

Another approach is to insert a sample into one of the two pathways, which influences the propagation of the electromagnetic wave by its optical properties. This type of interferometer is called a Mach–Zehnder arrangement in recent publications [139, 140], even though it differs slightly from the original set-up (Figure 12-34a) and is more similar to the Jamin interferometer (Figure 12-34b) [77]. Anyhow, this arrangement is used in two types of applications:

(a) either a sample can be placed in one of the two pathways, influencing the interferogram;
(b) or an integrated-optics arrangement can be used, as shown in Figure 12-35 [139]. One of the pathways of a waveguide can be covered with a chemical. The waveguide is intrinsically influenced by a gas or any liquid analyte, that comes in contact to give a sensor.

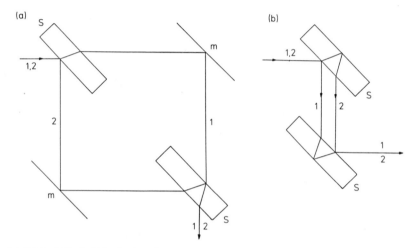

Figure 12-34. (a) Original Mach–Zehnder interferometer. The optical components cause a separation of the two interfering light beams as far apart as possible. (b) Yamin interferometer: similar arrangement, but shorter distance between the separated beams of radiation.

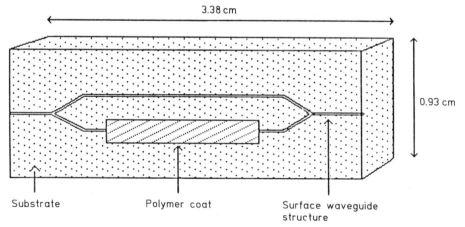

Figure 12-35. A Mach-Zehnder-like integrated optical setup, in which radiation is separated into two waveguides. One of the arms is coated by a polymer sensitive to gases or liquids.

Pressure and temperature effects can be used to change the refractive index in the fiber-optics setup [143, 144]. The influence of their effects causes a phase difference with respect to the other light beam. Such interferometric systems are promising for interesting applications [141, 142, 145].

In single-wavelength operation, interferometry causes a problem with respect to the phase. Starting with a certain phase-dependent signal level, the signal can increase or decrease with changing refractive index [141, 142]. To solve this problem, both arms of the surface waveguides are coupled to a third one, which yields an additional phase shift [146], or a Michelson-type interferometer with thermo-optic phase modulation is used in an integrated-optics arrangement [147].

Another approach is to emboss surface-relief gratings on planar $SiO_2 - TiO_2$ waveguides [148] and to use the gratings as input/output couplers (Figure 12-36). A laser beam passing the gratings excites guided modes in the waveguides. Let N be the effective index of the guided mode, a_I the incident angle of the laser beam with vacuum wavelength λ, l the mode order, and Λ the grating period, then the coupling condition is

$$\pm N = \sin a_I + l\,(\lambda/\Lambda). \tag{12-38}$$

According to this condition, the efficiency of the diffraction varies with N, which means such integrated-optics devices can be applied as sensors [149].

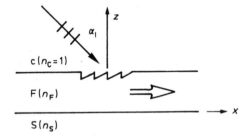

Figure 12-36.
Input grating couplex. In a waveguide film F (refractive index n_F) a surfacerelief grating is embossed. It is placed on a substrate S (refractive index n_S). At an angle a_I a laser beam is incoupled. Its resonance depends on the effective refractive index n_c of the cover C [148].

Plasmon surface polaritons are bound, nonradiative eletromagnetic modes propagating along a noble-metal/dielectric interface [150]. Their field intensity decays exponentially in normal direction to the interface. The propagation is influenced by dissipative losses. Therefore an attenuated total-reflectance arrangement (see Section 12.3.3.5) can be used to monitor the diffracted plasmon light at different angles. The measurements depend on wavelength, incident external angle a_i, and the properties of the substance in contact to the noble-metal surface. The set-up of a so-called surface plasmon microscope [151] is given schematically in Figure 12-37. According to Fresnel's conditions (see Section 12.3.3.1) and taking complex values of the refractive index instead of Equation (12-30), this instrument monitors the influence of the probe on surface plasmon resonance (SPR). Its first application was as a very sensitive tool for examination of interface structures and properties. Recently its commercial use as an immuno sensor has been introduced [152].

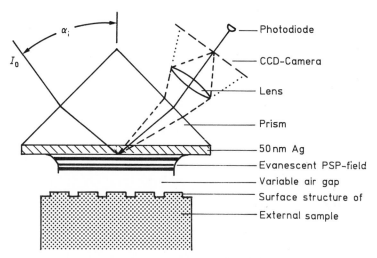

Figure 12-37. Surface plasmon microscope. A sample on the Ag layer influences the propagation of surface plasmons excited by a laser beam incident at an external angle α_i. The reflected intensity varies with angle of incidence, dielectric probe, and wawelength. Usually resonance condition is achieved by variation of α_i [151].

12.3.4.3 *Information Taken from Interferograms*

As mentioned above, two interfering light beams give an interferogram which depends on the factors d, α_i and n_{12}. By use of Fresnel's law and geometric considerations (perpendicular incidence ($\alpha_i \approx 0$)), the measured intensity of interference at a specific wavelength $I_m^{(\lambda)}$ is given by the superposition of beams a and b (see Figure 12-32) [78]. The result is a "modulated" reflected intensity, whereby $I_R^{(\lambda)}$ represents the mean value over all wavelengths:

$$I_m^{(\lambda)} \sim I_R^{(\lambda)} [1 + \cos (2\pi\Delta/\lambda)]. \tag{12-39}$$

The phase difference Δ between the two beams a and b is given by

$$\Delta = 2d \sqrt{n_2^2 - \sin^2\alpha} + \lambda/2 \tag{12-40}$$

whereby the phase shift occurs at the interface of the rarer to the denser medium. Assuming an angle of incidence α close to zero, $\sin^2\alpha$ can be approximated to zero. In this case,

$$I_m^{(\lambda)} \sim I_R^{(\lambda)} [1 + \cos ((2\pi/\lambda) (2n_2 d + \lambda/2))] \tag{12-41}$$

is obtained. Depending on the mentioned factors, a certain intensity results, which turns out to be small if the two beams show an extinguishing interference. In the case of amplification the measured intensity is high.

The use of polychromatic light causes an interference pattern in the observed wavelength range. One finds maxima and minima in the interferogram depending on the order of interference [136]. Taking into account the phase shift $\lambda/2$ for passing the interface from rarer to denser medium, one finds an order m' according to

$$2\pi/\lambda \, (2n_2 d + \lambda/2) = 2\pi m' \, . \tag{12-42}$$

Proximate maxima (or minima) in the interferogram can be measured according to the conditions

$$2n_2 d + \lambda_1/2 = m' \, \lambda_1 \tag{12-43 a}$$

and

$$2n_2 \, d + \lambda_2/2 = (m' + \delta m) \, \lambda_2 \, . \tag{12-43 b}$$

For $\delta m = 1$ (taking next extrema) one can combine the two equations above:

$$2n_2 d = \frac{\lambda_1 \lambda_2 \, \delta m}{\lambda_1 - \lambda_2} \, . \tag{12-44}$$

Taking the distances between certain extrema and knowing their wavelengths λ_1 and λ_2 exactly, one is able to determine the product of the ratio of refractive indices and the physical path length d by Equation (12-44). If one of these physical properties is known, the other can be determined by this means [136].

The physical path length does not depend on wavelength. Therefore, its knowledge allows one to determine the ratio of refractive indices depending on wavelength or frequency [153]. The advantage of a knowledge of this curve of dispersion has been discussed in Section 12.3.1 in detail. Applications of this principle of measurement are given in Section 12.3.7.

12.3.5 Photophysical and Photochemical Deactivation

In Section 12.3.1.3 the change in the photophysical properties of sensing material was mentioned as a tool for monitoring the variation in concentration of an analyte. For this reason, fluorescence sensors are widely used. Since the measured fluorescence intensity depends on the amount of light absorbed, this dependence will be discussed first.

12.3.5.1 *Amount of Light Absorbed*

The amount of light absorbed depends on the intensity of the excitation source, on the concentration of the fluorophore, and on its absorption coefficient. It is given by the difference between the incident ($I(0)$) and transmitted ($I(d)$) intensity at the wavelength λ' of irradiation (or excitation). Using Lambert–Beer's law one finds [91]

$$I'_{abs} = I(0) - I(d)) = I(0) \, [1 - 10^{-E'}] \, , \tag{12-45}$$

where $I(0)$ is the intensity of radiation at the front window and E' the total absorbance (of all the compounds) at wavelength λ' in decadic units. In fluorescence measurements small concentrations are normally used in order to avoid reabsorption of the emitted light on its way from inner volume elements to the outside. Therefore, Equation (12-45) can be simplified by extension of the exponential factor into a series [91], with a factor of 2.303:

$$I'_{abs} = I(0) \cdot 2.303 \ E' \ . \tag{12-46}$$

Sometimes additional components can absorb in addition to the fluorophore. Then Equation (12-45) has to be corrected by the ratio of the specific absorption of the fluorophore to the total absorption according to [91, 154]

$$I_{absA}(\lambda) = I(0) \ \varepsilon'_{\lambda A} \ a(t) \cdot \frac{1 - 10^{-E'}}{E'} \ . \tag{12-47}$$

At small absorptions ($E' \leqslant 0.01$) this equation approximates to Equation (12-46).

Further, a time dependence of the concentration $a(t)$ of the fluorophore is taken into account. This can be caused by a photodegradation, a process which competes with the photophysical deactivation at high irradiation intensities. Therefore, this amount of light absorbed determines the fluorescence intensities in addition to the photochemical reaction yields. The photodegradation products can influence the measured intensity by their absorption of $I(0)$ in addition to unexpected fluorescence at different wavelengths.

12.3.5.2 *Fluorescence*

The fluorescence intensity I_α^F of a compound A at an observation wavelength α is proportional to [90, 154]

$$I_\alpha^F \sim I_{absA}(\lambda) \ \eta_A \ f_{\alpha A} \ \Gamma \ , \tag{12-48}$$

where η is the fluorescence quantum yield ($\leqslant 1$), $f_{\alpha A}$ the spectral distribution of the fluorescence spectrum (comparable to the absorptivity ε_λ of the absorption spectrum), and Γ a geometric factor which takes into account that fluorescence is emitted into the whole cubic sphere from each volume element. By this means, absorption of the exciting radiation and reabsorption (inner filter effects) [90] can make I_α^F dependent on the volume element. Therefore, very low concentrations are generally taken. Thus, Equations (12-46) and (12-48) simplify to

$$I_\alpha^F \sim I(0) \ \varepsilon_{\lambda A} \ a(t) \ \eta_A \ f_{\alpha A} \ . \tag{12-49}$$

Any change in the fluorescence quantum yield η_A can be used to determine the influence of the analyte concentration on the fluorescence intensity of the indicator molecule. Since η is the ratio of fluorescence and the sum of all the deactivation processes from the excited state:

$$\eta = \frac{k_f}{k_f + k_r + k_q[Q] + k_p} \ , \tag{12-50}$$

where the rate constants k symbolize the rates of fluorescence (f), of radiationless deactivation (r), of photochemical reaction (p), and of the quenching process (q); [Q] is the quencher concentration [90, 155].

12.3.5.3   *Fluorescence Quenching*

The fluorescence quantum yield can be reduced either by association of fluorophore and quencher to give a non-fluorescent molecule *(static quenching)* or by an impact process with another molecule causing radiationless deactivation of the excited state *(dynamic quenching)*. Both types of quenching are discussed in great detail in the literature [90, 156]. They show a hyperbolic relationship and can be distinguished by measurement of time dependences of the fluorescence intensity and of the lifetimes τ [157]. The ratio of quenched (η) to normal (η_0) fluorescence quantum yield is given by (with k_p neglected)

$$\frac{\eta}{\eta_0} = \frac{1}{1 + K_{sv}\,[Q]}\; . \tag{12-51}$$

With

$$\frac{\eta_0}{\eta} = \frac{k_f}{k_f + k_r}\cdot\frac{k_f + k_r + k_q\,[Q]}{k_f} = \frac{\tau_0}{\tau} \tag{12-52}$$

and

$$\frac{\eta_0}{\eta} = \frac{k_f + k_r}{k_f + k_r} + \frac{k_q\,[Q]}{k_f + k_r} = 1 + k_q\,\tau_0\,[Q] \tag{12-53}$$

a linear relationship

$$\frac{\eta_0}{\eta} - 1 = k_q\,\tau_0\,[Q] = K_{sv}\,[Q] \tag{12-54}$$

is obtained, where K_{sv} is generally called the Stern–Volmer quenching constant [91]. A linear graph of the measured left-hand side of Equation (12-54) versus quencher concentration [Q] allows the determination of K_{sv}. An extended theory of dynamic quenching takes diffusion into account also [90].

12.3.5.4   *Photostability and Suitable Fluorescent Molecules*

According to the considerations above, only relatively photostable compounds have to be chosen as fluorescence indicator molecules in fluorosensors. Typical molecules are coumarins [158], which are also known as laser dyes. Even though these molecules are known to be very photostable, the intense excitation intensity causes photodegradation reactions [159, 160]. Only a few of those have been examined kinetically [161]. Non-fluorescent photoproducts cause a reduction of the fluorescence signal by inner filter effects and a decrease in concentration of the indicator molecule.

The fluorescent photoproducts can cause additional problems, since they emit a fluorescene spectrum themselves which can overlap with that of the indicator molecule. This makes a correlation of fluorescence intensity with the concentration of the indicator molecule difficult.

Normally coumarin dyes are either oxidized at the 4-methyl group to carboxylic acid, or degrade at the 7-amino group in deoxygenated media, where no photooxidation can take place. Oxazine and xanthene dyes are also potential indicator molecules [162]. They fluoresce in the long-wavelength region, but they are not totally photostable [163]. However, in some of these cases no fluorescent photodegradation products are formed.

Another group of interesting molecules are styryl derivatives. Their photokinetic behavior has been examined recently [164]. Fortunately, the photoproducts do not fluoresce. However, most of the styryls show a fast first step of photoreaction. This has been proved to be a photoisomerization at the $C = C$ double bond. For this reason attempts have been made to stabilize this double bond by increasing the viscosity of the solvent. Another approach was to embed them in polymers [151]. The problem with these compounds is the fixing in polymers. Under certain conditions some of them lose their fluorescence ability or at least their fluorescence quantum yield is reduced. The reasons may be either covalent bonding of the double bond of the styryl compound to the polymer during copolymerization or fluorescence quenching by the matrix.

12.3.6   Dynamic Measurements

It has been mentioned that the gas concentration changes the fluorescence intensity by quenching. Further, the lifetime of the excited state is also influenced [90, 91]. Therefore, the dynamic effect of the measurement of decays can be used for the determination of concentrations of gases [165]. Especially the interaction of the paramagnetic oxygen molecule with the excited state results in a change in fluorescence intensity in addition to a shorter lifetime of this excited state.

Modern dynamic methods of measurement of the lifetimes are therefore very sensitive to the effect of the quencher concentration on this fluorescence lifetime [155]. The decay rate of fluorescence intensity after an excitation pulse (not a continuous excitation) has to be measured to obtain the lifetime. This can be done

- by photographing the oscilloscope curve,
- recording a fast signal of a photomultiplier with time,
- or better by either so-called photon counting techniques or
- phase fluorimetry.

In general, the excitation pulse has to be very short (best in the pico- or even femtosecond time domain) and the detector and its amplification electronics have to be very fast. Hence the first-mentioned conventional methods run into problems. Therefore, single photon counting and phase fluorimetric techniques should be discussed in more detail.

12.3.6.1   Single Photon Counting

A relatively weak excitation light source is repetitively pulsed. These pulses are absorbed by the sample. The molecules are excited [166]. The emission of fluorescence is a statistical process. If a large number of molecules have been excited, their statistical degradation gives an exponential decrease in intensity after excitation.

In the case of photon counting only a few molecules should be excited. Therefore, the intensity of the pulse lamp is chosen to be so low that only one molecule emits per excitation pulse. This single photon has to be observed (counted). An electronic discriminator guarantees that onyl one photon is measured at a time. Other photons are neglected after the first one has reached the detector. The emitted photons are detected by a photomultiplier and electronically counted into a so-called multi-channel analyzer.

Further, it has to be determined at which time the photon was statistically emitted after excitation. To obtain the correlation with the moment of emission, a time window is electronically moved along the time axis. Only during this time window is the detector sensitive. The related time can be found via the electronic ramp by which the time window has been advanced.

It is evident that at short time intervals after the pulse statistically more photons are emitted. Consequently, in the multi-channel analyzer the related channels are more "filled" (show more counts) than the later ones. By firing more than 10000 pulses on the sample one finds a relative good signal-to-noise ratio. The problem is that the intensity of the excitation pulse also hits the detector. Therefore, its signal overlaps with the fluorescence intensity measured by the multi-channel analyzer (Figure 12-38) [157, 158].

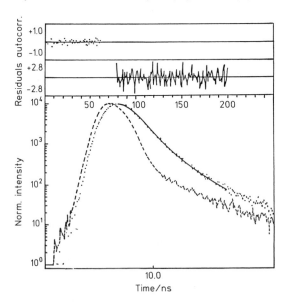

Figure 12-38.
Logarithmic plot of measured intensity versus time of a photon counting experiment. At the top the auto-correlation and the residuals between fitted decay and deconvoluted signal are shown. Dashed line: intensity of the excitation pulse; dotted line: measured fluorescence intensity.

For this reason, in a first step of evaluation deconvolution methods have to be used to separate the pulse intensity with its tailing and the beginning of the exponential decay of the fluorescence [169, 170]. In the second evaluation step, one has to try to fit this measured data set by exponential curves. The type of necessary exponential fit (one or more exponential functions) allows one to decide between one or more degradation pathways (for example, if more than one compound fluoresces). Next, lifetimes are calculated from the exponents in the evaluation formula.

12.3.6.2 Phase Fluorimetry

In phase fluorimetry, the sample of interest is illuminated by a source sinusoidally modulated at some frequency. The resulting fluorescence is also modulated. Then the phase shift of the modulated fluorescence relative to the exciting light is measured, ie, single points on the frequency response function of the sample are determined. Recording of the phase

$$\Phi\,(f) \,=\, \arctan\,(2\pi f\tau) \tag{12-55}$$

as a function of the modulation frequency f, the lifetime τ can be calculated [171].

12.3.6.3 Photokinetics

Kinetic examinations are another application of dynamic measurements [172]. These photochemical processes compete with the deactivation of the molecules by emission of radiation (fluorescence). The photochemical reaction taking place in the chromophore of an indicator molecule is influenced by the surroundings [173]. The measured rate of the reaction can be correlated with the increasing concentration of the product interfering with fluorescence. It is advisable to use spectral information to evaluate kinetic measurements. Parallel observations at many wavelengths result in increased information. Spectra are repetitively recorded during the advancement of the photoreaction and plotted "three-dimensionally" with the time axis to the back of the graph. The entirety of consecutive spectra is called a reaction spectrum [172, 174]. Such a spectrum is shown in Figure 12-39 for the thermally reversible photoreaction of a dihydroindolizine derivative recorded by a photodiode array.

In Section 12.3.1.6 it had been mentioned that a light-induced photoreaction causes physical and chemical changes of the compounds to be monitored optically. Therefore, it will reduce the dynamic range of a fluorophore and has to be avoided in sensor systems. Anyhow, in the past photochemical reactions have been used successfully as dosimetric systems. They allow either the radiation of non-visible light sources to be visualized by a color change or the number of photons to be counted quantitatively. They show some advantages in comparison with physical dosimeters. Nevertheless, it is preferable to remain with the usual terminology for such systems, which are called chemical actinometers [175], instead of renaming them as photochemical sensors.

In any case, photoreactions have to proceed in a suitable time domain in order to obtain a useful optical detection. The same is valid for any dynamic process, eg, the observation of the change in spectral fluorescence intensity by increasing the quencher concentration. This means that within 1 min the photometric or fluorometric change caused by the reaction or the quenching process should be so large that a good evaluation is possible. As a consequence, the requirements on the detection systems are high in order to achieve good spectra in this time domain.

Nowadays diode arrays allow repetitive scanning of a sample within milliseconds [116, 176]. Repetition rates are up to 75 Hz [136].

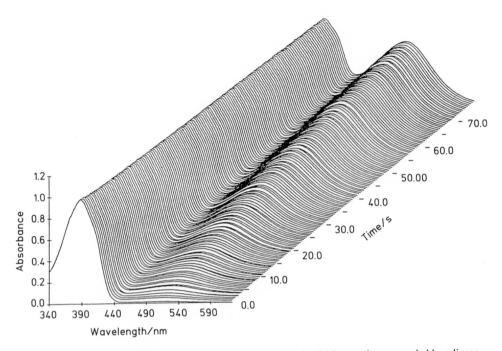

Figure 12-39. Reaction spectrum of dihydroindolizine (HD 137) in dichloromethane recorded by a linear diode array every 20 ms. The increase in absorbance at longer wavelength shows the advancement of the photoreaction. Only every fifth spectrum is plotted for clarity.

12.3.7   Applications

12.3.7.1   Optical Sensors for Monitoring Chemical Processes

Recently the process of microstructuring at either wafers or circuit boards has become extremely interesting. By this means the control of the radiation set-ups with respect to a homogeneous radiant flux and absolute intensity and also the process control of the photochemical reaction in the photoresists have been further studied. These photoresists are photochromic systems in polymers, coated on the surface of the circuit boards or the wafers. The photoreaction causes a change in the physical and/or chemical properties at areas of radiation. Further processing steps depend on the quality of the previous photoprocess [177].

The optical sensor arrangement demonstrated in Section 12.3.2.2 (see Figure 12-25) has turned out to be very useful in these examinations [178]. In Figure 12-40 the spectral change of such a photoresist during radiation is recorded as a reaction spectrum resolved with time. The combination of modern microprocessor equipment and the photodiode array allows a repetition rate of about 75 Hz [136, 178]. The combination of this photodiode array with fiber optics allows remote sensing in process control and gives the possibility of measuring with very good local spatial resolution.

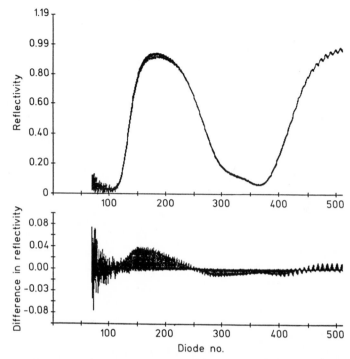

Figure 12-40. Top: reaction spectrum of a photoresist during the photoprocess. The spectral changes are small. Therefore the spectral differences of the individual spectra from the first spectrum are plotted at the bottom. Reflectivity or change in reflectivity are plotted against the number of the diode in the detection system. Diode 100 corresponds to 320 nm and diode 500 to 700 nm.

12.3.7.2 Refractive Index in Fluid Systems

This optical sensor device with remote sensing is used to monitor the optical properties of a flowing stream system in another application. High-performance liquid chromatography (HPLC) has become widely used in modern analytical chemistry. Generally UV–Vis detectors are used. However, various substances that are of interest in biological systems, such as amino acids, nucleotides, and saccharides, only absorb in a short wavelength range close to 200 nm [179]. Since photodiode arrays lack sensitivity in this wavelength range, frequently changes in refractive index are used as a detection principle. In addition, in microbore techniques [180, 181] transmission spectroscopy lacks sensitivity.

Under these conditions the path length in the cells has to be very small, since the cells themselves must have very small volumes. Otherwise the dead volume spoils the microbore separation [181]. On the other hand, measurements of refractive index can be made on biological materials according to the theoretical considerations in Section 12.3.1 even outside absorption bands, which means at longer wavelengths. By use of an interferometric method, as presented in Section 12.3.4.3, polychromatic radiation, multiple reflection, and remote sensing via fiber optics accompanied by subsequent recording of the obtained interferogram on the photodiode array allow the calculation of the dispersion curve [153]. Such a "corrected" interferogram is shown in Figure 12-41. The refractive index can be calculated for a known path length according to Equation (12-44) by use of the wavelength difference between prox-

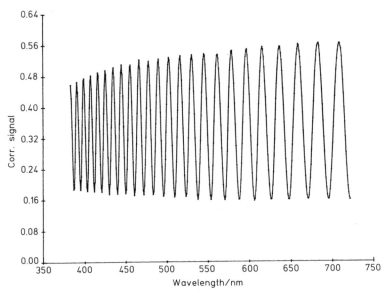

Figure 12-41. Corrected interference pattern of water plotted as signal amplitude versus wavelength. Correction with respect to intensity distribution of the light source and sensitivity of the photodiodes.

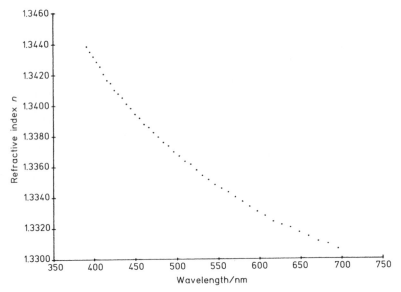

Figure 12-42. Calculated curve of dispersion of water by use of the pattern in Figure 12-41. Normal part of the curve of dispersion outside of bands of absorption.

imate maxima (minima). Since the information at two maxima has to be used, the optical path length obtained can only be approximated. Therefore, the equation

$$2nd = m\lambda$$

(12-56)

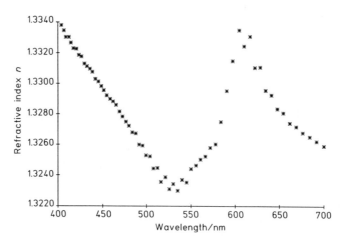

Figure 12-43. Calculated curve of dispersion of crystal violet in the wavelength range of absorption (anomalous part).

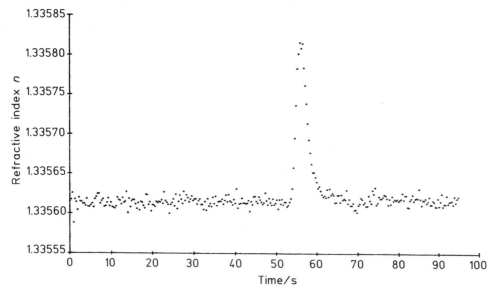

Figure 12-44. Result of an HPLC experiment with a 1-mm diameter microcolumn. The chromatogram shows the limit of detection of 5.0 µg of glucose injected in a 5-µl volume.

is used to determine in a first step the order *m* of the selected maximum (minimum). One knows by theory that *m* has to be an integer (minima) or a half-integer (maxima). Therefore, *m* is chosen accordingly. Then the correct refractive index can be calculated at the wavelength of each extremum (the path length *d* is independent of wavelength). The use of the whole interferogram allows one to calculate the refractive index as a function of wavelength. The result is shown in Figure 12-42. If absorption takes place in addition to dispersion in the measured wavelength range, one obtains an interferogram in which the anomalous dispersion becomes evident (Figure 12-43) [182].

By observing only one wavelength, the evaluation rate can be increased. An application to HPLC is shown in Figure 12-44 [136]. At present the method used is less sensitive than thermal crossed-beam arrangements [183, 184], but the set-up is less expensive and simpler. In addition to application in HPLC, this optical sensor system can be used for remote sensing in any flowing stream set-ups.

12.3.7.3 Biological Application

Biosensors using plant materials have been studied extensively [185]. Recent work has shown that plant materials (bananas, corn) are especially well suited for biocatalytic purposes. They specifically react in unique chemical pathways. They are usually combined with an electrochemical sensing element.

This principle can also be used in combination with optical sensing elements. For this reason, the optical equipment mentioned in Section 12.3.2.2 is used in combination with a He–Ne laser to excite fluorescence in a leaf and to measure it. The spectral measurement allows one to distinguish between different fluorescence spectra of the photosensitive units in the leaf. These spectra are influenced by the physiological state of the leaf caused either by light stress or by any change in the conditions of the natural surroundings [186].

12.3.7.4 Interferometric Gas and Liquid Sensors

Interferometric principles have frequently been used in optical sensing. Coded optical fibers in a Mach–Zehnder interferometer allow the detection of hydrogen in a concentration range from 20 ppb to 2% in 1 atm of nitrogen. When the coded fibers are exposed to hydrogen, a hydride is formed with an extended lattice constant which stretches the optical fiber [187]. This method can also be used in in-situ measurements of strain during electrodeposition of thin films [188]. The use of the air cavity of a Fabry–Perot interferometer as an optical sensing system has been described [189].

Temperature can be measured by a combined interferometric and polarimetric fiber-optic method. It uses a bifringent monomode fiber as a sensing element. The optical arrangement comprises a Mach–Zehnder and a Michelson interferometer set-up in tandem [190].

In Figure 12-35 a Mach-Zehnder-like arrangement was shown. One of the two waveguide arms of an integrated-optics device is covered with a thin polymer film (polysiloxane, thickness 1 μm) [142]. A gas stream of nitrogen passing over this film leads to a relative intensity for the superposition of the light waves guided in the fibers. If now toluene is injected in the gas stream for 1.5 min, this relative intensity varies with the relative gas concentration. The resulting intensity of the superimposed waves versus the injection time is shown in Figure 12-45. As in gas chromatography one peak is expected. Its height should correspond to the actual toluene concentration. Surprisingly two "negative" peaks are obtained. This fact can be explained by the theory of interferometry in Mach-Zehnder systems. The amplitude of the superposed waves depends on wavelength, refractive index of the polymer and "length" of the polymer window [140, 142, 191]. Using wavelength and length of the polymer window as parameters, simulation of the interferogram by applying wave-optics equations shows the dependence of intensity on refractive index, as plotted in Figure 12-46. Starting with the value of the refractive index of the pure polymer (start point), growing concentration of injected toluene leads to an increase of refractive index and next to a decrease of interference intensity (Figure 12-46). Continued gas injection yields a positive intensity peak. As soon as the maximal toluene concentration is reached (end point), the curve in the interferogram is followed backward to the start point.

In all these cases, interferometry at one wavelength has been used. To get more information about the sample, spectral interferometry must be applied [192]. This method allows the online measurement of changes in optical path length down to 1 Å (100 pm) [182]. Differently from monowavelength measurements, spectral observation also yields information about adsorbing substances. For example, the change in thickness of polymer (polysiloxane) films (caused by intruding of organic molecules passing over in a liquid or gas stream) can be observed with high repetition rates (75 Hz) [136].

The optical arrangement used is given in Figure 12-47 [193, 194, 197]. Within seconds, the polymer film (thickness of approximately 3 µm) is changed by the liquid or gas penetrating the polymer surface. As a result the optical path length increases. A 5% concentration of dichloromethane, ether, or toluene leads to a path-length change of approx. 300 nm. The detection limit of this integrated-optics device is 100 pm. That means, even low gas concentrations can be measured. The characteristics of the sensor depend on polymer constitution

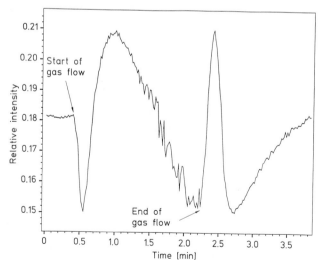

Figure 12-45. Relative intensity of a signal measured behind the Mach-Zehnder interferometer chip as a superposition of the propagating waves in the two arms.

and structure and on gas or liquid type. Figure 12-48 shows the changes in optical path length (expressed in relative sensitivity changes) for two polymers and for a large number of or organic solvents [194], 196]. In Figure 12-49 the response (increase of polymer thickness) to increasing concentration of ethanol in water is given [196].

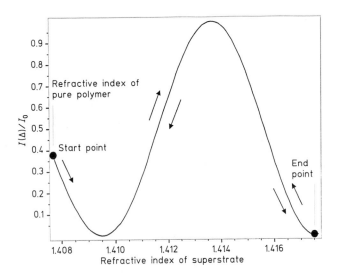

Figure 12-46. Calculated relative intensity in dependence on refractive index (for fixed wavelength and polymer window length). The sinusoidal form of the interferogram explains the appearance of a negative peak with increasing toluene concentration.

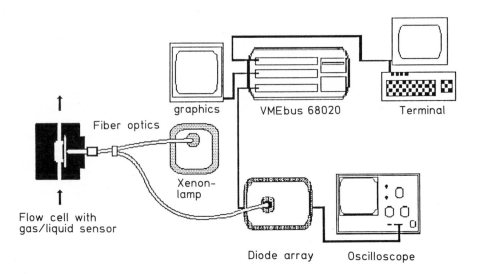

Figure 12-47. Experimental arrangement of an interferometric optical sensor in a flow cell for either gases or liquids.

A recent application of interferometric sensors is the investigation of antigen-antibody interaction at the surface of a polymer. According to Equation 3-10 additional "adsorption" influences the signal by two means. First the optical path length is affected. Second the reflectivity at the interface is changed. Both effects result in a change of the intensity of the multiply reflected beam. In other words, a signal is gained which correlates to the antigen-antibody interaction so that the device can be considered as a biosensor.

A typical ELIZA protocol is given in Figure 12-50. An antigen layer is coupled to a 1-μm polymer support, free sites are blocked, and the interferogram distinguishes between

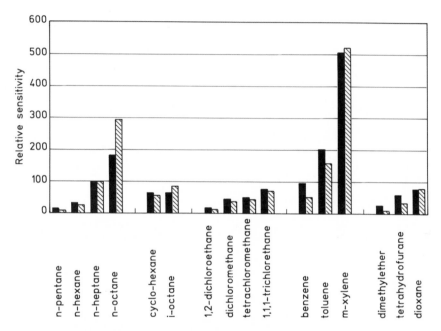

Figure 12-48. Relative response properties of two different siloxane polymers (marked by ■ and □) to some gasous organic solvents.

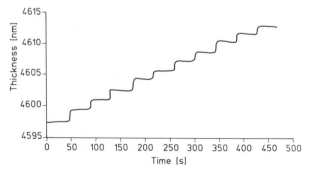

Figure 12-49. Determined increase in siloxane thickness by passing water containing ethanol. Each step of the curve correlates to 0.5% ethanol, starting at pure water at the left.

unspecific and specific antibody interaction. The different steps can also be controlled by ellipsometric measurements, a technique discussed in Section 12.3.3.2 [124]. The result is shown in Figure 12-51 [199].

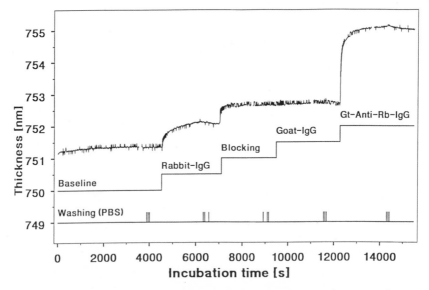

Figure 12-50. Interferometric observation of antigen-antibody interaction on a polystyrene/quartz substrate. Antigen: rabbit-IgG, blocking: ovalbumine; unspecific antibody (goat-IgG) shows no effect, specific goat-anti-rabbit-IgG yields label-free measurement of interaction via interferometry [199].

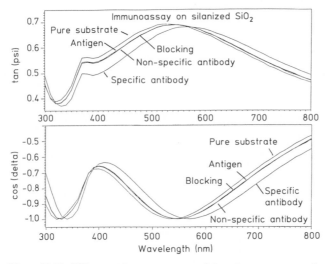

Figure 12-51. Ellipsometric measurement of the subsequent steps of a solid phase immunoassay (same procedure as in Figure 12-50). tan (psi) gives the ratio of the two reflected polarized partial beams, cos (delta) represents their relative phase shift by interaction with the layers [199].

12.4 References

[1] Chester, A. N., Martellucci, S., Verga-Scheggi, A. M. (eds.), *Optical Fiber Sensors;* Dordrecht: Nijhoof 1987.

[2] Wolfbeis, O. S., "Optic and Fiber Fluorosensors in Analytical and Clinical Chemistry" in: *Molecular Luminescence Spectrometry: Methods and Applications,* Vol. 2, Schulman, S. G. (ed.); New York: Wiley, 1988, Chap. 3, pp. 129–281.

[3] Wolfbeis, O. S. (ed.), *Fiber Optic Chemical Sensors and Biosensors;* Boca Raton, FL: CRC Press 1991.

[4] Wolfbeis, O. S., "Fiber Optic Sensors in Bioprocess Control" in: *Sensors in Bioprocess Control,* Twork, J. V., Yacynych, A. M. (eds.); New York: Dekker, 1990.

[5] Harmer, A. L., Narayanaswamy, R., "Spectroscopic and Fiber Optic Transducers", in: *Chemical Sensors,* Edmonds, T. E. (ed.), London: Blackie, 1988, Chap. 3, pp. 276–294.

[6] Peterson, J. I., *Encyclopedia of Medical Devices and Instrumentation;* New York: Wiley, 1987.

[7] Giallorenzi, T. G., Bucarro, J. A., Dandridge, A., Sigel, G. H., Cole, J. H., Rashleigh, S. C., Priest, R. G., *IEEE J. Quant. Electron.* **QE-18** (1982) 626.

[8] Hirschfeld, T., Dayton, T., Milanovich, F., Klainer, S., *Opt. Eng.* **22** (1983) 527.

[9] Seitz, W. R., *CRC Crit. Rev. Anal. Chem.* **19** (1988) 135.

[10] Lübbers, D. W., Opitz, N., *Sensors Actuators* **4** (1983) 641.

[11] Harmer, A. L., *Proc. Electrochem. Soc.* **87** (1987) 409.

[12] Peterson, J. I., Vurek, G. G., *Science* **224** (1984) 123.

[13] Hirschfeld, T., Callis, J. B., Kowalski, B. R., *Science* **226** (1984) 312.

[14] Boisdé, G., Perez, J. J., *C. R. Acad. Sci. Ser. Gen.* **5** (1988) 303.

[15] Arnold, M. A., (guest ed.), *Talanta* **35**, No. 2 (1988).

[16] Narayanaswamy, R., Sevilla, F., *J. Phys. E.* **21** (1988) 10.

[17] Opitz, N., Lübbers, D. W., *Eur. J. Physiol.* **355** (1975) R 120.

[18] Lübbers, D. W., Opitz, N., *Z. Naturforsch., Teil C* **30** (1975) 532.

[19] Hirschfeld, T., Deaton, T., Milanovich, F., Klainer, S., "The Feasibility of Using Fiber Optics for Monitoring Groundwater Contaminants", *EPA Report AD-89-F-2A 074* Environmental Protection Agency, 1983.

[20] O'Neil, R. A., Buja-Bijunas, L., Rayner, D. N., *Appl. Opt.* **19** (1980) 863, and Refs. 1–8 cited therein.

[21] Kapany, N. S., *Fiber Optics: Principles and Applications;* New York: Academic, 1967.

[22] Hopkins, H. H., Kapany, N. S., *Nature* (London) **173** (1954) 39–41.

[23] Capellaro, D. F., Kapany, N. S., Long, C., *Nature* (London) **191** (1961) 927–928.

[24] Gamble, W. J., Hugenholtz, P. G., Polanyi, M. L., Monroe, R. G., Nadas, A. S., *Circulation* **31** (1965) 328–343.

[25] Allan, W. B., *Fiber Optics: Theory and Practice;* New York: Plenum, 1973.

[26] Brown, R. G., *Appl. Opt.* **6** (1967) 1269–1270.

[27] Borman, S. A., *Anal. Chem.* **53** (1981) 1616A–1618A.

[28] Milanovich, F. P., Hirschfeld, T., *Adv. Instrum.* **38** (1983) 407–418.

[29] Seitz, W. R., *Anal. Chem.* **56** (1984) 16A–34A.

[30] Daly, J. C., *Fiber Optics;* Boca Raton, FL: CRC, 1984.

[31] Halley, P., *Les Systèmes à Fibres Optiques;* Paris: Eyrolles, 1985.

[32] Senior, J., *Optical Fiber Communications: Principles and Practice;* Englewood Cliffs, N. J.: Prentice-Hall, 1984.

[33] Howes, M. J., Morgan, D. V., *Optical Fiber Communications;* New York: Wiley, 1980.

[34] Geisler, J. A., Beaven, G. L., Boutruche, J. P., *Optical Fibers,* EPO Applied Technology Series, Vol. S; Oxford: Pergamon, 1986.

[35] Chabay, I., *Anal. Chem.* **54** (1982) 1071A–1080A.

[36] Whitten, W. B., *Appl. Spectrosc. Rev.* **19** (1983) 325–362.

[37] Payne, D. N., *Proc. SPIE Int. Soc. Opt. Eng.* **514** (1984) 353–360.

[38] Fuwa, K., Wei, L., Fujikara, K., *Anal. Chem.* **56** (1984) 1640–1644.

[39] Plaza, P., Nguyen, Quy Dao, Jouan, M., Fevrier, H., Saisse, H., *Analusis* **15** (1987) 504–507.

[40] Peterson, J. I., Goldstein, S. R., *US Pat. 4, 194, 877,* 1980.

[41] Boisdé, G., Rougeault, S., Perez, J. J., in: *Proceedings of Opto 86, ESI Publ.;* Paris: Masson, 1986, pp. 71–82; and *Microfilm CEA-Conf.,* 8572, 1986.

[42] Friebele, E. J., Long, K. J., Askins, C. G., Gingerich, M. E., Marrone, M. J., Griscom, D. L., *Proc. SPIE Int. Soc. Opt. Eng.* **541** (1985) 70–88.

[43] Boisdé, G., Bonnejean, C., Boucher, D., Neuman, V., Perez, J. J., Wurier, B., *Proc. SPIE Int. Soc. Opt. Eng.* **404** (1983) 17–24.

[44] Poulain, M., "New Glasses for Optical Fibers", *Endeavour, New Ser.* **11** (1987) 25–28.

[45] Ray, C. R., France, P. W., Carter, S. F., Moore, M. W., Williams, J. R., *Proc. SPIE Int. Soc. Opt. Eng.* **799** (1987) 94–100.

[46] Abgrall, A., Poulain, M., Boisdé, G., Cardin, V., Maze, G., *Proc. SPIE Int. Soc. Opt. Eng.* **618** (1986) 63–69.

[47] Le Sergent, C., *Proc. SPIE Int. Soc. Opt. Eng.* **799** (1987) 18–24.

[48] Dakin, J. P., King, A. J., *IEE Proc.* **131** (1984) 273–275.

[49] Boisdé, G., Carvalho, W., Dumas, P., Neuman, V., *Proc. SPIE Int. Soc. Opt. Eng.* **506** (1984) 196–201.

[50] Sutherland, R. M., Dähne, C., Place, J. F., Ringrose, A. S., *Clin. Chem.* **30** (1984) 1533–1538.

[51] Ives, J. T., Reichert, W. R., Lin, J. N., Reinecke, D., Suci, P. A., Van Wagenen, R. A., Newby, K., Herron, J., Dryden, P., Andrade, J. D., "Total Internal Reflection Fluorescence Surface Sensors" in: *Optical Fiber Sensors,* Chester, A. N., Martellucci, S., Verga-Scheggi, A. M. (eds.); Dordrecht: Nijhoff, 1987, pp. 391–397.

[52] Harrick, N. J., *Internal Reflection Spectroscopy;* New York: Wiley, 1967.

[53] Love, W. F., Slovacek, R. E., in: *Proceedings of OFS 86;* Tokyo: Instiute of Electronics and Communication Engineers of Japan, 1986, pp. 143–145.

[54] Smith, A. M., *Proc. SPIE Int. Soc. Opt. Eng.* **798** (1987) 206–213.

[55] Lew, A., Depeursinge, C., Cochet, F., Berthou, H., Parriaux, O., *Proc. SPIE Int. Soc. Opt. Eng.* **514** (1984) 71–74.

[56] Sutherland, R. M., Dähne, C., "IRS devices for optical immuno assays", in: *Biosensors: Fundamentals and Applications,* Turner, P. F., Karube, I., Wilson, G. S. (eds.); Oxford: Oxford University Press, 1987, Chap. 33, pp. 655–678.

[57] Fluitman, J., Popma, Th., *Sensors Actutators* **10** (1986) 25–46.

[58] Liedberg, B., Nylander, C., Lundstrom, I., *Sensors Actuators* **4** (1983) 299–304.

[59] Kreuwel, H. J. M., Lambeck, P. V., Gent, J. V., Popma, Th. J. A., *Proc. SPIE Int. Soc. Opt. Eng.* **798** (1987) 218–224.

[60] Boisdé, G., Kirsch, B., Mauchien, P., Rougeault, S., in: *Proceedings of Opto. 88, ESI Publ.;* Paris: Masson, 1988, pp. 294–299, and *Microfilm CEA-Conf.,* 9608, 1988.

[61] Malstrom, R. A., *Report DP-MS-85-76,* 1985, E. I. Du Pont de Nemours, Savannah River Laboratory.

[62] Chan, K., Ito, H., Inaba, H., Appl. Opt. **23** (1984) 3415–3420.

[63] Boisdé, G., Boissier, A., *US Pat. 4, 188, 126,* and *4, 225, 232,* 1980.

[64] Posch, H. E., Wolfbeis, O. S., Pusterhofer, J., *Talanta* **35** (1988) 89–94.

[65] Giulani, J. F., Wohltjen, H., Jarvis, N. L., *Opt. Lett.* **8** (1983) 54–56.

[66] Butler, M. A., *Appl. Phys. Lett.* **45** (1984) 1007–1009.

[67] Newman, D. H., Ritchie, S., *IEE Proc.* **133** (1986) 213–229.

[68] Kist, R., "Sources and Detectors for Fiber Optic Sensors" in: *Optical Fiber Sensors,* Chester, A. N., Martellucci, S., Verga-Scheggi, A. M. (eds.); Dordrecht: Nijhoff, 1987, pp. 267–298.

[69] Smardzewski, R. R., *Talanta* **35** (1988) 95–101.

[70] Boisdé, G., Chevalier, G., Perez, J. J., *Eur. Pat. 15 170,* 1983.

[71] Nakanishi, K., Imasaka, T., Ishibashi, N., *Anal. Chem.* **59** (1987) 1550–1554.

[72] Fuh, Ming-Ren, S., Burgess, L. W., *Anal. Chem.* **59** (1987) 1780–1783.

[73] Atkins, P. W., *Physical Chemistry,* 3rd ed.; Oxford: Oxford University Press, 1986.

[74] Gauglitz, G., et al., "Dynamische Untersuchungen photochemischer Prozeßschritte an dünnen Schichten von Photoresisten mit einem neuen optischen Sensorsystem", in: *Proceedings of Sensor 88;* Wunstorf: ACS Organisations GmbH, 1988.

[75] Gauglitz, G., *UV/Vis-Spektroskopie,* Weinheim: VCH, in preparation.

[76] *DIN-Norm 5030, Part 2:* Berlin: Beuth-Verlag, 1982.

[77] Bergmann, L., Schäfer, C., *Experimentalphysik, Vol. III, Optik;* Berlin: de Gruyter, 1978.

[78] Born, M., Wolf, E., *Principles of Optics;* New York: Pergamon, 1980.
[79] Hecht, E., Zajac, A., *Optics;* Reading, MA: Addison-Wesley, 1974.
[80] Gauglitz, G., *Praktische Spektroskopie im UV/Vis;* Tübingen: Attempto, 1983.
[81] Stuart, H. A., *Molekülstruktur;* Heidelberg: Springer, 1967.
[82] Böttcher, C. J. F., *Theory of Electric Polarization;* Amsterdam: Elsevier, 1952.
[83] Greschner, G. S., *Maxwell-Gleichungen, Vol. 1: Allgemeine Grundlagen;* Heidelberg: Hüthig und Wepf, 1981.
[84] Greschner, G. S., *Maxwell-Gleichungen, Vol. 2: Lichtstreuung an Molekülen;* Heidelberg: Hüthig und Wepf, 1981.
[85] Kerker, M., *The Scattering of Light;* New York: Academic, 1969.
[86] Stacey, K. A., *Light-Scattering in Physical Chemistry;* London: Butterworths, 1956.
[87] Van de Hulst, *Light Scattering by Small Particles;* New York: Dover, 1957.
[88] Chong, C. S., Colbow, K., *Biochim. Biophys. Acta* **436** (1976) 260–282.
[89] Bernstein, R. B., *Molecular Reaction Dynamics;* New York: Oxford University Press, 1974.
[90] Förster, Th., *Fluoreszenz organischer Verbindungen;* Göttingen: Vandenhoek, 1951.
[91] Parker, C. A., *Photoluminescence of Solutions;* Amsterdam: Elsevier, 1968.
[92] Raman, C. V., Krishnan, K. S., *Nature* (London) **121** (1928) 501.
[93] Freitag, C., *Infrarot- und Ramanspektroskopie, Handbuch der industriellen Meßtechnik;* Vulkan, pp. 731 ff.
[94] Kubelka, P., *J. Opt. Soc. Amr.* **38** (1948) 448.
[95] Kortüm, G., *Reflexionsspektroskopie;* Berlin: Springer, 1969.
[96] Debe, M. K., "Optical Probes of Organic Thin Films: Photons-in and Photons-out", *Prog. Surf. Sc.* **24** (1987) 1.
[97] Aizawa, M., *Photochemical Biosensors, International Workshop on Biosensors, June 1987;* Braunschweig: GBF, 1987, p. 217.
[98] Weber, L. G., *Appl. Spectrosc. Rev.* **21** (1985) 1.
[99] Goddu, R. F., *Anal. Chem.* **32** (1960) 140.
[100] Watson, C. A., *Anal. Chem.* **49** (1977) 835 A.
[101] Perkampus, H.-H., *UV–Vis Spektroskopie und ihre Anwendungen;* Heidelberg: Springer, 1986.
[102] Burgess, C., Knowles, A. (eds.), *Standards in Absorption Spectrometry: Techniques in Visible and Ultraviolet Spectrometry,* Vol. 1; London: Chapman and Hall, 1981.
[103] Kortüm, G., *Kolorimetrie, Photometrie und Spektrometrie;* Berlin: Springer, 1962.
[104] Kaye, W., Barber, D., Marasco, R., *Anal. Chem.* **52** (1980) 437 A.
[105] Schäfer, V., Heinrich, G., in: *Ultraviolette Strahlen,* Kiefer, J. (ed.); Berlin: de Gruyter, 1977, p. 47–171.
[106] Weber, H., Herziger, G., *Laser;* Weinheim: Physik-Verlag, 1972.
[107] Schäfer, F. P., in: *Dye Lasers: Topics in Applied Physics,* Vol. 1, Schäfer, F. P. (ed); Berlin: Springer, 1973, p. 1–90.
[108] Gauglitz, G., Goes, R., Stooss, W., Raue, R., *Z. Naturforsch. Teil A,* **40** (1985) 317.
[109] Rabek, J. F., *Experimental Methods in Photochemistry and Photophysics, Part 2;* New York: Wiley, 1982.
[110] Botez, D., *Laser Focus/Electro-Opt.* 68–79, 1987.
[111] Holmes, L., *Laser Focus/Electro-Opt.,* 90–94, 1986.
[112] Gordon, E. I., *Lasers Appl.,* 43–44, 1987.
[113] Fye, D. M., *Lasers Appl.,* 47–49, 1987.
[114] *Photomultiplier Tubes;* Hamamatsu Catalog, 1985.
[115] Svehla, G., *Comprehensive Analytical Chemistry,* Vol. XIX; Amsterdam: Elsevier, 1986.
[116] Talmi, Y., *Anal. Chem.* **47** (1975) 699 A.
[117] Marescaux, F., Hewitson, N., *Electro Opt.* **18** (1988) 15.
[118] *Lasers and Optronics, Buying Guide;* Dover, December, 1987.
[119] *Optoelectronics, Designer's Catalog;* Avondale, PA: Hewlett-Packard, 1988–89.
[120] Harrick, N. J., *J. Opt. Soc. Am.* **55** (1965) 851.
[121] Kane, J., Osterberg, H., *J. Opt. Soc. Am.* **54** (1964) 347.
[122] Hass, G., *J. Opt. Soc. Am.* **45** (1955) 945.
[123] Greenler, R. G., *J. Chem. Phys.* **44** (1966) 310.
[124] Azzam, R. M. A., Bashara, N. M., *Ellipsometry and Polarized Light;* North-Holland, 1977.

[125] Hauge, P. S., *Surf. Sci.* **96** (1980) 108.
[126] Hall, E. E., *Phys. Rev.* **15** (1902) 73.
[127] Goos, F., Lindberg-Hänchen, H., *Ann. Phys.* **5** (1949) 251.
[128] Schaefer, C., Gross, G., *Ann. Phys.* **30** (1937) 245.
[129] Court, I. N., von Willisen, F. K., *Appl. Opt.* **3** (1964) 719.
[130] *ASTM Nomenclature for Internal Reflection Spectroscopy, ASTM Book of Standards E131–66T, Part 31;* Philadelphia: American Society for Testing and Materials, 1967.
[131] Franz, W., *Z. Naturforsch. Teil A,* **13** (1958) 484.
[132] Keldys, L. V., Vavilov, V., Bricin, K. I., "The Influence of Strong Electric Field on the Optical Properties of Semiconductors", in: *Proceedings of International Conference on Semiconductor Physics, Prague, 1960;* 1961, p. 824.
[133] Hansen, W. N., *Spectrochim. Acta,* **21** (1965) 815.
[134] Hermann, T. S., *Anal. Biochem.* **12** (1965) 406.
[135] Okta, K., Iwamoto, R., *Anal. Chem.* **57** (1985) 2491.
[136] Gauglitz, G., Krause-Bonte, J., Schlemmer, H., Matthes, A., *Anal. Chem.* **60** (1988) 2609.
[137] Geick, R., *Chem. Labor Betr.* **23** (1972) 193, 250, 300.
[138] Genzel, L., *Fresenius' Z. Anal. Chem.,* **273** (1975) 391.
[139] Kist, R., Knoll, G., in: *Proceedings of AMA Seminar: Faser- und Integriert-optische Sensoren, Heidelberg, November, 1988;* Wunstorf: ACS Organisations GmbH, 1988, p. 157.
[140] Ross, L., *Glastechn. Ber.* **62** (1989) 285.
[141] Hollenbach, U., et al. *SPIE vol. 1014 Micro-Optics* (1988) 77.
[142] Ingenhoff, J., *Diploma Thesis,* Tübingen, 1990.
[143] Landolt-Börnstein, *Eigenschaften der Materie in ihren Aggregatzuständen, Part 8, Optische Konstanten,* Vol. 2; Berlin: Springer, 1962.
[144] Wolfbeis, O. S., *Fresenius' Z. Anal. Chem.* **325** (1986) 387.
[145] Wagner, E., Kist, R., in: *Proceedings of AMA Seminar: Faser- und Integriert-optische Sensoren, Heidelberg, November, 1988;* Wunstorf: ACS Organisations GmbH, p. 9.
[146] presented by LOT, Waghäusel, at Sensor 91' exhibition at Nürnberg, June 1991.
[147] Jestel, D., Baus, A., Vogel, E., *Electronics Lett.* **26** (15) (1988) 1144.
[148] Lukosz, W., Tiefenthaler, K., *Opt. Lett.* **8** (1983) 537.
[149] Lukosz, W., Tiefenthaler, K., *Sens. Actnat.* **15** (1988) 273.
[150] Burnstein, E., Chen, W. P., Hartstein, A., *J. Vac. Sci. Technol.* **11** (1974) 1004.
[151] Hickel, W., Rothenhäusler, B., Knoll, W., *J. Appl. Phys.* **66** (1989), 4832.
[152] BIA-Core (Biospecific Interaction Analysis) Pharmacia Biosenor, Uppsala, Sweden.
[153] Gauglitz, G., in: *Software-Entwicklung in der Chemie 1, Proceedings des Workshops Computer in der Chemie, Hochfilzen/Tirol, November, 1986,* Gasteiger, J. (ed.); Heidelberg: Springer, 1987, p. 165–200.
[154] Gauglitz, G., *Z. Phys. Chem., N. F.,* **88** (1974) 193.
[155] Guilbault, G. G., *Practical Fluorescence Theory, Methods/Techniques;* New York: Dekker, 1973.
[156] Wehry, E. L. (ed.), *Modern Fluorescence Spectroscopy* Vol. 3; New York: Plenum, 1981.
[157] Förster, Th., *Z. Elektrochem.* **53** (1949) 93.
[158] Wolfbeis, O. S., *Z. Naturforsch. Teil A* **32** (1979) 1065.
[159] Weber, J., *Phys. Lett. A* **57** (1976) 465.
[160] Drexhage, K. H., in: *Dye Lasers; Topics in Applied Physics* Vol. 1, Schäfer, F. P. (ed.); Berlin: Springer, 1973, p. 144–193.
[161] Stooss, W., *Dissertation,* Tübingen, 1987.
[162] Maeda, M., *Laser Dyes;* Tokyo: Academic, 1984.
[163] Lorch, A., *Diploma Thesis,* Tübingen, 1976.
[164] Gauglitz, G., Goes, R., Stooss, W., Raue, R., *Z. Naturforsch. Teil A* **40** (1985) 317.
[165] Lippitsch, M. E., Pusterhofer, J., Leiner, M. J. P., Wolfbeis, O. S., *Anal. Chim. Acta* **205** (1988) 1.
[166] Wehry, E. L. (ed.), *Modern Fluorescence Spectroscopy* Vol. 4, New York: Plenum, 1981, p. 9.
[167] Knight, A. E. W., Selinger, B. K., *Aust. J. Chem.* **26** (1973) 1.
[168] Gauglitz, G., Krabichler, G., Oelkrug, D., Stooss, W., in preparation.
[169] O'Connor, D. V., Ware, W. R., Andre, J. C., *J. Phys. Chem.* **83** (1979) 1333.
[170] O'Connor, D. V., Phillips, D., *Time-Correlated Single Photon Counting;* New York: Academic, 1984.

[171] Knight, A. E. W., Selinger, B. K., *Aust. J. Chem.* **26** (1973) 39.
[172] Gauglitz, G., Mauser, H., *Principles and Applications of Photokinetics;* Amsterdam: Elsevier, in press.
[173] Bär, R., Benz, R., Dürr, H., Gauglitz, G., Polster, J., Spang, P., *Z. Naturforsch. Teil A* **39** (1984) 662.
[174] Gauglitz, G., *GIT Fachz. Lab.* **26** (1982) 205; **26** (1982) 597; **29** (1982) 186.
[175] Gauglitz, G., Hubig, S., *Z. Phys. Chem., N. F.* **139** (1984) 237.
[176] Mächler, M., Sachse, R., Schlemmer, H., *Ger. Offen. DE 3414260,* Carl Zeiss, Oberkochen.
[177] Gauglitz, G., *Photochemie in der Leiterplattenfertigung,* Gomaringen: Simanowski, 1987.
[178] Gauglitz, G., Krause-Bonte, J., *Fresenius' Z. Anal. Chem.* **333** (1989) 518.
[179] Simpson, R. C., Brown, P. R., *J. Chromatogr.* **400** (1987) 297.
[180] Snyder, L. R., Kirkland, J. J., *Introduction to Modern Liquid Chromatography* 2nd ed.; New York: Wiley, 1979.
[181] Engelhardt, H., *Practice of High Performance Liquid Chromatography* 1st ed.; Heidelberg: Springer, 1986.
[182] Krause-Bonte, J., *Dissertation,* Tübingen, 1989.
[183] Bornhop, D. J., Nolan, T. G., Dovichi, N. J., *J. Chromatogr.* **384** (1987) 181.
[184] Dovichi, N. J., Nolan, T. G., *Anal. Chem.* **59** (1987) 2803.
[185] Rechnitz, G. A., Biosensors: Challenges for the 1990s in: *Biosensors, International Workshop, 1987,* Schmidt, R. D. (ed.), GBF Monographs, Vol. 10; Wallenheim: VCH, 1987, p. 3.
[186] Bigus, H. J., Krause-Bonte, J., Stransky, H., Gauglitz, G., Hager, A., in preparation.
[187] Butler, M. A., Ginley, D. S., *J. Appl. Phys.* **64** (1988) 3706.
[188] Butler, M. A., Ginley, D. S., *J. Electrochem. Soc.* **134** (1987) 510.
[189] Gerges, A. S., Newson, T. P., Farahi, F., Jones, J. D. C., Jackson, D. A., *Opt. Commun.* **68** (1988) 161.
[190] Newson, T. T., Farahi, F., Jones, J. D. C., Jackson, D. A., *Opt. Commun.* **68** (1988) 161.
[191] Fabricius, N., Gauglitz, G., Ingenhoff, J., *Ber. Bunsenges. Phys. Chem.,* in press.
[192] Gauglitz, G., Ingenhoff, J., Fabricius, N., *German Patent P 4033357.*
[193] Nahm, W., *Diploma Thesis,* Tübingen, 1990.
[194] Nahm, W., Gauglitz, G., *GIT Fachz. Lab.* **90** (7), (1990) 889.
[195] Gauglitz, G., Nahm, W., *Fresenius Z. Anal. Chem.,* in press.
[196] Nahm, W., Gauglitz, G., in: *Proceedings of Sensor 91, Vol. III;* Wunsdorf: ACS Organisations GmbH, 1991, p. 7.
[197] Brecht, A., Gauglitz, G., Ingenhoff, J., Nahm, W., Beck, W., Jung, G., Polster, J., in: *Proceedings of Sensor 91, Vol. III;* Wunsdorf: ACS Organisations GmbH, 1991, p. 43.
[198] Brecht, A., Ingenhoff, J., Gauglitz, G., in: *Proceedings of Eurosensors V, Rome, 1991;* announced for *Sens. Actuators,* in preparation.
[199] Gauglitz, G., *Optical Transducer Principles,* in: Schmid, R. D., Scheller, F. (eds.), *Biosensors-Fundamentals, Technologies, Applications;* Weinheim: VCH, in preparation.

13 Mass-Sensitive Devices

Maarten S. Nieuwenhuizen, Prins Maurits Laboratory TNO, Rijswijk,
The Netherlands
Adrian Venema, Delft University of Technology, Delft, The Netherlands

Contents

13.1   Introduction

Chemical sensors are required to control pre-set process conditions or they are needed for the protection of mankind, fauna, and flora by measuring the concentrations of explosive, corrosive, or toxic compounds by warning when pre-selected concentrations are exceeded.

In contrast to physical signals, there is always a variety of chemical signals present in the environment of the sensor. As a consequence, selectivity is very important. This requirement can be achieved by using a (bio)chemical interface which interacts with the analyte as selectively as possible, thereby modulating the flow of physical signals in the sensor device. This flow of physical signals is generated from electrical signals by using transducers. Actually the (bio)chemical interface is a transducer itself as it transfers signals from the chemical domain to the physical domain or directly to the electrical domain.

This chapter deals with acoustic sensing phenomena. Mass changes at the sensor surface can be measured very accurately. Especially two different acoustic phenomena will be treated: bulk acoustic waves (BAW) and surface acoustic waves (SAW). A large number of review papers dealing with BAW chemical sensors [1-11] and SAW chemical sensors [12-15] have been published, but so far (1989), they have not previously been reviewed in detail together.

13.2   Acoustic Chemical Sensors

Piezoelectricity was first observed by the Curie brothers in 1880 [16]. Compression of a quartz crystal produced an electric potential, whereas application of an electric potential caused mechanical deformations. By employing these properties wave phenomena can be generated.

The velocity of the waves and as a result their frequency are influenced by a large number of parameters, including mass effects at the surface of the piezoelectric material. This phenomenon has been used since the early days of radio to adjust the oscillating frequency of crystals.

A chemical sensor employing bulk acoustic waves was introduced by King in 1964 [17-19] and was used as a detector in gas chromatography [17-23]. Later also liquid chromatograhic detectors based on BAW were developed [24, 25].

Usually AT-cut quartz crystals with various resonance frequencies are used as the physical devices. They have the lowest temperature coefficient and the highest mass sensitivity. Small disks, squares, or rectangles of length 10-15 mm and thickness 0.1-0.2 mm are commonly used.

After the early work of King, a large number of papers were published concerning BAW sensors in both the gas and liquid phases. A representation of the output of publications on BAW sensors is shown in Figure 13-1.

Lord Rayleigh described waves that only occur in a relatively thin surface layer [26], among which are SAW. At first only seismologists studied SAW. SAW in solids could easily be generated after White and Volmer in 1965 [27] described the interdigital transducer (IDT).

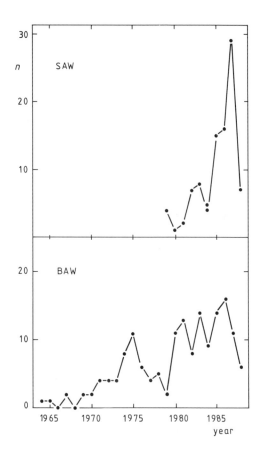

Figure 13-1.
Quantitative development of the output of
BAW and SAW chemical sensor research;
n = number of papers per year.

Basically the difference between BAW sensors and SAW sensors is the difference between a three-dimensional and a two-dimensional structure. As a result, SAW technology is compatible with planar silicon technology. Another advantage of SAW technology is concerned with sensitivity. Acoustic sensors in general are more sensitive at higher frequencies. In the case of BAW sensors, thinner an therefore mechanically weaker crystals are required to obtain higher frequencies, whereas in the case of SAW sensors only the IDT configuration and the electronics have to be adapted.

The first gas sensor based on SAW was reported by Wohltjen and Dessy in 1979 [28–30]. This SAW oscillator consisted of a delay line (two IDTs at some distance from each other) covered with a polymer [31]. As had happened during the development of the BAW sensors 15 years before, it was first applied as a detector in gas chromatography. In Figure 13-2 a number of possible SAW sensor configurations are shown. After Wohltjen's and Dessy's first papers, research on SAW chemical sensors exploded, as is also indicated in Figure 13-1.

Vetelino et al. [32, 33] reported on a SAW dual delay line gas sensor. One delay line is for measuring and the other acts as a reference (Figure 13-2). D'Amico et al. [34] reported on a three-transducer-type delay line (Figure 13-2). With both configurations improved temperature stability was demonstrated.

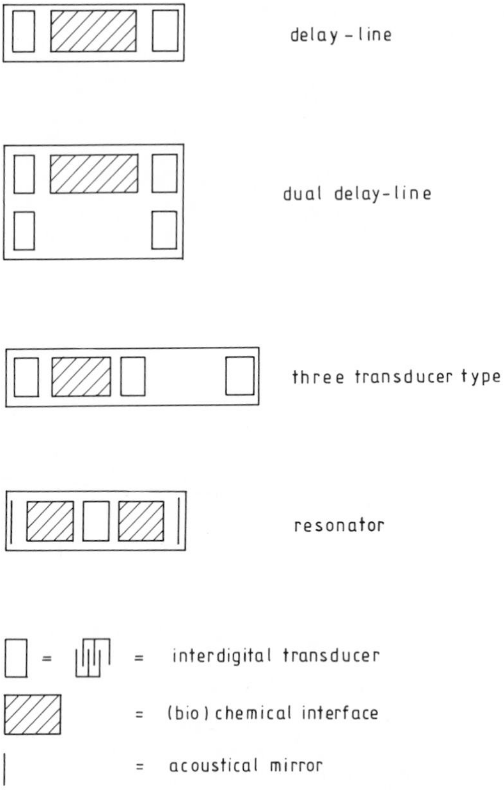

delay – line

dual delay-line

three transducer type

resonator

= interdigital transducer

= (bio) chemical interface

= acoustical mirror

Figure 13-2.
Different sensor configurations for SAW chemical sensors.

The first use of a SAW resonator sensor was reported by Martin et al. [35–37]. The resonator has one IDT which is located between two acoustically reflecting mirrors. The distance between the mirrors allows constructive interference between successive reflections, being maximum for one frequency (Figure 13-2).

The sensor materials usually applied for SAW sensors are STX-quartz or YZ-lithium niobate ($LiNbO_3$). Integration of SAW technology on silicon requires an additional layer. Since silicon is not a piezoelectric material, ZnO, CdS, or an electret have to be applied. D'Amico et al. [38] and Venema et al. [39, 40] first reported on such a system applying ZnO.

13.3   Physical and Electronic Aspects

13.3.1   Physics of Bulk Acoustic Wave Devices

The quartz crystal resonators used for BAW sensors can be easily excited by electrical means and sharp resonant frequencies are observed. The effects of changes in physical parameters are relatively small compared with those of mass changes. When putting electrodes onto the crystal and incorporating this system in a circuit with suitable gain and feedback, an oscillator is obtained with its resonant frequency being measured electronically in a very accurate way.

Several fundamental wave modes are obtained: longitudinal, lateral, and torsional as well as several overtones or combinations. One principle mode is selected when applying a particular cut of the crystal. The configuration of electrodes, the supporting structure, and the electronic circuit are also important parameters. The high-frequency thickness shear mode is most mass sensitive.

The theory of the physics and electronics of BAW chemical sensors has been well and thoroughly described [41]. Here only a brief outline will be given. For a piezoelectric material, there exists a connection between the electrical and mechanical properties. A voltage applied across the crystal gives rise to a mechanical deformation and vice versa. Since the mass of the crystal corresponds to an electrical inductance and the elastic properties to a capacitance, there will be a resonance frequency at which the crystal wants to oscillate causing a (high-frequency) voltage output from the crystal.

For so-called AT-cut quartz crystals operating in the thickness shear mode, the oscillation frequency (f_o) is inversely proportional to the thickness, d, of the crystal:

$$f_o = N/d \tag{13-1}$$

where N is a frequency constant; $N = 0.168$ MHz cm for AT-cut quartz at room temperature.

Since the mass of the crystal is $m = \rho A d$, where ρ is its density and A its cross-sectional area, Equation (13-1) indicates that the oscillation frequency is inversely proportional to the mass of the crystal. In pure shear mode vibrations, the stress is zero at the surfaces of the crystal. It is therefore assumed that the electrodes and extra layers on the electrodes (see Figure 13-3 a) only influence the oscillation frequency through the addition of an extra mass to the crystal. The frequency change obtained from Equation (13-1) is (as long as $\Delta m \ll m$)

$$\Delta f = -f_o \frac{\Delta m}{m} . \tag{13-2}$$

Equation (13-2) describes the physical basis for chemical sensors based on bulk acoustic waves. In this case, Δm is the increase in the mass of the sensing layer (see Figure 13-3) due to the interaction with the species to be detected. For AT-cut quartz one finds

$$\Delta f = -2.3 \times 10^6 f_o^2 \, \Delta m/A \tag{13-3}$$

where f_o is in MHz, Δm in g and A in cm^2. Since quartz crystals typically have $f_o = 10$ MHz and a frequency change of 0.1 Hz can readily be detected with modern electronics, it is possible to detect mass changes down to about 10^{-10} g/cm^2.

The physics of the BAW devices is relatively simple. The difficulty lies in finding a sensing layer with sufficient sensitivity (ie, absorption) and selectivity for the species to be detected and a sufficient speed of response and reversibility to be practical useful. Those problems are shared with the SAW devices and are further treated in Section 13.4. It should be pointed out, however, that the resonance frequency decrease due to the application of the sensor layers typically is of the order of 1–10 kHz.

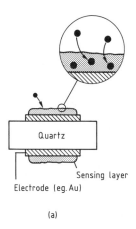

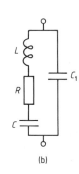

Quartz

Sensing layer

Electrode (eg. Au)

(a)

(b)

Figure 13-3.
(a) Schematic diagram of a quartz crystal with electrodes and sensing layers. It illustrates how the species to be detected are absorbed in the sensing layer. (b) Electrical equivalent of the quartz crystal. L represents the mass of the crystal (with electrodes and sensing layers), C the elasticity of the crystal and R losses in the crystal due to internal friction. C_1 represents the capacitance between the two electrodes. C_1 is normally much larger than C.

The quartz crystal oscillator is well documented and will not be treated here. Electrically, however, a quartz crystal can be described as in Figure 13-3 b. It is characterized by a high Q-value and a well defined resonance frequency. The inductance, L, is related to the total mass of the crystal and is thus the parameter that changes on absorption of molecules in the sensing layers. In several applications, two crystal oscillators are used, one as a passive reference and the other as the chemically sensitive device. Changes in the frequency difference between the two oscillators are monitored by the use of a frequency mixer and a low-pass filter.

Below the theory of physics and electronics of SAW chemical sensors will be treated in more detail. This theory follows the same lines as BAW physics and electronics, but is usually more complicated.

13.3.2   Surface Acoustic Wave Physics

Piezoelectric substrates are (mono)crystalline and therefore show an anisotropic behavior. The orientation dependence of the material properties implies that for efficient use of the waves certain precautions have to be taken [42].

One important precaution to ensure high performance of SAW sensors is the use of pure-mode Rayleigh waves. These waves have the mechanical and electrical components in one plane, the sagittal plane, which is normal to the substrate surface.

In homogeneous substrates the phase velocity and the amplitude of a SAW is determined by the elastic, piezoelectric, dielectric, and conductive properties and the mass of the substrates. If one of these parameters can be properly modulated by the quantity to be measured, the effect of sensing is created. This modulation can occur in the transducer and/or transmission region.

In layered substrates the physical properties per layer and additionally the thicknesses of the layers determine the phase velocity and amplitude of the SAW.

A very attractive phase-sensing device is the SAW oscillator employing IDTs [43]. An IDT [44, 45] is a planar interleaved metal electrode structure (Figure 13-4) whose adjacent electrodes are given equal but opposite potentials.

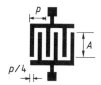

Figure 13-4.
Interdigital metal pattern of a uniform transducer with period p and
aperture A.

The resulting spatially periodic electric interdigital field produces a corresponding periodic
mechanical strain pattern employing the piezoelectric effect. This gives rise to the SAW, provided that the surface is stress free. The IDT behaves as a sequence of ultrasonic sources. For
an applied sinusoidal voltage, all vibrations interfere constructively only if the distance $p/2$
between two adjacent fingers is equal to half the elastic wavelength. The frequency $f_0 = v_R/p$
that corresponds to this cumulative effect is called the synchronous frequency or the resonance
frequency. If the frequency departs from this value, the interference between the elastic signals
is no longer constructive and the overall vibration is weaker. Thus, the bandwidth of an IDT
is narrower when there are more fingers. As a result of symmetry, SAW are emitted in opposite
directions, which results in an inherent minimum of 3 dB transducer conversion loss at f_0. A
minimum insertion loss of 6 dB is found for a delay line. The acoustic aperture A defines the
effective region of transduction between two adjacent electrodes. The IDT is uniform if a constant aperture is obtained.

Flexibility in SAW component design is obtained by the application of an equivalent circuit
model [46, 47]. It requires a knowledge of the piezoelectric coupling coefficient of the
material, the synchronous frequency f_0, and the static capacitance [48]. The piezoelectric
coupling coefficient is a measure of the efficiency of the conversion of electrical energy into
acoustic energy and vice versa. It is a constant in the case of homogeneous substrates and is
supplied together with other material data by the manufacturers of substrates.

Matching networks between the input and the output transducer and the source and load,
respectively, allow for the optimum transfer of signal energy. In many applications the matching networks are one inductor per electric port of the IDT, placed in series with the electric
port. The inductor impedance tunes out the reactive part of the IDT at the synchronous frequency and does so for a small frequency interval around the synchronous frequency. With
proper IDT design a conversion loss close to 3 dB can be obtained. If, however, a monolithic
integration of the SAW device and electronic circuits is pursued, the consequent absence of
the matching inductors can increase the conversion loss substantially. However, this is not
necessarily a restriction for proper sensor performance.

Several types of signal loss can occur in a SAW device, which can cause degradation of its
performance [42, 49]. The emitting and receiving IDTs of a delay line and the transmission
path between them are subjected to loss mechanisms. For example, in an IDT region the electrode resistivity and mass loading, internally caused acoustic reflection between the electrodes,
end effects, insufficient piezoelectric coupling, improper electrical matching, and bulk wave
excitation contribute to the conversion loss. The equivalent circuit mentioned above only deals
with the electrode resistivity, parasitic capacitance at the electric port, and loss in the tuning
inductor.

Size restrictions and impedance requirements limit the extent to which the acoustic aperture
A can be increased to reduce the overall device loss. The aperture determines the width of the
radiated beam and is typically 10–100 wavelengths in magnitude. The finite aperture of the

emitting IDT causes diffraction and a consequent signal loss because of the spreading of the beam and the departure of the transverse phase from a constant value. The magnitude of the electrical output signal of the receiving IDT depends on the fraction and profile of the beam intercepted, the IDT design and matching of the IDT to the load.

The diffraction losses are reduced if the crystal geometry is autocollimating. Relative misalignment of the two IDTs can also cause signal loss. In anisotropic media geometrically aligned IDTs can have an extra loss introduced by any departure of the axis of the combination from the pure-mode crystal axis, because then the ultrasonic beam does not follow the IDT axis (beam steering).

SAW propagating along the surface of an elastic solid are prone to attenuation. This is caused by scattering from surface imperfections (scratches, polishing defects), scattering from crystal defects (dislocations and grain boundaries in polycrystalline materials, eg, piezoelectric layers and (bio)chemical interfaces), interactions with thermal phonons (surface and bulk phonons), and interactions with charge carriers in metals and semiconductors.

Gas loading of the surface will lead to the emission of compressional waves and to a significant increase in the attenuation at microwave frequencies.

For sensor applications temperature effects are also very important. The ideal case would be a substrate with a zero temperature coefficient of delay combined with a high piezoelectric coupling coefficient. The present situation is, despite efforts in materials research, that a choice has to be made between these two parameters, eg, ST-quartz for a zero temperature coefficient or cuts of lithium niobate or lithium tantalate for high piezoelectric coupling.

13.3.3   The Surface Acoustic Wave Dual Delay Line Oscillator

The electronic system consists of two isolated identical delay line oscillators such as that shown in Figure 13-5. The loop gain and the loop phase of a delay line oscillator operating at an oscillating frequency f_c must satisfy the following conditions:

loop gain $A(f_c) > 1$ and loop phase $\varphi(f_c) = -2\pi n$.

An expression for the oscillation frequency f_c can be derived from the phase loop condition:

$$\varphi_{dll} + \varphi_A + 2\varphi_{tr} = -2\pi n$$

with $\varphi_{dll} = -2\pi f_c \tau$ hence

$$f_c = \frac{2\pi n + \varphi_A + 2\varphi_{tr}}{2\pi \tau} \tag{13-4}$$

where φ_A is the phase shift introduced by the amplifier and $2\varphi_{tr}$ is the sum of the phase shifts introduced by the two identical uniform IDTs. The phase shift introduced by each IDT is determined by its impedance and its load. This means that the phase shift of the transmitting and receiving IDT can be different. τ is the total delay time of the SAW between the IDTs (equal to twice the transit time).

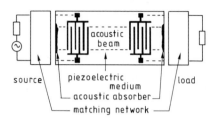

Figure 13-5.
Physical-electronic system of a SAW delay line
oscillator.

The equation for f_c shows that the oscillation frequency is inversely proportional to the acoustic signal delay. Changes in τ can thus be used for sensing purposes. It also shows that the influence of φ_A and φ_{tr} can be minimized by chosing φ_{dll} to be large, which implies a long delay line at the cost of an increase in insertion loss.

It should be noted that the amplifier gain or insertion loss of the delay line has no direct influence on the oscillator frequency. In practice, however, a change in the insertion loss or in the amplifier gain will have some influence on the loop phase and therefore on the oscillation frequency. When using a fixed gain amplifier, the loop gain must be chosen to be sufficiently larger than 1 to prevent any temperature, aging or other effect from decreasing the loop gain to less than 1, because this would stop oscillation. If the oscillator is switched on and if A is chosen to be larger than 1, the output signal will increase, until nonlinear effects limit the output amplitude. Then an increasing amplifier input signal will no longer cause a proportional increase in the amplifier output signal. Finally, the effective loop gain will reduce to 1, after which no further amplitude change will occur. In this way amplitude stabilization is caused by nonlinear effects such as clipping and/or saturation. These nonlinear effects may introduce a significant phase shift and consequently a frequency shift in the oscillator, which is dependent on temperature, amplifier gain setting, supply voltage, and delay line insertion loss.

Under the condition that the loop gain A is sufficiently larger than 1, a number of frequencies f_N will satisfy the loop phase condition. It was found that if oscillation was started at one of these frequencies, the oscillation remained stable [50]. This implies that after switching on the oscillator or after a momentary electrical disturbance, one cannot be sure at which frequency oscillation will occur.

One method of suppressing unwanted frequencies f_N is mode selection, in which a long but thinned array of finger pairs is used for one of the IDTs [51]. In this way an amplitude characteristic is obtained, for which only one frequency fulfils the loop phase condition.

A different approach provides the oscillator with automatic gain control (AGC). In this approach, the AGC circuit compares the output amplitude with a certain reference level. If there is any difference between the two levels the AGC circuit either increases or decreases the amplification until the output amplitude equals the reference level. In this way the loop gain is regulated to exactly 1 and the amplifier will operate in its linear region. When the ouput amplitude has reached its steady-state value (ie, the reference value), there will no longer be a frequency that has a loop gain larger than 1, as otherwise the output amplitude would increase. However, there will be a frequency for which the loop gain is exactly 1, as otherwise the output amplitude would decrease. For practical delay lines the insertion loss for different frequencies f_N will never be the same. This implies that the oscillator will always oscillate at only one frequency, namely that for which the delay line shows the lowest insertion loss.

Another advantage of using an AGC circuit is that the amplifier will never saturate and will always operate in its linear region. Hence no frequency dependence due to nonlinear effects as stated earlier will occur [52]. In addition, the harmonic content and other possible spurious signals due to the amplitude-limiting action of a saturating amplifier [53] is highly reduced or even suppressed.

The AGC circuit can be completely and easily integrated into a monolithic structure, which is important for the development of single-chip sensors.

13.4   The (Bio)chemical Interface

For the development of a suitable interface one can follow one of two routes, either chemical or biochemical [54]. A major advantage of the chemical route is that often more chemically and physically inert systems are obtained, which can withstand more aggressive environments, whereas biochemical interfaces make use of selectivity evolved during millions of years. Research on biochemical and chemical sensors is to be considered complementary rather than overlapping.

The interaction between a chemical interface and the analyte may vary from absorption/adsorption to chemisorption [13, 55]. The type of interaction and the morphology of the chemical interface material will determine the ultimate performance characteristics of the sensor. When applying absorption phenomena only slight differences in selectivity are obtained. In the case of adsorption only weak interactions occur. The energies involved range from van der Waals forces (0–10 kJ/mol) to acid-base interactions (<40 kJ/mol). In the case of chemisorption very strong interactions occur at the chemical interface. Chemical bonds are broken and formed (about 300 kJ/mol).

For selectivity, chemisorption is to be preferred. However, analytes are strongly and often irreversibly bound in that case. For reversibility, absorption or adsorption is to be preferred. Obviously a compromise is required: the area of coordination chemistry or charge-transfer complex formation. For example, metallophthalocyanines show a selective adsorption of NO_2 based on charge-transfer complexation. Table 13-1 shows the sensitivity towards NO_2 of different metallophthalocyanines on a SAW sensor and the selectivity towards a number of other gases [56].

Table 13-1. Sensitivities (Hz ppm^{-1} μm^{-1}) of 80-MHz SAW sensors with 0.4-μm thick layers of metallo- and H_2-phthalocyanines at 150 °C according to Nieuwenhuizen et al. [56].

Gas molecules	NO_2	CO	CO_2	CH_4	NH_3	SO_2	H_2O	C_7H_8
Metal in ppm	100	1200	200	400	200	200	8000	200
H_2	75	0	0	0	0	0	0.1	0
Mg	45	0.1	0	0	−2	0	0	0
Fe	67	−0.1	0	0	−10	0	−0.2	0
Co	133	0	0	0	−1	0	0	0
Ni	33	0.1	0	0	−3	0.3	0	0
Cu	80	0	0	0	−7	0	0	0

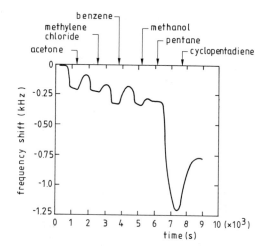

Figure 13-6.
Response to 2000 ppm of various vapors of a SAW device coated with poly(ethylene maleate) (after Snow and Wohltjen [57]).

An example from the work of Snow and Wohltjen [57] illustrates both absorption/adsorption and chemisorption phenomena (Figure 13-6). A SAW sensor was coated with poly(ethylene maleate). A number of organic solvents showed reversible responses due to adsorption. The response to cyclopentadiene proved to be partly irreversible due to a Diels-Alder reaction with the poly(ethylene maleate). The reversible part of the response was caused by desorption of unreacted cyclopentadiene.

For the biochemical interface one can make use of various biological systems such as enzymes, antibodies, polynucleotides, receptors, bacteria, organelles, or even pieces of an organism. In the past these systems could only be used in physiological or closely related liquid environments. At present there is a trend to apply these systems more and more in nonphysiological environments. In some cases even application in gaseous environments seems feasible [58].

Chemical sensors are often in open contact with the environment, whereas physical sensors usually can be encapsulated, thereby diminishing environmental, often detrimental, effects. Therefore, one has to minimize the destructive influence of the environment on the (bio)chemical interface and other parts of the sensor.

The stability of a (bio)chemical interface strongly depends on the attachment of the material to the surface of the sensor device. Several methods can be applied. In Table 13-2 an overview is given of methods described in the SAW chemical sensor literature. In many cases, however, information regarding the coating procedures was lacking. Very simple methods are smearing, spouting, spraying, and spin coating. More complicated techniques involve Langmuir-Blodgett films, physical vapor deposition (PVD) or chemical vapor deposition (CVD). In all cases the (bio)chemical interfaces are attached by weak physical bonds. For the sake of stability it is preferred to use a chemical immobilization technique employing the formation of chemical bonds [59–63]. As a result, sensitivity and response time will also be influenced. In 1988 the first experiments with a SAW sensor for NO_2 using chemically immobilized metallophthalocyanines as the chemical interface have been published [63]. When compared with the much thicker layers applied by PVD, two main differences were noted: faster responses and smaller sensitivities (see also Section 13.5.2, Table 13-5).

Table 13-2. Overview of coating techniques used with SAW sensors.

Coating technique	References
Smearing	[29]
Spin coating	[57, 187, 188]
Solvent evaporation	[33, 218, 219, 227]
Spraying	[193, 196, 197, 209, 210, 214, 228, 232, 233]
Langmuir-Blodgett film	[183, 222]
Physical vapor deposition	[56, 63, 184, 200–205, 211]
Sputtering	[33, 36, 37, 194–196]
Chemical immobilization	[17–21]

As in many cases it turned out to be very difficult to find a selective chemical interface for a certain analyte, arrays of sensors have been proposed. Many papers have dealt with this approach, involving both BAW [64–75] and SAW sensors [76–80]. Pattern recognition techniques employing increasingly powerful microprocessors are required to obtain the required signals. Still it is questionable whether reasonable selectivity will be achieved, especially over extended periods of time when aging phenomena occur.

The effect of changes in the relative humidity of the environment may be a general problem, wich requires a special chemical interface sensitive to water vapor.

13.5   Mass-Sensitive Sensors in the Gas Phase

In this section, an overview will be given of both BAW and SAW gas sensors. Many references claim sensor response for a certain analyte, but often the selectivity which is one of the most important criteria for a good sensor, is poorly studied or not studied at all. It is stressed that future reports on selective chemical sensors should provide information about the selectivity towards interferents relevant to the application. From several publications, therefore, no clear molecular view of selectivity can be obtained.

13.5.1   Bulk Acoustic Wave Gas Sensors

Carbon monoxide: Ho et al. [81] developed a sensor for 1–50 ppm CO by reacting CO with HgO at 210 °C to produce Hg, which was detected by a gold-coated sensor (see below). In a reference system CO was burned catalytically. In this way background levels of Hg could also be detected.

Carbon dioxide: Jordan [82] described a sensor for CO_2. 7, 10-Dioxa-3,4-diaza-1,5,12,16-hexadecatetrol was used as the chemical interface. The response to CO_2 was a function of the relative humidity (RH) and proved to be 898 Hz for 10% CO_2 at 90% RH. Response times were about 30 s.

To use a gas sensor in the liquid phase a gas-permeable membrane can be used [83]. Several of such membranes were tested to develop a sensor for CO_2 dissolved in water. Didodecyla-mine was used as the chemical interface.

Phosgene: Suleiman and Guilbault [84] reported on methyl trioctylphosphonium dimethylphosphonate as the chemical interface. Responses proved to be linear in the range 8–200 ppm. H_2O, NH_3, H_2S, and HCl were found to interfere.

Hydrogen: King [19] described a detector for H_2 using Pd or Pt films at 150 °C. Bucur [85] used Pd to detect both H_2 and D_2 in an inert gas at 50 °C. H_2 concentrations down to 10% could be detected. Mizutani et al. [86, 87] detected H_2 in N_2 in the 250–5000 ppm range with Pd. They also studied temperature effects (25–100 °C). First-order response rates were found, whereas no interferences from CH_4, SO_2, and CO_2 were observed. A French Patent [88] claimed Pd electrodes as the chemical interface. A Japanese group [89] reported on $LaNi_5$ as a chemical interface. This compound is known to have a very large capacity to ad-sorb H_2.

Hydrogen chloride: Hlavay and Guilbault [90] detected HCl in the ranges 10–100 ppb and 0.1–100 ppm using triphenylamine. Despite the strong interaction between the amine and HCl, reversible responses were obtained.

Hydrogen cyanide: Alder et al. [91] used bis(pentane-2,4-dionato)nickel (II) for the detec-tion of HCN in the range 13–93 pm.
Irreversible replacement of the ligand by HCN caused a mass effect. The response rate was related to the concentration. Responses were studied at different relative humidities (40–92 %). In Figure 13-7 the normalized response rate vs. the HCN concentration is plotted.

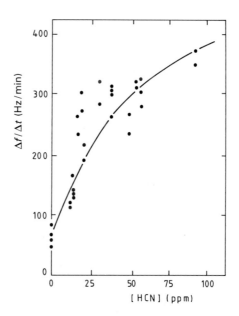

Figure 13-7.
Rate of frequency change versus HCN concentra-tion of a BAW sensor coated with bis(pentane-2,4-dionato)nickel(II) (after Alder et al. [91]).

Hydrogen sulfide: King [17] suggested several chemical interfaces for H_2S: Cu, Ag, and lead acetate.

Webber et al. [92] studied the use of various acetone extracts of half-burned organic chemicals as a chemical interface. With such a soot prepared from chlorobenzoic acid, 1–60 ppm H_2S could be detected.

Water: King [19] developed a sensor for water applying Au, Ni, and Al as the chemical interface. Since then many kinds of hygroscopic materials have been applied, such as evaporated films of SiO_x [93], gelatine [94], various kinds of polymers [95–101], and other materials [102, 103].

Tahara et al. [104] claimed a plasma-polymerized styrene film containing sulfonic acid groups as the chemical interface for water in the range 1–105 ppm. Hardly any interferences were measured from O_2, H_2, CO, CO_2, Cl_2, aromatic hydrocarbons, and fluorocarbons.

Mercury: Bristow [105] used a gold-coated sensor for Hg. A detection limit of 5 ng/L was found. Scheide et al. [106–109] extensively studied the same system at 150 °C. They detected Hg down to 1 ppb in an irreversible way. At 400 °C a reversible response was obtained.

Ammonia: Karmarkar and Guilbault [110] reported on Ucon LB-300X sensitized by NO_2 treatment as the chemical interface. Since then several other coatings have been proposed, such as glutamine. HCl [111, 112], pyridoxine. HCl [111–115], ascorbic acid [112, 116] and also nickel(II) dimethylglyoximate [117].

Poly(vinylpyrrolidone) was described by Edmonds et al. [118, 119]. A multi-sensor system was used with reference sensors for temperature and relative humidity using AgCl as the chemical interface. The detection range was 6–104 ppb, but the response curves levelled off at 7500 ppb. Also, aging of the chemical interface occurred, reducing the ammonia response to 50% after 40–50 days.

Nitrogen dioxide: Ucon LB-300X and Ucon 75-H-90000 have been tested as chemical interfaces for NO_2 [110] in the range 1–1000 ppb. At high concentrations the recovery was slow. Recently, Edmonds et al. [120] reported a study of the interaction between NO_2 and MnO_2 layers on a BAW sensor.

Oxygen: A U.S. Patent [121] described a coating for O_2 consisting of a Mn-thiocyanato-tributylphosphine-tetrahydrofuran-poly(vinyl-chloride) complex. A French Patent [88] claimed an In amalgam as the chemical interface.

Ozone: Fog and Rietz [122] described poly(butadiene) as a chemical interface for O_3 by the irreversible formation of ozonides. The influence of NO, NO_2, formaldehyde, phenol, and CO was found to be insignificant.

Sulfur dioxide: Guilbault and Lopez-Roman [123] reported on the use of tetrachloromercuriate, which was compared with many other kinds of materials. Later other groups reported on Carbowax 20 [124], various amines [112, 124–130], quadrol (N,N,N′,N′-tetra(2-hydroxypropyl)ethylenediamine) [131, 132], and ethylenedinitrotetraethanol [133] as chemical interfaces. In many cases NO_2 and H_2O were important interferences. Karmarkar et al. [132] showed that a hydrophobic membrane filter could easily reduce the response to water in the quadrol-coated sensor system.

Acetoin: Suleiman et al. [134] studied tetrabutylphosphonium chloride as a sensitive and reversible chemical interface for acetoin (3-hydroxy-2-butanone) in the range 8–120 ppb. Only acetone and ethanol interfered at high concentrations. The same group [135] reported on semicarbazide as a chemical interface, which was selected from a group of over 25 materials.

Amines: Guilbault et al. [136] studied many metal salts for the detection of methylamines, including Fe, Zn, Hg, Co, and Zn halides. Especially the Fe salts showed high sensitivity in a vacuum system. A Russian group [137] reported on a poly(methylsiloxane) as the chemical interface for amines and for ethanol.

Anaesthetics: Halogenated hydrocarbons such as halothane, isoforane, enflurane, trichloroethylene, and methoxyflurane are interesting analytes to monitor in inhalation systems during medical treatment. Two types of chemical interface were mentioned: various lipophilic materials [138, 139] and silicone rubber [140, 141, 144]. The silicone rubber is presently used in commercial equipment.

Dimethylhydrazine: Varga [142] used many kinds of poly(butadienes) as chemical interfaces.

Formaldehyde: Guilbault [143] reported on the detection of CH_2O using a formaldehyde dehydrogenase and a mixture of cofactors. In fact, this system is one of the first applications of a biochemical system in the gas phase. A linear response in the range $10–10^4$ ppb was obtained. A selectivity with respect to alcohols and other aldehydes was found.

Hydrocarbons: Many papers have reported on the detection of aliphatic and aromatic hydrocarbons. Chemical interfaces including GC stationary phases [141, 144–146], poly(isobutyl acrylate) [82], zeolites [147], cyclodextrin derivatives [148], and trans-Ir(II)Cl(CO)(PPh$_3$)$_2$ in Nujol [149] have been proposed. The last compound could easily detect aromatic compounds. The zeolites and the cyclodextrins employed shape selectivity: minor differences in size and shape gave rise to large differences in adsorption behavior and hence in sensor selectivity.

Nitro aromatics: Two papers reported on the detection of mononitrotoluene (MNT) and trinitrotoluene (TNT) [150, 151].

Tomita et al. [150] studied Carbowax 1000 as a sensitive, selective, and relatively fast chemical interface in the ppb-ppm range without serious interferences from organic solvents or perfumes.

Organophosphorus compounds: Organophosphorus compounds are very interesting analytes as many are very toxic (pesticides, chemical warfare agents; see below). Kristoff and Guilbault [152] reported on the use of a non-coated sensor where the Au, Ag, or Ni electrodes together with the quartz acted as a chemical interface.

Guilbault et al. [153–159] studied various salts of Fe, Cu, Ni, Cd, and Hg for the detection of diisopropyl methylphosphonate (DIMP) as a model compound. Especially HgBr$_2$ [153] and FeCl$_3$ [154] have been studied in detail, and also polymeric copper complexes (Cu-XAD-4 and Cu(II) tetramethylethylenediamine in poly(vinylpyrrolidone)).

Balog et al. [159] tested eight coating materials with 0.3 μg/L of the chemical warfare agent sarin. No responses were observed, but Cu-XAD-4, poly(ethylene maleate), and succinylcholine chloride gave responses of 59, 22, and 11 Hz, respectively, to 10 μg/L of DIMP.

Oxime systems have been reported: 3-pyridylaldoxime methiodide and isonitrobenzoyl acetone [160], 1-n-dodecyl-3-hydroxyiminomethylpyridinium iodide [161–165] and amidoximes [166, 167].

As the enzyme acetylcholinesterase is inhibited by organophosphorus compounds, this biochemical interface was studied by van Sant [168] and Guilbault et al. [169, 170]. The latter group found eel cholinesterase immobilized with glutaraldehyde to be the best biochemical interface for DIMP: 450 Hz for 4 ppm within 4–6 min with no important interferences. The responses increased at higher relative humidity values.

Guilbault et al. [170–172] reported on the use of antibodies for the detection of parathion in the gas phase. They claimed a response of 108 Hz for 36 ppb of parathion. Other organophosphorus compounds showed smaller responses. It should be mentioned that at present there are discussions on the interpretation of the experiments, as there are serious doubts about the biological activity of antibodies in the gas phase.

Propylene glycol dinitrate: Turnham et al. [173] reported on the detection of PGDN, a very toxic propellant in torpedoes. OV-175 and other GC stationary phases have been tested as chemical interfaces. A second sensor equipped with a trap for PGDN (cellulose acetate isobutyrate) acted as a reference. In Figure 13-8 the efficiency of this trap is illustrated.

Toluene diisocyanate: Alder et al. [91, 174–177] reported on the detection of TDI, especially by using poly(ethylene glycol) (PEG). In a dual sensor system one sensor contained PEG and the other a mixture of tri-n-octylphosphine oxide and $CoCl_2$. TDI could be detected in the range 0.1–15 ppm at 30–60% relative humidity without important interferences from common atmospheric components.

Morrison and Guilbault [178] detected TDI with silicone compounds at levels down to 10 ppb with minor interferences from alkylbenzenes and ketones. Selectivity originated from the irreversible reaction of the silicon compound and TDI.

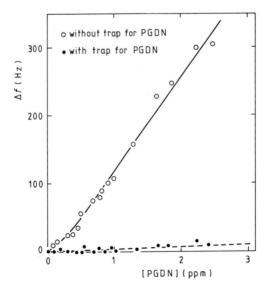

Figure 13-8.
Response as a function of the propylene glycol dinitrate concentration of a BAW sensor coated with OV-175 with and without a PGDN trap (after Turnham et al. [173]).

Vinyl chloride: This compound was detected by van Sant [168]. Vinyl chloride was decomposed catalytically to HCl by use of chromate and subsequently HCl was measured with an Amine 220-coated sensor. A linear response up to 80 ppm was reported without interferences from common atmospheric components. Only water had to be removed before the catalytic oxidation step. In Tables 13-3 and 13-4 an overview is given of the various BAW gas sensors for inorganic and organic analytes.

Table 13-3. Overview of BAW gas sensors for inorganic analytes.

Analyte	(Bio)chemical interface	References
CO	reaction to Hg[2]	[81]
CO_2	DDHT[1]	[82]
	didodecylamine and membrane	[83]
$COCl_2$	MPDP[3]	[84]
H_2	Pd, Pt	[19, 85–88]
	LaNi$_5$	[89]
HCl	tertiary amines	[90]
HCN	Bis (pentane-2,4-dionato)nickel(II)	[91]
H_2S	Various soots	[92]
H_2O	Au, Ni, Al	[19]
	SiO$_x$	[93]
	polymers and other materials	[95–104]
Hg	Au	[105–109]
NH_3	Ucon LB-300X, Ucon 75-H-90000	[110]
	glutamine.HCl	[111, 112]
	pyridoxine.HCl	[111–115]
	ascorbic acid	[112, 116]
	Ni(II) dimethylglyoximate	[117]
	poly(vinylpyrrolidone)	[118, 119]
NO_2	Ucon 75-H-90.000 etc.	[110]
	MnO$_2$	[120]
O_2	Mn complex	[121]
	In amalgam	[88]
O_3	Poly(butadiene)	[122]
SO_2	Sodium tetrachloromercuriate	[123]
	GC stationary phases	[123, 124]
	amines	[112, 125–130]
	quadrol	[131, 132]
	ethylenedinitrotetraethanol	[133]

1) 7,10-dioxa-3,4-diaza-1,5,12,16-hexadecatetrol.
2) see Hg.
3) methyltrioctylphosphonium dimethyl phosphate.

Table 13-4. Overview of BAW gas sensors for organic analytes.

Analyte	(Bio)chemical interface	References
Acetoin	Tetrabutylphosphonium chloride	[134]
	semicarbazide	[135]
Amines	Various metal salts	[136]
	poly(methylsiloxane)	[137]
Anesthetics	Various lipophilic layers	[138, 139]
	silicone rubber	[140, 141, 144]
Dimethylhydrazine	Butadiene copolymers	[142]
Ethanol	Poly(methylsiloxane)	[137]
Formaldehyde	Formaldehyde dehydrogenase	[143]
Hydrocarbons[1]	GC stationary phases	[144–146]
	poly(isobutyl acrylate)	[82]
	zeolites	[147]
	cyclodextrin derivatives	[148]
	trans-IrCl(CO)(PPh$_3$)$_2$	[149]
Nitroaromatics	Carbowax 100 etc.	[150]
	coal, quadrol, PEG	[151]
Organic P compounds	Au, Ag, Ni	[152]
	metal chlorides	[153, 154]
	Cu complexes	[155–159]
	oximes	[160–166]
	histidine.HCl	[167]
	succinyl choline chloride	[167]
	acetylcholinesterase	[168–170]
	antibody	[170–172]
PGDN[2]	OV 175 etc.	[173]
TDI[3]	poly(ethylene glycol) 400	[174–177]
	silicone fluid FS-1265	[178]
Vinyl chloride	Reaction to HCl	[168]

1) aliphatic and aromatic.
2) propylene glycol dinitrate.
3) toluene diisocyanate.

13.5.2   Surface Acoustic Wave Gas Sensors

Hydrogen: D'Amico et al. [34, 38, 179–181] have studied sensors for H_2 based on LiNbO$_3$, quartz, and the ZnO-SiO$_2$-Si system coated with a 0.2–0.4-μm Pd layer. The frequency change to 1% H_2 in N_2 was largest for the ZnO system, followed by quartz and LiNbO$_3$, with a response time of 10 min and a recovery time of about 2 min in all cases.

Oxygen: In a Russian Patent [182] a SAW sensor for oxygen was claimed. A metal oxide semiconductor material was used as the chemical interface.

Halogens and halogenated hydrocarbons: SAW sensors for I_2 were studied by Snow et al. [183], who used Langmuir-Blodgett films of metallophthalocyanine derivatives. They also measured conductivity changes of the films on exposure to I_2. The response times of SAW frequency and conductivity changes were different, indicating that the former are controlled by bulk adsorption and the latter by surface adsorption. Nieuwenhuizen and Nederlof [184] used metallophthalocyanines applied by PVD. Substantial destruction of the chemical interface was found to occur, owing to reaction with the halogens. When halogenated hydrocarbons (anesthetics) were decomposed by using a heated Pt wire, the resulting halogen vapors could be detected by a SAW sensor coated with phthalocyanines [185, 186].

Water: Brace et al. [187–190] studied the thermodynamics of different polymers using SAW devices. Martin et al. [191] used 97-MHz SAW devices for the characterization of the thin-film properties of poly(imide) films. They studied the transient behavior of different molecules diffusing into the polymer films and the adsorption of nitrogen at 77 K to determine the effective surface area of the films. Huang [192] reported on a non-coated $LiNbO_3$ delay line for high humidity sensing, making use of an amplitude-modulated SAW technique at 210–250 MHz.

In a search for a SAW sensor for CO_2, Nieuwenhuizen and Nederlof [193] found poly(imine) films to interact strongly with water using a 40-MHz quartz dual delay line sensor at 30–70 °C.

Hydrogen sulfide: Vetelino et al. [33, 194–196] described a system for H_2S with $LiNbO_3$ dual delay line sensors. On one delay line a sputtered film of WO_3 was applied, which was more stable than organic coatings. The sensor proved to be very sensitive (down to 10 ppb). Figure 13-9 shows the frequency change of the sensor versus the H_2S concentration. The sensor response was reversible at 200 °C. Sensitivity increases with temperature whereas from ordinary adsorption/desorption phenomena one would expect the opposite effect. It was suggested that the response was caused by the interaction of the SAW electrical field with the free charge carriers in the WO_3 film, which would also account for the peculiar temperature effects.

Sulfur dioxide: Following many papers dealing with BAW sensors for SO_2, Vetelino et al. [32, 33, 196–198] reported on a dual delay line SAW sensor using triethanolamine as the chemical interface. The sensor could detect down to 10 ppb SO_2, while reproducible results

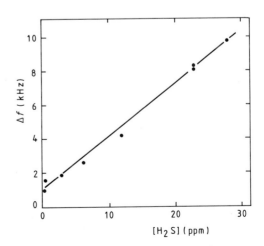

Figure 13-9.
Frequency change vs. H_2S concentration of a YZ-$LiNbO_3$ dual delay line SAW sensor with WO_3 as a chemical interface at 130 °C (after Vetelino et al. [195]).

were obtained down to 70 ppb. The response of the 60-MHz LiNbO$_3$ sensor was found to be 15 times higher than that of a comparitive 11-MHz AT-quartz BAW sensor. NO$_2$ was found to produce significant irreversible responses. Prolonged treatment with NO$_2$ made the film insensitive to NO$_2$ but it was still sensitive to SO$_2$, indicating different response mechanisms for the two gases.

Ammonia: D'Amico et al. [199] described a quartz SAW detector for NH$_3$ in the range 300–1000 ppm at 25 and 35 °C. On one delay line a Pt film was applied. It was found that even the dual delay line configuration showed considerable temperature dependence which was a function of the film thickness.

Nitrogen dioxide: Many papers have dealt with SAW NO$_2$ sensors with metallophthalo-cyanines as the chemical interface. Ricco et al. [200] found that when a lead phthalocyanine layer was deposited on a LiNbO$_3$ sensor (100 MHz), the interaction with NO$_2$ in N$_2$ increased the conductivity. Via acousto-electric coupling of the electric potential (associated with the wave because of the piezoelectric effect) to charge carriers in the phthalocyanine, a frequency change was obtained. When an underlying layer of Cr was applied, most of the response to NO$_2$ disappeared owing to short-circuiting of the electric field. Nieuwenhuizen et al. [56, 63, 184, 201–205] also extensively studied metallophthalocyanines as chemical interfaces. These

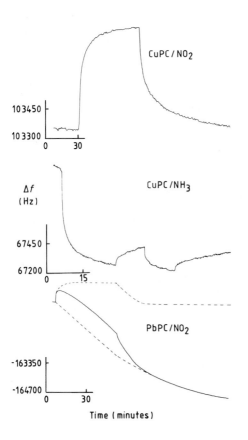

Figure 13-10.
Response curve of an 80-MHz copper phthalo-cyanine sensor to NO$_2$ (top) and to NH$_3$ (middle) and a response curve (bottom) of a lead phthalocyanine sensor to NO$_2$ (after Nieuwenhuizen et al. [56]).

layers were applied by PVD. Different central metal ions were involved in their studies (Table 13-1). In Figure 13-10 a typical response of an 80-MHz copper phthalocyanine sensor at 150 °C is shown. CO, CO_2, H_2O, SO_2, CH_4, and toluene hardly interfered. To illustrate that the response of other gases could originate from different response mechanisms, ie, from different physico-chemical interactions, the response to NH_3 is also depicted in Figure 13-10, indicating partial irreversible behavior. After a few responses complete insensitivity to NH_3 was obtained whereas the sensor still responded to NO_2. The response of lead phthalocyanine to NO_2 in air showed a peculiar behavior, as is also depicted in Figure 13-10. This was due to destruction of the phthalocyanine by NO_2. Studies concerning the response times indicated two response mechanisms with different activation energies. It is shown in Table 13-5 that different sensitivities were obtained when the chemical interface was applied on top of the sensor as a whole or only in between the IDTs. The $Si-SiO_2-ZnO/Cu$ phthalocyanine system was also studied in order to develop SAW technology on silicon. The sensitivity was found to be higher with the ZnO-based sensor. It was also found that ZnO requires a passivation layer because it interacts with gases itself.

Organophosphorus compounds: Detection (warning, monitoring) forms an integral part of a total defense system against chemical warfare agents. Several investigations in the USA and The Netherlands have concerned the search for new sensors to detect chemical warfare agents [206, 207]. The work was mostly related to finding suitable chemical interfaces for the detection of chemical warfare agents. Especially the knowledge obtained by Guilbault et al. [1, 11] in the field of BAW sensors could easily be translated to SAW sensor. A large number of studies by US defense groups [14, 208–226] used dimethyl methylphosphonate (DMMP), diisopropyl methylphosphonate (DIMP), or tributyl phosphate (TBP) as model compounds. Pyridinium salts [222] and phthalocyanine derivatives [220] have been studied as chemical interfaces. Almost all other coatings were polymers with which some kind of selectivity was obtained. However, as the groups concerned adhered to the concept of sensor arrays, they did not seek the ultimate selectivity. As it is known that amidoximes react with DMMP (nucleophilic phosphorylation), Wohltjen et al. [218, 219] studied the terpolymer butadiene-acrylonitrile-acrylamidoxime (PAOX). The polymer fluoropolyol (FPOL) showed interesting

Table 13-5. Sensitivity (S in Hz ppm^{-1}) and 80% response time (RT in min) of different sensor types with copper phthalocyanine as the chemical interface for NO_2, NH_3, and H_2O at 150 °C (from Nieuwenhuizen et al. [205]).

Substrate	Chemical interface	Frequency (MHz)	NO_2 S	NO_2 RT	NH_3 S	H_2O S
Quartz	In between IDTs	52	19.8	3	−1.6	0.0
Quartz	All over sensor	80	−74.0	7	3.7	0.0
Quartz	Immobilized	52	0.3	1	−0.6	0.003
Quartz	Quartz[1]	80	0.0	–	0.0	0.0
ZnO-SiO$_2$-Si	In between IDTs[1]	70	200.0	10	−1.0	0.0
ZnO-SiO$_2$-Si	All over sensor[1]	70	920.0	7	−5.2	0.0
ZnO-SiO$_2$-Si	ZnO[1]	70	40.0	10	−0.7	−0.1

1) single delay line measurements.

properties, as did derivatives of poly(ethylene maleate) (PEM), poly(vinylpyrrolidone) (PVP), poly(epichlorohydrin) (PECH), poly(butadiene-acrylonitrile) (PBAN) and many others. Some of these polymers are shown in Figure 13-11. In Table 13-6 the responses of both DMMP and TBP on a 112-MHz quartz sensor are given [210]. The responses were normalized with respect to the mass of the chemical interface. In Table 13-6 also the calculated responses are given for 1 µg/L DMMP or TBP on these coatings. As the noise level was 10–15 Hz it can be observed that only PEM and FPOL gave an acceptable response at the 1 µg/L level, which is still high from a chemical warfare point of view.

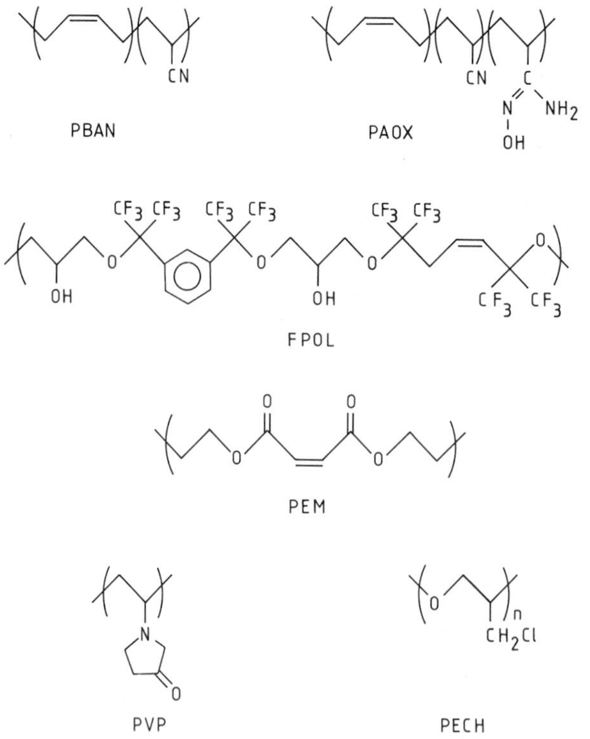

Figure 13-11.
Examples of polymeric chemical interfaces for SAW sensors for organophosphorus compounds as studied by U.S. defense research groups (for abbreviations, see text).

Styrene: A SAW sensor for styrene was described by Zellers et al. [227]. On one delay line of a dual delay line SAW sensor Pt(II) (ethylene)(pyridine)Cl$_2$ embedded in an amorphous poly(isobutlene) matrix, was applied. The other delay line was coated with a polymer. Styrene reacts with the complex as illustrated in Figure 13-12. The rate of the frequency change, due to the irreversible adsorption of styrene which replaces ethylene in the complex, was a function of the styrene concentration. The device could be regenerated at room temperature by exposure to large amounts of ethylene gas. Selectivity was claimed by measuring responses to mixtures of styrene and ethylbenzene and of toluene and 2-butanone. However, it would be interesting to see what the response to divinylbenzene, another relevant interferent, is.

Cyclopentadiene: For the description by Snow and Wohltjen [57] of a cyclopentadiene sensor, the reader is referred to Section 13-4.

Table 13-6. Normalized responses of dimethyl methylphosphonate (DMMP) and tributyl phosphate (TBP) on different coatings on a 112-MHz quartz SAW sensor and calculated responses to 1 µg/L of DMMP and TBP on 100-kHz coating (after Ballantine et al. [210]).

Coating[1]	Normalized responses		Calculated responses, 1 µg/L (Hz)	
	DMMP	TBP	DMMP	TBP
PBAN	450	6.4	8	0.05
PAOX	830	17	15	0.14
FPOL	15 800	13	285	0.11
PEM	6 500	24	117	0.20
PVP	40	33	0.7	0.28
PECH	550	7.2	10	0.06

1) for abbrevations, see text.

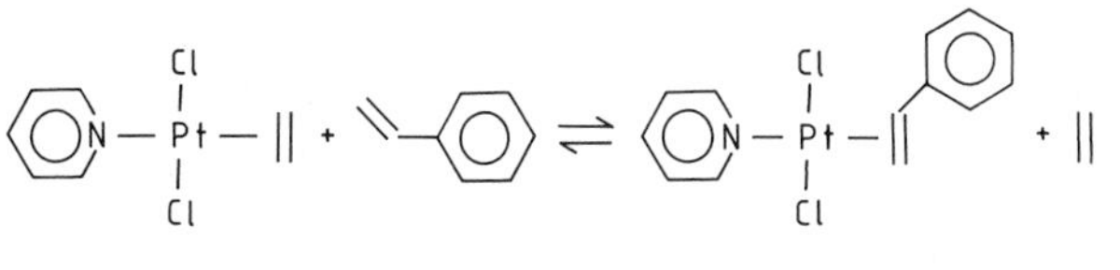

Figure 13-12. Reaction of irreversible adsorption of styrene on a platinum complex according to Zellers et al. [227].

Table 13-7. SAW sensors for inorganic gases.

Analyte	Sensor material	Frequency (MHz)	Temperature (°C)	Chemical interface	References
H_2	$LiNbO_3$	75	–	Pd	[34, 179, 181]
H_2	Quartz	23	–	Pd	[180]
H_2	ZnO	81	–	Pd	[38]
O_2	–	–	–	Metal oxide	[182]
I_2	Quartz	52	–	M-PC[1]	[183]
Halogens	Quartz	50–80	30–150	Cu-PC	[184]
H_2O	Quartz	97	–	Polyimide	[191]
H_2O	$LiNbO_3$	75	0–100	Polymer	[187, 188]
H_2O	$LiNbO_3$	75	–10–60	Polymer	[189, 190]
H_2O	$LiNbO_3$	210–250	25	$LiNbO_3$	[192]
H_2O	Quartz	40	30–70	Polyimine	[193]
H_2S	$LiNbO_3$	60	130	WO_3	[33, 194–196]
NH_3	Quartz	30	25	Pt	[199]
SO_2	$LiNbO_3$	60	–	TEA[2]	[32, 33, 196–198]
NO_2	Quartz	40–100	30–150	M-PC	[56, 63, 184, 201–204]
NO_2	$LiNbO_3$	110	80	Pb-PC	[200]
NO_2	ZnO	150	70	Cu-PC	[205]

1) PC = phthalocyanine; M = metal.
2) TEA = triethanolamine.

Table 13-8. SAW sensors for organic vapors.

Analyte	Sensor material	Frequency (MHz)	Temperature (°C)	Chemical interface	References
DMMP[1]	Quartz	31–112	–	PCs	[208, 220]
DMMP[1]	Quartz	52	–	Pyr salts[2]	[222]
DMMP[1]	Quartz	–	–	Amidoximes	[227, 219]
DMMP[1]	Quartz	31–300	20–80	Polymers	[14, 208–217, 220, 221, 223–226]
Styrene	Quartz	30	–	Pt complex	[227]
Cyclopentadiene	Quartz	31	25	PEM[3]	[57]
OC-comp.[4]	–	–	–	–	[185, 186]
Various	Quartz	31–290	–	Polymers	[229]
Various	LiNbO₃	–	–	Polymers	[29]
Various	ZnO	109	24	ZnO	[35–37]
Various	Quartz	97	–	Quartz	[229]
Various	Quartz	52, 112	–	Polymers	[230, 231]
Various	Quartz	158	35	Polymers	[232]
Various	Quartz	97	25	Polyimide	[233]

1) DMMP   = dimethyl methylphosphonate.
2) Pyr   = pyridinium.
3) PEM   = poly(ethylene maleate).
4) OC-comp. = organic chlorine-containing compounds.

Other SAW gas sensors: A number of papers [29, 35–37, 228–233] have been published dealing with SAW measurement of different gases. Usually attempts were made to correlate SAW sensor properties with physico-chemical properties of the gases in relation to the chemical interfaces (usually various kinds of polymers or unoccupied delay lines). In Tables 13-7 and 13-8 an overview is given of SAW gas sensors for inorganic and organic analytes published since Wohltjen introduced the sensor principle in 1979.

13.6   Mass-Sensitive Sensors in the Liquid Phase

13.6.1   Bulk Surface Wave Liquid Sensors

BAW chemical sensors can be applied in liquids (Table 13-9). Some examples are given below.

Electrodeposition of metals: It was found by Nomura et al. [234–237] that Ag, Cu, and Hg ions could by electrodeposited on one of the electrodes of a BAW sensor using a voltage of −0.2 V versus a Ag/AgCl electrode. The frequency change of this electrogravimetric system was proportional to the Ag concentration in the 0.1–10 μmol/L range after deposition for 5 min and in the 0.001–0.01 μmol/L range by recycling 20 ml of the solution for 3 h [235]. Selectivity could be obtained by masking other ions with EDTA.

Table 13-9. Overview of BAW liquid sensors for organic and inorganic analytes.

Analyte	(Bio)chemical interface	References
Ag	Electrodeposition	[234–237]
Cu	Poly(vinylpyridine)	[238]
Hg	Au	[239]
Pb	Cu(II) oleate	[240]
NH_3	Ni(II) dimethylglyoximate	[117]
CN^-	Electrodeposition	[241–243]
I^-	Electrodeposition	[244–246]
SO_2	Quadrol	[117]
	Reaction to Hg[1]	[247, 248]
SO_4^{2-}	Precipitation with Ba^{2+}	[249]
Alcohols	Lipid membranes	[250]
Odorants	Lipid membranes	[251]
Strychnine[2]	Lipid membranes	[252]
Enzyme[3]	Glucosamine	[253]
Immunoassay	Antibody	[254–258]
Polynucleotide	Other polynucleotide	[259]

1) see Hg.
2) and other bitter substances.
3) galactosyl transferase.

Copper: Poly(2-vinylpyridine) and poly(4-vinylpyridine) were used by Nomura and Sakai [238] to adsorb Cu^{2+} in the 5–35 µmol/L range in a maleate buffer at pH 6.6 with a contact time of 5 min. The bound Cu^{2+} could be removed with EDTA. Fe and Cd ions interfered.

Mercury: As with BAW gas sensors for Hg, Ho et al. [239] used a gold-coated sensor in the 5–100 ng range. Reversibility was achieved by thermal desorption at 170 °C.

Lead: Nomura [240] mentioned copper oleate as the chemical interface for the determination of Pb ions in solutions.

Ammonia: Webber and Guibault [117] used nickel(II) dimethylglyoxime for the detection of NH_3 in water up to 0.45 mol/L.

Cyanide: Silver-coated sensors were used for the detection of CN^- by Nomura et al. At first the sensor was dried before and after the electrodeposition procedure [241]. Later the frequency was monitored during passage of a CN^- solution [242]. Finally, a one-drop technique was developed [243]. Linear behavior was found in the ranges 0.1–10 µmol/L [241], 1–5 µmol/L [242] and 0.1–4 µmol/L [243], respectively. EDTA was used as a masking agent. In that case only Ag and Hg ions interfered.

Iodide: Nomura et al. [244, 245] studied the detection of I^- by an electrodeposition technique. Frequency changes could be measured in the range 0.5–7 µmol/L. $S_2O_3^{2-}$, CN^-, and Fe, Hg, and Ag ions interfered. Procedures for preventing these interferences were given. Yao et al. [246] employed the same technique. The only interference they found was from Br^-. Linearity was observed in the range 0.1–30 µmol/L. Reversibility of the detector was achieved by immersion in ammonia solution for 5–10 min.

Sulfur dioxide: Using quadrol as the chemical interface, SO_2 was measured in the ppb range by Webber and Guilbault [117]. Suleiman and Guilbault [247, 248] converted SO_2 to Hg and detected the latter by using a gold-coated BAW sensor.

Sulfate: Nomura [249] studied the detection of SO_4^{2-} ions in the range 0.5–10 ppm by precipitating them with Ba^{2+}. The sensor could be regenerated by treating it with an EDTA solution.

Alcohols, Odorants and Bitter substances: Okahata et al. [250–252] studied the adsorption of various compounds on a lipid bilayer membrane matrix. In that way alcohols, odorants, and bitter substances such as strychnine could be measured.

Enzymes: Grande et al. [253] detected an enzyme (galactosyl transferase) by using its substrate (glucosamine) as the chemical interface.

Immunological reactions: Several examples were found dealing with the detection of antigens using antibodies as the biochemical interface or the reverse [254–258]. Usually the sensor frequency was measured before and after immersion in the antigen solution. Thompson et al. [256] recorded frequency values for the reaction of human IgG and IgA with goat antihuman IgG immobilized on thin films of a polyacrylamide gel and directly to the crystal/ electrode surface. Muramatsu et al. [258] immobilized Protein A on a sensor. The frequency change resulting from the interaction between Protein A and human IgG correlated with the human IgG in the range 10^{-6}–10^{-2} g/L. In Figure 13-13 this correlation is shown after various reaction times at 30 °C.

Polynucleotides: The very selective interaction between a polynucleotide and a complementary other polynucleotide was observed by Fawcett and Evans [259]. Poly(butyl

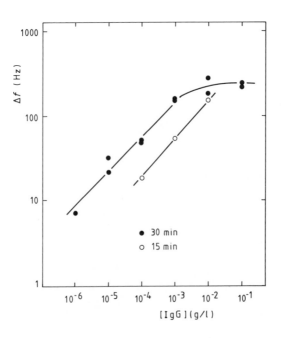

Figure 13-13.
Correlation between human IgG and frequency change of a BAW sensor coated with Protein A after 30 and 15 min contact time at 30 °C (after Muramatsu et al. [258]).

methacrylate) was applied to both sides of a BAW sensor followed by polynucleotide 1 using triazidotrinitrobenzene as a linker. Polynucleotide 2 (1 mg/L) was applied at 46 °C for 25 min for hybridization. A shift in resonance frequency of -613 Hz was observed.

13.6.2 Surface Acoustic Wave Liquid Sensors

Sensors applying pure SAW are not implementable in the liquid phase although some controversy exists in the literature as to whether this is true or not [59, 61, 260–262]. When the wave path is in contact with a liquid, acoustic energy is dissipated into the liquid, causing severe attenuation. In this way a so-called leaky-SAW sensor is obtained [262]. Sometimes, especially at low frequencies, the attenuation is acceptable. Bastiaans et al. [59–61] reported on the detection of influenza A virus employing an immunological interaction. Japanese workers [261] claimed to have sensed ions in solutions with valinomycin, a biological complexing agent.

So far, this paper has only dealt with pure SAW which work in a pseudo-semi-infinite space, ie, the decay length of the wave amplitude is very small in comparison with the thickness of the crystal. When the wave path is very thin other wave types are generated: Lamb waves. A very thin membrane was applied by White et al. [263–267] by underetching the wave-path area. Another sensor system used a very thin cantilever beam acting as a plate wave resonator [268–270]. These types of sensors can find application in both the gas phase and the liquid phase.

Martin et al. [271–273] generated horizontally polarized shear waves. As only viscous coupling between the wave and the liquid occurred and no energy was dissipated into the liquid by radiation of SAW, this sensor system could be employed as both a viscosity sensor and as a chemical sensor.

13.7 Conclusions and Outlook

Mass-sensitive devices and especially SAW sensors are very promising as chemical sensors as they employ the most basic physical effect which occur when one chemical substance interacts with another: a mass effect. No additional electrical or optical effects are required. One of the most critical phases in chemical sensors research, the search for a suitable chemical interface, is thus not influenced by these extra requirements. Any compound which is selective and which also meets other general chemical sensor requirements can be applied as a chemical interface.

Research on chemical sensors in general is multidisciplinary. In the future, research on mass-sensitive chemical sensor chemistry will focus on the (bio)chemical interface with respect to selectivity and stability. Also in the physical and electronic disciplines a lot of work still needs to be done, dealing with sensitivity (higher frequencies, other wave types), miniaturization, and integration on silicon.

13.8   References

[1] Hlavay, J., Guilbault, G. G., *Anal. Chem.* **49** (1977) 1890–1898.
[2] Webber, L. M., Hlavay, J., Guilbault, G. G., *Mikrochimica Acta* (1978) 351–358.
[3] Guilbault, G. G., *Ion Sel. Electrode Rev.* **2** (1980) 3–16.
[4] Guilbault, G. G., *Int. J. Environ Anal. Chem.* **10** (1981) 89–98.
[5] Guilbault, G. G., *Anal. Proc.* **19** (1982) 68–70.
[6] Alder, J. F., McCallum, J. J., *Analyst* **108** (1983) 1169–1189.
[7] Guilbault, G. G., *Anal. Chem. Symp. Ser.* **17** (1983) 637–643.
[8] Guilbault, G. G., *Proc. Int. Meeting on Chemical Sensors 1983, Fukuoka;* pp. 637–643.
[9] Guilbault, G. G., in: *Methods and Phenomena* Vol. 7, Lu, C., Czanderna, A. W. (eds.); New York: Elsevier, 1984, pp. 251–280.
[10] Ho, M. H., *Proc. 2nd Int. Meeting on Chemical Sensors 1986, Bordeaux;* pp. 639–643.
[11] Guilbault, G. G., Jordan, J. M., *CRC Critical Reviews Anal. Chem.* **19** (1988) 1–28.
[12] White, R. M., *Proc. 41st Annual Frequency Control Symp. 1987;* pp. 333–338.
[13] Nieuwenhuizen, M. S., Barendsz, A. W., *Sens. Actuators* **11** (1987) 45–62.
[14] Wohltjen, H., *Sens. Actuators* **5** (1984) 307–325.
[15] Nieuwenhuizen, M. S., Venema, A., *Sensors Materials,* **5** (1989) 261–300.
[16] Curie, J., Curie, P., *Bull. Soc. Min. Paris* **3** (1880) 90.
[17] King Jr., W. H., *Anal. Chem.* **36** (1964) 1735–1739.
[18] King Jr., W. H., *US Patent 3 164004,* 1965.
[19] King Jr., W. H., *Research and Development* **20**, No. 4 (1969) 28–34 and **20**, No. 5 28–33.
[20] Karasek, F. W., Gibbins, K. R., *J. Chrom. Sci.* **9** (1971) 535–540.
[21] Janghorbani, M., Freund, H., *Anal. Chem.* **45** (1973) 325–332.
[22] Karasek, F. W., Tiernay, J. M., *J. Chrom.* **89** (1974) 31–38.
[23] Karasek, F. W., Guy, P., Hill, H. H., Tiernay, J. M., *J. Chrom.* **124** (1976) 179–186.
[24] Schulz, W. W., King Jr., W. H., *J. Chrom.* **11** (1973) 343–348.
[25] Konash, P. L., Bastiaans, G. J., *Anal. Chem.* **52** (1980) 1928–1931.
[26] Lord Rayleigh, *Proc. London Math. Soc.* **17** (1885) 4–11.
[27] White, R. M., Volmer, F. W., *Appl. Phys. Lett.* **7** (1965) 314–316.
[28] Wohltjen, H., Dessy, R., *Anal. Chem.* **51** (1979) 1458–1464.
[29] Wohltjen, H., Dessy, R., *Anal. Chem.* **51** (1979) 1465–1470.
[30] Wohltjen, H., Dessy, R., *Anal. Chem.* **51** (1979) 1470–1478.
[31] Wohltjen, H., *US Patent 4 312 328,* 1979.
[32] Vetelino, J. F., Lee, D. L., *PCT Int. Appl. WO 83-1511,* 1983; Rijswijk, Holland: European Patent Office.
[33] Bryant, A., Lee, D. L., Vetelino, J. F., *Proc. IEEE Ultrasonics Symp. 1981, Chicago;* New York: IEEE, 1981, pp. 171–174.
[34] D'Amico, A., Palma, A., Verona, E., Sens. Actuators **3** (1982) 31–39.
[35] Martin, S. J., Schwartz, S. S., Gunshor, R. L., Pierret, R. F., *J. Appl. Phys.* **54** (1983) 561–569.
[36] Martin, S. J., Schweizer, K. S., Schwartz, S. K., Gunshor, R. L., *Proc. IEEE Ultrasonics Symp. 1984, Dallas;* New York: IEEE, 1984, pp. 207–212.
[37] Martin, S. J., Schweizer, K. S., Ricco, A. J., Zipperian, T. E., *Proc. 3rd Int. Conf. on Sensors and Actuators 1985, Philadelphia;* New York: IEEE, 1985, pp. 71–73.
[38] Caliendo, C., D'Amico, A., Verardi, P., Verona, E., *Proc. IEEE Ultrasonics Symp. 1988, Chicago;* New York: IEEE, 1988, pp. 569–574.
[39] Venema, A., Vellekoop, M. J., Nieuwkoop, E., Haartsen, J. C., Nieuwenhuizen, M. S., Nederlof, A. J., Barendsz, A. W., *Proc. 4rd Int. Conf. on Sensors and Actuators 1987,* Tokyo: IEE Japan, 1981, pp. 481–486.
[40] Vellekoop, M. J., Nieuwkoop, E., Haartsen, J. C., Venema, A., *Proc. IEEE Ultrasonics Symp. 1987, Denver;* New York: IEEE, 1987, pp. 641–644.
[41] *Applications of piezoelectric quartz crystal microbalances,* Lu, C., Czanderna, A. W. (eds.); Amsterdam: Elsevier, 1984.

[42] Slobodnik Jr., A. J., in: *Acoustic Surface Waves* Oliner, A. A. (ed.); Berlin: Springer Verlag, 1978, Chapter 6.

[43] Maines, J. D., Paige, E. G. S., Saunders, A. F., Young, A. S., *Electronics Lett.* **5** (1969) 678-680.

[44] Morgan, D. P., *Ultrasonics* **11** (1973) 121-131.

[45] Dieulesaint, E., Royer, D., in: *Elastic waves in solids (Applications to signal processing)*; Chichester: Wiley, 1980, Chapter 7.

[46] Smith, W. R., Gerard, H. M., Collings, J. H., Reeder, T. M., Shaw, H. J., *IEEE Trans. Microwave Theory Tech.* **17** (1969) 856-864.

[47] Smith, W. R., Gerard, H. M., Collins, J. H., Reeder, T. M., Shaw, H. J., *IEEE Trans. Microwave Theory Tech.* **17** (1969) 865-873.

[48] Farnell, G. W., Cermak, I. A., Silvester, P., Wong, S. K., *IEEE Trans. Sonics Ultrasonics* **17** (1970) 188-195.

[49] Farnell, G. W., in: *Surface wave filters,* Matthews, H. (ed.); New York: Wiley, 1977, Chapter 1.

[50] Lewis, M. F., *Proc. IEEE Ultrasonics Symp. 1973, Monterey;* New York: IEEE, 1973, pp. 344-347.

[51] Crabb, J., Lewis, M. F., Maines, J. D., *Electron. Lett.* **9** (1973) 195-197.

[52] Bale, R., Lewis, M. F., *Proc. IEEE Ultrasonics Symp. 1974, Milwaukee;* New York: IEEE, 1974, pp. 272-275.

[53] Browning, I., Lewis, M. F., *Proc. IEEE Ultrasonics Symp. 1976, Annapolis;* New York: IEEE, 1976, pp. 256-259.

[54] Dijk, C., van, Nieuwenhuizen, M. S., in: *Bioinstrumentation: Research, Developments and Applications*; Wise, D. K. (ed.); Boston: Butterworths, 1990, pp. 679-698.

[55] Janata, J., *Proc. 2nd Int. Meeting on Chemical Sensors 1986, Bordeaux*; pp. 25-37.

[56] Nieuwenhuizen, M. S., Nederlof, A. J., Barendsz, A. W., *Anal. Chem.* **60** (1988) 230-236.

[57] Snow, A. W., Wohltjen, H., *Anal. Chem.* **56** (1984) 1411-1416.

[58] Barzana, E., Klibanov, A. M., Karel, M., *Appl. Biochem. Biotechn.* **15** (1987) 25-34.

[59] Roederer, J. E., Bastiaans, G. J., *Anal. Chem.* **55** (1983) 2333-2336.

[60] Bastiaans, G. J., Good, C. M., *Proc. 2nd Int. Meeting on Chemical Sensors 1986, Bordeaux;* pp. 618-621.

[61] Bastiaans, G. J., *US Patent 4 735 906,* 1984.

[62] Barendsz, A. W., Nieuwenhuizen, M. S., *Dutch Patent 85-02705,* 1985, *European Patent 239 609,* 1987 and *PCT Int. Appl. WO 87-2135,* 1987.

[63] Nieuwenhuizen, M. S., Nederlof, A. J., Coomans, A., *Fresenius' Z. Anal. Chem.* **330** (1988) 123-124.

[64] Olness, D. U., T. Hirschfeld, *UCID Report 20047,* Livermore CA: Lawrence Livermore NL, 1984.

[65] Hager, H. E., *Sens. Actuators* **7** (1985) 271-283.

[66] Olness, D. U., Hirschfeld, T., Kishiyama, K. I., Lane, J. E., Steinhaus, R., *UCRL Report 93121,* Livermore CA: Lawrence Livermore NL, 1985.

[67] Olness, D. U., Hirschfeld, T., Kishiyama, K. I., Lane, J. E., Steinhaus, R., *Proc. US Army Conf. on Chemical Defense Research 1985,* Aberdeen PG: US Army CRDEC, 1986; pp. 897-902.

[68] Fraser, S. M., Edmonds, T. E., West, T. S., *Analyst* **111** (1986) 1183-1188.

[69] Carey, W. P., Kowalski, B. R., *Anal. Chem.* **58** (1986) 3077-3084.

[70] Carey, W. P., Beebe, K. R., Kowalski, B. R., Illman, D. L., Hirschfeld, T., *Anal. Chem.* **58** (1986) 149-153.

[71] Olness, D. U., Hirschfeld, T., Lane, J. E., Steinhaus, R., *UCID Report 20729,* Livermore/CA: Lawrence Livermore NL, 1986.

[72] Olness, D. U., Hirschfeld, T., Kishiyama, K., Steinhaus, R., Lane, J. E., Leonard, D., *Proc. US Army Conf. on Chemical Defense Research 1986,* Aberdeen PG: US Army CRDEC, 1987; pp. 429-434.

[73] Carey, W. P., Beebe, K. R., Kowalski, B. R., *Anal. Chem.* **59** (1987) 1529-1538.

[74] Olness, D. U., *UCID Report 21094,* Livermore CA: Lawrence Livermore NL, 1987.

[75] Carey, W. P., Beebe, K. R., Kowalski, B. R., *Report TR-35,* Univ. Washington, 1987.

[76] Ballato, A., *US Patent 4 598 224,* 1985.

[77] Miller, R. E., *Proc. US Army Conf. on Chemical Defense Research 1987,* Aberdeen PG: US Army CRDEC, 1988.

[78] Rose, S. L., Grate, J. W., Ballantine, D. S., *Proc. US Army Conf. on Chemical Defense Research 1986,* 421-428.

[79] Rose-Pehrsson, S. L., Grate, J. W., Ballantine, D. S., *Proc. US Army Conf. on Chemical Defense Research 1987,* Aberdeen PG: US Army CRDEC, 1988.

[80] Ballantine, D. S., Rose, S. L., Grate, J. W., Wohltjen, H., *Memorandum Report 5813,* US Naval Research Laboratory, 1986.

[81] Ho, M. H., Guilbault, G. G., Scheide, E. P., *Anal. Chem.* **52** (1982) 1998–2002.

[82] Jordan, J. M., Dissertation, Univ. New Oleans, 1985.

[83] Fogleman, W. W., Shuman, M. S., *Anal. Lett.* **9** (1976) 751–765.

[84] Suleiman, A., Guilbault, G. G., *Anal. Chim. Acta* **162** (1984) 97–102.

[85] Bucur, R. V., *Rev. Roum. Phys.* **19** (1971) 779–786.

[86] Mizutani, F., Abe, S., Yoshida, T., *Nippon Kagaku Kaishi* (1987) 472–476.

[87] Abe, S., Mizutani, F., Abe, I., *Adv. Hydrogen Energy* **5** (1986) 787–792.

[88] Thibault, M., Carballiera, A., *Fr. Patent 2 114 103,* 1972.

[89] Sakaguchi, H., Taniguchi, N., Nagai, H., Niki, G., Adachi, G., Shiokawa, J., *J. Phys. Chem.* **89** (1985) 5550–5552.

[90] Hlavay, J., Guilbault, G. G., *Anal. Chem.* **50** (1978) 965–967.

[91] Alder, J. F., Bentley, A. E., Drew, P. K. P., *Anal. Chim. Acta* **182** (1986) 123–131.

[92] Webber, L. M., Karmarkar, K. H., Guilbault, G. G., *Anal. Chim. Acta* **97** (1978) 29–35.

[93] Gjessing, D. T., Holm, C., Lanes, T., *Electron. Lett.* **3**, No. 4 (1967) 156–157.

[94] Lee, C. W., Fung, Y. S., Fung, K. W., *Anal. Chim. Acta* **147** (1982) 277–283.

[95] Ivashchenko, V. E., Popov, K. V., Rudykh, I. A., *USSR Patent 441 496,* 1974.

[96] Ivashchenko, V. E., Rudykh, I. A., Kolomyitsev, V. P., *USSR Patent 463 901,* 1975.

[97] Karaul'nik, A. E., Gruzinenko, V. B., Smirnov, Y. T., Lebedev, G. A., *USSR Patent 585 464,* 1977.

[98] Savchenko, V. E., Gribova, L. K., *Izv. Vyssh. Uchebn. Zaved. Tekhnol. Tekst. Promsti.* No. 1 (1977) 89–101.

[99] Shimadzu Comp., *European Patent 100 661,* 1983.

[100] Czanderna, A. W., Thomas, T. W., *J. Vac. Sci. Techn.* **A5** (1987) 2412–2417.

[101] Ito, H., *IEEE Trans. Ultrasonics* **34** (1987) 137–141.

[102] King Jr., W. H., *German Patent 1 901 845,* 1970.

[103] Randin, J. P., Zullig, F., *Sens. Actuators* **11** (1987) 319–328.

[104] Tahara, S., Kobayashi, J., Oka, S., *Proc. Int. Meeting on Chemical Sensors 1983, Fukuoka;* pp. 405–409.

[105] Bristow, Q., *J. Geochim. Explor.* **1** (1972) 55–76.

[106] Scheide, E. P., Taylor, J. K., *Environ Sci. Techn.* **8** (1974) 1097–1099.

[107] Scheide, E. P., Taylor, J. K., *Am. Chem. Soc. Div. Environm. Chem. Prep.* **14** (1974) 329–335.

[108] Scheide, E. P., Taylor, J. K., *Am. Ind. Hyg. Assoc. J.* **36** (1975) 897–901.

[109] Scheide, E. P., Warnar, R. B. J., *J. Am. Ind. Hyg. Assoc. J.* **39** (1978) 745–749.

[110] Karmarkar, K. H., Guilbault, G. G., *Anal. Chim. Acta* **75** (1975) 111–117.

[111] Hlavay, J., Guilbault, G. G., *Anal. Chem.* **50** (1978) 1044–1046.

[112] Beitness, H., Schroder, K., *Anal. Chim. Acta* **158** (1984) 57–65.

[113] Moody, G. J., Thomas, J. D. R., Yarmo, M. A., *Anal. Chim. Acta* **152** (1983) 225–229.

[114] Lai, C. S. I., Moody, G. J., Thomas, J. D. R., *Anal. Proc.* **22** (1985) 13.

[115] Lai, C. S. I., Moody, G. J., Thomas, J. D. R., *Analyst* **111** (1986) 511–515.

[116] Webber, L. M., Guilbault, G. G., *Anal. Chem.* **48** (1976) 2244–2247.

[117] Webber, L. M., Guilbault, G. G., *Anal. Chim. Acta* **93** (1977) 145–151.

[118] Edmonds, T. E., Fraser, S. M., West, T. M., *Proc. Int. Conf. on Detection and Measurement of Hazardous Substances in the Atmosphere 1982,* London.

[119] Edmonds, T. E., Fraser, S. M., West, T. M., *Anl. Proc.* **22** (1985) 366–367.

[120] Edmonds, T. E., Hepher, M. J., West, T. M., *Anal. Chim. Acta* **207** (1988) 67–75.

[121] Minten, K., Krug, W., *US Patent 4 637 987,* 1987.

[122] Fog, H. M., Rietz, B., *Anal. Chem.* **57** (1985) 2634–2638.

[123] Guilbault, G. G., Lopez-Roman, A., *Environm. Lett.* **2** (1971) 35–45.

[124] Lopez-Roman, A., Guilbault, G. G., *Anal. Lett.* **5** (1972) 225–235.

[125] Cheng, H. M., Wang, H. C., *Huan Ching K'O Hsueh* **2** (1981) 115–118.

[126] Cheney, J. L., Homolya, J. B., *Anal. Lett.* **8** (1975) 175–193.

[127] Cheney, J. L., Homolya, J. B., *Sci. Total Environ* **5** (1976) 69–77.

[128] Frechette, M. W., Fasching, J. L., Rosie, D. L., *Anal. Chem.* **45** (1973) 1765–1766.

[129] Karmarkar, K. H., Guilbault, G. G., *Anal. Chim. Acta* **71** (1974) 419-424.
[130] Frechette, M. W., Fasching, J. L., *Environ Sci. Techn.* **7** (1975) 1135-1137.
[131] Karmarkar, K. H., Webber, L. M., Guilbault, G. G., *Environm. Lett.* **8** (1975) 345-352.
[132] Karmarkar, K. H., Webber, L. M., Guilbault, G. G., *Anal. Chim. Acta* **81** (1976) 265-271.
[133] Cheney, J. L., Norwood, T., Homolya, J. B., *Anal. Lett.* **9** (1976) 361-377.
[134] Suleiman, A., Hahn, E. C., Guilbault, G. G., Cavanaugh, J. R., *Anal. Lett.* **17** (1984) 2205-2211.
[135] Hahn, E. C., Suleiman, A., Guilbault, G. G., Cananaugh, J. R., *Anal. Chim. Acta* **197** (1987) 195-202.
[136] Guilbault, G. G., Lopez-Roman, A., Billdedeau, S. M., *Anal. Chim. Acta* **52** (1972) 421-427.
[137] Chernenko, Z. V., Sinyakova, I. B., Samodumova, I. M., *Dopov. Akad. Nauk Ukr. RSR, Ser. B,* No. 4 (1983) 62-64.
[138] Hayes, J. K., Dwayne, R. W., Jordan, W. S., *Anaesthesiology* **59** (1983) 435-439.
[139] Hamilton, C., Gedeon, A., *J. Phys. Sci. E: Sci. Instrum.* **19** (1986) 271-274.
[140] Cooper, J. B., Edmondson, J. H., Joseph, D. M., Newbower, R. S., *IEEE Trans. Biomed. Eng.* **28** (1981) 459-466.
[141] Kindlund, A., Lundstroem, I., *Sens. Actuators* **3** (1982) 63-77.
[142] Varga Jr., G. M., *US NTIS AD Report 780171/5GA,* 1974.
[143] Guilbault, G. G., *Anal. Chem.* **55** (1983) 1682-1684.
[144] Kindlund, A., Lundstrom, I., *IEEE Trans. Biomed. Eng.* **27** (1980) 544.
[145] Kindlund, A., Sundgren, H., Lundstrom, I., *Sens. Actuators* **6** (1984) 1-17.
[146] Ho, M. H., Guilbault, G. G., Rietz, B., *Anal. Chem.* **55** (1983) 1830-1832.
[147] Bellomo, A., De Robertis, A., De Marco, D., *Inquinamento* **22** (1980) 47-50.
[148] Lai, C. S., Moody, G. J., Thomas, J. D. R., Mulligan, D. C., Stoddart, D. S., Zarzycki, R., *J. Chem. Soc. Perkin Trans.* **2** (1988) 319-324.
[149] Karmarkar, K. H., Guilbault, G. G., *Environm. Lett.* **10** (1975) 237-246.
[150] Tomita, Y., Ho, M. H., Guilbault, G. G., *Anal. Chem.* **51** (1979) 1475-1478.
[151] Sanchez-Pedreno, J. A. O., Drew, P. K. P., Alder, J. F., *Anal. Chim. Acta* **182** (1986) 285-291.
[152] Kristoff, J., Guilbault, G. G., *Anal. Chim. Acta* **149** (1983) 337-341.
[153] Guilbault, G. G., *Anal. Chim. Acta* **39** (1967) 260-264.
[154] Scheide, E. P., Guilbault, G. G., *Anal. Chem.* **44** (1972) 1764-1768.
[155] Guilbault, G. G., Affolter, J., Tomita, Y., Kolesar, E. S., *US Airforce Report SAM-TR-81-6;* Brocks AFB: Aerospace Med. Div., 1981.
[156] Guilbault, G. G., Affolter, J., Tomita, Y., Kolesar, E. S., *Anal. Chem.* **53** (1981) 2057-2060.
[157] Guilbault, G. G., Kristoff, J., Kolesar, E. S., *US Airforce Report SAM-TR-82-14;* Brocks AFB: Aerospace Med. Div., 1982.
[158] Guilbault, G. G., Kristoff, J., Owens, R., *Anal. Chem.* **57** (1985) 1754-1756.
[159] Balog, P. B., Stanford, T. B., Nordstrom, R. J., Burgener, R. C., *AMD Report TR-86-003;* Aberdeen PG: US Army CRDEC, 1986.
[160] Shackelford, W. M., Guilbault, G. G., *Anal. Chim. Acta* **73** (1974) 383-389.
[161] Guilbault, G. G., Tomita, Y., Kolesar, E. S., *US Airforce Report SAM-TR-80-21,* 1980.
[162] Guilbault, G. G., Tomita, Y., *Anal. Chem.* **52** (1980) 1484-1489.
[163] Kolesar, E. S., *Proc. Annu. Sci. Meeting Aerospace Med. Ass., 1981,* pp. 19-20.
[164] Olofsson, G., *FOA Report C 40142-Cl,* 1981.
[165] Kolesar, E. S., Guilbault, G. G., Affolter, J., Kristoff, J., Tomita, Y., Scheide, E. P., Warnar, R. B. J., *US Airforce Report SAM-TR-83-28,* 1983.
[166] Olofsson, G., Jonsson, U., Bergek, S., *Proc. IEEE Ultrasonics Symposium 1985, San Fransisco;* New York: IEEE, 1985, pp. 443-445.
[167] Guilbault, G. G., Tomita, Y., Kolesar, E. S., *Sens. Actuators* **2** (1981) 43-57.
[168] Sant, M. J., van, Dissertation, Univ. New Orleans, 1985.
[169] Guilbault, G. G., Lubrano, G., Ngeh-Ngwainbi, J., Jordan, J., Foley, P., *CRDEC Report CR-87056,* 1987.
[170] Guilbault, G. G., Ngeh-Ngwainbi, J., *NATO ASI Ser.* **C226** (1988) 187-194.
[171] Ngeh-Ngwainbi, J., Foley, P. H., Jordan, J. M., Guilbault, G. G., Palleschi, G., *Proc. 2nd Int. Meeting on Chemical Sensors 1986, Bordeaux;* pp. 515-518.
[172] Ngeh-Ngwainbi, J., Foley, P. H., Kuan, S. S., Guilbault, G. G., *J. Am. Chem. Soc.* **108** (1986) 5444-5447.

[173] Turnham, B. D., Yee, L. K., Luoma, G. A., *Anal. Chem.* **57** (1985) 2120–2124.
[174] Alder, J. F., Isaac, C. A., *Anal. Chim. Acta* **129** (1981) 163–174.
[175] Alder, J. F., Isaac, C. A., *Anal. Chim. Acta* **129** (1981) 175–188.
[176] Fielden, P. R., McGallum, J. J., Stanios, T., Alder, J. F., *Anal. Chim. Acta* **162** (1984) 84–96.
[177] McGallum, J. J., Fielden, P. R., Volkan, M., Alder, J. F., *Anal. Chim. Acta* **162** (1984) 75–83.
[178] Morrison, R. C., Guilbault, G. G., *Anal. Chem.* **57** (1985) 2342–2344.
[179] D'Amico, A., Palma, A., Verona, E., *Proc. IEEE Ultrasonics Symp. 1982, San Diego;* New York: IEEE, 1982, pp. 308–311.
[180] D'Amico, A., Gentilli, M., Verardi, P., Verona, E., *Proc. 2nd Int. Meeting on Chemical Sensors 1986, Bordeaux;* pp. 743–746.
[181] D'Amico, A., Palma, A., Verona, E., *Appl. Phys. Lett.* **41** (1982) 300–301.
[182] Koleshko, V. M., Meshkov, Y. V., *USSR Patent 1 191 817,* 1985.
[183] Snow, A. W., Barger, W. R., Klusty, M., Wohltjen, H., Jarvis, N. L., *Langmuir* **2** (1986) 513–519.
[184] Nieuwenhuizen, M. S., A. J. Nederlof, *Anal. Chem.* **60** (1988) 237–240.
[185] VEB Leipzig, *German Patent 3 220 854,* 1983.
[186] Forke, K., Zuerich, V., Clemens, H. G., *East German Patent 243 992,* 1985.
[187] Brace, J. G., Sanfelippo, T. S., *Proc. 4th Int. Conf. on Sensors and Actuators 1987,* Tokyo: IEE Japan, 467–470.
[188] Brace, J. G., Sanfelippo, T. S., Joshi, S., *Sens. Actuators* **14** (1988) 47–68.
[189] Joshi, S. G., Brace, J. G., *Proc. Int. Symp. on Moisture and Humidity 1985;* pp. 225–227.
[190] Joshi, S. G., Brace, J. G., *Proc. IEEE Ultrasonics Symp. 1985, San Francisco;* New York: IEEE, 1985, pp. 600–603.
[191] Martin, S. J., Frye, G. C., Ricco, A. J., Zipperian, T. E., *Proc. IEEE Ultrasonics Symp. 1987, Denver;* New York: IEEE, 1987, pp. 563–567.
[192] Huang, P. H., *Proc. 4th Int. Conf. on Sensors and Actuators 1987,* Tokyo: IEE Japan, pp. 462–466.
[193] Nieuwenhuizen, M. S., Nederlof, A. J., *Sens. Actuators, B2* (1990) 97–101.
[194] Vetelino, J. F., Lade, R. K., Falconer, R. S., *Proc. of the 2nd Int. Meeting on Chemical Sensors 1986, Bordeaux;* pp. 688–691.
[195] Vetelino, J. F., Lade, R. K., Falconer, R. S., *IEEE Trans. on Ultrasonics, Ferroelectrics and Frequency Control* **34** (1987) 157–161.
[196] Bryant, A., Poirier, M., Riley, G., Lee, D. L., Vetelino, J. F., *Sens. Actuators* **4** (1983) 105–111.
[197] Bryant, A., Poirier, M., Lee, D. L., Vetelino, J. F., *Proc. 36th Annual Frequency Control Symp. 1982;* pp. 276–283.
[198] Vetelino, J. F., *Proc. 36th Annual Frequency Control Symp. 1982,* pp. 284–289.
[199] D'Amico, A., Petri, A., Verardi, P., Verona, E., *Proc. IEEE Ultrasonics Symp. 1987, Denver;* New York: IEEE, 1987, pp. 633–636.
[200] Ricco, A. J., Martin, S. J., Zipperian, T. E., *Sens. Actuators* **8** (1985) 319–333.
[201] Barendsz, A. W., Vis, J. C., Nieuwenhuizen, M. S., Vellekoop, M. J., Chijsen, W. J., Venema, A., *Proc. IEEE Ultrasonics Symp. 1985, San Francisco;* New York: IEEE, 1985, pp. 586–590.
[202] Nieuwenhuizen, M. S., Barendsz, A. W., Nieuwkoop, E., Vellekoop, M. J., Venema, A., *Electronics Letters* **22** (1986) 184–185.
[203] Venema, A., Nieuwkoop, E., Vellekoop, M. J., Barendsz, A. W., Nieuwenhuizen, M. S., *Sensors Actuators* **10** (1986) 47–64.
[204] Venema, A., Nieuwkoop, E., Vellekoop, M. J., Chijsen, W. J., Barendsz, A. W., Nieuwenhuizen, M. S., *IEEE Trans. on Ultrasonics, Ferroelectrics and Frequency Control* **34** (1987) 148–155.
[205] Nieuwenhuizen, M. S., Nederlof, A. J., Vellekoop, M. J., Venema, A., *Sens. Actuators,* **19** (1989) 385–392.
[206] Barendsz, A. W., Nieuwenhuizen, M. S., *Proc. Int. Symp. Protection Against Chemical Warfare 1986, Stockholm;* pp. 165–172.
[207] Nieuwenhuizen, M. S., Barendsz, A. W., *Int. J. of Environm. Anal. Chem.* **29** (1987) 105–118.
[208] Barger, W. R., Giuliani, J., Jarvis, N. L., Snow, A. W., Wohltjen, H., *J. Environ Sci. Health* **B20** (1985) 359–371.
[209] Ballantine, D. S., Grate, J. W., Wohltjen, H., Snow, A. W., *Proc. US Army Conf. on Chemical Defense Research 1985;* Aberdeen PG: US Army CRDEC, 1986, pp. 107–112.
[210] Ballantine, D. S., Rose, S. L., Grate, J. W., Wohltjen, H., *Anal. Chem.* **58** (1986) 3058–3066.

[211] Barger, W. R., Wohltjen, H., Snow, A. W., Lint, J., Jarvis, N. L., *ACS Symposium Series* **309** (1986) 155–165.

[212] Grate, J. W., Snow, A. W., Ballantine, D. S., Wohltjen, H., Abraham, M., McGill, R. A., Sasson, P., *Proc. 4th Int. Conf. on Sensors and Actuators 1987,* Tokyo: IEE Japan, pp. 579–582.

[213] Grate, J. W., Snow, A. W., Ballantine, D. S., Wohltjen, H., Abraham, M., McGill, R. A., Sasson, P., *Proc. US Army Conf. on Chemical Defense Research 1987,* Aberdeen PG: US Army CRDEC, 1988.

[214] Wohltjen, H., Ballantine, D. S., Snow, A. W., Jarvis, N. L., *Proc. US Army Conf. on Chemical Defense Research 1984;* Aberdeen PG: US Army CRDEC, 1985, pp. 129–134.

[215] Wohltjen, H., Jarvis, N. L., Snow, A. W., Giuliani, J., Barger, W. R., *Proc. Int. Symp. Protection Against Chemical Warfare 1983, Stockholm;* pp. 51–59.

[216] Wohltjen, H., Snow, A. W., Lint, J., Jarvis, N. L., *Proc. US Army Conf. on Chemical Defense Research 1983,* Aberdeen PG: US Army CRDEC, *1984.*

[217] Wohltjen, H., Snow, A. W., Ballantine, D. S., *Proc. 3rd Conf. on Sensors and Actuators 1985, Philadelphia;* New York: IEEE, 1985, pp. 66–70.

[218] Jarvis, N. L., Lint, J., Snow, A. W., Wohltjen, H., *Proc. US Army Conf. on Chemical Defense Research 1983,* Aberdeen PG: US Army CRDEC, 1984.

[219] Lint, J., Wohltjen, H., Jarvis, N. L., Snow, A. W., *ACS Symposium Series* **309** (1986) 309–319.

[220] Ballantine, D. S., Snow, A. W., Klusty, J. W., Chimas, G., Wohltjen, H., *Memorandum Report 5865,* Naval Research Laboratory, 1986.

[221] Ballantine, D. S., Grate, J. W., Abraham, M., McGill, R. W., Sasson, P., *Proc. US Army Conf. on Chemical Defense Research 1987,* Aberdeen PG: US Army CRDEC, 1988.

[222] Katritzsky, A. R., Offerman, R. J., Wang, Z., *Proc. US Army Conf. on Chemical Defense Research 1987,* Aberdeen PG: US Army CRDEC, 1988.

[223] Miller, R. E., Davis, D., Parsons, J. A., *Proc. US Army Conf. on Chemical Defense Research 1987,* Aberdeen PG: US Army CRDEC, 1988.

[224] Sprague, L. G., Snow, A. W., Soulen, R. L., Grate, J. W., Ballantine, D. S., Wohltjen, H., *Proc. US Army Conf. on Chemical Defense Research 1987,* Aberdeen PG: US Army CRDEC, 1988.

[225] Snow, A. W., Ballantine, D. S., Whitney, T., Barger, W. R., Klusty, M., Wohltjen, H., Grate, J. W., Chaput, D., *Proc. US Army Conf. on Chemical Defense Research 1986;* Aberdeen PG: US Army CRDEC, 1987; pp. 695–703.

[226] Klusty, M., Barger, W. R., Snow, A. W., *Proc. US Army Conf. on Chemical Defense Research 1987,* Aberdeen PG: US Army CRDEC, 1988.

[227] Zellers, E. T., White, R. M., Rappaport, S. M., Wenzel, S. W., *Proc. 4rd Int. Conf. on Sensors and Actuators 1987,* Tokyo: IEE Japan, pp. 459–461.

[228] Wohltjen, H., Snow, A. W., Barger, W. R., Ballantine, D. S., *IEEE Trans. on Ultrasonics, Ferroelectrics and Frequency Control* **34** (1987) 172–178.

[229] Ricco, A. J., Ginley, D. S., Zipperian, T. E., Martin, S. J., *IEEE Trans. on Ultrasonics, Ferroelectrics and Frequency Control* **34** (1987) 143–147.

[230] Ballantine, D. S., Snow, A. W., Grate, J. W., *Proc. US Army Conf. on Chemical Defense Research 1986;* Aberdeen PG: US Army CRDEC, 1987, pp. 819–824.

[231] Ballantine, D. S., Wevv, D., Allen, D., Grate, J. W., *Proc. US Army Conf. on Chemical Defense Research 1987,* Aberdeen PG: US Army CRDEC, 1988.

[232] Grate, J. W., Snow, A. W., Ballantine, D. S., Wohltjen, H., Abraham, M. H., McGill, R. A., Sasson, V., *Anal. Chem.* **60** (1988) 869–875.

[233] Martin, S. J., Frye, G. C., Ricco, A. J., Zipperian, T. E., *Proc. IEEE Ultrasonics Symp. 1987, Denver;* New York: IEEE, 1987, pp. 563–567.

[234] Nomura, T., Maruyama, M., *Anal. Chim. Acta* **147** (1983) 365–369.

[235] Nomura, T., Iijima, L., *Anal. Chim. Acta* **131** (1981) 97–102.

[236] Nomura, T., Nagamune, T., *Anal. Chim. Acta* **155** (1983) 231–234.

[237] Nomura, T., Tsuge, K., *Anal. Chim. Acta* **169** (1985) 257–262.

[238] Nomura, T., Sakai, M., *Anal. Chim. Acta* **183** (1986) 301–305.

[239] Ho, M. H., Guilbault, G. G., Scheide, E. P., *Anal. Chim. Acta* **130** (1981) 141–147.

[240] Nomura, T., *Anal. Chim. Acta* **182** (1986) 261–265.

[241] Nomura, T., Hattori, O., *Anal. Chim. Acta* **115** (1980) 323–326.

[242] Nomura, T., Minemura, A., *Nippon Kagaku Kaishi* (1980) 1621–1625.

[243] Nomura, T., *Anal. Chim. Acta* **124** (1981) 81–84.
[244] Nomura, T., Mimatsu, T., *Anal. Chim. Acta 143* (1982) 237–241.
[245] Nomura, T., Watanabe, M., West, T. M., *Anal. Chim. Acta* **175** (1985) 107–116.
[246] Yao, S., Dan, S., Nie, L., *Anal. Chim. Acta* **209** (1988) 213–222.
[247] Suleiman, A., Guilbault, G. G., *Anal. Lett.* **17** (1984) 1927–1929.
[248] Suleiman, A., Guilbault, G. G., *Anal. Chem.* **56** (1984) 2964–2966.
[249] Nomura, T., *Buseki Kagaku* **36** (1987) 93–97.
[250] Okahata, Y., Ebato, H., Ye, X., *J. Chem. Soc. Chem. Comm.* (1988) 1037–1039.
[251] Okahata, Y., Shimizu, O., *Langmuir* **3** (1987) 1171–1173.
[252] Okahata, Y., Ebato, H., Taguchi, K., *J. Chem. Soc. Chem. Comm.* (1987) 1363–1365.
[253] Grande, L. H., Geren, C. R., Paul, D. W., *Sens. Actuators* **14** (1988) 387–403.
[254] Rice, T. K., *US Patent 4 314 821,* 1982.
[255] Roederer, J. E., Bastiaans, G. J., *Anal. Chem.* **55** (1983) 2333–2336.
[256] Thompson, M., Dhaliwal, G. K., Arthur, C. L., *Anal. Chem.* **58** (1986) 1206–1209.
[257] Thompson, M., Dhaliwal, G. K., Arthur, C. L., Calabrese, G. S., *IEEE Trans. Ultrasonics* **34** (1987) 127–135.
[258] Muramatsu, H., Dicks, J. M., Tamiya, M., Karube, I., *Anal. Chem.* **59** (1987) 2760–2763.
[259] Fawcett, N. C., Evans, J. A., *PCT Int. Appl. WO 87-2066,* 1987.
[260] Calabrese, G. S., Wohltjen, H., Roy, M. K., *Anal. Chem.* **59** (1987) 833–837.
[261] Hitachi Ltd., *Japanese Patent 60-73348,* 1985.
[262] Moriizumi, T., Unno, Y., Shiyokawa, S., *Proc. IEEE Ultrasonics Symp. 1987, Denver;* New York: IEEE, 1987, pp. 579–582.
[263] Muller, R. S., White, R. M., *US Patent 4 361 026,* 1980.
[264] Chuang, C., White, R. M., Bernstein, J. J., *Electronic Device Letters* **3** (1982) 145–148.
[265] Uozumi, K., Ohsone, K., White, R. M., *Anal. Chem.* **55** (1983) 2333–2336.
[266] White, R. M., Wicher, P. J., Wenzel, S. W., Zellers, E. T., *IEEE Trans. on Ultrasonics, Ferroelectrics and Frequency Control* **34** (1987) 162–171.
[267] Zellers, E. T., White, R. M., Wenzel, S. W., *Sens. Actuators* **14** (1988) 35–45.
[268] O'Connor, J. M., *European Patent 72 744,* 1982.
[269] Kolesar, E. S., *US Patent 4 549 427,* 1985.
[270] Hou, J., Vaart, H. van der, *Proc. IEEE Ultrasonics Symp. 1987, Denver;* New York: IEEE, 1987, pp. 573–578.
[271] Martin, S. J., Ricco, A. J., Hughes, R. C., *Proc. 4th Int. Conf. on Sensors and Actuators 1987,* Tokyo: IEE Japan, pp. 478–481.
[272] Ricco, A. J., Martin, S. J., *Appl. Phys. Letters* **50** (1987) 1474–1476.
[273] Ricco, A. J., Martin, S. J., Frye, G. C., Niemcyk, T. M., *Proc. Solid-State Sensors and Actuators Workshop 1988;* pp. 23–26.

Index

List of Symbols and Abbreviations

The following list contains the symbols most frequently used in this book. To avoid redundancy, subscripts are only noted in exceptional cases. References to chapters (where the quantities are explained in more detail) are only given for symbols with special meanings or in cases of uncommon use.

Symbol	Designation	Chapter
a	activity	
	sensitivity	6
	radius of microelectrode	5
a_0	initial surface area	11
a_i	activity of species i	
a_p	poisoned surface area	11
a_w	cross-sectional area of wire	11
A	area	
	aperture	13
A_k	vector of a_{ik} of normalized sensor responses	6
$A, A(f_c)$	loop gain	13
A_{eff}	catalytically active area	6
b	channel width	10
	distance between electrodes	4
b_A, b_B	adsorption equilibrium constant of species A, B	11
B	Biot number	
c	concentration	
c_i	concentration of ion i	
C	capacitance	
	bulk gas concentration	11
	elasticity of a crystal	13
$C(V)$	function of voltage	10
C_1	capacitance	13
C_f	mass sensitivity	4
C_s	function of the metal structure	10
C_V	heat capacity at constant volume	11
Cl	clorinity of sea water	7
d	thickness, distance	
d_i	insulator thickness	
D	diffusion coefficient	
$\tilde{D}$	diffusion coefficient	9
D_{12}	binary diffusion coefficient	11
D_K	Knudsen diffusion coefficient	8
D_s	standard diffusion coefficient	8

Symbol	Designation	Chapter
e	elementary charge	
e_0	elementary charge	7
emf	electromotive force	
E	electric field strength	
	activation energy	11
	open-circuit voltage, electromotive force (emf)	5
	potentials	7
	voltage	1, 8
E	functional or calibration matrix	6
ΔE	activation energy	6
E_a	activation energy	9, 11
$E_{Ag/AgCl}$	reference electrode potential	10
E_C	conduction band edge energy	
E_{CE}	voltage of the counter electrode	7
E_F	Fermi energy	
E_{FM}	Fermi energy of the metal	10
E_{FS}	Fermi energy of the semiconductor	10
E_g	band gap of the semiconductor	10
	activation energy of uncatalysed homogeneous reaction	11
E_i	energy	4
E_{pol}	polarization voltage	7
E_s	species	11
E_V	valence band edge energy	
E_{WE}	effective voltage at the working electrode	7
f	frequency	
	activation function	6
$f(P_{O_2})$	function of reactionscheme for water formation	10
f_0	oscillation frequency, resonance frequency	13
f_c	oscillation frequency	13
f_i	activity coefficient	7
f_N	oscillation frequencies	13
F	Faraday constant	
	electrical mobility	8
F_e	number of unoccupied adsorption sites on surface	10
F_i	number of unoccupied adsorption sites at the metal/insulator interface	10
g	geometric factor	11
G	conductance	
	total free enthalpy	4
ΔG_{cell}	Gibbs free energy of cell reaction	5
h	Thiele modulus	11
h^+	defect electrons	4
H	enthalpy	4
ΔH	heat of combustion, heat of adsorption	11
ΔH_i	oxidation heat of component i	4

Symbol	Designation	Chapter
i	electric current	
$i(t)$	time dependent current	6
i_0	exchange current density	8
I	electric current	
	ionic strength	7
	intensity of light	4
I_d	drain current	10
I_D	diffusion-controlled limiting current	7
I_h	hole current	8
I_i	current of the carriers i	8
I_1	plateau current of diffusion-limited regime	8
j	current density	
J	flux	
k	cell constant	7
	electrical bridge constant	11
	reaction rate constant	6, 11
	Boltzmann constant	8
k_B	Boltzmann constant	
k_f	intrinsic reaction rate constant	11
k_{ij}	selectivity constant	7
k_w	thermal conductivity of platinum wire	11
K	dissociation constant	7
	mass transfer coefficient	11
K, K_{a1}, K_{a2}	function of activity and differently occupied surface sites	10
K_g	thermal conductivity of the gas surrounding the element	11
K_i	accelerating force on ions	7
	characteristic parameters	9
K_{ij}	selectivity coefficients	1, 5
K_L	solubility product	7
K_m, K_1, K_2	net conductance of mixture, $-$ of component 1 or 2	11
K_{xy}	equilibrium constant of the surface ion exchange	8
l	length, distance	
l_i	distance of electron-hole recombination from electrode II	8
L	inductance	
	channel length	10
	mass of crystal	13
L_D	Debye-length	9
L_k	number of different partial pressures	6
m	mass	
	number of fragments upon dissociation	11
	electrical bridge constant	11
	characteristic parameter of defects	4
	intrinsic cell sensitivity	6

Symbol	Designation	Chapter
M	molecular weight	11
M_i	metal interstitial	4
n	counting number	
	exponent in the range 1–2	11
	number of electrons	4, 5, 7, 8
	number of fragments upon dissociation	11
n_c	number of moles of gas in volume V	6
n_e, n_i	number of absorbed hydrogen atoms	10
n_i	reaction orders	4
	characteristic parameter	4, 9
	number of moles of oxidized component i	4
n_i	number of hydrogen atoms at the inner interface	10
n_s	electron concentration at semiconductor surface	10
N	doping density in the semiconductor	10
	frequency constant	13
ΔN	excess concentration	4
N_A	Avogadro number	
	concentration of electrons at the semiconductor/insulator-interface	4
	bulk concentration of holes (acceptor concentration)	10
N_i	number of measuring signals	1
N_s	surface site density	10
O_L	lattice oxygen	4
p	partial pressure	
	dipole moment	
	pyroelectric coefficient	11
	period	13
p_{H_2}	partial pressure of hydrogen	7, 10
p_i	analytical information	4
	partial pressure of component i	1, 4
p_{O_2}	partial pressure of oxygen	4
p_R	reference pressure	8
P	pressure	
	electrical power	
	electrical power consumption	4
	statistical probability	1
P	analytical information	6
ΔP	excess concentration	4
$\Delta P(t)$	net flux of heat into the pyroelectric	11
P_a	electrical power	11
P_{cd}	power drain due to conduction	11
P_{CO_2}	partial pressure of carbon dioxide	10
P_{cv}	power drain due to convection	11
P_{cw}	power lost by conduction along the support wires	11
P_L	sum of heat losses	11
P_M	permeability	4

Symbol	Designation	Chapter
P_{O_2}	partial pressure of Oxygen	10
P_r	heat loss by radiation	11
q	elementary charge	
q_i, q_j	transported charge due to ions i, j	5
Q	quality factor	13
	heat of adsorption	11
Q_{ad}, Q_{react}	heat of adsorption, heat of reaction	4
Q_i	equivalent insulator charge per unit area	4, 10
r	reaction rate	4, 11
	radius	
	radial distance	
	catalytic activity	6
r_d	rate of bulk diffusion	11
R	resistance	
	gas constant	
	response	11
	losses in the crystal	13
ΔR	change of filament resistance	11
R_a	actual rate	11
R_{ct}	charge transfer resistance	8
R_d	rate of diffusion to the element	11
R_e	reaction rate on the bead in absence of EMTC	11
R_g	gas constant per mole	6
R_i	partial resistance	8
s_i	best estimate for a standard deviation	1
$sol(a_i)$	solution of activity	10
S	slope of potentiometric electrodes	7
	salinity of seawater	7
	entropy	4
	area of diffusion hole	8
S_0	sticking coefficient	4
S_i	sensitivity	1
S_{ij}	sensitivity	6
S_k	similarity	6
S_{MX}	solubility product of MX	8
t	time	
$\bar{t}_e$	average electronic transference number	8
t_{90}	90 % level of response time	11
t_K	transference number of the K^{z+} carriers	8
T	temperature	4
u_i	mobility of ion i	4, 5
	flow rate	6

Symbol	Designation	Chapter
$\tilde{u}_i$	electrochemical mobility of the carriers i	8
$\tilde{u}_K$	electrical mobility	8
U	voltage	7
	inner energy	4
v	AC voltage	10
V	voltage	
	DC voltage	10
	volume	
ΔV_{GS}	gate-source voltage	10
$V_O^{\cdot\cdot}$	positively charged oxygen vacancy	4
V_M	metal vacancy	4
V_O	oxygen vacancy	4
V_R	phase velocity	13
w_i	migration speed of ions	7
w_{ik}	statistical weight	6
W	reduced electro-chemical potential	8
W_d	width of depletion layer	10
W_{el}	electrical work	5
W_M, W_S	work funktion of metal, − of semiconductor	10
x	distance	
	species variable	11
	mole fraction	11
x'	measuring signal	1, 4, 6
x_0'	reference signal	1
X	impedance	7
X	measuring signal, sensor signal	6
X_{O_2}	relative concentration of O	8
y	amplitude of output signal	6
$\tilde{Y}$	complex admittance	9
z_i	charge number of ion i	
z_K	electrical mobility	8
Z	partition function	4
	number of electrons per val	8
	complex impedance	9
α	reaction order	11
	Seebeck coefficient	11
	transfer coefficient	5
α_i	real potential of species i	10
α_{ik}	normalized sensor response	6
β	dimensionless pH-sensitive parameter	10
	reaction order	11

Symbol	Designation	Chapter
β_{ik}	exponential non-linearity factor	6
γ	reaction order	11
	partial sensitivity	1
	activity coefficient	8
γ^*	normalization factor of sensor array	6
γ, γ'	proportionality constants	11
γ_{ik}	partial sensitivity	6
δ	reaction order	11
	partial charge	4
	thickness of the diffusion layers	5, 7
Δ	confidence range	1
Δ_k^*	Euclidean distance	6
ε	dielectric constant	
	emissivity of the sensing element	11
	optical constant	4
	electrode potential	5
	relative permittivity	9
	error, random variable	6
	error matrix	6
ε_i	dielectric permittivity of insulator	4, 10
ε_s	dielectric permittivity of semiconductor	4, 10
η	viscosity	7
	overpotential	5
ϑ	coverage	4, 11
κ	electrolytic conductivity	7
λ	empirically derived constant	11
	wavelength	
	mobility	7
Λ	molar conductivity	7
μ	mobility of electrons in the channel	10
	chemical potential	8
	electrochemical potential	5, 8
μ_i	chemical potential of ion, atom, molecule i	
	standard chemical potential of ion i	10
	statisitical average	1
ν	frequency	
ρ	density	
σ	magnitude of induced mobile charge	10
	force constant	11
	Stephan-Boltzmann constant	11
	conductivity	4
	specific conductivity	9
	diffusity of leak aperture	8
σ_e^0	free electron conductivity	8
σ_h^0	electron hole conductivity	8
σ_i	partial conductivity of ion i	4, 5
σ_{ik}	dispersion	6
σ_K	partial conductivity of species K^{z+}	8

Symbol	Designation	Chapter
τ	transit (or delay) time	13
τ_s	time constant of sensor	6
φ	electrical potential	
	phase angle between current and voltage	9
φ_A	phase shift	13
φ_B	difference between semiconductor mid-gap and bulk Fermi level	10
	voltage	10
φ_i	voltage drop accross the insulator	10
$\varphi(f_c)$	loop phase	13
Φ	work function	
Φ_{ms}	difference between the effective work function of the metal and that of the semiconductor	10
χ	surface dipole potential	10
	electron affinity	4, 9, 10
Ξ	selectivity	6
Ψ_0	surface potential	10
Ψ_i	partial specificity	6
ω	angular frequency	
	frequency of AC pumping current	8
Ω	collision integral	11

Abbreviation	Explanation
a.c.	alternating current
AAS	atomic absorption spectroscopy
AES	Auger electron spectroscopy
AES	atomic emission spectroscopy
AFS	atomic fluorescence spectroscopy
AGC	automatic gain control
AIM	adsorption isotherm measurements
AIS	atom inelastic scattering
ALE	atomic layer epitaxy
APS	appearance potential spectroscopy
ARIES	angular resolved ion and electron spectroscopy
ASEA	Allmaenna Svenska Elektriska Aktiebolaget
ASIA	atomizer, source, inductively coupled plasmas in atomic fluorescence spectroscopy
ATR	attenuated total reflection
BAW	bulk acoustic waves
BLM	bilayer lipid membrane
BMFT	Bundesministerium für Forschung und Technologie
CADI	computer assisted dispersive infrared
CARS	coherent anti-Stokes Raman spectroscopy
CCC	counter current chromatography

Abbreviation	Explanation
CD	circular dichroism
CDS	corona discharge spectroscopy
CE	counter electrode
CEA	Commissariat a ' l'Energie Atomique
CFS	constant final state spectroscopy
CHEMFET	chemically sensitive field effect transistor
CI	chemical ionization
CIR	cylindrical internal reflection
CIS	constant initial state spectroscopy
CM	conductance measurements
COMAS	concentration-modulated absorption spectroscopy
CPMAS	cross polarization magic angle spinning
CPAC	Center for Process Analysis and Control
CPD	contact potential difference measurements
CSNICP	conductive solids nebulizer inductively coupled plasma
CSP	chiral stationary phase
CSSD	chemically sensitive semiconductor devices
CV	calorific value
CVD	chemical vapor deposition
CWE	coated wire electrode
d.c.	direct current
DAC	diamond anvil cell
DCCC	droplet counter current chromatography
DCI	direct current ionization
DCPAES	direc current plasma atomic emission spectroscopy
DM	diffusion measurements
DMS	dynamic mass spectrometer
DR	diffuse reflectance
DRIFTS	diffuse reflectance infrared Fourier transform spectroscopy
DTA	differential thermal analysis
EBD	electron beam deposition
EBIC	electron beam induced current
EDS	energy dispersive spectroscopy
EDXRF	electron diffraction X-ray fluorescence
EDXS	energy dispersive X-ray spectroscopy
EEC	European Economic Community
EELS	electron energy loss spectroscopy
EGFET	extended gate field effect transistor
EI	electron ionization
EIE	easily ionized element
EIEIO	easily ionized element interface observation
ELL	ellipsometry
EM	electron microscopy
EMA	electron microprobe analysis
EMF	electromotive force
EMTC	external mass transfer control

Abbreviation	Explanation
ENDOR	electron nuclear double beam resonance
EnFET	enzyme-modified field effect transistor
EPR	electron paramagnetic resonance
ESA	electrostatic analyzer
ESCA	electron spectroscopy for chemical analysis
ESD	electron stimulated desorption
ESPRIT	European Strategic Program for Research in Information Technology
ESR	electron spin resonance
ETA	electrothermal atomization
EXAFS	extended X-ray absorption fine structure
FAB-MS	fast atom bombardment mass spectroscopy
FD	field desorption
FDIR	fast dispersive infrared
FDM	field desorption microscopy
FEC	field effect of conductance
FEM	field effect microscopy
FER	field effect of reflectance
FES	field emission spectroscopy
FET	field effect transistor
FIA	flow-injection analysis
FIAP	field ionisation atom probe
FIM	field ion microscopy
FIMS	field ion mass spectrometry
FIR	far infrared
FTIR	Fourier transform infrared
FTMS	Fourier transform mass spectrometry
FTR	frustrated total reflection
GASFET	gas sensitive field effect transistor
GC	gas chromatography
GCIR	gas chromatography infrared
GCMS	gas chromatography mass spectrometry
GCNMR	gas chromatography nuclear magnetic resonance
GDMS	glow discharge mass spectrometry
GFAAS	graphite furnace atomic absorption spectroscopy
GOD	glucose oxidase
GPC	gel permeation chromatography
GPMAS	gas phase molecular absorption spectroscopy
HAM	heat of adsorption measurements
HDC	hydrodynamic chromatography
HE	Hall effect
HEED	high energy electron diffraction
HID	hydrogen-induced drift
HIXSE	heavy ion induced X-ray satellite emission
hmds	hexamethyldisiloxane

Abbreviation	Explanation
HOL	holography
HPLC	high performance liquid chromatography
HPTLC	high perfomance thin layer chromatography
HREELS	high resolution electron energy loss spectrometer
HUP	hydrogen uranyl phosphate tetrahydrate
IC	integrated circuit
IC	ion chromatography
ICAP	inductively coupled argon plasma
ICB	ion cluster beam deposition
ICD	ion controlled diode
ICP	inductively coupled plasma
ICPAES	inductively coupled plasma atomic emission spectroscopy
ICPES	inductively coupled plasma emission spectroscopy
ICPMS	inductively coupled plasma mass spectrometry
ICRS	ion cyclotron resonance spectroscopy
IDT	interdigital transducer
IETS	inelastic electron tunneling spectroscopy
IEX	ion excited X-ray fluorescence
IID	ion impact desorption
ImFET	immuno-sensing field effect transistor
IMPA	ion microprobe analysis
IMXA	ion microprobe for X-ray analysis
INEPT	insensitive nuclei enhanced by polarization transfer
INS	ion neutralization spectroscopy
IPS	inverse photoelectron spectroscopy
IRE	internal reflection element
IRS	internal reflection spectroscopy
ISFET	ion sensitive field effect transistor
ISM	solvent polymeric membrane
ISS	ion scattering spectroscopy
LAMMA	laser microprobe mass analysis
LASER	light amplification by stimulated electromagnetic radiation
LB	Langmuir-Blodgett
LC	liquid chromatography
LCIR	liquid chromatography infrared
LCMS	liquid chromatography mass spectrometry
LEED	low energy electron diffraction
LEL	lower explosive limits
LETI	Laboratoire d'Electronique et des Technologies de l'Information
LIDAR	light detection and ranging
LIMA	laser ionization mass analysis
LIMS	laser ionization mass spetrometry
LM	light microscope
LPCVD	low-pressure chemical vapor deposition
LRS	laser Raman spectroscopy

Abbreviation	Explanation
MAK	maximal tolerable concentration in the ambient air which for a dayly exposition of 8 h is without any influence on the health of a person
MBAS	molecular beam atom scattering
MBE	molecular beam epitaxy
MBT	molecular beam techniques
MCD	magnetic circular dichroism
MIKE	mass selection followed by ion kinetic energy analysis
MIP	microwave induced plasma
MIR	multiple internal reflection
MIR	mid infrared
MIS CAP	metal insulator semiconductor capacitor
MISFET	metal insulator semiconductor field effect transistor
MOS	metal oxide semiconductor
MOSCAP	metal oxide semiconductor capacitor
MOSFET	metal oxide semiconductor field effect transistor
MPS	modulated photoconductivity spectroscopy
MRI	magnetic resonance imaging
NADH	nicotinamide adenine dinucleotide
NEC	National Electric Code
NHE	normal-hydrogen-electrode
NIR	near infrared
NIS	neutron inelastic scattering
NMR	nuclear magnetic resonance
NQR	nuclear quadrupole resonance
NTIS	National Technical Information Service
OAS	opto-acoustic spectroscopy
OES	optical emission spectroscopy
ORD	optical rational dispersion
ORP	oxidation reduced potential
OS	optical spectroscopy
OSCA	Optical Sensor Collaborative Association
PARUPS	polarization and angle resolved ultraviolet photoelectron spectroscopy
PAS	photoacoustical spectroscopy
PASCA	positron annihilation spectroscopy for chemical analysis
PC	photoconductivity
Pc	phatalocyanines
PD	photodesorption
PDMS	plasma desorption mass spectrometry
PDS	photodischarge spectroscopy
PEM	photoelastic modulator
PES	photoelectron spectroscopy
PIGME	particle induced gamma ion emission
PIS	Penning ionization spectroscopy
PIXE	proton/particle induced X-ray emission
PM	permeation measurements

Abbreviation	Explanation
PNMR	proton nuclear magnetic resonance
PVD	physical vapor deposition
PVS	photovoltage spectroscopy
r.h.	relativ humidity
RBS	Rutherford backscattering
RE	reference electrode
REFET	reference field effect transistor
RGA	residual gas analysis
RH	relative humidity
RHEED	reflection high energy electron diffraction
RI	refractive index
ROA	Raman optical activity
RRS	resonance Raman spectroscopy
RT	response time
RTP	room temperature phosphorescence
RTPL	room temperature phosphorescence in liquids
s.i.c.	solid ionic conductor
SAM	scanning Auger microscopy
SAM	scanning acoustic microscopy
SAW	surface acoustic wave
SAX	selected areas X-ray photo-electron spectroscopy
SDS	surface discharge spectroscopy
SEM	scanning electron microscopy
SERC	Science and Engineering Research Council
SERS	surface enhanced Raman spectroscopy
SES	spin echo spectroscopy
SES	secondary electron spectroscopy
SFC	supercritical fluid chromatography
SGFET	suspended gate field effect transistor
SIM	scanning ion microscopy
SIMS	secondary ion mass spectrometry
SIS	silicon-insulator-silicon
SNMS	secondary neutral mass spectrometry
SNMS	sputtered neutral mass spectrometry
SNR	signal-to-noise ratio
SOS	silicon-on-sapphire
SS	stain-less steel
SSIMS	scanning mode secondary ion mass spectrometry
SSMS	spark source mass spectrometer
STAT	slotted tube atom trap
STM	scanning tunneling microscope
TAB	tape automated bonding
TDS	thermal desorption spectroscopy
TEM	transmission electron microscopy
TIMS	thermal ionization mass spectrometry

Abbreviation	Explanation
TL	thermoluminescence
TLC	thin layer chromatography
TLV	threshold-limited value
TMOS	ultra-thin gate metal oxide semiconductor
TOF	time-of-flight mass spectrometer
TOSFET	ionsensitive Ta i 2 i/ O i 5 i/ -based field effect transistor
TPD	temperature-programmed desorption
UEL	upper explosive limits
UHV	ultra high vacuum
UPS	ultraviolet photoelectron spectroscopy
UV-VIS	ultraviolet-visible spectroscopy
VUV	vacuum ultraviolet
WDS	wavelength dispersive spectroscopy
WE	working electrode
X-AES	X-ray induced Auger electron spectroscopy
XPS	X-ray photoelectron spectroscopy
XRD	X-ray diffraction
XRF	X-ray fluorescence
YSZ	yttrium stabilized zirconia